Numerical Methods Using MathCad

Laurene Fausett
Georgia Southern University

Prentice Hall
Upper Saddle River, New Jersey 07458

Library of Congress Cataloging-in-Publication Data

Fausett, Laurene V.
　　Numerical methods using MathCAD / Laurene Fausett
　　p.　cm.
　　Includes bibliographical references and index.
　　ISBN 0-13-061081-X
　　1. Numerical analysis—Data processing.　2. MathCAD.　I. Title.
　　QA297.F384 2002
　　519.4'0285'5369—dc 21　　　　　　　　　　　2001021855

Vice Presidential and Editorial Director, ECS: *Marcia J. Horton*
Publisher: *Tom Robbins*
Acquisitions Editor: *Eric Frank*
Associate Editor: *Alice Dworkin*
Editorial Assistant: *Jessica Romeo*
Vice President and Director of Production
　and Manufacturing, ESM: *David W. Riccardi*
Executive Managing Editor: *Vince O'Brien*
Managing Editor: *David A. George*
Production Editor: *Patty Donovan*
Director of Creative Services: *Paul Belfanti*
Creative Director: *Carole Anson*
Art Director: *Jayne Conte*
Art Editor: *Adam Velthaus*
Cover Designer: *Bruce Kenselaar*
Manufacturing Manager: *Trudy Pisciotti*
Manufacturing Buyer: *Lisa McDowell*
Marketing Manager: *Holly Stark*
Marketing Assistant: *Karen Moon*
Cover Image: *Colors Dance*, Quilt by Susan Webb Lee, 1985, Weddington, NC. From the collection of the artist.

© 2002 by Prentice-Hall
Prentice-Hall, Inc.
Upper Saddle River, New Jersey 07458

All rights reserved. No part of this book may be reproduced in any form or by any means, without permission in writing from the publisher.

The author and publisher of this book have used their best efforts in preparing this book. These efforts include the development, research, and testing of the theories and programs to determine their effectiveness. The author and publisher make no warranty of any kind, expressed or implied, with regard to these programs or the documentation contained in this book. The author and publisher shall not be liable in any event for incidental or consequential damages in connection, or arising out of, the furnishing, performance, or use of these programs.

Mathcad® is a registered trademark of MathSoft Engineering & Education, Inc., 101 Main Street, Cambridge, MA 02142-1521.

Printed in the United States of America

10　9　8　7　6　5　4　3　2　1

ISBN 0-13-061081-X

Pearson Education LTD., *London*
Pearson Education Australia PTY, Limited, *Sydney*
Pearson Education Singapore, Pte. Ltd
Pearson Education North Asia Ltd, *Hong Kong*
Pearson Education Canada, Ltd, *Toronto*
Pearson Educación de Mexico, S.A. de C.V.
Pearson Education—Japan, *Tokyo*
Pearson Education Malaysia, Pte. Ltd
Pearson Education, Upper Saddle River, New Jersey

*Books extend our world through time and space.
In that spirit, I dedicate this work to all of my teachers
and all of my students.*

Contents

PREFACE XI

EXAMPLES/MATHCAD FUNCTIONS/ALGORITHMS XIII

1 FOUNDATIONS 1

1.1 Sample Problems and Numerical Methods 4
    1.1.1 Roots of Nonlinear Equations 4
    1.1.2 Fixed-Point Iteration 5
    1.1.3 Linear Systems 6
    1.1.4 Gaussian Elimination 7
    1.1.5 Numerical Integration 8
    1.1.6 Trapezoid Rule 8

1.2 Some Basic Issues 10
    1.2.1 Key Issues for Iterative Methods 10
    1.2.2 How Good Is the Result? 14
    1.2.3 Getting Better Results 22

1.3 Getting Started in Mathcad 28
    1.3.1 Overview of the Mathcad Workspace 28
    1.3.2 Mathematical Computations 31
    1.3.3 Operators on the Math Toolbars 32
    1.3.4 Built-In Functions 35
    1.3.5 Programming in Mathcad 38

2 SOLVING EQUATIONS OF ONE VARIABLE 47

2.1 Bisection Method 50
    2.1.1 Step-by-Step Computation 50
    2.1.2 Mathcad Function for Bisection 52
    2.1.3 Discussion 54

2.2 Regula Falsi and Secant Methods 55
    2.2.1 Step-by-Step Computation for Regula Falsi 56
    2.2.2 Mathcad Function for the Regula Falsi Method 58
    2.2.3 Step-by-Step Computation for the Secant Method 60
    2.2.4 Mathcad Function for the Secant Method 62
    2.2.5 Discussion 64

2.3 Newton's Method 68
    2.3.1 Step-by-Step Computation 68
    2.3.2 Mathcad Function for Newton's Method 70
    2.3.3 Discussion 72

2.4 Muller's Method 75
    2.4.1 Step-by-Step Computation for Muller's Method 76

2.4.2 Mathcad Function for Muller's Method   78
2.4.3 Discussion   80
2.5 Mathcad's Methods   81
2.5.1 Using the Built-In Functions   81
2.5.2 Understanding the Algorithms   84

3 SOLVING SYSTEMS OF LINEAR EQUATIONS: DIRECT METHODS   93

3.1 Gaussian Elimination   96
3.1.1 Using Matrix Notation   97
3.1.2 Step-by-Step Procedure   98
3.1.3 Mathcad Function for Basic Gaussian Elimination   101
3.1.4 Discussion   103
3.2 Gaussian Elimination with Row Pivoting   106
3.2.1 Step-by-Step Computation   106
3.2.2 Mathcad Function for Gaussian Elimination with Pivoting   110
3.2.3 Discussion   113
3.3 Gaussian Elimination for Tridiagonal Systems   113
3.3.1 Step-by-Step Procedure   116
3.3.2 Mathcad Function for the Thomas Method   118
3.3.3 Discussion   119
3.4 Mathcad's Methods   122
3.4.1 Using the Built-In Functions   122
3.4.2 Understanding the Algorithms   122

4 SOLVING SYSTEMS OF LINEAR EQUATIONS: ITERATIVE METHODS   131

4.1 Jacobi Method   135
4.1.1 Step-by-Step Procedure for Jacobi Iteration   136
4.1.2 Mathcad Function for the Jacobi Method   139
4.1.3 Discussion   142
4.2 Gauss-Seidel Method   144
4.2.1 Step-by-Step Computation for Gauss-Seidel Method   145
4.2.2 Mathcad Function for Gauss-Seidel Method   148
4.2.3 Discussion   150
4.3 Successive Overrelaxation   151
4.3.1 Step-by-Step Computation of SOR   152
4.3.2 Mathcad Function for SOR   154
4.3.3 Discussion   155
4.4 Mathcad's Methods   157
4.4.1 Using the Built-In Functions   157
4.4.2 Understanding the Algorithms   159

5 SYSTEMS OF EQUATIONS AND INEQUALITIES   171

5.1 Newton's Method for Systems of Equations   174
5.1.1 Matrix-Vector Notation   176
5.1.2 Mathcad Function for Newton's Method   177
5.2 Fixed-Point Iteration for Nonlinear Systems   181
5.2.1 Step-by-Step Computation   182

    5.2.2 Mathcad Function for Fixed-Point Iteration for Nonlinear Systems 182
    5.2.3 Discussion 186
 5.3 Minimum of a Nonlinear Function 187
    5.3.1 Step-by-Step Computation of Minimization by Gradient Descent 187
    5.3.2 Mathcad Function for Minimization by Gradient Descent 188
 5.4 Mathcad's Methods 192
    5.4.1 Using the Built-In Functions 192
    5.4.2 Understanding the Algorithms 193

6   LU Factorization 201

 6.1 LU Factorization from Gaussian Elimination 203
    6.1.1 A Step-by-Step Procedure for LU Factorization 204
    6.1.2 Mathcad Function for LU Factorization Using Gaussian Elimination 206
 6.2 LU Factorization of Tridiagonal Matrices 207
    6.2.1 Step-by-Step LU Factorizaiton of a Tridiagonal Matrix 207
    6.2.2 Mathcad Function for LU Factorization of a Tridiagonal Matrix 208
 6.3 LU Factorization with Pivoting 209
    6.3.1 Step-by-Step Computation 209
    6.3.2 Mathcad Function for LU Factorization with Row Pivoting 210
    6.3.3 Discussion 212
 6.4 Direct LU Factorization 215
    6.4.1 Direct LU Factorization of a General Matrix 215
    6.4.2 LU Factorization of a Symmetric Matrix 217
 6.5 Applications of LU Factorization 219
    6.5.2 Solving a Tridiagonal System Using LU Factorization 222
    6.5.3 Determinant of a Matrix 224
    6.5.4 Inverse of a Matrix 224
 6.6 Mathcad's Methods 226
    6.6.1 Using the Built-In Functions 226
    6.6.2 Understanding the Algorithms 226

7   Eigenvalues, Eigenvectors, and QR Factorization 233

 7.1 Power Method 236
    7.1.1 Basic Power Method 237
    7.1.2 Inverse Power Method 242
    7.1.3 Discussion 247
 7.2 QR Factorization 248
    7.2.1 Householder Transformations 248
    7.2.2 Givens Transformations 257
    7.2.3 Basic QR Factorization 261
 7.3 Finding Eigenvalues Using QR Factorization 267
    7.3.1 Basic QR Eigenvalue Method 267
    7.3.2 Better QR Eigenvalue Method 268
    7.3.3 Discussion 270
 7.4 Mathcad's Methods 270
    7.4.1 Using the Built-In Functions 270
    7.4.2 Understanding the Algorithms 271

8 INTERPOLATION 283

- 8.1 Polynomial Interpolation 286
  - 8.1.1 Lagrange Interpolation Polynomials 286
  - 8.1.2 Newton Interpolation Polynomials 295
  - 8.1.3 Difficulties with Polynomial Interpolation 306
- 8.2 Hermite Interpolation 310
- 8.3 Rational Function Interpolation 316
- 8.4 Spline Interpolation 320
  - 8.4.1 Piecewise Linear Interpolation 321
  - 8.4.2 Piecewise Quadratic Interpolation 322
  - 8.4.3 Piecewise Cubic Interpolation 325
- 8.5 Mathcad's Methods 334
  - 8.5.1 Using the Built-In Functions 334
  - 8.5.2 Understanding the Algorithms 335

9 FUNCTION APPROXIMATION 349

- 9.1 Least Squares Approximation 352
  - 9.1.1 Linear Least-Squares Approximation 352
  - 9.1.2 Quadratic Least-Squares Approximation 359
  - 9.1.3 Cubic Least-Squares Approximation 364
  - 9.1.4 Least-Squares Approximation for Other Functional Forms 369
- 9.2 Continuous Least-Squares Approximation 373
  - 9.2.1 Continuous Least-Squares with Orthogonal Polynomials 376
  - 9.2.2 Gram-Schmidt Process 376
  - 9.2.3 Legendre Polynomials 378
  - 9.2.4 Least-Squares Approximation with Legendre Polynomials 379
- 9.3 Function Approximation at a Point 381
  - 9.3.1 Taylor Approximation 381
  - 9.3.2 Padé Function approximation 382
- 9.4 Mathcad's Methods 385
  - 9.4.1 Using the Built-in Functions 385
  - 9.4.2 Understanding the Algorithms 386

10 FOURIER METHODS 393

- 10.1 Fourier Approximation and Interpolation 396
- 10.2 Fast Fourier Transforms for $n = 2^r$ 407
  - 10.2.1 Discrete Fourier Transform 407
  - 10.2.2 Fast Fourier Transform 408
- 10.3 Fast Fourier Transforms for General n 415
- 10.4 Mathcad's Methods 423
  - 10.4.1 Using the Built-In Functions 423
  - 10.4.2 Understanding the Algorithms 424

11 NUMERICAL DIFFERENTIATION AND INTEGRATION 436

- 11.1 Differentiation 436
  - 11.1.1 First Derivatives 436

    11.1.2 Higher Derivatives 440
    11.1.3 Partial Derivatives 441
    11.1.4 Richardson Extrapolation 442
 11.2 Basic Numerical Integration 445
    11.2.1 Trapezoid Rule 446
    11.2.2 Simpson Rule 448
    11.2.3 The Midpoint Formula 450
    11.2.4 Other Newton-Cotes Open Formulas 452
 11.3 Better Numerical Integration 452
    11.3.1 Composite Trapezoid Rule 453
    11.3.2 Composite Simpson's Rule 455
    11.3.3 Extrapolation Methods for Quadrature 458
 11.4 Gaussian Quadrature 462
    11.4.1 Gaussian Quadrature on [−1,1] 462
    11.4.2 Gaussian Quadrature on [a,b] 464
 11.5 Mathcad's Methods 468
    11.5.1 Using the Operators 468
    11.5.2 Understanding the Algorithms 469

12 Ordinary Differential Equations: Initial-Value Problems 477

 12.1 Taylor Methods 479
    12.1.1 Euler's Method 479
    12.1.2 Higher-Order Taylor Methods 484
 12.2 Runge-Kutta Methods 487
    12.2.1 Midpoint Method 487
    12.2.2 Other Second-Order Runge-Kutta Methods 492
    12.2.3 Third-Order Runge-Kutta Methods 494
    12.2.4 Classic Runge-Kutta Method 495
    12.2.5 Other Runge-Kutta Methods 499
    12.2.6 Runge-Kutta-Fehlberg Methods 501
 12.3 Multistep Methods 502
    12.3.1 Adams-Bashforth Methods 502
    12.3.2 Adams-Moulton Methods 508
    12.3.3 Predictor-Corrector Methods 509
 12.4 Stability 514
 12.5 Mathcad's Methods 517
    12.5.1 Using the Built-In Functions 517
    12.5.2 Understanding the Algorithms 520

13 Systems of Ordinary Differential Equations 529

 13.1 Higher-Order ODEs 532
 13.2 Systems of Two First-Order ODE 534
    13.2.1 Euler's Method for Solving Two ODE-IVPs 534
    13.2.2 Midpoint Method for Solving Two ODE-IVPs 537
 13.3 Systems of First-Order ODE-IVP 541
    13.3.1 Euler's Method for Solving Systems of ODEs 542

        13.3.2 Runge-Kutta Methods for Solving Systems of ODEs   544
        13.3.3 Multistep Methods for Systems   552
13.4 Stiff ODE and Ill-Conditioned Problems   557
13.5 Mathcad's Methods   559
        13.5.1 Using the Built-In Functions   559
        13.5.2 Understanding the Algorithms   562

14   Ordinary Differential Equations—Boundary Value Problems   575

14.1 Shooting Method for Solving Linear BVP   578
        14.1.1 Simple Boundary Conditions   578
        14.1.2 General Boundary Condition at $x = b$ 583
        14.1.3 General Boundary Conditions at Both Ends of the Interval   584
14.2 Shooting Method for Solving Nonlinear BVP   585
        14.2.1 Nonlinear Shooting Based on the Secant Method   585
        14.2.2 Nonlinear Shooting Using Newton's Method   588
14.3 Finite-Difference Method for Solving Linear BVP   592
14.4 Finite-Difference Method for Nonlinear BVP   599
14.5 Mathcad's Methods   602
        14.5.1 Using the Built-In Functions   602
        14.5.2 Understanding the Algorithms   604

15   Partial Differential Equations   609

15.1 Classification of PDE   613
15.2 Heat Equation: Parabolic PDE   614
        15.2.1 Explicit Method for Solving the Heat Equation   615
        15.2.2 Implicit Method for Solving the Heat Equation   623
        15.2.3 Crank-Nicolson Method for Solving the Heat Equation   628
        15.2.4 Heat Equation with Insulated Boundary   632
15.3 Wave Equation: Hyperbolic PDE   633
        15.3.1 Explicit Method for Solving Wave Equations   634
        15.3.2 Implicit Method for Solving Wave Equation   638
15.4 Poisson Equation: Elliptic PDE   640
15.5 Finite-Element Method for Solving an Elliptic PDE   645
15.6 Mathcad's Methods   658
        15.6.1 Using the Built-In Functions   658
        15.6.2 Understanding the Algorithms   659

Bibliography   667

Answers to Selected Problems   673

Index   695

Preface

The purpose of this text is to present the fundamental numerical techniques used in engineering, applied mathematics, computer science, and the physical and life sciences in a manner that is both interesting and understandable to undergraduate and beginning graduate students in those fields. The organization of the chapters, and of the material within each chapter, the use of Mathcad worksheets and functions to illustrate the methods, and the exercises provided are all designed with student learning as the primary objective.

The first chapter sets the stage for the material in the rest of the text, by giving a brief introduction to the long history of numerical techniques, and a "preview of coming attractions" for some of the recurring themes of the remainder of the text. It also presents enough description of Mathcad to allow students to use the Mathcad functions presented for each of the numerical methods discussed in the other chapters. An algorithmic statement of each method is also included; the algorithm may be used as the basis for computations using a variety of types of technological support, ranging from paper and pencil, to calculators, Mathcad worksheets or developing computer programs.

Each of the subsequent chapters begins with a one-page overview of the subject matter, together with an indication as to how the topics presented in the chapter are related to those in previous and subsequent chapters. Introductory examples are presented to suggest a few of the types of problems for which the topics of the chapter may be used. Following the sections in which the methods are presented, each chapter concludes with a summary of the most important formulas, a selection of suggestions for further reading, and an extensive set of exercises. The first group of problems provide fairly routine practice of the techniques; the second group are applications adapted from a variety of fields, and the final group of problems encourage students to extend their understanding of either the theoretical or the computational aspects of the methods.

The presentation of each numerical technique is based on the successful teaching methodology of providing examples and geometric motivation for a method, and a concise statement of the steps to carry out the computation, before giving a mathematical derivation of the process or a discussion of the more theoretical issues that are relevant to the use and understanding of the topic. Each topic is illustrated by examples that range in complexity from very simple to moderate.

Geometrical or graphical illustrations are included whenever they are appropriate. A simple Mathcad function is presented for each method, which also serves as a clear step-by-step description of the process; discussion of theoretical considerations is placed at the conclusion of the section. The last section of each chapter gives a brief discussion of Mathcad's built-in functions for solving the kinds of problems covered in the chapter.

The chapters are arranged according to the following general areas:

Chapters 2–5 deal with solving linear and nonlinear equations.
Chapters 6 and 7 treat topics from numerical linear algebra.
Chapters 8–10 cover numerical methods for data interpolation and approximation.
Chapters 11 presents numerical differentiation and integration.
Chapters 12–15 introduce numerical techniques for solving differential equations.

For much of the material, a calculus sequence that includes an introduction to differential equations and linear algebra provides adequate background. For more in depth coverage of the topics from linear algebra (especially the QR method for eigenvalues) a linear algebra course would be an appropriate prerequisite. The coverage of Fourier approximation and FFT (Chapter 10) and partial differential equations (Chapter 15) also assumes that the students have somewhat more mathematical maturity than the other chapters, since the material in intrinsically more challenging. The subject matter included is suitable for a two-semester sequence of classes, or for any of several different one-term courses, depending on the desired emphasis, student background, and selection of topics.

Many people have contributed to the development of this text. My colleagues at Florida Institute of Technology, the Naval Postgraduate School, the University of South Carolina Aiken, and Georgia Southern University have provided support, encouragement, and suggestions. I especially want to thank Jacalyn Huband for the development of the Mathcad functions and examples. I also wish to thank three other colleagues for their particular contibutions: Jane Lybrand for the data from classroom experiments used in several examples and exercises in Ch 9. Jack Leifer for providing data, as well as helpful discussions on engineering applications and the use of Mathcad in engineering; Pierre Larochelle for the example of robot motion in Ch 13. I also appreciate the many contributions my students have made to this text, which was after all written with them in mind. The comments made by the reviewers of the text have helped greatly in the fine-tuning of the final presentation. The editorial and production staff at Prentice Hall, as well as Patty Donovan, and the rest of the staff at Pinetree Composition, have my heartfelt gratitude for their efforts in insuring that the text is as accurate and as well designed as possible. And, saving the most important for last, I thank my husband and colleague, Don Fausett, for his patience and support.

Laurene Fausett

Examples/Mathcad Functions/Algorithms

EXAMPLES

Chapter 1

- **1.1** Fixed Point Iterations to Find Square Root of 3   5
- **1.2** Solving a Linear System   7
- **1.3** Approximating an Integral   9
- **1.4** Fixed Point Iteration   10
- **1.5** Bounds on Eigenvalues   12
- **1.6** Effect of Order of Operations   18
- **1.7** Cancellation Errors   19
- **1.8** Improving an Integral by Acceleration   24
- **1.9** Careful Application of Algorithm Reduces Error   26–27
- **1.10** Using Calculator Toolbar   29
- **1.11** Defining a Range Variable   30
- **1.12** A Simple Mathcad Program   39

Chapter 2

- **2-A** Floating Sphere   48
- **2-B** Planetary Orbits   49
- **2.1** Finding the Square Root of 3 Using Bisection   51
- **2.2** Approximating the Floating Depth for a Cork Ball   53
- **2.3** Finding the Cube Root of 2 Using Regula Falsi   57
- **2.4** Finding the Central Angle of an Elliptical Orbit   59
- **2.5** Finding the Square Root of 3 Using the Secant Method   61
- **2.6** Finding the Central Angle Using the Secant Method   63
- **2.7** A Challenging Problem   66–67
- **2.8** Finding the Square Root of 3/4 Using Newton's Method   69
- **2.9** Finding the Floating Depth for a Wooden Ball   71
- **2.10** Oscillations in Newton's Method   74
- **2.11** Finding the Sixth Root of 2 Using Muller's Method   77
- **2.12** Another Challenging Problem   79
- **2.13** Using polyroots   82
- **2.14** Using the Function Root without Bracketing the Solution   83
- **2.15** Using the Function Root with Bracketing the Solution   84

Chapter 3

- **3-A** Circuit Analysis Application   94
- **3-B** Forces on a Truss   95
- **3.1** Three-by-Three System   96
- **3.2** Circuit Analysis   99
- **3.3** Analysis of Forces on a Simple Truss   102
- **3.4** Difficult System
- **3.5** Solving a Three-by-Three System with Pivoting   108-109
- **3.6** Forces in a Simple Truss   111-112
- **3.7** Solving a Tridiagonal System   114-115
- **3.8** Using the Thomas Algorithm   117
- **3.9** Using the Mathcad Function Thomas   119
- **3.10** Thomas Method Recovery from a Zero-Pivot   120-121

Chapter 4

- **4-A** Finite Difference Solution of a PDE   132
- **4-B** Steady-State Concentrations   133
- **4.1** Illustrating the Jacobi Method Graphically   135
- **4.2** Solving a Three-by-Three System   136
- **4.3** Using the Jacobi Algorithm   138
- **4.4** A Linear System That Comes from a PDE   140–141
- **4.5** Jacobi Method is Sensitive to the Order of Equations   143
- **4.6** Graphical Illustration of the Gauss-Seidel Method   144

4.7 Using Gauss-Seidel Iteration 145
4.8 Using the Gauss-Seidel Algorithm 147
4.9 Using the Mathcad Function for Gauss-Seidel Method 149
4.10 Solving a three-by-three system using SOR 153
4.11 Solving Linear Systems that come from a PDE 155
4.12 Using Mathcad's Solve Block Structure 158

Chapter 5

5-A Nonlinear System from Geometry 172
5-B Position of a Two-link Robot Arm 173
5.1 Intersection of a Circle and an Ellipse 174–175
5.2 Intersection of a Circle and an Ellipse 178
5.3 Newton's Method for a System of Three Equations 179
5.4 Positioning a Robot Arm 180
5.5 Fixed-point Iteration for a System of Two Functions 181, 183
5.6 Fixed-point Iteration for a System of Three Equations 184–185
5.7 Finding a Zero of a nonlinear System by Minimization 189
5.8 Optimal Location of a Point in the Plane 190–191

Chapter 6

6-A Circuit-Analysis Application 202
6.1 Three-by-three System 203
6.2 Using the Algorithm for LU Factorization 205
6.3 Finding an LU Factorization 206
6.4 Finding the LU Factorization of a Tridiagonal Matrix 208
6.5 LU Factorization with Pivoting 211
6.6 Direct LU Factorization 216
6.7 Cholesky Factorization 218
6.8 Solving Electrical Circuit for Several Voltages 220
6.9 Solving a Tridiagonal System Using LU Factorization 223
6.10 Finding the Determinant of a Matrix 224
6.11 Finding a Matrix Inverse Using LU Factorization 225

Chapter 7

7-A Inertia 234
7-B Buckling and Breaking of a Beam 235
7.1 Using the Basic Power Method 237
7.2 Using the Mathcad Function for the Power Method 239
7.3 Dominant Eigenvalue for Symmetric Matrix 240
7.4 Dominant Eigenvalue of Shifted Matrix 241
7.5 Using the Inverse Power Method 243
7.6 Finding the Smallest Eigenvalue 245
7.7 Buckling Beam 245
7.8 Finding the Eigenvalue Closest to a Given Number 246
7.9 Using the Householder Algorithm 249
7.10 Using the Function Householder 251
7.11 Householder Transformations to Convert Matrix to Upper-Triangular Form 252-3
7.12 Similarity Transformation to Upper-Hessenberg Form 255
7.13 Transformation of Upper-Hessenberg Form to Upper-Triangular Form 259
7.14 QR Factorization Using Householder Transformations 262–263
7.15 QR Factorization of an Upper-Hessenberg Matrix 266
7.16 Finding Eigenvalues Using QR Factorization 269

Chapter 8

8-A Chemical Reaction Product 284
8-B Spline Interpolation 285
8.1 Lagrange Interpolation Parabola 288
8.2 Additional Data Points 290
8.3 Chemical Reaction Product Data 291
8.4 Higher Order Interpolation Polynomials 292
8.5 Newton Interpolation Parabola 296
8.6 Additional Data Points 297, 302
8.7 Higher Order Interpolation Polynomials 298–299
8.8 Humped and Flat Data 306
8.9 Noisy Straight Line 307
8.10 Runge Function 308–309
8.11 More Data for Product Concentration 311, 315
8.12 Difficult Data 314
8.13 Rational-Function Interpolation 320
8.14 Piecewise Linear Interpolation 321
8.15 Piecewise Quadratic Interpolation 324
8.16 Natural Cubic Spline Interpolation 326–327
8.17 Runge Function 330
8.18 Difficult Data 331
8.19 Chemical Reaction Product Data 332

Chapter 9

9-A Oil Reservoir Modeling 350
9-B Logistic Population Growth 351
9.1 Linear Approximation to Four Points 353
9.2 Least Squares Straight Line to Fit Four Data Points 356
9.3 Noisy Straight-Line Data 357
9.4 Chemical Reaction Data 361
9.5 Gompertz Growth Curve 362

- 9.6 Oil Reservoir   363
- 9.7 Cubic Least-Squares   367
- 9.8 Cubic Least-Squares, Continued   368
- 9.9 Oil Reservoir Data   370
- 9.10 Least-Squares Approximation of a Reciprocal Relation   371
- 9.11 Light Intensity   372
- 9.12 Continuous Least-Squares   374
- 9.13 Least-Squares Approximation Using Legendre Polynomials   380
- 9.14 Pade Approximation of the Runge Function   384

Chapter 10

- 10-A Waveforms of Different Instruments   394
- 10-B Geometric Figures   395
- 10.1 Trigonometric Interpolation   398
- 10.2 Trigonometric Approximation   399
- 10.3 A Step Function   401
- 10.4 Geometric Figures   406
- 10.5 FFT with four points   414
- 10.6 FFT for Six Data Points   418-9
- 10.7 FFT for Six Data Points   421
- 10.8 FFT for Six Data Points   422
- 10.9 Interpolation of a Triangle in the Plane   424
- 10.10 Waveform of a Clarinet   426
- 10.11 Rotation-Invariant Figures   427

Chapter 11

- 11-A Simple Functions That Do Not Have Simple Antiderivatives   434
- 11-B Length of an Elliptical Orbit   435
- 11.1 Forward, Backward, and Central Differences   437
- 11.2 Three-point Difference Formulas   438
- 11.3 Second Derivative   440
- 11.4 Improved Estimate of the Derivative   443
- 11.5 Integral Using the Trapezoid Rule   446
- 11.6 Integral Using Simpson's Rule   448
- 11.7 Integral Using Simpson's Rule   449
- 11.8 The Midpoint Rule   451
- 11.9 Integral of 1/x Using the Composite Trapezoid Rule   454–455
- 11.10 Integral Using Composite Simpson's Rule   456
- 11.11 Length of Elliptical Orbit   457
- 11.12 Integral of 1/x using Romberg Integration   459–460, 461
- 11.13 Integral on [-1, 1] Using Gaussian Quadrature   462
- 11.14 Gaussian Quadrature Using More Quadrature Points   463
- 11.15 Integral on [0, 2] Using Gaussian Quadrature   466

Chapter 12

- 12-A Motion of a Falling Body   478
- 12.1 Solving a simple ODE with Euler's Method   481
- 12.2 Another Example of the Use of Euler's Method   482
- 12.3 Solving a Simple ODE with Taylor's Method   484, 486
- 12.4 Solving a Simple ODE with the Midpoint Method   488–489
- 12.5 ODE for Dawson's Integral   491
- 12.6 Solving a Simple ODE with the Classic Runge-Kutta Method   497
- 12.7 Another Example of the Use of the Classic Runge-Kutta method   498
- 12.8 Solving a Simple ODE with an Adams-Bashforth Method Worksheet   505
- 12.9 Solving a Simple ODE with an Adams-Bashforth Method Function   507
- 12.10 Using the Adams Predictor Corrector Method   512
- 12.11 Velocity of Falling Parachutist   513
- 12.12 Weakly Stable Method   515–516
- 12.13 Using the Mathcad function rkfixed   518

Chapter 13

- 13-A Motion of a Nonlinear Pendulum   530
- 13-B Chemical Flow   531
- 13.1 Nonlinear Pendulum   532
- 13.2 Nonlinear Pendulum Using Euler's Method   535
- 13.3 Series Dilution Problem Using Euler's Method   536
- 13.4 Nonlinear Pendulum Using Runge-Kutta Method   539
- 13.5 Series Dilution Using the Midpoint Method   540
- 13.6 A Higher-Order System of ODEs   541
- 13.7 Solving a Higher-Order System Using Euler's Method   542
- 13.8 Solving Another Higher-Order System Using Euler's Method   543
- 13.9 Solving a Fourth Order System Using the Midpoint Method   546–547
- 13.10 Using Mathcad Function for Midpoint Method   548
- 13.11 Solving a Circular Chemical Reaction using a Runge-Kutta Method   550–551
- 13.12 Mass and Spring System   555
- 13.13 Motion of a Baseball   556
- 13.14 Using the Mathcad function rkfixed   560
- 13.15 Robot Motion   563–564

Chapter 14

- 14-A Deflection of a Beam   576
- 14-B A Well Hit Ball   577
- 14.1 Electrostatic Potential Between Two Spheres   580
- 14.2 A Simple Linear Shooting Problem   580–581
- 14.3 Deflection of a Simple Supported Beam   582
- 14.4 More General Boundary Conditions   583
- 14.5 Linear Shooting with Mixed Boundary Conditions   584
- 14.6 Nonlinear Shooting Method   586
- 14.7 Flight of a Baseball   588
- 14.8 Nonlinear Shooting with Newton's Method   590–591
- 14.9 A Finite Difference Problem   592–593
- 14.10 Mathcad Function for a Linear Finite-Difference Problem   596
- 14.11 Deflection of a Beam, Using Finite Differences   598
- 14.12 Solving a Nonlinear BVP by Using Finite Differences   600

Chapter 15

- 15-A Heat Equation   610
- 15-B Wave Equation   611
- 15-C Poisson's Equation   612
- 15.1a Temperature in a Rod, Explicit Method, Stable Solution   616-7
- 15.1b Temperature in a Rod, Explicit Method, Stable Solution   619
- 15.2 Temperature in a Rod, Explicit Method, Unstable Solution   622
- 15.3a Temperature in a Rod, Implicit Method   624
- 15.3b Solving the Hear Equation Using an Implicit Method   627
- 15.4 Temperature in a Rod, Crank-Nicolson Method   631
- 15.5 Vibrating String, Explicit Method, Stable Solution   636
- 15.6 Potential Equation   643
- 15.7 Poisson's Equation   644
- 15.8 Finding Basis Functions   647
- 15.9 A Finite-element Solution   650-51
- 15.10 Using Mathcad to Find the Basis Functions   652
- 15.11 Solving the Laplace Equation Using Finite Elements   656–657

MATHCAD FUNCTIONS

Chapter 2

Bisect   52
Falsi   58
Secant   62
Newton   70
Muller   78

Chapter 3

Gauss_B   101
Gauss_pivot   110
Thomas   118

Chapter 4

Jacobi   139
GaussSeidel   148
SOR   154

Chapter 5

Newton_sys   177
Fixed_pt_sys   182
GradDesc   88

Chapter 6

LU_factor_G   206
LU_factor_T   208
LU_pivot   210
Doolittle   216
Cholesky   218
LU_solve   221
LU_solve_T   222

Chapter 7

Power_m   238
InvPower   244
Householder   250
Hessenberg   256
QR_factor   263
QR_factor_H   265
QR_eigen_g   268

Chapter 8

Lagrange_coef   289
Lagrange_Eval   289
Newton_coef   301
Newton_Eval   301
Hermite_coef   313
Hermite_Eval   314
Rat_interp   319
Splline   329

Chapter 9

Lin_LS   55
Quad_LS   360

Cubic_LS 366
Ln_Linear_LS 369

Chapter 10

Trig_poly 400
FFT_4 413
FFT_rs 420

Chapter 11

Trap 454
Simp 456
Romberg 461
Gauss_quad 465

Chapter 12

Euler 480
Taylor 485
RK2 490
RK4 496
AB3 506
ABM3 511

Chapter 13

Euler_sys2 534
RK2_sys2 538
RK2_sys 547
RK4_sys 550
ABM3_sys 554
Robot2 565
S_robot_motion 566

Chapter 14

S_nonlinear_shoot 587
Nonlinear_shoot_Newton 589
S_linear_FD 597
S_nonlinear_FD 601

Chapter 15

Heat 618
Heat_imp 626
Heat_CN 630
S_15_5 637
Poisson 642
Basis_func 653
Coeff 655

ALGORITHMS

Chapter 2

Bisection Algorithm 50
Regula Falsi Algorithm 56
Secant Algorithm 60
Newton's Method Algorithm 68
Muller's Method Algorithm 76

Chapter 3

Algorithm for Gaussian Elimination 98
Algorithm for Gaussian Elimination with Row Pivoting 107
Algorithm for Thomas Method 116

Chapter 4

Algorithm for Jacobi Method 137
Algorithm for Gauss-Seidel Method 146
Algorithm for SOR Method 152
Algorithm for Conjugate Gradient Method 160

Chapter 5

Algorithm for Newton's Method 177
Algorithm for Fixed-point Iteration for Nonlinear Systems 182
Algorithm for Minimization by Gradient Descent 187
Levenberg-Marquardt Algorithm 194
Quasi-Newton Algorithm 195

Chapter 6

Algorithm for LU Factorization Using Gaussian Elimination 204
Algorithm for LU Factorization of a Tridiagonal Matrix 207
Algorithm for LU Factorization with Row Pivoting 209
Algorithm for Direct LU Factorization 215
Algorithm for Cholesky LU Factorization 217
Algorithm to Solve the Linear System LUx = b 219
Algorithm for Solving a Tridiagonal System Using LU Factorization 222

Chapter 7

Algorithm for the Inverse Power Method 242
Algorithm for Forming a Householder Matrix 249
Algorithm for Similarity Transformation o Hessenberg Form 254
Algorithm for Givens Transformation of Hessenberg Matrix to Triangular Matrix 60
Algorithm for QR Factorization Using Householder Matrices 261
Algorithm for QR Factorization Using Householder Transformations 262
Algorithm for QR Factorization of Upper-Hessenberg Matrix 264

Algorithm for Finding Eigenvalues by QR
  Factorization   267
Algorithm for Finding Eigenvalues Using QR-Improved
  Method   268

Chapter 8

Algorithm to Find Coefficients of Lagrange Interpolation
  Polynomial   287
Algorithm to Evaluate Lagrange Interpolation
  Polynomial   287
Algorithm to Find Coefficients of Newton Interpolation
  Polynomial   300
Algorithm to Evaluate Newton Interpolation
  Polynomial   300
Algorithm to Create Coefficients for Hermite
  Interpolation Polynomial   312
Algorithm to Evaluate Hermite Interpolation
  Polynomial   312
Algorithm for Bulirsch-Stoer Rational Function
  Interpolation   318
Algorithm for Natural Cubic Spline Interpolation   328

Chapter 9

Algorithm for Linear Least-Squares Approximation   354
Algorithm for Quadratic Least-Squares
  Approximation   359
Algorithm for Cubic Least-Squares
  Approximation   365
Algorithm for Exponential Approximation
  (Linear Fit to ln x)   369

Chapter 10

Algorithm for Fourier Approximation
  of Interpolation   397
Algorithm for FFT with n = 4   409
Algorithm for FFT with n = r s
  (data evenly space on [0, 2p]   418

Chapter 11

Algorithm for Trapezoid Rule   453
Algorithm for Simpson's Rule   455

Algorithm for Romberg Integration   458
Algorithm for Gaussian Quadrature   464–465

Chapter 12

Algorithm for Euler's Method   480
Algorithm for Second-order Taylor's Method   485
Algorithm for the Midpoint Method   487
Algorithm for Classic Runge-Kutta Method
  (Fourth-Order)   495
Algorithm for the Third-Order Adams-Bashforth
  Method   504
Algorithm for Third-Order Adams-Bashforth-Moulton
  Method   510

Chapter 13

Algorithm for the Midpoint Method for a System
  of Two First-Order ODE   537
Algorithm for the Midpoint Method for a System
  of First-Order ODE   545
Algorithm for Fourth-Order Runge-Kutta Method
  for System of ODE   549
Algorithm for Third Order Predictor-Corrector Method
  for a System of ODE   553

Chapter 14

Algorithm for Linear Shooting Method to Solve
  Boundary Value Problem   579
Algorithm for Nonlinear Shooting Using Newton's
  Method   589
Algorithm for Linear Finite-Difference Method   595

Chapter 15

Algorithm for Solving the Heat Equation, Explicit
  Method   616
Algorithm for Solving the Heat Equation, Implicit
  Method   625
Algorithm for Solving the Heat Equation, Crank-
  Nicolson Method   629
Algorithm to Solve Wave Equation, Finite
  Differences   635
Algorithm to Solve Poisson Equation, Finite
  Differences   641

1
Foundations

1.1 Sample Problems and Numerical Methods

1.2 Some Basic Issues

1.3 Getting Started in Mathcad

From the earliest times, the search for solutions of real-world problems has been an important aspect of mathematical study. In many interesting applications, an exact solution may be unattainable, or it may not give the answer in a convenient form. Useful answers may involve finding good approximate results with a reasonable amount of computational effort.

Many numerical methods have a very long history. There is evidence that the Babylonians (more than 3,700 years ago) knew how to find numerical solutions of quadratic equations and approximations to the square root of an integer. They also used linear interpolation to solve problems involving compound interest.

An example of the method of solving systems of linear equations that we know as Gaussian elimination appears in a Chinese manuscript (the *Nine Chapters*) from the Han Dynasty (approximately 2,000 years ago); matrix notation was used. The famous German mathematician Carl Friedrich Gauss (1777–1855) indicated that the method was well known.

Chinese mathematics during the Sung dynasty (960–1279) generalized the method of successive approximations from the *Nine Chapters* to find numerical solutions of higher-degree equations. Matrix solution techniques for linear systems were also extended to equations of higher degree (an approach similar to work in the West in the 19th century).

Greek mathematics included methods of calculating areas based on approximating the desired quantity by a large number of regions of known area. (A similar process was used for volumes.) A letter from Archimedes to Eratosthenes (c. 250 B.C.) describes one of these methods; the letter was discovered in 1906.

The continuation of the Greek mathematical traditions by Middle Eastern scholars tended to stress computational and practical aspects. Omar Khayyam, who lived 900 years ago, wrote a treatise on algebra that includes a systematic investigation of cubic equations. (He is perhaps better known as the author of the *Rubaiyat of Omar Khayyam*.) Jemshid Al-Kashi (who died about 1436), another Persian mathematician, solved cubic equations by iterative and trigonometric methods and also knew the method for solving general algebraic equations, which is now usually called Horner's method. W. G. Horner published the method in 1819, presumably without being aware of its previous history.

Leonardo Fibonacci (c. 1200) showed that certain cubic equations cannot be solved in terms of square roots, but that very accurate approximate solutions can be generated.

The first tables of logarithms were published by the Scotsman John Napier in 1614; they were revised by Henry Briggs (based on Napier's suggestions) in 1624 and provided a great tool for improving computation.

The connection between mathematics and astronomy has been extremely close throughout history. Leaders of the Copernican revolution ventured into a number of areas of mathematics. For example, Johannes Kepler published the *New Solid Geometry of Wine Barrels* in 1615, in which he used geometric approximations to calculate the volume of a solid of revolution.

A widely known method for approximating the roots of an equation called Newton's method (or the Newton-Raphson method) is a generalization of an iterative approach to finding the roots of polynomials published in the early 1700s. According to recent research, Thomas Simpson (well known for Simpson's rule for the numerical approximation of definite integrals) extended Newton's method to more general functions and published the results in 1740.

Taylor's formula is the theoretical basis for many numerical techniques. Taylor polynomials were introduced in an article by Brook Taylor published in 1715; the remainder term first appeared in a book by Joseph Louis Lagrange in 1797.

One of the most popular approaches to finding numerical approximations to ordinary differential equations, the Runge-Kutta method, was developed 100 years ago by the German applied mathematicians Carl Runge (1856–1927) and M. W. Kutta (1867–1944). Runge is also known for his work on the Zeeman effect and Kutta for his contributions to the theory of airfoil lift in aerodynamics.

Our modern-day calculators follow the basic design introduced in Blaise Pascal's adding machine (1642) and Gottfried Wilhelm Leibniz's multiplication machine (1671). The origin of computers, on the other hand, is usually traced to Charles Babbage's analytical engine (developed in the 1830s). However, essentially the entire development of the personal computer (PC or Macintosh) dates from about 1980.

Although the classical numerical methods mentioned above continue to be important in the development of computer software, many of the preferred methods of modern computing include relatively recent innovations and refinements. Some of these, with the date of publication of the primary reference for the method, are summarized here. They are discussed in more detail in the appropriate section of the text. For finding the root of a single equation, Brent's method (1973) and Ridders' method (1979) are the methods of choice in software packages such as Mathcad. For constrained linear optimization problems, the simplex method (1948) is the standard method. For general nonlinear least-squares problems, the Levenberg-Marquardt method (1963) provides an elegant algorithm for combining two more classical approaches. Other important methods for these problems are the conjugate gradient methods, such as the Polak-Ribiere method (1971), or the closely related Fletcher-Reeves method. The routines for finding eigenvalues and eigenvectors that are widely used in software libraries and packages are generally based on routines published by Wilkinson and Reinsch (1971). For function interpolation, the work by DeBoor on cubic splines (1978) is the standard. For rational function interpolation, the Bulirsch-Stoer algorithm (1980) is important. The existence of fast methods for computing discrete Fourier transforms (DFT) became widely known in the 1960s through the work of Cooley and Tukey; however, Danielson and Lancoz had also developed such methods (early 1940s). In fact, Gauss (1805) led the list of perhaps a dozen people who discovered efficient methods for the DFT independently over the years. There have also been numerous improvements in recent years in the numerical methods for finding derivatives, integrals, and solutions to ordinary and partial differential equations.

Many of the issues that confront scientists or engineers who use numerical methods are the same today as throughout the history of the subject, although the relative importance of the competing considerations may change, depending on the computational resources available. Two primary considerations are the computational effort required and the accuracy of the resulting solution. Numerical methods for solving a problem may be classified as either direct or iterative. A direct method, such as Gaussian elimination, produces an answer to a problem in a fixed number of computational steps. An iterative method produces a sequence of approximate answers (designed to converge ever closer to the true solution, under the proper conditions).

For a direct method that would give an exact result if the computations were carried out in exact arithmetic, such as Gaussian elimination, the effect of numerical round-off may be significant. Also, since the linear systems that occur in modern applications may be extremely large, efficiency of computation is a critical aspect of choosing a solution technique for these types of problems. Other direct methods, such as techniques for numerical integration, are developed by replacing the given function by an approximating function (such as a Taylor polynomial) for which the integral can be found. The accuracy of the method depends in part on the number of terms that are retained before the Taylor series is truncated.

For iterative techniques, it is imperative that questions of convergence be understood. Do the successive approximate answers actually approach the true answer? If so, how quickly? How should the decision be made to terminate the process?

In the next section, we present three problems that illustrate basic types of numerical methods and significant issues, such as convergence and computational effort, which form the recurring themes of subsequent chapters. We conclude this chapter with an introduction to the capabilities of Mathcad, the programming environment used throughout the text.

In the remainder of the text, numerical methods are grouped according to the types of problems for which they are intended. Techniques for solving nonlinear equations of a single variable and systems of linear or nonlinear equations are presented in Chapters 2 to 5. Some basic methods from numerical linear algebra are given in Chapters 6 and 7. Functional approximation, including interpolation, least squares approximation, and Fourier methods, are discussed in Chapters 8 to 10. Numerical approximation of differentiation and integration, the basic operations of calculus, are the subject of Chapter 11. Numerical solutions of ordinary and partial differential equations are considered in Chapters 12 to 15.

1.1 SAMPLE PROBLEMS AND NUMERICAL METHODS

To illustrate the types of problems for which a numerical solution may be desired, we consider three examples. The simple problems presented in these examples can be solved exactly by techniques that are well known in algebra or calculus. However, there are closely related problems for which no exact solution can be found. There are many numerical methods for solving problems such as these examples. We introduce one method for each problem presented, to illustrate some of the basic issues and themes that recur throughout numerical analysis.

1.1.1 Roots of Nonlinear Equations

Although the zeros of a quadratic function such as $f(x) = x^2 - 3$ can be found exactly by the quadratic formula, no such exact methods exist for most nonlinear functions. The formula for finding the zeros of a cubic function is much more complicated, and Niels Henrik Abel (1802–1829) proved that no formula exists for fifth-order polynomials. (A translation of Abel's paper appears in Smith [1959].)

There are many methods for finding approximate zeros of nonlinear functions. The simplest, the bisection method, is a systematic searching technique; the secant, false-position, and Newton's methods use a straight-line approximation to the function whose zero is sought. More powerful methods use a quadratic approximation to the function or a combination of these techniques. Each approach produces a succession of approximations. One consideration in choosing such a technique is whether, and how rapidly, these approximations approach the desired solution. The computational effort required for each iteration may also be important.

Finding the zeros of $f(x) = x^2 - 3$ is equivalent to the problem of finding the square root of 3. A simple iterative method for finding square roots is illustrated in Section 1.2. The graph of $y = x^2 - 3$ is shown in Fig. 1.1.

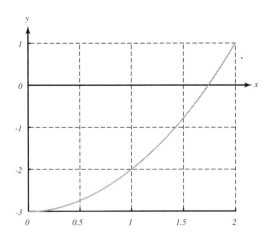

FIGURE 1.1 $y = x^2 - 3$ on the interval [0, 2].

1.1.2 Fixed-point Iteration

To find the square root of a positive number c, it is convenient to rewrite the equation $x^2 = c$ as the implicit equation

$$x = \frac{1}{2}\left(x + \frac{c}{x}\right);$$

this form provides the basis for an iterative solution technique by using the right-hand side of the equation to generate an updated estimate for the desired value of x.

A solution of an implicit equation of the form $x = g(x)$ is called a *fixed point*. In more detail, starting with an initial guess x_0, we evaluate

$$x_1 = \frac{1}{2}\left(x_0 + \frac{c}{x_0}\right); \quad x_2 = \frac{1}{2}\left(x_1 + \frac{c}{x_1}\right); \quad \ldots \quad x_k = \frac{1}{2}\left(x_{k-1} + \frac{c}{x_{k-1}}\right).$$

Geometrically this corresponds to finding the intersection of the line $y = x$ and the curve $y = \frac{1}{2}\left(x + \frac{c}{x}\right) = g(x)$. The method is known as fixed-point iteration, since we are seeking a value of x for which $x_k = x_{k+1} = g(x_k)$.

Example 1.1 Fixed Point Iterations to Find $\sqrt{3}$

The first two iterations in the procedure for finding a root of $x^2 = 3$ give

$$x_1 = \frac{1}{2}\left(1 + \frac{3}{1}\right) = 2; \quad x_2 = \frac{1}{2}\left(2 + \frac{3}{2}\right) = \frac{7}{4}.$$

The process of generating x_1 from $x_0 = 1$ is illustrated in Fig. 1.2.

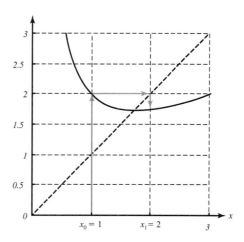

FIGURE 1.2 First step of fixed point iteration.

1.1.3 Linear Systems

If a linear system of equations has a unique solution, then a (nonzero) linear combination of two of the equations produces another linear equation that also passes through the same solution point. The well-known Gaussian elimination method systematically transforms the original system into an equivalent system (with the same solution) for which the solution point can be more easily identified. The process is illustrated in Example 1.2 for two equations in two unknowns. The graphs of the following two linear equations are illustrated in Fig. 1.3:

L_1: $\qquad 4x_1 + x_2 = 6,$

M_1: $\qquad -x_1 + 5x_2 = 9.$

Linear systems can be written more compactly in matrix-vector form. Thus, the foregoing system is written as $\mathbf{A}\,\mathbf{x} = \mathbf{b}$, where

$$\mathbf{A} = \begin{bmatrix} 4 & 1 \\ -1 & 5 \end{bmatrix}, \quad \mathbf{x} = \begin{bmatrix} x_1 \\ x_2 \end{bmatrix}, \quad \mathbf{b} = \begin{bmatrix} 6 \\ 9 \end{bmatrix}.$$

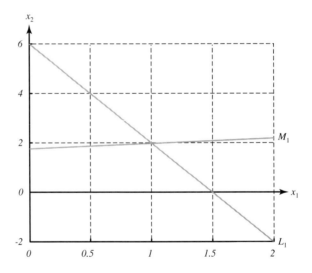

FIGURE 1.3 Graphs of the equations $4x_1 + x_2 = 6$ and $-x_1 + 5x_2 = 9$ on $[0, 2]$.

The linear systems for which numerical methods are required are frequently very large. In many important applications, the coefficient matrix has a particular structure that allows specialized solution techniques, which reduce computation and memory requirements.

1.1.4 Gaussian Elimination

Basic Gaussian elimination systematically transforms a system of linear equations into an equivalent system for which the solution is easier to find.

> ### Example 1.2 Solving a Linear System
>
> To illustrate this process, consider the following simple system:
>
> L_1: $\qquad 4x + y = 6,$
>
> M_1: $\qquad -x + 5y = 9.$
>
> Using basic Gaussian elimination, we multiply the first equation by 0.25 and add the result to the second equation to give the new (equivalent) system
>
> L_1: $\qquad 4x + y = 6,$
>
> M_2: $\qquad + 5.25y = 10.5.$
>
> Solving the second equation gives $y = 2$; substituting that value for y into the first equation yields $x = 1$.
>
> The original two equations are shown as lines L_1 and M_1 in Fig. 1.4. The modified second equation is shown as M_2. The solution for y is found from M_2; substituting into the first equation gives $x = 1$.
>
>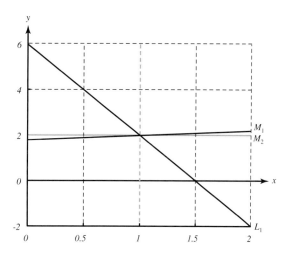
>
> **FIGURE 1.4** Geometric representation of Gaussian elimination for two equations.

If the computations in Gaussian elimination could be carried out exactly, then the main issue would be computational efficiency. However, not all numbers are represented in exact form in computer calculations; the extent of the difficulties this causes depends on certain characteristics of the coefficient matrix of the linear system.

1.1.5 Numerical Integration

The fundamental theorem of calculus states that the definite integral of a function may be found from the antiderivative of the function. However, for many functions, it is much easier to show that they have a definite integral than it is to find an expression for the antiderivative in terms of elementary functions. Several numerical techniques for finding definite integrals are based on approximating the function to be integrated by a simpler function whose antiderivative can be found exactly. The accuracy of the approximate integral depends on the form of the approximating function and the number of function evaluations. The basic numerical methods for integration may be improved by subdividing the interval of integration and using the fact that the integral over the entire interval is the sum of the integrals over the subintervals.

1.1.6 Trapezoid Rule

The basic trapezoid rule for integration is based on approximating the function to be integrated by a straight line connecting the points $(a, f(a))$ and $(b, f(b))$. We denote the length of the interval of integration as $h = b - a$. The formula for the basic trapezoid rule is

$$\int_a^b f(x)dx \approx \frac{h}{2}[f(a) + f(b)].$$

In order to obtain a useful formula for approximate integration based on trapezoidal regions, it is necessary to subdivide the interval of integration into relatively small subintervals and apply the basic trapezoid rule to each subinterval. We take n subintervals, each of the same length, so that each subinterval is of length $h = (b - a)/n$. For $n = 2$, the first subinterval is $[a, m]$ and the second subinterval is $[m, b]$ where $m = (a + b)/2$.

To apply the trapezoid rule to approximate a definite integral using two subintervals, write

$$\int_a^b f(x)\,dx = \int_a^m f(x)\,dx + \int_m^b f(x)\,dx$$

and apply the basic trapezoid to each of the integrals on the right-hand side.

$$\int_a^b f(x)\,dx \approx \frac{h}{2}[f(a) + f(m)] + \frac{h}{2}[f(m) + f(b)] = \frac{h}{2}[f(a) + 2f(m) + f(b)].$$

Note that h is the length of each subinterval, namely $(b - a)/2$.

In general, the subintervals are $[a, x_1], [x_1, x_2], ...[x_{n-1}, b]$, where $x_1 = a + h$, $x_2 = a + 2h$, etc. It is often convenient to set $x_0 = a$ and $x_n = b$. The trapezoid rule is

$$\int_a^b f(x)\,dx \approx \frac{h}{2}[f(x_0) + 2f(x_1) + + 2f(x_2) + ... + 2f(x_{n-1}) + f(x_n)].$$

Example 1.3 Approximating an Integral

Using the basic trapezoid rule approximation to the definite integral

$$I = \int_1^3 \frac{1}{x^3} dx$$

gives

$$I \approx \frac{h}{2}[f(a) + f(b)] = 1 + 1/27 = 28/27 \approx 1.037037$$

Taking two subintervals, we have $m = x_1 = 2$, $f(2) = 1/8$, $h = 1$, and

$$I \approx \frac{h}{2}[f(a) + 2f(m) + f(b)] = \frac{1}{2}(1 + 2(1/8) + 1/27) \approx 0.6435185.$$

With four subintervals, we have $x_1 = 3/2$, $x_2 = 2$, $x_3 = 5/2$, and $h = 1/2$,

$$I \approx \frac{h}{2}[f(a) + 2f(x_1) + 2f(x_2) + 2f(x_3) + f(b)]$$

$$= \frac{1}{4}(1 + 2(8/27) + 2(1/8) + 2(8/125) + 1/27) \approx 0.5019074$$

The exact value of this integral is $4/9 \approx 0.4444$

The graph of the function $f(x) = \frac{1}{x^3}$, and the straight-line approximations used in the trapezoid rule and the trapezoid rule with $n = 2$, are shown in Fig. 1.5. As the graphs suggest, the value of the integral given by the basic trapezoid rule is much larger than the true value of the integral, since the area under the straight line is larger than the area under the curve.

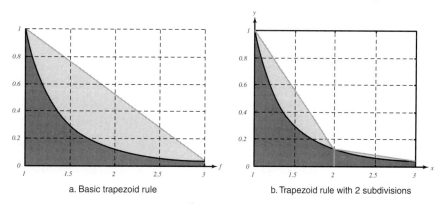

a. Basic trapezoid rule b. Trapezoid rule with 2 subdivisions

FIGURE 1.5 The areas given by $\int_1^3 \frac{1}{x^3} dx$ and by the trapezoid rule approximations.

1.2 SOME BASIC ISSUES

We now introduce some of the recurring themes in the analysis of numerical methods; these ideas are revisited in various settings in the remainder of the text. First we consider the two primary issues for iterative methods: "Does the process converge?" and "When do we stop?" We then discuss some issues related to the question of how good the result of a numerical method is, and how the result can be improved.

1.2.1 Key Issues for Iterative Methods

For iterative techniques, it is imperative to know whether the method converges, i.e., whether the sequence of approximate results approaches the true solution. If the method does converge, we must decide when to terminate the process. For an important class of methods, the convergence depends on the eigenvalues of the iteration matrix; a useful theorem for bounds on the eigenvalues is given below.

Convergence of Iterative Methods

For some iterative techniques, the convergence, or divergence of the method can be illustrated geometrically. The sequence of points generated by the fixed-point formula $x_k = g(x_{k-1})$ is shown in the next example.

Example 1.4 Fixed Point Iteration

Fig. 1.6a shows the first two steps of the convergent fixed-point iteration for $x = \cos(x)$, starting with $x_0 = 0.5$. Fig. 1.6b shows the first two steps of the divergent fixed-point iteration for $x = g(x) = 1 - x^3$ with $x_0 = 0.5$

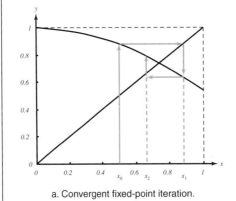
a. Convergent fixed-point iteration.

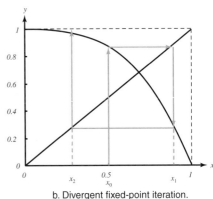
b. Divergent fixed-point iteration.

FIGURE 1.6 Fixed-point iterations.

Chapter 1 Foundations

It is very useful to be able to analyze algebraically whether a fixed-point formula will converge, and if so, to estimate how rapidly. The following theorem states conditions which guarantee that
  (i) the equation $x = g(x)$ has a fixed point in the interval $I = [a, b]$, and
  (ii) the iterative procedure $x_k = g(x_{k-1})$ will converge to that fixed point.

Fixed-point Convergence Theorem

If

1. $g(x)$ maps $[a, b]$ into $[a, b]$,
2. $g'(x)$ is continuous on $[a, b]$, and
3. there is a number $N < 1$ such that $|g'(x)| \leq N$ for all x in $[a, b]$;

then

1. $x = g(x)$ has exactly one solution (call it x^*) in the interval $[a, b]$, and
2. the fixed-point iteration $x_k = g(x_{k-1})$ converges to x^*, for any starting estimate in $[a, b]$.

Furthermore, the value of N gives an estimate of the error at any stage of the iteration, in that the error $e_k = x_k - x^*$ satisfies the inequalities

$$|e_{k+1}| \leq N|e_k| \quad \text{and} \quad |e_{k+1}| \leq N^{k+1}|e_0|.$$

It is easy to construct examples of functions $g(x)$ that do not map a given interval $I = [a, b]$ into I and for which the curves $y = x$ and $y = g(x)$ do not cross (at least for x in I). If g is continuous and does map I into I, then the curves will cross (at least once) in the interval. The guarantee of convergence of the iterative process hinges on the magnitude of $g'(x)$ being less than 1, at least near the fixed point.

To illustrate what this theorem says, consider the fixed-point iteration $x = \cos(x)$ illustrated in Example 1.4. For $0 \leq x \leq 1$, $\cos(x)$ is also between 0 and 1, so $g(x)$ does map $[0, 1]$ into $[0, 1]$. Furthermore, $g'(x) = -\sin(x)$ ranges between 0 and $-\sin(1)$, so $|g'(x)| \leq 0.85$ for all x in $[0, 1]$, and the theorem guarantees convergence of the iterations.

On the other hand, for $x = g(x) = 1 - x^3$, $g'(x) = -3x^2$, which is not bounded by a number less than 1 on $[0, 1]$. Although there is a fixed point in the interval (and $g(x)$ does map $[0, 1]$ into $[0, 1]$), the conditions of the theorem are not satisfied, and in fact, the iterations do not converge. The difficulty is that in any neighborhood of the fixed point, $|g'(x)| > 1$.

Estimating Eigenvalues

The convergence of iterative methods based on repeated multiplication by a matrix (call it **M**) often depends on the eigenvalues of **M**. A number λ is an eigenvalue of **M** (with a corresponding nonzero eigenvector **v**) if and only if $\mathbf{Mv} = \lambda\mathbf{v}$. Eigenvalues are extremely useful in many applications; methods for calculating them form the subject of Chapter 7. However, in some situations it is sufficient to be able to estimate the eigenvalues (especially if we can show that all of the eigenvalues are positive, for example). The following theorem gives bounds on the location of the eigenvalues of **M**; as in standard matrix notation, the element in row i, column j, is denoted m_{ij}.

Gerschgorin Circle Theorem

If, for $i = 1, \ldots, n$, C_i is the circle in the complex plane with center at $(m_{ii}, 0)$ and radius $r_i = \Sigma_{j \in I} |m_{ij}|$, where I indicates the index set $\{1, 2, \ldots, i-1, i+1, \ldots, n\}$, then all of the eigenvalues of **M** lie within the union of the disks bounded by these circles. Furthermore, if there are k disks, the union of which is disjoint from the other disks, then exactly k eigenvalues lie within that union.

Example 1.5 Bounds on Eigenvalues

To illustrate the Gerschgorin circle theorem, consider the problem of finding bounds on the eigenvalues of the matrix

$$M = \begin{bmatrix} 2 & -1/2 & 0 \\ -1/2 & 3 & 1/2 \\ 0 & 1/2 & 6 \end{bmatrix}.$$

We find that

C_1 has center $(2, 0)$ and radius $|-1/2| + |0| = 1/2$;
C_2 has center $(3, 0)$ and radius $|-1/2| + |1/2| = 1$;
C_3 has center $(6, 0)$ and radius $|0| + |1/2| = 1/2$.

The union of the interiors of circles C_1 and C_2 is disjoint from the disk bounded by circle C_3, so we know that there are exactly two eigenvalues in the union of the regions enclosed by circles C_1 and C_2. Furthermore, there is one eigenvalue in the region bounded by circle C_3. These regions are illustrated in Fig. 1.7.

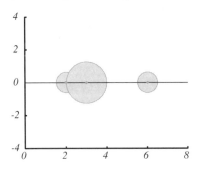

FIGURE 1.7   Gerschgorin disks.

Termination Conditions

A variety of conditions can be used for deciding when to stop an iterative procedure; however, there is no perfect test for a "stopping condition." The conditions may be characterized as being of three general types: the problem is "solved," the iteration has "converged," or the iteration process has continued "long enough." For the example of finding the zero of $f(x) = x^2 - 3$, with the true zero denoted x^*, possible convergence tests include the following:

The problem is "solved":

$$|f(x_k)| \leq f_{\text{tol}} \quad \text{(function value reduced to specified tolerance)}.$$

The iteration has "converged":

$|x_{k+1} - x_k| \leq \text{tol}$ (absolute change is within specified tolerance);
if $\text{tol} = 10^{-n}$, then x_{k+1} should approximate x^* to n decimal places.

$|x_{k+1} - x_k| \leq \text{tol} |x_{k+1}|$ (relative change is within specified tolerance);
if $\text{tol} = 10^{-n}$, then x_{k+1} should approximate x^* to n significant digits.

The iterations have gone on "long enough":

$$k \geq \text{max\_it} \quad \text{(iteration counter exceeds a specified limit)}.$$

It may also be desirable to check whether "the solution is looking bad":

$|f(x_k)| \geq f_{\text{big}}$ (function value exceeds a specified limit);
$|x_k| \geq x_{\text{big}}$ (value of iterated variable exceeds a specified limit).

It is important to realize that none of these tests guarantees the desired result, namely that $|x_k - x^*| < \text{tol}$. In addition, an iterative process could pass the successive iterates test at the same time that the iterates were diverging to ∞. As an example, consider a process in which $x_k = 1 + \frac{1}{2} + \frac{1}{3} + \cdots + \frac{1}{k}$. The difference between successive iterates, $|x_{k+1} - x_k| = \frac{1}{k+1} \to 0$ as $k \to \infty$, but $x_{k+1} \to \infty$ as $k \to \infty$.

The relative-change test is appropriate for problems in which the desired roots may be of greatly differing magnitudes. It is not suitable, however, if $x = 0$ is a desired root and the method is converging rapidly, since that would produce a relatively large change in the iterates.

1.2.2 How Good Is the Result?

There are several reasons that the results of a numerical solution to a problem from the "real world" may not be the exact answer. Simplifying assumptions made in modeling the original problem are one source of inaccuracies. Errors arising from data collection are another.

In this section, we illustrate several basic types of errors that are more directly linked to the numerical solution of the stated problem. We first define some standard terminology for describing errors that occur in numerical methods. We then summarize the fundamentals of computer arithmetic and illustrate two types of error that arise because computers do not represent most numbers in an exact form; i.e., they do not do exact arithmetic. The third example in this section illustrates one common form of error introduced in replacing a continuous process by a discrete approximation.

Measuring Error

If x is our approximate result, and the exact (but usually unknown) result is denoted x^*, then the error in using the approximate result is

$$\text{Error}(x) = x^* - x$$

However, especially for problems in which the magnitude of the true value may be very large, or very small, the relative error may be more important that the actual error.

$$\text{Rel Error}(x) = \frac{x^* - x}{x^*}$$

In computing the relative error, the approximate value is often used in the denominator in place of the unknown true value x^*.

Significant Digits

The number x is said to approximate x^* to t significant digits if t is the largest nonnegative integer for which $\left|\frac{x^* - x}{x}\right| < 5 \cdot 10^{-t}$.

Big Oh Approximation

For errors that come from using a finite step size, h, in approximating a continuous process by a discrete one, it is often useful to describe how the error depends on h, as h approaches zero. We say that a function $f(h)$ is "Big Oh" of h if $|f(h)| \leq c|h|$ for some constant c, when h is near 0. This is written $f(h) = O(h)$. Similarly, $f(h) = O(h^2)$ means that $|f(h)| \leq c|h^2|$ for some constant c, when h is near 0. If a method has an error term that is $O(h^k)$, the method is often called a kth order method.

For example, if we use a Taylor polynomial to approximate the function f at $x = a + h$, we have

$$f(x) = f(a + h) = f(a) + h f'(a) + \frac{h^2}{2!} f''(a) + \frac{h^3}{3!} f'''(\eta);$$

for some $\eta \in [a, a + h]$.

Assuming that f is sufficiently smooth, we can let M be the maximum of $f'''(x)$ for $a \le x \le a + h$. Then this approximation is $O(h^3)$, since the error, $\frac{h^3}{3!} f'''(\eta)$ satisfies

$$\left| \frac{h^3}{3!} f'''(\eta) \right| \le c |h^3|$$

where $c = \frac{1}{3!} M$.

Computer Representation of Real Numbers

Computers represent real numbers in a form, called *floating point*, that is similar to scientific notation. For example, a number N is stored as

$$N = \pm .d_1 d_2 d_3 \cdots d_p B^e,$$

where B is the base and the d_i's are the digits. For a computer, the base is usually 2, 8, or 16; in scientific notation, the base is 10. Each digit is an integer between 0 and $B - 1$. There are a fixed number of digits, p; the integer exponent, e, is restricted to a range of values; i.e., $e \in [\text{emin}, \text{emax}]$. If, as is generally the case, it is required that $d_1 \ne 0$, the system is called a *normalized floating-point system*. Note that for a binary system, this means that $d_1 = 1$, so there is in fact no need to store its value.

In a binary floating-point system, the decimals correspond to sums of negative powers of 2. To illustrate these ideas, consider the numbers that can be expressed exactly in this form, for very small values of p and e, for base 2. (The opposite of each number listed can also be represented.)

It is important to note that the numbers that can be represented exactly are not evenly distributed between the largest and smallest such numbers. The numbers that can be represented exactly, using base 2 with two digits and exponents of $-1, 0$, and 1 are illustrated in Fig. 1.8.

FIGURE 1.8   Exact numbers in base 2.

Base 2 numbers expressible using three digits (leading digit must be 1)

Exponent	Binary decimal	Expansion	Decimal
0	0.100_2	$(1)\frac{1}{2} + (0)\frac{1}{4} + (0)\frac{1}{8}$	0.5
	0.101_2	$(1)\frac{1}{2} + (0)\frac{1}{4} + (1)\frac{1}{8}$	0.625
	0.110_2	$(1)\frac{1}{2} + (1)\frac{1}{4} + (0)\frac{1}{8}$	0.75
	0.111_2	$(1)\frac{1}{2} + (1)\frac{1}{4} + (1)\frac{1}{8}$	0.875
1	1.00_2	$(1)1 + (0)\frac{1}{2} + (0)\frac{1}{4}$	1.0
	1.01_2	$(1)1 + (0)\frac{1}{2} + (1)\frac{1}{4}$	1.25
	1.10_2	$(1)1 + (1)\frac{1}{2} + (0)\frac{1}{4}$	1.5
	1.11_2	$(1)1 + (1)\frac{1}{2} + (1)\frac{1}{4}$	1.75
−1	0.0100_2	$(1)\frac{1}{4} + (0)\frac{1}{8} + (0)\frac{1}{16}$	0.25
	0.0101_2	$(1)\frac{1}{4} + (0)\frac{1}{8} + (1)\frac{1}{16}$	0.3125
	0.0110_2	$(1)\frac{1}{4} + (1)\frac{1}{8} + (0)\frac{1}{16}$	0.375
	0.0111_2	$(1)\frac{1}{4} + (1)\frac{1}{8} + (1)\frac{1}{16}$	0.4375

Precision

The precision with which numbers can be stored, and computations carried out, depends on the number of digits and the range of exponents used to represent a real number. In *single precision,* a real variable is stored in four words, or 32 bits. A *bit* is a binary digit (0 or 1); a *byte* is 4 bits (so a byte can have $2^4 = 16$ possible values); a word is 2 bytes (8 bits). Of the 32 bits, 23 are used for the digits, 8 for the exponent, and 1 for the sign. The 8 bits for the exponent can take on 256 possible values, from $0 = 00000000_2$ to $255 = 11111111_2$. In *double precision,* each floating-point number occupies eight words (64 bits, 11 for the exponent, 52 for the digits, and 1 for the sign.)

How Many?

In general, the total number of values, V, that can be represented (assuming that $d_1 \neq 0$) is given by

$$V = 2(B-1)(B^{p-1}) \text{ (total number of exponents)} + 1.$$

The factor of 2 corresponds to the sign bit; the factor of $(B-1)$ gives the number of possible values for the first digit. Each of the digits $d_2 \ldots d_p$ can take on B different values. The bit required to store zero accounts for the one additional value not counted in the product.

For a base-10 system, with two digits and exponents of 0 or 1, we have

$$V = 2(10-1)(10^{2-1})(2) + 1 = 2(9)(10)(2) + 1 = 361.$$

The maximum value is $.99 \times 10 = 9.9$ and the minimum value is -9.9. The positive numbers that can be represented with the exponent 1 are of the form 1.0, 1.1, 1.2, ... 2.0, 2.1, 2.3, ... 9.9; the positive numbers represented with the exponent 0 are 0.10, 0.11, 0.12, ..., 0.20, 0.21, 0.23, ..., 0.99. Thus, there are 91 values in the interval [0, 0.99], but only 10 values in the interval [1.0, 1.9].

In a binary floating-point system, with two digits and exponents of 0 or 1, we have

$$V = 2(2-1)(2^{2-1})(2) + 1 = 2(2-1)(2^1)(2) + 1 = 9.$$

The positive numbers (with their decimal equivalents in parentheses) that can be represented with exponent 0 are 0.10_2 (= 0.5) and 0.11_2 (= 0.75); those with exponent 1 are 1.0_2 (=1.0) and 1.1_2 (=1.5). Thus, the largest value is 1.5 and the minimum value is -1.5.

How Large and How Small?

The range of exponents that are available determines the smallest and largest numbers (in magnitude) that can be represented. For single precision, the binary numbers in the interval $[00000000_2, 11111111_2]$ are mapped to the interval $[-128, 127]$, so the smallest number is approximately $0.14693 \cdot 10^{-38}$; the largest number is approximately $0.9414 \cdot 10^{127}$. Numbers smaller than 10^{-38} cause *underflow* (which is often set to be zero). Numbers larger than 10^{127} (or 10^{1023} for double precision) cause *overflow* (which usually halts the program).

Errors from Inexact Representation

There are two approaches to shortening a number that has more digits than can be represented by the available floating-point system. The simplest is to *chop* the number, by discarding any digits beyond what the system can accommodate. The second method is to *round* the number; the result depends on the value of the first digit to be discarded. If the system allows for n digits, rounding produces the same result as chopping if the $(n + 1)$st digit is 0, 1, 2, 3, or 4. If the $(n + 1)$st digit is 6, 7, 8, or 9, the nth digit is increased by 1. If the $(n + 1)$st digit is 5, it is common to round so that the nth digit is even, rounding up about half of the time. The errors that occur from rounding are less likely to accumulate during repeated calculations, since the true value is larger than the rounded value about half of the time and smaller about half of the time. Furthermore, the largest absolute error that can occur is twice as large for chopping as for rounding. On the other hand, chopping requires no decisions as to whether to change the last retained digit. The inaccuracies that result from either rounding or chopping are known as round-off errors. We now consider some examples of the difficulties that can occur because of inexact computations.

Example 1.6 Effect of Order of Operations

As an example of the effect of round-off, consider the following addition problem:

$$0.99 + 0.0044 + 0.0042.$$

With exact arithmetic, the result is 0.9986, regardless of the order in which the additions are performed. However, if we have three-digit arithmetic, and the operations are nested from left to right, we find that

$$(0.99 + 0.0044) + 0.0042 = 0.994 + 0.0042 = 0.998.$$

On the other hand, if we change the nesting so that the small numbers are added together first, we have

$$0.99 + (0.0044 + 0.0042) = 0.99 + 0.0086 = 0.999.$$

Using the definition of x approximating x^* to t significant digits, we see that 0.998 approximates the true solution, $x^* = 0.9986$ to three significant digits, since

$$\left| \frac{0.9986 - 0.998}{0.998} \right| = 6.012 \cdot 10^{-4} < 5 \cdot 10^{-3}.$$

On the other hand, 0.999 approximates the true solution, $x^* = 0.9986$ to four significant digits, since $\left| \frac{0.9986 - 0.999}{0.999} \right| = 4.004 \cdot 10^{-4} < 5 \cdot 10^{-4}.$

For cases with greater difference in the sizes of the numbers, and with more terms, the loss of significance can be extreme.

Cancellation Error

A second example of the effect of inexact calculations occurs when a computation involves the subtraction of two nearly equal numbers. It is advisable to rewrite the formula to avoid the difficulty if possible. Consider the problem of using the quadratic formula to solve the quadratic equation

$$x^2 - bx + 1 = 0$$

The effect of rounding the discriminant $r = \sqrt{b^2 - 4}$ is illustrated in this example; note that for large b ($b \gg 4$), r is quite close to b.

The quadratic formula gives $x_1 = \dfrac{b + r}{2}$ and $x_2 = \dfrac{b - r}{2}$.

If b is positive, x_2 will involve the difference of two numbers that are very close to each other, a dangerous situation. This difficulty can be avoided by rationalizing the numerator in the quadratic formula,

$$x_2 = \frac{(b - r)}{2} \frac{(b + r)}{(b + r)} = \frac{(b^2 - r^2)}{2(b + r)} = \frac{4}{2(b + r)} = \frac{2}{(b + r)}.$$

If the same process is applied to the formula for x_1 (with which the standard quadratic formula does not have a problem), the rationalized numerator formula results in division by a quantity that is close to zero, which is a much worse situation.

Example 1.7 Cancellation Errors

To illustrate the effect of rounding, consider the problem of solving the quadratic equation $x^2 - 97x + 1 = 0$. The exact roots (shown to nine digits) and the approximations computed with rounding to five digits are summarized in Table 1.1.

Table 1.1 Effect of Rounding for Roots of Quadratic Equation

	x_1	x_2
Exact	96.9896896	0.0103103743
Standard quadratic formula rounded	96.990	0.01050
Rationalized quadratic formula rounded	95.238	0.01031

For the standard quadratic formula, rounding has a much larger effect on x_2 than on x_1, as expected. Using the rationalized quadratic formula for x_2 gives the correct result to the number of digits used in the rounded computation. On the other hand, if the rationalized formula is used for x_1, for which it is not appropriate, the results for the rounded computations approximate the true solution only to two digits.

Errors from Mathematical Approximations

Many numerical methods are based on a Taylor series expansion or on a Taylor polynomial approximation with remainder:

$$f(a + h) = f(a) + h f'(a) + \frac{h^2}{2!} f''(c), \text{ for some } c \in [a, a + h] \quad (***)$$

In analyzing the error associated with using the trapezoid rule, a Taylor polynomial approximation is used for both the function being integrated and for the function defined as the indefinite integral.

Local Truncation error

To introduce the analysis of the error produced by approximating a function by a simpler function, consider again the basic trapezoid rule for numerical integration, which we write now as

$$\int_a^{a+h} f(x) \, dx \approx \frac{h}{2} [f(a) + f(a + h)].$$

Define the function $F(t) = \int_a^t f(x) \, dx$.

We can represent F by its Taylor polynomial with remainder, as follows

$$F(a + h) = F(a) + h F'(a) + \frac{h^2}{2!} F''(a) + \frac{h^3}{3!} F'''(c_1), \text{ for some } c_1 \in [a, a + h].$$

Since $F' = f$, $F'' = f'$, $F''' = f''$, ... and $F(a) = 0$, we have

$$\int_b^{a+h} f(x) \, dx = F(a + h) = h f(a) + \frac{h^2}{2!} f'(a) + \frac{h^3}{3!} f''(c_1) \quad (1.1)$$

On the other hand, the Taylor polynomial with remainder for f is given by (***) which gives after a little algebra, with the remainder evaluated at c_2

$$\frac{h}{2} [f(a + h) + f(a)] = h f(a) + \frac{h^2}{2} f'(a) + \frac{h^3}{4} f''(c_2) \quad (1.2)$$

The error in the trapezoid rule is the difference of Eqs. (1.1) and (1.2), or

$$\int_a^{a+h} f(x) \, dx - \frac{h}{2} [f(a + h) + f(a)] = \frac{h^3}{6} f''(c_1) - \frac{h^3}{4} f''(c_2)$$

We now show that if f is sufficiently smooth; i.e., f'' is continuous and bounded on $[a, a+h]$, we can combine these two remainder terms. If

$$m \leq f''(x) \leq M, \text{ for all } x \text{ on } [a, a + h]$$

Chapter 1 Foundations

then
$$\frac{h^3}{6}m \le \frac{h^3}{6}f''(c_1) \le \frac{h^3}{6}M,$$

and
$$\frac{h^3}{4}m \le \frac{h^3}{4}f''(c_2) \le \frac{h^3}{4}M,$$

so
$$\frac{-h^3}{12}m \le \frac{h^3}{6}f''(c_1) - \frac{h^3}{4}f''(c_2) \le \frac{-h^3}{12}M.$$

By the Intermediate Value Theorem (applied to f'') there is some point η in the interval $[a, a + h]$ such that

$$f''(\eta) = \frac{-12}{h^3}\left[\frac{h^3}{6}f''(c_1) - \frac{h^3}{4}f''(c_2)\right].$$

Thus we have,

$$\int_a^{a+h} f(x)\, dx - \frac{h}{2}[f(a + h) + f(a)] = \frac{-h^3}{12}f''(\eta),$$

for some $\eta \in [a, a + h]$.

This is the local truncation error, which comes from truncating the Taylor series expansions, for one step of the trapezoid rule.

Global Truncation Error

To improve the results, it is useful to subdivide the interval into n equal subintervals $[a, x_1], [x_1, x_2], \ldots [x_{n-1}, b]$ and apply the method in each region. The length of each subinterval is $h = (b - a)/n$. This gives the more general (composite) trapezoid rule:

$$\int_a^b f(x)\, dx \approx \frac{h}{2}[f(a) + 2f(x_1) + \ldots + 2f(x_{n-1}) + f(b)] = T(h)$$

The global error is the result of adding the local error in each of the regions. If f is sufficiently smooth, the global error can represented as $\dfrac{-(b-a)h^2}{12}f''(\eta)$ for some $a \le \eta \le b$. Thus, the global error for the trapezoid rule is proportional to h^2, and the method is $O(h^2)$. This means that if the step size is cut in half, the bound on the global truncation error is reduced by a factor of one-fourth.

Using the Truncation Error

Representation of the approximation error as a term that depends on a derivative of the function, evaluated at some (unknown) point in the interval, and a power of the step size, is useful in several ways. It is sometimes not too difficult to find bounds on the derivative, so that the error term gives bounds on how large, or how small the error can be on the interval. Even without finding bounds on the derivative, the power of the step size that occurs in the error term (that is, the order of the method) gives some indication as to how different methods may be expected to compare. A higher order method does not always give better results than a lower order method, but in general one can at least expect to obtain more improvement in the results by reducing the step size in a higher order method.

For some analysis it is useful to represent the truncation error as a series, by retaining all of the terms in the Taylor series expansion used in deriving the method. For some methods, including the trapezoid rule, the series can be expressed as a power series in h with coefficients that do not depend on the derivatives of the function. Especially if that power series depends only on even powers of h, as is the case for the trapezoid rule, and several other important numerical methods, there is a technique, know as acceleration (or extrapolation) by which two applications of the method may be combined to obtain an even more accurate result. The series form for the error in the trapezoid rule can be written as

$$\int_a^b f(x)\,dx - T(h) = a_2 h^2 + a_4 h^4 + a_6 h^6 + \ldots$$

1.2.3 Getting Better Results

Two of the important issues in judging an numerical method are the accuracy of the results and the amount of computational effort required to achieve them. For many methods, the accuracy of the results can be improved by reducing the step size; e.g., taking more subdivisions of the interval of integration for the trapezoid method. However, the increased computational effort may be an unacceptably high price to pay for the improvement. Carried to an extreme, this approach also leads to round-off errors. We begin this section with a method, acceleration, which can be used to improve the results of certain basic approximation formulas (one of which is the Trapezoid Rule); it is considered further in Chapter 11. We also illustrate the effect on the results when an algorithm (such as Gaussian elimination) is applied carefully, or in an unwise manner. Finally, we introduce some of the ideas of computationally efficiency.

Acceleration

A technique known as acceleration provides a method of improving the accuracy of an approximation formula $A(h)$ whose error can be expressed as

$$A - A(h) = a_2 h^2 + a_4 h^4 + a_6 h^6 + \ldots,$$

where A is the true (unknown) value of the quantity being approximated by $A(h)$ and the coefficients of the error terms do not depend on the step size h. The trapezoid rule for integration is one such formula.

To apply acceleration, we form approximations to A using steps h and $h/2$. Let A_1 be the approximation using (the larger) step $h_1 = h$, and let A_2 be the approximation using step $h_2 = h/2$. Thus, we have

$$A - A_1 = a_2 h^2 + a_4 h^4 + a_6 h^6 + \ldots$$

and

$$A - A_2 = a_2 (h/2)^2 + a_4 (h/2)^4 + a_6 (h/2)^6 + \ldots$$
$$= \frac{1}{4} a_2 h^2 + \frac{1}{16} a_4 h^4 + \frac{1}{64} a_6 h^6 + \ldots$$

After some simple algebra, these equations can be written as

$$A = A_1 + a_2 h^2 + a_4 h^4 + a_6 h^6 + \ldots$$

and

$$4A = 4A_2 + a_2 h^2 + \frac{1}{4} a_4 h^4 + \frac{1}{16} a_6 h^6 + \ldots$$

Subtracting the first equation from the second gives

$$3A = 4A_2 - A_1 + \frac{-3}{4} a_4 h^4 + \text{higher order terms}$$

Solving for A yields

$$A = \frac{4A_2 - A_1}{3} - \frac{1}{4} a_4 h^4 + \text{higher order terms}$$

Thus, the linear combination

$$A \approx \frac{4A_2 - A_1}{3}$$

of two applications of the $O(h^2)$ approximation, using step sizes h and $h/2$, gives an $O(h^4)$ approximation to A.

Acceleration may give very good results with only a simple application, but in fact the process can be repeated for even more improvement. To see how the second stage works, call the approximation generated by the first stage of acceleration, with step sizes h and $h/2$, B_1. Furthermore, we apply the original formula with $h_3 = h/4$, and call the result, A_3. We call the result of combining A_2 and A_3 using first-stage acceleration B_2. Thus

$$B_1 = \frac{4A_2 - A_1}{3} \quad \text{and} \quad B_2 = \frac{4A_3 - A_2}{3}$$

The error in using B_1 is a power series in even powers of h, starting with the h^4 term. We do not need the explicit form of the coefficients, so we write

$$A = B_1 + b_4 h^4 + b_6 h^6 + \ldots$$

and

$$A = B_2 + b_4 (h/2)^4 + b_6 (h/2)^6 + \ldots$$
$$= B_2 + \frac{1}{16} b_4 h^4 + \frac{1}{64} b_6 h^6 + \ldots$$

since B_2 is produced by the same formula as B_1 but with the step sizes cut in half. Some simple algebra yields

$$A = \frac{16 B_2 - B_1}{15} + c_6 h^6 + \text{higher order terms}$$

Thus, three applications of the original formula, with error $O(h^2)$, together with three linear combinations of those results has produced an approximation with error $O(h^6)$.

Example 1.8 Improving an Integral by Acceleration

To illustrate the use of acceleration for the trapezoid rule (generally known as *Richardson extrapolation*), we use the results obtained in Example 1.3. It is often convenient to show the results of acceleration as additional columns in a table. For the first stage of acceleration, we obtain

step	approx. integral	1st-stage acceleration
$h = 2$	$A_1 = 1.037037$	
		$B_1 = 0.51233$
$h = 1$	$A_2 = 0.6435185$	
		$B_2 = 0.457037$
$h = 1/2$	$A_3 = 0.5019074$	

Continuing to the second stage, we extend the table

step	approx. integral	1st-stage acceleration	2nd-stage acceleration
$h = 2$	$A_1 = 1.037037$		
		$B_1 = 0.51233$	
$h = 1$	$A_2 = 0.6435185$		$C_1 = 0.4508619$
		$B_2 = 0.457037$	
$h = 1/2$	$A_3 = 0.5019074$		

Efficient Computations

Problems of interest for numerical methods often require many applications of certain computations that individually are not too time consuming. As computing capabilities have developed, the time required for basic operations, such as addition, subtraction, multiplication, and division, has been reduced dramatically. Not too many years ago, multiplication and division required significantly more computational effort than addition and subtraction; it was common then to analyze algorithms based on counting only multiplications and divisions. The differential between multiplication and addition is much less now, and effort is usually evaluated in terms of floating-point operations (flops). In this section, we consider the computational effort (flops) for polynomial evaluation.

The straightforward evaluation of each term in the polynomial

$$P(x) = a_n x^n + a_{n-1} x^{n-1} + \ldots + a_1 x + a_0$$

requires n multiplications for the highest (n^{th} degree) term and one less for each lower term. The total number of multiplications for the polynomial is

$$\sum_{k=1}^{n} k = \frac{n(n+1)}{2}.$$

There are n additions, giving the final count for the individual term evaluation as

$$\text{flops} = \frac{n(n+3)}{2}.$$

Of course, no one would evaluate each term independently if he or she were doing it by hand. Each power of x would utilize the value of the previous power; i.e., $x^k = x \cdot x^{k-1}$. This approach requires $n - 1$ multiplications by x, n multiplications by the coefficients, and n additions. The corresponding count of operations is flops $= 3n - 1$.

A more efficient algorithm, Horner's method for evaluating polynomials, is based on expressing $P(x)$ as $(x-c)Q(x) + P(c)$, where

$$P(x) = a_n x^n + a_{n-1} x^{n-1} + \ldots + a_1 x + a_0,$$

$$Q(x) = b_n x^{n-1} + b_{n-1} x^{n-2} + \ldots + b_2 x + b_1, \text{ and } b_0 = P(c).$$

Horner's Algorithm

    Define $b_n = a_n$.
    For $k = n - 1, \ldots 0$, compute
        $b_k = a_k + b_{k+1} c$.
    End   ($b_0 = P(c)$)

Each of the n stages requires one multiplication and one addition, giving flops $= 2n$. In some applications, the value of the derivative of P is also required. Since $P'(x) = (x - c)Q'(x) + Q(x)$, we have $P'(c) = Q(c)$, which can be found by applying Horner's algorithm to $Q(x)$.

Apply the Algorithm Carefully

The next example illustrates the fact that the way in which an algorithm is applied may affect the quality of the result.

Example 1.9 Careful Application of Algorithm Reduces Error

Consider again the system introduced in Example 1.2, but assume now that there is some error in the values on the right-hand side of the equation (either as a result of round-off in previous computations, or from inaccuracies in data measurement, or whatever):

L_1: $\quad\quad\quad\quad\quad\quad\quad\quad 4x + y = 6 \pm 0.4$ $\quad\quad\quad\quad\quad$ (1.3)

M_1: $\quad\quad\quad\quad\quad\quad\quad -x + 5y = 9 \pm 0.4.$ $\quad\quad\quad\quad$ (1.4)

After Gaussian elimination, the system becomes

L_1: $\quad\quad\quad\quad\quad\quad\quad\quad 4x + y = \quad 6 \pm 0.4,$ $\quad\quad\quad\quad$ (1.3)

M_2: $\quad\quad\quad\quad\quad\quad\quad\quad\quad\quad 5.25y = 10.5 \pm 0.5.$ $\quad\quad\quad$ (1.5)

Solving Eq. (1.5) for y gives $y = 2 \pm 0.0952...$; thus (to 4 decimal places) $y = 2.0952$ or 1.9048.

Backsubstitution gives the four points which determine the parallelogram containing the solution (as shown in Figure 1.9b):

$(1.0762, 2.0952), (0.8762, 2.0952), (1.1238, 1.9048), (0.9238, 1.9048)$

This region is very close to the region containing the actual solution, shown in Figure 1.9a. The true solution of this system is contained within the parallelogram determined by the points

$(1.0762, 2.0952), (0.8857, 2.057), (1.1143, 1.9420), (0.9238, 1.9048).$

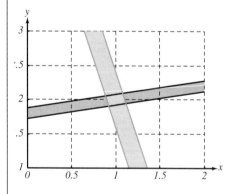
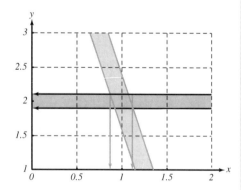

FIGURE 1.9A Graphical representation of a linear system with uncertainties.

FIGURE 1.9B Graph of Eqs. (1.3) and (1.5).

Chapter 1 Foundations

On the other hand, if the equations are listed in the opposite order, and the computations of Gaussian elimination are carried out without modification, the final result is much less accurate. The system is now

M_1: $\qquad -x + 5y = 9 \pm 0.4,$ (1.4)

L_1: $\qquad 4x + y = 6 \pm 0.4$ (1.3)

After Gaussian elimination, the system becomes

M_1: $\qquad -x + 5y = 9 \pm 0.4$ (1.4)

L_2: $\qquad 21y = 42 \pm 2$ (1.6)

Solving Eq. (1.6) for y gives $y = 2 \pm 0.0952$, so (to 4 decimal places) $y = 2.0952$ or 1.9048, as before. However, now backsubstitution results in the four points

$$(1.076, 2.0952), (0.124, 1.9048), (1.876, 2.0952), (0.9238, 1.9048).$$

As shown in Figure 1.10b, the region determined by these points is not very close to the region containing the actual solution, shown in Figure 1.10a. The true solution of this system is contained within the parallelogram determined by the points

$$(1.0762, 2.0952), (0.8857, 2.057), (1.1143, 1.9420), (0.9238, 1.9048).$$

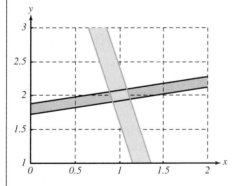 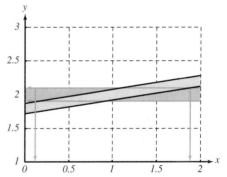

FIGURE 1.10A Graphical representation of a linear system with uncertainties

FIGURE 1.10B Graph of Eqs. (1.4) and (1.6).

Note that as in the previous solution, two of the points are the same for the true solution and the solution by Gaussian elimination. However the other two points are much more different in this solution.

1.3 GETTING STARTED IN MATHCAD

Mathcad has been chosen as the computational environment for the presentation of the numerical methods in this text because it combines the versatility and power of a programming language with the ease of use of a spreadsheet. The syntax for entering equations is very straight forward; text may be added for documentation, and two or three dimensional plots are easy to generate for illustration. Solutions may be either numeric or symbolic, however, we will not utilize Mathcad's symbolic capabilities. Unless otherwise noted, discussion is based on the standard version of Mathcad2000.

1.3.1 Overview of the Mathcad Workspace

Starting Mathcad opens a computational window known as a Worksheet. Worksheets may be printed, saved, opened, closed, etc. using the appropriate items under the File Menu. Several Worksheets may be open at the same time. Each Worksheet is composed of various *regions*. Each region is of one of three types: a math region, a text region, or a graph region. The default type is a math region. A region is placed wherever the crossbar cursor is positioned. Before discussing the creation of text and graph regions, we consider the basics of working in a math region.

Toolbars

The primary method of entering operators or symbols in a math region of the Worksheet is by means of the Math Toolbar. The Math Toolbar contains icons (or buttons) for nine toolbars, arranged in a three-by-three array (although the shape can be reconfigured if desired). If the Math Toolbar is not displayed, it can be opened by selecting the View menu, then selecting Toolbars, then Math (this sequence may be abbreviated as `View/Toolbar/Math`). The nine toolbars on the Math Toolbar are the Calculator (arithmetic), Graph, Vector&Matrix, Evaluation, Calculus, Boolean, Programming, Greek Symbol, and Symbolic Keyword Toolbars. These toolbars can also be accessed directly from the `View/Toolbars` menu. In future discussion, reference to these toolbars will not include specific mention of the means of accessing them.

Notation

Icons (buttons) will be denoted with brackets around either the symbol on the icon, or the name of its function.
    Special keys on the keyboard will also be indicated with an abbreviation of their function enclosed in brackets; for example

```
[Ctrl]    control key
[Retn]    return key
[Tab]     tab key
[Shft]    shift key
[Sp]      space
```

We introduce the toolbars by giving a brief example of the use of the Calculator Toolbar. The Calculator Toolbar contains icons (or buttons) for the standard arithmetic operators. The operations of addition, subtraction, multiplication, division, and exponentiation can also be entered using the standard keyboard symbols (+, −, *, /, and ^, respectively). Using the icon for an arithmetic operation produces a template with placeholders for each of the numbers to be entered. One can move from one placeholder to the next (especially to move out of an exponent or the denominator of a fraction) by using the tab key. However, one does not tab out of the last placeholder before evaluating the expression. Evaluation is achieved by pressing the = key immediately after the last number has been entered.

Example 1.10 Using the Calculator Toolbar

To enter 10^3 using the Calculator Toolbar,
    Type: 10 [x^y] 3   (followed by either space or tab)

The operators on the Calculator, Matrix, Calculus, Evaluation, Boolean and Programming Toolbars are discussed in more detail in Section 1.3.3.

Variables and Functions

In general, Mathcad is both case- and font-sensitive in naming variables, and is case-, but not font-sensitive for function names.

Mathcad evaluates mathematical expressions (and regions) from left to right and from top to bottom. This means that a value assigned to a variable is used in all expressions to the right of, and below where it is defined.

Define a Variable

To define a variable t with the value 10, you may use either the keyboard or the Calculator Toolbar.
    To use the keyboard,
        Type   t:10   (Mathcad displays t:=10)
    To use the Calculator Toolbar,
        Type   t [:=] 10

Define a Range Variable

A range variable is a variable that takes on a range of values. Any expression or equation in which a range variable is used will be evaluated for all values of the variable.

> ### Example 1.11 Defining a Range Variable
>
> To define the range variable r, taking all integer values from 2 to 10
> Type: `r:2,3;10`
> The difference between the first number listed and the second number is the step size. If the second number is omitted a step size of 1 is assumed (or -1 if the final number is smaller than the first number. As in defining an ordinary variable, the symbol ":" is displayed by Mathcad as :=. Similarly, the symbol ";" is displayed as .., so for this example, Mathcad will display
> `r:=2..10`
> this results in the variable r having the values 2, 3, 4, 5, 6, 7, 8, 9, 10.

Define a Function

A function is defined in the same manner as a variable, using the assignment equal symbol (typed as : , or entered from the Calculator Toolbar as the ▣ icon).

Creating Text Regions

There are two ways to create a text region, either by using the menu or by using keystrokes. To use the menu, select the Insert menu, followed by Text Region, i.e., Insert/Text Region. To use keystrokes, press the " key. Use the return key [Retn] to continue the text region to additional lines. To exit the Text Region, simply click anywhere in the Worksheet outside the text region.

Creating Graph Regions

Mathcad has several types of graphs available; we summarize the basic method of creating a graph using an x-y plot as the example. The first step is to define a range variable, and a function of the variable; for example, to graph a function $d(t)$, define t as a range variable over the interval of interest, and also define the function $d(t)$. The graph will be created using the menus (Insert/Graph/$x - y$ plot) or the Graph Toolbar.

Position the crossbar at the desired location for the graph.
Type $d(t)$, then (`Insert/Graph/x-y plot`); (with no space after $d(t)$)
Type t in the placeholder on the horizontal axis.

The graph can be resized by the following steps:
click to select the graph,
move the cursor to a handle on the edge of the graph,
where it becomes a double arrow symbol, and
hold mouse button and move as desired to resize graph.

The graph can be reformatted using options from the Format menu, or by double-clicking on the graph to see options.

1.3.2 Mathematical Computations

Numbers

Real numbers may be entered in either decimal or exponential notation. The decimal point is not required for an integer. To enter 314500 in exponential notation type `3.145*10^5`.

Mathcad's default is to display 3 decimal places in all results (although computations are made using the full precision available for the computer on which Mathcad is running, typically 14 or 15 decimal places). To change the number of decimals displayed, use `Format/Result`.

Complex numbers may be entered using either i or j for the imaginary unit; Mathcad's default display is to use i, but this can be changed using `Format/Result/DisplayOption` if desired. It is necessary to include a numerical coefficient preceding the imaginary unit, even if that coefficient is 1 (to show that the imaginary unit is intended rather than a variable named i or j); however Mathcad will not display the coefficient if it is 1.

Binary, octal, and hexadecimal numbers are indicated by the lowercase letter b, o, or h respectively immediately following the digits of the number.

Arrays

Vectors and matrices are rectangular arrays of numbers, which may be entered either from the menu using `Insert/Matrix` or from the Matrix Toolbar. A vector is a column of numbers, i.e. an m-by-1 matrix. A row vector may be created as an 1-by-n matrix.

To create a vector or matrix:

1) Menu: , or
   Matrix Toolbar: `[matrix]`
2) Specify number of rows and number of columns, then click OK
3) Fill in desired values in placeholders, using `[Tab]` to move from one placeholder to the next.

The operators on the Matrix Toolbar are discussed further in Section 1.3.3.

The default indexing for array elements begins with 0. This may be changed, if desired, by either defining `ORIGIN:1` (or the desired starting index) or using the menu (`Math/Options/Array Origin`).

One may also define individual elements of an array. For example, to define a 3-by-4 array **A** that is zero except for the element in the lower right corner, which is 1, type **A**`[2,3:1`. This defines **A**(2,3)=1, and all other elements will be set to 0.

Range variables are often useful in defining arrays. For example, one might define the range variable

`j:0;3`

and the array

`x[j:j^2 +3`

The result is an array

```
x=[3, 4, 7, 12]
```

To access an element in an array, type the array name followed by the subscript of the desired element. For example, in matrix **A,** the element in the *i*th row, *j*th column is $\mathbf{A}_{i,j}$, which can be typed as A[i,j (note that one does not type the closing bracket, it is supplied automatically by Mathcad).

Naming Variables and Functions

A name is a sequence of characters in a math region. Variables and functions are either built-in or user-defined. Built-in variables are defined to have the conventional value, but may be redefined if desired. The built-in value is set for all fonts (except Greek font); a redefined value of the variable applies only to the specific font used.

Literal subscripts (subscripts which are not numbers) are useful in naming variables. They are created by preceding the portion of the variable name which is to be a subscript by a period.

Greek Letters

    Keyboard: Type corresponding Roman letter followed by [Ctrl]g
    Menu: Select View/Toolbars/Greek

The letter π can also be typed by pressing [Ctrl][Shft]P.

Not All Equal Signs Are Equal

Mathcad has several different equal signs, which are not interchangeable.

Symbol	Use	Keyboard
=	evaluate (give previously defined value)	=
:=	assign (define value)	:
≡	global assignment	~
=	**bold equal**	[Ctrl]=

1.3.3   Operators on the Math Toolbars

There are nine toolbars available in Mathcad2000; these are accessed from the Math Toolbar or from the main menu (View/Toolbars). The Toolbars are

Calculator Toolbar	Evaluation Toolbar
Graph Toolbar	Boolean Toolbar
Matrix Toolbar	Greek Letters
Calculus Toolbar	Symbolic Toolbar
Programming Toolbar (Professional Mathcad only)	

Calculator Toolbar

The operators on the Calculator Toolbar, with their access by icon or keystroke:

Operator	icon	keystroke
Parentheses	()	'
Addition	+	+
Subtraction	−	−
Multiplication	×	*
Division	/	/
Range variable	m..n	;
Factorial	n!	!
Vector/matrix subscript	x_n	[
Absolute value	|x|	|
Square root	√	\
nth root	$\sqrt[n]{}$	[Ctrl]\
Exponentiation	x^y	^
Equals	=	=
Definition	:=	:

Matrix Toolbar

The operators on the Matrix Toolbar, together with their access by icon or keystroke:

Operator	icon	keystroke
Insert matrix	▦	[Ctrl]M
Vector/matrix subscript	x_n	[
Dot product	u·v	*
Cross product	x×y	[Ctrl]8
Vector sum	Σv	[Ctrl]4
Matrix inverse	x^{-1}	^-1
Magnitude/determinant	|x|	|
Matrix superscript	$M^{<>}$	[Ctrl]6
Matrix transpose	M^T	[Ctrl]1
Vectorize	$\vec{m}$	[Ctrl]-

(forces operations to be done element by element)

Calculus Toolbar

The operators on the Calculus Toolbar, with their icon and keystroke access:

Operator	icon	keystroke
Summation		[Ctrl][Shft]4
Product		[Ctrl][Shft]3
Range sum		$
Range product		#
Definite integral		&
Indefinite integral		[Ctrl]i
Derivative		?
Nth derivative		[Ctrl]?
Limit		[Ctrl]L
Right-hand limit		[Ctrl][Shft]A
Left-hand limit		[Ctrl][Shft]B

Boolean Toolbar

The operators on the Boolean Toolbar, with their icon and keystroke access:

Operator	icon	keystroke
Greater than		>
Less than		<
Greater than or equal		[Ctrl])
Less than or equal		[Ctrl](
Not equal		[Ctrl]3
Bold equals		[Ctrl]=
and		[Ctrl][Shft]7
or		[Ctrl][Shft]6
xor		[Ctrl][Shft]1
not		[Ctrl][Shft]5

Evaluation Toolbar

The operators on the Evaluation Toolbar, with their icon and keystroke access:

Operator	*icon*	*keystroke*
Equals	▭	=
Definition	▭	:
Global definition	▭	~
Symbolic equals	▭	[Ctrl].

(additional operators in Professional version)

1.3.4 Built-in Functions

Mathcad has built-in functions for many standard mathematical functions (Trigonometric, Exponential, Logarithmic, Hyperbolic, Bessel, and other special functions). There are also functions for interpolation and prediction, regression and smoothing, Fourier transforms, solving differential equations, optimization, and many other types of mathematical problems.

Function names are case-sensitive, but not font-sensitive (except that the Greek font is different).

Some of the standard mathematical functions that are available in Mathcad2000 are summarized below. The functions that are related to the topics covered in this text are discussed in more detail in the last section of the chapter of the text that deals with the corresponding material.

Notation

The letter used for a variable name is intended to indicate the variable type:

z	complex (or real)
x or y	real
m or n	integer
capitol letter	matrix (or vector)
M	square matrix, real or complex
v	vector
A	rectangular array

Math Functions

Bessel Functions

```
I0(x), I1(x), In(m,x)    modified Bessel function of the first kind
J0(x), J1(x), Jn(m,x)    Bessel function of the first kind
K0(x), K1(x), Kn(m,x)    modified Bessel function of the second kind
Y0(x), Y1(x), Yn(m,x)    Bessel function of the second kind
```

Hyperbolic Functions

`cosh(z),sinh(z),tanh(z)`	hyperbolic functions
`sech(z),csch(z),coth(z)`	co-functions
`acosh(z),asinh(z),atanh(z)`	inverse hyperbolic functions
`asech(z),acsch(z),acoth(z)`	

Logarithmic and Exponential Functions

`exp(z)`	natural exponential function
`ln(z)`	natural logarithm function
`log(z)`	logarithm base 10

Piecewise Continuous Functions

δ`(m,n)`	Kronecker delta function	(to type δ, press d`[Ctrl]`g)
	returns 1 if $m = n$, 0 otherwise	
Φ`(x)`	Heaviside step function	(to type Φ, press F`[Ctrl]`g)
	returns 1 if $x \geq 0$, 0 otherwise	
`sign(x)`	returns 0 if $x = 0$; returns 1 if $x > 0$; returns -1 if $x < 0$.	
`if(cond, x, y)`	returns x if cond is true (non-zero);	
	returns y if cond is false (zero)	

Special Functions

`erf(x)`	error function	
`cerf(x)`	complementary error function, $1 - \mathrm{erf}(x)$	
Γ`(z)`	Euler gamma function	(to type Γ, press G`[Ctrl]`g)

Trigonometric Functions

`cos(z), sin(z), tan(z)`	basic trig functions
`sec(z), csc(z), cot(z)`	co-functions
`acos(z), asin(z), atan(z)`	inverse trig functions
`asec(z), acsc(z), acot(z)`	
`atan2(x,y)`	returns angle in radians from positive x-axis to the point (x, y); result is between $-\pi$ and π.

Some Useful Functions for Numerical Manipulation

Complex Numbers

`arg(z)`	argument in radians from the positive real axis to the point z in the complex plane (result is between $-\pi$ and π)
`Im(z)`	imaginary part of complex number z
`Re(z)`	real part of complex number z

Truncation and Round-off

`ceil(x)`	returns least integer $\geq x$
`floor(x)`	returns greatest integer $\leq x$
`round(x)`	rounds real number x to nearest integer
`round(x, n)`	rounds real number x to n decimal places (if $n < 0$, rounds to left of decimal point)
`trunc(x)`	returns integer part of x (same as `floor(x)` for $x > 0$ and `ceil(x)` for $x < 0$)

Some Functions for Manipulation of Vectors and Matrices

Sorting

`csort(A,j)`	sorts rows of **A**; elements in column j are in ascending order.
`reverse(v)`	reverses order of elements of vector **v** for matrix, reverses order of rows of matrix
`rsort(A,j)`	sorts columns of **A**; elements in row i are in ascending order.
`sort(v)`	sorts elements of vector **v** into ascending order

Vector and Matrix

`augment(A,B,C,...)`	form new matrix, placing **A B C** ... from left to right matrices must have the same number of rows
`cols(A)`	number of columns in **A**
`eigenvals(M)`	vector of eigenvalues of **M**
`eigenvec(M,z)`	normalized eigenvector for eigenvalue z.
`identity(n)`	n-by-n identity matrix
`last(v)`	index of last element of vector **v**
`length(v)`	number of elements in vector **v**; (same as rows(v))
`matrix(m,n,f)`	$m_{ij} = f(i,j)$, for $i = 0, \ldots m-1;\ j = 0 \ldots n-1$.
`max(A)`	largest element in **A**
`min(A)`	smallest element in **A**
`rank(A)`	maximum number of linearly independent columns in **A**
`rows(A)`	number of rows in **A**
`rref(A)`	matrix representing row-reduced echelon form of **A**
`stack(A,B,C, . . .)`	forms new matrix by placing A B C .. top to bottom matrices must have the same number of columns
`submatrix(A,ia,ib,ja,jb)`	a_{ij}, for $ia \leq i \leq ib$ and $ja \leq j \leq jb$
`tr(M)`	trace (sum of diagonal elements) of **M**

Differential Equation Solving

There are several functions in Mathcad2000(Pro), but none in Standard Mathcad. See Chapter 12 for discussion. Mathcad2001 has the capabilities of Mathcad2000(Pro).

Fourier Transform

Mathcad has several functions for Fast Fourier Transforms and Inverse Fast Fourier Transforms. See Chapter 10 for discussion.

Interpolation

The built-in Mathcad functions for spline interpolation are discussed in Chapter 8.

Regression and Smoothing

Built-in Mathcad functions for function approximation are discussed in Chapter 9.

Solving

The built-in Mathcad functions for finding the zeros of a function of one variable are discussed in Chapter 2. The functions for finding the zeros of a function of several variables, or the maximum or minimum of such a function are discussed in Chapter 5.

1.3.5 Programming in Mathcad

Although the computations required to use the numerical methods discussed in this text can be carried out easily using the capabilities of Standard Mathcad2000, Professional Mathcad2000 has several additional features that are useful for solving problems of the types considered in the study of numerical analysis or numerical methods. The two features that are of primary interest are the built-in functions for solving differential equations (discussed in Chapters 12 to 15) and capability for creating mathematical programs. Although a simple program may be defined using the if function (available in Standard Mathcad2000), it becomes unwieldy if more than two branches are required.

In Mathcad a program is a compound expression comprised of (many) programming operators. Programming operators are accessed from the Programming Toolbar. As with other expressions, a program returns a value (when its name is typed followed by the = sign).

A function may be defined by a program (although a simple function may also be defined by an expression as discussed in Section 1.3.1).

> **Example 1.12 A Simple Mathcad Program**
>
> To define the function $f(x,w) = \log(x/w)$ using a simple program, carry out the following steps. All directions to click on an icon or button refer to the Programming Toolbar; alternative access to the desired operator from the keyboard are also given. Begin by opening the Programming Toolbar (from Math Toolbar or View menu).
>
> 1) Type $f(x,w)$:
> 2) click [add line] icon
> 3) Click top placeholder, type z
> 4) Click [¨] icon
> 5) Click right placeholder, type x/w
> 6) Click bottom placeholder, type $\log(z)$

A program may refer to variables and functions defined previously in the worksheet. Definitions of variables inside the program must be made with the local definition symbol, not the := operator. Variables defined inside a program are local to the program. To add more lines to a program, just click on the [add line] icon; to delete a placeholder, click on it and backspace (press [delete] key).

To access a function which is defined in a different worksheet, use `Insert/Reference` to specify the location (path name) for the function.

Conditional Statements

To define a function that has a different formula for different parts of its domain, begin as in the previous example for the first two steps. Then click on the [if] icon (or press the } key). In the right placeholder type a Boolean expression (using relational operators in Boolean Toolbar). In the left placeholder, type the value to be returned if the Boolean expression is true. Select the remaining placeholder (or add more lines and more if statements if necessary) and click the [otherwise]icon (or press [Ctrl]3). Type the value you want the function to return if the condition in the first statement is false.

Looping

Mathcad has two types of loops: the "for" loop for use when the number of times the loop is to be executed is known, and the "while" loop for use when the execution of the loop is to be continued as long as a condition is true.

Loops include automatic indentation to show the structure. If the loop consists of more than one statement, then a line is also place down the left edge of the loop. However, a loop with only one statement (even if that statement is another loop which itself contains many lines) does not have a line down its left edge.

For Loops

Click [for] icon (or press [Ctrl]").
    Type name of iteration variable in placeholder to the left of the ε symbol.
    Type the range of values for the iteration variable in placeholder to the right of ε symbol. (Typically the values are given as for a range variable, see Section 1.3.1.)
    Type expression to be evaluated in the second line placeholder (add lines if needed).

While Loops

Click [while] icon (or press [Ctrl]]).
    Click in top placeholder and type condition.
    Type expression to be evaluated in the second line placeholder (add lines if needed).
    Note: the condition is checked at the beginning of the loop.

Controlling Program Execution

Mathcad has three means of controlling program execution. These are the break, continue, and return statements. Their use is illustrated in functions in the text where appropriate.

Programs Within Programs

The power of a programming environment is greatly increased by the capability of a program to call another program or subroutine. There are also situations in which it is very convenient to be able to define a function recursively, in terms of previous values of the same function. Mathcad has both of these desirable capabilities.

A Few Comments

A function can only return a single result (number or matrix). To return more than one matrix (with the same number of rows in each) they may be concatenated (placed side by side using the augment function). The final results may then be separated (by extracting the desired columns) after the function returns the combined matrix.

SUMMARY

Fixed-point iteration for finding a solution $x_k = g(x_{k-1})$ of $x = g(x)$. To find a numerical approximation to $\sqrt{c}$ by fixed-point iteration, use

$$x = g(x) = \frac{1}{2}\left(x + \frac{c}{x}\right),$$

so that the iteration formula is

$$x_k = \frac{1}{2}\left(x_{k-1} + \frac{c}{x_{k-1}}\right).$$

The *trapezoid rule* approximates the definite integral:

$$\int_a^b f(x)\,dx \approx \frac{b-a}{2}[f(a) + f(b)].$$

A computationally efficient method of evaluating the polynomial

$$P(x) = a_n x^n + a_{n-1} x^{n-1} + \ldots + a_1 x + a_0,$$

at $x = c$ is given by Horner's algorithm:

Begin by setting $b_n = a_n$.
For $k = n - 1, \ldots, 0$,
$b_k = a_k + b_{k+1} c$.
The result is $b_0 = P(c)$.

The quantities $b_n, \ldots, b_1$ are the coefficients of $Q(x)$, with $P(x) = (x - c) Q(x) + P(c)$. $P'(c)$ can be found by evaluating $Q(c)$ using horner's method.

Gerschgorin circle theorem All eigenvalues of the matrix A lie within the union of the disks bounded by the n circles, $C_1, \ldots C_n$, where circle C_i has center at a_{ii} and radius

$$r_i = -|a_{ii}| + \sum_{j=1}^{n} |a_{ij}|.$$

Acceleration The order of convergence for some approximation formulas, including the trapezoid rule for numerical integration, can be improved by combining the approximate values obtained with step size h and step size $h/2$. If we denote these approximations as $T(h)$ and $T(h/2)$, respectively, a better estimate of the desired result is given by

$$T \approx \frac{1}{3}[4\,T(h/2) - T(h)]$$

The trapezoid rule is a formula of order $O(h^2)$, but the extrapolated value is an approximation of order $O(h^4)$.

Chapter 1 Foundations

SUGGESTIONS FOR FURTHER READING

Much of the historical discussion in the introduction to this chapter is based on material in the following excellent books:

> Struik, D. J., *A Concise History of Mathematics* (4th ed.), Dover, New York, 1987.
> Eves, H., *Great Moments in Mathematics* (V. 1, before 1650; v. 2, after 1650), The Mathematical Association of America, Washington, D.C., 1983.
> Smith, D. E., *A Source Book in Mathematics*, Dover, New York, 1959.

For historical notes related to calculus or differential equations, the following books are especially recommended:

> Boyer, C. B., *The History of the Calculus and its Conceptual Development,* Dover, New York, 1949.
> Simmons, G. F., *Calculus with Analytic Geometry*, McGraw-Hill, New York, 1985.
> Simmons, G. F., *Differential Equations with Applications and Historical Notes,* McGraw-Hill, New York, 1972.

Among the references for more theoretical and advanced treatments of numerical methods are:

> Atkinson, K. E., *An Introduction to Numerical Analysis* (2nd ed.), John Wiley, New York, 1989.
> Dahlquist. G., and A. Bjorck, *Numerical Methods*, (Translated by Ned Anderson), Prentice-Hall, Englewood Cliffs, NJ, 1974.
> Isaacson, E., and H. B. Keller, *Analysis of Numerical Methods*, Dover, New York, 1994 (originally published by John Wiley & Sons, 1966).
> Ralston, A., and P. Rabinowitz, *A First Course in Numerical Analysis* (2nd ed.), McGraw-Hill, New York, 1978.
> Press, W. H., S. A. Teukolsky, W. T. Vetterling, and B. P. Flannery, *Numerical Recipes in C, The Art of Scientific Computing,* (2d ed.), Cambridge Unversity Press, Cambridge, U.K., 1992.

For further discussion on the use of Mathcad2000, see

> *Mathcad2000 Reference Manual,* Mathsoft, Cambridge, MA, 1999.
> *Mathcad2000 User's Guide*, Mathsoft, Cambridge, MA, 1999.
> Larsen, R. W., *Introduction to Mathcad,* Prentice Hall, Upper Saddle River, NJ, 1999.

PRACTICE THE TECHNIQUES

For problems P1.1 to P1.5, investigate the use of Gaussian elimination to solve systems of two equations in two unknowns.
  a. *Solve the system as given.*
  b. *Graph the given equations, and graph the transformed second equation after the elimination step.*
  c. *Repeat Parts a and b for the system consisting of the same two equations, but given in reverse order.*

P1.1 $4x + y = 6$,
$-x + 2y = 3$.

P1.2 $4x + y = 9$,
$x + 2y = 4$.

P1.3 $3x + y = 4$,
$6x + 7y = 13$

P1.4 $5x + y = 17$,
$-10x + 17y = 4$.

P1.5 $3x + y = 6$,
$x + 5y = 16$.

For problems P1.6 to 1.10, investigate the convergence of the given fixed-point iteration formula.
  a. *Show the first three iterations graphically.*
  b. *Compute the first three iterates algebraically*
  c. *Determine whether the conditions of the fixed-point convergence theorem are satisfied.*

P1.6 $x = g(x) = 0.5\,x^3 + 0.3$.

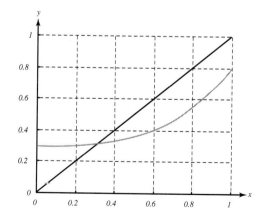

P1.7 $x = g(x) = \sin(2x)$.

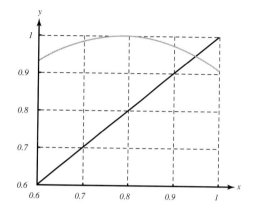

P1.8 $x = g(x) = -0.5x^3 + 0.8$.

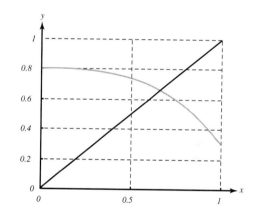

P1.9 $x = g(x) = 1 - x^2$.

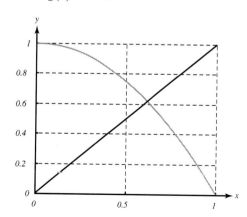

P1.10 $x = g(x) = \dfrac{1}{2} + \dfrac{1}{2} + \sin(3x)$

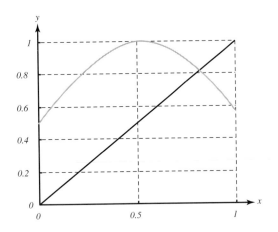

Problems P1.11 to P1.15 illustrate the use of the Gerschgorin theorem to find bounds on the eigenvalues of a matrix. Find the Gerschgorin circles for each row of the given matrix. Graph the regions, and give bounds on the eigenvalues.

P1.11 $\mathbf{A} = \begin{bmatrix} 1 & 1/8 & 1/4 \\ 1/2 & 2 & 0 \\ 0 & 0 & 3 \end{bmatrix}$.

P1.12 $\mathbf{A} = \begin{bmatrix} 1 & 3/4 & 0 \\ 1/2 & 2 & -1/8 \\ 0 & 1/8 & 3 \end{bmatrix}$.

P1.13 $\mathbf{A} = \begin{bmatrix} 1 & 1/4 & 0 \\ 1/4 & 2 & 1/4 \\ 0 & 1/4 & 3 \end{bmatrix}$.

P1.14 $\mathbf{A} = \begin{bmatrix} 1 & 1/3 & 1/3 \\ 1/4 & 2 & 1/4 \\ 1/2 & 1/4 & 3 \end{bmatrix}$.

P1.15 $\mathbf{A} = \begin{bmatrix} 1 & -1/2 & 0 \\ 1/2 & 2 & 1/8 \\ 0 & 1/8 & 3 \end{bmatrix}$.

P1.16 to P1.17 illustrate the effect of round-off error in adding numbers of differing magnitudes.
  a. Add from left to right, rounding to three digits at each step.
  b. Add from right to left, rounding to three digits at each step.
  c. Compare the relative error for the results from parts a and b.

P1.16 $100 + 0.49 + 0.49$

P1.17 $10.0 + 0.333 + 0.333 + 0.333$

Problems P1.18 to P1.20 illustrate the effect of round-off in the quadratic formula.
  a. Use the standard quadratic formula with rounding.
  b. Use the rationalized-numerator quadratic formula with rounding.
  c. Compare the results from parts a and b with the results found without rounding.

P1.18 $x^2 - 973\,x + 1 = 0$. (Round to three digits.)
P1.19 $x^2 - 57\,x + 1 = 0$. (Round to four digits.)
P1.20 $x^2 - 23\,x + 1 = 0$. (Round to four digits.)

Problems P1.21 to P1.25 illustrate the use of the trapezoid rule for numerical integration.
  a. Approximate the given integral, using the basic trapezoid rule with $h = b - a$.
  b. Approximate the integral, using the composite trapezoid rule with $h = \dfrac{b-a}{2}$.
  c. Improve the approximation by acceleration, using results from parts a and b.

P1.21 Find $\displaystyle\int_0^2 \dfrac{1}{1+x^2}\,dx$.

P1.22 Find $\displaystyle\int_0^{\pi/2} \sin(x)\,dx$.

P1.23 Find $\displaystyle\int_0^2 2^x\,dx$.

P1.24 Find $\displaystyle\int_0^2 e^{-x^2}\,dx$.

P1.25 Find $\displaystyle\int_0^{\pi/2} \dfrac{3}{1+\sin(x)}\,dx$.

Problems P1.26 to P1.30 illustrate the effect of the order of the equations in Gaussian elimination (with error).

a. Solve the equations in the order given; determine bounds on x and y.
Graph the given equations, and the transformed second equation
b. Solve the equations in reverse order; determine bounds on x and y.
Graph the given equations, and the transformed second equation
c. Compare the relative error in the values of x and y found in parts a and b.

P1.26 $4x + y = 6 \pm 0.1$,
$-x + 2y = 3 \pm 0.1$.

P1.27 $4x + y = 9 \pm 0.1$,
$x + 2y = 4 \pm 0.1$.

P1.28 $3x + y = 4 \pm 0.1$,
$6x + 7y = 13 \pm 0.1$.

P1.29 $5x + y = 17 \pm 0.2$,
$-10x + 17y = 4 \pm 0.2$.

P1.30 $3x + y = 6 \pm 0.2$,
$x + 5y = 16 \pm 0.2$.

EXTEND YOUR UNDERSTANDING

U1.1 For each of the Problems P1.1 to P1.5 for which the conditions of the theorem are not satisfied on [0, 1], investigate the following questions.

a. Is there a starting value of x for which the iterations do not converge? (Show this on the graph.)
b. Is there a subinterval on which the conditions are satisfied?

U1.2 Use the Gerschgorin circle theorem to show that, for any matrix

$$\mathbf{A} = \begin{bmatrix} 1+2r & -r & 0 \\ -r & 1+2r & -r \\ 0 & -r & 1+2r \end{bmatrix},$$

any eigenvalue m satisfies $|m| \geq 1$, regardless of the value of r. This result is used in Chapter 15. Extend the pattern to larger dimension matrices (e.g.,tridiagonal, with $1 + 2r$ on the diagonal, and $-r$ on the subdiagonal and superdiagonal).

U1.3 Show that Horner's method is equivalent to synthetic division.

U1.4 Show that Horner's method for evaluating polynomials can be viewed as rearranging the polynomial such that

$$P(x) = a_n x^n + a_{n-1} x^{n-1} + \ldots + a_1 x + a_0$$
$$= (\ldots((a_n x + a_{n-1})x + a_{n-1})x$$
$$+ \ldots + a_1)x + a_0$$

If only the value of P(c) is required, it is not necessary to save the intermediate quantities (the b_i in the statement of the algorithm in the text). Write an algorithm for Horner's method using a single temporary variable.

U1.5 Consider a very limited binary normalized floating point system in which there are four bits to store the positive numbers. What exponents can be represented if 2 bits are used? What numbers can be represented if 2 bits are used for digits, and 2 for exponents? What numbers can be represented if 3 bits are used for digits and 1 for the exponents?

U1.6 Find the positive numbers that can be represented with only 1 digit and exponents of 0 or 1 in a base-10 normalized floating-point system.

2
Solving Equations of One Variable

2.1 Bisection Method
2.2 *Regula Falsi* and Secant Methods
2.3 Newton's Method
2.4 Muller's Method
2.5 Mathcad's Methods

As discussed in Chapter 1, the problem of finding the zeros of a nonlinear function (or roots of a nonlinear equation) has a long history. Although quadratic equations of one variable can be solved analytically, numerical estimation of the zeros may be desired. For many other types of equations, it is either difficult or impossible to find an exact solution. In this chapter, we investigate several techniques for finding roots or zeros of nonlinear equations of a single variable.

The first technique presented, bisection, is a systematic approach to subdividing an interval on which we know the function has a zero. In addition to being simple and intuitive, this method can be used to obtain an adequate initial estimate of the zero that will be refined by more powerful methods. The second group of solution techniques—the *regula falsi*, secant, and Newton methods—are based on approximating the function whose zero is desired by a straight-line approximation, either a secant line through two points on the function or a tangent line to the function. We then investigate a technique, known as Muller's method, that is based on a quadratic approximation to the function.

The presentation of each technique includes several simple examples and a Mathcad function implementing the method.

Applications of zero-finding techniques occur throughout science and engineering. We begin with two examples, which we also use to illustrate the techniques throughout the chapter. Further examples from fields such as statics occur in the exercises.

In Chapters 3 and 4, we consider some methods for solving systems of linear equations; Chapter 5 presents techniques for dealing with systems of nonlinear equations.

Example 2-A Floating Sphere

According to Archimedes, if a solid that is lighter than a fluid is placed in the fluid, the solid will be immersed to such a depth that the weight of the solid is equal to the weight of the displaced fluid. For example, a spherical ball of unit radius will float in water at a depth x (the distance from the bottom of the ball to the water line) determined by ρ, the specific gravity of the ball ($\rho < 1$).

The volume of the submerged segment of the sphere is

$$V = \pi x(3r^2 + x^2)/6,$$

where r and x are related by the Pythagorean theorem, $r^2 + (1-x)^2 = 1$. (See Fig. 2.1.)

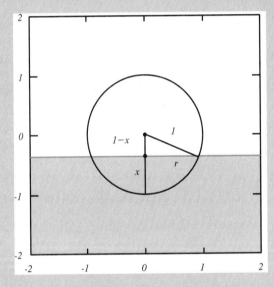

FIGURE 2.1 Floating sphere.

To find the depth at which the ball floats, we must solve the equation stating that the volume of the submerged segment is ρ times the volume of the entire sphere; i.e.,

$$\pi x(3r^2 + x^2)/6 = \rho(4\pi/3),$$

which simplifies to

$$x^3 - 3x^2 + 4\rho = 0.$$

In general, the zero that is of physical interest is between 0 and 2 (since the ball is of unit radius and x is measured up from the bottom of the ball).

Representative values of specific gravity include $\rho \approx 0.25$ for cork and $0.33 < \rho < 0.99$ for air-dried timber, depending on the type of wood. We investigate several methods of solving the preceding equation in this chapter.

Example 2-B Planetary Orbits

The position of a moon that revolves around a planet in an elliptical orbit can be described by Kepler's equation, which gives the central angle θ as a function of time. The relationship between time t and the central angle is given by

$$2\pi t = P(\theta - e \sin \theta),$$

where P is the period of revolution of the moon about the planet, e is the eccentricity of the moon's orbit, and the moon is at $(a, 0)$ at $t = 0$. The planet is located at a focus of the ellipse, $(a\,e, 0)$. To find the central angle for any given time t, we must find the root of the equation

$$2\pi t - P\theta + Pe \sin \theta = 0.$$

If the period of revolution is 100 days, and the eccentricity is 0.5, then, for any specified time, t, the central angle can be found from the relation

$$2\pi t - 100\theta + 50 \sin \theta = 0.$$

Kepler's law says that the orbit sweeps out equal areas in equal times. The position of the planet at 10-day intervals and the areas swept out between $t = 0$ and $t = 10$, and between $t = 60$ and $t = 70$ are illustrated in Fig. 2.2.

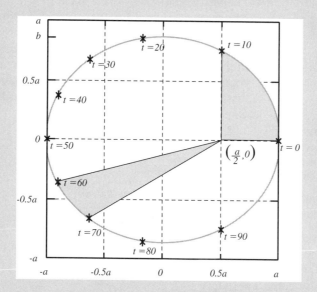

FIGURE 2.2 Position of the moon at 10-day intervals and areas swept out from day 0 to 10 and day 60 to 70.

The equation of the ellipse is $x(\theta) = a \cos(\theta)$, $y(\theta) = b \sin(\theta)$, where the coefficients a and b and the eccentricity e are related by the equation

$$b^2 = a^2(1 - e^2).$$

2.1 BISECTION METHOD

Bisection is a systematic search technique for finding a zero of a continuous function. The method is based on first finding an interval in which a zero is known to occur because the function has opposite signs at the ends of the interval, then dividing the interval into two equal subintervals, and determining which subinterval contains a zero, and continuing the computations on the subinterval that contains the zero.

Suppose that an interval $[a, b]$ has been located which is known to contain a zero, since the function changes signs between a and b. The midpoint of the interval is

$$m = \frac{a + b}{2},$$

and a zero must lie in either $[a, m]$ or $[m, b]$. The appropriate subinterval is determined by testing the function to see whether it changes sign on $[a, m]$. If so, the search continues on that interval; otherwise it continues on $[m, b]$.

2.1.1 Step-by-Step Computation

The algorithmic statement of the bisection method given here consists of an initialization step, and a block of steps which may be repeated several times depending on the accuracy desired for the result. The final step in the repeated block involves testing for the accuracy of the current approximation, recording the current values of the variables of interest, and updating the value of certain variables if computation is to continue. These steps can be carried out conveniently in a Mathcad Worksheet, as illustrated in Example 2.1. As long as Mathcad is in automatic computation mode, the remaining variables will be recomputed based on the updated values.

Bisection Algorithm

Initialization:
1) Define function, $f(x)$, whose zero is desired.

Begin computation block:
2) Define interval $[a, b]$ containing the desired zero.
   Compute $f(a)$ and $f(b)$; they must have opposite signs.
3) Compute midpoint of interval, m, and $f(m)$
4) Compare values of $f(a)$, $f(m)$ and $f(b)$
   to determine if zero is in $[a, m]$ or in $[m, b]$
5) Record current values of a, m, b, $f(a)$, $f(m)$, $f(b)$ in a table.
   Decide whether to continue computations.
   To continue, change definition of a or b in step 2 as follows:
      If zero is in $[a, m]$, redefine b to have the current value of m.
      If zero is in $[m, b]$, redefine a to have the current value of m.

End computation block

Example 2.1 Finding the Square Root of 3 Using Bisection

It is often helpful to graph the function whose zero is desired before proceeding with the computations. The graph in Figure 2.3 shows $y = x^2 - 3$ and the first two approximations to the zero starting with the interval $[1, 2]$.

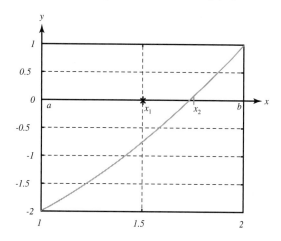

FIGURE 2.3 The graph of $y = x^2 - 3$, and the first two approximations to its zero on $[1, 2]$ using the bisection method.

This worksheet shows the first iteration of using the bisection algorithm to find the square root of 3.

Begin by defining the function whose zero is desired:

$$f(x) := x^2 - 3$$

Begin the computation by giving the initial bounds on the zero:

$a := 1 \qquad b := 2$

Evaluate the function at these points

$f(a) = -2 \qquad f(b) = 1$

Define the midpoint, evaluate it, and evaluate the function at the midpoint

$$m := \frac{a + b}{2} \qquad m = 1.5 \qquad f(m) = -0.75$$

Record the values of a, b, m, f(a), f(b), and f(m) in the Table below. Redefine a or b depending on the location of the root; Mathcad will recompute m.

Table of results for 3 iterations

a	b	m	f(a)	f(b)	f(m)
1	2	1.5	-2	1	-0.75
1.5	2	1.75	-0.75	1	0.063
1.5	1.75	1.625	-0.75	0.063	-0.359

Chapter 2 Solving Equations of One Variable

2.1.2 Mathcad Function for Bisection

The following Mathcad function for the bisection method finds an approximation to a zero of the function in the interval $[a,b]$. The function values $f(a)$ and $f(b)$ must have opposite signs (and the method terminates with an error message if they do not). The process stops when either

1) The magnitude of the function value at the approximate zero is less than "tol"

or

2) The maximum number of iterations "max" has been reached.

$$\text{Bisect}(f, a0, b0, \text{tol}, \text{max}) := \begin{array}{|l} a_0 \leftarrow a0 \\ b_0 \leftarrow b0 \\ ya_0 \leftarrow f(a0) \\ yb_0 \leftarrow f(b0) \\ \text{error("Function has same sign at endpoints")} \text{ if } ya_0 \cdot yb_0 > 0 \\ \text{for } i \in 0..\text{max} \\ \quad \begin{array}{|l} x_i \leftarrow \dfrac{(a_i + b_i)}{2} \\ y_i \leftarrow f(x_i) \\ \text{break if } |y_i| < \text{tol} \\ \text{if } y_i \cdot ya_i < 0 \\ \quad \begin{array}{|l} a_{i+1} \leftarrow a_i \\ b_{i+1} \leftarrow x_i \\ ya_{i+1} \leftarrow ya_i \\ yb_{i+1} \leftarrow y_i \end{array} \\ \text{otherwise} \\ \quad \begin{array}{|l} a_{i+1} \leftarrow x_i \\ b_{i+1} \leftarrow b_i \\ ya_{i+1} \leftarrow y_i \\ yb_{i+1} \leftarrow yb_i \end{array} \end{array} \\ x \end{array}$$

Example 2.2 Approximating the Floating Depth for a Cork Ball

To find the floating depth for a cork ball of radius 1 whose density is one-fourth that of water, we must find the zero (between 0 and 1) of

$$y = x^3 - 3x^2 + 1.$$

The function is shown in Fig. 2.4 for the region of interest. The computations using the Mathcad function Bisect are shown below. The function returns the approximate values of the desired zero, for interations 0, 1, . . . 5.

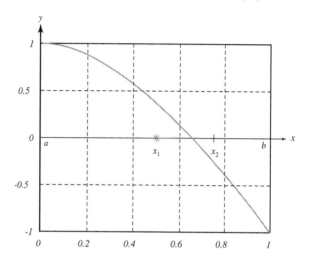

FIGURE 2.4 Graph of $y = x^3 - 3x^2 + 1$ and the first two approximations to its zero in [0, 1].

To find the floating depth of a cork ball, we seek the zero (between 0 and 1) of the function

$$g(x) := x^3 - (3 \cdot x^2) + 1$$

The initial bounds on the zero are

$$a0 := 0 \quad b0 := 1$$

We define the parameter values for the Bisect function as

$$tol := 0.0001 \qquad max := 5$$

The results are

$$\text{Bisect}(g, a0, b0, tol, max) = \begin{pmatrix} 0.5 \\ 0.75 \\ 0.625 \\ 0.688 \\ 0.656 \\ 0.641 \end{pmatrix}$$

2.1.3 Discussion

The bisection method is based on a well-known property of continuous functions, the intermediate value theorem. Applied specifically to the cases we are interested in, the theorem states that if $f(a) > 0$ and $f(b) < 0$, or if $f(a) < 0$ and $f(b) > 0$, then there is a number c between a and b such that $f(c) = 0$.

It is helpful to know how good our answers are for any numerical technique. In general, bisection is slow, but sure. At the first stage, the length of the interval is $b\text{-}a$. The furthest that our estimated zero (the midpoint of the interval) can be from the true solution is $(b\text{-}a)/2$. At each stage the length of the interval is halved, so the maximum possible error is reduced by a factor of one-half. The error at stage k is at most $(b\text{-}a)/2^k$.

One definition of linear convergence is that the inequality

$$|x^* - x_k| \leq r^k |x^* - x_1|$$

holds for some constant $r < 1$, where x^* is the true zero, r is the convergence rate, x_k is the approximate zero at the kth stage, and $x_1 = (a + b)/2$ is the first approximate zero. Thus, bisection is linearly convergent with rate $1/2$.

We have an estimate of the zero at each stage of the iteration, and we know that after k iterations, the error is at most

$$\frac{b_1 - a_1}{2^k},$$

where $[a_1, b_1]$ is the original interval within which the zero was bracketed. In many cases, we would not choose to retain the entire sequence of values of the endpoints of the interval bracketing the zero, but the values could be stored in vectors **a** and **b** if desired.

As indicated by the values of the function at the approximate zeros in Example 2.1, it is possible to be quite close to the true zero at some stage of the iteration and then move away from the zero before returning to a good approximation later (with a smaller error bracket for the zero). In computing the square root of 3, we had a better approximation at the second bisection step than we had at step 3 or step 4.

Bisection works well with problems that cause difficulties for other methods. It is also useful as a preprocessing algorithm for the methods we discuss in the remainder of the chapter.

2.2 REGULA FALSI AND SECANT METHODS

Bisection makes no use of information about the shape of the function $y = f(x)$ whose zero is desired. The first way to incorporate such information is to consider a straight-line approximation to the function. Since we know how to find the zero of a linear function, it is not much more work to find our approximation to the zero of $f(x)$ not as the midpoint of the interval, but as the point where the straight-line approximation to f crosses the x-axis. In this section, we consider two closely related methods based on straight-line approximations using two initial values of the independent variable that bracket the desired zero (as with the bisection method).

The *regula falsi* and secant methods start with two points, $(a, f(a))$ and $(b, f(b))$, satisfying the condition that $f(a) \cdot f(b) < 0$. The next approximation to the zero is the value of x where the straight line through the initial points crosses the x-axis; this approximate solution is

$$s = b - \frac{b - a}{f(b) - f(a)} f(b),$$

The *regula falsi* and secant methods may differ in the choice of the points to be used to define the next iteration.

The geometric basis for this new approximate zero is illustrated in Fig. 2.5.

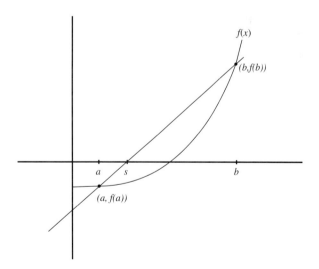

FIGURE 2.5 New approximate zero.

Chapter 2 Solving Equations of One Variable

2.2.1 Step-by-Step Computation for *Regula Falsi*

The *regula falsi* method, or the rule of false position, proceeds as in bisection to find the subinterval $[a, s]$ or $[s, b]$ that contains the zero by testing for a change of sign of the function, i.e., testing whether $f(a) \cdot f(s) < 0$ or $f(s) \cdot f(b) < 0$. If there is a zero in the interval $[a, s]$, we leave the value of a unchanged and set $b = s$. See Fig. 2.6. On the other hand, if there is no zero in $[a, s]$, the zero must be in the interval $[s, b]$; so we set $a = s$ and leave b unchanged.

The stopping condition may test the size of y, the amount by which the approximate solution s has changed on the last iteration, or whether the process has continued too long. Typically, a combination of these conditions is used.

Regula Falsi Algorithm

Initialization
  1) Define function, $f(x)$, whose zero is desired.
Begin computation block
  2) Define interval $[a, b]$ containing the desired zero.
     Compute $f(a)$ and $f(b)$
  3) Compute approximate solution,
$$s := b - f(b)*(b - a)/(f(b) - f(a))$$
     Compute $f(s)$
       If $f(s) = 0$, stop (s is the desired zero)
  4) Determine if zero is in $[a, s]$ or in $[s, b]$
       If $f(a)*f(s) < 0$, then zero is in $[a, s]$
       Otherwise, zero is in $[s, b]$
  5) Record current values of $a, s, b, f(a), f(s), f(b)$ in table.
     Decide whether to continue computation.
     If continue, change definition of a or b in step 2 as follows:
       If zero is in $[a, s]$, redefine b to have the current value of s.
       If zero is in $[s, b]$, redefine a to have the current value of s.
End computation block

Example 2.3 Finding the Cube Root of 2 Using *Regula Falsi*

To find a numerical approximation to, $\sqrt[3]{2}$ we seek the zero of $y = f(x) = x^3 - 2$, illustrated, with the first approximation, in Fig. 2.6.

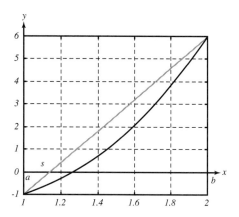

FIGURE 2.6 Graph of $y = x^3 - 2$ and approximation line in the interval $[1, 2]$.

This worksheet shows the first iteration of the Regula Falsi method to find the cube root of 2. Begin by defining the function:

$$f(x) := x^3 - 2$$

Begin the computation by giving the initial bounds on the zero:

$a := 1$ $\qquad$ $b := 2$

Evaluate the function at these points

$f(a) = -1$ $\qquad$ $f(b) = 6$

Define the approximate zero, s, evaluate it, and evaluate the function at s

$$s := b - f(b) \cdot \frac{(b-a)}{(f(b) - f(a))} \qquad s = 1.143 \qquad f(s) = -0.507$$

Record the values of a, b, s, f(a), f(b), and f(s) in the Table below.

Redefine a and b depending on the location of the zero; (must have f(a)*f(b) < 0) ; Mathcad will recompute f(a), f(b), s, and f(s).
Note that Format/Result has been set to show 4 decimal places.

Table of results for 3 iterations

a	b	s	f(a)	f(b)	f(s)
1	2	1.143	-1	6	-0.507
1.143	2	1.21	-0.507	6	-0.23
1.21	2	1.239	-0.23	6	-0.098

2.2.2 Mathcad Function for the *Regula Falsi* Method

The following Mathcad function finds a zero of a function in the interval $[a, b]$ using the *regula falsi* method. As with the bisection method, $f(a)$ and $f(b)$ must have opposite signs (and the method terminates with an error message if they do not).

The process stops when either:

1. The norm of the function value at the solution is less than "tol" or
2. The maximum number of iterations, "max", has been reached.

$$\text{Falsi}(f, a0, b0, \text{tol}, \text{max}) := \begin{array}{|l} a_0 \leftarrow a0 \\ b_0 \leftarrow b0 \\ ya_0 \leftarrow f(a_0) \\ yb_0 \leftarrow f(b_0) \\ \text{error("no root in interval")} \text{ if } ya_0 \cdot yb_0 > 0 \\ \text{for } i \in 0..\text{max} \\ \quad \begin{array}{|l} x_i \leftarrow b_i - yb_i \cdot \left(\dfrac{b_i - a_i}{yb_i - ya_i} \right) \\ y_i \leftarrow f(x_i) \\ \text{break if } |y_i| < \text{tol} \\ \text{if } ya_i \cdot y_i < 0 \\ \quad \begin{array}{|l} a_{i+1} \leftarrow a_i \\ ya_{i+1} \leftarrow ya_i \\ b_{i+1} \leftarrow x_i \\ yb_{i+1} \leftarrow y_i \end{array} \\ \text{otherwise} \\ \quad \begin{array}{|l} a_{i+1} \leftarrow x_i \\ ya_{i+1} \leftarrow y_i \\ b_{i+1} \leftarrow b_i \\ yb_{i+1} \leftarrow yb_i \end{array} \end{array} \\ x \end{array}$$

Example 2.4 Finding the Central Angle of an Elliptical Orbit

To find the central angle θ at $t = 10$ days (see Example 2-B), we must find the zero of the equation $y = 10\pi - 50\theta + 25\sin\theta$, which is illustrated in Fig. 2.7. The function `Falsi` returns the approximate value of the desired zero for iterations 0, ... 5.

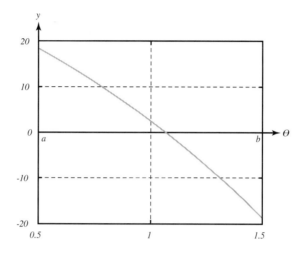

FIGURE 2.7 $y = 10\pi - 50\theta + 25\sin\theta$.

To find the central angle at day 10 of an elliptical orbit using the Regula Falsi method
Define the function whose zero is desired

$$f(\theta) := 10\cdot\pi - 50\cdot\theta + 25\cdot\sin(\theta)$$

Give the initial bounds on the zero $a := 0.5$ $b := 1.5$

and the parameter values $\text{tol} := 0.0001$ $\text{max} := 10$

The results show that the method has converged in 6 iterations.

$$\text{Falsi}(f, a, b, \text{tol}, \text{max}) = \begin{pmatrix} 0.9967 \\ 1.0577 \\ 1.065 \\ 1.0658 \\ 1.0659 \\ 1.0659 \end{pmatrix}$$

Chapter 2 Solving Equations of One Variable

2.2.3 Step-by-Step Computation for the Secant Method

The secant method, closely related to the *regula falsi* method, results from a slight modification of the latter. Instead of choosing the subinterval that must contain the zero, we form the next approximation from the two most recently generated points:

$$x_2 = x_1 - \frac{x_1 - x_0}{y_1 - y_0} y_1.$$

At the kth stage, the new approximation to the zero is

$$x_{k+1} = x_k - \frac{x_k - x_{k-1}}{y_k - y_{k-1}} y_k.$$

This process does not require any testing to determine the action to take at the next step. It also converges more rapidly than the *regula falsi* method, in general, even though the zero is not required to lie within the interval at each stage.

The following algorithm uses the same notation for the iterates as does the *regula falsi* algorithm. This facilitates Mathcad's recomputation of the appropriate values at each stage of the computation.

Secant Algorithm

Initialization

1) Define function, $f(x)$, whose zero is desired.

Begin computation block

2) Define first two approximations, a and b ($a \neq b$).
   It is not required that $a < b$, or that there actually be a zero between a and b.
   Evaluate $f(a)$ and $f(b)$.

3) Compute new approximate solution,

$$s = b - f(b)*(b - a)/(f(b) - f(a))$$

   Evaluate $f(s)$.
   If $f(s) = 0$, stop

4) Record current values of a, b, s, $f(a)$, $f(b)$, and $f(s)$ in a table.
   Decide whether to continue computations.
   If continue, update definitions of a and b in step 2, by setting

$$a = b, \qquad b = s.$$

End computation block

Example 2.5 Finding the Square Root of 3 Using the Secant Method

The graph of the function is shown in Figure 2.1. Comparing the results found here with the secant method to those found in Example 2.1, using the bisection method clearly illustrates the superior results that are typical for the secant method.

This worksheet shows the first iteration of using the secant method to find the square root of 3. Begin by defining the function:

$$f(x) := x^2 - 3$$

Begin the computation by giving the initial bounds on the zero:

$$a := 1 \qquad b := 2$$

Evaluate the function at these points

$$f(a) = -2 \qquad f(b) = 1$$

Define the approximate zero, s, evaluate it, and evaluate the function at s

$$s := b - f(b) \cdot \frac{(b - a)}{(f(b) - f(a))}$$

$$s = 1.6667$$

$$f(s) = -0.2222$$

Record the values of a, b, s, f(a), f(b), and f(s) in the Table below. Redefine a and b; Mathcad will recompute f(a), f(b), s, and f(s).
Note that Format/Result has been set to show 4 decimal places.

Table of results for 3 iterations

a	b	s	f(a)	f(b)	f(s)
1	2	1.6667	-2	1	-0.2222
2	1.6667	1.7273	1	-0.2222	-0.0165
1.6667	1.7273	1.7321	-0.75	0.063	3.1886×10^{-4}

2.2.4 Mathcad Function for the Secant Method

The following Mathcad function utilizes the secant method to find a zero of the given function, using the starting estimates $x_0 = a$ and $x_1 = b$. In contrast to the bisection and *regula falsi* methods, $f(a)$ and $f(b)$ need not have opposite signs, and there is no guarantee that there is a zero in the interval between two successive approximations, either initially or at any stage of the iteration. The process stops if any of the following conditions occur:

1. The magnitude of the function at the approximate zero is less than "tol".
2. The magnitude of the change in the approximate zero is less than "tol".
3. The maximum number of iterations, "max", has been reached.

$$\text{Secant}(f, a, b, \text{tol}, \text{max}) := \begin{vmatrix} x_0 \leftarrow a \\ y_0 \leftarrow f(a) \\ x_1 \leftarrow b \\ y_1 \leftarrow f(b) \\ \text{for } i \in 1..\text{max} \\ \quad \begin{vmatrix} \text{return } x \quad \text{if } |y_i| < \text{tol} \\ \text{return } x \quad \text{if } |x_i - x_{i-1}| < \text{tol} \\ x_{i+1} \leftarrow x_i - y_i \cdot \left(\dfrac{x_i - x_{i-1}}{y_i - y_{i-1}} \right) \\ y_{i+1} \leftarrow f(x_{i+1}) \end{vmatrix} \\ \text{return "Zero not found to desired tolerance"} \quad \text{if } i \geq \text{max} \\ x \end{vmatrix}$$

Example 2.6 Finding the Central Angle Using the Secant Method

To find the central angle at day 20 of an elliptical orbit

Define the function whose zero is desired

$$f(\theta) := 20 \cdot \pi - 50 \cdot \theta + 25 \cdot \sin(\theta)$$

Give the initial bounds on the zero

$a := 1 \qquad b := 2$

and the parameter values

$\text{tol} := 0.0001 \qquad \text{max} := 10$

The results show that the method has converged in 4 iterations; (the first two results listed are the user supplied initial bounds). Note that Format/Result has been set to show 6 decimal places.

$$\text{Secant}(f, a, b, \text{tol}, \text{max}) = \begin{pmatrix} 1 \\ 2 \\ 1.7012 \\ 1.7462 \\ 1.7488 \\ 1.7487 \end{pmatrix}$$

2.2.5 Discussion

The standard method of comparing how fast various iterative zero finding methods converge is to investigate the behavior of $\frac{|x_k - x^*|}{|x_{k-1} - x^*|^p}$ for large values of k; as in the discussion of the bisection method, we denote the true zero x^*.

Rate of convergence

If

$$\lim_{k \to \infty} \frac{|x_k - x^*|}{|x_{k-1} - x^*|^p} = \lambda \text{ for some } \lambda > 0,$$

we say that the sequence x_k converges to x^* with order $p > 0$; the number λ is called the *asymptotic error constant*. In general, higher values of p give faster convergence.

If a sequence converges with $p = 1$, we say it is linearly convergent;
If a sequence converges with $p = 2$, we say it converges quadratically.

It is not too hard to see that if a function is concave up on the interval $[a_k, b_k]$, the point $(b_k, y(b_k))$ will not change during the *regula falsi* iterations $(k + 1, \ldots.)$ (as illustrated in Fig. 2.8); similarly, if the function is concave down, the point $(a_k, y(a_k))$ does not change. At some stage, one or the other of these conditions will be met, and from that stage onward, the convergence is linear. The *regula falsi* method can be modified to improve the convergence order to 1.4 or 1.6. (See Ralston and Rabinowitz, 1978, for further discussion.)

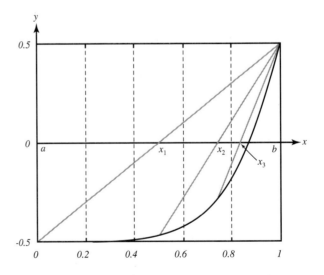

FIGURE 2.8 A function that is concave up near the zero.

Although it is appealing to know that the zero is bracketed at each step of the process, some effort is required to find the appropriate subinterval at each stage. Also, an iterative process in which the form of the function being iterated can change at each stage (as is the case for a process where a choice must be made as to which subinterval to pursue) is more difficult to analyze.

The calculation of the update formula for the secant method is the same as that for the *regula falsi* method; only the choice of which two of the x values are used for the next iteration differs. Because the secant method does not bracket the zero at each iteration (as do the bisection and *regula falsi* methods), the secant method is not guaranteed to converge. However, when it does, the convergence is usually more rapid than for either bisection or *regula falsi*. Typically, as the iterations progress, the secants become increasingly more accurate approximations to $f(x)$.

The rate of convergence is $p = \dfrac{1 + \sqrt{5}}{2} \approx 1.62$, so convergence is faster than linear, but less than quadratic. The asymptotic error constant is $\lambda = \left|\dfrac{f''(x^*)}{2f'(x^*)}\right|^{\beta}$, with $\beta = \dfrac{\sqrt{5} - 1}{2}$.

Convergence Theorem for the Secant Method

If

1. $f(x), f'(x)$ and $f''(x)$ are continuous on $I = [x^* - e, x^* + e]$
2. $f'(x^*) \neq 0$, and
3. the initial estimates x_0 and x_1 (in I) are sufficiently close to x^*, then the secant method will converge.

The requirements for "sufficiently close" can be made more precise by defining $M = \dfrac{\max |f''|}{2 \min |f''|}$, where max and min are for all x in the interval $I = [x^* - e, x^* + e]$. Then if $\max \{M | x^* - x_0|, M | x^* - x_1|\} < 1$, the secant method will converge. It may converge for starting values that do not satisfy this inequality, but in general, the larger the value of M (if it can be computed), the closer the starting values should be to the zero. (See Atkinson, 1989, pp. 65 to 73, for a development of these results.)

We conclude our discussion of the *regula falsi* and secant methods by considering the challenging problem of finding the zero of a function that is quite flat near the desired zero.

Example 2.7 A Challenging Problem

To illustrate the difficulties that may occur when a function is relatively flat near a zero, consider the simple example of finding the positive real zero of $y = x^5 - 0.5$. The function and the first three straight-line approximations are illustrated in Fig. 2.9. The results of the calculations using *regula falsi* are summarized in the table below. The *regula falsi* method converges in nine iterations with tol = 0.0001.

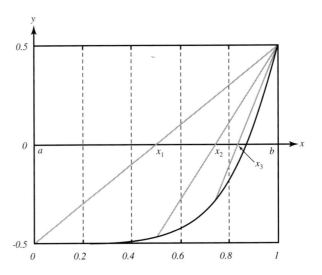

FIGURE 2.9 $y = x^5 - 0.5$ and first three approximations using *regula falsi*.

Calculation of $\sqrt[5]{0.5}$ using *regula falsi*.

step	a	b	x	y
1	0	1	0.5	−0.46875
2	0.5	1	0.74194	−0.27518
3	0.74194	1	0.83355	−0.09761
4	0.83355	1	0.86801	−0.0072543
6	0.86801	1	0.8699	−0.0018732
7	0.8699	1	0.87038	−0.00048132
8	0.87038	1	0.87051	−0.00012352
9	0.87051	1	0.87054	−3.1687e-05

Now consider the problem of finding the positive real zero of $y = x^5 - 0.5$ using the secant method (see Fig. 2.10). As shown in the table, the third approximation is outside the original interval that contains the root. Nevertheless, after seven iterations, the method has achieved essentially the same result as *regula falsi*.

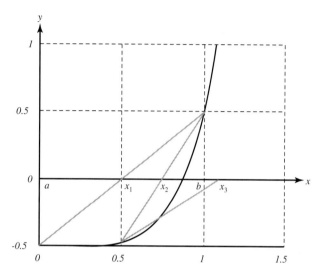

FIGURE 2.10 $y = x^5 - 0.5$ and first three approximations using secant method.

Calculation of $\sqrt[5]{0.5}$ using the secant method.

step	x	y
1	0.5	−0.46875
2	0.74194	−0.27518
3	1.0859	1.0098
4	0.81559	−0.13911
5	0.84832	−0.060656
6	0.87362	0.008891
7	0.87039	−0.00046064

On the other hand, if we give the starting values as $x_0 = 1$ and $x_1 = 0$, we again find $x_2 = 0.5$, but the calculation of x_3 is based on a straight line with almost zero slope (the line determined by the points $(0, -0.5)$ and $(0.5, -0.46875)$). There are extreme oscillations before the method converges to the correct value. Although it may seem contrived to give the starting estimates in this "nonnatural" order, completely analogous behavior occurs for $y = x^5 + 0.5$ when the starting estimates are $x_0 = -1$ and $x_1 = 0$.

Chapter 2 Solving Equations of One Variable

2.3 NEWTON'S METHOD

Like the *regula falsi* and secant methods, Newton's method uses a straight-line approximation to the function whose zero we wish to find, but in this case the line is the tangent to the curve. The next approximation to the zero is the value of x where the tangent crosses the x-axis. This requires additional information about the function (i.e., its derivative). The secant method discussed in the previous section can be viewed as Newton's method with the derivative approximated by a difference quotient.

2.3.1 Step-by-Step Computation

Given an initial estimate of the zero, x_0; the value of the function at x_0, $y_0 = f(x_0)$; and the value of the derivative at x_0, $y'_0 = f'(x_0)$; the x intercept of the tangent line, which is the new approximation to the zero, is

$$x_1 = x_0 - \frac{y_0}{y'_0}.$$

The process continues until the change in the approximations is sufficiently small or some other stopping condition is satisfied. At the kth stage, we have

$$x_{k+1} = x_k - \frac{y_k}{y'_k}.$$

Newton's Method Algorithm

Initialization

1) Define function, $f(x)$, whose zero is desired.
   Define $f'(x)$.

Begin computation block

2) Define first approximation of solution, x_0.
   Compute function value, $y_0 = f(x_0)$.
3) Compute new approximate solution,

$$x_s = x_0 - y_0/f'(x_0)$$

   Compute $y_s = f(x_s)$.
4) Record current approximate solution x_s, and function value y_s, in table.
   Decide whether to continue computations.
   If continue, update definition of x_0 in step 2: $x_0 = x_s$

End computation block

Example 2.8 Finding the Square Root of 3/4 Using Newton's Method

The function and the first tangent approximation are shown in Fig. 2.11.

This worksheet shows the first iteration of using Newton's method to find the square root of 3/4. Begin by defining the function and its derivative:

$$f(x) := 4x^2 - 3 \qquad g(x) := 8 \cdot x$$

Begin the computation by giving an initial estimate of the zero:

$$xold := 1$$

Define the approximate zero, xnew, evaluate it, and evaluate the function at xnew

$$xnew := xold - \frac{f(xold)}{g(xold)} \qquad xnew = 0.875 \qquad f(xnew) = 0.063$$

Record the values of xnew and f(xnew) in the Table below.
Redefine xold to be equal to xnew; Mathcad will recompute the required function values.
Note that Format/Result has been set to show 4 decimal places.

Table of results for 3 iterations, starting with initial estimate of xold = 1

xnew	f(xnew)
0.875	0.063
0.866	3.189×10^{-4}
0.866	2.581×10^{-9}

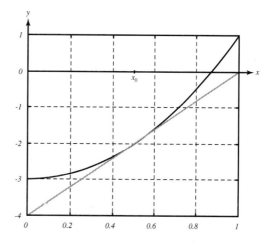

FIGURE 2.11 The graph of $y = 4x^2 - 3$ and the tangent line approximation at $x = 0.5$.

2.3.2 Mathcad Function for Newton's Method

The following Mathcad function utilizes Newton's method to find a zero of the given function, using the starting estimate x_0. The process stops when either:

1. The magnitude of the function at the approximate zero is less than "tol."
2. The maximum number of iterations, "max," has been reached.

The function for Newton's method given here does not protect against the difficulties that occur if the derivative value at an approximate solution is either zero, or nearly zero.

$$\text{Newton}(f, df, x0, tol, max) := \begin{vmatrix} x_0 \leftarrow x0 \\ y_0 \leftarrow f(x0) \\ dy_0 \leftarrow df(x0) \\ \text{for } i \in 1..max \\ \quad \begin{vmatrix} x_i \leftarrow x_{i-1} - \dfrac{y_{i-1}}{dy_{i-1}} \\ y_i \leftarrow f(x_i) \\ \text{return } x \text{ if } |y_i| < tol \\ dy_i \leftarrow df(x_i) \end{vmatrix} \\ x \end{vmatrix}$$

Example 2.9 Finding the Floating Depth for a Wooden Ball

The function is shown in Fig. 2.12.

To find the floating depth of a ball of relative density 1/3 using Newton's method, begin by defining the function whose zero is sought, and its derivative:

$$f(x) := x^3 - 3 \cdot x^2 + \frac{4}{3} \qquad\qquad g(x) := 3 \cdot x^2 - 6 \cdot x$$

Then define parameters and call the function, Newton

$\quad$ x0 := 0.5 $\qquad$ tol := 0.0001 $\qquad$ max := 8

The results show that the method has converged in 3 iterations (the first result is the initial estimate of the zero).

$$\text{Newton}(f, g, x0, \text{tol}, \text{max}) = \begin{pmatrix} 0.5 \\ 0.8148 \\ 0.7743 \\ 0.7739 \end{pmatrix}$$

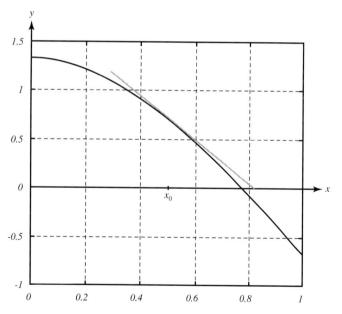

FIGURE 2.12 Floating depth of wooden ball

2.3.3 Discussion

Newton's method is based on using the tangent line to the function $y = f(x)$ at the current approximate zero to find the new approximation to the zero.

The tangent line at (x_0, y_0) has slope $m = f'(x_0)$, so the equation of the tangent line can be written as

$$y - y_0 = m(x - x_0).$$

To find the value of x where this line crosses the x-axis, set $y = 0$ and solve for x:

$$0 - y_0 = m x - m x_0$$

$$m x_0 - y_0 = m x$$

$$x = x_0 - \frac{y_0}{m}$$

This value of x is the new approximate zero, since $m = f'(x_0) = y'_0$.

We can obtain more information about the approximation to the zero x^* if we consider the Taylor series expansion for $f(x)$ near x_k, i.e.,

$$f(x) = f(x_k) + (x - x_k)f'(x_k) + 0.5(x - x_k)^2 f''(\eta)$$

where η is some unknown point between x and x_k. If $x = x^*$ and $f(x^*) = 0$, then

$$0 = f(x_k) + (x^* - x_k) f'(x_k) + 0.5(x^* - x_k)^2 f''(\eta),$$

and

$$x^* = x_k - \frac{f(x_k)}{f'(x_k)} - 0.5(x^* - x_k)^2 \frac{f''(\eta)}{f'(x_k)}.$$

If we now set

$$x_{k+1} = x_k - \frac{f(x_k)}{f'(x_k)}$$

and substitute this into the previous equation, we can solve for the error at the $(k + 1)$st approximation:

$$x^* - x_{k+1} = -0.5(x^* - x_k)^2 \frac{f''(\eta)}{f'(x_k)}.$$

Convergence Theorem for Newton's Method

If

1. $f(x), f'(x)$ and $f''(x)$ are continuous for all x in a neighborhood of x^*,
2. $f'(x^*) \neq 0$, and
3. x_0 is chosen sufficiently close to x^*,

then the iterates

$$x_{k+1} = x_k - \frac{f(x_k)}{f'x_k}$$

will converge to x^*.

Furthermore,

$$\lim_{k \to \infty} \frac{x_k - x^*}{(x_{k-1} - x^*)^2} = \frac{f''(x^*)}{2f'(x^*)}.$$

This result can be used to obtain some information about how close the initial estimate x_0 must be to the actual zero. Let I be an interval around x^* such that $f'(x) \neq 0$ on I, and let

$$M = \frac{\max |f''(x)|}{2 \min |f'(x)|},$$

where the max and min are taken over all x in I; then convergence is guaranteed if the initial estimate is chosen close enough to the true zero so that $|x^* - x_0| < 1/M$. (See Atkinson, 1989, p. 60, for further discussion.)

Newton's method is quadratically convergent (the order of convergence is $p = 2$), with asymptotic error constant $\left|\frac{f''(x^*)}{2f'(x^*)}\right|$.

Newton's method has a high order of convergence and a fairly simple statement; hence, it is often the first method people use. However, the method can encounter difficulties, as illustrated in the next example.

Furthermore, the derivative must not be zero at any approximation to the zero, or Newton's method will fail. The stopping condition should be a combination of a specified maximum number of iterations and minimum tolerance on the change in the computed zero. Because of the potential for divergence, it is also wise to test for large changes in the value of the computed zero, which might signal difficulties.

Example 2.10 Oscillations in Newton's Method

Newton's method can give oscillatory results for some functions and some initial estimates. For example, consider the cubic equation

$$y = x^3 - 3x^2 + x + 3.$$

The derivative is $y' = 3x^2 - 6x + 1$. If we don't consider the graph, we might guess $x_0 = 1$ as the initial point. The calculations for the first three iterations are summarized in the table. The tangent approximations for the first two iterations are illustrated in Fig. 2.13. It is easy to see that the process will oscillate between these two values.

Oscillations resulting from the use of Newton's method.

step	x_{i-1}	x_i	y_{i-1}	y_i
1	1	2	2	1
2	2	1	1	2
3	1	2	2	1

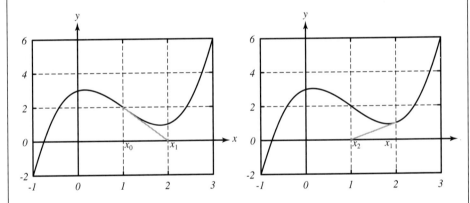

FIGURE 2.13 Oscillatory behavior of Newton's method.

2.4 MULLER'S METHOD

A logical extension of the methods based on linear approximations to the function whose zero we are seeking is to approximate the function by a quadratic function. This method is known as Muller's method. It has the advantage of being able to generate approximations to complex zeros even if the initial estimates are real. The idea is quite simple, although the formulas are more complicated than in the previous methods. Using three points on the function, we can find the equation of the quadratic that passes through those points and then find the zeros of that quadratic. It is also possible to start with two points that bracket the zero and use the midpoint of the interval between the points as the third point. In general, Muller's method is less sensitive to starting values than Newton's method is.

The parabola passing through the points (x_a, y_a), (x_b, y_b), and (x_c, y_c) can be written as $y = y_c + q(x - x_c) + r(x - x_c)(x - x_b)$, where $p = (y_b - y_a)/(x_b - x_a)$ and the coefficients are $q = (y_c - y_b)/(x_c - x_b)$ and $r = (q - p)/(x_c - x_a)$. This somewhat strange form is closely related to Newton's form of an interpolating polynomial, which is discussed in Chapter 8. Letting $s = c_2 + r(x_3 - x_2)$ and solving for the zero that is closest to x_c gives the next approximation to the desired zero:

$$x_n = x_c - \frac{2 y_c}{s + \text{sign}(s) \sqrt{s^2 - 4 y_c r}}.$$

Of course, once the formulas for the new approximation are derived, it is not necessary to actually find the equation of the quadratic at each step.

The first step is illustrated in Fig. 2.14.

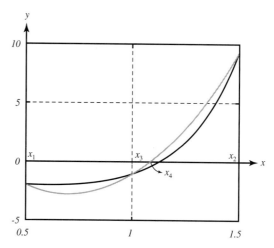

FIGURE 2.14 $y = x^6 - 2$ (solid line) and parabola (dashed line) for Muller's method.

2.4.1 Step-by-Step Computation for Muller's Method

Muller's Method Algorithm

Initialization

1) Define function, $f(x)$, whose zero is desired.

Begin computation block:

2) Define first three (distinct) approximations of solution, x_a, x_b, and x_c. Compute corresponding function values,

$$y_a = f(x_a),$$
$$y_b = f(x_b), \text{ and}$$
$$y_c = f(x_c).$$

3) Compute new approximate solution, x_r

$$p = (y_b - y_a)/(x_b - x_a)$$
$$q = (y_c - y_b)/(x_c - x_b)$$
$$r = (q - p)/(x_c - x_a)$$
$$s = q + r(x_c - x_b)$$
$$x_n = x_c - 2 y_c/(s + \text{sign}(s)*\sqrt{s^2 - 4 y_c r}.$$

Compute $y_n = f(x_n)$.

4) Record current approximate solution x_n, and function value y_n, in table. Decide whether to continue computations.
If continue, update definitions in step 2:

$$x_a = x_b,$$
$$x_b = x_c, \text{ and}$$
$$x_c = x_n.$$

(Also update function values to reduce computation)

End computation block

(If desired, notation can be adjusted to allow some parameter values to be reused from one step to the next.)

Example 2.11 Finding the Sixth Root of 2 Using Muller's Method

The worksheet shows the first iteration of the computations to approximate the real zero of $y = f(x) = x^6 - 2$. The graph of $f(x)$ is shown in Fig. 2.14. The table summarizes the results of the first few interations.

Begin by defining the function

$$f(x) := x^6 - 2$$

Give three starting estimates of the zero, xa, xb, and xc:

    xa := 0.5   xb := 1.5   xc := 1.0

Define the function values at these points as ya, yb, and yc

    ya := f(xa)   yb := f(xb)   yc := f(xc)

Define required parameters

$$p := \frac{(yb - ya)}{(xb - xa)} \qquad q := \frac{(yc - yb)}{(xc - xb)} \qquad r := \frac{(q - p)}{(xc - xa)} \qquad s := q + r \cdot (xc - xb)$$

Compute new approximate zero, xn, and display its value and the function value, yn

$$xn := xc - 2 \cdot \frac{yc}{\left[s + \text{sign}(s) \cdot (s^2 - 4 \cdot yc \cdot r)^{0.5}\right]} \qquad yn := f(xn)$$

xn = 1.078   yn = −0.432

Record values in table, and update values of xa, xb, and xc for next iteration

xa	xb	xc	xn	yn
0.5	1.5	1.0	1.078	-0.432
1.5	1.0	1.078	1.117	-0.057
1.0	1.078	1.117	1.123	7.603×10^{-4}

2.4.2 Mathcad Function for Muller's Method

The following Mathcad function for Muller's method requires two user-supplied starting estimates for the zero. (Typically these are chosen so that they define an interval that brackets the zero.) The midpoint of the interval is taken to be the third starting point. The function returns a vector that gives the approximate zero at each step of the process (including the starting values).

$$\text{Muller}(f, a, b, \text{tol}, \text{max}) := \begin{vmatrix} x_0 \leftarrow a \\ x_1 \leftarrow b \\ x_2 \leftarrow \dfrac{a+b}{2} \\ y_0 \leftarrow f(a) \\ y_1 \leftarrow f(b) \\ y_2 \leftarrow f(x_2) \\ c_0 \leftarrow \dfrac{y_1 - y_0}{x_1 - x_0} \\ \text{for } i \in 2 .. \text{max} \\ \quad \begin{vmatrix} c_{i-1} \leftarrow \dfrac{y_i - y_{i-1}}{x_i - x_{i-1}} \\ d_{i-2} \leftarrow \dfrac{c_{i-1} - c_{i-2}}{x_i - x_{i-2}} \\ s \leftarrow c_{i-1} + (x_i - x_{i-1}) \cdot d_{i-2} \\ x_{i+1} \leftarrow x_i - \dfrac{2 \cdot y_i}{s + \text{sign}(s) \cdot \sqrt{s^2 - 4 \cdot y_i \cdot d_{i-2}}} \\ y_{i+1} \leftarrow f(x_{i+1}) \\ \text{break if } |y_{i+1}| < \text{tol} \end{vmatrix} \\ x \end{vmatrix}$$

Example 2.12 Another Challenging Problem

Consider the problem of finding a zero of $y = x^{10} - 0.5$. The function and the first parabolic approximation are shown in Fig. 2.15.

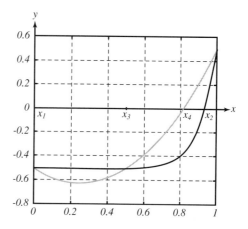

FIGURE 2.15 $y = x^{10} - 0.5$ (solid line) and parabola (dashed line) for Muller's method.

We begin by defining the function

$$f(x) := x^{10} - 0.5$$

the first two estimates of the zero

$a := 0 \qquad b := 1$

and the parameters

$\text{tol} := 0.0001 \qquad \text{max} := 10$

The results show that the method has converged in six iterations (the first three values listed are the initial estimates and the midpoint, which is taken as the third starting value). Format/Result has been set to show 6 decimal places.

$$\text{Muller}(f, a, b, \text{tol}, \text{max}) = \begin{pmatrix} 0 \\ 1 \\ 0.5 \\ 0.808747 \\ 0.908105 \\ 0.943252 \\ 0.932688 \\ 0.933032 \end{pmatrix}$$

2.4.3 Discussion

The parabola passing through the points (x_1, y_1), (x_2, y_2), and (x_3, y_3) can be written as

$$y = y_3 + c_2(x - x_3) + d_1(x - x_3)(x - x_2), \tag{2.1}$$

where the coefficients are found by first calculating

$$c_1 = \frac{y_2 - y_1}{x_2 - x_1}.$$

Then the coefficients are found to be

$$c_2 = \frac{y_3 - y_2}{x_3 - x_2}, \quad \text{and} \quad d_1 = \frac{c_2 - c_1}{x_3 - x_1}.$$

We write the equation for the zeros of eq. 2.1 in terms of powers of $(x - x_3)$:

$$0 = y_3 + c_2(x - x_3) + d_1(x - x_3)(x - x_2),$$
$$0 = y_3 + c_2(x - x_3) + d_1(x - x_3)(x - x_3 + x_3 - x_2),$$
$$0 = y_3 + c_2(x - x_3) + d_1(x - x_3)(x - x_3) + d_1(x - x_3)(x_3 - x_2),$$

Letting $s = c_2 + d_1(x_3 - x_2)$, we wish to find the zero of

$$0 = y_3 + s(x - x_3) + d_1(x - x_3)^2$$

that is closest to x_3; so we divide by $(x - x_3)^2$, set $z = (x - x_3)^{-1}$, and find the smallest zero of

$$0 = y_3 z^2 + s z + d_1.$$

Using the quadratic formula (and choosing the zero with the largest magnitude, so that $z^{-1} = x - x_3$ will be as small as possible) gives

$$z = \frac{-s - \text{sign}(s) \sqrt{s^2 - 4 y_3 d_1}}{2 y_3},$$

which, after some algebra, yields

$$x = x_3 - \frac{2 y_3}{s + \text{sign}(s) \sqrt{s^2 - 4 y_3 d_1}}.$$

The notation in the derivation above is designed to facilitate the reuse of the indexed parameters c_i from one step of the method to the next. The new approximate root becomes x_4, and the next stage is based on the points (x_2, y_2), (x_3, y_3) and (x_4, y_4); the parameters at the next stage are c_2, c_3 and d_2.

In contrast to Newton's method, Muller's method requires only function values; the derivative need not be calculated. Another advantage of Muller's method is that it may be used to find complex as well as real zeros. The method fails if $f(x_1) = f(x_2) = f(x_3)$, which can occur if x is a zero of multiplicity greater than 2.

The rate of convergence of Muller's method is slightly less than quadratic, since $p \approx 1.84$; the asymptotic error constant is $\left|\dfrac{f''(x^*)}{2f'(x^*)}\right|^\beta$, where p is the positive root of $f(x) = x^3 - x^2 - x - 1$, and $\beta = \dfrac{p-1}{2}$. (See Atkinson, 1989, p. 67, for a development of these results.)

2.5 MATHCAD'S METHODS

Mathcad has two built-in functions for finding roots of equations of the form $f(x) = 0$. One function, polyroots, finds all roots (real or complex) of a polynomial with real or complex coefficients. The second function, root, finds one root of the specified function, $f(x)$. Roots of such equations are also known as zeros of the function $f(x)$. We begin this section by describing the use of each of these functions. We then consider in more detail the methods implemented by each function.

2.5.1 Using the Built-in Functions

Using polyroots

There are two steps required to use the function polyroots to find the roots of a polynomial equation $p(x) = 0$.

1) Define the polynomial by defining the vector **v** of the coefficients of the polynomial $p(x)$, starting with the constant term and including 0's for any terms that do not explicitly appear. If the polynomial is of degree n, **v** has $n + 1$ components.

2) The function returns a vector of the n roots of the polynomial equation $p(x) = 0$. To obtain a vector of the roots, one can call the function as polyroots(v)=; on the other hand, to define a vector of the roots to use later, set r:= polyroots(v)).

Example 2.13 Using `polyroots`

To use the Mathcad polyroots function to find the roots of

$$p(x) := x^3 - 7x^2 + 14x - 7$$

Define the vector of coefficients

$$c := \begin{pmatrix} -7 \\ 14 \\ -7 \\ 1 \end{pmatrix}$$

Define a variable, X, to be the result of the call to the function polyroots, and display its value.

$$X := \text{polyroots}(c) \qquad X = \begin{pmatrix} 0.753 \\ 2.445 \\ 3.802 \end{pmatrix}$$

Define a range variable for plotting p(x)

$$x := 0.2, 0.3 .. 4$$

Plot p(x) and the roots, p(X), using Insert/Graph from the menu.

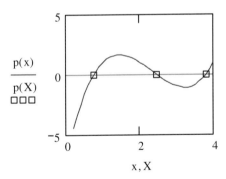

FIGURE 2.16 Polynomial and its roots.

Using roots

There are two ways of using the function root to find the roots of a general equation of the form $f(x) = 0$. These are the bracketed form and the unbracketed form. It may be helpful to create a graph of the function $f(x)$ before calling either of the forms of root, in order to have a good idea of the locations of the solutions. We begin by describing the use of the unbracketed form.

The *unbracketed form* of the function root requires an initial guess of the solution to begin its search for a zero of the function $f(x)$. The function whose zero is desired may be defined in advance, or its defining expression may be included in the call to root.

1) Set initial guess for value of the zero: x :=
2) Call function as root(f(x), x) =

Using different initial guesses may give different results. It is not permitted to put the numerical value of the initial guess in the list of calling parameters for the function root. The function root can find complex roots, but in order for it to do so the initial guess must be complex. The function root can solve equations of the form $f(x) = g(x)$ by calling the function as root(f(x)-g(x), x) =

Example 2.14 Using the Function root without Bracketing the Solution

To use the Mathcad function, root, in the unbracketed mode, one must give a starting estimate of the root.

x := 1

To find the 6th root of 2, one may call the function root as

$$\text{root}(x^6 - 2, x) = 1.122$$

or one may define the function f(x), and the initial estimate, x, and then call root(f(x),x)

$$f(x) := x^6 - 2 \qquad x := 1$$

$$\text{root}(f(x), x) = 1.122$$

The *bracketed form* of the function root does not require an initial guess for the solution, but as its name suggests, it does require two values of the independent variable that bracket the desired root. That is, the function is called as

```
root(f(x), x, a, b)
```

where $f(a)$ and $f(b)$ must have opposite signs, but x does not need to have a value assigned before the function is called.

The bracketed form of the function root is limited to real valued functions of a real variable, and only real roots are produced.

Example 2.15 Using the Function root **with Bracketing of the Solution**

To use the Mathcad function, root, in the bracketed mode, one must give two starting values that bracket the root.

To find the 6th root of 2, one may call the function root as

$$\text{root}\left(x^6 - 2, x, 0, 2\right) = 1.122$$

or one may define the function f(x), and then call root(f(x),x,a,b)

$$f(x) := x^6 - 2$$

$$\text{root}(f(x), x, 0, 2) = 1.122$$

2.5.2 Understanding the Algorithms

The function polyroots implements the Laguerre method with deflation and polishing as its default method. An alternative method, based on the eigenvalues of the companion matrix is also available (by right-clicking on the word polyroots and changing the method selection on the menu).

Laguerre's method is the most straightforward of a group of methods that are guaranteed to converge to all types of roots of a polynomial, whether real or complex, single or repeated. It requires that the computations be carried out in complex arithmetic. The motivation for the method is based on the following relationships.

If we write the polynomial as

$$P_n(x) = (x - x_1)(x - x_2) \ldots (x - x_n)$$

and take logarithms of both sides of the equation, we obtain

$$\ln |P_n(x)| = \ln |x - x_1| + \ln |x - x_2| + \ldots + \ln |x - x_n|$$

The first derivative is

$$\frac{d}{dx}(\ln|P_n(x)|) = \frac{1}{x-x_1} + \frac{1}{x-x_2} + \ldots + \frac{1}{x-x_n}$$

and the negative of the second derivative is

$$-\frac{d^2}{dx^2}(\ln|P_n(x)|) = \frac{1}{(x-x_1)^2} + \frac{1}{(x-x_2)^2} + \ldots + \frac{1}{(x-x_n)^2}$$

We define

$$G = \frac{1}{x-x_1} + \frac{1}{x-x_2} + \ldots + \frac{1}{x-x_n}$$

and

$$H = \frac{1}{(x-x_1)^2} + \frac{1}{(x-x_2)^2} + \ldots + \frac{1}{(x-x_n)^2}$$

Let a be the approximate distance of the desired root, x_1 from the current guess x, and let the distance of each of the other roots, $x_2, \ldots x_n$ from the current guess x be b. Expressing G and H in terms of these two variables, we get

$$G = \frac{1}{a} + \frac{n-1}{b} \quad \text{and} \quad H = \frac{1}{a^2} + \frac{n-1}{b^2}$$

Solving for a gives

$$a = \frac{n}{G + -\sqrt{(n-1)(nH - G^2)}}$$

where the sign in the denominator is chosen to give the largest possible magnitude for the denominator. Starting with an initial value for x, the next approximation to the root is $x + a$.

This motivation is based on discussion in [Press, et al., 1992]; for a more rigorous derivation see [Ralston and Rabinowitz, 1978].

The unbracketed form of the function `root` implements the secant and Muller methods, which have been discussed in Sections 2.2.2 and 2.4.

The bracketed form of the function `root` uses Ridders' method, or if that fails to find a root, the Brent method.

Ridders' method, a variation of the *regula falsi* method, uses three starting estimates of the desired root. The first two estimates must bracket the root; the third estimate is the midpoint between the first two. The derivation of Ridders' method utilizes (as an intermediate step) a quadratic equation involving an unknown exponential function. The new approximation to the root comes from applying the regula falsi method to a function which is a combination of the original function and this exponential function. The formula for the new approximation is independent of this unknown exponential function. For initial estimates $x_1 < x_2$ (and $x_3 = (x_1 + x_2)/2$), and the corresponding function values $y_1 = f(x_1)$, $y_2 = f(x_2)$ and $x_3 = f(x_3)$, the next estimate is

$$x_4 = x_3 + (x_3 - x_1) \frac{\text{sign}[y_1 - y_2] y_3}{\sqrt{y_3^2 - y_1 y_2}}$$

For the next step, one determines which of the previous estimates, x_1, x_2 or x_3 to retain, along with x_4, so that the root is still bracketed. The new midpoint, and the two new function values (for x_4 and the new midpoint) are computed, and the next approximate root is found from the formula above.

One of the nice properties of Ridders' method is that at each stage the new estimate is always within the bracketing interval (x_1, x_2) for that step. Another is that the convergence is quadratic; the number of significant digits in the solution approximately doubles with each step of the method.

For further details on this method, see [Ridders, 1979] or [Press et al., 1992, p. 358]

Brent's method, published in 1973, is an improved version of a general root-finding method developed at the mathematical Center in Amsterdam in the 1960s. The basic idea is to combine the guaranteed (but often slow) convergence of the bisection method with a more rapid (but not always convergent) method. In Brent's method, the more rapidly convergent method is inverse quadratic interpolation. That is, the method is based on using three points on the function whose root is desired to express x as a quadratic function of y (that is the inverse quadratic interpolation). Then by setting $y=0$, a new approximate root is obtained. Interpolation is discussed in Chapter 8. If the points to be interpolated are denoted

$$(x_a, y_a), (x_b, y_b) \text{ and } (x_c, y_c)$$

then it is straightforward to verify that they each satisfy the following quadratic function $(x = f(y))$

$$x = \frac{(y - y_b)(y - y_c)}{(y_a - y_b)(y_a - y_c)} x_a + \frac{(y - y_a)(y - y_c)}{(y_b - y_a)(y_b - y_c)} x_b + \frac{(y - y_a)(y - y_b)}{(y_c - y_a)(y_c - y_b)} x_c$$

Setting $y = 0$ gives the new approximate root x_n as

$$x_n = \frac{y_b y_c}{(y_a - y_b)(y_a - y_c)} x_a + \frac{y_a y_c}{(y_b - y_a)(y_b - y_c)} x_b + \frac{y_a y_b}{(y_c - y_a)(y_c - y_b)} x_c$$

The method must also perform various bookkeeping tasks in order to decide when bisection steps should be performed, and to make sure that the root remains bracketed by the appropriate approximations.

For further details, see [Brent, 1973] or [Press et al., 1992, p. 362]

SUMMARY

Bisection: To find a root in $[a, b]$, where $f(a) f(b) < 0$:
Find $m = (a+b)/2$; determine whether the root is in $[a, m]$ or in $[m, b]$ by testing whether $f(a) f(m) < 0$; continue the process on an appropriate subinterval.

***Regula Falsi* and Secant Methods:** Given two points (x_{k-1}, y_{k-1}) and (x_k, y_k), the next approximation to the root is

$$x_{k+1} = x_k - \frac{x_k - x_{k-1}}{y_k - y_{k-1}} y_k.$$

Regula falsi (false position) requires that the root is bracketed by the approximate roots at each stage, so the calculation of x_{k+1} uses x_k together with either x_{k-1} or x_{k-2}, and the corresponding y values.

The secant method uses x_k and x_{k-1}, and the corresponding y values to find x_{k+1}.

Newton's Method: Given the current estimate of the root, x_k, the value of the function at x_k, i.e., $y_k = f(x_k)$, and the value of the derivative at x_k, i.e., $y'_k = f'(x_k)$, it follows that

$$x_{k+1} = x_k - \frac{y_k}{y'_k}.$$

Muller's Method: Given three points (x_1, y_1), (x_2, y_2) and (x_3, y_3), compute

$$c_2 = \frac{y_3 - y_2}{x_3 - x_2}, \quad d_1 = \frac{c_2 - c_1}{x_3 - x_1}, \text{ and } s = c_2 + d_1(x_3 - x_2).$$

The next approximate root is then

$$x = x_3 - \frac{2 y_3}{s + \text{sign}(s) \sqrt{s^2 - 4 y_3 d_1}}.$$

SUGGESTIONS FOR FURTHER READING

For additional information on root finding techniques, the following references are recommended:

Atkinson, K. E., *An Introduction to Numerical Analysis* (2d ed.), John Wiley & Sons, New York, 1989.

Press, W. H., S. A. Teukolsky, W. T. Vetterling, and B. P. Flannery, *Numerical Recipes in C, The Art of Scientific Computing,* (2d ed.), Cambridge Unversity Press, Cambridge, U.K., 1992.

Ralston, A., and P. Rabinowitz, *A First Course in Numerical Analysis,* McGraw-Hill, New York, 1978.

Rice, J. R., *Numerical Methods, Software, and Analysis,* McGraw-Hill, New York, 1983.

For a particularly nice discussion of special methods for polynomials and why they are worthwhile, see

Hamming, R. W., *Numerical Methods for Scientists and Engineers* (2d ed.), McGraw-Hill, New York, 1973.

Examples of applications of root finding techniques occur throughout science and engineering. A few sources are suggested here.

Edwards, C. H. Jr., and D. E. Penney, *Calculus and Analytic Geometry* (5th ed.), Prentice Hall, Englewood Cliffs, NJ, 1998.

Greenberg, M. D., *Foundations of Applied Mathematics* (2d ed.), Prentice Hall, Englewood Cliffs, NJ, 1998.

Hibbeler, R. C., *Engineering Mechanics:Statics* (7th ed.), Prentice Hall, Englewood Cliffs, NJ, 1995.

Hibbeler, R. C., *Engineering Mechanics:Dynamics* (7th ed.), Prentice Hall, Englewood Cliffs, NJ, 1995.

Details of Brent's method and Ridder's method are given in

Brent, R., *Algorithms for Minimization Without Derivatives,* Prentice-Hall, Englewood Cliffs, NJ, 1973.

Forsythe, G. E., M. A. Malcolm, and C. B. Moler, *Computer Methods for Mathematical Computations,* Prentice-Hall, Englewood Cliffs, NJ, 1976.

Ridders, C. J. F., *IEEE Transaction on Circuits and Systems,* 1979, vol. CAS-26, pp. 979–980.

PRACTICE THE TECHNIQUES

For Problems P2.1 to P2.10, find the positive real zero of the functions that follow; find consecutive integers a and b that bracket the root to use as starting values for the bisection, regula falsi, or secant method. Use $\frac{a+b}{2}$ as the starting value for Newton's method and as the third starting value for Muller's method.

    a. Find the zero using bisection.
    b. Find the zero using regula falsi.
    c. Find the zero using the secant method.
    d. Find the zero using Newton's method.
    e. Find the zero using Muller's method.

P2.1 $f(x) = x^2 - 2$.
P2.2 $f(x) = x^2 - 5$.
P2.3 $f(x) = x^2 - 7$.
P2.4 $f(x) = x^3 - 3$.
P2.5 $f(x) = x^3 - 4$.
P2.6 $f(x) = x^3 - 6$.
P2.7 $f(x) = x^4 - 0.06$.
P2.8 $f(x) = x^4 - 0.25$.
P2.9 $f(x) = x^4 - 0.45$.
P2.10 $f(x) = x^4 - 0.65$.

For Problems P2.11 to P2.20, find all real zeros of the functions that follow; choose starting values as described for Problems P2.1 to P2.10.

    a. Find the zeros using bisection.
    b. Find the zeros using regula falsi.
    c. Find the zeros using the secant method.
    d. Find the zeros using Newton's method.
    e. Find the zeros using Muller's method.

P2.11 $f(x) = x^3 + 3x^2 - 1$.
P2.12 $f(x) = x^3 - 4x + 1$.
P2.13 $f(x) = x^3 - 9x + 2$.
P2.14 $f(x) = x^3 - 2x^2 - 5$.
P2.15 $f(x) = x^3 - x^2 - 4x - 3$.
P2.16 $f(x) = x^3 - 6x^2 + 11x - 5$.
P2.17 $f(x) = x^3 - x^2 - 24x - 32$.
P2.18 $f(x) = x^3 - 7x^2 + 14x - 7$.
P2.19 $f(x) = 6x^3 - 23x^2 + 20x$.
P2.20 $f(x) = 3x^3 - x^2 - 18x + 6$.

P2.21 For each of the following equations, use Newton's method with the specified starting value to find a root; discuss the source of the difficulty if Newton's method fails.

    a. $f(x) = -5x^4 + 11x^2 - 2$ $x_0 = 1$
    b. $f(x) = x^3 - 4x + 1$ $x_0 = 0$
    c. $f(x) = 5x^4 - 11x^2 + 2$
        $x_0 = 1; x_0 = 1/2;$ $x_0 = 0$
    d. $f(x) = x^5 - 0.5$ $x_0 = 1$

For Problems P2.22 to P2.29, use the method of your choice, including the appropriate built-in Mathcad functions.

P2.22 Find the zeros of the following Legendre polynomials:

    a. $P_2(x) = (3x^2 - 1)/2$.
    b. $P_3(x) = (5x^3 - 3x)/2$.
    c. $P_4(x) = (35x^4 - 30x^2 + 3)/8$.
    d. $P_5(x) = (63x^5 - 70x^3 + 15x)/8$.

P2.23 Find the first three positive zeros of $y = x \cos x + \sin x$.

P2.24 Find the intersection(s) of $y = e^x$ and $y = x^3$; i.e., find the zeros of $f(x) = e^x - x^3$.

P2.25 Find the intersection(s) of $y = e^x$ and $y = x^2$.

P2.26 Find all intersections of $y = 2^x$ and $y = x^2$.

P2.27 Find the point(s) of intersection of x^c and c^x for different values of c.

    a. $c = 3$ b. $c = 2.7$

(Note that for $c = e$, $x^c \le c^x$ for all x.)

P2.28 Find the intersection(s) of $y = -a + e^x$ and $y = b + \ln(x)$.

    a. $a = 5, b = 1$.
    b. $a = 3, b = 2$.
    c. $a = 1, b = 5$.

P2.29 Find the zeros of $y = f(x) = \ln(x + 0.1) + 1.5$.

EXPLORE SOME APPLICATIONS

A2.1 To determine the displacement d of a spring of stiffness 400 N/m and unstretched length 6 m when a force of 200 N is applied, as illustrated in the following figure, two expressions are found for the tension T in each half of the spring.

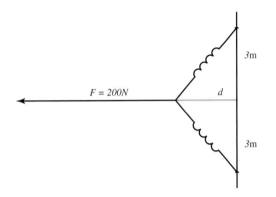

First, T is half the horizontal component of the applied force; i.e.,

$$T = 100\sqrt{9 + d^2}/d.$$

Second, T is the product of the spring constant and the amount by which the spring is stretched; i.e.,

$$T = 400(\sqrt{9 + d^2} - 3).$$

Find d by finding a root of the equation

$$4(\sqrt{9 + d^2} - 3) - \sqrt{9 + d^2}/d = 0.$$

(See Hibbeler, *Statics*, 1995 for a discussion of similar problems.)

A2.2 A boat can travel with a speed of $v_b = 20$ in still water. (See accompanying figure.) Determine the bearing angle θ of the boat in a river flowing at $v_w = -5$. The bearing angle is measured from the longitudinal axis along which the river flows. Let v be the velocity of the boat along the desired path, which is at 60° from the transverse axis (across the river).

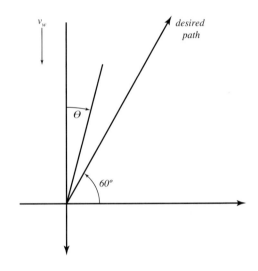

Equating the longitudinal and transverse components of the velocities, gives

$$v \cos(60°) = 20 \sin \theta$$
$$v \sin(60°) = -5 + 20 \cos \theta$$

Eliminating v, gives the equation for θ:

$$1.732 \sin \theta - \cos \theta + 0.25 = 0.$$

(See Hibbler, *Dynamics*, 1995, p. 88 for a discussion of a similar problem.)

A2.3 The van der Waals equation of state, a simple extension of the ideal gas law discovered in 1873 by the Dutch physicist Johanes Diderik van der Waals, is

$$\left(P + \frac{n^2 a}{V^2}\right)(V - nb) = nRT,$$

where the constants a and b, characteristic of the gas, are determined experimentally. For P in atmospheres, V in liters, n in moles, and T in kelvins, R is approximately 0.0820 liter atm deg^{-1} mole^{-1}. The volume of 1 mole of a perfect gas at standard conditions (1 atm, 273 K) is 22.415 liters. Find the volume occupied by 1 mole of the following gases, with given values of a and b:

Gas	a	b
O_2	1.36	0.0318
N_2O	3.78	0.0441
SO_2	6.71	0.0564

These parameter values come from Pauling, *General Chemistry*, 1989, p. 337. For further discussion of related problems, see also Himmelblau, *Basic Principles and Calculations in Chemical Engineering*, 1974, Sec 3.2.

A2.4 A simple model of oxygen diffusion around a capillary leads to an equation of the form

$$C(r) = \frac{R r^2}{4K} + B_1 \ln(r) + B_2,$$

where R, K, B_1, and B_2 depend on the geometry, reaction rates, and other specifics of the problem. As an example, without considering realistic values of these constants, find the value of r such that

$$C(r) = 2r^2 + 3\ln(r) + 1 = 2;$$

this corresponds to finding a zero of $y(r) = 2r^2 + 3\ln(r) - 1$.
(See Simon, 1986, pp. 185–189 for a discussion of the derivation of this equation.)

A2.5 The flow rate in a pipe system connecting two reservoirs (at different surface elevations) depends on the characteristics of the pump, the roughness of the pipe, the length and diameter of the pipe, and the specific gravity of the fluid. For an 800-ft section of 6-in. pipe connecting two reservoirs (with a 5-ft differential in elevation) containing oil of specific gravity 0.8, with a 6-hp pump, the equation for the flow rate Q is

$$12 Q^3 + 5 Q - 40 = 0.$$

Approximate the real root of the equation in the interval $0 \le Q \le 2$.
(For derivation of this equation, see Ayyub and McCuen, 1996, pp. 53–59.)

A2.6 The Peng-Robinson equation of state

$$P = \frac{RT}{V - b} - \frac{a}{V(V + b) + b(V - b)}$$

is a two-parameter extension of the ideal gas law. Find the volume of 1 mole of a gas at $P = 10^4$ kPa and $T = 340$ K; take as the parameter values $a = 364$ m^{6}kPa/(kg mole)^2, $b = 0.03$ m^3/kg mole and $R = 1.618$. Use $V = 0.055$ m^3/kg mole as an initial estimate (from the ideal gas law). (See Hanna and Sandall, 1995, pp. 161ff. for discussion)

A2.7 The Beattie-Bridgeman equation of state

$$P = \frac{RT}{V} + \frac{a}{V^2} + \frac{b}{V^3} + \frac{c}{V^4}$$

is a three-parameter extension of the ideal gas law. Using $a = -1.06$, $b = 0.057$, and $c = -0.0001$, find the volume of 1 mole of a gas at $P = 25$ atm, and $T = 293$ K. The constant $R = 0.082$ liter-atm/K-g mole. (See Ayyub and McCuen, 1996, p. 91).

A2.8 For given values of s (the length of the cable) and x (the distance between the support positions of the ends, the problem of finding the deflection of the hanging cable shown in the following diagram requires (as an intermediate step) the solution (for F) of the equation

$$s = F \sinh(x/F).$$

Find F for $s = 100$, and $x = 97$. (Adapted from Hibbeler, *Statics:*, 1996.)

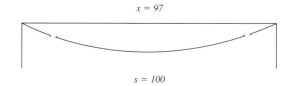

A2.9 The position of a ball, thrown up with a given initial velocity v_0, and initial position x_0, subject to air resistance proportional to its velocity is given by

$$x(t) = \rho^{-1}(v_0 + v_r)(1 - e^{-\rho t}) - v_r t + x_0$$

where ρ is the drag coefficient, g is the gravitational constant, and $v_r = g/\rho = mg/k$ is the terminal velocity. Find when the ball hits the ground, if $x_0 = 0$, $v_0 = 20$ m/s, $\rho = 0.35$, $g = 9.8$ m/s². See a differential equations text (e.g., Edwards and Penney, Boyce and DiPrima, etc.) for further discussion.

A2.10 Some techniques for solving the differential equation describing the deflection of a uniform beam with both ends fixed, subject to a load that is proportional to the distance from one end of the beam, require the positive roots of the function $f(x) = \cosh x \cos x - 1$. In order to keep the function values within a reasonable range, it is better to consider the equivalent problem, of finding the zeros of $g(x) = \cos x - 1/\cosh x$. Find the first 5 roots. (For more discussion of the deflection of a beam problem, see Edwards and Penney, *Differential Equations with Boundary Value Problems*, 1996, p. 624.)

A2.11 For a cantilever beam (one end fixed, the other free) the required parameters are the positive roots of $f(x) = \cosh x \cos x + 1$; in order to keep the function values within a reasonable range, consider zeros of $g(x) = \cos x + 1/\cosh x$. Find the first 5 roots.

EXTEND YOUR UNDERSTANDING

U2.1 Muller's method can also be given in the following form given three initial approximations to the root:

$$(x_0, y_0), (x_1, y_1) \text{ and } (x_2, y_2)$$

find a and b:

$$b = \frac{(x_0 - x_2)^2(y_1 - y_2) - (x_1 - x_2)^2(y_0 - y_2)}{(x_0 - x_2)(x_1 - x_2)(x_0 - x_1)}$$

$$a = \frac{(x_1 - x_2)(y_0 - y_2) - (x_0 - x_2)(y_1 - y_2)}{(x_0 - x_2)(x_1 - x_2)(x_0 - x_1)}$$

the next approximation is

$$x = x_2 + \frac{-2c}{b + \text{sign}(b)\sqrt{b^2 - 4ac}}$$

compare the computational effort for performing one step of this algorithm to the effort required for the algorithm given in the text.

U2.2 Use Newton's method, or the secant method, to find a root of

$$f(x) = -0.01 + \frac{1}{1 + x^2}$$

Compare the use of a convergence test of the form $|f(x_k)| < f\text{tol}$ with a test of the form

$$|x_{k+1} - x_k| < \text{tol}.$$

U2.3 Show that for the bisection method

$$\frac{|x_k - x^*|}{|x_{k-1} - x^*|} \leq \frac{1}{2}$$

is equivalent to $|x^* - x_k| \leq \frac{1}{2^k}|x^* - x_1|$, so that bisection is linearly convergent according to the general definition.

U2.4 Show that even without the initial estimates bracketing the zero, the secant method converges for Example 2.7; take $a = 2$, $b = 3$.

3

Solving Systems of Linear Equations: Direct Methods

3.1 Gaussian Elimination

3.2 Gaussian Elimination with Row Pivoting

3.3 Gaussian Elimination for Tridiagonal Systems

3.4 Mathcad's Methods

We now extend our investigation of numerical methods for solving equations to the consideration of linear systems. In this chapter, we present the method known as Gaussian elimination.

We begin with two examples that are used to illustrate basic Gaussian elimination and two important variations. Gaussian elimination is based on the fact that if two equations have a point in common, then that point also satisfies any linear combination of the two equations. If we can find linear combinations of suitably simple form, we will be able to find the solution of the original system more easily. The form that we desire is one in which certain of the variables have been eliminated from some of the equations.

Variations of Gaussian elimination are also presented for row pivoting and tridiagonal systems. These methods are direct techniques that require a single pass through the appropriate algorithm. The specific form of the resulting (equivalent) system of equations is illustrated in the examples that follow and in Mathcad worksheet computations and Mathcad functions. We restrict our inquiry to systems in which we have the same number of equations as unknowns.

Systems of linear equations occur in a wide variety of settings, including the analysis of electrical circuits, the determination of forces on a truss, balancing the reactants in a chemical reaction, economics, traffic flow, queuing theory, and calculating the equilibrium heat distribution in a plate.

In Chapters 4 and 5, we consider iterative techniques for linear and nonlinear systems. In Chapters 6 and 7, we investigate several topics from numerical linear algebra, including two important types of matrix factorization that can be useful in solving linear systems as well as in other applications. The first of these factorizations is closely related to the Gaussian elimination techniques presented here.

Example 3-A Circuit Analysis Application

Consider the problem of finding the currents in different parts of an electrical circuit, with resistors as shown in Fig. 3.1.

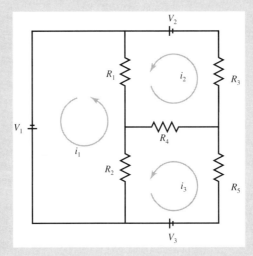

FIGURE 3.1 Simple electrical circuit.

The analysis tools come from elementary physics:

1. The sum of the voltage drops around a closed loop is zero;
2. The voltage drop across a resistor is the product of the current and the resistance.

We define each unknown current to be positive if it flows in the counterclockwise direction; if a computed current is negative, the flow is clockwise.

The analysis of the voltages around the three loops gives three equations, which we solve later in the chapter; if $R_1=20$, $R_2=0$, $R_3=25$, $R_4=10$, $R_5=30$, $V_1=0$, $V_2=0$ and $V_3=200$, we have

Flow around left loop

$$20(i_1 - i_2) + 10(i_1 - i_3) = 0.$$

Flow around upper right loop

$$25 i_2 + 10(i_2 - i_3) + 20(i_2 - i_1) = 0.$$

Flow around lower right loop

$$30 i_3 + 10(i_3 - i_2) + 10(i_3 - i_1) = 200.$$

Example 3-B Forces on a Truss

A lightweight structure constructed of triangular elements may be capable of supporting large weights. To analyze the forces on such a structure, known as a *truss*, a system of linear equations describing the equilibrium of the horizontal and vertical forces on each node (or joint) of the truss must be solved. A single triangular element is shown in Fig. 3.2.

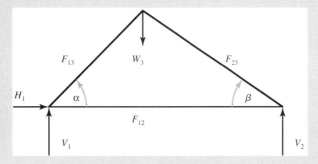

FIGURE 3.2 Element of a simple truss.

We assume that the forces in each of the members of the truss (F_{12}, F_{13}, and F_{23}) are acting to pull the structure together. V_1 and V_2 are unknown vertical forces supporting the structure (at nodes 1 and 2, respectively), H_1 is an unknown horizontal bracing force at node 1, and W_3 is a known force (at node 3) representing the weight of the structure. We define forces to be positive if they act to the right, or in an upward direction. If a computed quantity is negative, it indicates that the force acts in the opposite direction. The system of equations represents the conditions for vertical and horizontal equilibrium at the three nodes.

Node 1

$$V_1 \qquad\quad + F_{13} \sin \alpha \qquad\qquad = 0,$$
$$H_1 \quad + F_{12} + F_{13} \cos \alpha \qquad\qquad = 0.$$

Node 2

$$V_2 \qquad\qquad\quad + F_{23} \sin \beta = 0,$$
$$-F_{12} \qquad - F_{23} \cos \beta = 0.$$

Node 3

$$-F_{13} \sin \alpha - F_{23} \sin \beta = W_3,$$
$$-F_{13} \cos \alpha + F_{23} \cos \beta = 0.$$

3.1 GAUSSIAN ELIMINATION

The process of Gaussian elimination was introduced in Chapter 1 for a system of two equations. Example 3.1 illustrates the steps for a system of three equations in three unknowns.

Example 3.1 Three-by-Three System

The three-by-three system

$$x + 2y + 3z = 1,$$
$$2x + 6y + 10z = 0,$$
$$3x + 14y + 28z = -8,$$

can be solved by Gaussian elimination as described in the following steps:

Step 1

Use the first equation to eliminate x in the second and third equations.

Multiply the first equation by -2 and add it to the second equation to get a new second equation with the x variable eliminated.

Also, multiply the first equation by -3 and add it to the third equation to get a new third equation with the x variable eliminated. The resulting system is

$$x + 2y + 3z = 1,$$
$$2y + 4z = -2,$$
$$8y + 19z = -11.$$

Step 2

Use the second equation to eliminate the y term in the third equation.

Multiply the second equation by -4 and add it to the third equation to get a new third equation with the y variable eliminated. The system now looks like this:

$$x + 2y + 3z = 1,$$
$$2y + 4z = -2,$$
$$3z = -3.$$

This completes the "forward elimination" phase; we have an upper triangular system.

We now use "back substitution" to find the values of the unknowns:

$$3z = -3 \qquad \Rightarrow z = -1,$$
$$2y + 4(-1) = -2 \qquad \Rightarrow y = 1,$$
$$x + 2(1) + 3(-1) = 1 \Rightarrow x = 2.$$

3.1.1 Using Matrix Notation

We see in the preceding example that all of the computations are based on the coefficients and elements of the right hand side of the system of equations. This system is written in matrix-vector form as $\mathbf{A}\,\mathbf{x} = \mathbf{b}$ (see Chapter 1) with

$$\mathbf{A} = \begin{bmatrix} 1 & 2 & 3 \\ 2 & 6 & 10 \\ 3 & 14 & 28 \end{bmatrix}, \quad \mathbf{b} = \begin{bmatrix} 1 \\ 0 \\ -8 \end{bmatrix}$$

Since the same operations are performed on the matrix $\mathbf{A}$ and the vector $\mathbf{b}$, they are often combined in the augmented matrix:

$$\begin{bmatrix} 1 & 2 & 3 & | & 1 \\ 2 & 6 & 10 & | & 0 \\ 3 & 14 & 28 & | & -8 \end{bmatrix}$$

There are a variety of ways to transform a linear system of equations into an equivalent system having the same solution. However, the basic Gaussian elimination procedure follows a specific sequence of steps and only uses operations of the following form:

Add a multiple m of row R_i onto row R_j to form a new row R_j, or

$$R_j \leftarrow m R_i + R_j$$

At the kth stage of the basic Gaussian elimination procedure, the appropriate multiples of the kth equation are used to eliminate the kth variable from equations $k + 1, \ldots n$; in terms of the coefficient matrix $\mathbf{A}$, the appropriate multiple of the kth row is used to reduce each of the entries in the kth column below the kth row to zero. The kth row is called the *pivot row*, the kth column is the *pivot column*, and the element a_{kk} is the *pivot element*. Since finding the appropriate multiplier for each row requires dividing by the pivot element for that stage, the process fails if the pivot element is zero. Possible remedies for this situation are discussed later in the chapter; they allow interchanging the order of the rows of the augmented matrix.

In Mathcad, subscripts indicate an element in a vector or matrix. Vectors are denoted in bold, and with only a single subscript. For convenience, we give the rows different names. Note that Mathcad starts indices in vectors and matrices at 0 by default. Before using the following algorithm set ORIGIN:=1, so that the indices run from 1 to n. In the following discussion, the i,j element of matrix $\mathbf{M}$ is indicated as $\mathbf{M}[i,j]$, although in Mathcad it is not necessary to type the final].

3.1.2 Step-by-Step Procedure

Algorithm for Gaussian Elimination

The steps to solve the linear system $\mathbf{M}\,\mathbf{x} = \mathbf{b}$ using Gaussian elimination are illustrated here for a three-by-three system.

Initialize
1) Create row vectors $\mathbf{r}, \mathbf{s}, \mathbf{t}$, for each row of the augmented matrix $\mathbf{A} = [\mathbf{M}\ \mathbf{b}]$.

Begin computation
2) Transform $\mathbf{s}$ by adding a multiple of $\mathbf{r}$ to current $\mathbf{s}$

$$\mathbf{s}: = \mathbf{s} + \frac{-s_1}{r_1}\mathbf{r}$$

Transform $\mathbf{t}$ by adding a multiple of $\mathbf{r}$ to current $\mathbf{t}$

$$\mathbf{t}: = \mathbf{t} + \frac{-t_1}{r_1}\mathbf{r}$$

3) Transform $\mathbf{t}$ by adding a multiple of $\mathbf{s}$ to current $\mathbf{t}$

$$\mathbf{t}: = \mathbf{t} + \frac{-t_2}{s_2}\mathbf{s}$$

Back substitution

$$x_3: = \frac{t_4}{t_3}$$

$$x_2: = \frac{1}{s_2}(s_4 - x_3\, s_3)$$

$$x_1: = \frac{1}{r_1}(r_4 - x_3\, r_3 - x_2\, r_2)$$

The algorithm for Gaussian elimination is illustrated in the next example using the equations for the electrical circuit in Example 3-A.

Example 3.2 Circuit Analysis

The equations describing the electrical circuit in Example 3-A simplify to

$$+30i_1 - 20i_2 - 10i_3 = 0,$$
$$-20i_1 + 55i_2 - 10i_3 = 0,$$
$$-10i_1 - 10i_2 + 50i_3 = 200.$$

To illustrate the use of the algorithm for basic Gaussian elimination, define vectors for the rows of the augmented matrix for the electrical circuit introduced in Example 3-A.

$$r := (30 \ -20 \ -10 \ 0) \quad s := (-20 \ 55 \ -10 \ 0) \quad t := (-10 \ -10 \ 50 \ 200)$$

Transform vector s and t

$$s := s + r \cdot \left(\frac{-s_1}{r_1}\right) \qquad t := t + r \cdot \left(\frac{-t_1}{r_1}\right)$$

Display results if desired

$$s = (0 \ 41.667 \ -16.667 \ 0) \qquad t = (0 \ -16.667 \ 46.667 \ 200)$$

Transform vector t and display results

$$t := t + s \cdot \frac{(-t_{1,2})}{s_{1,2}} \qquad t = (0 \ 0 \ 40 \ 200)$$

Back substitution

$$x_3 := \frac{t_{1,4}}{t_{1,3}}$$

$$x_2 := \frac{[s_{1,4} - x_3 \cdot s_{1,3}]}{s_{1,2}}$$

$$x_1 := \frac{[r_{1,4} - x_3 \cdot r_{1,3} - x_2 \cdot r_{1,2}]}{r_{1,1}}$$

Display results

$$x_3 = 5 \qquad x_2 = 2 \qquad x_1 = 3$$

In order to implement the basic Gaussian elimination process in a computer program, it is helpful to describe the steps illustrated in Example 3.2 in somewhat more general terms. A general four-by-four system of equations can be represented in matrix form as $\mathbf{A}\,\mathbf{x} = \mathbf{b}$, where

$$\mathbf{A} = \begin{bmatrix} a_{11} & a_{12} & a_{13} & a_{14} \\ a_{21} & a_{22} & a_{23} & a_{24} \\ a_{31} & a_{32} & a_{33} & a_{34} \\ a_{41} & a_{42} & a_{43} & a_{44} \end{bmatrix} \quad \text{and} \quad \mathbf{b} = \begin{bmatrix} b_1 \\ b_2 \\ b_3 \\ b_4 \end{bmatrix}$$

Step 1: The pivot is a_{11}.

Multiply the first row by $m_{21} = -a_{21}/a_{11}$ and add the result to the second row

$a_{21} \Leftarrow 0;$ $\qquad\qquad a_{22} \Leftarrow a_{22} + m_{21}a_{12};$

$a_{23} \Leftarrow a_{23}; + m_{21}a_{13};$ $\quad a_{24} \Leftarrow a_{24} + m_{21}a_{14}.$

Transform the right-hand side: $b_2 \Leftarrow b_2 + m_{21}b_1$.

Multiply the first row by $m_{31} = -a_{31}/a_{11}$ and add the result to the third row

$a_{31} \Leftarrow 0;$ $\qquad\qquad a_{32} \Leftarrow a_{32} + m_{31}a_{12};$

$a_{33} \Leftarrow a_{33} + m_{31}a_{13};$ $\quad a_{34} \Leftarrow a_{34} + m_{31}a_{14}.$

Transform the right-hand side: $b_3 \Leftarrow b_3 + m_{31}b_1$.

Multiply the first row by $m_{41} = -a_{41}/a_{11}$ and add the result to the fourth row

$a_{41} \Leftarrow 0;$ $\qquad\qquad a_{42} \Leftarrow a_{42} + m_{41}a_{12};$

$a_{43} \Leftarrow a_{43} + m_{41}a_{13};$ $\quad a_{44} \Leftarrow a_{44} + m_{41}a_{14}.$

Transform the right-hand side: $b_4 \Leftarrow b_4 + m_{41}b_1$.

Step 2: The pivot is a_{22}.

Multiply the second row by $m_{32} = -a_{32}/a_{22}$ and add the result to the third row

$a_{31} = 0;$ $\quad a_{32} \Leftarrow 0;$ $\quad a_{33} \Leftarrow a_{33} + m_{32}a_{23};$ $\quad a_{34} \Leftarrow a_{34} + m_{32}a_{24}.$

Transform the right-hand side: $b_3 \Leftarrow b_3 + m_{32}b_2$.

Multiply the second row by $m_{42} = -a_{42}/a_{22}$ and add the result to the fourth row

$a_{41} = 0;$ $\quad a_{42} \Leftarrow 0;$ $\quad a_{43} \Leftarrow a_{43} + m_{42}a_{23};$ $\quad a_{44} \Leftarrow a_{44} + m_{42}a_{24}.$

Transform the right-hand side: $b_4 \Leftarrow b_4 + m_{42}b_2$.

Step 3: The pivot is a_{33}.

Multiply the third row by $m_{43} = -a_{43}/a_{33}$ and add the result to the fourth row

$a_{41} = 0;$ $\quad a_{42} = 0;$ $\quad a_{43} \Leftarrow 0;$ $\quad a_{44} \Leftarrow a_{44} + m_{43}a_{34}.$

Transform the right-hand side: $b_4 \Leftarrow b_4 + m_{43}b_3$.

Back substitution

$$x_4 = b_4/a_{44},$$
$$x_3 = (b_3 - a_{34}x_4)/a_{33},$$
$$x_2 = (b_2 - a_{23}x_3 - a_{24}x_4)/a_{22},$$
$$x_1 = (b_1 - a_{12}x_2 - a_{13}x_3 - a_{14}x_4)/a_{11}.$$

3.1.3 Mathcad Function for Basic Gaussian Elimination

We now consider a Mathcad function to solve an *n*-by-*n* system of linear equations using Gaussian elimination. The function allows us to solve *k* systems of the form $\mathbf{A}\mathbf{x} = \mathbf{b}_1, \ldots, \mathbf{A}\mathbf{x} = \mathbf{b}_k$, at the same time.

Note that this function requires that the indices on the matrix elements begin with 1; set ORIGIN:=1 before using the function.

$$\text{Gauss\_B}(A, B) := \begin{vmatrix} n \leftarrow \text{rows}(A) \\ k \leftarrow \text{cols}(B) \\ \text{for } i \in 1..n-1 \\ \quad \text{for } j \in (i+1)..n \\ \quad \quad \begin{vmatrix} m \leftarrow \dfrac{-A_{j,i}}{A_{i,i}} \\ \text{for } k1 \in i..n \\ \quad A_{j,k1} \leftarrow A_{j,k1} + m \cdot A_{i,k1} \\ \text{for } k2 \in 1..k \\ \quad B_{j,k2} \leftarrow B_{j,k2} + m \cdot B_{i,k2} \end{vmatrix} \\ \text{for } k1 \in 1..k \\ \quad x_{n,k1} \leftarrow \dfrac{B_{n,k1}}{A_{n,n}} \\ \text{for } i \in (n-1)..1 \\ \quad \text{for } k1 \in 1..k \\ \quad \quad \begin{vmatrix} x_{i,k1} \leftarrow \dfrac{B_{i,k1}}{A_{i,i}} \\ \text{for } j \in (i+1)..n \\ \quad x_{i,k1} \leftarrow x_{i,k1} - \dfrac{A_{i,j} \cdot x_{j,k1}}{A_{i,i}} \end{vmatrix} \\ x \end{vmatrix}$$

Example 3.3 Analysis of Forces on a Simple Truss

To illustrate the use of the preceding Gaussian elimination function, we investigate the effect of two different values of W_3 on the forces in the single-triangle truss shown in Fig. 3.2. Since varying W_3 affects only the right-hand side of the system, matrix **A** is unchanged as long as the geometry of the structure is not modified. We take $\alpha = \pi/6$ and $\beta = \pi/3$.

$$\alpha := \frac{\pi}{6} \qquad \beta := \frac{\pi}{3}$$

$$A := \begin{pmatrix} 1 & 0 & 0 & 0 & \sin(\alpha) & 0 \\ 0 & 1 & 0 & 1 & \cos(\alpha) & 0 \\ 0 & 0 & 1 & 0 & 0 & \sin(\beta) \\ 0 & 0 & 0 & -1 & 0 & -\cos(\beta) \\ 0 & 0 & 0 & 0 & -\sin(\alpha) & -\sin(\beta) \\ 0 & 0 & 0 & 0 & -\cos(\alpha) & \cos(\beta) \end{pmatrix} \qquad B := \begin{pmatrix} 0 & 0 \\ 0 & 0 \\ 0 & 0 \\ 0 & 0 \\ 100 & 75 \\ 0 & 0 \end{pmatrix}$$

$$x = \begin{pmatrix} 25 & 18.75 \\ 0 & 7.105 \times 10^{-15} \\ 75 & 56.25 \\ 43.301 & 32.476 \\ -50 & -37.5 \\ -86.603 & -64.952 \end{pmatrix}$$

Entering Gauss_B(A,B) produces the matrix **x** as its returned value; the first column of **x** gives the forces for a vertical force of 100 units at node 3; the second column gives the results for the second scenario. The smaller vertical force gives correspondingly smaller forces in each node.

3.1.4 Discussion

There are two key aspects to understanding why Gaussian elimination works. The first is to see why a linear combination of two equations passes through the point of intersection of the two equations. The second is to consider why (or when) the sequence of steps for Gaussian elimination produces a system that can be solved by backsubstitution. Analyzing these two questions suggests when Gaussian elimination works well, when it works poorly or fails, and when it can be improved.

If two equations have a point in common, then that point is also a solution of any equation formed as a linear combination of the equations. This result can be shown by simple algebra. Consider two linear equations,

$$S_1: \qquad a_0 + a_1 x_1 + a_2 x_2 + \cdots + a_n x_n = 0$$

and

$$T_1: \qquad b_0 + b_1 x_1 + b_2 x_2 + \cdots + b_n x_n = 0$$

and assume that $\mathbf{r} = (r_1, r_2, \ldots r_n)$ satisfies both equations. Then $\mathbf{r}$ also satisfies the linear combination $C = m_1 S_1 + m_2 T_1$, or

$$m_1 a_0 + m_1 a_1 x_1 + m_1 a_2 x_2 + \cdots + m_1 a_n x_n$$
$$+ m_2 b_0 + m_2 b_1 x_1 + m_2 b_2 x_2 + \cdots + m_2 b_n x_n = 0.$$

Substituting $(r_1, r_2, \ldots r_n)$ in the equation C and using the fact that $\mathbf{r}$ satisfies S_1 and T_1 gives the desired result.

Now consider the specific sequence of transformations on the linear system given in the basic Gaussian elimination algorithm presented earlier. The description assumes that it is possible to find the necessary multiplier to reduce each column to zero as indicated. Two situations can arise if a zero pivot element is encountered, depending on whether or not there are any nonzero elements in the pivot column below the pivot row.

If a zero pivot occurs (in an n-by-n linear system), and the entire pivot column below the pivot row is also zero, then the system of equations does not have a unique solution. The equations are either inconsistent or redundant.

On the other hand, if a zero element is encountered in a pivot position, but there is a nonzero element in the pivot column below the pivot element, the Gaussian elimination process can be modified to allow for interchanging the row whose pivot element is zero with a row below it. This process is called (partial) pivoting and is the subject of the next section.

Several considerations are operative in determining how well a particular numerical method works. Among the most important are questions dealing with the quality of the solution, the sensitivity of the method to errors (including inexact arithmetic), and the computational effort required.

Computational effort is usually measured in terms of the number of multiplications and divisions ($m + d$) or in terms of the number of floating-point operations (flops). On early computers, multiplication and division were much more time intensive than addition and subtraction, which led researchers to analyze algorithms in terms of multiplication and division. It is also typical for the number of additions and subtractions to be directly related to the number of multiplications and divisions. Today, because the difference in effort for different operations has been reduced, analysis in terms of flops has become more common. Since the linear systems that arise in practice are often very large, it is important to see how the computational effort required for Gaussian elimination is related to the size of the coefficient matrix **A** (assumed to be n-by-n).

At the first stage of Gaussian elimination, one division is required to find the multiplier for the second row ($m_{12} = -a_{12}/a_{11}$). Then n multiplications and n additions are required to form the new second row. Note that we must also multiply the right-hand side, but we do not have to multiply the first element in the row, since we know that the new first element in the second row will be zero. This process must be performed for each of the rows below the first row. Thus the first stage requires $(n + 1)(n - 1)$ multiplications and divisions, as well as $n(n - 1)$ additions.

At the kth stage of elimination, there is one division to form the multiplier m_{ki} and $(n - k)$ multiplications to generate the new i^{th} row (for $i = k + 1, \ldots, n$). There are also $n - k$ additions required for each new row.

The total number of multiplications and divisions is

$$\sum_{k=1}^{n-1} (n - k + 1)(n - k) = \sum_{k=1}^{n-1} (n^2 - 2nk + k^2 + n - k)$$

$$= \sum_{k=1}^{n-1} (n^2 + n) - \sum_{k=1}^{n-1} (2n + 1)k + \sum_{k=1}^{n-1} k^2$$

which simplifies to

$$= \frac{n^3}{3} - \frac{n}{3}.$$

If there is a unique solution, if computations are exact, and if the pivot element is not zero at any stage, Gaussian elimination gives the solution. However, since computer computations are not exact, we may be faced with errors in the calculations that result from round-off. We illustrate here two types of difficulties that can occur. The first can be avoided by a suitable reordering of the rows of the augmented matrix, which is the subject of the next section. The second example is indicative of a more serious problem.

Gaussian elimination works well for systems with coefficient matrices with special properties. For example, if **A** is strictly diagonally dominant, i.e., for each i, $|a_{ii}| > \sum_{j \neq 1} |a_{ij}|$, Gaussian elimination will work well [Golub and Van Loan, p. 120].

Some Difficulties Are Solvable with Pivoting

Consider the following system of two equations in two unknowns, and suppose that we have arithmetic with rounding to two digits at each stage of the Gaussian elimination process:

$$0.001 x_1 + x_2 = 3,$$
$$x_1 + 2x_2 = 5.$$

Proceeding according to the basic Gaussian elimination procedure, we multiply the first equation by -1000 and add it to the second equation to give (if arithmetic were exact) $-998\, x_2 = -2995$. After rounding, this equation becomes $-1000 x_2 = -3000$, which gives $x_2 = 3$. Substitution of this value in the first equation then yields $x_1 = 0$. Clearly, this is not a good approximation to a solution of the second equation. In the next section, we discuss an enhancement to Gaussian elimination to avoid such a dilemma.

An Ill-Conditioned Matrix Causes More Serious Difficulties

Consider the linear system

$$x_1 + \frac{1}{2} x_2 = \frac{3}{2},$$

$$\frac{1}{2} x_1 + \frac{1}{3} x_2 = \frac{5}{6}.$$

Using exact arithmetic gives the exact solution $x_1 = x_2 = 1$. However, if the right-hand side is modified slightly, to

$$x_1 + \frac{1}{2} x_2 = \frac{3}{2}$$

$$\frac{1}{2} x_1 + \frac{1}{3} x_2 = 1$$

the exact solution becomes $x_1 = 0$ and $x_2 = 3$.

This extreme sensitivity to small changes in the right hand side is evidence of the fact that the coefficient matrix is "ill-conditioned." The difficulties arising from ill-conditioning cannot be solved by simple refinements in the Gaussian elimination procedure. If the *condition number* of a matrix is defined as the ratio of the largest eigenvalue to the smallest eigenvalue, a matrix with a large condition number is ill-conditioned. Eigenvalues are discussed in Chapter 7.

The condition number of a matrix can also be found by using any of four built-in functions in Mathcad2000(Pro). These functions differ in the matrix norm used for computing the condition number. The function `cond1(M)` uses the L_1 norm; `cond2(M)` uses the L_2 norm, `conde(M)` uses the Euclidean norm, and `condi(M)` uses the infinity norm. The Hilbert matrix is a well known example of a matrix that is very ill-conditioned. For the 2-by-2 Hilbert matrix

$$H = \begin{bmatrix} 1 & 1/2 \\ 1/2 & 1/3 \end{bmatrix}$$

we have `cond1(H)` = 27, `cond2(H)` = 19.281, `conde(H)` = 19.333 and `condi(H)` = 27.

3.2 GAUSSIAN ELIMINATION WITH ROW PIVOTING

Basic Gaussian elimination, as presented in the previous section, fails if the pivot element at any stage of the elimination process is zero, because division by zero is not possible. In addition, difficulties that are not as easy to detect arise if the pivot element is significantly smaller than the coefficients it is being used to eliminate. In this section, we investigate an enhancement to Gaussian elimination that prevents or alleviates some of these shortcomings of the basic procedure.

In order to reduce the inaccuracies that occur in solutions computed with Gaussian elimination and avoid (if possible) the failure of the method resulting from a zero coefficient in the pivot position at some stage of the process, we may need to interchange selected rows of the augmented matrix. We illustrate this process, known as *row pivoting*, in Example 3.4.

Example 3.4 Difficult System

Consider again the following simple system of two equations in two unknowns, to be solved by Gaussian elimination with rounding to two significant digits at each stage of the process:

$$0.001 x_1 + x_2 = 3,$$

$$x_1 + 2x_2 = 5.$$

With basic Gaussian elimination (with rounding) we found that $x_2 = 3$ and $x_1 = 0$. This is not a good approximation to a solution of the second equation.

If, instead of solving the system with the equations in the order given, we recognize that the very small coefficient of x_1 in the first equation is dangerous (because we would be dividing by something that is close to zero), we can interchange the order of the equations as follows.

$$x_1 + 2x_2 = 5$$

$$0.001 x_1 + x_2 = 3$$

Now we multiply the first equation by -0.001 and add it to the second equation to give (if arithmetic were exact)

$$0.998 x_2 = 2.995$$

After rounding, this equation becomes $x_2 = 3$. Substitution of this value into the first equation yields $x_1 = -1$, a much better approximation to a solution of the system.

3.2.1 Step-by-Step Computation

The small pivot element shown in the previous example could occur at any stage of elimination. Gaussian elimination with row pivoting checks all entries in the pivot column (from the current diagonal element to the bottom of the column) and chooses the largest element as the pivot. The current row and the selected pivot row are interchanged. The process is summarized in the following algorithm, and illustrated in Example 3.5.

Algorithm for Gaussian Elimination with Row Pivoting

The steps to solve the linear system $\mathbf{M}\,\mathbf{x} = \mathbf{b}$ using Gaussian elimination are illustrated here for a three-by-three system.

Initialize
  1) Create row vectors $\mathbf{r}, \mathbf{s}, \mathbf{t}$, for each row of the augmented matrix $\mathbf{A} = [\mathbf{M}\ \mathbf{b}]$.

Begin computation
  2a) Pivot if necessary
    Consider the magnitude of the first elements of each of the vectors.
    If s_1 is larger than either r_1 or t_1, interchange vectors $\mathbf{r}$ and $\mathbf{s}$
      (rename vectors so that first vector is still called $\mathbf{r}$, second vector is $\mathbf{s}$).
    If t_1 is larger than either r_1 or s_1, interchange vectors $\mathbf{r}$ and $\mathbf{t}$
      (rename vectors so that first vector is still called $\mathbf{r}$, third vector is $\mathbf{t}$).
  2b) Transform $\mathbf{s}$ by adding a multiple of $\mathbf{r}$ to current $\mathbf{s}$

$$\mathbf{s} := \mathbf{s} + \frac{-s_1}{r_1}\mathbf{r}$$

  Transform $\mathbf{t}$ by adding a multiple of $\mathbf{r}$ to current $\mathbf{t}$

$$\mathbf{t} := \mathbf{t} + \frac{-t_1}{r_1}\mathbf{r}$$

  3a) Pivot if necessary
    Consider the magnitude of the second elements of each of the vectors $\mathbf{s}$ and $\mathbf{t}$.
    If s_2 is larger than t_2, interchange vectors $\mathbf{s}$ and $\mathbf{t}$
      (rename vectors so that second vector is still called $\mathbf{s}$, third vector is $\mathbf{t}$).
  3b) Transform $\mathbf{t}$ by adding a multiple of $\mathbf{s}$ to current $\mathbf{t}$

$$\mathbf{t} := \mathbf{t} + \frac{-t_2}{s_2}\mathbf{s}$$

Back substitution

$$x_3 := \frac{t_4}{t_3}$$

$$x_2 := \frac{1}{s_2}(s_4 - x_3 s_3)$$

$$x_1 := \frac{1}{r_1}(r_4 - x_3 r_3 - x_2 r_2)$$

Example 3.5 Solving a Three-By-Three System With Pivoting

Solve $\mathbf{M}\mathbf{x} = \mathbf{b}$, for

$$\mathbf{M} = \begin{bmatrix} 2 & 6 & 10 \\ 1 & 3 & 3 \\ 3 & 14 & 28 \end{bmatrix} \quad \mathbf{b} = \begin{bmatrix} 0 \\ 2 \\ -8 \end{bmatrix}$$

Initialize

1) Create row vectors $\mathbf{r}$, $\mathbf{s}$, and $\mathbf{t}$, for the rows of the augmented matrix $[\mathbf{M}\ \mathbf{b}]$.

$$\mathbf{r} := \begin{bmatrix} 2 & 6 & 10 & 0 \end{bmatrix}$$
$$\mathbf{s} := \begin{bmatrix} 1 & 3 & 3 & 2 \end{bmatrix}$$
$$\mathbf{t} := \begin{bmatrix} 3 & 14 & 28 & -8 \end{bmatrix}$$

Begin computation

2) Pivot on elements of the first column. Interchange rows 1 and 3.

$$\mathbf{r} := \begin{bmatrix} 3 & 14 & 28 & -8 \end{bmatrix}$$
$$\mathbf{s} := \begin{bmatrix} 1 & 3 & 3 & 2 \end{bmatrix}$$
$$\mathbf{t} := \begin{bmatrix} 2 & 6 & 10 & 0 \end{bmatrix}$$

Transform row 2 by adding a multiple of vector $\mathbf{r}$ to current vector $\mathbf{s}$

$$\mathbf{s} := \mathbf{s} + \frac{-s_1}{r_1}\mathbf{r}$$

$$\mathbf{s} = \begin{bmatrix} 0 & -5/3 & -19/3 & 14/3 \end{bmatrix}$$

Transform row 3 by adding a multiple of vector $\mathbf{r}$ to current vector $\mathbf{t}$

$$\mathbf{t} := \mathbf{t} + \frac{-t_1}{r_1}\mathbf{r}$$

$$\mathbf{t} = \begin{bmatrix} 0 & -10/3 & -26/3 & 16/3 \end{bmatrix}$$

3) Display the current system, and decide on pivoting, based on elements in second column, rows 2 and 3.

$$\mathbf{r} = \begin{bmatrix} 3 & 14 & 28 & -8 \end{bmatrix}$$
$$\mathbf{s} = \begin{bmatrix} 0 & -5/3 & -19/3 & 14/3 \end{bmatrix}$$
$$\mathbf{t} = \begin{bmatrix} 0 & -10/3 & -26/3 & 16/3 \end{bmatrix}$$

Largest magnitude element is in row 3, so interchange rows 2 and 3.

$$\mathbf{r} = \begin{bmatrix} 3 & 14 & 28 & -8 \end{bmatrix}$$
$$\mathbf{s} = \begin{bmatrix} 0 & -10/3 & -26/3 & 16/3 \end{bmatrix}$$
$$\mathbf{t} = \begin{bmatrix} 0 & -5/3 & -19/3 & 14/3 \end{bmatrix}$$

Transform row 3 by adding a multiple of vector **s** onto vector **t**.

$$\mathbf{t} := \mathbf{t} + \frac{-t_2}{s_2}\mathbf{s}$$

$$\mathbf{t} = \begin{bmatrix} 0 & 0 & -2 & 2 \end{bmatrix}$$

Display the current system, if desired.

$$\mathbf{r} = \begin{bmatrix} 3 & 14 & 28 & -8 \end{bmatrix}$$
$$\mathbf{s} = \begin{bmatrix} 0 & -10/3 & -26/3 & 16/3 \end{bmatrix}$$
$$\mathbf{t} = \begin{bmatrix} 0 & 0 & -2 & 2 \end{bmatrix}$$

Back substitution

$$x_3 := \frac{t_4}{t_3} \qquad\qquad x_3 = -1$$

$$x_2 := \frac{1}{s_2}(s_4 - x_3 s_3) \qquad x_2 = 1$$

$$x_1 := \frac{1}{r_1}(r_4 - x_3 r_3 - x_2 r_2) \qquad x_1 = 2$$

3.2.2 Mathcad Function for Gaussian Elimination with Pivoting

The following function, Gauss_pivot requires that the indexing of matrix and vector elements begin with 1. Define ORIGIN:=1 before using the function.

$$
\begin{aligned}
&\text{Gauss\_pivot}(A, b) := \\
&\quad n \leftarrow \text{rows}(A) \\
&\quad \text{for } i \in 1 \mathbin{..} n - 1 \\
&\qquad \text{pivot} \leftarrow |A_{i,i}| \\
&\qquad p \leftarrow i \\
&\qquad \text{for } j \in (i+1) \mathbin{..} n \\
&\qquad\quad \text{if } |A_{j,i}| > \text{pivot} \\
&\qquad\qquad \text{pivot} \leftarrow |A_{j,i}| \\
&\qquad\qquad p \leftarrow j \\
&\qquad \text{if } p > i \\
&\qquad\quad \text{for } j \in 1 \mathbin{..} n \\
&\qquad\qquad \text{temp1} \leftarrow A_{i,j} \\
&\qquad\qquad A_{i,j} \leftarrow A_{p,j} \\
&\qquad\qquad A_{p,j} \leftarrow \text{temp1} \\
&\qquad\quad \text{temp2} \leftarrow b_i \\
&\qquad\quad b_i \leftarrow b_p \\
&\qquad\quad b_p \leftarrow \text{temp2} \\
&\qquad \text{for } j \in (i+1) \mathbin{..} n \\
&\qquad\quad m \leftarrow \dfrac{-A_{j,i}}{A_{i,i}} \\
&\qquad\quad \text{for } k \in i \mathbin{..} n \\
&\qquad\qquad A_{j,k} \leftarrow A_{j,k} + m \cdot A_{i,k} \\
&\qquad\quad b_j \leftarrow b_j + m \cdot b_i \\
&\quad x_n \leftarrow \dfrac{b_n}{A_{n,n}} \\
&\quad \text{for } i \in (n-1) \mathbin{..} 1 \\
&\qquad x_i \leftarrow \dfrac{b_i}{A_{i,i}} \\
&\qquad \text{for } j \in (i+1) \mathbin{..} n \\
&\qquad\quad x_i \leftarrow x_i - \dfrac{A_{i,j} \cdot x_{j,k1}}{A_{i,i}} \\
&\quad x
\end{aligned}
$$

Example 3.6 Forces in a Simple Truss

Assuming that the weight of the structure (100 kg) is localized at node 2 (bottom center of Fig. 3.3), the following linear system describes the truss:

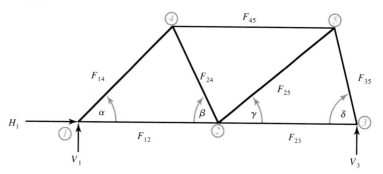

FIGURE 3.3 Triangular truss

$$
\begin{aligned}
V_1 + F_{14} \sin \alpha &= 0 \\
H_1 + F_{12} + F_{14} \cos \alpha &= 0, \\
+ F_{24} \sin \beta + F_{25} \sin \gamma &= 100, \\
- F_{12} + F_{23} - F_{24} \cos \beta + F_{25} \cos \gamma &= 0, \\
V_3 + F_{35} \sin \delta &= 0, \\
- F_{23} - F_{35} \cos \delta &= 0, \\
- F_{14} \sin \alpha - F_{24} \sin \beta &= 0, \\
- F_{14} \cos \alpha + F_{24} \cos \beta + F_{45} &= 0, \\
- F_{25} \sin \gamma - F_{35} \sin \delta &= 0, \\
- F_{25} \sin \gamma + F_{35} \sin \delta - F_{45} &= 0,
\end{aligned}
$$

The equations, in the order written, cannot be solved without pivoting.

The results show that for

$$\alpha = \beta = \gamma = \delta = \pi/4,$$

we have

$$V_1 = 50, \quad H_1 = 0, \quad V_3 = 50,$$
$$F_{12} = 50, \quad F_{14} = -7.07, \quad F_{23} = 50, \quad F_{24} = 70.7,$$
$$F_{25} = 70.7, \quad F_{35} = -70.7, \quad \text{and } F_{45} = -100.$$

To solve the truss problem illustrated in Figure 3.3, with a weight of 100 at node 2, we must solve the following linear system. Pivoting is required.

$$a := \frac{\pi}{4} \qquad b := \frac{\pi}{4} \qquad c := \frac{\pi}{4} \qquad d := \frac{\pi}{4}$$

$$M := \begin{pmatrix} 1 & 0 & 0 & 0 & \sin(a) & 0 & 0 & 0 & 0 & 0 \\ 0 & 1 & 0 & 1 & \cos(a) & 0 & 0 & 0 & 0 & 0 \\ 0 & 0 & 0 & 0 & 0 & 0 & \sin(b) & \sin(c) & 0 & 0 \\ 0 & 0 & 0 & -1 & 0 & 1 & -\cos(b) & \cos(c) & 0 & 0 \\ 0 & 0 & 1 & 0 & 0 & 0 & 0 & 0 & \sin(d) & 0 \\ 0 & 0 & 0 & 0 & 0 & -1 & 0 & 0 & -\cos(d) & 0 \\ 0 & 0 & 0 & 0 & -\sin(a) & 0 & -\sin(b) & 0 & 0 & 0 \\ 0 & 0 & 0 & 0 & -\cos(a) & 0 & \cos(b) & 0 & 0 & 1 \\ 0 & 0 & 0 & 0 & 0 & 0 & 0 & -\sin(c) & -\sin(d) & 0 \\ 0 & 0 & 0 & 0 & 0 & 0 & 0 & -\cos(c) & \cos(d) & -1 \end{pmatrix} \qquad b := \begin{pmatrix} 0 \\ 0 \\ 100 \\ 0 \\ 0 \\ 0 \\ 0 \\ 0 \\ 0 \\ 0 \end{pmatrix}$$

x := Gauss_pivot (M, b)

x =

	1
1	50
2	0
3	50
4	50
5	-70.711
6	50
7	70.711
8	70.711
9	-70.711
10	-100

3.2.3 Discussion

For situations in which row pivoting is desirable because some pivot elements are significantly smaller than others, some form of row scaling should also be used, since, of course, the difficulties illustrated in Example 3.4 could be masked by multiplying the first equation by 1000 to give

$$x_1 + 1000\, x_2 = 3000,$$
$$x_1 + 2\, x_2 = 5,$$

Now no row interchange occurs, because the elements in the first column are equal. However, the computational difficulties described in Example 3.5 are still present. Those difficulties were more apparent with the first equation in its original form.

Scaling strategies vary and are difficult to include in general purpose computer codes. One possible approach is to scale each row by the appropriate power of 10 (including, of course, the corresponding right hand side if it has been retained in a separate vector) so that the magnitude of the largest element in each row of the coefficient matrix is between 0.1 and 1. Dividing by a power of 2 is actually better, since it avoids any round-off error. An even easier method to incorporate into the foregoing functions for Gaussian elimination is to divide each row of **A** by the largest element in the row, making the corresponding scaling on **b** also. However, this may introduce additional round-off error.

The Mathcad function max can be used to find the largest element in a vector. Scaling is sometime suggested on the columns of the coefficient matrix also; this corresponds to changing the units in which the corresponding unknown is measured. Scaling does not generally mitigate problems caused by illconditioning.

The computational effort for Gaussian elimination is proportional to n^3, so, for large systems of equations more efficient methods may be desirable. Improved methods have been developed in recent years that reduce the exponent to values less than 2.5. (See Hager, 1988, or Strang, 1988.) When the system has the appropriate structure, iterative techniques, which we examine in Chapter 4, may be appropriate. In the next section, we consider Gaussian elimination for the special case of a tridiagonal matrix.

3.3 GAUSSIAN ELIMINATION FOR TRIDIAGONAL SYSTEMS

In many applications, the linear system to be solved has a banded structure. For a tridiagonal system, the only nonzero entries in the coefficient matrix are the diagonal, the subdiagonal, and the superdiagonal. To take advantage of this special structure, we apply a modified form of Gaussian elimination in which, following each elimination step, the pivot row is scaled so that the diagonal element is 1. We begin by illustrating the process for a four-by-four system.

Example 3.7 Solving a Tridiagonal System

Consider the following system of equations:

$$\begin{aligned} 2x_1 - x_2 &= 1, \\ -x_1 + 2x_2 - x_3 &= 0, \\ -x_2 + 2x_3 - x_4 &= 0, \\ -x_3 + 2x_4 &= 1, \end{aligned}$$

First, scale the first equation by dividing through by a_{11}, so that the new first equation has one on the diagonal:

$$\begin{aligned} x_1 - \tfrac{1}{2}x_2 &= \tfrac{1}{2}, \\ -x_1 + 2x_2 - x_3 &= 0, \\ -x_2 + 2x_3 - x_4 &= 0, \\ -x_3 + 2x_4 &= 1. \end{aligned}$$

This modifies two other elements also, the element on the upper diagonal and the one on the right-hand side.

Second, use the first equation to eliminate the x_1 term in the second equation (because of the tridiagonal structure, that is the only equation below the first in which x_1 appears):

$$\begin{aligned} x_1 - \tfrac{1}{2}x_2 &= \tfrac{1}{2}, \\ + \tfrac{3}{2}x_2 - x_3 &= \tfrac{1}{2}, \\ -x_2 + 2x_3 - x_4 &= 0, \\ -x_3 + 2x_4 &= 1. \end{aligned}$$

Complete this step by scaling the second equation

$$\begin{aligned} x_1 - \tfrac{1}{2}x_2 &= \tfrac{1}{2}, \\ + x_2 - \tfrac{2}{3}x_3 &= \tfrac{1}{3}, \\ - x_2 + 2x_3 - x_4 &= 0, \\ - x_3 + 2x_4 &= 1. \end{aligned}$$

Next, use the second equation to eliminate the x_2 term in the third equation:

$$x_1 - \frac{1}{2}x_2 \qquad\qquad\qquad = \frac{1}{2},$$
$$+x_2 - \frac{2}{3}x_3 \qquad\qquad = \frac{1}{3},$$
$$+ \frac{4}{3}x_3 - x_4 = \frac{1}{3},$$
$$-x_3 + 2x_4 = 1.$$

Now scale the third equation:

$$x_1 - \frac{1}{2}x_2 \qquad\qquad\qquad = \frac{1}{2},$$
$$+x_2 - \frac{2}{3}x_3 \qquad\qquad = \frac{1}{3},$$
$$+x_3 - \frac{3}{4}x_4 = \frac{1}{4},$$
$$-x_3 + 2x_4 = 1.$$

Finally, use the third equation to eliminate x_3 in the last equation:

$$x_1 - \frac{1}{2}x_2 \qquad\qquad\qquad = \frac{1}{2},$$
$$+x_2 - \frac{2}{3}x_3 \qquad\qquad = \frac{1}{3},$$
$$+x_3 - \frac{3}{4}x_4 = \frac{1}{4},$$
$$+ \frac{5}{4}x_4 = \frac{5}{4}.$$

And scale the last equation:

$$x_1 - \frac{1}{2}x_2 \qquad\qquad\qquad = \frac{1}{2},$$
$$+x_2 - \frac{2}{3}x_3 \qquad\qquad = \frac{1}{3},$$
$$+x_3 - \frac{3}{4}x_4 = \frac{1}{4},$$
$$+x_4 = 1.$$

Now solve by back substitution to obtain

$$x_4 = 1; \qquad\qquad x_3 = 1/4 - (-3/4)(1) = 1;$$
$$x_2 = 1/3 - (-2/3)(1) = 1; \quad x_1 = 1/2 - (-1/2)(1) = 1.$$

Chapter 3 Solving Systems of Linear Equations: Direct Methods

3.3.1 Step-by-Step Procedure

For a tridiagonal matrix **A**, we can reduce the storage requirements from n^2 to $3n$ by storing only the vector **d** containing the **d**iagonal elements, the vector **a** containing the elements **a**bove the diagonal, and the vector **b** containing the elements **b**elow the diagonal. Note that elements b_1 and a_n are zero. The right-hand side is stored as the vector **r**. In this notation, the general tridiagonal system of equations can be written as

$$d_1 x_1 + a_1 x_2 = r_1,$$
$$b_2 x_1 + d_2 x_2 + a_2 x_3 = r_2,$$
$$\cdots .$$
$$b_{n-1} x_{n-2} + d_{n-1} x_{n-1} + a_{n-1} x_n = r_{n-1},$$
$$b_n x_{n-1} + d_n x_n = r_n.$$

An efficient algorithm for the solution of a tridiagonal system is based on Gaussian elimination with the coefficients of the diagonal elements scaled to 1 at each stage. This algorithm takes advantage of the zero elements that are already present in the coefficient matrix and avoids unnecessary arithmetic operations. Thus, we need to store only the new vectors **a** and **r**. This procedure is known in the engineering literature as the *Thomas method*.

Algorithm for Thomas Method

Initialize
1) Create row vectors **a**, **d**, **b**, and **r** that define the tridiagonal system.
   $a_n = 0$ and $b_1 = 0$ (the above and below diagonals, respectively)

Update vector entries
2) Entries corresponding to first equation in the system
   $$a_1 = a_1/d_1$$
   $$r_1 = r_1/d_1$$
3) Entries corresponding to equations $j = 2$ to $n - 1$
   $$t = d_j - b_j a_{j-1}$$
   $$a_j = a_j/t$$
   $$r_j = (r_j - b_j r_{j-1})/t$$
4) Entries corresponding to equation n
   $$r_n = (r_n - b_n r_{n-1})/(d_n - b_n a_{n-1})$$

Form solution
5) First: $x_n = r_n$
   Then, for $j = n - 1 \cdots 1$
   $$x_j = r_j - a_j x_{j+1}$$

Example 3.8 Using the Thomas Algorithm

Use the Thomas algorithm to solve a tridiagonal system, $M x = r$, with vectors a, d, and b:

$$M := \begin{pmatrix} 3 & 1 & 0 & 0 \\ 1 & 2 & -1 & 0 \\ 0 & -1 & 4 & 1 \\ 0 & 0 & -1 & 3 \end{pmatrix} \quad r := \begin{pmatrix} 6 \\ 5 \\ 11 \\ 16 \end{pmatrix} \quad a := \begin{pmatrix} 1 \\ -1 \\ 1 \\ 0 \end{pmatrix} \quad d := \begin{pmatrix} 3 \\ 2 \\ 4 \\ 3 \end{pmatrix} \quad b := \begin{pmatrix} 0 \\ 1 \\ -1 \\ -1 \end{pmatrix}$$

The first step in the elimination phase of the solution

$$a_{0,0} := \frac{a_{0,0}}{d_{0,0}} \qquad r_{0,0} := \frac{r_{0,0}}{d_{0,0}}$$

The second step

$$t := d_{1,0} - b_{1,0} \cdot a_{0,0} \qquad a_{1,0} := \frac{a_{1,0}}{t} \qquad r_{1,0} := \frac{(r_{1,0} - b_{1,0} \cdot r_{0,0})}{t}$$

$$t = 1.667 \qquad a_{1,0} = -0.6 \qquad r_{1,0} = 1.8$$

The third step

$$t := d_{2,0} - b_{2,0} \cdot a_{1,0} \qquad a_{2,0} := \frac{a_{2,0}}{t} \qquad r_{2,0} := \frac{(r_{2,0} - b_{2,0} \cdot r_{1,0})}{t}$$

$$t = 3.4 \qquad a_{2,0} = 0.294 \qquad r_{2,0} = 3.765$$

The fourth step

$$t := d_{3,0} - b_{3,0} \cdot a_{2,0} \qquad r_{3,0} := \frac{(r_{3,0} - b_{3,0} \cdot r_{2,0})}{t} \qquad r_{3,0} = 6$$

Form the solution

$$x_{3,0} := r_{3,0} \qquad\qquad x_{3,0} = 6$$

$$x_{2,0} := r_{2,0} - a_{2,0} \cdot x_{3,0} \qquad x_{2,0} = 2$$

$$x_{1,0} := r_{1,0} - a_{1,0} \cdot x_{2,0} \qquad x_{1,0} = 3$$

$$x_{0,0} := r_{0,0} - a_{0,0} \cdot x_{1,0} \qquad x_{0,0} = 1$$

3.3.2 Mathcad Function for the Thomas Method

The following function for the Thomas method uses the Mathcad default for indexing vectors, i.e., a vector of n elements is indexed from 0 to $n - 1$.

$$\text{Thomas}(a, d, b, r) := \begin{vmatrix} n \leftarrow \text{rows}(a) - 1 \\ a_0 \leftarrow \dfrac{a_0}{d_0} \\ r_0 \leftarrow \dfrac{r_0}{d_0} \\ \text{for } i \in 1 .. n - 1 \\ \quad \begin{vmatrix} \text{denom} \leftarrow d_i - b_i \cdot a_{i-1} \\ \text{break if denom} = 0 \\ a_i \leftarrow \dfrac{a_i}{\text{denom}} \\ r_i \leftarrow \dfrac{r_i - b_i \cdot r_{i-1}}{\text{denom}} \end{vmatrix} \\ r_n \leftarrow \dfrac{r_n - b_n \cdot r_{n-1}}{d_n - b_n \cdot a_{n-1}} \\ x_n \leftarrow r_n \\ \text{for } i \in n - 1 .. 0 \\ \quad \begin{vmatrix} x_i \leftarrow r_i - a_i \cdot x_{i+1} \end{vmatrix} \\ x \end{vmatrix}$$

The next example illustrates the use of the Mathcad function `Thomas` for solving a tridiagonal system, such as those that arise in using spline interpolation (which is discussed in Chapter 8).

Example 3.9 Using the Mathcad function THOMAS

To solve the tridiagonal linear system M x = r using the Thomas method, define the diagonal, above diagonal, and below diagonal vectors that characterize M. For the system

$$M := \begin{pmatrix} 6 & 1 & 0 & 0 \\ 1 & 4 & 1 & 0 \\ 0 & 1 & 6 & 2 \\ 0 & 0 & 2 & 6 \end{pmatrix} \qquad r := \begin{pmatrix} 3 \\ -24 \\ 1 \\ 2 \end{pmatrix}$$

we have the vectors

$$d := \begin{pmatrix} 6 \\ 4 \\ 6 \\ 6 \end{pmatrix} \qquad a := \begin{pmatrix} 1 \\ 1 \\ 2 \\ 0 \end{pmatrix} \qquad b := \begin{pmatrix} 0 \\ 1 \\ 1 \\ 2 \end{pmatrix}$$

The solution from the function Thomas is

$$x := \text{Thomas}(a, d, b, r)$$

$$x = \begin{pmatrix} 1.623 \\ -6.737 \\ 1.326 \\ -0.109 \end{pmatrix} \qquad M \cdot x = \begin{pmatrix} 3 \\ -24 \\ 1 \\ 2 \end{pmatrix}$$

3.3.3 Discussion

In addition to the greatly reduced storage requirements achieved by taking advantage of the special structure of a tridiagonal matrix, the computational effort is much less for the Thomas method than for the general form of Gaussian elimination. The required multiplications and divisions for the Thomas method are as follows:

For the first equation, 2 divisions are needed.
For each of the next $n - 2$ equations, 2 multiplications and 2 divisions are needed.
For the last equation, 2 multiplications and 1 division are required.
The total for elimination is $5 + 4(n - 2)$.
For the backsubstitution, $n - 1$ multiplications are needed.

The Thomas algorithm requires that $d_1 \neq 0$ and that $d_i - b_i a_{i-1} \neq 0$ for each i. For many applications, the structure of the tridiagonal matrix guarantees that these quantities will not be zero. In other cases, if we do encounter a zero value (but the system is, in fact, nonsingular) we can solve for the appropriate variable directly, reduce the size of the system, and solve the new reduced system, as illustrated in the next example.

In general, the Thomas method works well when the system is diagonally dominant.

Example 3.10 Thomas Method Recovery from a Zero-Pivot

To illustrate the possibility of continuing the solution process with the Thomas method when a division by zero is encountered, consider the following system:

$$
\begin{aligned}
2x_1 - x_2 &= 1, \\
-x_1 + 2x_2 - x_3 &= 0, \\
-x_2 + \tfrac{2}{3}x_3 - x_4 &= -\tfrac{4}{3}, \\
-x_3 + 2x_4 - x_5 &= 0, \\
-x_4 + 2x_5 - x_6 &= 0, \\
-x_5 + 2x_6 &= 1.
\end{aligned}
$$

The solution begins in the normal manner by scaling the first equation and using the result to eliminate the x_1 term in the second equation. Continuing by scaling the second equation and using it to eliminate the x_2 term in the third equation, we obtain

$$
\begin{aligned}
x_1 - \tfrac{1}{2}x_2 &= \tfrac{1}{2}, \\
+ x_2 - \tfrac{2}{3}x_3 &= \tfrac{1}{3}, \\
- x_4 &= -1, \\
-x_3 + 2x_4 - x_5 &= 0, \\
-x_4 + 2x_5 - x_6 &= 0, \\
-x_5 + 2x_6 &= 1.
\end{aligned}
$$

However, we are unable to scale the third equation so as to have 1 on the diagonal, since the coefficient of x_3 is now 0. But because the third row does have a nonzero coefficient (for variable x_4), we can solve for that variable and proceed. Thus, the third equation is solved for x_4, giving $x_4 = 1$. The fourth equation is skipped for now, and the computed value of x_4 is substituted into the fifth equation. The elimination proceeds, using the fifth equation to eliminate x_5 from the final equation:

$$
\begin{aligned}
x_1 - \tfrac{1}{2}x_2 &= \tfrac{1}{2}, \\
x_2 - \tfrac{2}{3}x_3 &= \tfrac{1}{3}, \\
-x_4 &= -1 \quad \text{(solve)}, \\
-x_3 + 2x_4 - x_5 &= 0 \quad \text{(skip for now)}, \\
2x_5 - x_6 &= 1 \quad \text{(using } x_4 = 1\text{)}, \\
-x_5 + 2x_6 &= 1.
\end{aligned}
$$

Finally, we scale the fifth equation, use it to eliminate x_5 in the last equation, and scale the last equation:

$$
\begin{aligned}
x_1 - \tfrac{1}{2}x_2 &= \tfrac{1}{2}, \\
x_2 - \tfrac{2}{3}x_3 &= \tfrac{1}{3}, \\
-x_4 &= -1, \\
-x_3 + 2x_4 - x_5 &= 0, \\
x_5 - \tfrac{1}{2}x_6 &= \tfrac{1}{2}, \\
x_6 &= 1.
\end{aligned}
$$

Solving by back substitution yields

$$
\begin{aligned}
x_6 &= 1, \\
x_5 &= \tfrac{1}{2} + \tfrac{1}{2}x_6 = 1, \\
x_4 &= 1 \quad \text{(computed previously)}, \\
x_3 &= 2x_4 - x_5 = 1 \quad \text{(skipped previously)}, \\
x_2 &= \tfrac{1}{3} + \tfrac{2}{3}x_3 = 1, \\
x_1 &= \tfrac{1}{2} + \tfrac{1}{2}x_2 = 1.
\end{aligned}
$$

3.4 MATHCAD'S METHODS

It is possible to solve a linear system using Mathcad's built-in capabilities by either using a matrix function, or by forming a Solve Block. Since the Solve Block approach can also be used for more general systems of equations, we delay its discussion until Chapter 5.

3.4.1 Using the Built-in Functions

The Mathcad function `rref` transforms a matrix to its reduced row echelon form, that is a form in which elements above or below the diagonal are (as much as possible) reduced to 0 and the elements on the diagonal are equal to 1. To use this function to solve a linear system, **M x** = **b**, form the augmented matrix of the system, **A** = [**M b**] and transform **A** to reduced row echelon form. The last column of the reduced matrix is the solution vector **x**. The following steps summarize the process.

1) Create matrix **M** (an n-by-n matrix) and vector **v** (an n-by-1 matrix) using the operators on the Matrix Toolbar.
   Create matrix **A** by setting `A := augment(M, v)`.
   (Or create **A** directly using the operators on the Matrix Toolbar.)
2) Use `rref(A)` to row reduce matrix **A**.

$$S := \text{rref}(A)$$

3) Extract the last column of S using the operator M on the Matrix Toolbar

$$\mathbf{x:} = \mathbf{S}_M \, n + 1.$$

4) Verify solution if desired

$$\mathbf{M}^*\mathbf{x} =$$

It is also very straightforward to use matrix inversion to solve a small linear system written in matrix-vector form. To solve

$$\mathbf{M \, x} = \mathbf{b}$$

define $\mathbf{N} := \mathbf{M}^{-1}$ (find $\mathbf{M}^{-1}$ using the Matrix Toolbar), and compute

$$\mathbf{x} = \mathbf{N \, b}$$

This is not compuationally as efficient as reduction to reduced row echelon form.

3.4.2 Understanding the Algorithms

The function `rref` uses standard row reduction as discussed in Section 3.1, along with scaling of the diagonal elements to 1, and row reduction applied to rows above the pivot row as well as below the pivot row to also reduce the above diagonal elements to 0. This transforms the matrix into what is generally known as reduced row echelon form (RREF).

One well-known algorithm for reduction to RREF is Gauss-Jordan elimination. In this method the elimination process is carried out on all of the rows above and below the pivot row at each step. It is interesting to note that the operation count for Gauss-Jordan elimination is somewhat higher than for Gaussian elimination with back substitution ($O(\frac{1}{2} n^3)$ as compared to $O(\frac{1}{3} n^3)$). The reason for the higher computational load is that elimination is more expensive than back substitution, and at each stage the elimination must be applied to all of the rows other than the pivot row.

A small modification to the standard Gauss-Jordan scheme provides a computationally more efficient process; this is the standard RREF algorithm. By performing row reduction to transform the below diagonal elements first (forward elimination, as in standard Gauss elimination) together with scaling of the diagonal elements, followed by the reduction of the above diagonal elements (backward elimination), the operation count for RREF is of the same order as for Gaussian elimination. The reason that this small change in the algorithm (applying all of the backward elimination after the forward elimination stages are completed) makes a significant improvement on the operation count is that now each forward elimination step only involves elimination of the elements in one column. The resulting computation for backward elimination is of the same order as that for back-substitution.

SUMMARY

Basic Gaussian Elimination: To solve $\mathbf{Ax} = \mathbf{b}$, transform matrix $\mathbf{A}$ into an upper triangular matrix by systematically applying the following row transformations to the augmented matrix consisting of $\mathbf{A}$ together with $\mathbf{b}$, i.e., $[\mathbf{A} : \mathbf{b}]$.

Add a multiple m of row R_i onto row R_j to form a new row R_j:

$$R_j = \leftarrow mR_i + R_j.$$

Solve the resulting linear system by back substitution:

$$x_n = b_n/a_{nn},$$
$$x_{n-1} = (b_{n-1} - a_{n-1,n} x_n)/a_{n-1,n-1},$$
$$\ldots$$
$$x_2 = (b_2 - a_{23}x_3 - \ldots - a_{2,n}x_n)/a_{22},$$
$$x_1 = (b_1 - a_{12}x_2 - \ldots - a_{1,n-1}x_{n-1} - a_{1,n}x_n)/a_{11}.$$

Gaussian Elimination with Row Pivoting: To solve $\mathbf{Ax} = \mathbf{b}$, transform matrix $\mathbf{A}$ into an upper triangular matrix by systematically performing row interchanges on the augmented matrix so that at each stage the pivot element is as large as possible, then performing the same row transformations as for basic Gaussian elimination.

Thomas method (Modified Gaussian Elimination for Tridiagonal Systems): Let

$$d_1 x_1 + a_1 x_2 = r_1,$$
$$b_2 x_1 + d_2 x_2 + a_2 x_3 = r_2,$$
$$b_3 x_2 + d_3 x_3 + a_3 x_4 = r_3,$$
$$\ldots$$
$$b_n x_{n-1} + d_n x_n = r_n.$$

Step 1

For the first equation, form the new elements a_1 and r_1,

$$a_1 = \frac{a_1}{d_1}, \qquad r_1 = \frac{r_1}{d_1}.$$

Step 2

For each of the equations from $i = 2, \ldots, n-1$,

$$a_i = \frac{a_i}{d_i - b_i a_{i-1}}, \qquad r_i = \frac{r_i - b_i r_{i-1}}{d_i - b_i a_{i-1}}.$$

Step 3

For the last equation,

$$r_n = \frac{r_n - b_n r_{n-1}}{d_n - b_n a_{n-1}}.$$

Step 4

Solve by back substitution, yielding

$$x_n = r_n,$$
$$x_i = r_i - a_i x_{i+1}, \qquad i = n-1, n-2, n-3, \ldots, 2, 1.$$

SUGGESTIONS FOR FURTHER READING

The topics introduced in this chapter are part of the field of numerical linear algebra. We consider other topics from this area in Chapters 6 and 7. For more in depth treatments of these techniques, the following are a few suggested sources:

Hager, W. W., *Applied Numerical Linear Algebra*, Prentice-Hall, Englewood Cliffs, NJ, 1988. This book includes a discussion of the relative merits of row and column pivoting.

Strang, G., *Linear Algebra and Its Applications*, 3d ed., Harcourt Brace Jovanovich, San Diego, 1988.

For further discussion of the topics introduced in this chapter, see any of the following texts:

Atkinson, K. E., *An Introduction to Numerical Analysis,* 2d ed., John Wiley, New York, 1989.

Coleman, T. F. and C. Van Loan, *Handbook for Matrix Computations*, SIAM, Philadelphia, 1988.

Dongarra, J. J. et al., *LINPACK User's Guide*, SIAM, Philadelphia, 1979.

Forsythe, G. E. and C. B. Moler, *Computer Solution of Linear Algebraic Systems*, Prentice-Hall, Englewood Cliffs, NJ, 1967.

Gill, P. E., W. Murray, and M. H. Wright, *Numerical Linear Algebra and Optimization*, Addison-Wesley, Redwood City, CA, 1991.

Golub, G. H. and C. F. Van Loan, *Matrix Computations*, 3d ed., Johns Hopkins University Press, Baltimore, 1996.

Isaacson, E. and H. B. Keller, *Analysis of Numerical Methods*, Wiley, New York, 1966.

Jensen, J. A. and J. H. Rowland, *Methods of Computation*, Scott, Foresman and Company, Glenview, IL, 1975.

Press, W. H., S. A. Teukolsky, W. T. Vetterling, and B. P. Flannery, *Numerical Recipes in C, The Art of Scientific Computing*, (2d ed.), Cambridge Unversity Press, 1992. (Discussion of Gauss-Jordan elimination, pp. 36–41)

Ralston, A. and P. Rabinowitz, *A First Course in Numerical Analysis*, 2nd ed, McGraw Hill, New York, 1978.

Stoer, J. and R. Bulirsch, *Introduction to Numerical Analysis*, Springer-Verlag, New York, 1980.

Wilkinson, J. H., *The Algebraic Eigenvalue Problem*, Oxford University Press, New York, 1965.

Wilkinson, J. H. and C. Reinsch, *Linear Algebra,* vol II of *Handbook for Automatic Computation*, Springer-Verlag, New York, 1971.

PRACTICE THE TECHNIQUES

For problems P3.1 to P3. 5, solve the linear system $Ax = b$ using basic Gaussian elimination.

P3.1 $\mathbf{A} = \begin{bmatrix} 3 & -1 & 2 \\ 1 & 2 & 3 \\ 2 & -2 & -1 \end{bmatrix}$, $\mathbf{b} = \begin{bmatrix} 1 \\ 1 \\ 1 \end{bmatrix}$

P3.2 $\mathbf{A} = \begin{bmatrix} 2 & 5 & 3 \\ 8 & 9 & 7 \\ 4 & 6 & 1 \end{bmatrix}$ $\mathbf{b} = \begin{bmatrix} 2 \\ 14 \\ 3 \end{bmatrix}$

P3.3 $\mathbf{A} = \begin{bmatrix} 10 & -2 & 1 \\ -2 & 10 & -2 \\ -2 & -5 & 10 \end{bmatrix}$, $\mathbf{b} = \begin{bmatrix} 9 \\ 12 \\ 18 \end{bmatrix}$

P3.4 $\mathbf{A} = \begin{bmatrix} -1 & 5 & 2 \\ 2 & 3 & 1 \\ 3 & 2 & 1 \end{bmatrix}$, $\mathbf{b} = \begin{bmatrix} -2 \\ 7 \\ 3 \end{bmatrix}$

P3.5 $\mathbf{A} = \begin{bmatrix} 8 & 1 & -1 \\ -1 & 7 & -2 \\ 2 & 1 & 9 \end{bmatrix}$, $\mathbf{b} = \begin{bmatrix} 8 \\ 4 \\ 12 \end{bmatrix}$

For problems P3.6 to P3.10, solve the linear system $Ax = b$

a. *using basic Gaussian elimination.*
b. *using the Mathcad function* Gauss_B *given in the text.*

P3.6 $\mathbf{A} = \begin{bmatrix} 2 & 0 & -2 \\ 3 & -4 & -4 \\ -2 & 2 & -1 \end{bmatrix}$

i) $\mathbf{b} = [-10 \quad -8 \quad 3]^T$
ii) $\mathbf{b} = [6 \quad 2 \quad 15]^T$

P3.7 $\mathbf{A} = \begin{bmatrix} 3 & -5 & -5 \\ 5 & -5 & -2 \\ 2 & 3 & 4 \end{bmatrix}$

i) $\mathbf{b} = [-36 \quad -15 \quad 33]^T$
ii) $\mathbf{b} = [-14 \quad -32 \quad -7]^T$

P3.8 $\mathbf{A} = \begin{bmatrix} 4 & 12 & 8 & 4 \\ 1 & 7 & 18 & 9 \\ 2 & 9 & 20 & 20 \\ 3 & 11 & 15 & 14 \end{bmatrix} \quad \mathbf{b} = \begin{bmatrix} -4 \\ -5 \\ -25 \\ -18 \end{bmatrix}$

P3.9 $\mathbf{A} = \begin{bmatrix} 1 & 1 & 0 & 3 \\ 2 & 1 & -1 & 1 \\ 3 & -1 & -1 & 2 \\ -1 & 2 & 3 & -1 \end{bmatrix} \quad \mathbf{b} = \begin{bmatrix} 4 \\ 1 \\ -3 \\ 4 \end{bmatrix}$

P3.10 $\mathbf{A} = \begin{bmatrix} 1 & 1 & 1 & 1 \\ 2 & 4 & 4 & 4 \\ 3 & 11 & 14 & 14 \\ 5 & 17 & 38 & 42 \end{bmatrix} \quad \mathbf{b} = \begin{bmatrix} 0 \\ -2 \\ -8 \\ -20 \end{bmatrix}$

For problems P3.11 to P3.15, solve the linear system, $\mathbf{Ax} = \mathbf{b}$, using Gaussian elimination with row pivoting

P3.11 $\mathbf{A} = \begin{bmatrix} 6 & 2 & 2 \\ 6 & 2 & 1 \\ 1 & 2 & -1 \end{bmatrix}, \quad \mathbf{b} = \begin{bmatrix} 0 \\ 5 \\ 0 \end{bmatrix}$

P3.12 $\mathbf{A} = \begin{bmatrix} 1 & 2 & 3 \\ 2 & 4 & 10 \\ 3 & 14 & 28 \end{bmatrix} \quad \mathbf{b} = \begin{bmatrix} 0 \\ -2 \\ -8 \end{bmatrix}$

P3.13 $\mathbf{A} = \begin{bmatrix} 2 & 6 & 10 \\ 1 & 3 & 3 \\ 3 & 14 & 28 \end{bmatrix} \quad \mathbf{b} = \begin{bmatrix} 0 \\ 2 \\ -8 \end{bmatrix}$

P3.14 $\mathbf{A} = \begin{bmatrix} -1 & 1 & 0 & 0 \\ 1 & -1 & 1 & 0 \\ 0 & 1 & -1 & 1 \\ 0 & 0 & 1 & -1 \end{bmatrix} \quad \mathbf{b} = \begin{bmatrix} 1 \\ 1 \\ -1 \\ -1 \end{bmatrix}$

P3.15 $\mathbf{A} = \begin{bmatrix} 2 & -1 & 0 & 0 & 0 & 0 \\ -1 & 2 & -1 & 0 & 0 & 0 \\ 0 & -1 & 2/3 & -1 & 0 & 0 \\ 0 & 0 & -1 & 2 & -1 & 0 \\ 0 & 0 & 0 & -1 & 2 & -1 \\ 0 & 0 & 0 & 0 & -1 & 2 \end{bmatrix}$

$\mathbf{b} = [\,1 \quad 0 \quad -4/3 \quad 0 \quad 0 \quad -1\,]^T$

For problems P3.16 to P3.20, solve the linear system $\mathbf{Ax} = \mathbf{b}$ with rounding to two digits.
a. using Gaussian elimination.
b. using Gaussian elimination with row pivoting.
c. Compare the results from Parts a and b.

P3.16 $\quad 0.001\,x_1 + 2x_2 = 4$
$\quad\quad\quad x_1 + 2x_2 = 5$

P3.17 $\quad 0.001\,x_1 + 10x_2 = 30$
$\quad\quad\quad x_1 + x_2 = 2$

P3.18 $\quad 0.001\,x_1 + 2x_2 = 6$
$\quad\quad\quad x_1 + 3x_2 = 8$

P3.19 $\quad 0.001\,x_1 + x_2 + x_3 = 5$
$\quad\quad\quad x_1 + x_2 \phantom{{}+x_3} = 3$
$\quad\quad\quad x_1 \phantom{{}+x_2} + x_3 = 4$

P3.20 $\quad 6x_1 - 2.2x_2 + 3x_3 = 20$
$\quad\quad\quad -3x_1 + x_2 - 1.1x_3 = -8.1$
$\quad\quad\quad -1x_1 - 3\,x_2 + 0.9x_3 = 6$

For problems P3.21 to 3.25, solve the linear system $\mathbf{Ax} = \mathbf{b}$; round to 2 digits.
a. Use basic Gaussian elimination.
b. Use Gaussian elimination with row pivoting.
c. Use Gaussian elimination with scaling and pivoting; scale each row so that the largest element is 1 before choosing the pivot element.

P3.21 $\quad x_1 + 2000x_2 = 4000$
$\quad\quad\quad x_1 + 2x_2 = 5$

P3.22 $\quad x_1 + 1000x_2 = 3000$
$\quad\quad\quad x_1 + x_2 = 2$

P3.23 $\quad x_1 + 2000x_2 = 6000$
$\quad\quad\quad x_1 + 3x_2 = 8$

P3.24 $x_1 + 1000x_2 + 1000x_3 = 5000$
$x_1 + x_2 = 3$
$x_1 + + x_3 = 4$

P3.25 $600x_1 - 220x_2 + 300x_3 = 2000$
$-300x_1 + 100x_2 - 110x_3 = -810$
$-1x_1 - 3x_2 + 0.9x_3 = 6$

For problems P3.26 to 3.35, solve the linear system $Ax = r$
  a. using Gaussian elimination.
  b. using the Thomas method for tridiagonal systems.

P3.26 $A = \begin{bmatrix} 1 & 2 & 0 \\ 1 & 3 & 3 \\ 0 & 3 & 10 \end{bmatrix}$ $r = \begin{bmatrix} 10 & 17 & 22 \end{bmatrix}^T$

P3.27 $A = \begin{bmatrix} 1 & 2 & 0 \\ 1 & 3 & 4 \\ 0 & 3 & 13 \end{bmatrix}$ $r = \begin{bmatrix} 3 & 8 & 16 \end{bmatrix}^T$

P3.28 $A = \begin{bmatrix} -2 & 1 & 0 & 0 \\ 1 & -2 & 1 & 0 \\ 0 & 1 & -2 & 1 \\ 0 & 0 & 1 & -2 \end{bmatrix}$, $r = \begin{bmatrix} -1 \\ 0 \\ 0 \\ 0 \end{bmatrix}$.

P3.29 $A = \begin{bmatrix} 2 & 1 & 0 & 0 \\ -1 & 2 & 1 & 0 \\ 0 & 1 & 2 & -1 \\ 0 & 0 & -1 & 2 \end{bmatrix}$, $r = \begin{bmatrix} 0 \\ 0 \\ 0 \\ 11 \end{bmatrix}$.

P3.30 $A = \begin{bmatrix} 5 & 1 & 0 & 0 \\ 1 & 5 & 1 & 0 \\ 0 & 1 & 5 & 1 \\ 0 & 0 & 1 & 5 \end{bmatrix}$, $r = \begin{bmatrix} 33 \\ 26 \\ 30 \\ 15 \end{bmatrix}$

P3.31 $A = \begin{bmatrix} -3 & -4 & 0 & 0 & 0 & 0 \\ -3 & 4 & 5 & 0 & 0 & 0 \\ 0 & 1 & -1 & -3 & 0 & 0 \\ 0 & 0 & 0 & 4 & -5 & 0 \\ 0 & 0 & 0 & 3 & 1 & -5 \\ 0 & 0 & 0 & 0 & -1 & 2 \end{bmatrix}$
$r = \begin{bmatrix} 14 & -36 & -6 & 14 & 9 & 6 \end{bmatrix}^T$

P3.32 $A = \begin{bmatrix} 1 & 3 & 0 & 0 & 0 & 0 & 0 \\ 5 & -4 & -1 & 0 & 0 & 0 & 0 \\ 0 & 5 & -2 & -1 & 0 & 0 & 0 \\ 0 & 0 & 2 & 3 & 1 & 0 & 0 \\ 0 & 0 & 0 & 5 & -3 & -1 & 0 \\ 0 & 0 & 0 & 0 & 1 & -1 & 0 \\ 0 & 0 & 0 & 0 & 0 & -2 & 4 \end{bmatrix}$
$r = \begin{bmatrix} 19 & 1 & 28 & 0 & -25 & 0 & 2 \end{bmatrix}^T$

P3.33 $A = \begin{bmatrix} -1 & 1 & 0 & 0 & 0 & 0 & 0 & 0 \\ -1 & 4 & 1 & 0 & 0 & 0 & 0 & 0 \\ 0 & 4 & 1 & 3 & 0 & 0 & 0 & 0 \\ 0 & 0 & 0 & -1 & -2 & 0 & 0 & 0 \\ 0 & 0 & 0 & -2 & -2 & -2 & 0 & 0 \\ 0 & 0 & 0 & 0 & -4 & -2 & -2 & 0 \\ 0 & 0 & 0 & 0 & 0 & 2 & 4 & 0 \\ 0 & 0 & 0 & 0 & 0 & 0 & 2 & 0 \end{bmatrix}$
$r = \begin{bmatrix} 7 & 13 & -3 & -2 & -4 & -28 & 26 & 10 \end{bmatrix}^T$

P3.34
$A = \begin{bmatrix} -1 & 1 & 0 & 0 & 0 & 0 & 0 & 0 & 0 \\ 2 & 3 & 1 & 0 & 0 & 0 & 0 & 0 & 0 \\ 0 & 3 & -3 & -1 & 0 & 0 & 0 & 0 & 0 \\ 0 & 0 & -4 & 3 & 4 & 0 & 0 & 0 & 0 \\ 0 & 0 & 0 & 3 & 3 & 5 & 0 & 0 & 0 \\ 0 & 0 & 0 & 0 & -1 & -5 & 0 & 0 & 0 \\ 0 & 0 & 0 & 0 & 0 & -5 & 1 & -4 & 0 \\ 0 & 0 & 0 & 0 & 0 & 0 & -2 & 2 & -4 \\ 0 & 0 & 0 & 0 & 0 & 0 & 0 & -4 & 2 \end{bmatrix}$
$r = \begin{bmatrix} -1 & 19 & 20 & -1 & -19 & 14 & 0 & -4 & -2 \end{bmatrix}^T$

P3.35
$A = \begin{bmatrix} 3 & -4 & 0 & 0 & 0 & 0 & 0 & 0 & 0 & 0 \\ 3 & 3 & 5 & 0 & 0 & 0 & 0 & 0 & 0 & 0 \\ 0 & -1 & 1 & 2 & 0 & 0 & 0 & 0 & 0 & 0 \\ 0 & 0 & -2 & -4 & 5 & 0 & 0 & 0 & 0 & 0 \\ 0 & 0 & 0 & 1 & 0 & -2 & 0 & 0 & 0 & 0 \\ 0 & 0 & 0 & 0 & 5 & -3 & -2 & 0 & 0 & 0 \\ 0 & 0 & 0 & 0 & 0 & 1 & 0 & -5 & 0 & 0 \\ 0 & 0 & 0 & 0 & 0 & 0 & -3 & 0 & -1 & 0 \\ 0 & 0 & 0 & 0 & 0 & 0 & 0 & -3 & 0 & 1 \\ 0 & 0 & 0 & 0 & 0 & 0 & 0 & 0 & -4 & 1 \end{bmatrix}$
$r = \begin{bmatrix} -13 & -11 & -6 & 25 & 6 & 29 & 1 & 0 & 3 & -12 \end{bmatrix}^T$

For Problems P3.36 to P3.40, solve the linear system $\mathbf{A}\mathbf{x} = \mathbf{b}$ using the Mathcad function in the text or using Mathcad's built-in methods.

P3.36 $\mathbf{A} = \begin{bmatrix} -5 & 0 & -4 & 1 & 4 & 5 \\ 3 & 5 & -2 & -4 & 3 & -2 \\ -1 & -3 & 3 & 4 & 3 & 1 \\ 0 & 1 & 1 & 1 & -1 & -4 \\ -4 & -1 & -4 & -3 & 2 & 0 \\ -3 & -3 & -4 & 5 & 3 & 1 \end{bmatrix}$

a. $\mathbf{b} = [\;14 \quad -26 \quad 0 \quad -16 \quad 0 \quad -17\;]^T$
b. $\mathbf{b} = [\;19 \quad -3 \quad -15 \quad -4 \quad 15 \quad 10\;]^T$
c. $\mathbf{b} = [-16 \quad 29 \quad -4 \quad 2 \quad -13 \quad -31\;]^T$
d. $\mathbf{b} = [-44 \quad 3 \quad -23 \quad 3 \quad -10 \quad -40\;]^T$

P3.37

$\mathbf{A} = \begin{bmatrix} -4 & -2 & -1 & 4 & 5 & -2 & -1 & 5 \\ -4 & 1 & 3 & -4 & -4 & -4 & -2 & -3 \\ 2 & 1 & -2 & 3 & -4 & -1 & -3 & 0 \\ 1 & 2 & -2 & 5 & 2 & -5 & 1 & 3 \\ 0 & -4 & 4 & -2 & -1 & -3 & 1 & 4 \\ 4 & -1 & 5 & 0 & 3 & 3 & 0 & 3 \\ 4 & 3 & 2 & 2 & -1 & -3 & -1 & 1 \\ 0 & -5 & 4 & 1 & 2 & 3 & 3 & 1 \end{bmatrix}$

a. $\mathbf{b} = [\;10 \quad -30 \quad 0 \quad -16 \quad 23 \quad 61 \quad 9 \quad 40\;]^T$
b. $\mathbf{b} = [\;46 \quad -13 \quad -14 \quad 18 \quad 16 \quad 20 \quad 0 \quad 7\;]^T$
c. $\mathbf{b} = [\;48 \quad -10 \quad -27 \quad 24 \quad 16 \quad -13 \quad -22 \quad 16\;]^T$
d. $\mathbf{b} = [-19 \quad -7 \quad 7 \quad 23 \quad -12 \quad -11 \quad 21 \quad -38\;]^T$

P3.38

$\mathbf{A} = \begin{bmatrix} 0 & -3 & -3 & -2 & 3 & 3 & 1 & 2 \\ 3 & 1 & 0 & -4 & 4 & -4 & 2 & -1 \\ -3 & 2 & 3 & -5 & 3 & 5 & -2 & 0 \\ -1 & 0 & -1 & 0 & 4 & -1 & -4 & -5 \\ 5 & -1 & -2 & -1 & 4 & 0 & 2 & 3 \\ -4 & -3 & -2 & -3 & 0 & -2 & -1 & 0 \\ -3 & -3 & -4 & 1 & -3 & 2 & 4 & 0 \\ -4 & 0 & -3 & 0 & 2 & 3 & 3 & -3 \end{bmatrix}$

a. $\mathbf{b} = [-18 \quad -13 \quad -77 \quad 25 \quad -8 \quad 4 \quad 13 \quad -15\;]^T$
b. $\mathbf{b} = [\;3 \quad 14 \quad -53 \quad 3 \quad 14 \quad -7 \quad 36 \quad 25\;]^T$
c. $\mathbf{b} = [\;14 \quad -30 \quad 36 \quad -7 \quad -18 \quad 26 \quad 3 \quad -7\;]^T$
d. $\mathbf{b} = [-28 \quad -3 \quad -36 \quad -49 \quad 5 \quad -24 \quad -6 \quad -37\;]^T$

P3.39

$\mathbf{A} = \begin{bmatrix} 1 & -4 & 4 & 1 & 2 & 1 & 3 & -1 & -2 & 4 \\ 2 & 2 & -3 & -4 & 5 & 5 & -4 & -2 & 0 & -2 \\ 2 & 3 & 1 & 0 & -2 & 1 & -1 & 1 & -3 & 1 \\ 1 & -4 & -4 & -3 & -2 & -3 & -1 & 1 & 0 & 4 \\ 1 & 3 & -2 & 0 & -5 & -3 & 3 & -2 & 3 & -1 \\ -1 & -1 & -2 & 0 & -1 & -4 & 0 & -3 & -2 & -3 \\ 4 & -5 & 2 & 0 & 1 & -3 & -4 & 1 & -5 & -2 \\ 5 & -2 & 3 & 1 & 3 & 1 & 2 & 1 & -3 & -5 \\ 0 & -4 & 3 & -4 & -1 & -1 & -2 & -1 & -2 & -5 \\ -4 & 5 & 3 & 5 & -2 & -2 & 1 & -1 & 0 & 4 \end{bmatrix}$

a. $\mathbf{b} = [\;14 \quad 42 \quad -3 \quad -22 \quad -28 \quad 21 \quad 38 \quad 34 \quad 31 \quad -16\;]^T$
b. $\mathbf{b} = [-12 \quad -47 \quad 15 \quad 28 \quad 17 \quad 28 \quad 37 \quad -2 \quad -5 \quad 15\;]^T$
c. $\mathbf{b} = [-16 \quad 6 \quad 14 \quad 20 \quad -16 \quad -10 \quad -10 \quad -44 \quad -46 \quad 26\;]^T$
d. $\mathbf{b} = [\;-9 \quad 46 \quad 16 \quad -21 \quad -20 \quad 6 \quad 27 \quad 21 \quad 36 \quad -25\;]^T$

P3.40

$\mathbf{A} = \begin{bmatrix} 2 & -2 & -3 & 1 & 3 & 5 & 3 & 3 & -1 & -5 \\ 1 & 2 & 3 & 2 & -2 & -3 & 0 & 5 & -4 & 3 \\ 0 & 2 & 1 & -4 & 0 & -4 & 4 & -1 & 2 & 0 \\ -2 & -2 & 0 & -3 & -4 & 2 & -4 & 4 & -4 & 2 \\ 2 & -3 & -2 & -4 & -2 & 2 & -1 & 0 & 4 & -2 \\ 4 & -5 & -1 & -1 & 3 & -4 & 2 & 4 & 4 & -5 \\ 1 & 0 & 0 & 5 & 1 & 4 & 1 & -3 & 3 & -3 \\ 0 & -4 & 1 & -1 & 3 & -4 & -4 & 1 & -4 & -2 \\ 5 & 0 & 4 & -3 & -5 & -2 & 1 & -3 & 4 & 4 \\ -2 & -3 & 1 & -5 & 1 & -3 & 4 & 5 & 2 & 4 \end{bmatrix}$

a. $\mathbf{b} = [\;16 \quad -37 \quad 36 \quad -43 \quad 33 \quad 66 \quad -7 \quad 18 \quad 11 \quad 5\;]^T$
b. $\mathbf{b} = [\;0 \quad 29 \quad -29 \quad 25 \quad -3 \quad 12 \quad -11 \quad 17 \quad -3 \quad 7\;]^T$
c. $\mathbf{b} = [\;34 \quad -54 \quad -1 \quad -2 \quad 47 \quad -3 \quad 35 \quad -39 \quad 11 \quad -27\;]^T$
d. $\mathbf{b} = [-15 \quad 16 \quad -8 \quad -35 \quad -54 \quad -40 \quad 29 \quad -6 \quad -22 \quad -38\;]^T$

EXPLORE SOME APPLICATIONS

A3.1 Linear systems occur in many other problems involving numerical methods. The following system is a small example of the type of system encountered as part of quadratic spline interpolation problem (see Chapter 8):

$$a_1 - a_2 \qquad\qquad + b_1 \qquad\qquad = 4$$
$$a_2 - a_3 \qquad + b_1 + b_2 = 12$$
$$+ a_3 - a_4 \qquad + b_2 = -4$$
$$a_1 + a_2 \qquad\qquad - b_1 \qquad\qquad = 0$$
$$a_2 + a_3 \qquad + b_1 - b_2 = 0$$
$$a_3 + a_4 \qquad\qquad + b_2 = 0$$

Solve for the unknowns, a_1, a_2, a_3, b_1, b_2, and b_3.

A3.2 This is a small example of the type of system encountered as part of the cubic spline interpolation technique (discussed later in section 8.4.3). Solve for the unknowns, $a_1, a_2, \ldots, a_7$.

$$\tfrac{2}{3}a_1 + \tfrac{1}{6}a_2 = 1.2$$
$$\tfrac{1}{6}a_1 + \tfrac{2}{3}a_2 + \tfrac{1}{6}a_3 = 1.97$$
$$+ \tfrac{1}{6}a_2 + \tfrac{2}{3}a_3 + \tfrac{1}{6}a_4 = 2$$
$$+ \tfrac{1}{6}a_3 + \tfrac{2}{3}a_4 + \tfrac{1}{6}a_5 = 0$$
$$+ \tfrac{1}{6}a_4 + \tfrac{2}{3}a_5 + \tfrac{1}{6}a_6 = -2$$
$$+ \tfrac{1}{6}a_5 + \tfrac{2}{3}a_6 + \tfrac{1}{6}a_7 = -1.97$$
$$+ \tfrac{1}{6}a_6 + \tfrac{2}{3}a_7 = -1.2$$

A3.3 An 10-stage equilibrium process such as liquid extraction or gas absorption can be modeled by a tridiagonal system of linear equations; solve using $rr = 0.9$ (this depends on the ratio of flow rates (left to right and right to left) and the ratio of weight fractions of the two components that are flowing). Assume that the flow in is 0.05, and that the flow out is 0.5.

$$-(1 + rr)x_1 + rr\, x_2 = -(\text{flow into compartment 1})$$
$$x_{i-1} - (1 + rr)x_i + rr\, x_{i+1} = 0 \qquad i = 2, \ldots n - 1$$
$$x_{n-1} - (1 + rr)x_n = -(\text{flow out of compartment } n)$$

(See Hanna and Sandall, p. 58 for a discussion of similar problems.)

A3.4 Consider the single triangle truss shown in Figure 3.2, but take $\alpha = \beta = \pi/4$, so that $\cos \alpha = \sin \alpha = \cos \beta = \sin \beta = \sqrt{2}/2$. Take $W_3 = 100$.

$$V_1 \qquad\qquad + \sqrt{2}/2\, F_{13} \qquad\qquad = 0$$
$$H_1 \; + F_{12} + \sqrt{2}/2\, F_{13} \qquad\qquad = 0$$
$$V_2 \qquad\qquad\qquad\qquad + \sqrt{2}/2\, F_{23} = 0$$
$$F_{12} \qquad\qquad\qquad\qquad - \sqrt{2}/2\, F_{23} = 0$$
$$- \sqrt{2}/2\, F_{13} - \sqrt{2}/2\, F_{23} = 100$$
$$- \sqrt{2}/2\, F_{13} + \sqrt{2}/2\, F_{23} = 0$$

A3.5 Tridiagonal systems also occur in solutions of ordinary differential equations with boundary conditions when the equations are solved by means of finite differences. For this example, solve the system $A\mathbf{x} = \mathbf{b}$

$$A = \begin{bmatrix}
-1.99 & 1.00 & 0 & 0 & 0 & 0 & 0 & 0 & 0 \\
1.00 & -1.99 & 1.00 & 0 & 0 & 0 & 0 & 0 & 0 \\
0 & 1.00 & -1.99 & 1.00 & 0 & 0 & 0 & 0 & 0 \\
0 & 0 & 1.00 & -1.99 & 1.00 & 0 & 0 & 0 & 0 \\
0 & 0 & 0 & 1.00 & -1.99 & 1.00 & 0 & 0 & 0 \\
0 & 0 & 0 & 0 & 1.00 & -1.99 & 1.00 & 0 & 0 \\
0 & 0 & 0 & 0 & 0 & 1.00 & -1.99 & 1.00 & 0 \\
0 & 0 & 0 & 0 & 0 & 0 & 1.00 & -1.99 & 1.00 \\
0 & 0 & 0 & 0 & 0 & 0 & 0 & 1.00 & -1.99
\end{bmatrix}$$

$\mathbf{b} = [-0.99 \; 0.002 \; 0.0031 \; 0.0042 \; 0.0055 \; 0.0068 \; 0.0084 \; 0.0103 \; -0.6874]^T$

A3.6 The equilibrium positions of a system of blocks coupled by springs is described by a linear system of equations. Suppose that there are 4 blocks arranged as shown below. The unstretched length of the ith spring is L_i, and its spring constant is k_i. The distance between the two walls is l. The equation stating that the ith block is in equilibrium states that the forces on the block (from the springs on either side of it) are equal. For example, for B_2, this gives $k_2(x_1 + L_2 - x_2) = k_3(x_2 + L_3 - x_3)$

Chapter 3 Solving Systems of Linear Equations: Direct Methods

$$\begin{bmatrix} -k_1 - k_2 & k_2 & 0 & 0 \\ k_2 & -k_2 - k_3 & k_3 & 0 \\ 0 & k_3 & -k_3 - k_4 & k_4 \\ 0 & 0 & k_4 & -k_4 - k_5 \end{bmatrix} \begin{bmatrix} x_1 \\ x_2 \\ x_3 \\ x_4 \end{bmatrix} = \begin{bmatrix} -k_1 L_1 + k_2 L_2 \\ -k_2 L_2 + k_3 L_3 \\ -k_3 L_3 + k_4 L_4 \\ -k_4 L_4 + k_5(L_5 - L) \end{bmatrix}$$

Solve the system for $L_1 = 2$; $L_2 = 2$; $L_3 = 2$; $L_4 = 2$; $L_5 = 2$; $L_w = 8$. $k_1 = 1$; $k_2 = 1$; $k_3 = 1$; $k_4 = 1$; $k_5 = 5$. See (Garcia, p. 103) for discussion of a similar problem.

A3.7 Consider the simple truss shown in Figure 3.3, but take the angles to be $\alpha = \pi/6$, $\beta = \pi/3$, $\gamma = \pi/6$, $\delta = \pi/3$. Solve the linear system, and compare the forces to those found in Example 3.6.

A3.8 Consider the same truss as in A 3.7, but assume that half of the load is applied at node 4, and half at node 5, instead of the entire load being localized at node 2.

A3.9 Find the currents in the electrical circuit shown in Fig. 3.1 if the resistances are $R_1 = 12$, $R_2 = 4$, $R_3 = 5$, $R_4 = 2$, $R_5 = 10$, and the voltages are $V_1 = 100$, $V_2 = 0$, $V_3 = 0$.

A3.10 Find the currents in the electrical circuit shown in Fig. 3.1 if the resistances are $R_1 = 15$, $R_2 = 5$, $R_3 = 20$, $R_4 = 0$, $R_5 = 10$, and the voltages are $V_1 = 0$, $V_2 = 200$, $V_3 = 100$.

A3.11 Find the currents in the electrical circuit shown in Fig. 3.1 if the resistances are $R_1 = 10$, $R_2 = 5$, $R_3 = 0$, $R_4 = 10$, $R_5 = 5$, and the voltages are $V_1 = -200$, $V_2 = 0$, $V_3 = 35$.

EXTEND YOUR UNDERSTANDING

U3.1 In Gauss-Jordan elimination, elements above the diagonal are eliminated in the same manner as are elements below the diagonal, thus avoiding the back-substitution phase of the solution process. In a modified Gauss-Jordan technique, the elimination of elements below the diagonal is done in a first phase (as with basic Gaussian elimination), but the second phase eliminates elements above the diagonal (rather than using backsubstitution). Compare the number of operations (flops) required for the ordinary Gauss-Jordan and the modified Gauss-Jordan techniques to the number of flops required for basic Gaussian elimination.

U3.2 Consider the problem of solving the system $\mathbf{Ax} = \mathbf{b}$, where

$$\mathbf{A} = \begin{bmatrix} 1 & 1/2 & 1/3 \\ 1/2 & 1/3 & 1/4 \\ 1/3 & 1/4 & 1/5 \end{bmatrix}$$

and **b** is four different, but similar, right-hand sides:

$$\mathbf{b} = \begin{bmatrix} 3.0000 & 2.9000 & 3.1000 & 3.0000 \\ 1.9000 & 2.0000 & 1.8000 & 2.0000 \\ 1.4330 & 1.5000 & 1.4000 & 1.4000 \end{bmatrix}$$

Compare the solutions to $x = [\,1\ 2\ 3\,]^T$, the solution to

$$\mathbf{Ax} = [3.0000 \quad 1.9167 \quad 1.4333].$$

U3.3 Solve the following tridiagonal system, performing steps of the Thomas algorithm by hand (note that the process can be separated into two parts when division by zero would occur in the basic process):

$$\begin{aligned}
1x_1 + 1x_2 &= 2 \\
2x_1 + 3x_2 + 1x_3 &= 5 \\
+ 2x_2 + 3x_3 + 2x_4 &= 3 \\
+ 1x_3 + 2x_4 + 1x_5 &= 2 \\
+ 1x_4 + 1x_5 + 1x_6 &= 3 \\
+ 2x_5 + 1x_6 + 2x_7 &= 4 \\
+ 2x_6 + 5x_7 + 2x_8 &= 6 \\
+ 1x_7 + 1x_8 &= 3
\end{aligned}$$

4

Solving Systems of Linear Equations: Iterative Methods

4.1 Jacobi Method
4.2 Gauss-Seidel Method
4.3 Successive Overrelaxation
4.4 Mathcad's Methods

Systems of linear equations for which numerical solutions are needed are often very large, making the computational effort of general, direct methods, such as Gaussian elimination, prohibitively expensive. For systems that have coefficient matrices with the appropriate structure—especially large, sparse systems (i.e., systems with many coefficients whose value is zero)—iterative techniques may be preferable.

We begin this chapter with an example of a linear system that occurs in the solution of Poisson's equation by a finite-difference method. (See Chapter 15.) For applications such as this, in which the elements of the coefficient matrix can be generated as needed from a simple formula, it is beneficial to use a method that does not require storing (and modifying) the entire coefficient matrix, as was the case with Gaussian elimination.

In this chapter we consider the three most common classical iterative techniques for linear systems: the Jacobi, Gauss-Seidel, and successive overrelaxation (SOR) methods. The performance of each technique is illustrated for several examples. For each method, an algorithm is given which can be used as the basis for computing the solution of a small linear system in a Mathcad worksheet. A Mathcad function is also provided for each technique. Some theoretical results are presented to give guidance in determining when the method may be useful.

The convergence of each of the iterative techniques examined in this chapter depends on results from linear algebra. The eigenvalues of the iteration matrix play a crucial role in determining whether these methods will converge; numerical methods for finding eigenvalues are discussed in Chapter 7, but Mathcad's built-in function `eigenvals` provides an efficient way of checking the conditions of the theorems given in this chapter.

Example 4-A Finite Difference Solution of a PDE

The two-dimensional potential equation can be solved numerically by defining a mesh of points in the region of interest and approximating the derivatives by finite differences. For example, to solve the parital differential equation (PDE)

$$u_{xx} + u_{yy} = 0, \qquad 0 \le x \le 1, \ 0 \le y \le 1,$$

with the values of the unknown function $u(x,y)$ given along the boundaries of the region (at $b_1 \ldots b_{12}$) and a mesh of $\Delta x = \Delta y = 0.25$ (so that there are nine points interior to the region where the value of u must be computed), the grid looks like

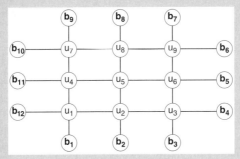

For each point there is an equation that states that four times the value at that point, minus the values of the unknowns at the four adjacent grid points (right, above, left, and below the point) is equal to 0. For some of the grid points, one or two of the adjacent points are on the boundary of the region, and the value of the solution of the PDE is given at those points. The linear system for the unknowns is:

$$\begin{array}{rrrrrrrrrl}
4u_1 & -u_2 & & -u_4 & & & & & & = b_1 + b_{12} \\
-u_1 & +4u_2 & -u_3 & & -u_5 & & & & & = b_2 \\
& -u_2 & +4u_3 & & & -u_6 & & & & = b_3 + b_4 \\
-u_1 & & & +4u_4 & -u_5 & & -u_7 & & & = b_{11} \\
& -u_2 & & -u_4 & +4u_5 & -u_6 & & -u_8 & & = 0 \\
& & -u_3 & & -u_5 & +4u_6 & & & -u_9 & = b_5 \\
& & & -u_4 & & & +4u_7 & -u_8 & & = b_9 + b_{10} \\
& & & & -u_5 & & -u_7 & +4u_8 & -u_9 & = b_8 \\
& & & & & -u_6 & & -u_8 & -4u_9 & = b_6 + b_7 \\
\end{array}$$

This system is well suited for an iterative solution scheme, such as any of the methods which we consider in this chapter (since it is diagonally dominant and quite sparse). We solve this system in Examples 4.4, 4.9 and 4.11. In practice, the linear system would often be very large, and the coefficient matrix would not actually be generated. The finite-difference method used to form the linear system is discussed in Chapter 15.

Example 4-B Steady State Concentrations

The modeling of an interconnected system of reservoirs, reactors, or any type of tank containing a chemical whose concentration may vary from one tank to another, may lead to a system of linear equations for which iterative solution techniques are appropriate. Each tank is characterized by known inflow and outflow rates. The contents of each tank are assumed to be well mixed, so that the concentration of the chemical is uniform throughout the tank.

The mass-balance equations for each tank state that the rate at which the chemical enters the tank must equal the rate at which it leaves. The rate of transfer of the chemical is the product of the concentration of the chemical in the flow stream (mass/volume) and the flow rate (volume/time). The diagram below illustrates an interconnected system of 9 tanks, labeled $T_1 \ldots T_9$. All tanks except 5 and 9 receive inflow from an external source, as well as from some other tanks. The concentration of the chemical in tank T_i is denoted c_i, and the flow rate from T_i to T_j is denoted r_{ij}. The concentration of chemical in flow into a tank from an external source is assumed known; the concentration into tank T_i is denoted c_{0i} and the corresponding flow rate is denoted r_{0i}. The rate of flow out from tank 9 is denoted r; for steady-state operation r equals the total flow into the system. Further more the total flow into each tank must equal the flow out.

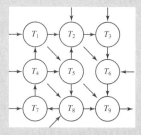

For the system illustrated in the diagram, the mass-balance equations are

$$r_{01} c_{01} + r_{41} c_4 = (r_{12} + r_{15}) c_1$$
$$r_{02} c_{02} + r_{12} c_1 + r_{52} c_5 = (r_{23} + r_{26}) c_2$$
$$r_{03} c_{03} + r_{23} c_2 = r_{36} c_3$$
$$r_{04} c_{04} + r_{74} c_7 = (r_{41} + r_{48} + r_{45}) c_4$$
$$r_{15} c_1 + r_{45} c_4 = (r_{52} + r_{58} + r_{59}) c_5$$
$$r_{06} c_{06} + r_{26} c_2 + r_{36} c_3 = r_{69} c_6$$
$$r_{07} c_{07} + r_{87} c_8 = r_{74} c_7$$
$$r_{08} c_{08} + r_{58} c_5 + r_{48} c_4 = (r_{87} + r_{89}) c_8$$
$$r_{59} c_5 + r_{69} c_6 + r_{89} c_8 = r c_9$$

As an example, if we take the flow rates into the tanks as $r_{01} = r_{02} = r_{03} = r_{04} = r_{08} = 3$ and $r_{06} = r_{07} = 2$, then the flow rate out from tank 9 is $r = 19$. One set of internal flow rates that are compatible with these external values are $r_{12} = 6$; $r_{15} = 2$; $r_{23} = 5$; $r_{26} = 5$; $r_{36} = 8$; $r_{41} = 5$; $r_{45} = 1$; $r_{48} = 1$; $r_{52} = 1$; $r_{58} = 1$; $r_{59} = 1$; $r_{69} = 15$; $r_{74} = 4$; $r_{87} = 2$; $r_{89} = 3$. The distribution of concentrations of the chemical in the various tanks depends on the concentration in the different inflow streams. As would be expected, unit concentration in each inflow stream leads to unit concentration in each tank. Doubling the concentration in the inflow to tank 2 (while leaving all others at unit value) gives steady-state concentrations in tanks 1 – 9 as follows:

| 1 | 1.3 | 1.1875 | 1 | 1 | 1.2 | 1 | 1 | 1.1579 |

An Overview of Iterative Methods for Linear Systems

Classical iterative methods for solving linear systems are based on converting the system $\mathbf{A}\mathbf{x} = \mathbf{b}$ into the equivalent system $\mathbf{x} = \mathbf{C}\mathbf{x} + \mathbf{d}$ and generating a sequence of approximations $\mathbf{x}^{(1)}, \mathbf{x}^{(2)}, \ldots$, where

$$\mathbf{x}^{(k)} = \mathbf{C}\mathbf{x}^{(k-1)} + \mathbf{d}.$$

This methodology is similar to the fixed-point iteration method introduced in Chapter 1 for nonlinear functions of a single variable.

In this chapter we consider three common iterative techniques for solving linear systems: the Jacobi, Gauss-Seidel, and SOR methods. The basic idea is to solve the ith equation in the system for the ith variable, in order to convert the given system (using a four-by-four system for illustration)

$$a_{11}x_1 + a_{12}x_2 + a_{13}x_3 + a_{14}x_4 = b_1,$$
$$a_{21}x_1 + a_{22}x_2 + a_{23}x_3 + a_{24}x_4 = b_2,$$
$$a_{31}x_1 + a_{32}x_2 + a_{33}x_3 + a_{34}x_4 = b_3,$$
$$a_{41}x_1 + a_{42}x_2 + a_{43}x_3 + a_{44}x_4 = b_4,$$

into the system

$$x_1 = \phantom{-\frac{a_{21}}{a_{22}}x_1} - \frac{a_{12}}{a_{11}}x_2 - \frac{a_{13}}{a_{11}}x_3 - \frac{a_{14}}{a_{11}}x_4 + \frac{b_1}{a_{11}},$$

$$x_2 = -\frac{a_{21}}{a_{22}}x_1 \phantom{- \frac{a_{12}}{a_{11}}x_2} - \frac{a_{23}}{a_{22}}x_3 - \frac{a_{24}}{a_{22}}x_4 + \frac{b_2}{a_{22}},$$

$$x_3 = -\frac{a_{31}}{a_{33}}x_1 - \frac{a_{32}}{a_{33}}x_2 \phantom{- \frac{a_{13}}{a_{11}}x_3} - \frac{a_{34}}{a_{33}}x_4 + \frac{b_3}{a_{33}},$$

$$x_4 = -\frac{a_{41}}{a_{44}}x_1 - \frac{a_{42}}{a_{44}}x_2 - \frac{a_{43}}{a_{44}}x_3 \phantom{- \frac{a_{14}}{a_{11}}x_4} + \frac{b_4}{a_{44}}.$$

The Jacobi and Gauss-Seidel methods differ in the manner in which they update the values of the variables on the right-hand side of the equations. The SOR update is a convex combination of the previous solution vector and the Gauss-Seidel update.

Stopping conditions must be specified for any iterative process. Two possibilities are to stop the iterations when the norm of the change in the solution vector $\mathbf{x}$ from one iteration to the next is sufficiently small or to stop the iterations when the norm of the residual vector, $\|\mathbf{A}\mathbf{x} - \mathbf{b}\|$, is below a specified tolerance.

4.1 JACOBI METHOD

The Jacobi method is based on the transformation of the linear system $\mathbf{A}\mathbf{x} = \mathbf{b}$ into the system $\mathbf{x} = \mathbf{C}\mathbf{x} + \mathbf{d}$, in which the matrix $\mathbf{C}$ has zeros on the diagonal. The vector $\mathbf{x}$ is updated using the previous estimate for all components of $\mathbf{x}$ to evaluate the right-hand side of the equation.

> **Example 4.1 Illustrating the Jacobi Method Graphically**
>
> To visualize the Jacobi iterations, consider the two-by-two system
> $$2x + y = 6,$$
> $$x + 2y = 6.$$
> For the Jacobi method, the equations are written as
> $$x = -\frac{1}{2}y + 3,$$
> $$y = -\frac{1}{2}x + 3.$$
> Starting with $x^{(1)} = 1/2$ and $y^{(1)} = 1/2$, the first equation produces the next estimate for x (using $y^{(1)}$), and the second equation gives the next value of y (using $x^{(1)}$):
> $$x^{(2)} = -\frac{1}{2}y^{(1)} + 3 = -\frac{1}{4} + 3 = \frac{11}{4},$$
> $$y^{(2)} = -\frac{1}{2}x^{(1)} + 3 = -\frac{1}{4} + 3 = \frac{11}{4}.$$
> New values of the variables are not used until a new iteration step is begun; this is called *simultaneous updating*. The first iteration is illustrated in Fig. 4.1.
>
>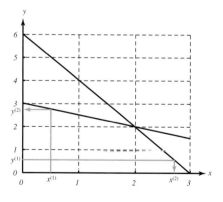
>
> **FIGURE 4.1** One iteration of the Jacobi method.

4.1.1 Step-by-Step Procedure for Jacobi Iteration

Example 4.2 Solving a Three-by-Three System

Consider the three-by-three system of equations

$$2x_1 - x_2 + x_3 = -1,$$
$$x_1 + 2x_2 - x_3 = 6,$$
$$x_1 - x_2 + 2x_3 = -3,$$

which are converted to

$$x_1 = + 0.5\, x_2 - 0.5\, x_3 + (-0.5),$$
$$x_2 = -0.5\, x_1 + 0.5\, x_3 + (3.0),$$
$$x_3 = -0.5\, x_1 + 0.5\, x_2 + (-1.5).$$

In matrix notation, the original system, $\mathbf{A}\,\mathbf{x} = \mathbf{b}$, i.e.,

$$\begin{bmatrix} 2 & -1 & 1 \\ 1 & 2 & -1 \\ 1 & -1 & 2 \end{bmatrix} \begin{bmatrix} x_1 \\ x_2 \\ x_3 \end{bmatrix} = \begin{bmatrix} -1 \\ 6 \\ -3 \end{bmatrix},$$

is transformed to

$$\begin{bmatrix} x_1^{(k)} \\ x_2^{(k)} \\ x_3^{(k)} \end{bmatrix} = \begin{bmatrix} 0.0 & 0.5 & -0.5 \\ -0.5 & 0.0 & 0.5 \\ -0.5 & 0.5 & 0.0 \end{bmatrix} \begin{bmatrix} x_1^{(k-1)} \\ x_2^{(k-1)} \\ x_3^{(k-1)} \end{bmatrix} + \begin{bmatrix} -0.5 \\ 3.0 \\ -1.5 \end{bmatrix},$$

(where the iteration counter is indicated by a superscript).
Starting with $x^{(0)} = (0, 0, 0)$ we find

$$\begin{bmatrix} x_1^{(1)} \\ x_2^{(1)} \\ x_3^{(1)} \end{bmatrix} = \begin{bmatrix} 0.0 & 0.5 & -0.5 \\ -0.5 & 0.0 & 0.5 \\ -0.5 & 0.5 & 0.0 \end{bmatrix} \begin{bmatrix} 0 \\ 0 \\ 0 \end{bmatrix} + \begin{bmatrix} -0.5 \\ 3.0 \\ -1.5 \end{bmatrix} = \begin{bmatrix} -0.5 \\ 3.0 \\ -1.5 \end{bmatrix}.$$

For the second iteration, we have

$$\begin{bmatrix} x_1^{(2)} \\ x_2^{(2)} \\ x_3^{(2)} \end{bmatrix} = \begin{bmatrix} 0.0 & 0.5 & -0.5 \\ -0.5 & 0.0 & 0.5 \\ -0.5 & 0.5 & 0.0 \end{bmatrix} \begin{bmatrix} -0.5 \\ 3.0 \\ -1.5 \end{bmatrix} + \begin{bmatrix} -0.5 \\ 3.0 \\ -1.5 \end{bmatrix} = \begin{bmatrix} 1.75 \\ 2.50 \\ 0.25 \end{bmatrix}.$$

Continuing these iterations, or using the Jacobi function given following the algorithm, we find that solution vector after 10 iterations is

$$\mathbf{x} = [0.99854 \quad 2.00098 \quad -1.00049]^T.$$

The true solution is $\mathbf{x}^* = \begin{bmatrix} 1 & 2 & -1 \end{bmatrix}^T$.

We summarize the steps in using the Jacobi Method in the following algorithm.

Algorithm for Jacobi Method

To solve $\mathbf{M}\mathbf{x} = \mathbf{b}$

Initialize

1) Define the matrix $\mathbf{M}$ (n-by-n), vector $\mathbf{b}$, and initial guess vector $\mathbf{x}_o$.

   Create iteration matrix $\mathbf{C}$ as follows
   set the diagonal elements
   $c(i, i) = 0$
   set the off diagonal elements
   $c(i,j) = -m(i,j)/m(i,i)$
   (opposite of corresponding element of M, divided by diag element for that row.)

   Create the vector $\mathbf{d}$
   $d(i) = b(i)/m(i,i)$.

Begin iterations

2) Form new solution vector, $\mathbf{x}_n$ from the old solution vector $\mathbf{x}_o$:

   $\mathbf{x}_n = \mathbf{C}\,\mathbf{x}_o + \mathbf{d}$

   Record new solution in table if desired

   Test for convergence

   if the test for convergence is not passed,
   update solution vector

   $\mathbf{x}_o = \mathbf{x}_n$

   and continue iterations

   Appropriate tests for convergence typically include the following:

   a check on the norm of the change in the solution vector, $|\mathbf{x}_n - \mathbf{x}_{n-1}|$,
   or
   a check on the norm of the residual, $|\mathbf{M}\,\mathbf{x}_n - \mathbf{b}|$
   together with
   a check on the maximum number of iterations.

Example 4.3 Using the Jacobi Algorithm

To illustrate the use of the Jacobi algorithm, we seek to solve the linear system M x = b, for

$$M := \begin{pmatrix} 25 & 1 & 5 \\ 1 & 20 & -1 \\ -5 & 2 & 50 \end{pmatrix} \qquad b := \begin{pmatrix} 20 \\ 2 \\ -55 \end{pmatrix} \qquad x0 := \begin{pmatrix} 0 \\ 0 \\ 0 \end{pmatrix}$$

Transform the matrix M to the iteration matrix C, by
    dividing each row of M by the opposite of the element on the diagonal,
    and then setting the diagonal elements of C to 0
Transform the vector b to the vector d, by
    dividing by the corresponding diagonal element of M

$$C := \begin{pmatrix} 0 & \frac{-1}{25} & \frac{-5}{25} \\ \frac{-1}{20} & 0 & \frac{1}{20} \\ \frac{5}{50} & \frac{-2}{50} & 0 \end{pmatrix} \qquad d := \begin{pmatrix} \frac{20}{25} \\ \frac{2}{20} \\ \frac{-55}{50} \end{pmatrix}$$

Compute new solution, and display the transpose of the new solution (to save space)

$$x1 := C \cdot x0 + d \qquad\qquad x1^T = (0.8 \quad 0.1 \quad -1.1)$$

Continue iterations

$$x2 := C \cdot x1 + d \qquad\qquad x2^T = (1.016 \quad 5 \times 10^{-3} \quad -1.024)$$

$$x3 := C \cdot x2 + d \qquad\qquad x3^T = (1.0046 \quad -2 \times 10^{-3} \quad -0.9986)$$

4.1.2 Mathcad Function for the Jacobi Method

The following function for the Jacobi method tests convergence using a combination of user specified tolerance on the change in the solution vector and a user specified maximum number of iterations allowed. The function returns a vector containing the approximate solution vector.

$$\text{Jacobi}(A, b, x0, \text{tol}, \text{max}) := \begin{vmatrix} n \leftarrow \text{rows}(A) - 1 \\ C \leftarrow -A \\ \text{for } i \in 0..n \\ \quad C_{i,i} \leftarrow 0 \\ \text{for } i \in 0..n \\ \quad \text{for } j \in 0..n \\ \quad \quad C_{i,j} \leftarrow \dfrac{C_{i,j}}{A_{i,i}} \\ \text{for } i \in 0..n \\ \quad d_i \leftarrow \dfrac{b_i}{A_{i,i}} \\ i \leftarrow 0 \\ \text{xold} \leftarrow x0 \\ \text{while } i \leq \text{max} \\ \quad \begin{vmatrix} \text{xnew} \leftarrow C \cdot \text{xold} + d \\ (\text{break}) \text{ if } |\text{xold} - \text{xnew}| < \text{tol} \\ \text{xold} \leftarrow \text{xnew} \\ i \leftarrow i + 1 \end{vmatrix} \\ \text{xold} \end{vmatrix}$$

In the very basic form of the function given here, the return vector does not indicate the iteration on which the process terminated, or whether the convergence criterion for change in the solution vector was satisfied. The iteration number when the process stopped could be returned together with the solution vector by utilizing Mathcad's `stack` function, if desired. A Mathcad function can return only a single entity (scalar, vector, or matrix).

The use of the function `Jacobi` is illustrated in the next example.

Example 4.4 A Linear System That Comes from a PDE

To illustrate the use of the Jacobi method for solving a system of linear equations that comes from a finite-difference approach to solving a partial differential equation (PDE), we consider the linear system introduced in Example 4-A. The finite difference method for solving Poisson's PDE is presented in Chapter 15. We take the values of the function on the boundary to be as shown on the grid below.

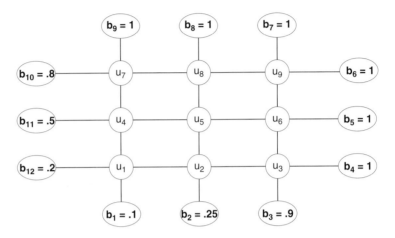

The nine equations that state that four times the value at that point, minus the values of the unknowns at the four adjacent grid points (right, above, left, and below the point) is equal to 0 are as follows.

$$4u_1 \quad -u_2 \quad -u_4 \quad -b_{12} \quad -b_1 = 0$$
$$4u_2 \quad -u_3 \quad -u_5 \quad -u_1 \quad -b_2 = 0$$
$$4u_3 \quad -b_4 \quad -u_6 \quad -u_2 \quad -b_3 = 0$$
$$4u_4 \quad -u_5 \quad -u_7 \quad -b_{11} \quad -u_1 = 0$$
$$4u_5 \quad -u_6 \quad -u_8 \quad -u_4 \quad -u_2 = 0$$
$$4u_6 \quad -b_5 \quad -u_9 \quad -u_5 \quad -u_3 = 0$$
$$4u_7 \quad -u_8 \quad -b_9 \quad -b_{10} \quad -u_4 = 0$$
$$4u_8 \quad -u_9 \quad -b_8 \quad -u_7 \quad -u_5 = 0$$
$$4u_9 \quad -b_6 \quad -b_7 \quad -u_8 \quad -u_6 = 0$$

Putting the known values on the right, and writing the unknowns in order we have

Chapter 4 Solving Systems of Linear Equations: Iterative Methods

$$\begin{vmatrix} 4u_1 & -u_2 & & -u_4 & & & & & & = .3 \\ -u_1 & +4u_2 & -u_3 & & -u_5 & & & & & = .25 \\ & -u_2 & +4u_3 & & & -u_6 & & & & = 1.9 \\ -u_1 & & & +4u_4 & -u_5 & & -u_7 & & & = .5 \\ & -u_2 & & -u_4 & +4u_5 & -u_6 & & -u_8 & & = 0 \\ & & -u_3 & & -u_5 & +4u_6 & & & -u_9 & = 1 \\ & & & -u_4 & & & +4u_7 & -u_8 & & = 1.8 \\ & & & & -u_5 & & -u_7 & +4u_8 & -u_9 & = 1 \\ & & & & & -u_6 & & -u_8 & +4u_9 & = 2 \end{vmatrix}$$

To illustrate the use of the Mathcad function Jacobi, consider the 9-by-9 system corresponding to a finite-difference solution of a small PDE. To check the convergence, we show the error at each step.

tol := 0.00001

$$M := \begin{pmatrix} 4 & -1 & 0 & -1 & 0 & 0 & 0 & 0 & 0 \\ -1 & 4 & -1 & 0 & -1 & 0 & 0 & 0 & 0 \\ 0 & -1 & 4 & 0 & 0 & -1 & 0 & 0 & 0 \\ -1 & 0 & 0 & 4 & -1 & 0 & -1 & 0 & 0 \\ 0 & -1 & 0 & -1 & 4 & -1 & 0 & -1 & 0 \\ 0 & 0 & -1 & 0 & -1 & 4 & 0 & 0 & -1 \\ 0 & 0 & 0 & -1 & 0 & 0 & 4 & -1 & 0 \\ 0 & 0 & 0 & 0 & -1 & 0 & -1 & 4 & -1 \\ 0 & 0 & 0 & 0 & 0 & -1 & 0 & -1 & 4 \end{pmatrix} \qquad b := \begin{pmatrix} .3 \\ .25 \\ 1.9 \\ .5 \\ 0 \\ 1 \\ 1.8 \\ 1 \\ 2 \end{pmatrix} \qquad x0 := \begin{pmatrix} 0 \\ 0 \\ 0 \\ 0 \\ 0 \\ 0 \\ 0 \\ 0 \\ 0 \end{pmatrix}$$

$x_a := \text{Jacobi}(M, b, x0, \text{tol}, 10)$ \qquad $|M \cdot x_a - b| = 0.055861$

$x_a^T = (0.348147 \ 0.521752 \ 0.816002 \ 0.581573 \ 0.696289 \ 0.852997 \ 0.805287 \ 0.850319 \ 0.923142)$

$x_b := \text{Jacobi}(M, b, x0, \text{tol}, 20)$ \qquad $|M \cdot x_b - b| = 1.745648 \times 10^{-3}$

$x_b^T = (0.359024 \ 0.538305 \ 0.826881 \ 0.598127 \ 0.718048 \ 0.869555 \ 0.816167 \ 0.866877 \ 0.934024)$

$x_c := \text{Jacobi}(M, b, x0, \text{tol}, 30)$ \qquad $|M \cdot x_c - b| = 5.45515 \times 10^{-5}$

$x_c^T = (0.359364 \ 0.538823 \ 0.827221 \ 0.598644 \ 0.718728 \ 0.870073 \ 0.816507 \ 0.867394 \ 0.934364)$

$x_d := \text{Jacobi}(M, b, x0, \text{tol}, 40)$ \qquad $|M \cdot x_d - b| = 3.857374 \times 10^{-5}$

$x_d^T = (0.359367 \ 0.538828 \ 0.827224 \ 0.59865 \ 0.718733 \ 0.870078 \ 0.81651 \ 0.8674 \ 0.934367)$

$x_e := \text{Jacobi}(M, b, x0, \text{tol}, 50)$ \qquad $|M \cdot x_e - b| = 3.857374 \times 10^{-5}$

$x_e^T = (0.359367 \ 0.538828 \ 0.827224 \ 0.59865 \ 0.718733 \ 0.870078 \ 0.81651 \ 0.8674 \ 0.934367)$

There is no change after 40 iterations.

4.1.3 Discussion

The Jacobi method may be derived by converting the original system, $\mathbf{A}\mathbf{x} = \mathbf{b}$, into a decomposed form $(\mathbf{L} + \mathbf{D} + \mathbf{U})\mathbf{x} = \mathbf{b}$, where $\mathbf{D}$ is a diagonal matrix, $\mathbf{L}$ is a lower triangular matrix, and $\mathbf{U}$ is an upper triangular matrix. The decomposed form is expressed as $\mathbf{x} = \mathbf{C}\mathbf{x} + \mathbf{d}$ in the following way:

$$\mathbf{D}\mathbf{x} = (-\mathbf{L} - \mathbf{U})\mathbf{x} + \mathbf{b},$$
$$\mathbf{x} = \mathbf{D}^{-1}(-\mathbf{L} - \mathbf{U})\mathbf{x} + \mathbf{D}^{-1}\mathbf{b},$$
$$\mathbf{x} = \mathbf{C}\mathbf{x} + \mathbf{d}.$$

This formulation assumes that the diagonal of $\mathbf{A}$ contains no element that is zero. If $\mathbf{A}$ is nonsingular, but has a zero element on the diagonal, rows and columns may be permuted to obtain a form with nonsingular $\mathbf{D}$. It is desirable to have the diagonal elements large relative to the off-diagonal elements.

The primary considerations in determining when the Jacobi method (or some other iterative method) works well are convergence and computational effort.

Some Useful Theoretical Results on Convergence

There are several useful theoretical results concerning the relationships between the characteristics of the matrix $\mathbf{A}$ (or the iteration matrix $\mathbf{C}$) and the convergence of Jacobi iteration for the system $\mathbf{A}\mathbf{x} = \mathbf{b}$. Two such results are summarized next.

A *sufficient* condition for Jacobi iteration to converge to the solution of the system $\mathbf{A}\mathbf{x} = \mathbf{b}$ is that the original matrix $\mathbf{A}$ be strictly diagonally dominant. This means that, for each row, the magnitude of the diagonal element is greater than the sum of the magnitudes of the other elements in the row.

A *necessary and sufficient* condition for the convergence of the Jacobi method is that the magnitude of the largest eigenvalue of the iteration matrix $\mathbf{C}$ be less than 1. Eigenvalues and eigenvectors are discussed further in Chapter 7.

Using the Mathcad function to compute eigenvalues of the iteration matrix $\mathbf{C}$ for the system in Example 4.5 (which follows) we find values of

$$3.4641 \quad \text{and} \quad -3.4641.$$

Clearly, Jacobi iteration is not suitable for this problem. The eigenvalues of the iteration matrix for the original ordering of the equations (in Example 4.1) are

$$0.28868 \quad \text{and} \quad -0.28868.$$

Of course, we would not normally use an iterative method on such a small system, but the point is that the method will not converge for this simple example. (For further discussion of these results, see Atkinson, 1989, pp. 546–7.)

Example 4.5 Jacobi Method is Sensitive to the Order of Equations

To illustrate the sensitivity of the Jacobi method to the form of the coefficient matrix **A** (or equivalently, to the form of the iteration matrix **C**), consider the simple system from the beginning of this section, but with the order of the equations reversed, so that the system is no longer diagonally dominant:

$$x + 2y = 6,$$
$$2x + y = 6.$$

In the Jacobi iterative form, we have (see Fig. 4.2)

$$L_1: x = -2y + 6,$$
$$L_2: y = -2x + 6.$$

Starting with $x^{(1)} = 1/2$ and $y^{(1)} = 1/2$, we obtain

$$x^{(2)} = -2y^{(1)} + 6 = 5,$$
$$y^{(2)} = -2x^{(1)} + 6 = 5.$$

Further iterations show that the solution diverges, with x and y alternating between large positive and large negative values.

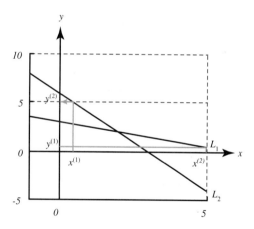

FIGURE 4.2 Divergent Jacobi iterations.

Computational Effort and Other Considerations

Each iteration of the Jacobi method requires one matrix-vector multiplication, or $(n - 1)^2$ scalar multiplications. Thus, if the method converges in a reasonable number of iterations, the computational effort could be significantly less than the $O(n^3)$ multiplications required for Gaussian elimination.

The Jacobi method is particularly convenient for parallel computation, because each component of the solution can be updated independently of the other components.

4.2 GAUSS-SEIDEL METHOD

The Gauss-Seidel method of solving linear systems is a simple iterative technique obtained by transforming the linear system $\mathbf{A}\mathbf{x} = \mathbf{b}$ into the system $\mathbf{x} = \mathbf{C}\mathbf{x} + \mathbf{d}$, in which the matrix $\mathbf{C}$ has zeros on the diagonal. However, in contrast to the Jacobi method, each component of the vector $\mathbf{x}$ on the right-hand side of the transformed equation is updated immediately as each iteration progresses. This procedure is called *sequential updating*.

Example 4.6 Graphical Illustration of the Gauss-Seidel Method

To visualize the Gauss-Seidel iterations, consider the two-by-two system introduced in Example 4.1; the equations are written as

$$x = -\frac{1}{2}y + 3$$

$$y = -\frac{1}{2}x + 3$$

The first iteration is illustrated in Fig. 4.3 for the starting values of $x^{(1)} = 1/2$ and $y^{(1)} = 1/2$. The first equation then produces the next estimate for x (using $y^{(1)}$), and the second equation is used to find the next value of y (using the newly computed value, $x^{(2)}$). The result is

$$x^{(2)} = -\frac{1}{2}y^{(1)} + 3 = -\frac{1}{4} + 3 = \frac{11}{4},$$

$$y^{(2)} = -\frac{1}{2}x^{(2)} + 3 = -\frac{11}{8} + 3 = \frac{13}{8}.$$

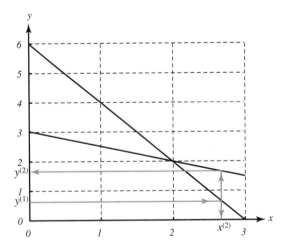

FIGURE 4.3 First step of Gauss-Seidel iteration.

4.2.1 Step-by-Step Computation for Gauss-Seidel Method

Example 4.7 Using Gauss-Seidel Iteration

Consider the three-by-three system of equations (as in Example 4.2)
$$2x_1 - x_2 + x_3 = -1,$$
$$x_1 + 2x_2 - x_3 = 6,$$
$$x_1 - x_2 + 2x_3 = -3,$$

which are converted to the iterative system
$$x_1^{(new)} = \quad\quad\quad + 0.5\, x_2^{(old)} - 0.5\, x_3^{(old)} + (-0.5),$$
$$x_2^{(new)} = -0.5\, x_1^{(new)} \quad\quad\quad + 0.5\, x_3^{(old)} + (\ 3.0),$$
$$x_3^{(new)} = -0.5\, x_1^{(new)} + 0.5\, x_2^{(new)} \quad\quad\quad + (-1.5).$$

Note that the most recent estimate for each of the unknowns is used in evaluating the right-hand side of each equation.

Starting with $x^{(0)} = (0, 0, 0)$, we find
$$x_1 = \quad\quad\quad (0.5)(0) + (-0.5)(0) + -0.5 = -0.5,$$
$$x_2 = (-0.5)(-0.5) \quad\quad\quad + (0.5)(0) + 3.0 = 3.25,$$
$$x_3 = (-0.5)(-0.5) + (0.5)(3.25) \quad\quad\quad + -1.5 = 0.375.$$

For the second iteration,
$$x_1 = \quad\quad\quad (0.5)(3.25) + (-0.5)(0.375) + -0.5 = 0.9375,$$
$$x_2 = (-0.5)(0.9375) \quad\quad\quad + (0.5)(0.375) + 3.0 = 2.7188,$$
$$x_3 = (-0.5)(0.9375) + (0.5)(2.7188) \quad\quad\quad + -1.5 = -0.6094.$$

For the third iteration we find
$$\mathbf{x} = \begin{bmatrix} 1.16406 & 2.11328 & -1.02539 \end{bmatrix}^T.$$

For the fourth iteration
$$\mathbf{x} = \begin{bmatrix} 1.06934 & 1.95264 & -1.05835 \end{bmatrix}^T.$$

For the fifth iteration
$$\mathbf{x} = \begin{bmatrix} 1.00549 & 1.96808 & -1.01871 \end{bmatrix}^T.$$

And, for the sixth iteration
$$\mathbf{x} = \begin{bmatrix} 0.99339 & 1.99395 & -0.99972 \end{bmatrix}^T.$$

Algorithm for Gauss-Seidel Method

To solve $\mathbf{M}\,\mathbf{x} = \mathbf{b}$, with initial estimate of solution $\mathbf{x}_0$.
Initialize

1) Create matrix $\mathbf{M}$ (n-by-n), vector $\mathbf{b}$, and initial guess vector $\mathbf{x}_0$.
2) Create iteration matrix $\mathbf{C}$ as for Jacobi method
    diagonal elements of $\mathbf{C}$ are
        $c(i, i) = 0$
    off-diagonal elements are
        $c(i,j) = -m(i,j)/m(i,i)$
            (opposite of corresponding element of $\mathbf{M}$,
            divided by diag element for that row.)
3) Create vector $\mathbf{d}$ as follows
    $d(i) = b(i)/m(i,i)$.
4) Set $\mathbf{x}$ equal to the initial guess for solution the solution $\mathbf{x}_0$
    $\mathbf{x} := \mathbf{x}_0$

Begin iterations (illustrated for a three-by-three system)

1) Update each component of solution vector $\mathbf{x}$ sequentially

$$x_1 := C_{1,1}\,x_1 + C_{1,2}\,x_2 + C_{1,3}\,x_3 + d_1$$
$$x_2 := C_{2,1}\,x_1 + C_{2,2}\,x_2 + C_{2,3}\,x_3 + d_2$$
$$x_3 := C_{3,1}\,x_1 + C_{3,2}\,x_2 + C_{3,3}\,x_3 + d_3$$

Display transpose of solution vector
    $\mathbf{x}^T =$

2) To continue iterations, copy the lines from step 1; Mathcad will recompute all values.
3) Check for convergence by considering the change in the solution vector, or the difference between $\mathbf{M}$, $\mathbf{x}$ and $\mathbf{b}$, (or some other convergence test) if desired.

We illustrate the use of the Gauss-Seidel Algorithm in the next example. Note that the algorithm assumes that the vectors and matrices are indexed starting from 1, rather than the Mathcad default of 0.

Example 4.8 Using the Gauss-Seidel Algorithm

To illustrate the use of the Gauss-Seidel algorithm, we seek to solve the linear system M x = b, for

ORIGIN:= 1

$$M := \begin{pmatrix} 25 & 1 & 5 \\ 1 & 20 & -1 \\ -5 & 2 & 50 \end{pmatrix} \qquad b := \begin{pmatrix} 20 \\ 2 \\ -55 \end{pmatrix} \qquad x0 := \begin{pmatrix} 0 \\ 0 \\ 0 \end{pmatrix} \qquad x := x0$$

Transform the matrix M to the iteration matrix C, by dividing each row of M by the opposite of the element on the diagonal, and then setting the diagonal elements of C to 0

Transform the vector b to the vector d, by dividing by the corresponding diagonal element of M

$$C := \begin{pmatrix} 0 & \frac{-1}{25} & \frac{-5}{25} \\ \frac{-1}{20} & 0 & \frac{1}{20} \\ \frac{5}{50} & \frac{-2}{50} & 0 \end{pmatrix} \qquad d := \begin{pmatrix} \frac{20}{25} \\ \frac{2}{20} \\ \frac{-55}{50} \end{pmatrix}$$

Compute each component of the new solution sequentially

$x_1 := (C_{1,2} \cdot x_2 + C_{1,3} \cdot x_3 + d_1)$ $x^T = (0.8 \quad 0 \quad 0)$

$x_2 := (C_{2,1} \cdot x_1 + C_{2,3} \cdot x_3 + d_2)$ $x^T = (0.8 \quad 0.06 \quad 0)$

$x_3 := (C_{3,1} \cdot x_1 + C_{3,2} \cdot x_2 + d_3)$ $x^T = (0.8 \quad 0.06 \quad -1.0224)$

Second iteration

$x_1 := (C_{1,2} \cdot x_2 + C_{1,3} \cdot x_3 + d_1)$ $x^T = (1.0021 \quad 0.06 \quad -1.0224)$

$x_2 := (C_{2,1} \cdot x_1 + C_{2,3} \cdot x_3 + d_2)$ $x^T = (1.0021 \quad -1.224 \times 10^{-3} \quad -1.0224)$

$x_3 := (C_{3,1} \cdot x_1 + C_{3,2} \cdot x_2 + d_3)$ $x^T = (1.0021 \quad -1.224 \times 10^{-3} \quad -0.9997)$

Third iteration

$x_1 := (C_{1,2} \cdot x_2 + C_{1,3} \cdot x_3 + d_1)$ $x^T = (1 \quad -1.224 \times 10^{-3} \quad -0.9997)$

$x_2 := (C_{2,1} \cdot x_1 + C_{2,3} \cdot x_3 + d_2)$ $x^T = (1 \quad 1.297 \times 10^{-5} \quad -0.9997)$

$x_3 := (C_{3,1} \cdot x_1 + C_{3,2} \cdot x_2 + d_3)$ $x^T = (1 \quad 1.297 \times 10^{-5} \quad -1)$

Chapter 4 Solving Systems of Linear Equations: Iterative Methods

4.2.2 Mathcad Function for Gauss-Seidel Method

The Mathcad function for the Gauss-Seidel method given here uses the Mathcad default of zero-offset for vectors and matrices (indices start at 0). The function returns a vector containing the approximate solution. The computations terminate if the norm of the change in the solution vector is less than the user-specified tolerance, or the maximum number of iterations has been performed.

$$
\begin{aligned}
&\text{GaussSeidel}(A, b, x0, tol, max) := \Bigg| \\
&\quad C \leftarrow -A \\
&\quad n \leftarrow \text{cols}(A) \\
&\quad x \leftarrow x0 \\
&\quad \text{for } i \in 0..n-1 \\
&\quad\quad C_{i,i} \leftarrow 0 \\
&\quad \text{for } i \in 0..n-1 \\
&\quad\quad \text{for } j \in 0..n-1 \\
&\quad\quad\quad C_{i,j} \leftarrow \dfrac{C_{i,j}}{A_{i,i}} \\
&\quad \text{for } i \in 0..n-1 \\
&\quad\quad r_i \leftarrow \dfrac{b_i}{A_{i,i}} \\
&\quad i \leftarrow 0 \\
&\quad \text{while } i \le max - 1 \\
&\quad\quad xold \leftarrow x \\
&\quad\quad \text{for } j \in 0..n-1 \\
&\quad\quad\quad sum \leftarrow 0 \\
&\quad\quad\quad \text{for } k \in 0..n-1 \\
&\quad\quad\quad\quad sum \leftarrow sum + C_{j,k} \cdot x_k \\
&\quad\quad\quad x_j \leftarrow \left(sum + r_j \right) \\
&\quad\quad \text{break if } |xold - x| \le tol \\
&\quad\quad i \leftarrow i + 1 \\
&\quad x
\end{aligned}
$$

The use of the Mathcad function `GaussSeidel` is illustrated in the next example.

Example 4.9 Using the Mathcad Function for Gauss-Seidel Method

To illustrate the use of the Mathcad function GaussSeidel, consider the 9-by-9 system introduced in Example 4.4

$$M := \begin{pmatrix} 4 & -1 & 0 & -1 & 0 & 0 & 0 & 0 & 0 \\ -1 & 4 & -1 & 0 & -1 & 0 & 0 & 0 & 0 \\ 0 & -1 & 4 & 0 & 0 & -1 & 0 & 0 & 0 \\ -1 & 0 & 0 & 4 & -1 & 0 & -1 & 0 & 0 \\ 0 & -1 & 0 & -1 & 4 & -1 & 0 & -1 & 0 \\ 0 & 0 & -1 & 0 & -1 & 4 & 0 & 0 & -1 \\ 0 & 0 & 0 & -1 & 0 & 0 & 4 & -1 & 0 \\ 0 & 0 & 0 & 0 & -1 & 0 & -1 & 4 & -1 \\ 0 & 0 & 0 & 0 & 0 & -1 & 0 & -1 & 4 \end{pmatrix} \quad b := \begin{pmatrix} .3 \\ .25 \\ 1.9 \\ .5 \\ 0 \\ 1 \\ 1.8 \\ 1 \\ 2 \end{pmatrix} \quad x0 := \begin{pmatrix} 0 \\ 0 \\ 0 \\ 0 \\ 0 \\ 0 \\ 0 \\ 0 \\ 0 \end{pmatrix}$$

tol := 0.00001

x_a := GaussSeidel(M, b, x0, tol, 10)

x_a^T = (0.35784 0.5373 0.82646 0.59712 0.71721 0.86932 0.81575 0.86664 0.93399)

$|M \cdot x_a - b| = 4.89412 \times 10^{-3}$

x_b := GaussSeidel(M, b, x0, tol, 20)

x_b^T = (0.35937 0.53884 0.82723 0.59866 0.71875 0.87009 0.81652 0.86741 0.93437)

$|M \cdot x_b - b| = 9.55891 \times 10^{-6}$

Comparing these results to those in Example 4.4, we see that the norm of the residual (the error) after 20 iterations of GaussSeidel is less than it is after 50 iterations of Jacobi.

4.2.3 Discussion

The Gauss-Seidel method is usually derived by converting the original system, $\mathbf{A}\mathbf{x} = \mathbf{b}$, into a decomposed form $(\mathbf{L} + \mathbf{D} + \mathbf{U})\mathbf{x} = \mathbf{b}$ where $\mathbf{D}$ is a diagonal matrix, $\mathbf{L}$ is a lower-triangular matrix, and $\mathbf{U}$ is an upper-triangular matrix. The decomposed form is expressed as $\mathbf{x} = \mathbf{T}\mathbf{x} + \mathbf{c}$ in the following way:

$$(\mathbf{D} + \mathbf{L})\mathbf{x} = -\mathbf{U}\mathbf{x} + \mathbf{b}$$

$$\mathbf{x} = (\mathbf{D} + \mathbf{L})^{-1}(-\mathbf{U})\mathbf{x} + (\mathbf{D} + \mathbf{L})^{-1}\mathbf{b}.$$

$$\mathbf{x} = \mathbf{T}\mathbf{x} + \mathbf{c}$$

Like the Jacobi method, the Gauss-Seidel method is sensitive to the form of the coefficient matrix $\mathbf{A}$ (or equivalently, to the form of the iteration matrix $\mathbf{T}$); reversing the order of the equations in Example 4.6 shows a divergence similar to that for the Jacobi method.

Some Useful Theoretical Results on Convergence

There are several useful theoretical results concerning the relationships between the characteristics of the matrix $\mathbf{A}$ (or the iteration matrix $\mathbf{T}$) and the convergence of Gauss-Seidel iteration for the system $\mathbf{A}\mathbf{x} = \mathbf{b}$. The matrix form of the Gauss-Seidel method is used primarily for analyzing the convergence of the method; however, general theoretical results that guarantee convergence (which depend on the eigenvalues of the matrix $\mathbf{T}$) are not necessarily the most convenient for important applications.

The numerical solution of partial differential equations (see Chapter 15) leads to linear systems in which the matrix $\mathbf{A}$ is real and symmetric, with positive diagonal elements. For such matrices, the Gauss-Seidel method will converge (for any $\mathbf{x}^{(0)}$) if, and only if, all of the eigenvalues of $\mathbf{A}$ are real and positive. (See Atkinson, 1989, pp. 499, 551; a proof is given in Isaacson and Keller, 1966, pp. 70–71.) For an example that does not satisfy the symmetry requirement, theoretical analysis is more difficult to carry out.

If $\mathbf{A}$ is positive definite, Gauss-Seidel iteration converges for any initial vector. (For a proof of this result, see Ralston and Rabinowitz, 1978, p. 445.) We discuss some tests for positive definite matrices in the next section.

If the iteration matrix $\mathbf{C}$ is nonnegative (a situation that often occurs in the numerical solution of partial differential equations), then the Jacobi and Gauss-Seidel methods either both converge or both diverge; when they both converge, Gauss-Seidel iteration converges more rapidly (except in the trivial case when the largest eigenvalue of the iteration matrix for both methods is zero; see Ralston and Rabinowitz, 1978, p. 446).

Computational Effort and Other Considerations

The Gauss-Seidel method typically converges more rapidly than the Jacobi method, although it is more difficult to use for parallel computation. In particular, for a linear system arising in the finite-difference solution of Poisson's PDE, with a J-by-J grid (so there are J^2 equations), it can be shown that the number of iterations

r required to reduce the error by a factor of 10^{-p} is $r \approx \frac{1}{4} p J^2$. The Jacobi method requires approximately twice as many iterations; $r \approx \frac{1}{2} p J^2$. Thus, although Gauss-Seidel converges more rapidly, it is still impractical for very large problems. Finite-difference problems often have $J = 100$ or more. In the next section, we consider a method of accelerating the convergence of Gauss-Seidel. See [Press et al., 1992] for further discussion of these issues.

4.3 SUCCESSIVE OVERRELAXATION

It is possible to modify the Gauss-Seidel method by introducing an additional parameter, ω (omega), that may accelerate the convergence of the iterations. The idea is to take a combination of the previous value of **x** and the current update (from the Gauss-Seidel method). The parameter ω controls the proportion of the update that comes from the previous solution and the proportion that comes from the current calculation. For $0 < \omega < 1$, the method is called successive underrelaxation; for $1 < \omega < 2$, the method is called successive overrelaxation (SOR). For $\omega = 1$, SOR reduces to the Gauss-Seidel method.

Consider the three-by-three system

$$a_{11}x_1 + a_{12}x_2 + a_{13}x_3 = b_1,$$
$$a_{21}x_1 + a_{22}x_2 + a_{23}x_3 = b_2,$$
$$a_{31}x_1 + a_{32}x_2 + a_{33}x_3 = b_3.$$

The SOR equations are

$$x_1^{(new)} = (1 - \omega)x_1^{(old)} + \frac{\omega}{a_{11}}(b_1 \qquad\qquad - a_{12}\,x_2^{(old)} - a_{13}\,x_3^{(old)})$$

$$x_2^{(new)} = (1 - \omega)x_2^{(old)} + \frac{\omega}{a_{22}}(b_2 - a_{21}\,x_1^{(new)} \qquad\qquad - a_{23}\,x_3^{(old)})$$

$$x_3^{(new)} = (1 - \omega)x_3^{(old)} + \frac{\omega}{a_{33}}(b_3 - a_{31}\,x_1^{(new)} - a_{32}\,x_2^{(new)} \qquad\qquad)$$

Although underrelaxation can sometimes be used to convert a nonconvergent problem into a convergent one, the primary use of relaxation methods is in applications where the coefficient matrix has a structure that allows overrelaxation to accelerate the convergence of an already convergent process. In particular, SOR is important for solving the linear systems that arise in the finite difference approach to solving certain partial differential equations (described in Chapter 15). Such problems often involve systems of $N = 10^4$ (or more) equations. The number of iterations required to achieve a specified accuracy using Jacobi or Gauss-Seidel is proportional to N^2. (The proportionality constant is smaller for Gauss-Seidel than for Jacobi.) With the proper choice of the relaxation parameter, the number of iterations for SOR is proportional to N rather than N^2. However, the proper choice of the relaxation parameter is a crucial consideration in the effective use of SOR; optimal values have been determined for some systems that occur in the solution of PDE. For systems for which the optimal value of ω has not already been determined, this can be a major difficulty in using the SOR method. We consider the choice of ω further after illustrating the use of SOR for some small problems.

4.3.1 Step-by-Step Computation of SOR

Algorithm for SOR Method

To solve $\mathbf{M}\,\mathbf{x} = \mathbf{b}$:
Initialize:

>Define relaxation parameter ω
>Form matrix $\mathbf{C}$ and vector $\mathbf{d}$ as for Jacobi and Gauss-Seidel methods
>Define initial guess for solution $\mathbf{x}_{old}$
>
>$\mathbf{x} = \mathbf{x}_{old}$

Begin iterations (illustrated for a 3-by-3 system):

1) Update each component of solution vector $\mathbf{x}$ sequentially

$$\mathbf{x}_1 := (\mathbf{C}_{1,1}\,\mathbf{x}_1 + \mathbf{C}_{1,2}\,\mathbf{x}_2 + \mathbf{C}_{1,3}\,\mathbf{x}_3 + \mathbf{d}_1)\,\omega + (1 - \omega)\mathbf{x}_1$$
$$\mathbf{x}_2 := (\mathbf{C}_{2,1}\,\mathbf{x}_1 + \mathbf{C}_{2,2}\,\mathbf{x}_2 + \mathbf{C}_{2,3}\,\mathbf{x}_3 + \mathbf{d}_2)\,\omega + (1 - \omega)\mathbf{x}_2$$
$$\mathbf{x}_3 := (\mathbf{C}_{3,1}\,\mathbf{x}_1 + \mathbf{C}_{3,2}\,\mathbf{x}_2 + \mathbf{C}_{3,3}\,\mathbf{x}_3 + \mathbf{d}_3)\,\omega + (1 - \omega)\mathbf{x}_3$$

Display transpose of solution vector

$$\mathbf{x}^T =$$

2) Check for convergence by considering the change in the solution vector, or the difference between $\mathbf{M}\,\mathbf{x}$ and $\mathbf{b}$, (or some other convergence test) if desired.
3) To continue iterations, copy the lines from step 1;
Mathcad will recompute all values.

The use of this algorithm is illustrated in the next example.

Example 4.10 Solving a three-by-three system using SOR

To illustrate the use of the SOR algorithm, we seek to solve the linear system M x = b, with
ORIGIN:= 1

$$M := \begin{pmatrix} 25 & 1 & 5 \\ 1 & 20 & -1 \\ -5 & 2 & 50 \end{pmatrix} \quad b := \begin{pmatrix} 20 \\ 2 \\ -55 \end{pmatrix} \quad x0 := \begin{pmatrix} 0 \\ 0 \\ 0 \end{pmatrix}$$

Transform the matrix M to the iteration matrix C, by dividing each row of M by the opposite of the element on the diagonal, and then setting the diagonal elements of C to 0

Transform the vector b to the vector d, by dividing by the corresponding diagonal element of M

$$C := \begin{pmatrix} 0 & \frac{-1}{25} & \frac{-5}{25} \\ \frac{-1}{20} & 0 & \frac{1}{20} \\ \frac{5}{50} & \frac{-2}{50} & 0 \end{pmatrix} \quad d := \begin{pmatrix} \frac{20}{25} \\ \frac{2}{20} \\ \frac{-55}{50} \end{pmatrix} \quad w := 1.2$$

$$x := x0$$

Compute each component of the new solution sequentially

$$x_1 := (C_{1,2} \cdot x_2 + C_{1,3} \cdot x_3 + d_1) \cdot w + (1 - w) \cdot x_1 \qquad x^T = (0.96 \ 0 \ 0)$$

$$x_2 := (C_{2,1} \cdot x_1 + C_{2,3} \cdot x_3 + d_2) \cdot w + (1 - w) \cdot x_2 \qquad x^T = (0.96 \ 0.0624 \ 0)$$

$$x_3 := (C_{3,1} \cdot x_1 + C_{3,2} \cdot x_2 + d_3) \cdot w + (1 - w) \cdot x_3 \qquad x^T = (0.96 \ 0.0624 \ -1.2078)$$

Second iteration

$$x_1 := (C_{1,2} \cdot x_2 + C_{1,3} \cdot x_3 + d_1) \cdot w + (1 - w) \cdot x_1 \qquad x^T = (1.0549 \ 0.0624 \ -1.2078)$$

$$x_2 := (C_{2,1} \cdot x_1 + C_{2,3} \cdot x_3 + d_2) \cdot w + (1 - w) \cdot x_2 \qquad x^T = (1.0549 \ -0.0282 \ -1.2078)$$

$$x_3 := (C_{3,1} \cdot x_1 + C_{3,2} \cdot x_2 + d_3) \cdot w + (1 - w) \cdot x_3 \qquad x^T = (1.0549 \ -0.0282 \ -0.9505)$$

Third iteration

$$x_1 := (C_{1,2} \cdot x_2 + C_{1,3} \cdot x_3 + d_1) \cdot w + (1 - w) \cdot x_1 \qquad x^T = (0.9785 \ -0.0282 \ -0.9505)$$

$$x_2 := (C_{2,1} \cdot x_1 + C_{2,3} \cdot x_3 + d_2) \cdot w + (1 - w) \cdot x_2 \qquad x^T = (0.9785 \ 9.908 \times 10^{-3} \ -0.9505)$$

$$x_3 := (C_{3,1} \cdot x_1 + C_{3,2} \cdot x_2 + d_3) \cdot w + (1 - w) \cdot x_3 \qquad x^T = (0.9785 \ 9.908 \times 10^{-3} \ -1.013)$$

4.3.2 Mathcad Function for SOR

The Mathcad function given here returns a vector containing as its first component the iteration number on which the process terminated, followed by the solution vector.

$$\text{SOR}(A, b, x0, w, \text{tol}, \text{max}) := \begin{vmatrix} C \leftarrow -A \\ n \leftarrow \text{cols}(A) \\ x \leftarrow x0 \\ \text{for } i \in 0..n-1 \\ \quad C_{i,i} \leftarrow 0 \\ \text{for } i \in 0..n-1 \\ \quad \text{for } j \in 0..n-1 \\ \quad\quad C_{i,j} \leftarrow \dfrac{C_{i,j}}{A_{i,i}} \\ \text{for } i \in 0..n-1 \\ \quad r_i \leftarrow \dfrac{b_i}{A_{i,i}} \\ i \leftarrow 0 \\ \text{while } i \leq \text{max} - 1 \\ \quad \begin{vmatrix} \text{xold} \leftarrow x \\ \text{for } j \in 0..n-1 \\ \quad \begin{vmatrix} \text{sum} \leftarrow r_j \\ \text{for } k \in 0..n-1 \\ \quad \text{sum} \leftarrow \text{sum} + C_{j,k} \cdot x_k \\ x_j \leftarrow (1-w) \cdot \text{xold}_j + w \cdot (\text{sum}) \end{vmatrix} \\ \text{break if } |\text{xold} - x| \leq \text{tol} \\ i \leftarrow i + 1 \end{vmatrix} \\ r \leftarrow \text{augment}\left(i, x^T\right) \\ r \end{vmatrix}$$

> **Example 4.11 Solving Linear Systems that come from a PDE**
>
> The SOR function given above allows us to investigate the effect of the choice of ω on the convergence of the solution to the linear system introduced in Example 4-A. The following table shows the number of iterations, N, required for SOR to converge, with tolerance 0.00001, for several values of ω.
>
> | ω | 1.0 | 1.1 | 1.2 | 1.3 |
> |---|---|---|---|---|
> | N | 18 | 14 | 9 | 12 |

4.3.3 Discussion

The SOR method can be derived by multiplying the decomposed system obtained from the Gauss-Seidel method by the relaxation parameter ω, i.e.,

$$\omega(\mathbf{D} + \mathbf{L})\mathbf{x} = -\omega \mathbf{U}\mathbf{x} + \omega\mathbf{b},$$

and adding $(1 - \omega)\mathbf{D}\mathbf{x}$ to each side of the equation. This gives

$$(\mathbf{D} - \omega\mathbf{L})\mathbf{x} = ((1 - \omega)\mathbf{D} - \omega\mathbf{U})\mathbf{x} + \omega\mathbf{b},$$

which can be solved for **x** (for purposes of analysis) to obtain

$$\mathbf{x} = (\mathbf{D} - \omega\mathbf{L})^{-1}((1 - \omega)\mathbf{D} + \omega\mathbf{U})\mathbf{x} + \omega(\mathbf{D} - \omega\mathbf{L})^{-1}\mathbf{b}.$$

The SOR iteration matrix is $\mathbf{C} = (\mathbf{D} - \omega\mathbf{L})^{-1}((1 - \omega)\mathbf{D} + \omega\mathbf{U})$. It can be shown that $\det \mathbf{C} = (1 - \omega)^n$. Since the determinant of a matrix is equal to the product of its eigenvalues, at least one eigenvalue of **C** must be greater than or equal to 1 in absolute value if either $\omega \leq 0$ or $\omega \geq 2$. Therefore, the iteration parameter ω should always be chosen such that $0 < \omega < 2$.

Given an approximate solution $\mathbf{x}^{(k)}$ to the linear system $\mathbf{A}\mathbf{x} = \mathbf{b}$, the residual vector **r** is defined to be $\mathbf{r} = \mathbf{b} - \mathbf{A}\mathbf{x}^{(k)}$. The SOR method is designed to reduce the residual vector more rapidly than the Gauss-Seidel method does. (See Golub and Ortega, 1992, p. 299.)

Some Useful Theoretical Results

It is always important to consider whether an iterative process will converge to a solution. The Ostrowski theorem gives some information on the convergence of the SOR method for an important class of matrices, namely, positive definite matrices.

Ostrowski Theorem

If **A** is a positive definite matrix and $0 < \omega < 2$, then the SOR method will converge for any initial vector **x**. (See Golub and Ortega, 1992, p. 298.)

The matrix **A** is *positive definite* if $\mathbf{x}^T \mathbf{A} \mathbf{x} > 0$ for any vector **x** that is not the zero vector. Unfortunately, this definition does not provide a convenient way to check whether a particular matrix is positive definite. Since symmetric matrices occur in many applications for which SOR is important, we summarize a few useful results for testing whether a symmetric matrix is positive definite.

Necessary and Sufficient Tests for a Positive Definite Matrix

For a real, symmetric matrix **A**, each of the following tests is a necessary and sufficient condition for **A** to be positive definite (See Strang, 1988, p. 331.):

$\mathbf{x}^T \mathbf{A} \mathbf{x} > 0$ for all $\mathbf{x} \neq \mathbf{0}$.
All eigenvalues of **A** are positive.
All upper left submatrices of **A** have positive determinants.
All pivots of **A** (without row interchanges) are positive.

Another Test for a Positive Definite Matrix

A symmetric matrix is positive definite if it is diagonally dominant and each diagonal element is positive. (Golub and Van Loan, 1996, p. 141.)

Test to exclude Matrix from being Positive Definite

If **A** is symmetric and positive definite, then (Faires and Burden, 1998, p. 276.)

1) **A** is non-singular
2) $a_{ii} > 0$ for $i = 1, \ldots n$
3) $a_{ii} a_{jj} > (a_{ij})^2$ for $i \neq j$

Tridiagonal coefficient matrices occur in connection with finite-difference methods for solving boundary value problems (cf. Chapter 14), some finite-difference methods for solving partial differential equations (cf. Chapter 15) and cubic spline interpolation (cf. Chapter 8). A useful result pertaining to such matrices is the following.

Optimal ω for Tridiagonal, Positive Definite Matrix

If **A** is positive definite and tridiagonal, then the optimal ω is

$$\omega = \frac{2}{1 + \sqrt{1 - (\rho_J)^2}}$$

where ρ_J is spectral radius (the maximum of the absolute values the eigenvalues) of the iteration matrix for the Jacobi method.

Results for Systems that Come from Finite-Difference Methods

For a linear system that arises in the finite difference method solution of Poisson's PDE, with a *J*-by-*J* grid, the optimal value of the relaxation parameter is

$$\omega \approx \frac{2}{1 + \pi/J}$$

This value is optimal in an asymptotic sense, and does not necessarily give the best results for the first few (on the order of *J*) iterations.

The spectral radius for the iteration matrix for the Jacobi method for a rectangular grid (*J*-by-*L*) with $s = \Delta x/\Delta y$, is

$$\rho_J = \frac{\cos(\pi/J) + s^2 \cos(\pi/L)}{1 + s^2}.$$

For $J = L$ and $\Delta x = \Delta y$, this simplifies to $\rho_J = \cos(\pi/J)$.

4.4 MATHCAD'S METHODS

The methods presented in this chapter are the classical methods for solving large sparse systems of linear equations. Mathcad uses the SOR method (with automatic selection of the relaxation parameter) in one of its functions for solving PDEs. This is discussed in Chapter 15.

In this section, we introduce Mathcad's Solve Block structure, which is useful for several types of linear and nonlinear problems. There are four functions which can be used in a Solve Block. These are `Find`, `Maximize`, `Minimize` and `Minerr`. The function `Find` is used to solve a system of linear or nonlinear equations. The functions `Maximize` and `Minimize` are used for constrained optimization problems. The function `Minerr` is appropriate for problems which may not have an exact solution, but for which the solution vector that comes closest to solving the given set of equations, subject to the given set of inequality constraints, is desired. The functions `Find`, `Maximize` and `Minimize` may be used for either linear or nonlinear system; their use for nonlinear systems, and the use of `Minerr`, is discussed at the end of the next chapter.

4.4.1 Using the Built-in Functions

Mathcad's function `Find` can be used to solve a system of linear (on nonlinear) equations and inequalities. In general, there are four steps to using the `Find` function to solve a system of equations and inequaltities. These are:

1) Provide initial values for all unknowns. For example

$$\text{var}_1: = 0, \ \text{var}_2: = 0, \ldots \text{var}_n: = 0$$

These values may all be set to 0 for a linear problem, since they are not actually used unless the problem is nonlinear.

2) Type the word `Given` (or `given`). This may be in any style, as long it is not in a text region.

3) Type the equations or inequalities to be soved. Use [Ctrl]= to type the equal sign in the equations. Inequalities may use $<, >, \leq,$ or $\geq$, but not $\neq$.

4) Type `Find(var`$_1$`, var`$_2$`, ... var`$_n$`)=`

This group of instructions forms a Solve Block (actually starting with `Given` and ending with `Find`).

The functions `Maximize` and `Minimize` are used to solve constrained optimization problems, in much the same way as the `Find` function. A linear constrained maximization or minimization problem has the form:

Maximize (or minimize) $f(x)$, where $f(x)$ is a linear function, subject to a set of linear inequalities.

1) Define the function to be maximized (or minimized)
   Provide initial values for all unknowns.
   If desired, define matrix and vector to be used in stating the constraints.
2) Type the word `Given` (or `given`).
3) Type the inequality constraints. Inequalities may use $<$, $>$, $\leq$, or $\geq$, but not $\neq$.
4) Type `Maximize`$(f, \text{var}_1, \text{var}_2, \ldots \text{var}_n)=$

Example 4.12 Using Mathcad's Solve Block Structure

To maximize
$$z = x_0 + 3 x_1$$

Subject to the inequality constraints
$$-x_0 + x_1 \leq 1$$
$$x_0 + 2x_1 \leq 5$$
$$2x_0 + x_1 \leq 8$$

1) Initializations: Define the function to be maximized
$$f(x) := x_0 + 3 x_1$$

define the matrix and vector for the constraints
$$M = \begin{bmatrix} -1 & 1 \\ 1 & 2 \\ 2 & 1 \end{bmatrix} \quad v = \begin{bmatrix} 1 \\ 5 \\ 8 \end{bmatrix}$$

initialize the solution vector
$$x_0 := 0, \quad x_1 := 1$$

2) Given
3) $M x \leq v$
   $x \geq 0$
4) r:= `Maximize`(f, x)
   $r^T = (1, 2)$

Chapter 4 Solving Systems of Linear Equations: Iterative Methods

4.4.2 Understanding the Algorithms

Mathcad uses a combination of methods for solving problems involving the `Find`, `Maximize`, `Minimize`, and `Minerr` functions. These include the simplex, conjugate gradient, Levenberg-Marquardt, and quazi-Netwon methods. Without pursuing the question of which method Mathcad selects for a particular problem, we present a brief overview of two of these methods that are appropriate for linear problems. We first consider the conjugate gradient method, which is useful for solving certain types of linear systems. We then introduce the simplex method, which is important for solving linear constrained optimization problems.

Conjugate Gradient Method

The simplest conjugate gradient method can be used to solve the *n*-by-*n* linear system

$$\mathbf{A}\,\mathbf{x} = \mathbf{b}$$

in the case where $\mathbf{A}$ is symmetric and positive definite. The solution is based on the fact that the function

$$f(\mathbf{x}) = \frac{1}{2}\mathbf{x}\,\mathbf{A}\,\mathbf{x} - \mathbf{b}\,\mathbf{x}$$

is minimized when its gradient (which is $\mathbf{A}\,\mathbf{x} - \mathbf{b}$) is zero.

Without exploring the details of how to minimize $f(\mathbf{x})$, we see that for matrices with very nice characteristics (symmetric and positive definite), there is a direct connection between solving a linear system and minimizing a linear function.

A generalization of the conjugate gradient method for use in solving an *n*-by-*n* linear system that is not necessarily either symmetric or positive definite, gives the biconjugate gradient method. This method does not have a direct connection with function minimzation.

For an iterative solution of a linear system, the residual at step k, $\mathbf{r}_k$, is an important measure of the quality of the solution at that stage.

$$\mathbf{r}_k = \mathbf{b} - \mathbf{A}\,\mathbf{x}_k.$$

The biconjugate gradient method works with four sequences of vectors, which we denote as $\mathbf{r}_k$, $\mathbf{s}_k$, $\mathbf{p}_k$ and $\mathbf{q}_k$. The sequence $\mathbf{r}_k$ is actually the sequence of residuals. The process is described in the following algorithm.

Algorithm for Conjugate Gradient Method

1) Define initial estimate of solution and compute the residual

    $\mathbf{x}_1$

    $\mathbf{r}_1 = \mathbf{b} - \mathbf{A}\mathbf{x}_1$

    Set

    $\mathbf{s}_1 = \mathbf{r}_1,$
    $\mathbf{p}_1 = \mathbf{r}_1,$
    $\mathbf{q}_1 = \mathbf{s}_1$

2) For $k = 1, \ldots n$

    Compute $\mathbf{r}_{k+1}, \mathbf{s}_{k+1}, \mathbf{p}_{k+1}$ and $\mathbf{q}_{k+1}$ as follows:

    $$\alpha_k = \frac{\mathbf{s}_k \mathbf{r}_k}{\mathbf{q}_k \mathbf{A} \mathbf{p}_k}$$

    $\mathbf{r}_{k+1} = \mathbf{r}_k - \alpha_k \mathbf{A} \mathbf{p}_k$

    $\mathbf{s}_{k+1} = \mathbf{r}_k - \alpha_k \mathbf{A}^T \mathbf{q}_k$

    $$\beta_k = \frac{\mathbf{s}_{k+1} \mathbf{r}_{k+1}}{\mathbf{s}_k \mathbf{r}_k}$$

    $\mathbf{p}_{k+1} = \mathbf{r}_k + \beta_k \mathbf{p}_k$

    $\mathbf{q}_{k+1} = \mathbf{s}_k + \beta_k \mathbf{q}_k$

    Compute new solution

    $\mathbf{x}_{k+1} = \mathbf{x}_k + \alpha_k \mathbf{p}_k$

    Check for termination
      if $\mathbf{r}_{k+1} = \mathbf{s}_{k+1} = 0$, stop; $\mathbf{x}_{k+1}$ is the solution,
      otherwise continue.

It can be shown that the vector $\mathbf{r}_{k+1}$ is, in fact, the residual corresponding to the solution vector $\mathbf{x}_{k+1}$, and that the process will terminate in at most n steps. There are several important relationships that are satisfied by combinations of the vectors formed in this process; see [Press et al., 1992] for more details on this and other conjugate gradient methods.

Simplex Method

A linear programming problem is a problem in which we seek to find the maximum or minimum of a linear function, subject to linear constraints (equalities and inequalities). The standard solution method for such problems is known as the simplex method. Mathcad's implementation also includes certain further refinements based on branch and bound techniques. We give a brief introduction to the basic simplex method here.

The problem of linear programming is to maximize a given linear function (called the objective function)

$$z = b_1 x_1 + b_2 x_2 + \ldots b_n x_n$$

subject to the constraints that each of the variables be non-negative, and also subject to m other linear inequality or equality constraints. A vector that satisfies the constraints is called a feasible solution.

To illustrate the process, consider the following simple problem:
Maximize:

$$z = x_1 + 3 x_2$$

Subject to the constraints that

$$1 + x_1 - x_2 \geq 0$$
$$5 - x_1 - 2x_2 \geq 0$$
$$8 - 2x_1 - x_2 \geq 0$$

The foundation of the simplex method is that the region containing all of the feasible solutions is an n-dimensional convex polyhedron or simplex (since the boundaries are hyperplanes), and that the optimal feasible solution must occur at a vertex of this simplex. At a vertex of the feasible region some of the inequality constraints are satisfied as equalities. There are a total of $n + m$ constraints; n of these will be exactly satisfied by the optimal feasible solution. The simplex method gives us a systematic procedure in which the objective function increases at each step and the number of steps is (almost always) no more than the larger of m and n.

The simplest form of linear programming problem is one that is in "restricted normal form." Although this form may seem very limited in its usefulness, in fact general linear programming problems may be converted to this form by introducing additional variables. We limit our discussion of linear programming to problems in restricted normal form. The requirements for this form are that all constraints, other than the non-negativity of the variables, are equality constraints. Furthermore, we require that each constraint equation has at least one variable that appears only in that equation. We solve each constraint for one such variable. The variables that are thus selected are called *basic variables;* the other variables (which appear on the right hand side of the equations) are *non-basic variables*. The solution of the system is found by setting the non-basic variables to zero. The method determines which of the variables should ultimately be basic, and which non-basic, by investigating, in a systematic manner, the effect on the objective function of interchanging one basic and one non-basic variable at each step.

Since the restricted normal form can only be achieved for $m \leq n$, we limit the discussion to that case. We write the objective function so that it depends only on non-basic variables (substituting for any basic variables that may appear). A feasible solution can be found by setting all non-basic variables to zero. A problem with inequality constraints can be transformed to the restricted normal form by the introduction of additional non-negative variables. The simplex method is usually presented in a tabular format (tableau). However, we only intend to sketch the general ideas of the method here, with as little special notation as is necessary for clarity.

To illustrate the process, consider the following simple problem, which is the problem introduced above, transformed to restricted normal form.

Maximize:

$$z = x_1 + 3 x_2$$

Subject to

$$x_3 = 1 + x_1 - x_2$$
$$x_4 = 5 - x_1 - 2x_2$$
$$x_5 = 8 - 2x_1 - x_2$$

The simplex method begins by considering the effect of setting all of the non-basic variables to 0. The non-basic variables in the example above are the variables x_1 and x_2. With that choice for x_1 and x_2, we would have $z = 0$ and $x_3 = 1$, $x_4 = 5$, and $x_5 = 8$. This is a feasible solution, in that all of the equations are satisfied, but is not a particularly good solution in terms of maximizing the objective function. We then consider the effect of increasing either x_1 or x_2, to improve the value of z. That is, we consider the effect of making either x_1 or x_2 into a basic variable. Increasing either x_1 or x_2 will increase (improve) z, since the coefficient of each of these variables in the z equation is positive; we now consider which one to increase. Whichever we choose, we would like to increase as much as possible, without causing any of the variables, x_3, x_4, or x_5 to become negative.

If we increase x_2, (and remember that x_1 is still 0), then:

The first equation limits the increase to $x_2 \rightarrow 1$ (so that $x_3 \geq 0$);
The second equation limits the increase to $x_2 \rightarrow 5/2$ (so that $x_4 \geq 0$);
The third equation limits the increase to $x_2 \rightarrow 8$ (so that $x_5 \geq 0$).

Since we require that *all* of the variables stay positive, we could only let x_2 increase to 1. This would change the value of z from 0 to 3, an increase of 3.

On the other hand, if we increase x_1, (with x_2 still equal to 0), then:

The first equation places no limits on the increase (since the coefficient of x_1 is positive);
The second equation limits the increase to $x_1 \to 5$ (so that $x_4 \geq 0$);
The third equation limits the increase to $x_1 \to 8/2 = 4$ (so that $x_5 \geq 0$).

Since we require that all of the variables stay positive, we can take $x_1 = 4$. Since this would increase the value of z by 4, which is greater than the increase in z that we can achieve by increasing x_2, this is the preferred change.

To accomplish the change (to make x_1 into a basic variable), we solve the third equation for x_1, since it is the equation that limits the acceptable change in x_1

$$x_3 = 1 + x_1 - x_2,$$

$$x_4 = 5 - x_1 - 2x_2,$$

$$x_5 = 8 - 2x_1 - x_2 \to x_1 = 4 - \frac{1}{2}x_2 - \frac{1}{2}x_5.$$

and substitute into the other equations

$$z = x_1 + 3x_2 \quad \to \quad z = \left(4 - \frac{1}{2}x_2 - \frac{1}{2}x_5\right) + 3x_2,$$

$$= 4 + \frac{5}{2}x_2 - \frac{1}{2}x_5,$$

$$x_3 = 1 + x_1 - x_2 \quad \to \quad x_3 = 1 + \left(4 - \frac{1}{2}x_2 - \frac{1}{2}x_5\right) - x_2,$$

$$x_3 = 5 - \frac{3}{2}x_2 - \frac{1}{2}x_5,$$

$$x_4 = 5 - x_1 - 2x_2 \to x_4 = 5 - \left(4 - \frac{1}{2}x_2 - \frac{1}{2}x_5\right) - 2x_2,$$

$$x_4 = 1 - \frac{3}{2}x_2 + \frac{1}{2}x_5,$$

(already done) $x_5 = 8 - 2x_1 - x_2 \to x_1 = 4 - \frac{1}{2}x_2 - \frac{1}{2}x_5.$

The basic variables are now x_1, x_3, and x_4. A couple of observations are in order before we proceed. The first is that only variables that appear in the z-equation with positive coefficients are candidates for selection as basic variables, as they are the only variables that are currently 0 that could increase the value of z if they were increased to a positive value. The second observation is that only equations (the constraint equations) in which the candidate variable appears with a negative coefficient place a restriction on how much the variable can be increased.

To proceed, we see that since

$$z = 4 + \frac{5}{2}x_2 - \frac{1}{2}x_5,$$

the only one of the variables on the right-hand side of these equations that could be increased (from its present value of 0) to increase the value of z, is x_2. To determine the maximum allowable increase, we see that:

The first constraint equation limits the increase of x_2 to at most 10/3;
The second constraint equation limits the increase of x_2 to at most 2/3; and
The third equation limits the increase of x_2 to at most 8.

Thus, we will make x_2 into a basic variable by solving the second equation for x_2:

$$x_3 = 5 - \frac{3}{2}x_2 - \frac{1}{2}x_5,$$

$$x_4 = 1 - \frac{3}{2}x_2 + \frac{1}{2}x_5, \rightarrow x_2 = \frac{2}{3} - \frac{2}{3}x_4 + \frac{1}{3}x_5,$$

$$x_1 = 4 - \frac{1}{2}x_2 - \frac{1}{2}x_5,$$

and substituting into each of the other equations.
New equations

$$z = 4 + \frac{5}{2}x_2 - \frac{1}{2}x_5 \rightarrow z = 4 + \frac{5}{2}\left(\frac{2}{3} - \frac{2}{3}x_4 + \frac{1}{3}x_5\right) - \frac{1}{2}x_5,$$

$$z = \frac{17}{3} - \frac{5}{3}x_4 + \frac{1}{3}x_5,$$

$$x_3 = 5 - \frac{3}{2}x_2 - \frac{1}{2}x_5 \rightarrow x_3 = 5 - \frac{3}{2}\left(\frac{2}{3} - \frac{2}{3}x_4 + \frac{1}{3}x_5\right) - \frac{1}{2}x_5,$$

$$x_3 = 4 + x_4 - x_5$$

(already transformed) $\quad x_2 = \frac{2}{3} - \frac{2}{3}x_4 + \frac{1}{3}x_5,$

$$x_1 = 4 - \frac{1}{2}x_2 - \frac{1}{2}x_5 \rightarrow x_1 = 4 - \frac{1}{2}\left(\frac{2}{3} - \frac{2}{3}x_4 + \frac{1}{3}x_5\right) - \frac{1}{2}x_5,$$

$$x_1 = \frac{11}{3} + \frac{1}{3}x_4 - \frac{2}{3}x_5.$$

Since the coefficient of x_5 in the objective function

$$z = \frac{17}{3} - \frac{5}{3}x_4 + \frac{1}{3}x_5$$

is positive, we consider making x_5 into a basic variable. The constraint equations are currently:

$$x_3 = 4 + x_4 - x_5,$$

$$x_2 = \frac{2}{3} - \frac{2}{3}x_4 + \frac{1}{3}x_5,$$

$$x_1 = \frac{11}{3} + \frac{1}{3}x_4 - \frac{2}{3}x_5.$$

Since the coefficient of x_5 is positive in the second constraint equation, only the first and third equations limit the amount by which x_5 could be increased. The first equation limits the increase to $x_5 \to 4$, whereas the third equation limits the increase to $x_5 \to 11/2$. Thus, we convert x_5 to a basic variable, by solving the first equation for x_5,

$$x_3 = 4 + x_4 - x_5 \quad \to \quad x_5 = 4 + x_4 - x_3$$

and substituting the resulting expression into the objective function and the other constraint equations.

$$z = \frac{17}{3} - \frac{5}{3}x_4 + \frac{1}{3}x_5 \to z = \frac{17}{3} - \frac{5}{3}x_4 + \frac{1}{3}(4 + x_4 - x_3),$$

$$z = 7 - \frac{4}{3}x_4 - \frac{1}{3}x_3,$$

$$x_2 = \frac{2}{3} - \frac{2}{3}x_4 + \frac{1}{3}x_5 \to x_2 = \frac{2}{3} - \frac{2}{3}x_4 + \frac{1}{3}(4 + x_4 - x_3),$$

$$x_2 = 2 - \frac{1}{3}x_4 - \frac{1}{3}x_3,$$

$$x_1 = \frac{11}{3} + \frac{1}{3}x_4 - \frac{2}{3}x_5 \to x_1 = \frac{11}{3} + \frac{1}{3}x_4 - \frac{2}{3}(4 + x_4 - x_3),$$

$$x_1 = 1 - \frac{1}{3}x_4 + \frac{2}{3}x_3.$$

Since the objective function, $z = 7 - \frac{4}{3}x_4 - \frac{1}{3}x_3$, cannot be increased by increasing either of the non-basic variables, the process is finished. The final solution is found by setting the non-basic variables to 0, which gives:

$$x_3 = x_4 = 0, x_1 = 1, x_2 = 2, x_5 = 4, \text{ and } z = 7.$$

SUMMARY

In this chapter, we solve the linear system $\mathbf{Ax} = \mathbf{b}$ iteratively by converting it into an equivalent system $\mathbf{x} = \mathbf{Cx} + \mathbf{d}$ to generate a sequence of approximations $\mathbf{x}^{(1)}, \mathbf{x}^{(2)}, \ldots$, according to the formula $\mathbf{x}^{(k)} = \mathbf{C}\mathbf{x}^{(k-1)} + \mathbf{d}$. The ith equation is solved explicitly for the ith component of $\mathbf{x}$.

The Jacobi method, illustrated for $n = 4$, is as follows:

$$x_1^{(k)} = \phantom{-\frac{a_{21}}{a_{22}}x_1^{(k-1)}} - \frac{a_{12}}{a_{11}}x_2^{(k-1)} - \frac{a_{13}}{a_{11}}x_3^{(k-1)} - \frac{a_{14}}{a_{11}}x_4^{(k-1)} + \frac{b_1}{a_{11}},$$

$$x_2^{(k)} = -\frac{a_{21}}{a_{22}}x_1^{(k-1)} \phantom{- \frac{a_{12}}{a_{22}}x_2^{(k-1)}} - \frac{a_{23}}{a_{22}}x_3^{(k-1)} - \frac{a_{24}}{a_{22}}x_4^{(k-1)} + \frac{b_2}{a_{22}},$$

$$x_3^{(k)} = -\frac{a_{31}}{a_{33}}x_1^{(k-1)} - \frac{a_{32}}{a_{33}}x_2^{(k-1)} \phantom{- \frac{a_{33}}{a_{33}}x_3^{(k-1)}} - \frac{a_{34}}{a_{33}}x_4^{(k-1)} + \frac{b_3}{a_{33}},$$

$$x_4^{(k)} = -\frac{a_{41}}{a_{44}}x_1^{(k-1)} - \frac{a_{42}}{a_{44}}x_2^{(k-1)} - \frac{a_{43}}{a_{44}}x_3^{(k-1)} \phantom{- \frac{a_{44}}{a_{44}}x_4^{(k-1)}} + \frac{b_4}{a_{44}}.$$

The Gauss-Seidel method, also illustrated for $n = 4$, is characterized by the following equations:

$$x_1^{(k)} = \phantom{-\frac{a_{21}}{a_{22}}x_1^{(k)}} - \frac{a_{12}}{a_{11}}x_2^{(k-1)} - \frac{a_{13}}{a_{11}}x_3^{(k-1)} - \frac{a_{14}}{a_{11}}x_4^{(k-1)} + \frac{b_1}{a_{11}},$$

$$x_2^{(k)} = -\frac{a_{21}}{a_{22}}x_1^{(k)} \phantom{- \frac{a_{12}}{a_{22}}x_2^{(k)}} - \frac{a_{23}}{a_{22}}x_3^{(k-1)} - \frac{a_{24}}{a_{22}}x_4^{(k-1)} + \frac{b_2}{a_{22}},$$

$$x_3^{(k)} = -\frac{a_{31}}{a_{33}}x_1^{(k)} - \frac{a_{32}}{a_{33}}x_2^{(k)} \phantom{- \frac{a_{33}}{a_{33}}x_3^{(k)}} - \frac{a_{34}}{a_{33}}x_4^{(k-1)} + \frac{b_3}{a_{33}},$$

$$x_4^{(k)} = -\frac{a_{41}}{a_{44}}x_1^{(k)} - \frac{a_{42}}{a_{44}}x_2^{(k)} - \frac{a_{43}}{a_{44}}x_3^{(k)} \phantom{- \frac{a_{44}}{a_{44}}x_4^{(k)}} + \frac{b_4}{a_{44}}.$$

Finally, the SOR method, illustrated as well for $n = 4$ ($1 < \omega < 2$), is described as follows:

$$x_1^{(k)} = (1 - \omega)x_1^{(k-1)} + \frac{\omega}{a_{11}}(b_1 \phantom{- a_{21}x_1^{(k)}} - a_{12}x_2^{(k-1)} - a_{13}x_3^{(k-1)} - a_{14}x_4^{(k-1)}),$$

$$x_2^{(k)} = (1 - \omega)x_2^{(k-1)} + \frac{\omega}{a_{22}}(b_2 - a_{21}x_1^{(k)} \phantom{- a_{32}x_2^{(k)}} - a_{23}x_3^{(k-1)} - a_{24}x_4^{(k-1)}),$$

$$x_3^{(k)} = (1 - \omega)x_3^{(k-1)} + \frac{\omega}{a_{33}}(b_3 - a_{31}x_1^{(k)} - a_{32}x_2^{(k)} \phantom{- a_{43}x_3^{(k)}} - a_{34}x_4^{(k-1)}),$$

$$x_4^{(k)} = (1 - \omega)x_4^{(k-1)} + \frac{\omega}{a_{44}}(b_4 - a_{41}x_1^{(k)} - a_{42}x_2^{(k)} - a_{43}x_3^{(k)} \phantom{- a_{44}x_4^{(k)}}).$$

SUGGESTIONS FOR FURTHER READING

The discussion of the conjugate gradient and simplex methods presented in this chapter is based on material in:

> Press, W. H., S. A. Teukolsky, W. T. Vetterling, and B. P. Flannery, *Numerical Recipes in C, The Art of Scientific Computing* (2d ed.), Cambridge Unversity Press, 1992.

For further discussion of conjugate gradient methods, see the following:

> Golub, G. H. and C. F. Van Loan, *Matrix Computations* (3d ed.), Johns Hopkins University Press, Baltimore, 1996.
>
> Freund, R. W., G. H. Golub, and N. M. Nachtigal, "Iterative Solution of Linear Systems," *Acta Numerica I*, 1992, pp. 57–100.
>
> Golub, G. H. and J. M. Ortega, *Scientific Computing and Differential Equations: An Introduction to Numerical Methods*, Academic Press, Boston, 1992.
>
> Stoer, J. and R. Bulirsch, *Introduction to Numerical Analysis*, Springer-Verlag, New York, 1980 (Chapter 8).

For further discussion of linear programming and the simplex method, see the following texts:

> Taha, H. A, *Operations Research, An Introduction* (6th ed.), Prentice Hall, Upper Saddle River, NJ, 1997.
>
> Winston, W. L., *Operations Research, Aplications and Algorithms* (3rd ed.), Duxbury Press (Wadsworth), Belmont, CA,1994.
>
> Danzig, G. B., *Linear Programming and Extensions*, Princeton University Press, Princeton, NJ, 1963.

The following are excellent references for applied linear algebra:

> Strang, G., *Linear Algebra and Its Applications* (3d ed.), Harcourt Brace Jovanovich, San Diego, CA, 1988.
>
> Hager, W. W., *Applied Numerical Linear Algebra*, Prentice-Hall, Englewood Cliffs, NJ, 1988.
>
> Datta, B. N., *Numerical Linear Algebra and Applications*, Brooks Cole, Pacific Grove, CA, 1995.

For more advanced discussion of iterative methods for linear systems, see the following texts:

> Greenbaum, A., *Iterative Methods for Solving Linear Systems*, SIAM, Philadelphia, 1997.
>
> Barrett, R., J. Donato, J. Dongarra, V. Eijkhout, R. Pozo, C. Romine, and H. van der Vorst, *Templates for the Solution of Linear Systems: Building Blocks for Iterative Methods*, SIAM, Philadelphia, 1993.

PRACTICE THE TECHNIQUES

For Problems P4.1 to P4.5 approximate the solution of the system $\mathbf{A}\,\mathbf{x} = \mathbf{b}$, *iteratively (if possible); use* $\mathbf{x}^{(0)} = [\,0,\,0,\,0\,]^T$.
  a. Use Jacobi iteration and perform three iterations by hand.
  b. Use Gauss-Seidel iteration and perform three iterations by hand.
  c. Use Jacobi iteration and perform 10 iterations with a Mathcad function.
  d. Use Gauss-Seidel iteration and perform 10 iterations with a Mathcad function.

P4.1 $\mathbf{A} = \begin{bmatrix} 10 & -2 & 1 \\ -2 & 10 & -2 \\ -2 & -5 & 10 \end{bmatrix},\ \mathbf{b} = \begin{bmatrix} 9 \\ 12 \\ 18 \end{bmatrix}.$

P4.2 $\mathbf{A} = \begin{bmatrix} 4 & 1 & 0 \\ 1 & 3 & -1 \\ 1 & 0 & 2 \end{bmatrix},\ \mathbf{b} = \begin{bmatrix} 3 \\ -4 \\ 5 \end{bmatrix}.$

P4.3 $\mathbf{A} = \begin{bmatrix} 5 & -1 & 0 \\ -1 & 5 & -1 \\ 0 & -1 & 5 \end{bmatrix},\ \mathbf{b} = \begin{bmatrix} 9 \\ 4 \\ -6 \end{bmatrix}.$

P4.4 $\mathbf{A} = \begin{bmatrix} 8 & 1 & -1 \\ -1 & 7 & -2 \\ 2 & 1 & 9 \end{bmatrix},\ \mathbf{b} = \begin{bmatrix} 8 \\ 4 \\ 12 \end{bmatrix}.$

P4.5 $\mathbf{A} = \begin{bmatrix} 4 & 1 & 0 \\ 1 & 3 & -1 \\ 0 & -1 & 4 \end{bmatrix},\ \mathbf{b} = \begin{bmatrix} 3 \\ 4 \\ 5 \end{bmatrix}.$

For Problems P4.6 to P4.18 approximate the solution of the system $\mathbf{A}\,\mathbf{x} = \mathbf{b}$, *iteratively; use* $\mathbf{x}^{(0)} = [\,0,\,0,\,\ldots\,0\,]^T$.
  a. Use Jacobi iteration, perform 10 iterations using a Mathcad function.
  b. Use Gauss-Seidel iteration, perform 10 iterations using a Mathcad function.
  c. Use SOR with $\omega = 1.25$, $\omega = 1.5$, $\omega = 1.75$, $\omega = 1.9$.

P4.6 $\mathbf{A} = \begin{bmatrix} -2 & 1 & 0 & 0 \\ 1 & -2 & 1 & 0 \\ 0 & 1 & -2 & 1 \\ 0 & 0 & 1 & -2 \end{bmatrix},\ \mathbf{b} = \begin{bmatrix} -1 \\ 0 \\ 0 \\ 0 \end{bmatrix}.$

P4.7 $\mathbf{A} = \begin{bmatrix} 5 & 1 & 0 & 0 \\ 1 & 5 & 1 & 0 \\ 0 & 1 & 5 & 1 \\ 0 & 0 & 1 & 5 \end{bmatrix},\ \mathbf{b} = \begin{bmatrix} 33 \\ 26 \\ 30 \\ 15 \end{bmatrix}.$

P4.8 $\mathbf{A} = \begin{bmatrix} 1 & 2 & 0 & 0 \\ 2 & 6 & 8 & 0 \\ 0 & 8 & 35 & 18 \\ 0 & 0 & 18 & 112 \end{bmatrix},\ \mathbf{b} = \begin{bmatrix} 2 \\ 6 \\ -10 \\ -112 \end{bmatrix};$
(the optimal value is $\omega = 1.9387$.)

P4.9 $\mathbf{A} = \begin{bmatrix} 4 & 8 & 0 & 0 \\ 8 & 18 & 2 & 0 \\ 0 & 2 & 5 & 1.5 \\ 0 & 0 & 1.5 & 1.75 \end{bmatrix},$
$\mathbf{b} = [8 \quad 18 \quad 0.50 \quad -1.75]^T;$
(the optimal value is $\omega = 1.634$.)

P4.10 $\mathbf{A} \begin{bmatrix} 4 & -8 & 0 & 0 \\ -8 & 18 & -2 & 0 \\ 0 & -2 & 5 & -1.5 \\ 0 & 0 & -1.5 & 1.75 \end{bmatrix},$
$\mathbf{b} = [-12 \quad 22 \quad 5 \quad 2]^T;$
(the optimal value is $\omega = 1.634$.)

P4.11 $\mathbf{A} = \begin{bmatrix} 1 & -2 & 0 & 0 \\ -2 & 5 & -1 & 0 \\ 0 & -1 & 2 & -0.5 \\ 0 & 0 & -0.5 & 1.25 \end{bmatrix},$
$\mathbf{b} = [-3 \quad 5 \quad 2 \quad 3.5]^T;$
(the optimal value is $\omega = 1.5431$.)

P4.12 $\mathbf{A} = \begin{bmatrix} 1 & -2 & 0 & 0 & 0 \\ -2 & 5 & 1 & 0 & 0 \\ 0 & 1 & 2 & -2 & 0 \\ 0 & 0 & -2 & 5 & 1 \\ 0 & 0 & 0 & 1 & 2 \end{bmatrix},$
$\mathbf{b} = [5 \quad -9 \quad 0 \quad 3 \quad 0]^T;$
(the optimal value is $\omega = 1.7684$.)

P4.13 $A = \begin{bmatrix} 1 & -2 & 0 & 0 & 0 \\ -2 & 6 & 4 & 0 & 0 \\ 0 & 4 & 9 & -0.5 & 0 \\ 0 & 0 & -0.5 & 1.25 & 0.5 \\ 0 & 0 & 0 & 0.5 & 3.25 \end{bmatrix}$,

$b = \begin{bmatrix} 5 & -2 & 18 & 0.5 & -2.25 \end{bmatrix}^T$;

(the optimal value is $\omega = 1.7064$.)

P4.14

$A = \begin{bmatrix} 1 & -2 & 0 & 0 & 0 & 0 \\ -2 & 6 & 4 & 0 & 0 & 0 \\ 0 & 4 & 9 & -0.5 & 0 & 0 \\ 0 & 0 & -0.5 & 3.25 & 1.5 & 0 \\ 0 & 0 & 0 & 1.5 & 1.75 & -3 \\ 0 & 0 & 0 & 0 & -3 & 13 \end{bmatrix}$

$b = \begin{bmatrix} -3 & 22 & 35.5 & -7.75 & 4 & -33 \end{bmatrix}^T$;

(the optimal value is $\omega = 1.7113$.)

P4.15

$A = \begin{bmatrix} 7.63 & 0.3 & 0.15 & 0.5 & 0.34 & 0.84 \\ 0.38 & 6.4 & 0.7 & 0.9 & 0.29 & 0.57 \\ 0.83 & 0.19 & 8.33 & 0.82 & 0.34 & 0.37 \\ 0.5 & 0.68 & 0.86 & 10.21 & 0.53 & 0.7 \\ 0.71 & 0.3 & 0.85 & 0.82 & 5.95 & 0.55 \\ 0.43 & 0.54 & 0.59 & 0.66 & 0.31 & 9.25 \end{bmatrix}$

$b = \begin{bmatrix} -9.44 & 25.27 & -48.01 & 19.76 & -23.63 & 62.59 \end{bmatrix}^T$;

P4.16

$A = \begin{bmatrix} 85.57 & 0.46 & 0.92 & 0.41 & 0.14 & 0.02 \\ 0.23 & 52.53 & 0.74 & 0.89 & 0.2 & 0.75 \\ 0.61 & 0.82 & 20.44 & 0.06 & 0.2 & 0.45 \\ 0.49 & 0.44 & 0.41 & 67.57 & 0.6 & 0.93 \\ 0.89 & 0.62 & 0.94 & 0.81 & 84.08 & 0.47 \\ 0.76 & 0.79 & 0.92 & 0.01 & 0.2 & 2.38 \end{bmatrix}$,

$b = \begin{bmatrix} 85.61 & -267.18 & 54.91 & -140.66 & 331.55 & -18.69 \end{bmatrix}^T$

P4.17

$A = \begin{bmatrix} 14.38 & 0.59 & 0.44 & 0.12 & 0.8 & 0.84 & 0.39 & 0.16 \\ 0.09 & 81.93 & 0.35 & 0.45 & 0.91 & 0.17 & 0.59 & 0.87 \\ 0.04 & 0.37 & 43.17 & 0.72 & 0.23 & 0.17 & 0.12 & 0.24 \\ 0.61 & 0.63 & 0.68 & 89.93 & 0.24 & 0.99 & 0.04 & 0.65 \\ 0.61 & 0.72 & 0.7 & 0.27 & 73.54 & 0.44 & 0.46 & 0.97 \\ 0.02 & 0.69 & 0.73 & 0.25 & 0.08 & 69.07 & 0.87 & 0.66 \\ 0.02 & 0.08 & 0.48 & 0.87 & 0.64 & 0.31 & 35.55 & 0.87 \\ 0.19 & 0.45 & 0.55 & 0.23 & 0.19 & 0.37 & 0.26 & 16.61 \end{bmatrix}$

$b = \begin{bmatrix} -23.49 & 87.25 & 170.88 & -530.36 & 227.13 & 141.59 & -136.83 & 117.43 \end{bmatrix}^T$

P4.18

$A = \begin{bmatrix} 10 & 0 & 1 & 0 & 0 & 0 & 0 & 0 \\ 0 & 10 & 0 & 0 & 0 & 0 & -1 & 0 \\ 0 & 0 & 10 & 0 & 0 & -2 & 0 & 0 \\ 2 & 0 & 0 & 10 & 0 & 0 & 0 & 0 \\ 0 & 0 & 1 & 0 & 10 & 0 & 0 & 0 \\ 0 & 0 & 0 & -3 & 0 & 10 & 0 & 0 \\ 0 & 3 & 0 & 0 & 0 & 0 & 10 & 0 \\ 0 & 0 & 0 & 0 & 1 & 0 & 0 & 10 \end{bmatrix}$

$b = \begin{bmatrix} 13 & 13 & 18 & 42 & 53 & 48 & 76 & 85 \end{bmatrix}^T$

EXPLORE SOME APPLICATIONS

A4.1 To find the forces in the single triangle truss with angles $a = \pi/6$ and $b = \pi/3$, solve the linear system $T x = b$, for T and b as given here.

$T = \begin{bmatrix} 1 & 0 & 0 & 0 & \sin(a) & 0 \\ 0 & 1 & 0 & 1 & \cos(a) & 0 \\ 0 & 0 & 1 & 0 & 0 & \sin(b) \\ 0 & 0 & 0 & -1 & 0 & -\cos(b) \\ 0 & 0 & 0 & 0 & -\cos(a) & \cos(b) \\ 0 & 0 & 0 & 0 & -\sin(a) & -\sin(b) \end{bmatrix}$

$b = \begin{bmatrix} 0 & 0 & 0 & 0 & 0 & 10 \end{bmatrix}^T$

A4.2 Iterative methods may be useful for systems that have a few small non-zero elements outside of a fairly pronounced banded structure (such elements prevent the application of the Thomas method and cause extensive fill in the lower triangular portion of the coefficient matrix if Gaussian elimination is used). For example, solve $A x = b$:

$$\mathbf{A} = \begin{bmatrix} -2.9 & 0.9 & 0 & 0 & 0 & 0.01 & 0 & 0 & 0 & 0 \\ 1.0 & -2.9 & 0.9 & 0 & 0 & 0 & 0 & 0 & 0.01 & 0 \\ 0 & 1.0 & -2.9 & 0.9 & 0 & 0 & 0 & 0 & 0 & 0 \\ 0 & 0 & 1.0 & -2.9 & 0.9 & 0 & 0 & 0 & 0 & 0 \\ 0.01 & 0 & 0 & 1.0 & -2.9 & 0.9 & 0 & 0 & 0 & 0 \\ 0 & 0 & 0 & 0 & 1.0 & -2.9 & 0.9 & 0 & 0 & 0 \\ 0 & 0 & 0 & 0 & 0 & 1.0 & -2.9 & 0.9 & 0 & 0 \\ 0 & 0 & 0.01 & 0 & 0 & 0 & 1.0 & -2.9 & 0.9 & 0 \\ 0.01 & 0 & 0 & 0 & 0 & 0 & 0 & 1.0 & -2.9 & 0.9 \\ 0 & 0 & 0 & 0 & 0 & 0 & 0 & 0 & 1.0 & -2.9 \end{bmatrix}$$

$\mathbf{b} = $
$[-0.1395\ -0.080\ -0.153\ -0.173\ -0.1842\ -0.255\ -0.296\ -0.3826\ -0.3492\ -1.05]^T$

A4.5 The following linear system corresponds to the finite difference solution to a PDE, with unequal spacing in the x and y directions.

$$\mathbf{A} = \begin{bmatrix} -68 & 9 & 25 & 0 & 0 & 0 & 0 & 0 \\ 9 & -68 & 0 & 25 & 0 & 0 & 0 & 0 \\ 25 & 0 & -68 & 9 & 25 & 0 & 0 & 0 \\ 0 & 25 & 9 & -68 & 0 & 25 & 0 & 0 \\ 0 & 0 & 25 & 0 & -68 & 9 & 25 & 0 \\ 0 & 0 & 0 & 25 & 9 & -68 & 0 & 25 \\ 0 & 0 & 0 & 0 & 25 & 0 & -68 & 9 \\ 0 & 0 & 0 & 0 & 0 & 25 & 9 & -68 \end{bmatrix}$$

$\mathbf{b} = [-25.6667\ -8.6933\ -8.0000\ -1.4400\ -9.0000\ -3.2400\ -34.0000\ -30.7600]^T$

EXPAND YOUR UNDERSTANDING

U4.1 For each of the linear systems in problems P3.1 to P3.5, determine whether $\mathbf{A}$ is strictly diagonally dominant. Also solve (if possible) using the iterative methods presented in this chapter.

For problems U4.2 to U4.6 write the linear system $\mathbf{A}\,x = \mathbf{b}$ in the explicit iterative form for the Jacobi method, i.e. $x = \mathbf{T}\,x + \mathbf{c}$, where $\mathbf{A} = \mathbf{D} - \mathbf{L} - \mathbf{U}$, and $\mathbf{T} = \mathbf{D}^{-1}(\mathbf{L} + \mathbf{U})$; find the eigenvalues of $\mathbf{T}$ using the Mathcad function `eigenvals`. *For those problems for which the theorems apply, find the optimal SOR parameter.*

U4.2 $\quad \mathbf{A} = \begin{bmatrix} 10 & -2 & 1 \\ -2 & 10 & -2 \\ -2 & -5 & 10 \end{bmatrix} \quad \mathbf{b} = \begin{bmatrix} 9 \\ 12 \\ 18 \end{bmatrix}$

U4.3 $\quad \mathbf{A} = \begin{bmatrix} 5 & -1 & 0 \\ -1 & 5 & -1 \\ 0 & -1 & 5 \end{bmatrix} \quad \mathbf{b} = \begin{bmatrix} 9 \\ 4 \\ -6 \end{bmatrix}$

U4.4 $\quad \mathbf{A} = \begin{bmatrix} 4 & 1 & 0 \\ 1 & 3 & -1 \\ 0 & -1 & 4 \end{bmatrix} \quad \mathbf{b} = \begin{bmatrix} 3 \\ 4 \\ 5 \end{bmatrix}$

U4.5 $\quad \mathbf{A} = \begin{bmatrix} -2 & 1 & 0 & 0 \\ 1 & -2 & 1 & 0 \\ 0 & 1 & -2 & 1 \\ 0 & 0 & 1 & -2 \end{bmatrix} \quad \mathbf{b} = \begin{bmatrix} -1 \\ 0 \\ 0 \\ 0 \end{bmatrix}$

U4.6 $\quad \mathbf{A} = \begin{bmatrix} 5 & 1 & 0 & 0 \\ 1 & 5 & 1 & 0 \\ 0 & 1 & 5 & 1 \\ 0 & 0 & 1 & 5 \end{bmatrix} \quad \mathbf{b} = \begin{bmatrix} 33 \\ 26 \\ 30 \\ 15 \end{bmatrix}$

U4.7 Suppose Jacobi's method is used for the linear system $\mathbf{A}\,x = \mathbf{b}$, with

$$\mathbf{A} = \begin{bmatrix} 4 & 12 & 8 & 4 \\ 1 & 7 & 18 & 9 \\ 2 & 9 & 20 & 20 \\ 3 & 11 & 15 & 14 \end{bmatrix} \quad \mathbf{b} = \begin{bmatrix} -4 \\ -5 \\ -25 \\ -18 \end{bmatrix}$$

Perform 2 iterations and discuss the expected results.

U4.8 Other decompositions of the matrix $\mathbf{A}$ may be appropriate for solving the system $\mathbf{A}\,x = \mathbf{b}$ iteratively when A has special structure. In particular, if $\mathbf{A} = \mathbf{B} + \mathbf{C}$, where $\mathbf{B}$ is tridiagonal and the elements of C are small, consider the iterative scheme

$$\mathbf{A}\,x = \mathbf{b}$$
$$(\mathbf{B} + \mathbf{C})\,x = \mathbf{b}$$
$$\mathbf{B}\,x = -\mathbf{C}\,x + \mathbf{b}$$

The method will converge if the elements of $\mathbf{C}$ are sufficiently small so that $\|\mathbf{B}^{-1}\mathbf{C}\| < 1$. The usefulness of this method depends on the fact that the tridiagonal system $\mathbf{B}\,x$ is relatively easy to solve at each stage. See Hager, p. 334.

5
Systems of Equations and Inequalities

5.1 Newton's Method for Systems of Equations

5.2 Fixed-Point Iteration for Nonlinear Systems

5.3 Minimum of a Nonlinear Function

5.4 Mathcad's Methods

Several of the methods introduced in Chapter 2 for finding a zero of a nonlinear function of a single variable can be extended to nonlinear functions of several variables. However, the problem is much more difficult in several variables. For two nonlinear functions of two variables, $z = f(x, y)$ and $z = g(x, y)$, finding a zero of the system requires finding the intersection of the curves $f(x, y) = 0$ and $g(x, y) = 0$. It is very important to use any information available about the specific problem to identify the region where these curves may intersect, i.e., the possible location of a common root of the equations $f(x, y) = 0$ and $g(x, y) = 0$.

We begin this chapter by considering the extension of Newton's method to systems of nonlinear equations. Starting with an initial estimate of the solution, each nonlinear function is approximated by its tangent plane (at the current approximate solution); the common root of the resulting linear equations provides the next approximation to the desired zero. Forming the linear system requires either the computation of the Jacobian matrix (the matrix of partial derivatives) for the nonlinear system or approximation of the Jacobian numerically.

Fixed-point iteration, introduced for functions of a single variable in Chapter 1, provides an approach to solving a system of nonlinear equations that does not require the computation of partial derivatives. However, the transformation of the original problem into fixed-point form is more straightforward for some problems than for others.

We also consider the problem of finding the minimum of a scalar function of several variables. In addition to many other applications of minimization, root-finding problems may be solved by constructing a function whose minimum corresponds to the desired root. A basic gradient-descent minimization function is presented in Section 5.3. More powerful approaches, implemented in Mathcad's built-in function Minerr, are discussed briefly in Section 5.4. Mathcad's built-in function for finding the solutions of nonlinear systems of equations, Find, is also presented in Section 5.4.

Nonlinear systems of equations occur in many settings, including the solution of nonlinear partial differential equations by means of finite differences.

Example 5-A Nonlinear System from Geometry

The problem of finding the points of intersection of two curves in the *xy*-plane may require solving a nonlinear system of equations. For example, suppose we would like to know where the ellipse of eccentricity 0.5 (which described the orbit in an example in Chapter 2) would intersect a circle with the same area and the same center. (See Fig. 5.1.) The area of an ellipse is $\pi a b$; for this example, with an eccentricity of 0.5, we take $a = 1$ and $b = \sqrt{3}/2$, so the equation of the ellipse is

$$x^2 + 4y^2/3 = 1.$$

Since the area of a circle is πr^2, the equation of a circle with the same area as the ellipse, is

$$x^2 + y^2 = r^2 = \sqrt{3}/2.$$

The points of intersection are solutions of the nonlinear system

$$3x^2 + 4y^2 - 3 = 0,$$
$$x^2 + y^2 - \sqrt{3}/2 = 0.$$

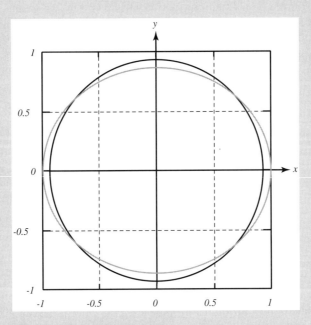

FIGURE 5.1 Intersections of a circle and an ellipse with the same area.

Example 5-B Position of a Two-link Robot Arm

The position of a two-link robot arm can be described in terms of the angle that the first link makes with the horizontal axis and the angle that the second link makes with the first link. In this example, we assume that the lengths of the two links are d_1 and d_2; the first link makes an angle α with the horizontal axis, and the second link makes an angle β with the direction defined by the first link. Our problem is to find the angles α and β that allow the end of the second link to be at a specified point, with coordinates (p_1, p_2). The arrangement is illustrated in Fig. 5.2. The equations for these requirements are as follows:

Location of the end of the first link (x_1, y_1)

$$x_1 = d_1 \cos(\alpha),$$
$$y_1 = d_1 \sin(\alpha);$$

Location of the end of the second link (x_2, y_2)

$$x_2 = x_1 + d_2 \cos(\alpha + \beta),$$
$$y_2 = y_1 + d_2 \sin(\alpha + \beta).$$

Thus, we need to solve

$$p_1 = d_1 \cos(\alpha) + d_2 \cos(\alpha + \beta),$$
$$p_2 = d_1 \sin(\alpha) + d_2 \sin(\alpha + \beta),$$

for the unknown angles α and β.

We solve a problem of this type in Example 5.4.

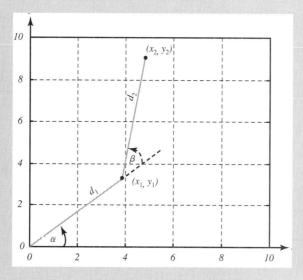

FIGURE 5.2 Two-link robot arm, initial position.

5.1 NEWTON'S METHOD FOR SYSTEMS OF EQUATIONS

Newton's method for finding the root of a nonlinear function, discussed in Chapter 2, can be extended to solving a system of nonlinear equations. In the following examples, we show iterations as superscripts (in parentheses), since we often use vector notation, denoting components of the unknown vector by subscripts. In cases where we do not wish to retain all iterates, we denote the current estimate as x_old, and the next estimate as x_new.

Example 5.1 Intersection of a Circle and an Ellipse

We first consider a system of two equations describing the intersection of the unit circle (centered at the origin) and a given parabola (with vertex at the origin). The curves are illustrated in Fig. 5.3. We are looking for the common zeros of the functions

$$f(x, y) = x^2 + y^2 - 1 \quad \text{and} \quad g(x, y) = x^2 - y.$$

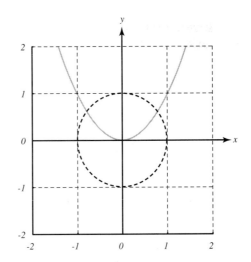

FIGURE 5.3 Intersections of a circle and a parabola.

We start with an initial estimate of a common solution, (x_0, y_0). The plane that is tangent to the function $z = f(x, y)$ at $(x_0, y_0, f(x_0, y_0))$ has the equation:

$$z - f(x_0, y_0) = f_x(x_0, y_0)(x - x_0) + f_y(x_0, y_0)(y - y_0),$$

where $f_x(x_0, y_0)$ is the partial derivative of $f(x,y)$ with respect to x, and $f_y(x_0, y_0)$ is the partial derivative of $f(x,y)$ with respect to y, each evaluated at (x_0, y_0).

The equation of the plane tangent to $z = g(x, y)$ at $(x_0, y_0, g(x_0, y_0))$ is

$$z - g(x_0, y_0) = g_x(x_0, y_0)(x - x_0) + g_y(x_0, y_0)(y - y_0).$$

To find the next approximation to the desired solution, we find the intersection of these two tangent planes with the xy-plane (i.e., with $z = 0$). We define $r = (x - x_0)$ and $s = (y - y_0)$ and solve the linear system:

$$f_x(x_0, y_0)\, r + f_y(x_0, y_0)\, s = -f(x_0, y_0),$$
$$g_x(x_0, y_0)\, r + g_y(x_0, y_0)\, s = -g(x_0, y_0).$$

Note that r and s give the location of the intersection point in terms of its displacement from the point (x_0, y_0); i.e., $(x, y) = (r + x_0, s + y_0)$.

For this example, $f_x = 2x$, $f_y = 2y$, $g_x = 2x$, and $g_y = -1$.

Choosing the initial estimate as $(x_0, y_0) = (1/2, 1/2)$, we have

$$f_x = 1,\ f_y = 1,\ g_x = 1,\ g_y = -1,\ f(1/2, 1/2) = -1/2,\ \text{and}\ g(1/2, 1/2) = -1/4.$$

The resulting linear system to be solved is

$$r + s = 1/2,$$
$$r - s = 1/4.$$

Its solution is $r = 3/8$, $s = 1/8$.

The second approximate solution, (x_1, y_1) is the intersection point in terms of its (x, y) coordinates:

$$x_1 = x_0 + r = 1/2 + 3/8 = 7/8,$$
$$y_1 = y_0 + s = 1/2 + 1/8 = 5/8.$$

The process is repeated, using the new approximate solution to evaluate the partial derivatives f_x, f_y, g_x, and g_y, and the functions, f and g. The true solution is $(x, y) = (0.78615, 0.61803)$. (See Fig. 5.4.)

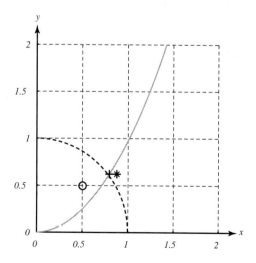

FIGURE 5.4 Initial estimate and two iterations of Newton's method.

Chapter 5 Systems of Equations and Inequalities

5.1.1 Matrix-vector Notation

The general nonlinear system

$$f_1(x_1, x_2, x_3, \ldots, x_n) = 0,$$
$$f_2(x_1, x_2, x_3, \ldots, x_n) = 0,$$
$$f_3(x_1, x_2, x_3, \ldots, x_n) = 0,$$
$$\ldots$$
$$f_n(x_1, x_2, x_3, \ldots, x_n) = 0,$$

can be written in a more compact vector form as $F(\mathbf{x}) = \mathbf{0}$, where

$$\mathbf{x} = [x_1, x_2, x_3, \ldots, x_n],$$

$$F(\mathbf{x}) = \begin{bmatrix} f_1(\mathbf{x}) \\ f_2(\mathbf{x}) \\ \vdots \\ f_n(\mathbf{x}) \end{bmatrix}, \text{ and } \mathbf{0} = \begin{bmatrix} 0 \\ 0 \\ \vdots \\ 0 \end{bmatrix}.$$

Newton's method uses the Jacobian (matrix of partial derivatives) of the system:

$$J(x_1, \ldots x_n) = \begin{bmatrix} \partial f_1/\partial x_1 & \cdots & \partial f_1/\partial x_n \\ \vdots & \cdots & \vdots \\ \partial f_n/\partial x_1 & \cdots & \partial f_n/\partial x_n \end{bmatrix}.$$

At each stage of the iterative process, an updated approximate solution vector **x_new** is found from the current approximate solution **x** according to the equation

$$\mathbf{x\_new} = \mathbf{x} - J^{-1}(\mathbf{x})\, F(\mathbf{x}).$$

In the case of a single equation and a single variable, this reduces to the Newton iteration introduced in Section 2.3:

$$x_{k+1} = x_k - \frac{y_k}{y'_k}, \text{ or } x_{(\text{new})} = x_{(\text{old})} - \frac{f(x_{(\text{old})})}{f'(x_{(\text{old})})}.$$

However, evaluating the inverse of the Jacobian matrix is an expensive computation, so in practice the equivalent system of linear equations

$$J(\mathbf{x})\, \mathbf{y} = -F(\mathbf{x}),$$

where $\mathbf{y} = \mathbf{x\_new} - \mathbf{x}$, is solved for the vector **y**, which is used to update **x**. That is,

$$\mathbf{x\_new} = \mathbf{x} + \mathbf{y}.$$

Algorithm for Newton's Method

$$\text{Define } \mathbf{F}(\mathbf{x}) = \begin{bmatrix} f_1(\mathbf{x}) \\ \cdot \\ \cdot \\ \cdot \\ \cdots f_n(\mathbf{x}) \end{bmatrix} \text{ and } J(\mathbf{x}) = \begin{bmatrix} \partial f_1/\partial x_1 & \cdots & \partial f_1/\partial x_n \\ \cdot & \cdots & \cdot \\ \cdot & \cdots & \cdot \\ \cdot & \cdots & \cdot \\ \partial f_n/\partial x_1 & \cdots & \partial f_n/\partial x_n \end{bmatrix}$$

1) Define initial solution vector $\mathbf{x}_{old}$.
2) Define $\mathbf{b} := -\mathbf{F}(\mathbf{x}_{old})$ and $\mathbf{M} := J(\mathbf{x}_{old})$
3) Solve the linear system $\mathbf{M}\,\mathbf{u} = \mathbf{b}$
4) Update solution
   $\mathbf{x}_{new} = \mathbf{x}_{old} + \mathbf{u}$
5) Save current solution in table if desired
   Test for convergence
      check norm of change in solution vector, $\mathbf{u}$, or
      check norm of $\mathbf{F}(\mathbf{x}_{new})$
      maximum number of iterations
   if test for convergence is passed, stop,
      otherwise continue (set $\mathbf{x}_{old}$ = current solution vector $\mathbf{x}_{new}$ in step 1)

5.1.2 Mathcad Function for Newton's Method

$$\text{Newton\_sys}(F, JF, x0, tol, max\_it) := \left| \begin{array}{l} x \leftarrow x0 \\ iter \leftarrow 1 \\ \text{while } iter \leq max\_it \\ \quad \left| \begin{array}{l} y \leftarrow -(JF(x))^{-1} \cdot F(x) \\ xnew \leftarrow x + y \\ (break) \text{ if } |x - xnew| \leq tol \\ x \leftarrow xnew \\ iter \leftarrow iter + 1 \end{array} \right. \\ x \leftarrow xnew \\ x \end{array} \right.$$

Example 5.2 Intersection of a Circle and an Ellipse

To illustrate the use of the algorithm for Newton's method, consider the system of equations introduced in Example 5.1.

ORIGIN := 1

$$F(x) := \begin{bmatrix} (x_1)^2 + (x_2)^2 - 1 \\ (x_1)^2 - x_2 \end{bmatrix} \qquad J(x) := \begin{pmatrix} 2 \cdot x_1 & 2 \cdot x_2 \\ 2 \cdot x_1 & -1 \end{pmatrix} \qquad x0 := \begin{pmatrix} 0.5 \\ 0.5 \end{pmatrix}$$

$$b := -F(x0) \qquad M := J(x0) \qquad y := M^{-1} \cdot b \qquad xn := x0 + y$$

Display result

$$xn = \begin{pmatrix} 0.875 \\ 0.625 \end{pmatrix}$$

To continue iterations, set x0:= xn and copy the line in which b, M, y, and xn are computed.

$$x0 := xn \qquad b := -F(x0) \qquad M := J(x0) \qquad y := M^{-1} \cdot b \qquad xn := x0 + y$$

Display result

$$xn = \begin{pmatrix} 0.79067 \\ 0.61806 \end{pmatrix}$$

Perform one more iteration

$$x0 := xn \qquad b := -F(x0) \qquad M := J(x0) \qquad y := M^{-1} \cdot b \qquad xn := x0 + y$$

Display result

$$xn = \begin{pmatrix} 0.78616 \\ 0.61803 \end{pmatrix}$$

Example 5.3 Newton's Method for a System of Three Equations

To illustrate the use of the Mathcad function Newton_sys, we seek a common solution of the following three equations, which represent the unit sphere centered at the origin, a cylinder with radius 1/2 and axis along the x_2-axis, and a paraboloid of revolution around the x_3-axis.

$$f_1(x_1, x_2, x_3) = x_1^2 + x_2^2 + x_3^2 - 1 = 0,$$
$$f_2(x_1, x_2, x_3) = x_1^2 + x_3^2 - 1/4 = 0,$$
$$f_3(x_1, x_2, x_3) = x_1^2 + x_2^2 - 4x_3 = 0.$$

$$F(x) := \begin{bmatrix} (x_0)^2 + (x_1)^2 + (x_2)^2 - 1 \\ (x_0)^2 + (x_2)^2 - \frac{1}{4} \\ (x_0)^2 + (x_1)^2 - 4x_2 \end{bmatrix} \quad JF(x) := \begin{pmatrix} 2x_0 & 2x_1 & 2x_2 \\ 2x_0 & 0 & 2x_2 \\ 2x_0 & 2x_1 & -4 \end{pmatrix} \quad x0 := \begin{pmatrix} 1 \\ 1 \\ 1 \end{pmatrix}$$

The following results show that the method converges rapidly; there is very little change after the first two iterations.

tol := 0.00001

$$\text{Newton\_sys}(F, JF, x0, tol, 1) = \begin{pmatrix} 0.79167 \\ 0.875 \\ 0.33333 \end{pmatrix}$$

$$\text{Newton\_sys}(F, JF, x0, tol, 2) = \begin{pmatrix} 0.52365 \\ 0.86607 \\ 0.2381 \end{pmatrix}$$

$$\text{Newton\_sys}(F, JF, x0, tol, 3) = \begin{pmatrix} 0.44733 \\ 0.86603 \\ 0.23607 \end{pmatrix}$$

$$\text{Newton\_sys}(F, JF, x0, tol, 4) = \begin{pmatrix} 0.44081 \\ 0.86603 \\ 0.23607 \end{pmatrix}$$

Example 5.4 Positioning a Robot Arm

Consider a two-link robot arm, as introduced in Example 5-B. Let the length of the first link be 5 and the second link of length 6. We wish to find the angles so that the arm will move to the point (10, 4), starting from initial angles of $\alpha = 0.7$ and $\beta = 0.7$.

The system of equations in this case is

$$5\cos(\alpha) + 6\cos(\alpha + \beta) - 10 = 0,$$
$$5\sin(\alpha) + 6\sin(\alpha + \beta) - 4 = 0.$$

Iterations to find position of robot arm using Newton's method.

step	α	β	$\|\Delta x\|$
0	0.7	0.7	
1	−0.59855	1.8339	1.724
2	−0.10782	0.89987	1.0551
3	0.086882	0.53893	0.4101
4	0.14791	0.426	0.12837
5	0.15585	0.41139	0.016621
6	0.15598	0.41114	0.00029053

The initial position and final position of the arm are illustrated in Fig. 5.5.

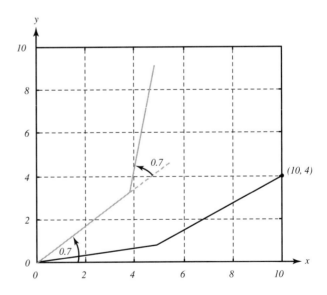

FIGURE 5.5 Initial and final positions of two-link robot arm.

5.2 FIXED-POINT ITERATION FOR NONLINEAR SYSTEMS

It is sometimes convenient to solve a system of nonlinear equations by an iterative process that does not require the computation of partial derivatives. An example of the use of fixed-point iteration for finding a zero of a nonlinear function of a single variable appears in Chapter 1. The extension of this idea to systems is straightforward.

Example 5.5 Fixed-point Iteration for a System of Two Functions

To introduce the use of fixed-point iteration for nonlinear systems, consider the problem of finding a zero of the system

$$f_1(x_1, x_2) = x_1^3 + 10\,x_1 - x_2 - 5 = 0,$$
$$f_2(x_1, x_2) = x_1 + x_2^3 - 10\,x_2 + 1 = 0,$$

by converting these equations to the form $x_1 = g_1(x_1, x_2)$, $x_2 = g_2(x_1, x_2)$, in the following manner:

$$x_1 = -0.1\,x_1^3 + 0.1\,x_2 + 0.5,$$
$$x_2 = 0.1\,x_1 + 0.1\,x_2^3 + 0.1.$$

The graphs of the equations $f_1 = 0$ and $f_2 = 0$ are shown in Fig. 5.6.

The computations in a Mathcad worksheet to solve this problem are shown following a statement of the algorithm.

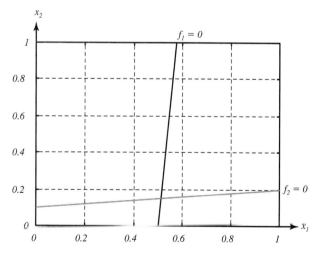

FIGURE 5.6 System of two nonlinear equations.

5.2.1 Step-by-Step Computation

Algorithm for Fixed-point Iteration for Nonlinear Systems

Define $\mathbf{G}(\mathbf{x}) = [g_1(\mathbf{x}), \ldots g_n(\mathbf{x})]^T$
and initial solution vector $\mathbf{x}_0$.

$i = 0$

Begin iterations

$\quad \mathbf{y} = \mathbf{G}(\mathbf{x}_i)$

$\quad \mathbf{x}_{i+1} = \mathbf{y}$

$\quad$ test for convergence

$\quad\quad$ Norm of change in solution vector, $\mathbf{x}_{i+1} - \mathbf{x}_i$;
$\quad\quad$ Norm of $\mathbf{G}(\mathbf{x}_{i+1}) - \mathbf{x}_{i+1}$;
$\quad\quad$ maximum number of iterations;
$\quad\quad$ if test for convergence is passed, exit iteration loop;
$\quad\quad$ otherwise, increment i and continue.

Continue iterations.

Display results.

5.2.2 Mathcad Function for Fixed-point Iteration for Nonlinear Systems

$$\text{Fixed\_pt\_sys}(G, x0, \text{tol}, \text{max\_it}) := \begin{array}{|l} x \leftarrow x0 \\ \text{iter} \leftarrow 1 \\ \text{while iter} \leq \text{max\_it} \\ \quad \begin{array}{|l} \text{xnew} \leftarrow G(x) \\ \text{if } |x - \text{xnew}| \leq \text{tol} \\ \quad \begin{array}{|l} x \leftarrow \text{xnew} \\ \text{break} \end{array} \\ x \leftarrow \text{xnew} \\ \text{iter} \leftarrow \text{iter} + 1 \end{array} \\ \text{xnew} \end{array}$$

Example 5.5 Fixed-point Iteration, continued.

To illustrate the use of the fixed-point algorithm, consider the problem of finding the zero of the system F(x) = 0, for

$$F(x) := \begin{bmatrix} (x_0)^3 + 10 \cdot x_0 - x_1 - 5 \\ x_0 + (x_1)^3 - 10 \cdot x_1 + 1 \end{bmatrix}$$

by finding the fixed-point of G(x) = x, for G(x) and the initial solution vector x0 given below

$$G(x) := \begin{bmatrix} -0.1(x_0)^3 + 0.1 \cdot x_1 + 0.5 \\ 0.1 \cdot x_0 + 0.1 \cdot (x_1)^3 + 0.1 \end{bmatrix} \qquad x0 := \begin{pmatrix} 0 \\ 0 \end{pmatrix}$$

First set x_old=x0 and x_new:=G(x_old)

$$x_{old} := x0 \qquad x_{new} := G(x_{old})$$

$$x_{new} = \begin{pmatrix} 0.5 \\ 0.1 \end{pmatrix} \qquad G(x_{new}) = \begin{pmatrix} 0.4975 \\ 0.1501 \end{pmatrix}$$

Consider the change in the solution values, and the difference between xn and G(xn) to decide whether to continue iterations. For the next iteration, set x_old=x_new and continue.

$$x_{old} := x_{new} \qquad x_{new} := G(x_{old})$$

$$x_{new} = \begin{pmatrix} 0.4975 \\ 0.1501 \end{pmatrix} \qquad G(x_{new}) = \begin{pmatrix} 0.5027 \\ 0.15009 \end{pmatrix}$$

$$x_{old} := x_{new} \qquad x_{new} := G(x_{old})$$

$$x_{new} = \begin{pmatrix} 0.5027 \\ 0.15009 \end{pmatrix} \qquad G(x_{new}) = \begin{pmatrix} 0.50231 \\ 0.15061 \end{pmatrix}$$

Example 5.6 Fixed-point Iteration for a System of Three Equations

As another illustration of the use of fixed-point iteration, we seek a common solution of the following three equations.

$$f_1(x_1, x_2, x_3) = x_1^2 + 50 x_1 + x_2^2 + x_3^2 - 200 = 0,$$
$$f_2(x_1, x_2, x_3) = x_1^2 + 20 x_2 + x_3^2 - 50 = 0,$$
$$f_3(x_1, x_2, x_3) = -x_1^2 - x_2^2 + 40 x_3 + 75 = 0.$$

These equations can be written compactly in vector form as $\mathbf{f}(\mathbf{x}) = 0$, where $\mathbf{x}$, $\mathbf{f}$ and $\mathbf{0}$ are vectors in R^3. There are many possibilities for converting this system to a fixed-point iteration form $\mathbf{x} = \mathbf{g}(\mathbf{x})$, but convergence depends on the magnitude of the partial derivatives of $\mathbf{g}$ being sufficiently small, so we choose to rewrite the system as

$$50 x_1 = -x_1^2 - x_2^2 - x_3^2 + 200,$$
$$20 x_2 = -x_1^2 \qquad - x_3^2 + 50,$$
$$40 x_3 = x_1^2 + x_2^2 \qquad - 75.$$

The desired iterative form is obtained by dividing each equation by the coefficient of the variable on the left-hand side:

$$x_1 = g_1(x_1, x_2, x_3) = -0.02\, x_1^2 - 0.02\, x_2^2 - 0.02\, x_3^2 + 4,$$
$$x_2 = g_2(x_1, x_2, x_3) = -0.05\, x_1^2 \qquad - 0.05\, x_3^2 + 2.5,$$
$$x_3 = g_3(x_1, x_2, x_3) = 0.025\, x_1^2 + 0.025\, x_2^2 \qquad - 1.875.$$

To solve the system F(x) = 0 using fixed point iteration,
i.e. using G(x) = x, for

$$F(x) := \begin{bmatrix} (x_0)^2 + 50 x_0 + (x_1)^2 + (x_2)^2 - 200 \\ (x_0)^2 + 20 x_1 + (x_2)^2 - 50 \\ -(x_0)^2 - (x_1)^2 + 40 x_2 + 75 \end{bmatrix}$$

$$G(x) := \begin{bmatrix} -0.02 \cdot (x_0)^2 + -0.02 \cdot (x_1)^2 - 0.02 \cdot (x_2)^2 + 4 \\ -0.05 \cdot (x_0)^2 + -0.05 \cdot (x_2)^2 + 2.5 \\ 0.025(x_0)^2 + 0.025(x_1)^2 - 1.875 \end{bmatrix}$$

with starting value, x0 and convergence parameters tol and max_it

$$x0 := \begin{pmatrix} 1 \\ 1 \\ 1 \end{pmatrix} \qquad \text{tol} := 0.000001 \qquad \text{max\_it} := 5$$

we define

y := Fixed_pt_sys(G, x0, tol, max_it)

display the result, and check the accuracy of the result by computing F(y), which should be the zero vector, and G(y) which should be equal to y

$$y = \begin{pmatrix} 3.63657 \\ 1.73818 \\ -1.47384 \end{pmatrix} \qquad F(y) = \begin{pmatrix} 0.2465 \\ 0.16053 \\ -0.1997 \end{pmatrix} \qquad G(y) = \begin{pmatrix} 3.63164 \\ 1.73016 \\ -1.46885 \end{pmatrix}$$

Extending the computations to 10 iterations, we find much better results:

y := Fixed_pt_sys(G, x0, tol, 10)

$$y = \begin{pmatrix} 3.63282 \\ 1.73205 \\ -1.47005 \end{pmatrix} \qquad F(y) = \begin{pmatrix} -7.90547 \times 10^{-4} \\ -5.11169 \times 10^{-4} \\ 6.41758 \times 10^{-4} \end{pmatrix} \qquad G(y) = \begin{pmatrix} 3.63283 \\ 1.73208 \\ -1.47007 \end{pmatrix}$$

Using the preceding Mathcad function and a starting estimate of (1, 1, 1), we find that the fixed-point iteration converges to the solution (3.633, 1.732, −1.47).

Chapter 5 Systems of Equations and Inequalities

5.2.3 Discussion

The conditions that guarantee a fixed-point for the vector function $\mathbf{g}(\mathbf{x})$ are similar to those presented in Chapter 1 for a fixed-point of a function of 1 variable. We consider the extension of the interval $a \le x \le b$ in R^1 into R^n by assuming that $\mathbf{g}(\mathbf{x})$ is defined on an n-dimensional rectangle D, the set of all points $(x_1, x_2, \ldots, x_n)$ such that $a_i \le x_i \le b_i$ for some constants $a_1 \ldots a_n$ and $b_1 \ldots b_n$. We assume further that all components of the Jacobian matrix $\mathbf{G}(\mathbf{x})$ are continuous on D.

Fixed-point Convergence Theorem for R^n

If $\mathbf{g}(\mathbf{x})$ maps D into D, then $\mathbf{g}$ has a fixed-point in D. In other words, if $\mathbf{g}(\mathbf{x})$ is in D whenever $\mathbf{x}$ is in D, then there is some point $\mathbf{p}$ in D such that $\mathbf{p} = \mathbf{g}(\mathbf{p})$.

If $\|\mathbf{G}(\mathbf{p})\|_\infty < 1$ then the sequence of approximations to the fixed-point, defined by

$$\mathbf{x}^{(k+1)} = \mathbf{g}(\mathbf{x}^{(k)}) \qquad (***)$$

converges, as long as the initial point $\mathbf{x}^{(0)}$ is sufficiently close to the fixed-point $\mathbf{p}$. The matrix norm $\|\mathbf{G}(\mathbf{p})\|_\infty$ is the maximum of the row sums of $\mathbf{G}$. (For a proof of this theorem, see Atkinson, 1989.)

The following result is a corollary to the fixed-point convergence theorem.

If there is a constant $K < 1$ such that for every $\mathbf{x}$ in D,

$$\left|\frac{\partial g_i(\mathbf{x})}{\partial x_j}\right| \le \frac{K}{n} \quad \text{for each } i = 1 \ldots n, \text{ and each } j = 1 \ldots n,$$

then, for any initial point $\mathbf{x}^{(0)}$ in D, the sequence of approximations to the fixed-point defined by (***) converges.

A bound on the error at the mth step is given by

$$\|\mathbf{x}^{(m)} - \mathbf{p}\|_\infty \le \frac{K^m}{1-K}\|\mathbf{x}^{(1)} - \mathbf{x}^{(0)}\|_\infty$$

These results, a special case of the contraction mapping theorem, are proved in more advanced numerical analysis texts, e.g., Ortega (1972) and Atkinson (1989).

To apply the fixed-point convergence theorem to Example 5.5, i.e.,

$$x_1 = -0.1\, x_1^3 + 0.1\, x_2 + 0.5 = g_1(x_1, x_2),$$

$$x_2 = 0.1\, x_1 + 0.1\, x_2^3 + 0.1 = g_2(x_1, x_2),$$

we first check that $\mathbf{g}(\mathbf{x})$ maps the rectangle $0 \le x_1 \le 1, 0 \le x_2 \le 1$ into itself; that is, for $0 \le x_1, x_2 \le 1$, we have $0 \le g_1, g_2 \le 1$. In fact $0.4 \le g_1 \le 0.6$ and $0.1 \le g_2 \le 0.3$. We also investigate the partial derivatives

$$\frac{\partial g_1(\mathbf{x})}{\partial x_1} = -0.3\, x_1^2, \quad \frac{\partial g_1(\mathbf{x})}{\partial x_2} = 0.1, \quad \frac{\partial g_2(\mathbf{x})}{\partial x_1} = 0.1, \text{ and } \frac{\partial g_2(\mathbf{x})}{\partial x_2} = 0.3\, x_2^2,$$

which are all less than 0.5 (for $0 \le x_1, x_2 \le 1$), as is required for the corollary.

5.3 MINIMUM OF A NONLINEAR FUNCTION

In this section, we turn our attention to the problem of finding a minimum of a scalar function of several variables. Minimization problems are important in many applications. Also, the problem of finding the common zeros of several nonlinear functions can be converted into a minimization problem, as illustrated in Example 5.7. An interesting geometric problem is presented in Example 5.8.

5.3.1 Step-by-Step Computation of Minimization by Gradient Descent

The process is described in the following algorithm and implemented in the Mathcad function, GradDesc, presented in the next section. The motivation for the method is described following the algorithm.

Algorithm for Minimization by Gradient Descent

Define $f(\mathbf{x}) = f(x_1, x_2, \ldots x_n)$
$\mathbf{g}(\mathbf{x}) = $ negative gradient vector of $f(\mathbf{x}) = (g_1(\mathbf{x}), \ldots, g_n(\mathbf{x}))$

$$g_1 = -\frac{\partial f}{\partial x_1}, \ldots, g_n = -\frac{\partial f}{\partial x_n}$$

$\mathbf{x}_0 = $ initial solution vector;
$i = 0$.
Begin iterations
    $\mathbf{dx} = \mathbf{g}(\mathbf{x}_i)$
    $\mathbf{x}_{i+1} = \mathbf{x}_i + \mathbf{dx}$
    $z_0 = f(\mathbf{x}_i)$
    $z_1 = f(\mathbf{x}_{i+1})$
    ch $= z_1 - z_0$
    while ch > 0
        $\mathbf{dx} = 0.5\,\mathbf{dx}$
        if $\|\mathbf{dx}\| < $ tol
            exit the iteration loop (process has stalled out)
        otherwise
            $\mathbf{x}_{i+1} = \mathbf{x}_i + \mathbf{dx}$
            $z_1 = f(\mathbf{x}_{i+1})$
            ch $= z_1 - z_0$
        end
    end
    if $|\mathrm{ch}| < $ tol,   exit iteration loop
    otherwise, increment i and continue
Continue interations

The gradient of a function of several variables is a vector in the direction of most rapid increase, and the negative of the gradient gives the direction of most rapid decrease. A simple gradient search technique starts with an initial estimate of the location of a minimum and determines the direction in which the function is decreasing most rapidly. A new approximate solution is found by moving a specified distance in the direction of the negative gradient. If the function value is lower at the new point, it is accepted as the new solution; if not, a smaller step in the direction of the negative gradient is taken. The step size is reduced until the new point is an improvement (i.e., the function value is smaller at the new point than at the previous point) or until the step size is less than a given tolerance, in which case convergence has been achieved.

5.3.2 Mathcad Function for Minimization by Gradient Descent

$$\text{GradDesc}(myfunc, myfuncg, x0, tol, max\_it) := \begin{vmatrix} x \leftarrow x0 \\ iter \leftarrow 1 \\ \text{while } iter \leq max\_it \\ \quad \begin{vmatrix} dx \leftarrow myfuncg(x) \\ xnew \leftarrow x + dx \\ z0 \leftarrow myfunc(x) \\ z1 \leftarrow myfunc(xnew) \\ ch \leftarrow z1 - z0 \\ \text{while } ch \geq 0 \\ \quad \begin{vmatrix} dx \leftarrow \dfrac{dx}{2} \\ \text{return } xnew \text{ if } |dx| < 0.0000 \\ xnew \leftarrow x + \dfrac{dx}{2} \\ z1 \leftarrow myfunc(xnew) \\ ch \leftarrow z1 - z0 \end{vmatrix} \\ \text{return } xnew \text{ if } |ch| < tol \\ x \leftarrow xnew \\ iter \leftarrow iter + 1 \end{vmatrix} \\ xnew \end{vmatrix}$$

Example 5.7 Finding a Zero of a Nonlinear System by Minimization

To use minimization techniques to find a zero of a system of nonlinear functions, we define a new function that is the sum of the squares of the functions whose common zero is desired. For example, if we want a zero of the functions

$$f(x, y) = x^2 + y^2 - 1 \quad \text{and} \quad g(x, y) = x^2 - y,$$

we define

$$h(x, y) = (x^2 + y^2 - 1)^2 + (x^2 - y)^2.$$

The minimum value of $h(x,y)$ is 0, which occurs when $f(x, y) = 0$ and $g(x, y) = 0$. The derivatives for the gradient are

$$h_x(x, y) = 2(x^2 + y^2 - 1)\,2x + 2(x^2 - y)\,2x,$$
$$h_y(x, y) = 2(x^2 + y^2 - 1)\,2y + 2(x^2 - y)(-1).$$

To find a zero of the nonlinear system introduced in Example 5.1, we seek a minimum of the function f(x) which is the sum of the squares of the two functions who zeros are sought.

$$f(x) := \left[\left(x_0\right)^2 + \left(x_1\right)^2 - 1 \right]^2 + \left[\left(x_0\right)^2 - x_1 \right]^2$$

To use the method of gradient descent, we also define the g(x), which is the opposite of the gradient vector, and the starting estimate of the solution, x0.

$$g(x) := \begin{bmatrix} -\left[\left[4 \cdot x_0 \cdot \left[\left(x_0\right)^2 + \left(x_1\right)^2 - 1\right]\right] + 4 \cdot x_0 \cdot \left[\left(x_0\right)^2 - x_1\right]\right] \\ -\left[4 \cdot x_1 \cdot \left[\left(x_0\right)^2 + \left(x_1\right)^2 - 1\right] - 2 \cdot \left[\left(x_0\right)^2 - x_1\right]\right] \end{bmatrix} \quad x0 := \begin{pmatrix} 0.5 \\ 0.5 \end{pmatrix}$$

With a convergence tolerance of tol

$$\text{tol} := 0.00001$$

the results after 5, 10 and 15 iterations are as follows

$$\text{GradDesc}(f, g, x0, \text{tol}, 5) = \begin{pmatrix} 0.78657 \\ 0.6181 \end{pmatrix}$$

$$\text{GradDesc}(f, g, x0, \text{tol}, 10) = \begin{pmatrix} 0.78605 \\ 0.61802 \end{pmatrix}$$

$$\text{GradDesc}(f, g, x0, \text{tol}, 15) = \begin{pmatrix} 0.78605 \\ 0.61802 \end{pmatrix}$$

We now consider an example in which finding the minimum of a function is the primary objective.

Example 5.8 Optimal Location of a Point in the Plane

Given three points in the plane, we wish to find the location of the point $P = (x, y)$ so that the sum of the squares of the distances from P to the three given points, (x_1, y_1), (x_2, y_2) and (x_3, y_3), is as small as possible. In other words, we need to find the minimum of:

$$f(x, y) = (x - x_1)^2 + (y - y_1)^2 + (x - x_2)^2 + (y - y_2)^2 + (x - x_3)^2 + (y - y_3)^2.$$

The necessary derivatives are

$$f_x(x, y) = 2(x - x_1) + 2(x - x_2) + 2(x - x_3) = 6x - 2x_1 - 2x_2 - 2x_3,$$
$$f_y(x, y) = 2(y - y_1) + 2(y - y_2) + 2(y - y_3) = 6y - 2y_1 - 2y_2 - 2y_3.$$

The initial estimate of $(0.2, 0.2)$ and the results of the first two updates are shown in Fig. 5.7, together with the three given points, $(0, 0)$, $(0, 1)$, and $(1/2, 1)$.

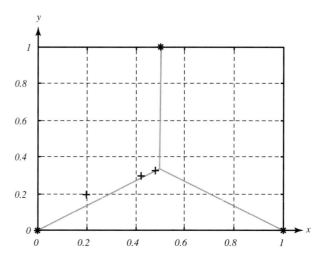

FIGURE 5.7 First three iterations to find optimal location of P.

To find the minimum of the sum of the squares of the distances of a point P from three given points, (xa,ya), (xb,yb) and (xc,yc)

$x_a := 0$ $y_a := 0$ $x_b := 1$ $y_b := 0$ $x_c := 0.5$ $y_c := 1$

we can apply gradient descent to the function.

$$f(x) := \left[(x_0 - x_a)^2 + (x_1 - y_a)^2 + (x_0 - x_b)^2 + (x_1 - y_b)^2 + (x_0 - x_c)^2 + (x_1 - y_c)^2 \right]$$

To use the method of gradient descent, we also define the g(x), which is the opposite of the gradient vector, and the starting estimate of the solution, x0.

$$g(x) := \begin{bmatrix} -(6 \cdot x_0 - 2 \cdot x_a - 2 \cdot x_b - 2 \cdot x_c) \\ -(6 \cdot x_1 - 2 \cdot y_a - 2 \cdot y_b - 2 \cdot y_c) \end{bmatrix} \qquad x0 := \begin{pmatrix} 0.2 \\ 0.2 \end{pmatrix}$$

With a convergence tolerance of tol

$$tol := 0.00001$$

the results after 1, 2, 3, 4, and 5 iterations are as follows

$$\text{GradDesc}(f, g, x0, tol, 1) = \begin{pmatrix} 0.65 \\ 0.4 \end{pmatrix}$$

$$\text{GradDesc}(f, g, x0, tol, 2) = \begin{pmatrix} 0.425 \\ 0.3 \end{pmatrix}$$

$$\text{GradDesc}(f, g, x0, tol, 3) = \begin{pmatrix} 0.5375 \\ 0.35 \end{pmatrix}$$

$$\text{GradDesc}(f, g, x0, tol, 4) = \begin{pmatrix} 0.48125 \\ 0.325 \end{pmatrix}$$

$$\text{GradDesc}(f, g, x0, tol, 5) = \begin{pmatrix} 0.50938 \\ 0.3375 \end{pmatrix}$$

5.4 MATHCAD'S METHODS

The problems discussed in this chapter may be solved using Mathcad's Solve Block structure, which was introduced in Section 4.4. In addition, problems discussed in previous chapters, such as finding the root of a single nonlinear equation (Chapter 2) and solving linear systems (Chapters 3 and 4) may be solved by the same methods. The usc of the functions Find, Maximize, and Minimize is essentially the same for nonlinear systems as for linear system. The function Minerr seeks an approximate solution to a system that may not have an exact solution.

5.4.1 Using the Built-in Functions

A Mathcad Solve Block is created by the following four steps:

1) Provide an initial guess for each of the unknowns in the problem.

$$x_0 := \qquad x_1 := \qquad \cdots \qquad x_{n-1} :=$$

Define the function to be maximized or minimized, if there is one.

2) Type the word "Given" (any style, but in a math region, not in a text region)
3) Type the equations and inequalities to be solved, use icons on Boolean Toolbar, or type symbols as described in Chapter 1. Not equals [≠] and assignment statements are not allowed
4) Call the desired function (Find, Maximize, Minimize, or Minerr); type function name in any style; do not supply numerical values in the call to the function.

To solve a system of equalities and inequalities

$$\text{Find }(x_0, x_1, \ldots x_{n-1})=$$

To solve a system of equalities and inequalities such that function f takes on maximum value

$$\text{Maximize }(f, x_0, x_1, \ldots x_{n-1})=$$

To solve a system of equalities and inequalities such that function f takes on minimum value

$$\text{Minimize }(f, x_0, x_1, \ldots x_{n-1})=$$

To approximately solve a system of equalities and inequalities that may not have an exact solution

$$\text{Minerr }(x_0, x_1, \ldots x_{n-1})=$$

The user can change the default values of the convergence tolerance TOL and constraint tolerance CTOL by setting the desired values somewhere above the Solve Block. Smaller values of TOL require that the difference between successive approximations to the solution be smaller before convergence is achieved. The constraint tolerance determines how closely a constraint must be to be being satisfied for the solution to be acceptable. The default value of CTOL is 10^{-3}.

5.4.2 Understanding the Algorithms

As in many of Mathcad's methods, the functions `Find`, `Maximize`, `Minimize`, or `Minerr` will AutoSelect an appropriate algorithm depending on the form of the equations and inequalities that appear in the problem, or if desired the user may choose a specific method from the menu which is accessed by right-clicking on the function name. Two of the methods used by Mathcad, the simplex method and the conjugate gradient method, were described in Section 4.4. We now introduce the other two methods which are used by these functions, namely the Levenberg-Marquardt and quasi-Newton methods.

Levenberg-Marquardt Method

The Levenberg-Marquardt algorithm is a method for solving nonlinear least-squares problems. It is also the basis for some globally-convergent methods for minimization and zero-finding. These methods are known collectively as the model-trust region approach; individual algorithms include the hook step and dogleg step methods. For a discussion of these methods, see [Dennis and Schnabel, 1983].

The general setting for the Levenberg-Marquardt (L-M) algorithm is the problem of fitting a model (which depends on some parameters) to a set of data. To find the best-fit values for the parameters, we minimize a function which measures the agreement of the data and the model for any particular choice of parameter values. This function, often called a merit function, is designed so that its value is small for parameter values that give good agreement between the model and the data. Since the merit function that is most widely used is the sum of the squares of the differences between the data values and the corresponding model value, these problems are generally known as least-squares problems. We consider the basics of least-squares problems in Chapter 9. The following description of the L-M algorithm is based on the discussion in [Press et al., 1992, pp. 683–685].

The L-M method provides a smooth transition between two approaches to the minimization of the merit function: the steepest descent method (used when the approximate solution is far from the desired value) and the inverse-Hessian method (used when the solution is close to the value that minimizes the merit function). The Hessian matrix of a function is the matrix of the second partial derivatives, i.e., $H(f) = [h_{ij}]$ where $h_{ij} = \dfrac{\partial^2 f}{\partial x_i \partial x_j}$. The L-M method uses a modification of the Hessian matrix, which we denote as $\mathbf{M} = [m_{ij}]$, where

$$m_{ij} = \begin{cases} \dfrac{1}{2} \dfrac{\partial^2 f}{\partial x_i \partial x_j} & \text{for } i \neq j \\[2ex] \dfrac{1}{2} \dfrac{\partial^2 f}{\partial x_i \partial x_j}(1 + \lambda) & \text{for } i = j \end{cases}$$

We also define the vector $\mathbf{b}$, with $b_i = \dfrac{-1}{2} \dfrac{\partial f}{\partial x_i}$.

The basic idea is to start with an initial estimate of the vector **x**, compute a correction vector **d** by solving the linear system **M d = b**, and then increase or decrease λ depending on whether $f(\mathbf{x} + \mathbf{d})$ is larger or smaller than $f(\mathbf{x})$. We continue iterating in this manner until the value of f decreases by an amount that is significantly less than 1.

Levenberg-Marquardt Algorithm

Define the merit function $f(\mathbf{x})$ and the initial estimate **x** of the vector that will minimize f. Compute the matrix **M** and vector **b** as defined above.

    Compute $f(\mathbf{x})$
    Set $\lambda = 0.0001$
**   Solve **M d = b**
    Compute $f(\mathbf{x} + \mathbf{d})$
    If $f(\mathbf{x} + \mathbf{d}) \geq f(\mathbf{x})$
        Set $\lambda = 10\,\lambda$
        return to ** and continue iterations
    If $f(\mathbf{x} + \mathbf{d}) \leq f(\mathbf{x})$
        If $f(\mathbf{x}) - f(\mathbf{x+d}) \leq 0.01$, stop
        Otherwise, set $\lambda = 0.1\,\lambda$
        return to ** and continue iterations

Quasi-Newton Methods

The goal of the quasi-Newton methods, which are also known as variable metric methods, is the same as the goal of a conjugate gradient method, namely to minimize a function of several variables by performing a sequence of minimizations along various well chosen lines. Both conjugate gradient and quasi-Newton methods require that the gradient of the function being minimized can be computed. The two main quasi-Newton methods are the Davidon-Fletcher-Powell (DFP) method (sometimes called just Fletcher-Powell) and the Broyden-Fletcher-Goldfarb-Shanno (BFGS) method. The methods differ only in some details; the following discussion is based on the presentation in [Press et al., 1992, pp. 425–428] and [Acton, 1990, pp. 467–469]. The assumption is that the function $f(\mathbf{x})$ whose minimum is desired can be approximated by a quadratic form, i.e., $f(\mathbf{x}) \approx c - \mathbf{b}\,\mathbf{x} + \frac{1}{2}\mathbf{x}\,\mathbf{A}\,\mathbf{x}$. However, we do not know the matrix **A**. The plan is to construct a sequence of approximations $\mathbf{H}_i \to \mathbf{A}^{-1}$, the inverse Hessian matrix.

If $\mathbf{x}_i$ is the approximation at step i to the vector **x** that minimizes f, and $\mathbf{g}(\mathbf{x})$ is the gradient of f, then we can write (to second order)

$$f(\mathbf{x}) = f(\mathbf{x}_i) + (\mathbf{x} - \mathbf{x}_i)\,g(\mathbf{x}_i) + \frac{1}{2}(\mathbf{x} - \mathbf{x}_i)\,\mathbf{A}\,(\mathbf{x} - \mathbf{x}_i);$$

which gives

$$g(\mathbf{x}) = g(\mathbf{x}_i) + \mathbf{A}\,(\mathbf{x} - \mathbf{x}_i).$$

If we knew $\mathbf{A}^{-1}$, we could set $g(\mathbf{x}) = 0$ and solve for $(\mathbf{x} - \mathbf{x}_i)$, as in standard Newton's method. Instead, we use our current approximation to $\mathbf{A}^{-1}$, which remarkably is often better than using the true Hessian. This paradox is a result of the fact that if we are far from a minimum there is no guarantee that the Hessian is positive definite, as it must be to be assured that Newton's method takes us in a direction where the function is decreasing. The sequence of approximations $\mathbf{H}_i$ are constructed so that they are always positive definite.

Quasi-Newton Algorithm

(subscripts denote iteration, not component of vector)
(vectors are column vectors, so that vector transpose is a row vector)
    Initialize
        $\mathbf{H}_1$ = Identity matrix
        $\mathbf{x}_1$ = initial estimate of minimum point
    Define the required functions
        $f(\mathbf{x})$ (function to be minimized)
        $\mathbf{g}(\mathbf{x})$ (gradient of $f(\mathbf{x})$)
    Compute
        $f_1 = f(\mathbf{x}_1)$
        $\mathbf{g}_1 = \mathbf{g}(\mathbf{x}_1)$
    For $i = 1, \ldots n$
        $\mathbf{s}_i = -\mathbf{H}_i \mathbf{g}_i$ (downhill direction for search)
        Find α_i so that $f(\mathbf{x}_i + \alpha_i \mathbf{s}_i)$ is minimal (along the direction $\mathbf{s}_i$)
        $\mathbf{d}_i = \alpha_i \mathbf{s}_i$
        $\mathbf{x}_{i+1} = \mathbf{x}_i + \mathbf{d}_i$
        $f_{i+1} = f(\mathbf{x}_{i+1})$
        $\mathbf{g}_{i+1} = \mathbf{g}(\mathbf{x}_{i+1})$
        $\mathbf{y}_i = \mathbf{g}_{i+1} - \mathbf{g}_i$

$$\mathbf{A}_i = \frac{\mathbf{d}_i \mathbf{d}_i^T}{\mathbf{d}_i^T \mathbf{y}_i} \quad \text{(numerator is matrix, denominator is scalar)}$$

$$\mathbf{B}_i = \frac{(\mathbf{H}_i \mathbf{y}_i)(\mathbf{H}_i \mathbf{y}_i)^T}{\mathbf{y}_i^T \mathbf{H}_i \mathbf{y}_i} \quad \text{(numerator is matrix, denominator is scalar)}$$

        $\mathbf{H}_{i+1} = \mathbf{H}_i + \mathbf{A}_i - \mathbf{B}_i$
    Stop if either $i > n$ (too many iterations)
        or if $\mathbf{d}_i$ and $\mathbf{s}_i$ are sufficiently small
    Otherwise
        Increment index i and continue

Efficient line search procedures (minimization along a ray) are discussed extensively in the literature. See [Acton, 1990] and [Press et al., 1992].

SUMMARY

The general nonlinear system

$$f_1(x_1, x_2, x_3, \ldots, x_n) = 0,$$
$$f_2(x_1, x_2, x_3, \ldots, x_n) = 0,$$
$$f_3(x_1, x_2, x_3, \ldots, x_n) = 0,$$
$$\ldots$$
$$f_n(x_1, x_2, x_3, \ldots, x_n) = 0,$$

is written in vector form as $F(\mathbf{x}) = \mathbf{0}$.

Newton's Method The update for Newton's method is given by

$$\mathbf{x\_new} = \mathbf{x} - \mathbf{J}^{-1}(\mathbf{x})\,\mathbf{F}(\mathbf{x}).$$

To avoid computing the inverse of the Jacobian, the equivalent system of linear equations

$$J(\mathbf{x})\,\mathbf{y} = -\mathbf{F}(\mathbf{x}), \text{ where } \mathbf{y} = \mathbf{x\_new} - \mathbf{x},$$

is solved for the vector $\mathbf{y}$, which is used to update $\mathbf{x}$. That is,

$$\mathbf{x\_new} = \mathbf{x} + \mathbf{y}.$$

Fixed-point Iteration The system $F(\mathbf{x}) = \mathbf{0}$ may be converted to fixed-point form $\mathbf{x} = \mathbf{g}(\mathbf{x})$ in many different ways.

The fixed-point iteration $\mathbf{x}^{(k+1)} = \mathbf{g}(\mathbf{x}^{(k)})$ will converge if there is a region D, such that $\mathbf{g}(D)$ is in D, and the Jacobian of g satisfies $\|\mathbf{G}(\mathbf{p})\|_\infty < 1$.

Minimization of a Scalar Function of Several Variables The problem of solving the nonlinear system $F(\mathbf{x}) = \mathbf{0}$ may be converted to a minimization problem by defining

$$h(\mathbf{x}) = (F_1(\mathbf{x}))^2 + (F_2(\mathbf{x}))^2 + \ldots + (F_n(\mathbf{x}))^2$$

The minimum value of $h(\mathbf{x})$ is 0, which occurs when each of the component functions of $F(\mathbf{x})$ is zero.

Minimization Problems Occur in Many Areas of Application A simple gradient search for the minimum of $h(\mathbf{x})$ begins with an initial estimate $\mathbf{x}^{(0)}$; at each iteration the approximate solution is moved a small distance in the direction of the negative gradient of h. Care must be taken in determining the distance the solution should be moved.

SUGGESTIONS FOR FURTHER READING

The following texts examine numerical methods applied to nonlinear functions of several variables:

Acton, F. S., *Numerical Methods That (usually) Work*, Harper and Row, New York, 1970.

Ortega, J. M. *Numerical Analysis—A Second Course*, Academic Press, New York, 1972.

Press, W. H., S. A. Teukolsky, W. T. Vetterling, and B. P. Flannery, *Numerical Recipes in C, The Art of Scientific Computing*, (2d ed.), Cambridge University Press, Cambridge 1992.

Reinboldt, W. C., *Methods for Solving Systems of Nonlinear Equations*, SIAM, Philadelphia, 1974.

For further reading on the Levenberg-Marquardt method, see

Marquardt, D. W., *Journal of the Society for Industrial and Applied Mathematics,* vol. 11, pp. 431–441, 1963.

Dennis, J. E. and Schnabel, R. B, *Numerical Methods for Unconstrained Optimization and Nonlinear Equations,* Prentice-Hall, Englewood Cliffs, NJ, 1983.

For discussion of quasi-Newton methods, see

Polak, E., *Computational Methods in Optimization*, Academic Press, New York, 1971, section 2.3.

Jacobs, D. A. H. (ed.) *The State of the Art in Numerical Analysis*, Academic Press, London, 1977, Chapter 3.1.7 (by K. W. Brodie).

The following are among the well-respected texts that provide a more in-depth treatment of multivariable analysis:

Dillon, W. R. and M. Goldstein, *Multivariate Analysis: Methods and Applications*, John Wiley and Sons, New York, 1984.

Hair, J. F., R. E. Anderson, R. L. Tatham, and W. Black, *Multivariate Data Analysis*, (5th ed.), Prentice Hall, Englewood Cliffs, NJ, 1998.

Johnson, R. A. and D. W. Wichern, *Applied Multivariate Statistical Analysis*, (4th ed.), Prentice Hall, Englewood Cliffs, NJ, 1998.

Kachigan, S. K., *Multivariate Statistical Analysis: A Conceptual Introduction*, (2d ed.), Radius Press, New York, 1991.

Mardia, K. V., *Multivariate Analysis*, Academic Press, London, 1980.

Morrison, D. F., *Multivariate Statistical Methods*, (3d ed.), McGraw-Hill, New York, 1990.

PRACTICE THE TECHNIQUES

For Problems P5.1 to P5.10, solve the nonlinear system
    a. using Newton's method
    b. by finding the minimum of the function $f^2 + g^2$ (or $f^2 + g^2 + h^2$)

P5.1 $f(x, y) = x^2 - \sqrt{3}\,xy + 2y^2 - 10 = 0$
$g(x, y) = 4x^2 + 3\sqrt{3}\,xy + y^2 - 22 = 0$

P5.2 $f(x, y) = x^2 - \sqrt{3}xy + 2y^2 - 10 = 0$
$g(x, y) = x^2 - \sqrt{3}\,xy + 2 = 0$

P5.3 $f(x,y) = 4x^2 + 3\sqrt{3}\,xy + y^2 - 22 = 0$
$g(x,y) = x^2 - \sqrt{3}\,xy + 2 = 0$

P5.4 $f(x,y) = -x^2 + xy - y + 7x - 11 = 0$
$g(x,y) = x^2 + y^2 - 9 = 0$

P5.5 $f(x,y) - x^2 + 4y^2 - 16 = 0$
$g(x,y) = xy^2 - 4 = 0$

P5.6 $f(x,y) = x^3 y = y = 2x^3 + 16 = 0$
$g(x,y) = x - y^2 + 1 = 0$

P5.7 $f(x,y,z) = xyz - 1 = 0$
$g(x, y, z) = x^2 + y^2 + z^2 - 4 = 0$
$h(x,y,z) = x^2 + 2y^2 - 3 = 0$

P5.8 $f(x,y,z) = x^2 + 4y^2 + 9z^2 - 36 = 0$
$g(x,y,z) = x^2 + 9y^2 - 47 = 0$
$h(x,y,z) = x^2 z - 11 = 0$

P5.9 $f(x,y,z) = x^2 + 2y^2 + 4z^2 - 7 = 0$
$g(x,y,z) = 2x^2 + y^3 + 6z - 10 = 0$
$h(x,y,z) = xyz + 1 = 0$

P5.10 $f(x,y,z) = x^2 + y^2 + z^2 - 14 = 0$
$g(x,y,z) = x^2 + 2y^2 - 9 = 0$
$h(x,y,z) = x - 3y^2 + z^2 = 0$

For Problems P5.11 to P5.15, solve the nonlinear system
    a. using Newton's method
    b. by finding the minimum of the function $f^2 + g^2$ (or $f^2 + g^2 + h^2$)
    c. using fixed point iteration

P5.11 $f(x,y,z) = x^3 - 10x + y - x + 3 = 0$
$g(x,y,z) = y^3 + 10y - 2x - 2z - 5 = 0$
$h(x,y,z) = x + y - 10z + 2\sin(z) + 5 = 0$

P5.12 $f(x,y,z) = x^2 + 20x + y^2 + z^2 - 20 = 0$
$g(x,y,z) = x^2 + 20y + z^2 - 20 = 0$
$h(x,y,z) = x^2 + y^2 - 40z = 0$

P5.13 $f(x,y,z) = x^2 + y^2 + z^2 + 10x - 4 = 0$
$g(x,y,z) = x^2 - y^2 + z^2 + 10y - 5 = 0$
$h(x,y,z) = x^2 + y^2 - z^2 + 10z - 6 = 0$

P5.14 $f(x,y,z) = 10x - x^2 - y - z^2 - 4 = 0$
$g(x,y,z) = 10y - x^2 - y^2 - z - 5 = 0$
$h(x,y,z) = 10z - x - y^2 - z^2 - 6 = 0$

P5.15 $f(x,y,z) = 10x - x^3 + y^2 - z - 5 = 0$
$g(x,y,z) = 10y + 0.5x^2 - y^2 - z - 3 = 0$
$h(x,y,z) = 10z - x - y^2 - z^3 - 6 = 0$

EXPLORE SOME APPLICATIONS

A5.1 Find the minimum of the following function:

$$f(x,y) = -y + x^{-1} + \frac{y^2 - y + 1}{(y - y^2)(1 - x)}$$

Find the minimum by setting the partial derivatives, f_x and f_y equal to zero and solving the resulting nonlinear system using Newton's method. Investigate the effect of different starting estimates; $x = y = 0.5$ is one suitable choice.

A5.2 The equations for the optimal two-stage finite-difference scheme for parabolic PDE are

$$12\,r\,q = 6\,r - 1$$
$$60\,r^2 - 180\,r^2\,q - 30\,r\,q = 1.$$

Compare your result to the exact solution, which is $r = \dfrac{\sqrt{5}}{10}$ and $q = \dfrac{3 - \sqrt{5}}{6}$. (See Ames, 1992, p. 65.)

A5.3 Solve the following nonlinear system using Newton's method (this gives the coefficients and evaluation points for Gauss-Legendre quadrature, $n = 2$, which are discussed in Chapter 11).

$$2 = a_1 + a_2 \qquad 0 = a_1 x_1 + a_2 x_2$$
$$\frac{2}{3} = a_1 x_1^2 + a_2 x_2^2 \qquad 0 = a_1 x_1^3 + a_2 x_2^3$$

A5.4 Find the roots of the steady-state of the Lorenz equations.

$$y - x = 0$$
$$5x - y - xz = 0$$
$$xy - 16z = 0$$

(See Garcia, 1994, p. 109 for discussion of a similar problem.)

A5.5 The steady-state of the concentration of two chemical species in an oscillatory chemical system described by the Brusselator model is given by the nonlinear system:

$$0 = A + x^2 y - (B + 1)x$$
$$0 = Bx - x^2 y$$

Find the solution for the following values of the parameters, A and B:
 a) $B = 1, A = 1$
 b) $B = 3, A = 1$
 c) $B = 2, A = 1$.

(See Garcia, 1994, p. 111 for discussion of a similar problem.)

A5.6 Find the angles of a two-link robot arm that enable it to reach the point (3, 4); (the lengths of the two links are $d_1 = 5$, $d_2 = 6$). Use the starting values for the angles of $a = 1, b = 1$.

A5.7 Find the location of two points, (x_1, x_2) and (x_3, x_4) so that the sum of the distances squared required to link the 4 given points by paths as shown in the figure is minimal.

p1 = [0 0]; p2 = [1.8 0];
p3 = [1.5 1]; p4 = [0.3 1.6];

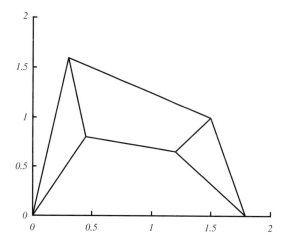

A5.8 Find the location of two points, (x_1, x_2) and (x_3, x_4) so that the sum of the distances (not squared) required to link the 4 given points by paths as shown in the figure is minimal.

p1 = [0 0]; p2 = [1.8 0];
p3 = [1.5 1]; p4 = [0.3 1.6];

EXTEND YOUR UNDERSTANDING

U5.1 to U5.5 For each of the problems P5.11 to P5.15, determine whether the conditions of the fixed point convergence theorem or the corollary are satisfied.

U5.6 Compare the sum of the distances found in A5.8 for the optimal placement of the two points with the sum of the distances if only one interior point is used.

U5.7 Repeat A5.7 and A5.8 using the four given points $(0, 0)$, $(1, 0)$, $(1, 1)$ and $(0, 1)$. Compare your results with those found using only one interior point.

6
LU Factorization

6.1 LU Factorization from Gaussian Elimination

6.2 LU Factorization of Tridiagonal Matrices

6.3 LU Factorization with Pivoting

6.4 Direct LU Factorization

6.5 Applications of LU Factorization

6.6 Mathcad's Methods

In Chapters 3 and 4 we used vector-matrix notation and operations to solve systems of linear equations. In this chapter and the next, we investigate numerical methods for carrying out several important tasks from linear algebra. In this chapter, we consider the factorization of a matrix into the product of a lower triangular matrix **L** and an upper-triangular matrix **U**. For a three-by-three matrix **A**, the problem is to find **L** and **U** so that **LU = A**, i.e.,

$$\begin{bmatrix} \ell_{11} & 0 & 0 \\ \ell_{21} & \ell_{22} & 0 \\ \ell_{31} & \ell_{32} & \ell_{33} \end{bmatrix} \cdot \begin{bmatrix} u_{11} & u_{12} & u_{13} \\ 0 & u_{22} & u_{23} \\ 0 & 0 & u_{33} \end{bmatrix} = \begin{bmatrix} a_{11} & a_{12} & a_{13} \\ a_{21} & a_{22} & a_{23} \\ a_{31} & a_{32} & a_{33} \end{bmatrix}.$$

There are two common methods of finding an LU factorization: Gaussian elimination and direct computation. The first creates an upper triangular matrix during the elimination process; the corresponding lower triangular matrix, with 1's on the diagonal, can be constructed from the multipliers used during the elimination. In the special case of a tridiagonal matrix, an LU factorization can be found very efficiently from the vectors containing the elements on the main diagonal, the elements above the main diagonal, and the elements below the main diagonal.

The second method, direct decomposition, is somewhat more general, in that it allows for the fact that the LU factorization of a given matrix is not unique; the three most common forms correspond to three different assumptions about the diagonal elements of **L** and **U**. The form that places 1's on the diagonal of **L** gives the same factors as are produced by Gaussian elimination. For symmetric matrices, the Cholesky form, which requires that the corresponding diagonal elements of **L** and **U** be equal, i.e., $u_{ii} = \ell_{ii}$, is especially useful, since it preserves the symmetry and produces a factorization with $\mathbf{L} = \mathbf{U}^T$. It can be shown that for a real, symmetric, positive definite matrix, the factorization can be accomplished without complex arithmetic.

An LU factorization of **A** can be used to solve linear systems of equations efficiently, especially when the system **A x = b** must be solved repeatedly using a given matrix **A** with different values of **b** that are not known in advance. LU factorization can also be used to find the inverse and the determinant of **A**.

Example 6-A Circuit Analysis Application

Consider the problem of finding the currents in different parts of an electrical circuit, as shown originally in Figure 3.1. Suppose now that we want to investigate the effect of changing the size of the voltage drop and even its position, placing it perhaps in one of the other two subloops of the circuit. (See Fig. 6.1.) This corresponds to changing the right-hand side of the system of equations developed in Chapter 3.

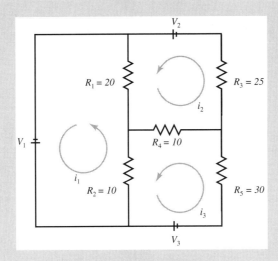

FIGURE 6.1 Simple electrical circuit.

The equations for the three loops can be written in a more general form as follows:

Flow around left loop

$$20(i_1 - i_2) + 10(i_1 - i_3) = V_1,$$

Flow around upper-right loop:

$$25 i_2 + 10(i_2 - i_3) + 20(i_2 - i_1) = V_2,$$

Flow around lower-right loop

$$30 i_3 + 10(i_3 - i_2) + 10(i_3 - i_1) = V_3.$$

In Figure 3.1 we had $V_1 = 0$, $V_2 = 0$, and $V_3 = 200$.
We can solve the new circuit equations efficiently by finding the LU factorization of the coefficient matrix (which does not change as long as we do not modify the sizes or positions of the resistors).

6.1 LU FACTORIZATION FROM GAUSSIAN ELIMINATION

The process of Gaussian elimination forms the basis for finding a very useful representation of a matrix **A** known as an LU factorization. A lower-triangular matrix **L**, with 1's on the diagonal, can be constructed from the multipliers used in Gaussian elimination. (See Chapter 3.) The elimination process transforms the original matrix **A** into an upper-triangular matrix **U**. The lower-triangular matrix **L** is formed by placing the negatives of the multipliers (as used in the multiply-and-add procedure) in the appropriate positions as shown in the following examples. In the following illustrative example, we create new matrices **L** and **U** so that we can verify the result; for efficient use of computer memory, **L** and **U** can be stored in the locations originally occupied by matrix **A**.

Example 6.1 Three-by-three System

We first illustrate the process for the coefficient matrix from Example 3.1; the **L** and **U** matrices are initialized as **L** = **I** and **U** = **A**. The matrices are

$$\mathbf{A} = \begin{bmatrix} 1 & 2 & 3 \\ 2 & 6 & 10 \\ 3 & 14 & 28 \end{bmatrix}, \quad \mathbf{L} = \begin{bmatrix} 1 & 0 & 0 \\ 0 & 1 & 0 \\ 0 & 0 & 1 \end{bmatrix}, \quad \mathbf{U} = \begin{bmatrix} 1 & 2 & 3 \\ 2 & 6 & 10 \\ 3 & 14 & 28 \end{bmatrix}.$$

Step 1: The first row of **U** remains unchanged.
Multiply the first row by −2, and add the result to the second row. Store the negative of the multiplier in the first column, second row, of **L**. Also, multiply the first row by −3, and add the result to the third row. Store the negative of the multiplier in the first column, third row, of **L**. The resulting matrices are

$$\mathbf{L} = \begin{bmatrix} 1 & 0 & 0 \\ 2 & 1 & 0 \\ 3 & 0 & 1 \end{bmatrix}, \quad \mathbf{U} = \begin{bmatrix} 1 & 2 & 3 \\ 0 & 2 & 4 \\ 0 & 8 & 19 \end{bmatrix}.$$

Step 2: The first and second rows of **U** remain unchanged.
Multiply the second row by −4, and add the result to the third row. Store the negative of the multiplier in the second column, third row, of **L**. The matrices that result are

$$\mathbf{L} = \begin{bmatrix} 1 & 0 & 0 \\ 2 & 1 & 0 \\ 3 & 4 & 1 \end{bmatrix}, \quad \mathbf{U} = \begin{bmatrix} 1 & 2 & 3 \\ 0 & 2 & 4 \\ 0 & 0 & 3 \end{bmatrix}.$$

Multiply **L** by **U** and verify the result:

$$\mathbf{LU} = \begin{bmatrix} 1 & 0 & 0 \\ 2 & 1 & 0 \\ 3 & 4 & 1 \end{bmatrix} \cdot \begin{bmatrix} 1 & 2 & 3 \\ 0 & 2 & 4 \\ 0 & 0 & 3 \end{bmatrix} = \begin{bmatrix} 1 & 2 & 3 \\ 2 & 6 & 10 \\ 3 & 14 & 28 \end{bmatrix} = \mathbf{A}.$$

6.1.1 A Step-by-Step Procedure for LU Factorization

Algorithm for LU Factorization Using Gaussian Elimination

Initialize:
  $\mathbf{L} = \mathbf{identity}(n)$ (n-by-n identity matrix)
  $\mathbf{U} = \mathbf{A}$.
Begin computation

1) For $i = 2, \ldots n$ (each row of matrix $\mathbf{U}$ after the first row)
    Form multiplier: $m_{i,1} = -U_{i,1}/U_{1,1}$
    Transform row i by adding $m_{i,1}$ times row 1 onto to current row i
    For $j = 1, \ldots n$
    $$U_{i,j} = U_{i,j} + m_{i,1} U_{1,j}$$
    end j loop
    Update $\mathbf{L}$ matrix: $L_{i,1} = -m_{i,1}$
  end i loop

2) For $i = 3, \ldots n$ (each row of matrix $\mathbf{U}$ after the second row)
    Form multiplier: $m_{i,2} = -U_{i,2}/U_{2,2}$
    Transform row i by adding $m_{i,2}$ times row 2 onto to current row i
    For $j = 2, \ldots n$
    $$U_{i,j} = U_{i,j} + m_{i,2} U_{2,j}$$
    end j loop
    Update $\mathbf{L}$ matrix: $L_{i,2} = -m_{i,2}$
  end i loop

k^{th} stage:
  For $i = k+1, \ldots n$ (each row of matrix $\mathbf{U}$ below the kth row)
    Form multiplier: $m_{i,k} = -U_{i,k}/U_{k,k}$
    Transform row i by adding $m_{i,k}$ times row k onto to current row i
    For $j = k, \ldots n$
    $$U_{i,j} = U_{i,j} + m_{i,k} U_{k,j}$$
    end j loop
    Update $\mathbf{L}$ matrix: $L_{i,k} = -m_{i,k}$
  end i loop
  Continue to form multiplier, transform rows below k^{th} row, and update the $\mathbf{L}$ matrix through the $k = (n-1)$ st stage.
  The matrix $\mathbf{L}$ is the desired lower-triangular matrix; the transformed matrix $\mathbf{U}$ is the desired upper triangular matrix.

The use of the algorithm for Gaussian elimination is illustrated in the next example.

Example 6.2 Using the Algorithm for LU Factorization

We illustrate the use of the algorithm for LU factorization for the following 4-by-4 matrix A
ORIGIN := 1

$$A := \begin{pmatrix} 4 & 12 & 8 & 4 \\ 1 & 7 & 18 & 9 \\ 2 & 9 & 20 & 20 \\ 3 & 11 & 15 & 14 \end{pmatrix} \qquad L := \begin{pmatrix} 1 & 0 & 0 & 0 \\ 0 & 1 & 0 & 0 \\ 0 & 0 & 1 & 0 \\ 0 & 0 & 0 & 1 \end{pmatrix} \qquad U := A$$

Step 1 (use range variable j to carry out computations for elements in each row)

$j := 1..4$

$$m_{2,1} := \frac{-U_{2,1}}{U_{1,1}} \qquad U_{2,j} := U_{2,j} + m_{2,1} \cdot U_{1,j} \qquad L_{2,1} := -m_{2,1}$$

$$m_{3,1} := \frac{-U_{3,1}}{U_{1,1}} \qquad U_{3,j} := U_{3,j} + m_{3,1} \cdot U_{1,j} \qquad L_{3,1} := -m_{3,1}$$

$$m_{4,1} := \frac{-U_{4,1}}{U_{1,1}} \qquad U_{4,j} := U_{4,j} + m_{4,1} \cdot U_{1,j} \qquad L_{4,1} := -m_{4,1}$$

Step 2 (use range variable j to carry out computations for elements in each row)

$j := 2..4$

$$m_{3,2} := \frac{-U_{3,2}}{U_{2,2}} \qquad U_{3,j} := U_{3,j} + m_{3,2} \cdot U_{2,j} \qquad L_{3,2} := -m_{3,2}$$

$$m_{4,2} := \frac{-U_{4,2}}{U_{2,2}} \qquad U_{4,j} := U_{4,j} + m_{4,2} \cdot U_{2,j} \qquad L_{4,2} := -m_{4,2}$$

Step 3 (use range variable j to carry out computations for elements in each row)

$j := 3..4 \qquad m_{4,3} := \frac{-U_{4,3}}{U_{3,3}} \qquad U_{4,j} := U_{4,j} + m_{4,3} \cdot U_{3,j} \qquad L_{4,3} := -m_{4,3}$

$$L = \begin{pmatrix} 1 & 0 & 0 & 0 \\ 0.25 & 1 & 0 & 0 \\ 0.5 & 0.75 & 1 & 0 \\ 0.75 & 0.5 & 0.25 & 1 \end{pmatrix} \qquad U = \begin{pmatrix} 4 & 12 & 8 & 4 \\ 0 & 4 & 16 & 8 \\ 0 & 0 & 4 & 12 \\ 0 & 0 & 0 & 4 \end{pmatrix} \qquad L \cdot U = \begin{pmatrix} 4 & 12 & 8 & 4 \\ 1 & 7 & 18 & 9 \\ 2 & 9 & 20 & 20 \\ 3 & 11 & 15 & 14 \end{pmatrix}$$

6.1.2 Mathcad Function for LU Factorization Using Gaussian Elimination

The returned value for the following function is a *n*-by-2*n* matrix with **L** in the left-most *n* columns and **U** in the right-most *n* columns. We are assuming that **M** is square.

$$\text{LU\_factor\_G}(M) := \begin{vmatrix} n \leftarrow \text{cols}(M) \\ U \leftarrow M \\ L \leftarrow \text{identity}(n) \\ \text{for } j \in 0..n-2 \\ \quad \text{for } i \in j+1..n-1 \\ \quad \quad L_{i,j} \leftarrow \dfrac{U_{i,j}}{U_{j,j}} \\ \quad \quad \text{for } k \in j..n-1 \\ \quad \quad \quad U_{i,k} \leftarrow U_{i,k} - L_{i,j} \cdot U_{j,k} \\ x \leftarrow \text{augment}(L, U) \\ x \end{vmatrix}$$

Example 6.3 Finding an LU Factorization

To factor the coefficient matrix M of the circuit analysis example in Chapter 3, we first define M:

$$M := \begin{pmatrix} 30 & -20 & -10 \\ -20 & 55 & -10 \\ -10 & -10 & 50 \end{pmatrix}$$

and then call the function LU_factor_G:

$$B := \text{LU\_factor\_G}(M) \qquad B = \begin{pmatrix} 1 & 0 & 0 & 30 & -20 & -10 \\ -0.667 & 1 & 0 & 0 & 41.667 & -16.667 \\ -0.333 & -0.4 & 1 & 0 & 0 & 40 \end{pmatrix}$$

$L := \text{submatrix}(B, 0, \text{rows}(M) - 1, 0, \text{cols}(M) - 1) \qquad U := \text{submatrix}(B, 0, \text{rows}(M) - 1, \text{cols}(M), 5)$

$$L = \begin{pmatrix} 1 & 0 & 0 \\ -0.667 & 1 & 0 \\ -0.333 & -0.4 & 1 \end{pmatrix} \qquad U = \begin{pmatrix} 30 & -20 & -10 \\ 0 & 41.667 & -16.667 \\ 0 & 0 & 40 \end{pmatrix}$$

6.2 LU FACTORIZATION OF TRIDIAGONAL MATRICES

As we found with Gaussian elimination, the LU factorization of a tridiagonal matrix **T** can be accomplished using much less computation (and less computer memory for the storage of the matrices) than for a full matrix **M** of the same size.

6.2.1 Step-by-Step LU Factorizaiton of a Tridiagonal Matrix

Mathcad default indexing on arrays is to start with 0. The following algorithm assumes that the indexing has been changed to start from 1, by defining ORIGIN:=1.

Algorithm for LU Factorization of a Tridiagonal Matrix

> The n-by-n matrix **M** is given by the (1-by-n) vectors **a**, **d**, **b**, where
> **a** is the upper diagonal of matrix **M**, with $a_n = 0$
> **d** is the diagonal of matrix **M**
> **b** is the lower diagonal of matrix **M**, with $b_1 = 0$
> The factorization of **M** consists of
> Lower bidiagonal matrix,
>   main diagonal = $[\,1, 1, \ldots, 1\,]$
>   lower diagonal = **bb**
> Upper bidiagonal matrix
>   main diagonal is **dd**
>   upper diagonal is **a**
> Begin computations
> $bb_1 = 0$
> $dd_1 = d_1$
> for $i = 2$ to n
> $$bb_i = \frac{b_i}{dd_{i-1}}$$
> $$dd_i = d_i - bb_i\, a_{i-1}$$
> end
> Display results

If desired, the multipliers that are stored in vector **bb** and that form the lower diagonal of the matrix **L**, could be written directly into the original vector **L**, and the modified main diagonal of the upper triangular matrix, **dd**, could be written over the original diagonal **d**. We have chosen to use somewhat more vectors than necessary to try to make the process as clear as possible and to allow verification of the result if necessary.

Note that this factorization follows the basic Gaussian elimination process and does not scale the diagonal elements to 1, as in the Thomas method. It is not, in general, possible to obtain an LU factorization with all of the diagonal elements of both **L** and **U** set equal to 1.

6.2.2 Mathcad Function for LU Factorization of a Tridiagonal Matrix

The Mathcad function LU_factor_T given here returns a matrix with the lower diagonal of **L** in the left column and the main diagonal of **U** in the right column. The main diagonal of **L** is all 1's. The upper diagonal of **U** is unchanged.

$$\text{LU\_factor\_T}(a,d,b) := \begin{vmatrix} n \leftarrow \text{length}(d) \\ bb_0 \leftarrow 0 \\ dd_0 \leftarrow d_0 \\ \text{for } i \in 1..n-1 \\ \quad \begin{vmatrix} bb_i \leftarrow \dfrac{b_i}{dd_{i-1}} \\ dd_i \leftarrow d_i - bb_i \cdot a_{i-1} \end{vmatrix} \\ x \leftarrow \text{augment}(bb, dd) \\ x \end{vmatrix}$$

Example 6.4 Finding the LU Factorization of a Tridiagonal Matrix

Find the LU factorization of the tridiagonal matrix defined by the above, below, and main diagonals:

$$a := \begin{pmatrix} 4 \\ 1 \\ 2 \\ 5 \\ 0 \end{pmatrix} \quad b := \begin{pmatrix} 0 \\ 4 \\ -2 \\ 3 \\ 4 \end{pmatrix} \quad d := \begin{pmatrix} 1 \\ 15 \\ 3 \\ 7 \\ 21 \end{pmatrix} \quad B := \text{LU\_factor\_T}(a,d,b) \quad B = \begin{pmatrix} 0 & 1 \\ 4 & -1 \\ 2 & 1 \\ 3 & 1 \\ 4 & 1 \end{pmatrix}$$

Create the L and U matrices and verify result, if desired.

$$L := \begin{pmatrix} 1 & 0 & 0 & 0 & 0 \\ 4 & 1 & 0 & 0 & 0 \\ 0 & 2 & 1 & 0 & 0 \\ 0 & 0 & 3 & 1 & 0 \\ 0 & 0 & 0 & 4 & 1 \end{pmatrix} \quad U := \begin{pmatrix} 1 & 4 & 0 & 0 & 0 \\ 0 & -1 & 1 & 0 & 0 \\ 0 & 0 & 1 & 2 & 0 \\ 0 & 0 & 0 & 1 & 5 \\ 0 & 0 & 0 & 0 & 1 \end{pmatrix} \quad L \cdot U = \begin{pmatrix} 1 & 4 & 0 & 0 & 0 \\ 4 & 15 & 1 & 0 & 0 \\ 0 & -2 & 3 & 2 & 0 \\ 0 & 0 & 3 & 7 & 5 \\ 0 & 0 & 0 & 4 & 21 \end{pmatrix}$$

6.3 LU FACTORIZATION WITH PIVOTING

For problems in which row pivoting must be performed on the coefficient matrix **A** during Gaussian elimination, we can obtain an LU factorization of the permuted matrix **PA**, where **P** is the permutation matrix that represents the row interchanges which occurred during the pivoting.

6.3.1 Step-by-Step Computation

Algorithm for LU Factorization with Row Pivoting

> Initialize
>
> $\quad$ **L** = **identity**(n) $\quad$ *n*-by-*n* identity matrix
> $\quad$ **P** = **identity**(n) $\quad$ *n*-by-*n* identity matrix
>
> Create row vectors $R1, R2, \ldots Rn$, for each row of the matrix **M**.
> Stage $k = 1$.
> $\quad$ Find row that has first element of largest magnitude, call it Rp.
> $\quad$ If $Rp \neq R1$, interchange first and *p*th rows of matrix **M** and matrix **P**.
> $\quad$ For $i = 2, \ldots n$, transform Ri by adding multiple of $R1$ to current Ri.
> $\quad$ Define multiplier: $M_{i,1} = \dfrac{-Ri_1}{R1_1}$
> $\quad$ Transform row i: $\ Ri := Ri + M_{i,1} R1$;
> $\quad$ Update **L** matrix: $\ L_{i,1} = -M_{i,1}$.
> Stage $k = 2 \ldots n-1$.
> $\quad$ Of rows $Rk \ldots Rn$, find the row that has the *k*th element of largest magnitude, call the row Rp.
> $\quad$ If $Rp \neq Rk$, interchange rows k and p in matrix **M** and matrix **P**, also interchange columns $1 \ldots k-1$ of rows k and p in matrix **L**
> $\quad$ For $i = k + 1, \ldots n$, transform Ri by adding multiple of Rk to current Ri.
> $\quad$ Define multiplier: $M_{i,k} = \dfrac{-Ri_k}{Rk_k}$.
> $\quad$ Transform row i: $\ Ri := Ri + M_{i,k} Rk$;
> $\quad$ Update **L** matrix: $\ L_{i,k} = -M_{i,k}$.
> Form matrix **U** from the rows $R1, R2, \ldots Rn$ as
>
> $\quad$ **U** = stack($R1, R2, \ldots Rn$)
>
> Lower-triangular matrix is **L**.

6.3.2 Mathcad Function for LU Factorization with Row Pivoting

The Mathcad function LU_factor_P given here returns a n-by-$3n$ matrix consisting of (from left to right) the columns of **L**, the columns of **U**, and the columns of **P**.

$$\text{LU\_pivot}(M) := \begin{vmatrix} n \leftarrow \text{cols}(M) \\ U \leftarrow M \\ P \leftarrow \text{identity}(n) \\ L \leftarrow \text{identity}(n) \\ \text{for } i \in 0..n-2 \\ \quad \begin{vmatrix} \text{pivot} \leftarrow |U_{i,i}| \\ p \leftarrow i \\ \text{for } j \in i+1..n-1 \\ \quad \begin{vmatrix} \text{if } |U_{j,i}| > \text{pivot} \\ \quad \begin{vmatrix} \text{pivot} \leftarrow |U_{j,i}| \\ p \leftarrow j \end{vmatrix} \end{vmatrix} \\ \text{if } p > i \\ \quad \begin{vmatrix} \text{for } j \in 0..n-1 \\ \quad \begin{vmatrix} \text{temp1} \leftarrow U_{i,j} \\ \text{temp2} \leftarrow P_{i,j} \\ U_{i,j} \leftarrow U_{p,j} \\ P_{i,j} \leftarrow P_{p,j} \\ U_{p,j} \leftarrow \text{temp1} \\ P_{p,j} \leftarrow \text{temp2} \end{vmatrix} \\ \text{for } j \in 0..i-1 \quad \text{if } i \geq 1 \\ \quad \begin{vmatrix} \text{temp3} \leftarrow L_{i,j} \\ L_{i,j} \leftarrow L_{p,j} \\ L_{p,j} \leftarrow \text{temp3} \end{vmatrix} \end{vmatrix} \\ \text{for } m \in i+1..n-1 \\ \quad \begin{vmatrix} L_{m,i} \leftarrow \dfrac{U_{m,i}}{U_{i,i}} \\ \text{for } k \in i..n-1 \\ \quad \begin{vmatrix} U_{m,k} \leftarrow U_{m,k} - L_{m,i} \cdot U_{i,k} \end{vmatrix} \end{vmatrix} \end{vmatrix} \\ x \leftarrow \text{augment}(L, U, P) \\ x \end{vmatrix}$$

Example 6.5 LU Factorization with Pivoting

$$M := \begin{pmatrix} 2 & 6 & 10 \\ 1 & 3 & 3 \\ 3 & 14 & 28 \end{pmatrix}$$

B := LU_pivot(M)

$$B = \begin{pmatrix} 1 & 0 & 0 & 3 & 14 & 28 & 0 & 0 & 1 \\ 0.667 & 1 & 0 & 0 & -3.333 & -8.667 & 1 & 0 & 0 \\ 0.333 & 0.5 & 1 & 0 & 0 & -2 & 0 & 1 & 0 \end{pmatrix}$$

L := submatrix(B,0,2,0,2)   U := submatrix(B,0,2,3,5)

$$L = \begin{pmatrix} 1 & 0 & 0 \\ 0.667 & 1 & 0 \\ 0.333 & 0.5 & 1 \end{pmatrix} \qquad U = \begin{pmatrix} 3 & 14 & 28 \\ 0 & -3.333 & -8.667 \\ 0 & 0 & -2 \end{pmatrix}$$

$$L \cdot U = \begin{pmatrix} 3 & 14 & 28 \\ 2 & 6 & 10 \\ 1 & 3 & 3 \end{pmatrix}$$

P := submatrix(B,0,2,6,8)

$$P \cdot M = \begin{pmatrix} 3 & 14 & 28 \\ 2 & 6 & 10 \\ 1 & 3 & 3 \end{pmatrix}$$

It is worth noting that if we store the elements of **L** below the diagonal in the positions of **A** that have been reduced to zero by the elimination process, the elements of **L** will appear in the correct positions, even when pivoting is used. However, a record of the row interchanges that are performed must be kept so that one can determine which permutation of **A** will be produced by the product **LU**, in order to permute the right-hand side of a system of equations if the permuted LU factorization is used to solve the system. On the other hand, if we store **L** as a separate matrix, then, whenever pivoting occurs, the elements of **L** that have been already been computed must be permuted along with matrix **A**. However, the diagonal elements of **L** are not permuted.

6.3.3 Discussion

The operations of Gaussian elimination can be expressed as actions performed on matrix **A** by certain elementary matrices. First consider the effect of multiplying a matrix by several simple matrices. (We use three-by-three matrices for purposes of illustration.) Multiply the first row of **A** by c, and add the result to the second row to form the new second row:

$$\begin{bmatrix} 1 & 0 & 0 \\ c & 1 & 0 \\ 0 & 0 & 1 \end{bmatrix} \cdot \begin{bmatrix} a_{11} & a_{12} & a_{13} \\ a_{21} & a_{22} & a_{23} \\ a_{31} & a_{32} & a_{33} \end{bmatrix} = \begin{bmatrix} a_{11} & a_{12} & a_{13} \\ d_{21} & d_{22} & d_{23} \\ a_{31} & a_{32} & a_{33} \end{bmatrix}.$$

Here $d_{21} = c\,a_{11} + a_{21}$, $d_{22} = c\,a_{12} + a_{22}$, and $d_{23} = c\,a_{13} + a_{23}$.
For convenience, we write this as **C A = D**.

We also need the inverse of the matrix **C**; i.e., we want $\mathbf{C}^{-1}\mathbf{C} = \mathbf{I}$. By direct calculation it is evident that

$$\begin{bmatrix} 1 & 0 & 0 \\ -c & 1 & 0 \\ 0 & 0 & 1 \end{bmatrix} \cdot \begin{bmatrix} 1 & 0 & 0 \\ c & 1 & 0 \\ 0 & 0 & 1 \end{bmatrix} = \begin{bmatrix} 1 & 0 & 0 \\ 0 & 1 & 0 \\ 0 & 0 & 1 \end{bmatrix},$$

or $\qquad\qquad \mathbf{C}^{-1} \quad\cdot\quad \mathbf{C} \quad=\quad \mathbf{I}.$

We can combine several such operations in a single matrix. For instance, the operations that perform the first stage of Gaussian elimination can be written as

$$\begin{bmatrix} 1 & 0 & 0 \\ m_{21} & 1 & 0 \\ m_{31} & 0 & 1 \end{bmatrix} \cdot \begin{bmatrix} a_{11} & a_{12} & a_{13} \\ a_{21} & a_{22} & a_{23} \\ a_{31} & a_{32} & a_{33} \end{bmatrix} = \begin{bmatrix} a_{11} & a_{12} & a_{13} \\ 0 & a'_{22} & a'_{23} \\ 0 & a'_{32} & a'_{33} \end{bmatrix},$$

where the elements a'_{22}, a'_{23}, a'_{32}, and a'_{33} have been modified by the actions needed to reduce the first column of **A** to zero (below the diagonal element).

To show that we can obtain the LU factorization of **A** as previously illustrated, we define the matrices

$$\mathbf{M}_1 = \begin{bmatrix} 1 & 0 & 0 \\ m_{21} & 1 & 0 \\ m_{31} & 0 & 1 \end{bmatrix} \text{ and } \mathbf{M}_1^{-1} = \begin{bmatrix} 1 & 0 & 0 \\ -m_{21} & 1 & 0 \\ -m_{31} & 0 & 1 \end{bmatrix},$$

where m_{21} and m_{31} are determined as before. In a similar manner, we define

$$\mathbf{M}_2 = \begin{bmatrix} 1 & 0 & 0 \\ 0 & 1 & 0 \\ 0 & m_{32} & 1 \end{bmatrix} \text{ and } \mathbf{M}_2^{-1} = \begin{bmatrix} 1 & 0 & 0 \\ 0 & 1 & 0 \\ 0 & -m_{32} & 1 \end{bmatrix},$$

where m_{32} is found in the manner required to reduce to zero the second column of the transformed matrix **A**, below the diagonal. Thus, $\mathbf{U} = \mathbf{M}_2 \cdot (\mathbf{M}_1 \cdot \mathbf{A})$.

The following identities provide the justification for the LU factorization of **A**

$$\mathbf{M}_1^{-1} \cdot (\mathbf{M}_1 \cdot \mathbf{A}) = \mathbf{A},$$
$$(\mathbf{M}_1^{-1} \cdot \mathbf{M}_2^{-1}) \cdot \{\mathbf{M}_2 \cdot (\mathbf{M}_1 \cdot \mathbf{A})\} = \mathbf{A}.$$

Now consider the elements of $\mathbf{M}_1^{-1} \cdot \mathbf{M}_2^{-1} = \mathbf{L}$,

or
$$\begin{bmatrix} 1 & 0 & 0 \\ -m_{21} & 1 & 0 \\ -m_{31} & 0 & 1 \end{bmatrix} \cdot \begin{bmatrix} 1 & 0 & 0 \\ 0 & 1 & 0 \\ 0 & -m_{32} & 1 \end{bmatrix} = \begin{bmatrix} 1 & 0 & 0 \\ -m_{21} & 1 & 0 \\ -m_{31} & -m_{32} & 1 \end{bmatrix}.$$

Thus, the lower-triangular matrix with 1's on the diagonal and the negatives of the Gaussian elimination multipliers in the appropriate lower triangular positions, together with the upper triangular matrix formed during the elimination process, provides an LU factorization of matrix $\mathbf{A}$.

When row pivoting is used in Gaussian elimination, the matrices $\mathbf{L}$ and $\mathbf{U}$ that are obtained give a factorization of a *permutation* of $\mathbf{A}$, corresponding to the row interchanges performed during pivoting. Row interchanges on matrix $\mathbf{A}$ correspond to multiplying $\mathbf{A}$ on the left by a permutation matrix—that is, a matrix with elements that are either 0 or 1 and with the further property that there is exactly one 1 in each row and in each column. In other words, the matrix $\mathbf{P}$ is a permutation of the identity matrix. Column interchanges are accomplished by multiplying $\mathbf{A}$ on the right by a permutation matrix. The inverse of a permutation matrix is the matrix itself.

For example, to interchange the second and third rows of $\mathbf{A}$, we perform the following multiplication:

$$\begin{bmatrix} 1 & 0 & 0 \\ 0 & 0 & 1 \\ 0 & 1 & 0 \end{bmatrix} \cdot \begin{bmatrix} a_{11} & a_{12} & a_{13} \\ a_{21} & a_{22} & a_{23} \\ a_{31} & a_{32} & a_{33} \end{bmatrix} = \begin{bmatrix} a_{11} & a_{12} & a_{13} \\ a_{31} & a_{32} & a_{33} \\ a_{21} & a_{22} & a_{23} \end{bmatrix}.$$

Now, consider a simple three-by-three example. The operations that perform the first stage of Gaussian elimination can be written as

$$\begin{bmatrix} 1 & 0 & 0 \\ m_{21} & 1 & 0 \\ m_{31} & 0 & 1 \end{bmatrix} \cdot \begin{bmatrix} a_{11} & a_{12} & a_{13} \\ a_{21} & a_{22} & a_{23} \\ a_{31} & a_{32} & a_{33} \end{bmatrix} = \begin{bmatrix} a_{11} & a_{12} & a_{13} \\ 0 & a'_{22} & a'_{23} \\ 0 & a'_{32} & a'_{33} \end{bmatrix}$$

where the new elements $a'_{22}, a'_{23}, a'_{32}$, and a'_{33} are the result of the actions needed to reduce the first column of $\mathbf{A}$ to zero, below the diagonal.

To show that we can obtain the LU factorization of $\mathbf{A}$ as illustrated, we define the matrices $\mathbf{M}_1$ and $\mathbf{M}_1^{-1}$ as above.

For the second stage of Gaussian elimination, we work with the transformed matrix, which is actually $\mathbf{M}_1\mathbf{A}$. If we interchange the second and third rows of $\mathbf{M}_1\mathbf{A}$, we will find $\mathbf{M}_2$ on the basis of the permuted matrix $\mathbf{PM}_1\mathbf{A}$. Accordingly, we define

$$\mathbf{M}_2 = \begin{bmatrix} 1 & 0 & 0 \\ 0 & 1 & 0 \\ 0 & m_{32} & 1 \end{bmatrix} \text{ and } \mathbf{M}_2^{-1} = \begin{bmatrix} 1 & 0 & 0 \\ 0 & 1 & 0 \\ 0 & -m_{32} & 1 \end{bmatrix}$$

where m_{32} is found to reduce the second column of $\mathbf{PM}_1\mathbf{A}$ to zero, below the diagonal.

As above, we first consider the identity

$$\mathbf{M}_1^{-1} \cdot (\mathbf{M}_1 \cdot \mathbf{A}) = \mathbf{A}.$$

The form of $\mathbf{M}_1 \cdot \mathbf{A}$ and $\mathbf{M}_1^{-1}$ are:

$$\mathbf{M}_1 \cdot \mathbf{A} \approx \begin{bmatrix} x & x & x \\ 0 & x & x \\ 0 & x & x \end{bmatrix}, \quad \mathbf{M}_1^{-1} \approx \begin{bmatrix} \# & 0 & 0 \\ \# & \# & 0 \\ \# & 0 & \# \end{bmatrix}$$

Before $\mathbf{M}_2$ is constructed, the second and third rows of the matrix $\mathbf{M}_1 \cdot \mathbf{A}$ are interchanged; we express this interchange as multiplication by the permutation matrix $\mathbf{P}$. Making use of the fact that $\mathbf{P}$ is its own inverse, we have the identity

$$\mathbf{M}_1^{-1} \cdot \mathbf{P} \cdot \mathbf{P} \cdot (\mathbf{M}_1 \cdot \mathbf{A}) = \mathbf{A}.$$

Now, $\mathbf{M}_2$ is found to reduce the (3,2) element in the matrix $\mathbf{P} \cdot (\mathbf{M}_1 \cdot \mathbf{A})$ to zero; we thus have the identity

$$\mathbf{M}_1^{-1} \cdot \mathbf{P} \cdot \mathbf{M}_2^{-1} \cdot \{\mathbf{M}_2 \cdot \mathbf{P} \cdot \mathbf{M}_1 \cdot \mathbf{A}\} = \mathbf{A}.$$

Hence, $\mathbf{M}_2 \cdot \mathbf{P} \cdot \mathbf{M}_1 \cdot \mathbf{A}$ is the upper-triangular matrix formed by the Gaussian elimination. However, $\mathbf{M}_1^{-1} \cdot \mathbf{P} \cdot \mathbf{M}_2^{-1}$ is not lower triangular. In order to construct the lower triangular matrix with the negatives of the multipliers in the right places, we need $\mathbf{P} \cdot \mathbf{M}_1^{-1} \cdot \mathbf{P} \cdot \mathbf{M}_2^{-1}$. This gives

$$\mathbf{P} \cdot \mathbf{M}_1^{-1} \cdot \mathbf{P} \cdot \mathbf{M}_2^{-1} \cdot \{\mathbf{M}_2 \cdot \mathbf{P} \cdot \mathbf{M}_1 \cdot \mathbf{A}\} = \mathbf{P}\mathbf{A}$$

or $\mathbf{L}\mathbf{U} = \mathbf{P}\mathbf{A}$ when row interchanges are performed during elimination. The permutation matrix $\mathbf{P}$ performs the bookkeeping associated with the row interchanges.

Finally, consider the actual elements of $\mathbf{B} = \mathbf{P} \cdot \mathbf{M}_1^{-1} \cdot \mathbf{P}$

$$\begin{bmatrix} 1 & 0 & 0 \\ 0 & 0 & 1 \\ 0 & 1 & 0 \end{bmatrix} \cdot \begin{bmatrix} 1 & 0 & 0 \\ -m_{21} & 1 & 0 \\ -m_{31} & 0 & 1 \end{bmatrix} \cdot \begin{bmatrix} 1 & 0 & 0 \\ 0 & 0 & 1 \\ 0 & 1 & 0 \end{bmatrix} =$$

$$\begin{bmatrix} 1 & 0 & 0 \\ -m_{31} & 0 & 1 \\ -m_{32} & 1 & 0 \end{bmatrix} \cdot \begin{bmatrix} 1 & 0 & 0 \\ 0 & 0 & 1 \\ 0 & 1 & 0 \end{bmatrix} = \begin{bmatrix} 1 & 0 & 0 \\ -m_{31} & 1 & 0 \\ -m_{21} & 0 & 1 \end{bmatrix}$$

We now verify that $\mathbf{P} \cdot \mathbf{M}_1^{-1} \cdot \mathbf{P} \cdot \mathbf{M}_2^{-1} = \mathbf{B} \cdot \mathbf{M}_2^{-1} = \mathbf{L}$:

$$\begin{bmatrix} 1 & 0 & 0 \\ -m_{31} & 1 & 0 \\ -m_{21} & 0 & 1 \end{bmatrix} \cdot \begin{bmatrix} 1 & 0 & 0 \\ 0 & 1 & 0 \\ 0 & -m_{32} & 1 \end{bmatrix} = \begin{bmatrix} 1 & 0 & 0 \\ -m_{31} & 1 & 0 \\ -m_{21} & -m_{32} & 1 \end{bmatrix}$$

Thus, the positions of the elements of $\mathbf{L}$ reflect the interchanges that occurred during elimination; the matrix formed from the product of $\mathbf{L}$ and $\mathbf{U}$ is a permuted form of the original matrix $\mathbf{A}$.

6.4 DIRECT LU FACTORIZATION

An alternative approach to Gaussian elimination for finding the LU factorization of matrix **A** is based on equating the elements of the product **LU** with the corresponding elements of **A**, in a systematic manner. The three most common forms of LU factorization correspond to three different choices for the form of the diagonal elements in **L** and **U**: the diagonal elements of **L** are 1, the diagonal elements of **U** are 1, or the diagonal elements of **L** and **U** are equal.

6.4.1 Direct LU Factorization of a General Matrix

The form of LU factorization that assumes that the diagonal elements of matrix **L** are 1's is known in the literature as Doolittle's method (or sometimes Crout's method). For a three-by-three matrix **A**, the problem is to find matrices **L** and **U** so that **L U = A**:

$$\begin{bmatrix} 1 & 0 & 0 \\ \ell_{21} & 1 & 0 \\ \ell_{31} & \ell_{32} & 1 \end{bmatrix} \cdot \begin{bmatrix} u_{11} & u_{12} & u_{13} \\ 0 & u_{22} & u_{23} \\ 0 & 0 & u_{33} \end{bmatrix} = \begin{bmatrix} a_{11} & a_{12} & a_{13} \\ a_{21} & a_{22} & a_{23} \\ a_{31} & a_{32} & a_{33} \end{bmatrix}.$$

We begin by finding $u_{11} = a_{11}$ and then solving for the remaining elements in the first row of **U** and the first column of **L**. At the second stage, we find u_{22} and then the remainder of the second row of **U** and the second column of **L**. Continuing in this manner, we determine all of the elements of **U** and **L**. This process is implemented in the following Mathcad function and illustrated in Example 6.6.

A similar procedure can be carried out assuming that there are 1's on the diagonal of **U** rather than **L**. This variation is sometimes called Crout's method. We limit our investigation to direct factorization with 1s on the diagonal of **L**.

Algorithm for Direct LU Factorization

Initialize   **U** = **L** = **identity**(n) n-by-n identity matrix
for $k = 1$ to n

$$U_{k,k} = A_{k,k} - \sum_{i=1}^{k-1} L_{k,i} U_{i,k}$$

for $j = k+1$ to n
    for $i = j+1$ to n (for each row below the jth row)

$$U_{k,j} = A_{k,j} - \sum_{i=1}^{k-1} L_{k,i} U_{i,j} \qquad L_{j,k} = \frac{1}{U_{k,k}} \left(A_{j,k} - \sum_{i=1}^{k-1} L_{j,i} U_{i,k} \right)$$

    end
end
end

Mathcad Function for Direct LU Factorization

$$\text{Doolittle}(M) := \begin{array}{|l} n \leftarrow \text{cols}(M) \\ U \leftarrow \text{identity}(n) \\ L \leftarrow \text{identity}(n) \\ \text{for } k \in 0..n-1 \\ \quad \begin{array}{|l} \text{for } j \in k..n-1 \\ \quad \begin{array}{|l} \text{dotu} \leftarrow 0 \\ \text{if } k > 0 \\ \quad \begin{array}{|l} \text{dotl} \leftarrow 0 \\ \text{for } i \in 0..k-1 \\ \quad \begin{array}{|l} \text{dotu} \leftarrow \text{dotu} + L_{k,i} \cdot U_{i,j} \\ \text{dotl} \leftarrow \text{dotl} + L_{j,i} \cdot U_{i,k} \end{array} \end{array} \\ U_{k,j} \leftarrow M_{k,j} - \text{dotu} \\ L_{j,k} \leftarrow \dfrac{M_{j,k} - \text{dotl}}{U_{k,k}} \quad \text{if } k < j \end{array} \end{array} \\ x \leftarrow \text{augment}(L, U) \\ x \end{array}$$

Example 6.6 Direct LU Factorization

$$M := \begin{pmatrix} 1 & 4 & 5 \\ 4 & 20 & 32 \\ 5 & 32 & 64 \end{pmatrix} \qquad B := \text{Doolittle}(M) \qquad B = \begin{pmatrix} 1 & 0 & 0 & 1 & 4 & 5 \\ 4 & 1 & 0 & 0 & 4 & 12 \\ 5 & 3 & 1 & 0 & 0 & 3 \end{pmatrix}$$

$L := \text{submatrix}(B, 0, \text{rows}(M) - 1, 0, \text{cols}(M) - 1) \quad U := \text{submatrix}(B, 0, \text{rows}(M) - 1, \text{cols}(M), 5)$

$$L = \begin{pmatrix} 1 & 0 & 0 \\ 4 & 1 & 0 \\ 5 & 3 & 1 \end{pmatrix} \qquad U = \begin{pmatrix} 1 & 4 & 5 \\ 0 & 4 & 12 \\ 0 & 0 & 3 \end{pmatrix} \qquad L \cdot U = \begin{pmatrix} 1 & 4 & 5 \\ 4 & 20 & 32 \\ 5 & 32 & 64 \end{pmatrix}$$

6.4.2 LU Factorization of a Symmetric Matrix

If the matrix **A** is symmetric, there is a convenient form of LU factorization, called Cholesky factorization, for which the upper-triangular matrix **U** is the transpose of the lower-triangular matrix **L**; i.e., $\mathbf{A} = \mathbf{L}\mathbf{L}^T$. Furthermore, only $n(n+1)/2$ storage locations are required, rather than the usual n^2. For a three-by-three matrix **A**, the problem is to find matrices **L** and **U** so that $\mathbf{L}\mathbf{U} = \mathbf{A}$; that is, we seek

$$\begin{bmatrix} x_{11} & 0 & 0 \\ \ell_{21} & x_{22} & 0 \\ \ell_{31} & \ell_{32} & x_{33} \end{bmatrix} \cdot \begin{bmatrix} x_{11} & u_{12} & u_{13} \\ 0 & x_{22} & u_{23} \\ 0 & 0 & x_{33} \end{bmatrix} = \begin{bmatrix} a_{11} & a_{12} & a_{13} \\ a_{21} & a_{22} & a_{23} \\ a_{31} & a_{32} & a_{33} \end{bmatrix}.$$

where corresponding diagonal elements of **L** and **U** are equal ($u_{ii} = \ell_{ii} = x_{ii}$.) If **A** is symmetric positive definite, then the square roots that occur will be real.

Algorithm for Cholesky LU Factorization

Define symmetric n-by-n matrix **A**
Initialize $\mathbf{L} = \mathbf{I}_n$ n-by-n identity matrix
Results   **L** is computed, $\mathbf{U} = \mathbf{L}^T$

Begin computation
for $k = 1$ to n
    $\mathbf{x} = \mathbf{L}(k, 1:k-1)$ columns 1 to $k-1$ of the kth row of L
    $\mathbf{L}_{kk} = \sqrt{\mathbf{A}_{kk} - \mathbf{x} \cdot \mathbf{x}^T}$
    for $j = k + 1$ to n
       $\mathbf{y} = \mathbf{L}(j, 1:k-1)$ columns 1 to $k-1$ of the jth row of L
       $\mathbf{L}_{jk} = \sqrt{\mathbf{A}_{jk} - \mathbf{y} \cdot \mathbf{x}^T}$
    end
end
$\mathbf{U} = \mathbf{L}^T$

The following function for Cholesky LU factorization assumes that **A** is symmetric. A check for the symmetry of **A** could easily be added, by testing to see whether $\mathbf{A} = \mathbf{A}^T$. The function does not exploit the storage savings that can be accomplished by writing **L** and **U** over the original matrix **A**; instead, **A** is saved to allow verification of the factorization.

Mathcad Function for Cholesky LU Factorization

If the matrix **M** is symmetric and positive definite, the Cholesky factorization can be carried out without pivoting or scaling. If **M** is not positive definite, the procedure may encounter the square root of a negative number at some stage of the computation. However, since all real symmetric matrices have real eigenvalues, and all eigenvalues may shifted by the amount k, simply by adding *k* to each element on the principal diagonal, one can apply Cholesky factorization to $\mathbf{M} + k\,\mathbf{I}$ as long as **M** is symmetric. For further discussion, see [Atkinson, 1989].

The Mathcad function Cholesky given here returns an n by $2n$ matrix containing the columns of **L** on the left and the columns of **U** on the right.

$$\text{Cholesky(M)} := \begin{vmatrix} n \leftarrow \text{cols(M)} \\ L \leftarrow \text{identity}(n) \\ \text{for } i \in 0..n-1 \\ \quad L_{i,i} \leftarrow 0 \\ \text{for } k \in 0..n-1 \\ \quad \begin{vmatrix} \text{dot} \leftarrow 0 \\ \text{for } i \in 0..k-1 \qquad \text{if } k > 0 \\ \quad \text{dot} \leftarrow \text{dot} + L_{k,i} \cdot L_{k,i} \\ L_{k,k} \leftarrow \sqrt{M_{k,k} - \text{dot}} \\ \text{for } j \in k+1..n-1 \qquad \text{if } k < n-1 \\ \quad \begin{vmatrix} \text{dot} \leftarrow 0 \\ \text{for } i \in 0..k-1 \qquad \text{if } k > 0 \\ \quad \text{dot} \leftarrow \text{dot} + L_{j,i} \cdot L_{k,i} \\ L_{j,k} \leftarrow \dfrac{M_{j,k} - \text{dot}}{L_{k,k}} \end{vmatrix} \end{vmatrix} \\ x \leftarrow \text{augment}(L, L^T) \\ x \end{vmatrix}$$

Example 6.7 Cholesky Factorization

$M := \begin{pmatrix} 1 & 4 & 5 \\ 4 & 20 & 32 \\ 5 & 32 & 64 \end{pmatrix}$ $\quad$ B := Cholesky(M) $\quad$ $B = \begin{pmatrix} 1 & 0 & 0 & 1 & 4 & 5 \\ 4 & 2 & 0 & 0 & 2 & 6 \\ 5 & 6 & 1.732 & 0 & 0 & 1.732 \end{pmatrix}$

L := submatrix(B, 0, rows(M) − 1, 0, cols(M) − 1) $\quad$ U := submatrix(B, 0, rows(M) − 1, cols(M), 5)

$L = \begin{pmatrix} 1 & 0 & 0 \\ 4 & 2 & 0 \\ 5 & 6 & 1.732 \end{pmatrix}$ $\quad$ $U = \begin{pmatrix} 1 & 4 & 5 \\ 0 & 2 & 6 \\ 0 & 0 & 1.732 \end{pmatrix}$ $\quad$ $L \cdot U = \begin{pmatrix} 1 & 4 & 5 \\ 4 & 20 & 32 \\ 5 & 32 & 64 \end{pmatrix}$

6.5 APPLICATIONS OF LU FACTORIZATION

6.5.1 Solving Systems of Linear Equations

One advantage of saving the multipliers so that the **L** matrix can be formed is evident from the situation in which the same system of equations must be solved again later, with a different right-hand side. If the coefficient matrix **A** of a linear system of equations is written as the product of a lower-triangular matrix **L** and an upper-triangular matrix **U**, the linear system can be solved easily in two steps.

The original matrix-vector equation, **A x = b**, is written in terms of the LU factorization of **A** as **L U x = b**. We introduce the unknown vector **y**, defined as **U x = y**. We first solve the system **L y = b** for **y** by "forward substitution"; i.e., we solve for y_1 first, y_2 next, etc. We then solve the system **U x = y** for **x** by "backward substitution" finding x_n first, then x_{n-1} next, etc.

Algorithm to Solve the Linear System L U x = b

Define

    L   Lower-triangular matrix (with 1's on diagonal);
    U   Upper-triangular matrix;
    B   Right-hand side matrix (n-by-m).

Solve **L Z = B** using forward substitution.

for $j = 1$ to m
    $\mathbf{Z}_{1j} = \mathbf{B}_{1j}$
    for $i = 2$ to n
        $\mathbf{Z}_{ij} = \mathbf{B}_{ij} - \mathbf{L}_{i,:} * \mathbf{Z}_{:j}$ (dot product of ith row of **L** with j^{th} column of **Z**)
    end
end

Solve **U X = Z** using back substitution

for $j = 1$ to m

    $\mathbf{X}_{n:} = \dfrac{\mathbf{Z}_{n:}}{\mathbf{U}_{nn}}$

    for $i = n-1$ to 1
        $\mathbf{X}_{ij} = \dfrac{1}{\mathbf{U}_{ii}}(\mathbf{Z}_{ij} - \mathbf{U}_{i,:} * \mathbf{X}_{:j})$ (dot product of ith row of **U** with j^{th} column of **Z**)
    end
end

Example 6.8 Solving Electrical Circuit for Several Voltages

Consider again the electric circuit of Figure 3.1 with resistances as shown, but with $V_1 = 0$, $V_2 = 80$, and $V_3 = 0$. The system to be solved is $\mathbf{A}\,\mathbf{x} = \mathbf{b}$, where

$$\mathbf{A} = \begin{bmatrix} 30 & -20 & -10 \\ -20 & 55 & -10 \\ -10 & -10 & 50 \end{bmatrix}, \text{ and } \mathbf{b} = \begin{bmatrix} 0 \\ 80 \\ 0 \end{bmatrix}.$$

As found in Example 6.3, $\mathbf{A} = \mathbf{L}\,\mathbf{U}$, for

$$\mathbf{L} = \begin{bmatrix} 1 & 0 & 0 \\ -2/3 & 1 & 0 \\ -1/3 & -2/5 & 1 \end{bmatrix} \text{ and } \mathbf{U} = \begin{bmatrix} 30 & -20 & -10 \\ 0 & 125/3 & -50/3 \\ 0 & 0 & 40 \end{bmatrix}.$$

We first solve $\mathbf{L}\,\mathbf{y} = \mathbf{b}$, i.e.,

$$\begin{bmatrix} 1 & 0 & 0 \\ -2/3 & 1 & 0 \\ -1/3 & -2/5 & 1 \end{bmatrix} \begin{bmatrix} y_1 \\ y_2 \\ y_3 \end{bmatrix} = \begin{bmatrix} 0 \\ 80 \\ 0 \end{bmatrix},$$

and find, by forward substitution, that

$$y_1 = 0, \ y_2 = 80, \text{ and } y_3 = (2/5)(80) = 32.$$

Next, we solve $\mathbf{U}\,\mathbf{x} = \mathbf{y}$, or

$$\begin{bmatrix} 30 & -20 & -10 \\ 0 & 125/3 & -50/3 \\ 0 & 0 & 40 \end{bmatrix} \begin{bmatrix} x_1 \\ x_2 \\ x_3 \end{bmatrix} = \begin{bmatrix} 0 \\ 80 \\ 32 \end{bmatrix},$$

and find, by backward substitution, that

$$x_3 = 32/40 = 4/5,$$
$$x_2 = (80 + 40/3)(3/125) = 56/25,$$

and

$$x_1 = (1/30)[-20(56/25) - 10(4/5)] = 44/25.$$

The resulting electrical currents in each of the three loops are

$$i_1 = 1.76, \ i_2 = 2.24, \text{ and } i_3 = 0.80.$$

Mathcad Function to Solve the Linear System L U x = b

$\text{zero}(i, j) := 0$

$\text{LU\_solve}(L, U, b) := \begin{vmatrix} n \leftarrow \text{cols}(L) \\ z \leftarrow \text{matrix}(n, 1, \text{zero}) \\ x \leftarrow \text{matrix}(n, 1, \text{zero}) \\ z_0 \leftarrow b_0 \\ \text{for } i \in 1..n-1 \\ \quad \begin{vmatrix} \text{dot} \leftarrow 0 \\ \text{for } k \in 0..i-1 \\ \quad \begin{vmatrix} \text{dot} \leftarrow \text{dot} + L_{i,k} \cdot z_k \end{vmatrix} \\ z_i \leftarrow b_i - \text{dot} \end{vmatrix} \\ x_{n-1} \leftarrow \dfrac{z_{n-1}}{U_{n-1,n-1}} \\ \text{for } i \in n-2, n-3..0 \\ \quad \begin{vmatrix} \text{dot} \leftarrow 0 \\ \text{for } k \in i+1..n-1 \\ \quad \begin{vmatrix} \text{dot} \leftarrow \text{dot} + U_{i,k} \cdot x_k \end{vmatrix} \\ x_i \leftarrow \dfrac{(z_i - \text{dot})}{U_{i,i}} \end{vmatrix} \\ x \end{vmatrix}$

Note that if pivoting had been used in decomposing **A**, then the system to be solved no longer would be **A x = b**, but rather would be **P A x = P b**, where **P A = L U**. In such a case, the original right-hand side must be transformed to **c = P b**.

As described for Gaussian elimination in Chapter 3, this routine can be generalized easily to the case of multiple right-hand sides. For each column of the right-hand-side matrix **b**, there will be a corresponding column in the solution matrices **z** and **x**.

6.5.2 Solving a Tridiagonal System Using LU Factorization

A tridiagonal system whose coefficient matrix has been factored as described in Section 6.2 can also be solved as described in the algorithm given here, or by the Mathcad function below.

Algorithm for Solving a Tridiagonal System Using LU Factorization

The n-by-n matrix $\mathbf{A}$ is given by its $\mathbf{LU}$ factorization:

Solve $\mathbf{L\,z = b}$ using forward substitution:
$z_1 = b_1$
for $i = 2$ to n
   $z_i = r_i - b_i\,z_{i-1}$
end

Solve $\mathbf{U\,x = z}$ using back substitution:

$$x_n = \frac{z_n}{d_n}$$

for $i = n-1$ to 1

$$x_i = \frac{z_i - a_i\,x_{i+1}}{d_i}$$

end

Mathcad Function for Solving a Tridiagonal System Using LU Factorization

$$\text{LU\_solve\_T}(a,d,b,r) := \begin{vmatrix} n \leftarrow \text{rows}(a) \\ z_0 \leftarrow r_0 \\ \text{for } i \in 1\,..\,n-1 \\ \quad z_i \leftarrow r_i - b_i\cdot z_{i-1} \\ x_{n-1} \leftarrow \dfrac{z_{n-1}}{d_{n-1}} \\ \text{for } i \in n-2, n-3\,..\,0 \\ \quad x_i \leftarrow \dfrac{\left(z_i - a_i\cdot x_{i+1}\right)}{d_i} \\ x \end{vmatrix}$$

The input to the function LU_Solve_T is a tridiagonal matrix in LU factored form. The lower matrix is bidiagonal with 1's on the diagonal and lower diagonal given in vector **b**. The upper matrix is also bidiagonal, the main diagonal is given in vector **d**, the upper diagonal in vector **a**. The right hand side of the system to be solved is given in vector **r**.

Example 6.9 Solving a Tridiagonal System Using LU Factorization.

To solve a tridiagonal linear system, with above diagonal a, below diagonal b, and diagonal d, and right hand side given by vector r,

$$a := \begin{pmatrix} 4 \\ 1 \\ 2 \\ 5 \\ 0 \end{pmatrix} \quad b := \begin{pmatrix} 0 \\ 4 \\ -2 \\ 3 \\ 4 \end{pmatrix} \quad d := \begin{pmatrix} 1 \\ 15 \\ 3 \\ 7 \\ 21 \end{pmatrix} \quad r := \begin{pmatrix} -7 \\ -23 \\ 5 \\ 6 \\ 89 \end{pmatrix}$$

we first find the LU decomposition (which was done in Example 6.5); the modified below diagonal, bb, and main diagonal, dd, are

$$bb := \begin{pmatrix} 0 \\ 4 \\ 2 \\ 3 \\ 4 \end{pmatrix} \quad dd := \begin{pmatrix} 1 \\ -1 \\ 1 \\ 1 \\ 1 \end{pmatrix}$$

The solution is

$$x := LU\_solve\_T(a, dd, bb, r) \quad x = \begin{pmatrix} 1 \\ -2 \\ 3 \\ -4 \\ 5 \end{pmatrix}$$

6.5.3 Determinant of a Matrix

The determinant of a matrix is useful in a variety of situations, including solving eigenvalue problems and making a change of variable in a multiple integral. Cramer's rule for solving systems of linear equations is based on the ratio of two determinants, but it is not an efficient numerical method for solving large systems.

The determinant of $\mathbf{A}$ can be found as follows:

$$\text{If } \mathbf{A} = \mathbf{LU}, \text{ then } \det(\mathbf{A}) = \prod_{i=1}^{n} \ell_{ii} \prod_{i=1}^{n} u_{ii}.$$

If pivoting is used, so that $\mathbf{A} = \mathbf{P}^{-1}\mathbf{L}\mathbf{U}$, we have

$$\det(\mathbf{A}) = (-1)^k \prod_{i=1}^{n} \ell_{ii} \prod_{i=1}^{n} u_{ii},$$

where k is the number of row interchanges that occurred during the LU factorization.

Example 6.10 Finding the Determinant of a Matrix

To find the determinant of the three-by-three matrix

$$\mathbf{A} = \begin{bmatrix} 1 & -1 & 2 \\ -2 & 1 & 1 \\ -1 & 2 & 1 \end{bmatrix},$$

we begin by finding its L U factorization, where

$$\mathbf{L} = \begin{bmatrix} 1 & 0 & 0 \\ -2 & 1 & 0 \\ -1 & -1 & 1 \end{bmatrix} \text{ and } \mathbf{U} = \begin{bmatrix} 1 & -1 & 2 \\ 0 & -1 & 5 \\ 0 & 0 & 8 \end{bmatrix}.$$

Since pivoting was not required, we obtain

$$\det(\mathbf{A}) = u_{11} u_{22} u_{33} = (1)(-1)(8) = -8.$$

6.5.4 Inverse of a Matrix

The inverse of an n-by-n matrix $\mathbf{A}$ can be found by solving the system of equations, $\mathbf{A}\mathbf{x}_i = \mathbf{e}_i$ ($i = 1, \ldots, n$) for the vectors $\mathbf{e}_i = [\,0\;0\;\ldots\;1\;\ldots\;0\;0\,]^T$ where the 1 appears in the ith position. The matrix $\mathbf{X}$ whose columns are the solution vectors $\mathbf{x}_1, \ldots, \mathbf{x}_n$ is $\mathbf{A}^{-1}$. The process is illustrated in Example 6.11.

Example 6.11 Finding a Matrix Inverse Using LU Factorization

Find the inverse of the three-by-three matrix $\mathbf{A} = \mathbf{L}\mathbf{U}$, where

$$\mathbf{A} = \begin{bmatrix} 1 & -1 & 2 \\ -2 & 1 & 1 \\ -1 & 2 & 1 \end{bmatrix}, \quad \mathbf{L} = \begin{bmatrix} 1 & 0 & 0 \\ -2 & 1 & 0 \\ -1 & -1 & 1 \end{bmatrix}, \quad \mathbf{U} = \begin{bmatrix} 1 & -1 & 2 \\ 0 & -1 & 5 \\ 0 & 0 & 8 \end{bmatrix}.$$

First we solve $\mathbf{L}\mathbf{Y} = \mathbf{I}$ for $\mathbf{Y}$, i.e.,

$$\begin{bmatrix} 1 & 0 & 0 \\ -2 & 1 & 0 \\ -1 & -1 & 1 \end{bmatrix} \begin{bmatrix} y_{11} & y_{12} & y_{13} \\ y_{21} & y_{22} & y_{23} \\ y_{31} & y_{32} & y_{33} \end{bmatrix} = \begin{bmatrix} 1 & 0 & 0 \\ 0 & 1 & 0 \\ 0 & 0 & 1 \end{bmatrix}.$$

The first column of $\mathbf{Y}$ is the solution vector for the system

$$\begin{bmatrix} 1 & 0 & 0 \\ -2 & 1 & 0 \\ -1 & -1 & 1 \end{bmatrix} \begin{bmatrix} y_{11} \\ y_{21} \\ y_{31} \end{bmatrix} = \begin{bmatrix} 1 \\ 0 \\ 0 \end{bmatrix}.$$

The second and third columns of $\mathbf{Y}$ are found in a similar manner, using the second and third columns of $\mathbf{I}$. Each column of $\mathbf{Y}$ is found by forward substitution, using the corresponding column of $\mathbf{I}$. The solution that is obtained is

$$Y = \begin{bmatrix} 1 & 0 & 0 \\ 2 & 1 & 0 \\ 3 & 1 & 1 \end{bmatrix}.$$

Finally, we solve $\mathbf{U}\mathbf{X} = \mathbf{Y}$ for $\mathbf{X}$; then $\mathbf{A}^{-1} = \mathbf{X}$. We have

$$\begin{bmatrix} 1 & -1 & 2 \\ 0 & -1 & 5 \\ 0 & 0 & 8 \end{bmatrix} \begin{bmatrix} x_{11} & x_{12} & x_{13} \\ x_{21} & x_{22} & x_{23} \\ x_{31} & x_{32} & x_{33} \end{bmatrix} = \begin{bmatrix} 1 & 0 & 0 \\ 2 & 1 & 0 \\ 3 & 1 & 1 \end{bmatrix}.$$

The solution is easily found for each column of $\mathbf{X}$, using back substitution and the corresponding column of $\mathbf{Y}$. The solution is

$$\mathbf{X} = \mathbf{A}^{-1} = \begin{bmatrix} 1/8 & -5/8 & 3/8 \\ -1/8 & -3/8 & 5/8 \\ 3/8 & 1/8 & 1/8 \end{bmatrix}.$$

6.6 MATHCAD'S METHODS

6.6.1 Using the Built-in Functions

Mathcad has matrix operators on the Matrix Toolbar that can be used to find the inverse or the determinant of a matrix **M**; **M** is a real or complex square matrix. **M** must also be nonsingular for the inverse to exist. There are also keystroke shortcuts for finding the inverse or the determinant.

	toolbar icon	keystroke
Matrix inverse	x^{-1}	M^-1
Matrix determinant	|x|	| M

In addition, Mathcad2000Pro has two built-in functions for LU factorization, `lu(M)` and `cholesky(M)`. The function `lsolve(M, v)` solves the linear system **M x = v** using LU decomposition of **M** and forward/backward substitution.

The function `lu(M)` is described in the Mathcad 2000 Reference Manual as using Crout's method with partial pivoting, with references to [Press et al., 1992] and [Golub and Van Loan, 1989].

6.6.2 Understanding the Algorithms

The computation of the determinant of **M** is based on the LU decomposition of **M** as discussed in section 6.5.3.

The function `lu(M)` is described in the Mathcad 2000 Reference Manual as using Crout's method with partial pivoting. Details of this method (which also includes some implicit row scaling in the pivoting strategy) are given in [Press et al., 1992, pp. 44–47]. This method places 1's on the diagonal of the lower triangular matrix, and thus will normally give the same results as the LU factorization with pivoting described in Section 6.3, although the computations in Section 6.3 are based on Gaussian elimination. Differences in the results could occur because of differences in the details of the pivoting strategies. Golub and Van Loan [1996, p. 104] observe that the Doolittle/Crout factorizations, which are based on inner products rather than the outer-products that occur in Gaussian elimination, are popular because there are fewer intermediate results required than in Gaussian elimination. For discussion of many methods of implementing the computational loops for Gaussian elimination (and the corresponding LU factorization), see [Golub and Van Loan, 1996, Ch. 3].

The function `cholesky(M)` assumes that **M** is symmetric, and only uses the upper-triangular part of **M** in the computations.

SUMMARY

LU Factorization from Gaussian Elimination
$\mathbf{U}$ is the upper-triangular matrix obtained by Gaussian elimination is formed from the multipliers used in the elimination:

$$\mathbf{L} = \begin{bmatrix} 1 & 0 & 0 & 0 \\ -m_{21} & 1 & 0 & 0 \\ -m_{31} & -m_{32} & 1 & 0 \end{bmatrix}$$

LU Factorization of Tridiagonal System (illustrated for $n = 4$)

$$\begin{aligned} d_1 x_1 + a_1 x_2 &= r_1, \\ b_2 x_1 + d_2 x_2 + a_2 x_3 &= r_2, \\ b_3 x_2 + d_3 x_3 + a_3 x_4 &= r_3, \\ b_4 x_3 + d_4 x_4 &= r_4. \end{aligned}$$

Compute:
$dd_1 = d_1,$
$bb_2 = b_2/dd_1, \quad dd_2 = d_2 - bb_2 a_1,$
$bb_3 = b_3/dd_2, \quad dd_3 = d_3 - bb_3 a_2,$
$bb_4 = b_4/dd_3, \quad dd_4 = d_4 - bb_4 a_3.$

The factorization is

$$\mathbf{L} = \begin{bmatrix} 1 & 0 & 0 & 0 \\ bb_2 & 1 & 0 & 0 \\ 0 & bb_3 & 1 & 0 \\ 0 & 0 & bb_4 & 1 \end{bmatrix} \quad \mathbf{U} = \begin{bmatrix} dd_1 & a_1 & 0 & 0 \\ 0 & dd_2 & a_2 & 0 \\ 0 & 0 & dd_3 & a_3 \\ 0 & 0 & 0 & dd_4 \end{bmatrix}$$

Direct LU Factorization

1's on the diagonal of $\mathbf{L}$:

$$\begin{bmatrix} 1 & 0 & 0 \\ \ell_{21} & 1 & 0 \\ \ell_{31} & \ell_{32} & 1 \end{bmatrix} \cdot \begin{bmatrix} u_{11} & u_{12} & u_{13} \\ 0 & u_{22} & u_{23} \\ 0 & 0 & u_{33} \end{bmatrix} = \begin{bmatrix} a_{11} & a_{12} & a_{13} \\ a_{21} & a_{22} & a_{23} \\ a_{31} & a_{32} & a_{33} \end{bmatrix}.$$

1's on the diagonal of $\mathbf{U}$:

$$\begin{bmatrix} \ell_{11} & 0 & 0 \\ \ell_{21} & \ell_{22} & 0 \\ \ell_{31} & \ell_{32} & \ell_{33} \end{bmatrix} \cdot \begin{bmatrix} 1 & u_{12} & u_{13} \\ 0 & 1 & u_{23} \\ 0 & 0 & 1 \end{bmatrix} = \begin{bmatrix} a_{11} & a_{12} & a_{13} \\ a_{21} & a_{22} & a_{23} \\ a_{31} & a_{32} & a_{33} \end{bmatrix}.$$

Cholesky form makes the diagonals of $\mathbf{L}$ and $\mathbf{U}$ equal:

$$\begin{bmatrix} x_{11} & 0 & 0 \\ \ell_{21} & x_{22} & 0 \\ \ell_{31} & \ell_{32} & x_{33} \end{bmatrix} \cdot \begin{bmatrix} x_{11} & u_{12} & u_{13} \\ 0 & x_{22} & u_{23} \\ 0 & 0 & x_{33} \end{bmatrix} = \begin{bmatrix} a_{11} & a_{12} & a_{13} \\ a_{21} & a_{22} & a_{23} \\ a_{31} & a_{32} & a_{33} \end{bmatrix}$$

SUGGESTIONS FOR FURTHER READING

The following are a few of the many excellent undergraduate texts on linear algebra:

> Kolman, B., *Introductory Linear Algebra with Applications* (6th ed.), Prentice Hall, Upper Saddle River, NJ, 1997.
>
> Leon, S. J., *Linear Algebra with Applications* (5th ed.), Prentice Hall, Upper Saddle River, NJ, 1998.
>
> Strang, G., *Linear Algebra and Its Applications* (3rd ed.), Harcourt Brace Jovanovich, San Diego, CA, 1988.

Some references at a somewhat more advanced level include:

> Dongarra, J. J. et al., *LINPACK User's Guide*, SIAM, Philadelphia, 1979.
>
> Forsythe, G. E., M. A. Malcolm and C. B. Moler, *Computer Methods for Mathematical Computations*, Prentice-Hall, Englewood Cliffs, 1977.
>
> Fox, L., *An Introduction to Numerical Linear Algebra*, Oxford University Press, New York, 1965. This classic work also contains many bibliographic entries.
>
> Golub, G. H. and C. F. Van Loan, *Matrix Computations* (3d ed.), Johns Hopkins University Press, Baltimore, 1996.
>
> Horn, R. A. and C. R. Johnson, *Matrix Analysis*, Cambridge University Press, Cambridge, 1985.

PRACTICE THE TECHNIQUES

For problems P6.1 to P6.10 find

    a. the LU factorization of the matrix, using Gaussian elimination.
    b. the inverse
    c. the determinant
    d. solve $\mathbf{A}\,\mathbf{x} = \mathbf{b}$ with $\mathbf{b} = [\,1\ 1\ \ldots\ 1]^T$; then solve $\mathbf{A}\,\mathbf{y} = \mathbf{x}$.

P6.1 $\mathbf{A} = \begin{bmatrix} 1 & 2 & 3 \\ 2 & 8 & 11 \\ 3 & 22 & 35 \end{bmatrix}$

P6.2 $\mathbf{A} = \begin{bmatrix} 3 & 6 & 12 \\ -1 & 0 & 2 \\ 3 & 2 & 1 \end{bmatrix}$

P6.3 $\mathbf{A} = \begin{bmatrix} 2 & 1 & -2 \\ 4 & -1 & 2 \\ 2 & -1 & 1 \end{bmatrix}$

P6.4 $\mathbf{A} = \begin{bmatrix} 3/2 & -1 & 1/2 \\ -1/2 & 1/2 & -1/4 \\ 1/2 & -1/2 & 1/2 \end{bmatrix}$

P6.5 $\mathbf{A} = \begin{bmatrix} 1 & 1/2 & 1/3 \\ 1/2 & 1/3 & 1/4 \\ 1/3 & 1/4 & 1/5 \end{bmatrix}$

P6.6 $\mathbf{A} = \begin{bmatrix} 1/5 & 0 & 0 & 0 \\ -26 & 13 & -4 & 2 \\ 12 & -6 & 2 & -1 \\ -3/2 & 3/4 & -1/4 & 1/4 \end{bmatrix}$

P6.7 $\mathbf{A} = \begin{bmatrix} 1 & 1 & 0 & 3 \\ 2 & 1 & -1 & 1 \\ 3 & -1 & -1 & 2 \\ -1 & 2 & 3 & -1 \end{bmatrix}$

P6.8 $\mathbf{A} = \begin{bmatrix} 1 & -2 & 6 & -12 \\ 0 & 1 & -3 & 6 \\ 0 & -5 & 16 & -32 \\ 0 & 15 & -48 & 97 \end{bmatrix}$

P6.9 $\mathbf{A} = \begin{bmatrix} 3 & 7 & 4 & 0 \\ 0 & 3 & 13 & 3 \\ 0 & 0 & 1 & 4 \\ 1 & 2 & 0 & 0 \end{bmatrix}$

P6.10 $\mathbf{A} = \begin{bmatrix} 1 & 2 & 3 & -1 \\ 1 & 1 & -1 & 2 \\ 0 & -1 & -1 & 3 \\ 3 & 1 & 2 & -1 \end{bmatrix}$

P6.11 $\mathbf{A} = \begin{bmatrix} 1 & 2 & 3 & 1 & 0 & 0 \\ 2 & 6 & 10 & 2 & 1 & 0 \\ 3 & 14 & 28 & 3 & 4 & 1 \\ 1 & 2 & 3 & 4 & 6 & 12 \\ 0 & 2 & 4 & -1 & 1 & 2 \\ 0 & 0 & 3 & 3 & 2 & 2 \end{bmatrix}$

P6.12 $\mathbf{A} = \begin{bmatrix} 2 & 1 & -2 & 1 & 0 & 0 \\ 4 & -1 & 2 & 2 & 1 & 0 \\ 2 & -1 & 1 & 1 & 2/3 & 1 \\ 2 & 1 & -2 & 5/2 & -1 & 1/2 \\ 0 & -3 & 6 & -1/2 & 10/3 & -7/6 \\ 0 & 0 & -1 & 1/2 & -7/3 & 29/12 \end{bmatrix}$

P6.13 $\mathbf{A} = \begin{bmatrix} 1 & 2 & 1 & 1 & 0 & 0 \\ 2 & 6 & 4 & 2 & 1 & 0 \\ 3 & 14 & 12 & 3 & 4 & 1 \\ 1 & 2 & 1 & 2 & 6 & 12 \\ 0 & 2 & 2 & -1 & -3 & -6 \\ 2 & 4 & 3 & 3 & 2 & 2 \end{bmatrix}$

P6.14

$\mathbf{A} = \begin{bmatrix} 1 & 2 & 1 & 1 & 0 & 0 & 1 & 0 & -1 & 0 \\ 2 & 6 & 4 & 2 & 1 & 0 & 2 & 2 & 0 & 1 \\ 3 & 14 & 12 & 3 & 4 & 1 & 4 & 9 & 6 & 5 \\ 1 & 2 & 1 & 2 & 2 & 3 & 1 & 1 & 1 & 3 \\ 0 & 2 & 2 & -1 & 1 & 3 & 2 & 0 & 1 & -1 \\ 2 & 4 & 3 & 3 & -2 & -7 & -1 & 4 & 0 & 2 \\ 1 & 2 & 0 & 1 & 0 & 1 & 1 & 0 & -1 & 0 \\ 0 & 2 & 4 & 0 & -1 & -4 & 3 & 9 & 0 & 7 \\ 1 & 0 & -3 & 2 & 1 & 2 & -1 & -3 & -1 & 0 \\ 0 & 2 & 2 & 1 & 3 & 4 & 0 & 4 & 5 & 7 \end{bmatrix}$

P6.15

$\mathbf{A} = \begin{bmatrix} 1 & 1 & 1 & 1 & 0 & 0 & 1 & 0 & -1 & 0 \\ 1 & 3 & 2 & 1 & 1 & 0 & 1 & 1 & 0 & 1 \\ 1 & 3 & 3 & 1 & 1 & 1 & 2 & 2 & 1 & 2 \\ 1 & 1 & 1 & 3 & 1 & 1 & 1 & 1 & 1 & 1 \\ 0 & 2 & 1 & -2 & 2 & 0 & 1 & -1 & 0 & 1 \\ 1 & 1 & 2 & 3 & -1 & 2 & 1 & 3 & 2 & 1 \\ 1 & 1 & 0 & 1 & 0 & 0 & 2 & 0 & -2 & 0 \\ 0 & 2 & 2 & 0 & -1 & 0 & 2 & 6 & 0 & 3 \\ 1 & -1 & -1 & 3 & 0 & 1 & 0 & -1 & 2 & -1 \\ 0 & 2 & 1 & 2 & 2 & 2 & 0 & 4 & 4 & 4 \end{bmatrix}$

For Problems P6.16 to P6.25, find an LU factorization of the given matrix
    a. *Using the method for tridiagonal matrices.*
    b. *Using Gaussian elimination.*

P6.16 $\mathbf{A} = \begin{bmatrix} 5 & 0 & 0 \\ 5 & 3 & 1 \\ 0 & 9 & 7 \end{bmatrix}$

P6.17 $\mathbf{A} = \begin{bmatrix} 4 & 2 & 0 \\ 16 & 12 & 1 \\ 0 & 8 & 5 \end{bmatrix}$

P6.18 $\mathbf{A} = \begin{bmatrix} 2 & 4 & 0 \\ 2 & 8 & 2 \\ 0 & 4 & 4 \end{bmatrix}$

P6.19 $\mathbf{A} = \begin{bmatrix} 5 & 1 & 0 \\ 15 & 6 & 3 \\ 0 & 9 & 11 \end{bmatrix}$

P6.20 $\mathbf{A} = \begin{bmatrix} 3 & 5 & 0 \\ 12 & 24 & 3 \\ 0 & 8 & 7 \end{bmatrix}$

P6.21 $\mathbf{A} = \begin{bmatrix} 1 & 3 & 0 & 0 \\ 2 & 7 & 2 & 0 \\ 0 & 2 & 5 & 5 \\ 0 & 0 & 1 & 6 \end{bmatrix}$

P6.22 $\mathbf{A} = \begin{bmatrix} 3 & 1 & 0 & 0 \\ 12 & 5 & 2 & 0 \\ 0 & 2 & 8 & 5 \\ 0 & 0 & 16 & 23 \end{bmatrix}$

P6.23 $\mathbf{A} = \begin{bmatrix} 1 & 4 & 0 & 0 \\ 1 & 6 & 2 & 0 \\ 0 & 6 & 10 & 3 \\ 0 & 0 & 12 & 13 \end{bmatrix}$

P6.24 $\mathbf{A} = \begin{bmatrix} 2 & 1 & 0 & 0 \\ 6 & 5 & 1 & 0 \\ 0 & 2 & 3 & 5 \\ 0 & 0 & 8 & 22 \end{bmatrix}$

P6.25 $\mathbf{A} = \begin{bmatrix} 5 & 0 & 0 & 0 \\ 10 & 1 & 2 & 0 \\ 0 & 3 & 7 & 4 \\ 0 & 0 & 1 & 8 \end{bmatrix}$

P6.26 $\mathbf{A} = \begin{bmatrix} 1 & 2 & 0 & 0 & 0 & 0 \\ 2 & 6 & 2 & 0 & 0 & 0 \\ 0 & 8 & 9 & 0 & 0 & 0 \\ 0 & 0 & 0 & 1 & 6 & 0 \\ 0 & 0 & 0 & -1 & -4 & 6 \\ 0 & 0 & 0 & 0 & -4 & -11 \end{bmatrix}$

P6.27 $\mathbf{A} = \begin{bmatrix} 1 & -5 & 0 & 0 & 0 & 0 \\ 2 & -8 & -4 & 0 & 0 & 0 \\ 0 & 6 & -9 & -3 & 0 & 0 \\ 0 & 0 & 12 & -8 & -2 & 0 \\ 0 & 0 & 0 & 20 & -5 & -1 \\ 0 & 0 & 0 & 0 & 30 & 0 \end{bmatrix}$

P6.28 $\mathbf{A} = \begin{bmatrix} 4 & 4 & 0 & 0 & 0 & 0 \\ 16 & 20 & 0 & 0 & 0 & 0 \\ 0 & 4 & 3 & 1 & 0 & 0 \\ 0 & 0 & 9 & 5 & 4 & 0 \\ 0 & 0 & 0 & 10 & 21 & 2 \\ 0 & 0 & 0 & 0 & 3 & 7 \end{bmatrix}$

P6.29 $\mathbf{A} = \begin{bmatrix} 5 & 2 & 0 & 0 & 0 & 0 \\ 15 & 10 & 0 & 0 & 0 & 0 \\ 0 & 8 & 2 & 2 & 0 & 0 \\ 0 & 0 & 2 & 5 & 4 & 0 \\ 0 & 0 & 0 & 9 & 16 & 4 \\ 0 & 0 & 0 & 0 & 16 & 17 \end{bmatrix}$

P6.30

$\mathbf{A} = \begin{bmatrix} 1 & 1 & 0 & 0 & 0 & 0 & 0 & 0 & 0 & 0 \\ 1 & 3 & 1 & 0 & 0 & 0 & 0 & 0 & 0 & 0 \\ 0 & 4 & 3 & 1 & 0 & 0 & 0 & 0 & 0 & 0 \\ 0 & 0 & 3 & 5 & 1 & 0 & 0 & 0 & 0 & 0 \\ 0 & 0 & 0 & -2 & 1 & 1 & 0 & 0 & 0 & 0 \\ 0 & 0 & 0 & 0 & -4 & -1 & 0 & 0 & 0 & 0 \\ 0 & 0 & 0 & 0 & 0 & -3 & 2 & 1 & 0 & 0 \\ 0 & 0 & 0 & 0 & 0 & 0 & 2 & 3 & 0 & 0 \\ 0 & 0 & 0 & 2 & 1 & 0 & 0 & 4 & 2 & 0 \\ 0 & 0 & 0 & 2 & 1 & 0 & 0 & 0 & 6 & 1 \end{bmatrix}$

*For Problems P6.31 to P6.35 find an LU factorization of a permutation of **A**, using Gaussian elimination with row pivoting.*

P6.31 $\mathbf{A} = \begin{bmatrix} 6 & 2 & 2 \\ 6 & 2 & 1 \\ 1 & 2 & -1 \end{bmatrix}$

P6.32 $\mathbf{A} = \begin{bmatrix} 1 & 2 & 3 \\ 2 & 4 & 10 \\ 3 & 14 & 28 \end{bmatrix}$

P6.33 $\mathbf{A} = \begin{bmatrix} -1 & 1 & 0 & 0 \\ 1 & -1 & 1 & 0 \\ 0 & 1 & -1 & 1 \\ 0 & 0 & 1 & -1 \end{bmatrix}$

P6.34 $\mathbf{A} = \begin{bmatrix} 2 & -1 & 0 & 0 & 0 & 0 \\ -1 & 2 & -1 & 0 & 0 & 0 \\ 0 & -1 & 2/3 & -1 & 0 & 0 \\ 0 & 0 & -1 & 2 & -1 & 0 \\ 0 & 0 & 0 & -1 & 2 & -1 \\ 0 & 0 & 0 & 0 & -1 & 2 \end{bmatrix}$

P6.35 $\mathbf{A} = \begin{bmatrix} 1 & -2 & 0 & 0 & 0 & 0 \\ -2 & 6 & 4 & 0 & 0 & 0 \\ 0 & 4 & 8 & -1/2 & 0 & 0 \\ 0 & 0 & -1/2 & 13/4 & 3/2 & 0 \\ 0 & 0 & 0 & 3/2 & 7/4 & -3 \\ 0 & 0 & 0 & 0 & -3 & 13 \end{bmatrix}$;

For Problems P6.36 to P6.40 find the LU factorization of the given matrix
  a. *Doolittle form*
  b. *Cholesky form*
  c. *use the LU factorization from part a or b to solve the linear system* $\mathbf{Ax} = \mathbf{b}$

P6.36 $\mathbf{A} = \begin{bmatrix} 1 & 2 & 3 \\ 2 & 20 & 26 \\ 3 & 26 & 70 \end{bmatrix}$

  i. $\mathbf{b} = \begin{bmatrix} 14 & 120 & 265 \end{bmatrix}^T$
  ii. $\mathbf{b} = \begin{bmatrix} 7 & 38 & 123 \end{bmatrix}^T$
  iii. $\mathbf{b} = \begin{bmatrix} 10 & 48 & 173 \end{bmatrix}^T$

P6.37 $\mathbf{A} = \begin{bmatrix} 9 & 18 & 36 \\ 18 & 40 & 84 \\ 36 & 84 & 181 \end{bmatrix}$

  i. $\mathbf{b} = \begin{bmatrix} 63 & 146 & 313 \end{bmatrix}^T$
  ii. $\mathbf{b} = \begin{bmatrix} 0 & -8 & -25 \end{bmatrix}^T$
  iii. $\mathbf{b} = \begin{bmatrix} -18 & -32 & -59 \end{bmatrix}^T$

P6.38 $\mathbf{A} = \begin{bmatrix} 9 & 18 & 36 \\ 18 & 52 & 116 \\ 36 & 116 & 265.25 \end{bmatrix}$

  i. $\mathbf{b} = \begin{bmatrix} 63.00 & 186.00 & 417.25 \end{bmatrix}^T$
  ii. $\mathbf{b} = \begin{bmatrix} -9.00 & -46.00 & -113.25 \end{bmatrix}^T$
  iii. $\mathbf{b} = \begin{bmatrix} 81 & 306 & 721 \end{bmatrix}^T$

P6.39 $\mathbf{A} = \begin{bmatrix} 2 & -1 & 0 & 0 & 0 & 0 \\ -1 & 2 & -1 & 0 & 0 & 0 \\ 0 & -1 & 2/3 & -1 & 0 & 0 \\ 0 & 0 & -1 & 2 & -1 & 0 \\ 0 & 0 & 0 & -1 & 2 & -1 \\ 0 & 0 & 0 & 0 & -1 & 2 \end{bmatrix}$

  i. $\mathbf{b} = \begin{bmatrix} 3 & -3 & 2 & -2 & -3 & -1 \end{bmatrix}^T$
  ii. $\mathbf{b} = \begin{bmatrix} 0 & 0 & 1 & -2 & -3 & -1 \end{bmatrix}^T$
  iii. $\mathbf{b} = \begin{bmatrix} 3 & -5 & 13/3 & -9 & 11 & -9 \end{bmatrix}^T$

P6.40

$\mathbf{A} = \begin{bmatrix} 1 & 1 & 1 & 1 & 0 & 0 & 1 & 0 & -1 & 0 \\ 1 & 5 & 3 & 1 & 2 & 0 & 1 & 2 & 1 & 2 \\ 1 & 3 & 3 & 1 & 1 & 1 & 2 & 2 & 1 & 2 \\ 1 & 1 & 1 & 5 & 2 & 2 & 1 & 2 & 3 & 2 \\ 0 & 2 & 1 & 2 & 6 & 3 & 2 & 0 & 5 & 4 \\ 0 & 0 & 1 & 2 & 3 & 4 & 2 & 1 & 5 & 3 \\ 1 & 1 & 2 & 1 & 2 & 2 & 7 & 2 & -1 & 4 \\ 0 & 2 & 2 & 2 & 0 & 1 & 2 & 9 & 2 & 5 \\ -1 & 1 & 1 & 3 & 5 & 5 & -1 & 2 & 14 & 4 \\ 0 & 2 & 2 & 2 & 4 & 3 & 4 & 5 & 4 & 7 \end{bmatrix}$

  i. $\mathbf{b} = \begin{bmatrix} 0 & 8 & 3 & -14 & -39 & -32 & -1 & 45 & -56 & -1 \end{bmatrix}^T$
  ii. $\mathbf{b} = \begin{bmatrix} -15 & -43 & -43 & -27 & -12 & -14 & -59 & -79 & 2 & -63 \end{bmatrix}^T$
  iii. $\mathbf{b} = \begin{bmatrix} -6 & -10 & -8 & -12 & -7 & 0 & -29 & -19 & 24 & -21 \end{bmatrix}^T$

EXPLORE SOME APPLICATIONS

LU factorization is especially useful for solving linear systems that must be solved for several different right-hand sides which are not known in advance. One place that this occurs is in using the inverse power method, discussed in the next chapter.

For each of the following matrices, find the LU decomposition, and use it to solve the sequence of linear systems. Start with $\mathbf{b} = [1\ 1\ 1]^T$, *and solve* $\mathbf{A x} = \mathbf{b}$. *Then find* $\mathbf{y}$ *such that* $\mathbf{A y} = \mathbf{x}$.

A6.1 $\mathbf{A} = \begin{bmatrix} 1 & 0 & 0 \\ 2 & -1 & 2 \\ 4 & -4 & 5 \end{bmatrix}$.

A6.2 $\mathbf{A} = \begin{bmatrix} 5 & -2 & 1 \\ 3 & 0 & 1 \\ 0 & 0 & 2 \end{bmatrix}$.

A6.3 $\mathbf{A} = \begin{bmatrix} 2 & 2 & -1 \\ -5 & 9 & -3 \\ -4 & 4 & 1 \end{bmatrix}$.

A6.4 $\mathbf{A} = \begin{bmatrix} -19 & 20 & -6 \\ -12 & 13 & -3 \\ 30 & -30 & 12 \end{bmatrix}$.

Problems A6.5 to A6.7 make use of a matrix formulation of quadratic equations. The quadratic form $a x^2 + b xy + c y^2 = F(x,y)$ *can be written in matrix form as* $[\mathbf{A v}]\mathbf{v} = F(x,y)$, *with* $\mathbf{v} = [x, y]^T$ *and* $\mathbf{A} = \begin{bmatrix} a & b/2 \\ b/2 & c \end{bmatrix}$

The form of the graph of the quadratic equation

$$a x^2 + b xy + c y^2 = d, \quad (d \neq 0)$$

is given by the sign of the determinant of the matrix $\mathbf{A}$:

If $det(A) < 0$, *the graph is a hyperbola;*

if $det(A) > 0$, *the graph is an ellipse, or a circle, (or is degenerate);*

if $det(A) = 0$, *the graph is a pair of straight lines (or is degenerate).*

(For further discussion of these ideas, see Grossman and Derrick, 1988, p. 490; Fraleigh and Beauregard, 1987, or Leon, 1998)

Write the following equations in matrix form, and determine whether the equation is a hyperbola, an ellipse, a circle, or a pair of straight lines.

A6.5 $x^2 - \sqrt{3}\, x y + 2 y^2 = 10$

A6.6 $4 x^2 + 3 \sqrt{3}\, xy + y^2 = 22$

A6.7 $x^2 - \sqrt{3}\, xy = -2$

7

Eigenvalues, Eigenvectors, and QR Factorization

7.1 Power Method
7.2 QR Factorization
7.3 Finding Eigenvalues Using QR Factorization

The factorization of a matrix **A** into the product of lower-triangular and upper-triangular matrices **L** and **U**, discussed in Chapter 6, provides a method for finding several quantities associated with **A**, including the determinant of **A**, the inverse of **A**, and the solution of the linear system **A x** = **b**. In this chapter, we conclude our investigation into topics from numerical linear algebra by considering techniques for finding the eigenvalues and eigenvectors of a matrix. We already have encountered eigenvalues in Chapter 4, where they appear in the analysis of the convergence characteristics of iterative methods for solving linear systems. The estimation of eigenvalues based on the Gerschgorin theorem was presented in Chapter 1.

We first consider a method, known as the power method, for finding a specific eigenvalue and its associated eigenvector for a given matrix **A**. The basic power method finds the dominant eigenvalue, i.e., the eigenvalue of largest magnitude. Variations of the power method can be used to find the eigenvalue of smallest magnitude or the eigenvalue closest to a specified value.

We next consider the factorization of **A** into the product of an orthogonal matrix **Q** and a right (upper) triangular matrix **R**. *QR factorization* can be used for a number of types of problems, including solving a linear system and finding an orthonormal basis for the space spanned by a collection of vectors. We restrict our application of QR factorization to its use in finding the eigenvalues of **A**.

Eigenvalues and eigenvectors are important in many areas of science and engineering, including solving differential equations and finding physical characteristics of a structure, such as the principal stress, moments of inertia, etc.

The ratio of the largest to the smallest eigenvalue of a matrix is a useful measure of the "condition" of a matrix; a matrix with a large condition number is called ill conditioned. The Hilbert matrix, introduced in Chapter 1, is a famous example of an ill-conditioned matrix.

Example 7-A Inertia

The principal inertias and principal axes of a three-dimensional object can be found from the eigenvalues and eigenvectors, respectively, of its inertial matrix. For example, consider a body consisting of unit point masses at (1, 0, 0), (1, 2, 0) and (0, 0, 1). (See Fig. 7.1.) Its inertial matrix is

$$G = \begin{bmatrix} 5 & -2 & 0 \\ -2 & 3 & 0 \\ 0 & 0 & 6 \end{bmatrix}.$$

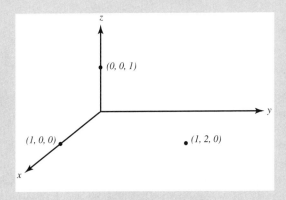

FIGURE 7.1 A body consisting of three unit point masses.

In general, the inertial matrix is

$$G = \begin{bmatrix} I_{xx} & -I_{xy} & -I_{xz} \\ -I_{yx} & I_{yy} & -I_{yz} \\ -I_{zx} & -I_{zy} & I_{zz} \end{bmatrix},$$

where the moments of inertia of the body around the x-, y-, and z-axes are, respectively,

$$I_{xx} = \int (y^2 + z^2)\, dm, \quad I_{yy} = \int (x^2 + z^2)\, dm, \quad I_{zz} = \int (x^2 + y^2)\, dm,$$

and the corresponding products of inertia are

$$I_{xy} = I_{yx} = \int xy\, dm, \quad I_{xz} = I_{zx} = \int xz\, dm, \quad I_{yz} = I_{zy} = \int yz\, dm.$$

(See Greenberg, 1998, p. 581, or Thomson, 1986, p. 103 for further discussion.)

Example 7-B Buckling and Breaking of a Beam

The bending moment of a simply supported beam (see Fig. 7.2) is described by the differential equation

$$-x'' = \lambda x \qquad 0 \leq x \leq 1$$

with boundary conditions

$$x(0) = x(1) = 0,$$

where λ is the applied load. When λ reaches a critical value, the beam buckles (and may break soon after the load exceeds that value).

This smallest eigenvalue of the differential equation can be approximated by the smallest eigenvalue of the linear system obtained by the finite difference techniques we consider in Chapter 14. Thus, for large n, we are interested in the smallest eigenvalue of the n-by-n matrix

$$\mathbf{A} = (n+1)^2 \begin{bmatrix} 2 & -1 & & & & \\ -1 & 2 & -1 & & & \\ & -1 & 2 & -1 & & \\ & & \cdot & \cdot & \cdot & \\ & & & -1 & 2 & -1 \\ & & & & -1 & 2 \end{bmatrix},$$

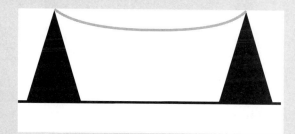

FIGURE 7.2 Simply supported beam.

The eigenvalues of the boundary value problem can be shown to be

$$\mu_j = (j\pi)^2$$

Using the Mathcad functions for finding eigenvalues discussed in this chapter, one can experiment to investigate how large n should be to achieve a reasonably accurate estimate of the smallest eigenvalue of the differential equation, namely $\mu_1 = \pi^2$, by using the eigenvalue of the discrete system. One might also compare the next smallest eigenvalue of the discrete system with the corresponding eigenvalue for the continuous problem, i.e., $\mu_2 = 4\pi^2$. (See Hager, 1988, for further discussion.)

7.1 POWER METHOD

A (real or complex) number λ is an eigenvalue of a matrix $\mathbf{A}$, and a nonzero vector $\mathbf{x}$ is a corresponding eigenvector of $\mathbf{A}$, if and only if $\mathbf{A}\mathbf{x} = \lambda\mathbf{x}$.

The power method is an iterative procedure for determining the dominant eigenvalue of a matrix $\mathbf{A}$. The basic idea behind the power method is that the defining relationship which is satisfied if $\mathbf{x}$ is an eigenvector of $\mathbf{A}$, namely, $\lambda\mathbf{x} = \mathbf{A}\mathbf{x}$, can be converted into a sequence of approximations to λ and $\mathbf{x}$. We start with an initial guess $\mathbf{z}$ for the true eigenvector $\mathbf{x}$ and compute $\mathbf{w} = \mathbf{A}\mathbf{z}$. If $\mathbf{z}$ is an eigenvector, then for any component of $\mathbf{z}$ and $\mathbf{w}$, we would have $\lambda z_k = w_k$. If $\mathbf{z}$ is not an eigenvector, we would like to use $\mathbf{w}$ as the next approximation, and iterate until the process converges (if it does). However, because an eigenvector is determined only up to a scale factor, we normalize $\mathbf{w}$ before using it as the next approximation $\mathbf{z}$; the normalization is chosen so that the largest component of $\mathbf{z}$ is 1 at each stage of the iteration.

The method will converge if the initial estimate of the eigenvector has a (nonzero) component in the direction of the eigenvector corresponding to the dominant eigenvalue. For that reason, a starting vector with all components equal to 1 is used in the computations of the power method.

Briefly, the power method approach is based on the following observations. We iterate according to the equation $\mathbf{w} = \mathbf{A}\mathbf{z}$. From the basic definition of an eigenvalue, we also have the approximate relationship $\lambda\mathbf{z} \approx \mathbf{A}\mathbf{z}$. Therefore $\mathbf{w} \approx \lambda\mathbf{z}$.

Denoting the dominant component of $\mathbf{w}$ as w_k, we have

$$w_k \approx \lambda z_k \Rightarrow \lambda \approx \frac{w_k}{z_k}.$$

Since we scale the vector $\mathbf{z}$ at each stage so that $z_k = 1$, we have the approximation

$$\lambda \approx w_k.$$

Carrying out the iterations has the effect of multiplying the original estimate by successively higher powers of $\mathbf{A}$—hence the name of the method. The first two iterations, shown here, illustrate the basic pattern:

$$\mathbf{w}^{(1)} = \mathbf{A}\,\mathbf{z}^{(1)};$$

$$\mathbf{z}^{(2)} = \frac{1}{w^{(1)}_k}\mathbf{w}^{(1)} = \frac{1}{w^{(1)}_k}\mathbf{A}\,\mathbf{z}^{(1)};$$

$$\mathbf{w}^{(2)} = \mathbf{A}\,\mathbf{z}^{(2)} = \mathbf{A}\frac{1}{w^{(1)}_k}\mathbf{A}\,\mathbf{z}^{(1)} = \frac{1}{w^{(1)}_k}\mathbf{A}^2\,\mathbf{z}^{(1)};$$

$$\mathbf{z}^{(3)} = \frac{1}{w^{(2)}_k}\mathbf{w}^{(2)} = \frac{1}{w^{(2)}_k}\frac{1}{w^{(1)}_k}\mathbf{A}^2\,\mathbf{z}^{(1)}.$$

7.1.1 Basic Power Method

The basic power method is illustrated in the following example.

Example 7.1 Using the Basic Power Method

Find the dominant eigenvalue λ, and a corresponding eigenvector, of A, starting with the initial vector $\mathbf{z} = [1, 1, 1]^T$, for

$$\mathbf{A} = \begin{bmatrix} 21 & 7 & -1 \\ 5 & 7 & 7 \\ 4 & -4 & 20 \end{bmatrix},$$

Step 1: $\mathbf{w} = \mathbf{A}\,\mathbf{z} = [\,27, 19, 20\,]^T$.

Since the first component of $\mathbf{w}$ has the largest magnitude, the first estimate of λ is $w_1 = 27$. We use that component to scale the approximate eigenvector:

$$\mathbf{z} = \mathbf{w}/w_1 = [1, 19/27, 20/27]^T = [1.0000, 0.7307, 0.7407]^T.$$

Step 2: $\mathbf{w} = \mathbf{A}\,\mathbf{z} = [\,25.1852, 15.1111, 16.0000\,]^T$.

Since the first component of $\mathbf{w}$ has the largest magnitude, the estimate of λ is $w_1 = 25.1852$, and

$$\mathbf{z} = \mathbf{w}/w_1 = [1.0000, 0.6000, 0.6353]^T.$$

Step 3: $\mathbf{w} = \mathbf{A}\,\mathbf{z} = [\,24.5647, 13.6471, 14.3059\,]^T$.
The estimate of λ is $w_1 = 24.5647$, and

$$\mathbf{z} = \mathbf{w}/w_1 = [1.0000, 0.5556, 0.5824]^T.$$

Step 4: $\mathbf{w} = \mathbf{A}\,\mathbf{z} = [\,24.3065, 12.9655, 13.4253\,]^T$.
The estimate of λ is $w_1 = 24.3065$, and

$$\mathbf{z} = \mathbf{w}/w_1 = [1.0000, 0.5334, 0.5523]^T.$$

Checking the accuracy of our estimate, $\lambda = 24.3065$ and

$$\mathbf{z} = [1.0000, 0.5334, 0.5523]^T,$$

we compute $\mathbf{A}\,\mathbf{z} - \lambda\,\mathbf{z} = [-0.1249, -0.3653, -0.5123\,]^T$.

The maximum norm of this vector gives us a bound on the accuracy of the estimate after four steps, namely, $\|\mathbf{A}\,\mathbf{z} - \lambda\,\mathbf{z}\|_\infty = 0.5123$.

Using Mathcad's built-in function `eigenvals` to compute the eigenvalues of $\mathbf{A}$ gives:

$$\text{eigenvals}(\mathbf{A}) = [24, 8, 16]^T.$$

From this, we see that the largest eigenvalue is 24. The successive approximations of the power method converge toward that value.

Mathcad Function for the Basic Power Method

The Mathcad function power_m uses a simple function one(i,j) = 1, to initialize the vector **z** to be all 1's. The function power_m returns a table showing the eigenvalue (column 0), and the eigenvector at each iteration.

$\text{one}(i, j) := 1$

$\text{power\_m}(A, \text{max\_it}, \text{tol}) :=$
$\quad | \; n \leftarrow \text{rows}(A)$
$\quad | \; z \leftarrow \text{matrix}(n, 1, \text{one})$
$\quad | \; \text{it} \leftarrow 0$
$\quad | \; \text{error} \leftarrow 100$
$\quad | \; \text{while } (\text{it} < \text{max\_it} \wedge \text{error} > \text{tol})$
$\quad\quad | \; w \leftarrow A \cdot z$
$\quad\quad | \; ww \leftarrow |w|$
$\quad\quad | \; mx \leftarrow \max(w)$
$\quad\quad | \; mn \leftarrow \min(w)$
$\quad\quad | \; m \leftarrow \max(|mx|, |mn|)$
$\quad\quad | \; \text{for } k \in 0..n-1$
$\quad\quad\quad | \; \text{if } m = |w_k|$
$\quad\quad\quad\quad | \; kk \leftarrow k$
$\quad\quad\quad\quad | \; \text{continue}$
$\quad\quad | \; m \leftarrow w_{kk}$
$\quad\quad | \; z \leftarrow \dfrac{w}{w_{kk}}$
$\quad\quad | \; \text{error} \leftarrow |A \cdot z - m \cdot z|$
$\quad\quad | \; \text{Table}_{\text{it}, 0} \leftarrow \text{it}$
$\quad\quad | \; \text{Table}_{\text{it}, 1} \leftarrow m$
$\quad\quad | \; \text{for } k \in 0..n-1$
$\quad\quad\quad | \; \text{Table}_{\text{it}, k+2} \leftarrow z_k$
$\quad\quad | \; \text{it} \leftarrow \text{it} + 1$
$\quad | \; \text{eigens} \leftarrow \text{stack}(m, z)$
$\quad | \; \text{Table}$

Example 7.2 Using the Mathcad Function for the Power Method

To find the dominant eigenvalue of matrix **A**, we begin by defining the matrix and the parameters max_it (the maximum number of iterations) and tol (the tolerance on the error. The process terminates when either the number of iteration exceeds max_it or the error |**A z** − *m* **z**| is less than tol.

$$A := \begin{pmatrix} -44 & 9 & -6 \\ -280 & 57 & -40 \\ -75 & 15 & -13 \end{pmatrix} \quad \text{max\_it} := 15 \quad \text{tol} := 0.005 \quad C := \text{power\_m}(A, \text{max\_it}, \text{tol})$$

C =

	0	1	2	3	4
0	0	-263	0.156	1	0.278
1	1	2.247	0.212	1	-0.134
2	2	3.125	0.159	1	0.28
3	3	1.385	0.246	1	-0.387
4	4	3.667	0.139	1	0.436
5	5	-1.07	-0.263	-0.68	1
6	6	-5.013	0.106	1	0.687
7	7	-1.884	-0.113	0.089	1
8	8	-3.283	0.069	1	0.971
9	9	-2.799	-0.049	0.415	1
10	10	-3.082	0.032	0.83	1
11	11	-2.947	-0.022	0.556	1
12	12	-3.036	0.014	0.74	1
13	13	-2.976	-9.622·10⁻³	0.617	1
14	14	-3.016	6.38·10⁻³	0.699	1

To extract the eigenvalue and eigenvector:

$r := \text{rows}(C) - 1 \qquad m := C_{r,1} \qquad m = -3.016$

$z := \text{submatrix}(C, r, r, 2, 4) \qquad z = \begin{pmatrix} 6.38 \times 10^{-3} & 0.699 & 1 \end{pmatrix}$

Given additional iterations, the dominant eigenvalue is found to be −3; the corresponding eigenvector is $[0 \ 2/3 \ 1]^T$.

Accelerated Power Method

In some cases, it is possible to accelerate the convergence of the power method by using an estimate of λ that does not rely on a single component of the current vectors **w** and **z**. The improved estimate of λ is called the *Rayleigh quotient*. It is given by:

$$\lambda = (\mathbf{z}^T \mathbf{w})/(\mathbf{z}^T \mathbf{z}).$$

When **A** is symmetric, the power method with the Rayleigh quotient converges more rapidly than the basic power method. The accelerated power method can be implemented by replacing the definition of λ in the Mathcad function for the basic power method.

Example 7.3 Dominant Eigenvalue for Symmetric Matrix

To illustrate the acceleration in convergence of the dominant eigenvalue of a symmetric matrix that generally results when using the Rayleigh quotient, consider the following four-by-four matrix:

$$\mathbf{A} = \begin{bmatrix} 4 & 2/3 & -4/3 & 4/3 \\ 2/3 & 4 & 0 & 0 \\ -4/3 & 0 & 6 & 2 \\ 4/3 & 0 & 2 & 6 \end{bmatrix}.$$

The estimate of the dominant eigenvalue from the basic power method λ_b, the estimate from the Rayleigh quotient λ_r, and the eigenvector **z** are shown in the table below. After 10 iterations, the error from the basic method is 0.1328; that from the Rayleigh quotient approximation is 0.1036.

Iterations of basic and accelerated power method

Step	λ_b	λ_r	z(1)	z(2)	z(3)	z(4)
1	9.3333	6.3333	0.5000	0.5000	0.7143	1.0000
2	8.0952	7.2792	0.3353	0.2882	0.6941	1.0000
3	7.8353	7.6621	0.2477	0.1757	0.7297	1.0000
4	7.7898	7.8286	0.1885	0.1114	0.7764	1.0000
5	7.8042	7.9083	0.1443	0.0732	0.8210	1.0000
6	7.8344	7.9495	0.1104	0.0497	0.8595	1.0000
7	7.8661	7.9718	0.0842	0.0346	0.8911	1.0000
8	7.8945	7.9842	0.0640	0.0246	0.9164	1.0000
9	7.9181	7.9911	0.0485	0.0178	0.9362	1.0000
10	7.9371	7.9950	0.0366	0.0131	0.9516	1.0000

Shifted Power Method

We may need to find eigenvalues besides (or instead of) the eigenvalue of largest magnitude. Some simple properties of eigenvalues can help us do this. In particular, we can make use of the fact that if a matrix **A** has eigenvalues $\lambda_1, \lambda_2, \ldots, \lambda_n$, with corresponding eigenvectors $\mathbf{v}_1, \mathbf{v}_2, \ldots \mathbf{v}_n$, then the eigenvalues of $\mathbf{A} - b\mathbf{I}$ are $\mu_1 = \lambda_1 - b, \mu_2 = \lambda_2 - b, \ldots \mu_n = \lambda_n - b$; the eigenvectors are unchanged by the shift.

If we already know an eigenvalue λ of a matrix **A**, we can find another eigenvalue of **A** by applying the power method to the matrix $\mathbf{B} = \mathbf{A} - \lambda \mathbf{I}$. We denote the dominant eigenvalue of the shifted matrix **B** as μ. No additional algorithm or function is required for the shifted power method, since it is simply the basic (or accelerated basic) power method applied to the shifted matrix.

Example 7.4 Dominant Eigenvalue of Shifted Matrix

Consider the inertial matrix introduced in Example 7-A. One eigenvalue of

$$\mathbf{G} = \begin{bmatrix} 5 & -2 & 0 \\ -2 & 3 & 0 \\ 0 & 0 & 6 \end{bmatrix}$$

is 6; to find another eigenvalue, we apply the power method to the shifted matrix

$$\mathbf{B} = \mathbf{G} - 6\mathbf{I} = \begin{bmatrix} -1 & -2 & 0 \\ -2 & -3 & 0 \\ 0 & 0 & 0 \end{bmatrix}.$$

We start with $\mathbf{z} = [\,1, 1, 1\,]^T$ and use the Rayleigh quotient approximation.

Step 1: $\mathbf{w} = \mathbf{B}\,\mathbf{z} = [-3, -5, 0]^T$, $\quad \mu = \dfrac{\mathbf{z}^T\mathbf{w}}{\mathbf{z}^T\mathbf{z}} = -\dfrac{8}{3}$; $\quad \mathbf{z} = \dfrac{\mathbf{w}}{w_2} = \left[\dfrac{3}{5}, 1, 0\right]^T$.

Step 2: $\mathbf{w} = \mathbf{B}\,\mathbf{z} = [-\dfrac{13}{5}, -\dfrac{21}{5}, 0]^T$, $\quad \mu = \dfrac{\mathbf{z}^T\mathbf{w}}{\mathbf{z}^T\mathbf{z}} = -\dfrac{72}{17}$; $\quad \mathbf{z} = \dfrac{\mathbf{w}}{w_2} = \left[\dfrac{13}{21}, 1, 0\right]^T$.

The results of the first three iterations using a Mathcad function for the power method with Rayleigh quotient to estimate the eigenvalue are summarized below. The stopping condition is $\|\mathbf{A}\,\mathbf{z} - \mu\,\mathbf{z}\| < 0.0001$. The eigenvalue of **G** corresponding to $\mu = -4.2361$ is $\lambda = 6 - 4.2361 = 1.7639$.

Iterations of shifted power method

Step	μ	z(1)	z(2)	z(3)
1	−2.6667	0.6000	1.0000	0
2	−4.2353	0.6190	1.0000	0
3	−4.2361	0.6180	1.0000	0

7.1.2 Inverse Power Method

The inverse power method provides an estimate of the eigenvalue of $\mathbf{A}$ that is of smallest magnitude. It is based on the fact that eigenvalues of $\mathbf{B} = \mathbf{A}^{-1}$ are the reciprocals of the eigenvalues of $\mathbf{A}$. Therefore, we apply the power method to $\mathbf{B} = \mathbf{A}^{-1}$ to find its dominant eigenvalue, μ. The reciprocal of μ will give the eigenvalue of $\mathbf{A}$ with the smallest magnitude; we denote this eigenvalue as λ. However, it is not desirable to actually compute $\mathbf{A}^{-1}$; instead, at the stage where the power method would compute $\mathbf{A}^{-1}\mathbf{z} = \mathbf{w}$ to find the next approximation to the eigenvector $\mathbf{w}$, we solve the system $\mathbf{A}\mathbf{w} = \mathbf{z}$ for $\mathbf{w}$. This is an example of a situation in which the LU factorization is useful, since we must solve a linear system with the same coefficient matrix, but different right-hand sides, at each stage.

The inverse power method can also be generalized to find the eigenvalue that is closest to a given number. This procedure is discussed further in the next section.

Algorithm for the Inverse Power Method

Find eigenvalue of smallest magnitude, m, and corresponding eigenvector, $\mathbf{z}$, for matrix $\mathbf{M}$.

Initialize $\mathbf{z}$ (n-by-1 vector, all components = 1)

Begin iterations
     solve $\mathbf{M}\mathbf{w} = \mathbf{z}$ (this corresponds to setting $\mathbf{w} = \mathbf{M}^{-1}\mathbf{z}$)
     approximate eigenvalue if desired at each step

         set $m = \dfrac{\mathbf{z}^T \mathbf{z}}{\mathbf{z}^T \mathbf{w}}$ (approx. eigenvalue)

     (reciprocal of estimate of dominant eigenvalue, using Rayleigh quotient)
     find k such that $|\mathbf{w}_k| \geq |\mathbf{w}_j|$ for $1 \leq j < n$

         set $\mathbf{z} = \dfrac{\mathbf{w}}{\mathbf{w}_k}$ (approx. eigenvector)

         (scaled so that largest component is 1)

     test for convergence
         A quick test is to compute change in approximate eigenvector, or norm of the change in the eigenvector. This would require saving the previous eigenvector at each step.

         A better test would be to compute the norm of the residual, $|\mathbf{M}\mathbf{z} - m\mathbf{z}|$, which would be zero for the exact eigenvalue and eigenvector.

End iterations

The next example illustrates the first three steps of the inverse power method.

Example 7.5 Using the Inverse Power Method

To find the smallest eigenvalue of

$$\mathbf{A} = \begin{bmatrix} 21 & 7 & -1 \\ 5 & 7 & 7 \\ 4 & -4 & 20 \end{bmatrix} = \begin{bmatrix} 1 & 0 & 0 \\ 0.24 & 1 & 0 \\ 0.19 & -1 & 1 \end{bmatrix} \begin{bmatrix} 21 & 7 & -1 \\ 0 & 5.33 & 7.24 \\ 0 & -0 & 27.43 \end{bmatrix},$$

we apply the inverse power method with initial vector $\mathbf{z} = [\,1, 1, 1\,]^T$.

Step 1: Solve $\mathbf{A}\,\mathbf{w} = \mathbf{z} \Rightarrow \mathbf{w} = [\,0.0286, 0.0651, 0.0573\,]^T$.

Since the largest component of $\mathbf{w}$ is w_2, the estimate of the largest eigenvalue of $\mathbf{A}^{-1}$ is $\mu = \dfrac{w_2}{z_2} = 0.0651$.

The estimate of the smallest eigenvalue of $\mathbf{A}$ is $\lambda = 1/\mu = \dfrac{z_2}{w_2} = 15.3610$, and

$$\mathbf{z} = \dfrac{\mathbf{w}}{w_2} = [0.4400, 1.0001, 0.8801]^T$$

Step 2: Solve $\mathbf{A}\,\mathbf{w} = \mathbf{z} \Rightarrow \mathbf{w} = [-0.0042, 0.0842, 0.0617]^T$.
Since the largest component of $\mathbf{w}$ is w_2, we have

$$\mu = \dfrac{w_2}{z_2} = 0.0842, \quad \lambda = 1/\mu = \dfrac{z_2}{w_2} = 11.8777,$$

$$\mathbf{z} = \dfrac{\mathbf{w}}{w_2} = [-0.0495, 0.9997, 0.7324]^T.$$

Step 3: Solve $\mathbf{A}\,\mathbf{w} = \mathbf{z} \Rightarrow \mathbf{w} = [-0.0336, 0.1029, 0.0639\,]^T$.
The estimate of λ is $\dfrac{z_2}{w_2} = 9.7153$.

As indicated in Example 7.1, using Mathcad to compute the eigenvalues of $\mathbf{A}$ gives eigenvalues of 24, 8, and 16.

Mathcad Function for the Inverse Power Method

The function `InvPower` uses the function `lu_factor` and `lu_solve`. These functions may be repeated preceding the function `InvPower` (in the same worksheet where `InvPower` is given and used) or a reference to the function may be given before `InvPower`, using Insert Reference on the Mathcad toolbar to specify the location of the required functions (using the relative path option).

$$\text{one}(i,j) := 1$$

$$\text{InvPower}(A, \text{max\_it}, \text{tol}) := \begin{vmatrix} n \leftarrow \text{rows}(A) \\ z \leftarrow \text{matrix}(n, 1, \text{one}) \\ \text{it} \leftarrow 0 \\ \text{error} \leftarrow 100 \\ B \leftarrow \text{lu\_factor}(A) \\ L \leftarrow \text{submatrix}(B, 0, n-1, 0, n-1) \\ U \leftarrow \text{submatrix}(B, 0, n-1, n, 2 \cdot n - 1) \\ \text{while } (\text{it} < \text{max\_it} \wedge \text{error} > \text{tol}) \\ \quad \begin{vmatrix} w \leftarrow \text{lu\_solve}(L, U, z) \\ ww \leftarrow |w| \\ \text{wmax} \leftarrow \max(w) \\ m \leftarrow \dfrac{\left(z^T \cdot z\right)_{0,0}}{\left(z^T \cdot w\right)_{0,0}} \\ z \leftarrow \dfrac{w}{\text{wmax}} \\ \text{out}_{\text{it}, 0} \leftarrow m \\ \text{for } i \in 1..n \\ \quad \text{out}_{\text{it}, i} \leftarrow z_1 \\ \text{error} \leftarrow |A \cdot z - m \cdot z| \\ \text{it} \leftarrow \text{it} + 1 \end{vmatrix} \\ \text{eigens} \leftarrow \text{stack}(m, z) \\ \text{eigens} \end{vmatrix}$$

The next example shows the results of 5, 10, 15, and 20 iterations of the inverse power method to find the smallest eigenvalue, and corresponding eigenvector for the given matrix **A**.

Example 7.6 Finding the Smallest Eigenvalue

We find the smallest eigenvalue, and corresponding eigenvector of matrix A, using the Inverse Power method with the stated tolerance.

$$A := \begin{pmatrix} 66 & -21 & 9 \\ 228 & -73 & 33 \\ 84 & -28 & 16 \end{pmatrix} \qquad \text{tol} := 0.0001$$

The first component of the returned vector is the eigenvalue, the remainder is the eigenvector

$$\text{InvPower}(A, 5, \text{tol}) = \begin{pmatrix} 1.858 \\ 0.296 \\ 1 \\ 0.221 \end{pmatrix} \qquad \text{InvPower}(A, 10, \text{tol}) = \begin{pmatrix} 1.984 \\ 0.299 \\ 1 \\ 0.203 \end{pmatrix}$$

$$\text{InvPower}(A, 15, \text{tol}) = \begin{pmatrix} 1.998 \\ 0.3 \\ 1 \\ 0.2 \end{pmatrix} \qquad \text{InvPower}(A, 20, \text{tol}) = \begin{pmatrix} 2 \\ 0.3 \\ 1 \\ 0.2 \end{pmatrix}$$

The next example illustrates the results of using the inverse power method for the application introduced in Example 7-B.

Example 7.7 Buckling Beam

The buckling and subsequent breaking of a simply supported beam is approximated by the smallest eigenvalue of the matrix A given in Example 7-B.

A better approximation is obtained for larger values of n.

For $n = 6$, with five iterations of the inverse power method and a tolerance equal to 10^{-5}, we find that $m = 0.1981$, and therefore, the first buckling mode is approximately $(6 + 1)^2(0.198) = 9.7069$.

For $n = 10$, performing five iterations of the inverse power method with a tolerance equal to 10^{-5}, we find that $m = 0.0810$, and the estimate of the first buckling mode is $(10 + 1)^2(0.081) = 9.8027$.

General Inverse Power Method

The ideas of the shifted power method and the simple inverse power method can be combined to enable us to find the eigenvalue of a matrix $\mathbf{A}$ that is closest to a given number b. This method is based on the fact that if the eigenvalues of a matrix $\mathbf{C}$ are $\lambda_1, \ldots, \lambda_n$, then the eigenvalues of $\mathbf{D} = \mathbf{C} - b\mathbf{I}$ are $\lambda_1 - b, \ldots, \lambda_n - b$, and the eigenvalues of $\mathbf{D}^{-1}$ are $(\lambda_1 - b)^{-1}, \ldots, (\lambda_n - b)^{-1}$. Thus, μ_1, the dominant eigenvalue of $\mathbf{D}^{-1}$, corresponds to the eigenvalue of $\mathbf{C}$ that is closest to b, according to the relation $\mu_1 = (\lambda_1 - b)^{-1}$. We illustrate the method in the next example. We do not include a separate algorithm or Mathcad function, since the method is simply the inverse power method applied to a shifted matrix.

Example 7.8 Finding the Eigenvalue Closest to a Given Number

Find the eigenvalue of $\mathbf{A}$ that is closest to $b = 15$ by applying the inverse power method to $\mathbf{B} = \mathbf{A} - 15\mathbf{I}$, where

$$\mathbf{A} = \begin{bmatrix} 21 & 7 & -1 \\ 5 & 7 & 7 \\ 4 & -4 & 20 \end{bmatrix}, \text{ and } \mathbf{B} = \begin{bmatrix} 6 & 7 & -1 \\ 5 & -8 & 7 \\ 4 & -4 & 5 \end{bmatrix}.$$

Start with $\mathbf{z} = [1, 1, 1]^T$
Step 1: Solve $\mathbf{B}\mathbf{w} = \mathbf{z} \Rightarrow \mathbf{w} = [0.0317, 0.1587, 0.3016]^T$;
estimate of largest eigenvalue of $\mathbf{B}^{-1}$: $\mu = \dfrac{w_3}{z_3} = 0.3016$;
estimate of smallest eigenvalue of $\mathbf{B}$: $1/\mu = \dfrac{z_3}{w_3} = 3.3158$;

$$\mathbf{z} = \dfrac{\mathbf{w}}{w_3} = [0.1053, 0.5263, 1.000]^T.$$

Step 2: Solve $\mathbf{B}\mathbf{w} = \mathbf{z} \Rightarrow \mathbf{w} = [0.3718, 0.4570, 0.8630]^T$;

$$\mu = \dfrac{w_3}{z_3} = 0.8630; \quad 1/\mu = \dfrac{z_3}{w_3} = 1.1588;$$

$$\mathbf{z} = \dfrac{\mathbf{w}}{w_3} = [-0.4308, 0.5295, 1.000]^T.$$

Step 3: Solve $\mathbf{B}\mathbf{w} = \mathbf{z} \Rightarrow \mathbf{w} = [-0.4723, 0.4808, 0.9624]^T$;
estimate of smallest eigenvalue of $\mathbf{B}$: $1/\mu = \dfrac{z_3}{w_3} = 1.0390$.

The estimate of the desired eigenvalue of $\mathbf{A}$ is $b + 1/\mu$. Thus, the estimated eigenvalue of $\mathbf{A}$ that is closest to 15 is approximately 16.

7.1.3 Discussion

The proof of the convergence of the basic power method assumes that the dominant eigenvalue is real and not a repeated eigenvalue, so that the eigenvalues can be ordered as $|\lambda_1| > |\lambda_2| \geq |\lambda_3| \geq \ldots \geq |\lambda_n|$. Often, however, the method is applicable even if these assumptions are not met. (See Ralston and Rabinowitz, 1978.)

The proof is based on the fact that if matrix $\mathbf{A}$ is diagonalizable (i.e., if $\mathbf{A}$ has n linearly independent eigenvectors $\mathbf{v}^{(1)}, \ldots, \mathbf{v}^{(n)}$), then any vector $\mathbf{z}$ can be written as a unique linear combination of the eigenvectors:

$$\mathbf{z} = \sum_{i=1}^{n} c_i \mathbf{v}^{(i)} \Rightarrow \mathbf{A}^j \mathbf{z} = \sum_{i=1}^{n} c_i \mathbf{A}^j \mathbf{v}^{(i)}.$$

Since λ_i is the eigenvalue corresponding to eigenvector $\mathbf{v}^{(i)}$, it follows that

$$\mathbf{A} \mathbf{v}^{(i)} = \lambda_i \mathbf{v}^{(i)}, \text{ and } \mathbf{A}^j \mathbf{v}^{(i)} = \lambda_i^j \mathbf{v}^{(i)}.$$

Therefore,

$$\mathbf{A}^j \mathbf{z} = \sum_{i=1}^{n} c_i \lambda_i^j \mathbf{v}^{(i)} = \lambda_1^j \left[c_1 \mathbf{v}^{(1)} + \sum_{i=2}^{n} c_i \frac{\lambda_i^j}{\lambda_1^j} \mathbf{v}^{(i)} \right]$$

If $\mathbf{A}$ has a single dominant eigenvalue, then $\dfrac{\lambda_i^j}{\lambda_1^j} \to 0$ as $j \to \infty$.

Without some control over the iterative process that this suggests, we cannot distinguish the unknown quantities λ_1^j, c_1, and $\mathbf{v}^{(1)}$ on the right-hand side of the equation for $\mathbf{A}^j \mathbf{z}$. It is also quite likely that all components of the limiting vector will grow without bound. However, by an appropriate scaling of the vectors produced during iteration, we can assure that this limit is finite and nonzero.

The convergence of the estimate at the jth iteration depends on $|\lambda_2 \lambda_1|^j$, so the method will converge more rapidly when the dominant eigenvalue is much larger than the next most dominant eigenvalue.

Convergence may be improved for symmetric matrices by using the Rayleigh quotient to estimate the eigenvalue. In this case, the estimate at the jth iteration depends on $|\lambda_2 \lambda_1|^{2j}$ (because the eigenvectors can be taken to be orthonormal).

The *trace* of a square matrix $\mathbf{A}$, $\text{tr}(\mathbf{A})$, is defined as the sum of the diagonal elements of $\mathbf{A}$. The trace is also equal to the sum of the eigenvalues of $\mathbf{A}$ (Cf. Golub and Van Loan, 1996, p. 310). This fact can be used to help determine values of the shift parameter b to use for finding other eigenvalues. It is especially useful when all diagonal element are nonnegative. In such a case, after the dominant eigenvalue λ_1 has been found, one might try a value of b given by $b = \dfrac{\text{tr}(\mathbf{A}) - \lambda_1}{n - 1}$.

7.2 QR FACTORIZATION

In this section we consider the problem of factoring a square matrix **A** into the product of an orthogonal matrix **Q** and an upper (right) triangular matrix **R**. Thus we seek to write **A** = **QR**, where **R** is right triangular, **Q** is orthogonal (so that $\mathbf{Q}^{-1} = \mathbf{Q}^T$), and both are real. The QR factorization is more computationally intensive (about twice as many operations) to obtain than an LU factorization, but has superior stability properties. In the next section we present a method, based on QR factorization, for finding all of the eigenvalues of a matrix.

The process of finding the QR factorization of a matrix requires that we be able to transform a matrix **A** so that certain elements of the resulting matrix **B** are zero. The transformation may be of the form **B** = **M A** or **B** = **M A M**$^{-1}$. If the matrix **M** is orthogonal, the transformation, **B** = **M A M**$^{-1}$, is a similarity transformation, which preserves eigenvalues. However, in some situations, the desired form of the resulting matrix **B** is such that we cannot achieve it with similarity transformations.

We begin by introducing two important transformations, the Householder and Givens transformations. Each is based on the creation of an orthogonal matrix with certain desirable properties, although in practice the equivalent operations are usually carried out more efficiently without explicitly forming the matrix or performing the matrix multiplication. The use of Householder transformations is illustrated for two important applications, transforming a matrix to upper triangular form, and transforming a matrix to upper Hessenberg form. Transformation to upper Hessenberg form can be achieved with a similarity transformation, but upper triangular form cannot. The use of Givens transformations is illustrated for transforming an upper Hessenberg matrix to upper triangular form.

The remainder of this section deals with the process of finding the QR factorization of a matrix, using Householder or Givens transformations.

7.2.1 Householder Transformations

A Householder matrix is an orthogonal matrix **H** (n-by-n) with the property that for a given vector **x** and a specified index k ($1 \leq k \leq n - 1$), the vector **z**, defined as **z** = **Hx**, has the form $[z_1, \ldots z_k, 0 \ldots 0]^T$. That is, the elements $x_{k+1}, \ldots x_n$ have been "zeroed-out". In applications it is often the case that vector **x** is a column of a given matrix and we need to zero-out all of the elements of the column below a certain point. A Householder matrix has the form $\mathbf{H} = \mathbf{I} - 2\,\mathbf{w}\,\mathbf{w}^T$, where **w** is a column vector with $\|\mathbf{w}\|_2 = 1$ and **I** is the n-by-n identity matrix. Because **H** is orthogonal, the transformation **B** = **H A H** is a similarity transformation.

The formation of the Householder matrix, which reduces to zero positions $k + 1$ through n of a vector **x**, is summarized in the following algorithm. Note that the algorithm begins by setting the first $k - 1$ components of **x** to 0; this simplifies the computation of the parameter g, and the vector **w**.

Algorithm for Forming a Householder Matrix

Initialize index k and vector $\mathbf{x}$ (to be transformed)
$\mathbf{x} = [0, \ldots, 0, x_k, \ldots, x_n]^T$ (set $x_1 \ldots x_{k-1}$ to 0)
$g = \text{norm}(\mathbf{x})$ (if $g = 0$, stop)
$s = \sqrt{2\,g\,(g + |x_k|)}$
$\mathbf{w} = \mathbf{x}/s$
$\mathbf{w}_k = \dfrac{1}{s}(x_k + \text{sign}(x_k)\,g)$
$\mathbf{H} = \mathbf{I} - 2*\mathbf{w}*\mathbf{w}^T$

Example 7.9 Using the Householder Algorithm

ORIGIN := 1

We illustrate the use of the Mathcad algorithm for creating a Householder matrix by using it to create the matrices needed to perform a similarity transform of matrix A to Hessenberg form

$$A := \begin{pmatrix} 4 & \frac{2}{3} & \frac{-4}{3} & \frac{4}{3} \\ \frac{2}{3} & 4 & 0 & 0 \\ \frac{-4}{3} & 0 & 6 & 2 \\ \frac{4}{3} & 0 & 2 & 6 \end{pmatrix}$$

The first column of A is to be zeroed-out below the subdiagonal.

$x := \text{submatrix}(A, 1, 4, 1, 1)$ $\quad x^T = (4 \quad 0.667 \quad -1.333 \quad 1.333)$

Define parameters n and k $\quad n := 4 \quad k := 1$

Perform computations

$x_1 := 0 \quad g := |x| \quad s := \sqrt{2 \cdot g \cdot (g + |x_{k+1}|)} \quad w := \dfrac{x}{s} \quad w_{k+1} := \dfrac{(x_{k+1} + \text{sign}(x_{k+1}) \cdot g)}{s}$

$w^T = (0 \quad 0.816 \quad -0.408 \quad 0.408) \qquad H := \text{identity}(4) - 2 \cdot w \cdot w^T$

$H = \begin{pmatrix} 1 & 0 & 0 & 0 \\ 0 & -0.333 & 0.667 & -0.667 \\ 0 & 0.667 & 0.667 & 0.333 \\ 0 & -0.667 & 0.333 & 0.667 \end{pmatrix} \quad B := H \cdot A \cdot H \quad B = \begin{pmatrix} 4 & -2 & 0 & 0 \\ -2 & 4 & 0 & 0 \\ 0 & 0 & 6 & 2 \\ 0 & 0 & 2 & 6 \end{pmatrix} \quad \text{eigenvals}(B) = \begin{pmatrix} 6 \\ 2 \\ 4 \\ 8 \end{pmatrix}$

Define vector to be transformed: the second column of B is to be zeroed-out below the diagonal.

$x := \text{submatrix}(B, 1, 4, 2, 2) \quad x_1 := 0 \quad x_2 := 0 \quad k := 2 \quad g := |x| \quad g = 0$

Since $g = 0$, no transformation should be performed.

Chapter 7 Eigenvalues, Eigenvectors, and QR Factorization

The following Mathcad function for the construction of a Householder matrix **H** is presented for instructional purposes. In practice, the matrix **B** that would result from using the matrix transformation **B** = **H A** is usually produced without explicitly forming the Householder matrix; the procedure is described following Example 7.10. To prevent the difficulties that would occur if one performs division by a quantity that is essentially 0, a test should be added to this function. One possibility would be to set **w** = **y** if $s = 0$, since in that case **y** would be the zero vector.

Mathcad Function for Householder Matrix

In using this function in applications, note that the vector x is an n-tuple, the first k components of x are not used in the computation.

$$\text{zero}(i,j) := 0$$

$$\text{Householder}(x,k) := \begin{vmatrix} n \leftarrow \text{rows}(x) \\ y \leftarrow \text{matrix}(n,1,\text{zero}) \\ \text{for } i \in k-1..n-1 \\ \quad y_i \leftarrow x_i \\ g \leftarrow |y| \\ p \leftarrow \text{sign}(x_{k-1}) \\ s \leftarrow \sqrt{2 \cdot g \cdot (g + p \cdot x_{k-1})} \\ w \leftarrow \dfrac{y}{s} \\ w_{k-1} \leftarrow \dfrac{x_{k-1} + p \cdot g}{s} \\ H \leftarrow \text{identity}(n) - 2 \cdot w \cdot w^T \\ H \end{vmatrix}$$

The next example illustrates the use of Householder matrices as part of the process of transforming a matrix **A** into an upper-triangular matrix. In this setting, Householder matrices are determined for each column of **A**, except the last, in order to reduce the elements below the diagonal of the matrix to zero.

Example 7.10 Using the Function `Householder`

To illustrate the use of the Mathcad function Householder, we use it to create the matrices needed to transform a given matrix A to upper triangular form.

$$A := \begin{pmatrix} 21 & 7 & -1 \\ 5 & 7 & 7 \\ 4 & -4 & 20 \end{pmatrix}$$

Define vector to be transformed: the first column of A is to be zeroed-out below the diagonal.

$x := \text{submatrix}(A, 0, 2, 0, 0)$ $\qquad x^T = (21 \ 5 \ 4)$

Apply Householder function to create matrix H1

$H1 := \text{Householder}(x, 1)$

Display result, and verify that H1 A has required form

$$H1 = \begin{pmatrix} -0.957 & -0.228 & -0.182 \\ -0.228 & 0.973 & -0.021 \\ -0.182 & -0.021 & 0.983 \end{pmatrix} \qquad H1 \cdot A = \begin{pmatrix} -21.954 & -7.561 & -4.282 \\ 0 & 5.305 & 6.618 \\ 0 & -5.356 & 19.694 \end{pmatrix}$$

Define B = H1 A, and apply Householder function to transform second column of B

$B := H1 \cdot A \qquad x := \text{submatrix}(B, 0, 2, 1, 1) \qquad x^T = (-7.561 \ 5.305 \ -5.356)$

$H2 := \text{Householder}(x, 2)$

$$H2 = \begin{pmatrix} 1 & 0 & 0 \\ 0 & -0.704 & 0.71 \\ 0 & 0.71 & 0.704 \end{pmatrix} \qquad H2 \cdot B = \begin{pmatrix} -21.954 & -7.561 & -4.282 \\ 0 & -7.539 & 9.335 \\ 0 & 0 & 18.561 \end{pmatrix}$$

For future reference, we note that while the transformation C=H2*H1*A produces an upper triangular matrix, it is not a similarity transform.

$C := H2 \cdot B \qquad \text{eigenvals}(C) = \begin{pmatrix} -21.954 \\ 18.561 \\ -7.539 \end{pmatrix} \qquad \text{eigenvals}(A) = \begin{pmatrix} 24 \\ 8 \\ 16 \end{pmatrix}$

The similarity transformation D = H2*H1*A*H1*H2 does not preserve the triangular form.

$D := H2 \cdot H1 \cdot A \cdot H1 \cdot H2 \qquad \text{eigenvals}(D) = \begin{pmatrix} 24 \\ 16 \\ 8 \end{pmatrix} \qquad D = \begin{pmatrix} 23.502 & 1.563 & -1.647 \\ 0.016 & 11.937 & 1.216 \\ -3.382 & 13.241 & 12.561 \end{pmatrix}$

Householder Transformations in Practice

In applications, Householder matrices are not explicitly constructed. In applications, we typically need to find either $\mathbf{B} = \mathbf{H}\,\mathbf{A}$ or $\mathbf{B} = \mathbf{H}\,\mathbf{A}\,\mathbf{H}$. These products can be found without actually constructing $\mathbf{H}$. The vector $\mathbf{w}$ is as described in the algorithm above.

To find $\mathbf{H}\,\mathbf{A}$, we observe that

$$\begin{aligned}\mathbf{H}\,\mathbf{A} &= (\mathbf{I} - 2\,\mathbf{w}\,\mathbf{w}^T)\,\mathbf{A} \\ &= \mathbf{A} - 2\,\mathbf{w}\,\mathbf{w}^T\,\mathbf{A} \\ &= \mathbf{A} - \mathbf{w}\,\mathbf{u}^T,\end{aligned}$$

where $\mathbf{u} = 2\,\mathbf{A}^T\,\mathbf{w}$.

To find $\mathbf{H}\,\mathbf{A}\,\mathbf{H}$, we see that

$$\begin{aligned}\mathbf{H}\,\mathbf{A}\,\mathbf{H} &= (\mathbf{I} - 2\,\mathbf{w}\,\mathbf{w}^T)\,\mathbf{A}\,(\mathbf{I} - 2\,\mathbf{w}\,\mathbf{w}^T) \\ &= (\mathbf{I} - 2\,\mathbf{w}\,\mathbf{w}^T)\,\mathbf{A} - 2\,(\mathbf{I} - 2\,\mathbf{w}\,\mathbf{w}^T)\,\mathbf{A}\,\mathbf{w}\,\mathbf{w}^T \\ &= \mathbf{A} - 2\,\mathbf{w}\,\mathbf{w}^T\,\mathbf{A} - 2\,\mathbf{A}\,\mathbf{w}\,\mathbf{w}^T + 4\,\mathbf{w}\,\mathbf{w}^T\,\mathbf{A}\,\mathbf{w}\,\mathbf{w}^T \\ &= \mathbf{A} - 2\,\mathbf{C}\,\mathbf{A} - 2\,\mathbf{A}\,\mathbf{C} + 4\,\alpha\,\mathbf{C}\end{aligned}$$

where $\mathbf{C} = \mathbf{w}\,\mathbf{w}^T$ and $\alpha = \mathbf{w}^T\,\mathbf{A}\,\mathbf{w}$.

For further discussion, see [Hager, 1988] or [Press et al., 1992].

The procedure for finding $\mathbf{B} = \mathbf{H}\,\mathbf{A}$ described above is used in the transformation of $\mathbf{A}$ to upper triangular form, described below. As noted above, tridiagonal form cannot, in general, be achieved with a similarity transformation. The procedure for computing the similarity transformation $\mathbf{B} = \mathbf{H}\,\mathbf{A}\,\mathbf{H}$ is used to transform $\mathbf{A}$ to the slightly less restrictive form known as upper Hessenberg; this allows for non-zero elements on the first subdiagonal, as well as in the upper-triangular portion of the matrix.

Transformation to Tridiagonal Form

The next example illustrates the use of Householder transformations without the formation of the Householder matrices.

Example 7.11 Householder Transformations to Convert Matrix to Upper-Triangular Form

To transform the matrix $\mathbf{A}$ to upper-triangular form, we use Householder transformations to zero-out the subdiagonal elements, column by column.

$$\mathbf{A} = \begin{bmatrix} 21 & 7 & -1 \\ 5 & 7 & 7 \\ 4 & -4 & 20 \end{bmatrix}$$

To introduce zeros below the diagonal in the first column of $\mathbf{A}$:

$$\mathbf{x} = [21, 5, 4]^T$$

$$g = \text{norm}(\mathbf{x}) = 21.954,$$

$$s = \sqrt{2g(g + |\mathbf{x}_k|)} = 43.429,$$

$$\mathbf{w} = \mathbf{x}/s$$

$$w_1 = \frac{1}{s}(\mathbf{x}_1 + \text{sign}(\mathbf{x}_1)\, g)$$

$$\mathbf{w} = = [0.98907, 0.11513, 0.092104]^T.$$

This gives

$$\mathbf{u} = 2\,\mathbf{A}^T\mathbf{w} = [43.429, 14.733, 3.3178]^T.$$

$$\mathbf{w}\,\mathbf{u}^T = \begin{bmatrix} 42.954 & 14.561 & 3.2816 \\ 5 & 1.6949 & 0.38198 \\ 4 & 1.356 & 0.30559 \end{bmatrix},$$

and

$$\mathbf{B} = \mathbf{A} - \mathbf{w}\,\mathbf{u}^T = \begin{bmatrix} -21.954 & -7.5611 & -4.2816 \\ 0 & 5.3051 & 6.618 \\ 0 & -5.3560 & 19.694 \end{bmatrix}.$$

To introduce zeros below the diagonal in the second column of $\mathbf{B}$:

$$\mathbf{x} = [0, 5.3051, -5.3560]^T$$

$$\mathbf{w} = [0, 0.92296, -0.38489]^T$$

$$\mathbf{u} = 2\,\mathbf{B}^T\mathbf{w} = [0, 13.916, -2.9436]^T$$

$$\mathbf{w}\,\mathbf{u}^T = \begin{bmatrix} 0 & 0 & 0 \\ 0 & 12.844 & -2.7169 \\ 0 & -5.356 & 1.133 \end{bmatrix},$$

and

$$\mathbf{C} = \mathbf{B} - \mathbf{w}\,\mathbf{u}^T = \begin{bmatrix} -21.954 & -7.5611 & -4.2816 \\ 0 & -7.5386 & 9.3349 \\ 0 & 0 & 18.561 \end{bmatrix}.$$

Similarity Transformation to Hessenberg Form

An application of QR factorization, which we consider in the next section, is the problem of finding the eigenvalues of a matrix. The solution to this problem requires forming a sequence of matrices, each of which must be factored. If the matrix to be factored is Hessenberg, then there is only one element below the diagonal in each column that must be zeroed out; Givens rotations (discussed in Section 7.2.2) are usually preferable to Householder transformations in this case. A matrix may be transformed to Hessenberg form using Householder transformations, and that form is preserved during the iterative process used in the QR eigenvalue routine. We now investigate how the matrix **A** (which is n-by-n) can be transformed to an upper-Hessenberg form, without the construction of the Householder matrices used in the process.

We need to transform each column of **A**, from column 1 to column $n - 2$; for column k we require that elements in rows $k + 2 \ldots n$ be 0. In order that the transformations preserve eigenvalues, we must compute **H A H** at each stage. We use the fact that $\mathbf{H A H} = \mathbf{A} - 2\,\mathbf{C A} - 2\,\mathbf{A C} + 4\,\alpha\,\mathbf{C}$, where $\mathbf{C} = \mathbf{w w}^T$ and $\alpha = \mathbf{w}^T \mathbf{A w}$.

Algorithm for Similarity Transformation to Hessenberg Form

Initialize
    matrix **M** (n-by-n)
For $k = 1$ to $n - 2$
    $\mathbf{x} = \mathbf{M}^{<k>}$ (k^{th} column of **M**)
    for $i = 1$ to k
        $x_i = 0$
    end
    $g = \text{norm }\mathbf{x}$
    $s = \sqrt{2\,g\,(g + |\mathbf{x}_{k+1}|)}$
    $\mathbf{w} = \dfrac{\mathbf{x}}{s}$
    $\mathbf{w}_{k+1} = \dfrac{1}{s}(\mathbf{x}_{k+1} + \text{sign}(\mathbf{x}_{k+1})\,g)$
    $\mathbf{A} = \mathbf{M}$ (save previous matrix)
    $\alpha = \mathbf{w}^T \mathbf{A w}$
    $\mathbf{C} = \mathbf{w w}^T$
    $\mathbf{M} = \mathbf{A} - 2\,\mathbf{C A} - 2\,\mathbf{A C} + 4\,\alpha\,\mathbf{C}$ (form new matrix)
End

We illustrate the use of the algorithm to transform a matrix to upper-Hessenberg form in the next example.

Example 7.12 Similarity Transformation to Upper-Hessenberg Form

$$\text{ORIGIN1} :=$$

$$M := \begin{pmatrix} 4 & \frac{2}{3} & \frac{-4}{3} & \frac{4}{3} \\ \frac{2}{3} & 4 & 0 & 0 \\ \frac{-4}{3} & 0 & 6 & 2 \\ \frac{4}{3} & 0 & 2 & 6 \end{pmatrix}$$

$k := 1 \quad x := M^{\langle k \rangle} \quad\quad x_1 := 0 \quad\quad x^T = (0 \ \ 0.667 \ \ -1.333 \ \ 1.333)$

$g := |x| \quad s := \sqrt{2 \cdot g \cdot (g + |x_{k+1}|)} \quad\quad w := \frac{x}{s} \quad w_{k+1} := \frac{(x_{k+1} + \text{sign}(x_{k+1}) \cdot g)}{s}$

$A := M \quad\quad B := w^T \cdot A \cdot w \quad\quad C := w \cdot w^T \quad\quad d := 4 \cdot B_{1,1}$

$$M := A - 2 \cdot C \cdot A - 2 \cdot A \cdot C + d \cdot C \quad\quad M = \begin{pmatrix} 4 & -2 & 0 & 0 \\ -2 & 4 & 0 & 0 \\ 0 & 0 & 6 & 2 \\ 0 & 0 & 2 & 6 \end{pmatrix}$$

Step 2

$k := 2 \quad x := M^{\langle k \rangle} \quad\quad x_1 := 0 \quad x_2 := 0 \quad\quad x^T = (0 \ \ 0 \ \ 0 \ \ 0)$

Note that computations stop when **x** is the zero vector to avoid division by zero in forming vector **w**.

A similarity transformation to upper-Hessenberg form can be accomplished using the following Mathcad function.

$$
\text{Hessenberg}(M) := \begin{vmatrix}
n \leftarrow \text{rows}(M) \\
\text{for } k \in 0..n-3 \\
\quad \begin{vmatrix}
x \leftarrow M^{\langle k \rangle} \\
\text{for } i \in 0..k \\
\quad x_i \leftarrow 0 \\
g \leftarrow |x| \\
s \leftarrow \sqrt{2 \cdot g \cdot (g + |x_{k+1}|)} \\
w \leftarrow \dfrac{x}{s} \\
w_{k+1} \leftarrow \dfrac{x_{k+1} + \text{sign}(x_{k+1}) \cdot g}{s} \\
A \leftarrow M \\
\alpha \leftarrow (w^T \cdot A \cdot w)_0 \\
C \leftarrow w \cdot w^T \\
M \leftarrow A - 2 \cdot C \cdot A - 2 \cdot A \cdot C + 4 \cdot \alpha \cdot C
\end{vmatrix} \\
M
\end{vmatrix}
$$

Analysis of the Householder Transformation

To understand how the Householder transformation works, we must do a little analysis. The basic ideas are as follows. Any matrix formed according to the equation $\mathbf{H} = \mathbf{I} - 2\,\mathbf{w}\,\mathbf{w}^T$ is symmetric. If $\mathbf{w}$ is a unit vector ($\|\mathbf{w}\|_2 = 1$), then $\mathbf{H}$ is orthogonal, since $\mathbf{H}^T \mathbf{H} = (\mathbf{I} - 2\mathbf{w}\mathbf{w}^T)(\mathbf{I} - 2\mathbf{w}\mathbf{w}^T) = \mathbf{I} - 4\mathbf{w}\mathbf{w}^T + 4\mathbf{w}\mathbf{w}^T\mathbf{w}\mathbf{w}^T = \mathbf{I}$. Because it is both symmetric and orthogonal, $\mathbf{H}$ is its own inverse! Any vector $\mathbf{v} \neq 0$ that is not a unit vector can be scaled to produce a unit vector $\mathbf{w}$ in the same direction by defining $\mathbf{w} = \dfrac{\mathbf{v}}{\|\mathbf{v}\|_2}$.

Now we outline how we can determine the vector $\mathbf{v}$ that will provide a Householder matrix of the form $\mathbf{H} = \mathbf{I} - \dfrac{2}{\mathbf{v}^T \mathbf{v}} \mathbf{v}\mathbf{v}^T$ such that multiplication of an n-dimensional vector $\mathbf{x}$ by $\mathbf{H}$ will reduce all the components of $\mathbf{x}$ to zero except for the kth component. Let $\mathbf{e}_k$ denote the n-dimensional vector with a 1 as its kth component and 0's for all other components. Then we want to determine $\mathbf{v}$ such that $\mathbf{H}\,\mathbf{x} = \alpha\,\mathbf{e}_k$, for some scalar α. Some matrix algebra tells us that $\mathbf{v}$ must be a linear combination of **x** and $\mathbf{e}_k$. We set $\mathbf{v} = \mathbf{x} + \alpha \, \mathbf{e}_k$ and find (after more algebra) that $\mathbf{v} = \mathbf{x} \pm \|\mathbf{x}\|_2 \, \mathbf{e}_k$ and $\mathbf{H}\mathbf{x} \pm \|\mathbf{x}\|_2 \, \mathbf{e}_k = 0$, as desired.

It is good practice to choose the sign of $\alpha = \pm \|\mathbf{x}\|_2$ to agree with the sign of x_k, in order to avoid cancellation in the kth component of **v**. It is for this reason that we compute $w_k = (x_k + \text{sign}(x_k) \, g)/s$ in step 3 of the algorithm for the Householder transformation.

To apply the Householder transformation to the first column of a matrix **A**, we compute the vector **v** with $k = 1$, using the first column of **A** as the vector **x**. For columns other than the first, we have $k > 1$, and we want to reduce to zero only elements after the kth component of column k. Therefore, we set the components of **v** before the kth component equal to zero; i.e., $v_i = 0$ for $i = 1, \ldots k - 1$. We use only components k through n to compute the norm of the **x** term, denoted as g in step 2. We ignore the first $k - 1$ components and treat the remaining components as an $(n - k + 1)$-dimensional vector. The norm of **v** is denoted as s in step 2. The vector **w** is obtained by dividing the component of **v** by the norm of **v**. That is the reason for division by s in step 3 of the algorithm.

Computational Effort

To see that this computation of **H A** is preferable to actually forming **H** and performing the matrix multiplication, consider the following analysis of the computational effort of the two approaches. Computation of **u** requires the multiplication of a matrix and a vector, that of $\mathbf{w}\,\mathbf{u}^T$ requires the outer product of two vectors, and that of $\mathbf{A} - \mathbf{w}\,\mathbf{u}^T$ requires the subtraction of one matrix from another. Altogether these steps require $4n^2$ flops if **A** is an n-by-n matrix and **w** is n-by-1. On the other hand, explicit formation of the matrix $\mathbf{H} = \mathbf{I} - 2\,\mathbf{w}\,\mathbf{w}^T$ requires the outer product of a vector with itself and multiplication of the matrices **H** and **A**, for a total of $2n^3$ flops. This is an order of magnitude more computational effort than it takes to compute $\mathbf{A} - \mathbf{w}\,\mathbf{u}^T$. For that reason, Householder matrices are rarely explicitly formed in practice.

7.2.2 Givens Transformations

A Givens rotation is a matrix of the form

$$G = \begin{bmatrix} 1 & 0 & 0 & 0 & 0 \\ 0 & c & 0 & -s & 0 \\ 0 & 0 & 1 & 0 & 0 \\ 0 & s & 0 & c & 0 \\ 0 & 0 & 0 & 0 & 1 \end{bmatrix} \begin{matrix} \\ \leftarrow \text{row } i \\ \\ \leftarrow \text{row } j \\ \\ \end{matrix}$$

where $c^2 + s^2 = 1$.

In general, the c's occur on the diagonal in row i and row j.

By selecting the appropriate values for c and s, a Givens rotation can be used to reduce a specific matrix element to zero. In particular, if

$$c = \frac{x_1}{\sqrt{x_1^2 + x_2^2}} \quad \text{and} \quad s = \frac{-x_2}{\sqrt{x_1^2 + x_2^2}}$$

then $c^2 + s^2 = 1$ and

$$\begin{bmatrix} c & -s \\ s & c \end{bmatrix} \begin{bmatrix} x_1 \\ x_2 \end{bmatrix} = \begin{bmatrix} \sqrt{x_1^2 + x_2^2} \\ 0 \end{bmatrix}$$

If **G** is a Givens matrix with $g_{ii} = g_{jj} = c$, and $g_{ij} = -s$, $g_{ji} = s$, (with $i \leq j$), and all other elements as for the identity matrix, then multiplication of a matrix **A** by **G** affects only rows i and j of matrix **A**; i.e., if **C** = **G A** then

$$\mathbf{C}(i,1{:}n) = c\,\mathbf{A}(i,1{:}n) - s\,\mathbf{A}(j,1{:}n),$$

$$\mathbf{C}(j,1{:}n) = s\,\mathbf{A}(i,1{:}n) + c\,\mathbf{A}(j,1{:}n),$$

and

$$\mathbf{C}(k,1{:}n) = \mathbf{A}(k,1{:}n) \qquad \text{for } k \neq i, j.$$

A Givens rotation matrix is orthogonal, since $\mathbf{G}^T \mathbf{G} = I$. This means that the product $\mathbf{B} = \mathbf{G} \mathbf{A} \mathbf{G}^T$ is a similarity transformation of matrix **A**. Multiplication on the right by $\mathbf{G}^T$ affects only columns i and j, that is, if $\mathbf{D} = \mathbf{C}\,\mathbf{G}^T$, then

$$\mathbf{D}(1{:}n,i) = c\,\mathbf{C}(1{:}n,i) - s\,\mathbf{C}(1{:}n,j),$$

$$\mathbf{D}(1{:}n,j) = s\,\mathbf{C}(1{:}n,i) + c\,\mathbf{C}(1{:}n,j),$$

and

$$\mathbf{D}(1{:}n,k) = \mathbf{C}(1{:}n,k) \qquad \text{for } k \neq i,j.$$

Hence, the actions of the multiplication on the left or the right by **G** or $\mathbf{G}^T$ can be accomplished without explicitly forming **G** or $\mathbf{G}^T$ and without performing the matrix multiplication.

Our primary interest in Givens transformations is for the transformation of an upper Hessenberg matrix into an upper triangular matrix. In this case, when we are working on the kth column of the Hessenberg matrix **B**, we know that the only element that must be zeroed out is the element on the subdiagonal, namely $b_{k+1,k}$.

We illustrate the process in the next example. An algorithmic statement of the method follows the example.

Example 7.13 Transformation of Upper-Hessenberg Form to Upper-Triangular Form

Transformation of upper-Hessenberg matrix B to upper-triangular matrix T, using Givens rotation; we initialize B to be the matrix found in Example 7.12.

$B := M$

$$B = \begin{pmatrix} 4 & -2 & 0 & 0 \\ -2 & 4 & 0 & 0 \\ 0 & 0 & 6 & 2 \\ 0 & 0 & 2 & 6 \end{pmatrix}$$

Step 1

$k := 1 \quad y_1 := B_{k,k} \quad y_2 := B_{k+1,k} \quad c := \dfrac{y_1}{\sqrt{(y_1)^2 + (y_2)^2}} \quad s := \dfrac{-y_2}{\sqrt{(y_1)^2 + (y_2)^2}}$

$j := 1..4$

$T := B \quad T_{k,j} := c \cdot B_{k,j} - s \cdot B_{k+1,j} \quad T_{k+1,j} := s \cdot B_{k,j} + c \cdot B_{k+1,j} \quad B := T$

Step 2

$k := 2 \quad y_1 := B_{k,k} \quad y_2 := B_{k+1,k} \quad c := \dfrac{y_1}{\sqrt{(y_1)^2 + (y_2)^2}} \quad s := \dfrac{-y_2}{\sqrt{(y_1)^2 + (y_2)^2}}$

$j := 1..4$

$T := B \quad T_{k,j} := c \cdot B_{k,j} - s \cdot B_{k+1,j} \quad T_{k+1,j} := s \cdot B_{k,j} + c \cdot B_{k+1,j} \quad B := T$

Step 3

$k := 3 \quad y_1 := B_{k,k} \quad y_2 := B_{k+1,k} \quad c := \dfrac{y_1}{\sqrt{(y_1)^2 + (y_2)^2}} \quad s := \dfrac{-y_2}{\sqrt{(y_1)^2 + (y_2)^2}}$

$j := 1..4$

$T := B \quad T_{k,j} := c \cdot B_{k,j} - s \cdot B_{k+1,j} \quad T_{k+1,j} := s \cdot B_{k,j} + c \cdot B_{k+1,j}$

$$T = \begin{pmatrix} 4.472 & -3.578 & 0 & 0 \\ 0 & 2.683 & 0 & 0 \\ 0 & 0 & 6.325 & 3.795 \\ 0 & 0 & 0 & 5.06 \end{pmatrix}$$

Algorithm for Givens Transformation of Hessenberg Matrix to Triangular Matrix

Given (upper) Hessenberg matrix **B** (n-by-n)
For $k = 1, \ldots n - 1$ (for each column of B except the last)
    $x_1 = \mathbf{B}_{kk}$ (diagonal element)
    $x_2 = \mathbf{B}_{k+1,k}$ (subdiagonal element, to be zeroed-out)
    $c = \dfrac{x_1}{\sqrt{x_1^2 + x_2^2}}$
    $s = \dfrac{-x_2}{\sqrt{x_1^2 + x_2^2}}$
    T = **B**
    for $j = 1$ to n
        $\mathbf{T}(k,j) = c\,\mathbf{B}(k,j) - s\,\mathbf{B}(k+1,j)$ (**T**=**GB**, changes all columns of
        $\mathbf{T}(k+1,j) = s\,\mathbf{B}(k,j) + c\,\mathbf{B}(k+1,j)$ row k and row $k+1$)
    end
    B = **T**
End

Discussion of Givens Rotations

If the parameters c and s that occur in a Givens matrix are interpreted as $\cos(\theta)$ and $\sin(\theta)$, then multiplication by the two-by-two Givens matrix

$$G = \begin{bmatrix} c & -s \\ s & c \end{bmatrix}$$

corresponds to rotation by the angle θ in R^2.

    Givens rotations require approximately twice as many multiplications as Householder reflections. However, in the case of matrices with only a few elements that must be reduced to zero, Givens rotations are the method of choice. More efficient implementations of the Givens scheme are also available. (See [Hager, 1988] for further discussion.)

7.2.3 Basic QR Factorization

A real matrix $\mathbf{A}$ (n-by-n) can be factored into the form $\mathbf{A} = \mathbf{Q}\mathbf{R}$, where $\mathbf{Q}$ is orthogonal, $\mathbf{R}$ is upper-triangular, and both are real. In this section, we consider several ways of achieving this factorization using Householder and/or Givens transformations. For a single factorization of a matrix, Householder transformations are the more efficient method. For this case, we present an algorithm using Householder matrices, and a more efficient algorithm using Householder transformations (without forming the matrices explicitly). A Mathcad function is also presented for finding the QR factorization using Householder transformations.

In the application of QR factorization to the problem of finding the eigenvalues of a matrix $\mathbf{A}$, a sequence of matrices must be factored. In this setting, it is more efficient to perform a similarity transformation to matrix $\mathbf{A}$ to produce a matrix $\mathbf{B}$, which is in Hessenberg form and has the same eigenvalues as $\mathbf{A}$; the iterative procedure to find the eigenvalues is then carried out for matrix $\mathbf{B}$. Since the sequence of matrices that are produced (and must be factored) are all of Hessenberg form, it is preferable to use Givens rotations in the factorizations. An algorithm for QR factorization of an upper-Hessenberg matrix using Givens transformation is also presented.

QR Factorization Using Householder Matrices

In order to perform the factorization using Householder matrices, we form a sequence of matrices $\mathbf{H}^{(1)}, \mathbf{H}^{(2)}, \ldots \mathbf{H}^{(n-1)}$ and a sequence of matrices $\mathbf{R}^{(1)}, \mathbf{R}^{(2)}, \ldots \mathbf{R}^{(n-1)}$. The matrix $\mathbf{H}^{(1)}$ is found to zero out all elements of the first column of $\mathbf{A}$ below the diagonal. $\mathbf{R}^{(1)} = \mathbf{H}^{(1)} \mathbf{A}$. The matrix $\mathbf{H}^{(2)}$ is found to zero out all elements of the second column of $\mathbf{R}^{(1)}$ below the diagonal (zero out rows $3 \ldots n$). $\mathbf{R}^{(2)} = \mathbf{H}^{(2)} \mathbf{R}^{(1)}$. The process continues until $\mathbf{A}$ has been transformed to the upper triangular matrix $\mathbf{R}^{(n-1)}$ by successive multiplications by the orthogonal matrices $\mathbf{H}^{(k)}$. The desired factorization is $\mathbf{A} = \mathbf{Q}\mathbf{R}$ where $\mathbf{R} = \mathbf{R}^{(n-1)}$ and $\mathbf{Q} = \mathbf{H}^{(1)} \mathbf{H}^{(2)} \ldots \mathbf{H}^{(n-1)}$.

Algorithm for QR Factorization Using Householder Matrices

Begin computations
Define $\mathbf{R}^{(0)} = \mathbf{A}$
Define $\mathbf{Q}^{(0)} = \mathbf{I}$
For $k = 1$ to $n - 1$
    Find Householder matrix, $\mathbf{H}^{(k)}$, with $\mathbf{x} = k^{\text{th}}$ column of $\mathbf{R}^{(k-1)}$
    Define $\mathbf{R}^{(k)} = \mathbf{H}^{(k)} \mathbf{R}^{(k-1)}$
    Define $\mathbf{Q}^{(k)} = \mathbf{Q}^{(k-1)} \mathbf{H}^{(k)}$
End
Define $\mathbf{R} = \mathbf{R}^{(n-1)}$
Define $\mathbf{Q} = \mathbf{Q}^{(n-1)}$

QR Factorization Using Householder Transformations

As indicated previously, it is more efficient not to actually form the Householder matrices used in a Householder transformation. The following algorithm restates the QR factorization process in terms of Householder transformations, without the formation of the matrices.

Algorithm for QR Factorization Using Householder Transformations

Initialize
  $R = A$
  $Q = I$
Begin computation
for $k = 1$ to $n-1$
  for $j = 1$ to $k-1$ (apply Householder transformation to
    $x_j = 0$ kth column of R, to zero out elements
  end   in rows $k+1 \ldots n$)
  for $j = k$ to n
    $x_j = R_{j,k}$
  end
  $g = \text{norm}(x)$
  $s = \sqrt{2g(g + |x_k|)}$
  $w = \dfrac{x}{s}$
  $w_k = \dfrac{1}{s}(x_k + \text{sign}(x_k)\, g)$
  $u = 2\, R^T w$
  $R = R - w\, u^T$ ($R_{\text{new}} = H\, R_{\text{old}}$)
  $Q = Q - 2\, Q\, w\, w^T$ ($Q_{\text{new}} = Q_{\text{old}}\, H$)
end

In the next example, we illustrate the use of Householder matrices to find the QR factorization of the matrix introduced in Example 7.9. The function QR_factor follows the example.

Example 7.14 QR Factorization Using Householder Transformations

$$A := \begin{pmatrix} 21 & 7 & -1 \\ 5 & 7 & 7 \\ 4 & -4 & 20 \end{pmatrix} \qquad B := \text{QR\_factor}(A) \qquad r := \text{rows}(A) - 1$$

$$Q := \text{submatrix}(B,0,r,0,r) \qquad R := \text{submatrix}(B,0,r,r+1,2r+1)$$

$$Q = \begin{pmatrix} -0.957 & 0.031 & -0.29 \\ -0.228 & -0.7 & 0.677 \\ -0.182 & 0.713 & 0.677 \end{pmatrix} \qquad R = \begin{pmatrix} -21.954 & -7.561 & -4.282 \\ 0 & -7.539 & 9.335 \\ 0 & 0 & 18.561 \end{pmatrix}$$

Mathcad Function to Find QR Factorization by Householder Transformations

The Mathcad function that follows illustrates the use of Householder transformations, without explicitly forming the Householder matrices, to obtain the QR factorization of a matrix **A**.

$$\text{QR\_factor}(A) := \begin{vmatrix} n \leftarrow \text{rows}(A) \\ Q \leftarrow \text{identity}(n) \\ R \leftarrow A \\ \text{for } k \in 0..n-2 \\ \quad \begin{vmatrix} \text{for } i \in 0..n-1 \\ \quad \begin{vmatrix} x_i \leftarrow 0 \text{ if } i < k \\ x_i \leftarrow R_{i,k} \text{ otherwise} \end{vmatrix} \\ g \leftarrow |x| \\ v \leftarrow x \\ v_k \leftarrow x_k + g \\ s \leftarrow |v| \\ w \leftarrow \dfrac{v}{s} \\ u \leftarrow 2 \cdot R^T \cdot w \\ R \leftarrow R - w \cdot u^T \\ Q \leftarrow Q - 2Q \cdot w \cdot w^T \end{vmatrix} \\ B \leftarrow \text{augment}(Q,R) \\ B \end{vmatrix}$$

Chapter 7 Eigenvalues, Eigenvectors, and QR Factorization

QR Factorization of Hessenberg Matrix Using Givens Transformations

The following algorithm gives an alternative method of finding the QR factorization of an upper Hessenberg matrix. It is computationally more efficient than the Householder transformations for factoring a matrix in which there is only one nonzero element below the diagonal in each column that must be zeroed out.

Algorithm for QR Factorization of Upper-Hessenberg Matrix

Initialize
$\quad \mathbf{Q} = \mathbf{I}$
$\quad \mathbf{R} = \mathbf{B}$
Begin computaion
for $k = 1$ to $n - 1$
$\quad x_1 = \mathbf{R}(k,k)$
$\quad x_2 = \mathbf{R}(k+1,k)$
$\quad c = \dfrac{x_1}{\sqrt{x_1^2 + x_2^2}}$
$\quad s = \dfrac{-x_2}{\sqrt{x_1^2 + x_2^2}}$

$\quad \mathbf{R}_n = \mathbf{R}$
$\quad \mathbf{Q}_n = \mathbf{Q}$
$\quad$ for $j = 1$ to n
$\quad\quad \mathbf{R}_n(k,j) = c\,\mathbf{R}(k,j) - s\,\mathbf{R}(k+1,j)$
$\quad\quad \mathbf{R}_n(k+1,j) = s\,\mathbf{R}(k,j) + c\,\mathbf{R}(k+1,j)$
$\quad\quad \mathbf{Q}_n(j,k) = c\,\mathbf{Q}(j,k) - s\,\mathbf{Q}(j,k+1)$
$\quad\quad \mathbf{Q}_n(j,k+1) = s\,\mathbf{Q}(j,k) + c\,\mathbf{Q}(j,k+1)$
$\quad$ end
$\quad \mathbf{R} = \mathbf{R}_n$
$\quad \mathbf{Q} = \mathbf{Q}_n$
end

Mathcad Function for QR Factorization of Upper-Hessenberg Matrix

$\text{QR\_factor\_H}(A) := \bigg|$ $n \leftarrow \text{rows}(A)$
$Q \leftarrow \text{identity}(n)$
$R \leftarrow A$
for $k \in 0 .. n-2$
$\quad\bigg|$ $x1 \leftarrow R_{k,k}$
$x2 \leftarrow R_{k+1,k}$
$rr \leftarrow \sqrt{x1^2 + x2^2}$
$c \leftarrow \dfrac{x1}{rr}$
$s \leftarrow \dfrac{-x2}{rr}$
$R \leftarrow R^T$
$Rnew \leftarrow R$
$Qnew \leftarrow Q$
$Rnew^{\langle k \rangle} \leftarrow c \cdot R^{\langle k \rangle} - s \cdot R^{\langle k+1 \rangle}$
$Rnew^{\langle k+1 \rangle} \leftarrow s \cdot R^{\langle k \rangle} + c \cdot R^{\langle k+1 \rangle}$
$Qnew^{\langle k \rangle} \leftarrow c \cdot Q^{\langle k \rangle} - s \cdot Q^{\langle k+1 \rangle}$
$Qnew^{\langle k+1 \rangle} \leftarrow s \cdot Q^{\langle k \rangle} + c \cdot Q^{\langle k+1 \rangle}$
$R \leftarrow Rnew^T$
$Q \leftarrow Qnew$
$B \leftarrow \text{augment}(Q, R)$
B

Example 7.15 QR Factorization of an Upper-Hessenberg Matrix

$$M := \begin{pmatrix} 4 & -2 & 0 & 0 \\ -2 & -17.333 & -15.085 & 0 \\ 0 & -15.085 & 24.19 & -12.781 \\ 0 & 0 & -12.781 & 13.294 \end{pmatrix} \qquad B := \text{QR\_factor\_H}(M) \qquad n := \text{rows}(M)$$

$$B = \begin{pmatrix} 0.894 & -0.329 & -0.274 & 0.13 & 4.472 & 5.963 & 6.746 & 0 \\ -0.447 & -0.658 & -0.547 & 0.26 & 0 & 22.281 & -6.448 & 8.653 \\ 0 & -0.677 & 0.665 & -0.315 & 0 & 0 & 29.816 & -14.197 \\ 0 & 0 & -0.429 & -0.903 & 0 & 0 & 0 & -7.979 \end{pmatrix}$$

$$Q := \text{submatrix}(B, 0, n-1, 0, n-1) \qquad R := \text{submatrix}(B, 0, n-1, n, 2n-1)$$

$$Q = \begin{pmatrix} 0.894 & -0.329 & -0.274 & 0.13 \\ -0.447 & -0.658 & -0.547 & 0.26 \\ 0 & -0.677 & 0.665 & -0.315 \\ 0 & 0 & -0.429 & -0.903 \end{pmatrix} \qquad R = \begin{pmatrix} 4.472 & 5.963 & 6.746 & 0 \\ 0 & 22.281 & -6.448 & 8.653 \\ 0 & 0 & 29.816 & -14.197 \\ 0 & 0 & 0 & -7.979 \end{pmatrix}$$

$$Q \cdot R = \begin{pmatrix} 4 & -2 & 0 & 0 \\ -2 & -17.333 & -15.085 & 0 \\ 0 & -15.085 & 24.19 & -12.781 \\ 0 & 0 & -12.781 & 13.294 \end{pmatrix}$$

7.3 FINDING EIGENVALUES USING QR FACTORIZATION

To find the eigenvalues of a real matrix **A** using QR factorization, we generate a sequence of matrices $\mathbf{A}^{(m)}$ that are orthogonally similar to **A** (and thus have the same eigenvalues as **A**). A similarity transformation is a transformation of the form $\mathbf{H}^{-1}\mathbf{A}\mathbf{H}$, where **H** is any nonsingular matrix. The sequence $\mathbf{A}^{(m)}$ converges to a matrix from which the eigenvalues can be found easily. If the eigenvalues of **A** satisfy $|\lambda_1|>|\lambda_2|> \ldots > |\lambda_n|$, the iterates converge to an upper-triangular matrix with the eigenvalues on the diagonal.

We first consider the basic QR-eigenvalue algorithm and a Mathcad function to implement it. However, since the QR-eigenvalue method involves a sequence of factorizations, we conclude this section with a more efficient QR factorization method for finding eigenvalues. This better QR-eigenvalue algorithm is based on a preliminary transformation of **A** into Hessenberg form—i.e., upper-triangular, except for the first subdiagonal. The QR factorization of a matrix that is in Hessenberg form is computationally more efficient, since there is only one element that must be transformed to zero in each column. Furthermore the upper Hessenberg form is preserved through the iterations of the algorithm. The QR factorization in this case is usually based on Givens rotations.

7.3.1 Basic QR Eigenvalue Method

The following is the basic procedure for finding eigenvalues using QR factorization. The matrices formed at each iteration are indicated by superscripts. If **M** has n distinct real eigenvalues, the method will converge to an upper triangular matrix with the eigenvalues on the diagonal. In other cases, the final result may contain small blocks on the diagonal.

Algorithm for Finding Eigenvalues by QR Factorization

Given a real n-by-n matrix **A**:
Define $\mathbf{A}^{(1)} = \mathbf{A}$.
For $k = 1$ to k_max,
 Factor $\mathbf{A}^{(k)} = \mathbf{Q}^{(k)} \mathbf{R}^{(k)}$ (so that $\mathbf{Q}^{(k)T} \mathbf{A}^{(k)} = \mathbf{R}^{(k)}$).
 Define $\mathbf{A}^{(k+1)} = \mathbf{R}^{(k)} \mathbf{Q}^{(k)}$ (so that $\mathbf{A}^{(k+1)} = \mathbf{Q}^{(k)T} \mathbf{A}^{(k)} \mathbf{Q}^{(k)}$).
End .

7.3.2 Better QR Eigenvalue Method

As mentioned in the previous section, in employing QR factorizations to find the eigenvalues of matrix **A**, it is preferable to first use a similarity transformation to convert **A** to Hessenberg form. As described earlier, the required computations can be performed more efficiently without explicitly forming the Householder matrices.

Algorithm for Finding Eigenvalues Using QR—Improved Method

> convert **M** to Hessenberg form
> begin iteratations
>   find **QR** factorization of **M**
>   set **M = R*Q**
>   test for convergence
> end iterations

Mathcad Function for Finding Eigenvalues Using QR—Improved Method

The function QR_eigen_g calls the functions QR_factor_H and Hessenberg. These functions must be included in the worksheet before QR_eigen_g, or a reference inserted.

$$\text{QR\_eigen\_g}(A, \text{max}) := \begin{vmatrix} n \leftarrow \text{rows}(A) \\ A \leftarrow \text{Hessenberg}(A) \\ \text{for } i \in 1..\text{max} \\ \quad \begin{vmatrix} B \leftarrow \text{QR\_factor\_g}(A) \\ R \leftarrow \text{submatrix}(B, 0, n-1, n, 2n-1) \\ Q \leftarrow \text{submatrix}(B, 0, n-1, 0, n-1) \\ A \leftarrow R \cdot Q \end{vmatrix} \\ \text{for } i \in 0..n-1 \\ \quad e_i \leftarrow A_{i,i} \\ e \end{vmatrix}$$

Example 7.16 Finding Eigenvalues Using QR Factorization

To illustrate the use of the QR method for finding eigenvalues, consider again the matrix A:

$$A := \begin{pmatrix} 4 & \frac{2}{3} & \frac{-4}{3} & \frac{4}{3} \\ \frac{2}{3} & 4 & 0 & 0 \\ \frac{-4}{3} & 0 & 6 & 2 \\ \frac{4}{3} & 0 & 2 & 6 \end{pmatrix} \qquad A = \begin{pmatrix} 4 & 0.66666667 & -1.33333333 & 1.33333333 \\ 0.66666667 & 4 & 0 & 0 \\ -1.33333333 & 0 & 6 & 2 \\ 1.33333333 & 0 & 2 & 6 \end{pmatrix}$$

$$\text{QR\_eigen\_g}(A, 1) = \begin{pmatrix} 5.6 \\ 2.4 \\ 7.2 \\ 4.8 \end{pmatrix} \qquad \text{QR\_eigen\_g}(A, 2) = \begin{pmatrix} 5.95121951 \\ 2.04878049 \\ 7.76470588 \\ 4.23529412 \end{pmatrix}$$

$$\text{QR\_eigen\_g}(A, 3) = \begin{pmatrix} 5.99452055 \\ 2.00547945 \\ 7.93846154 \\ 4.06153846 \end{pmatrix} \qquad \text{QR\_eigen\_g}(A, 4) = \begin{pmatrix} 5.99939043 \\ 2.00060957 \\ 7.9844358 \\ 4.0155642 \end{pmatrix}$$

$$\text{QR\_eigen\_g}(A, 5) = \begin{pmatrix} 5.99993226 \\ 2.00006774 \\ 7.99609756 \\ 4.00390244 \end{pmatrix} \qquad \text{QR\_eigen\_g}(A, 6) = \begin{pmatrix} 5.99999247 \\ 2.00000753 \\ 7.99902368 \\ 4.00097632 \end{pmatrix}$$

7.3.3 Discussion

Especially for a general matrix, which may have real or complex eigenvalues, setting good termination conditions is not an easy task. If the matrix has distinct, real eigenvalues then the QR eigenvalue method will converge to an upper triangular matrix with the eigenvalues of the original matrix on the diagonal. In other cases the final form may have one or more two-by-two blocks on the diagonal. The eigenvalues of these submatrices are also eigenvalues of the original matrix. It is suggested [Press et al., p. 490] that the process terminate in the following manner

1) if $a_{n,n-1}$ is "negligible", then $a_{n,n}$ is an eigenvalue (delete n^{th} row, n^{th} col and continue)
2) if $a_{n-1,n-2}$ is "negligible", then the eigenvalues of the 2-by-2 submatrix in lower-right corner are eigenvalues, (delete n^{th} and $(n-1)^{st}$ rows and columns and procede).

For proof of fundamental theorem that sequence of iterates of QR method converges to an upper (or lower) triangular form with eigenvalues on the diagonal, see [Stoer and Bulirsch, 1980].

7.4 MATHCAD'S METHODS

Mathcad2000 has built-in functions for finding eigenvalues and eigenvectors; Mathcad2000Pro has an additional function for eigenvectors and also a function for performing QR factorization.

7.4.1 Using the Built-in Functions

The function `eigenvals(M)` has been used in the previous chapters to find the eigenvalues of a matrix. The matrix **M** is a real or complex square matrix. The return from the function is a vector containing the eigenvalues of **M**.

The function `eigenvec(M, z)` returns the normalized eigenvector associated with the eigenvalue z of the square matrix **M**. The matrix **M** is a real or complex square matrix; the eigen value may be real or complex.

Mathcad2000Pro has an additional fucntion for finding eigenvectors of a real or complex, square matrix **M**. The function `eigenvecs(M)` returns a matrix whose columns contain the normalized eigenvectors corresponding to the eigenvalues returned by `eigenvalues(M)`.

Mathcad2000Pro also has a built-in function for QR decomposition, `qr(A)`, where **A** is a real m-by-n matrix. The function returns an m-by-$(m+n)$ matrix. The first m columns contain the matrix **Q**, the remaining n columns contain **R**, such that **A = Q R**.

7.4.2 Understanding the Algorithms

For the function `eigenvals(M)`, the computations are based on reduction to Hessenberg form followed by QR decomposition. This is also the method used by the function `eigenvecs(M)` (in Mathcad2000Pro). The procedure is an efficient implementation of the ideas discussed in Section 7.2. See [Press et al., 1992, Ch 11] for a detailed discussion of the QR and related methods.

The function `eigenvec(M, z)` uses inverse iteration, as described in [Press et al., 1992, sect. 11.7]. This is essentially the same method as described in 7.1.4 as the general inverse power method.

Efficient conversion to Hessenberg form

An alternative to the algorithm for conversion to Hessenberg form by Householder transformations is given in the following method based on Gaussian elimination; this description follows the presentation in [Press et al., 1992, pp. 484–486]. In general, elimination is more efficient (half the number of flops [Golub and Van Loan, 1989, p. 370]) than the Householder method. For some matrices, the Householder method is stable and elimination is not, but such matrices are not common in actual practice. Pivoting is included in the following algorithm to improve stability.

Gaussian elmination must be modified to make the transformations into *similiarity* transformations, by performing the indicated operations on columns to correspond to the operations on rows that are required to achieve the desired upper Hessenberg form. This process requires $n - 2$ stages for an n-by-n matrix.

First stage:
    Find element of maximum magnitude in first column, below the diagonal
        If this element is zero, this stage is finished.
        Otherwise, denote the element as $a_{p,1}$ (that is, the element is in the pth row)
            If $p > 2$, interchange rows 2 and p; also interchange columns 2 and p.
            For rows $i = 3, 4, \ldots n$
                compute multiplier: $m_{i,2} = a_{i,1}/a_{2,1}$
                subtract multiple of row 2 from row i
                add multiple of column i to column 2

Stage k:
    Find element of max. magnitude in kth column, rows $k+1, \ldots n$.
        If this element is zero, this stage is finished.
        Otherwise, denote the element as $a_{p,k}$ (that is, the element is in the pth row)
            If $p > k+1$, interchange rows $k+1$ and p;
                also interchange columns $k+1$ and p
            For rows $i = k+2, \ldots n$
                compute multiplier: $m_{i,k+1} = a_{i,k}/a_{k+1,k}$
                subtract multiple of row $k+1$ from row i
                add multiple of column i to column $k+1$

SUMMARY

The basic power method to find the dominant eigenvalue of matrix $\mathbf{A}$:

$$\mathbf{z}^{(1)} = \begin{bmatrix} 1 & 1 & 1 & \ldots & 1 \end{bmatrix}^T;$$

$$\mathbf{w}^{(1)} = \mathbf{A}\,\mathbf{z}^{(1)}; \qquad \mathbf{z}^{(2)} = \frac{1}{w^{(1)}_k}\mathbf{w}^{(1)},$$

$w^{(1)}_k$ is the component of $\mathbf{w}$ that is of largest magnitude;
$w^{(1)}_k$ is the first estimate of the dominant eigenvalue.
$\mathbf{z}^{(2)}$ is the corresponding (estimate of the) eigenvector.

$$\mathbf{w}^{(2)} = \mathbf{A}\,\mathbf{z}^{(2)}; \qquad \mathbf{z}^{(3)} = \frac{1}{w^{(2)}_k}\mathbf{w}^{(2)}.$$

Continue until the estimates have converged.

The Rayleigh quotient gives an improved estimate of the eigenvalue, especially for symmetric matrices.

$$\lambda = (\mathbf{z}^T\mathbf{w})/(\mathbf{z}^T\mathbf{z}).$$

The inverse power method finds the eigenvalue of $\mathbf{A}$ that is of smallest magnitude.

Apply the power method to $\mathbf{B} = \mathbf{A}^{-1}$ to find its dominant eigenvalue, μ.
The reciprocal of μ gives the smallest magnitude eigenvalue λ of $\mathbf{A}$.
To avoid computing $\mathbf{A}^{-1}$, instead of finding $\mathbf{A}^{-1}\mathbf{z} = \mathbf{w}$ we solve the system $\mathbf{A}\mathbf{w} = \mathbf{z}$ for $\mathbf{w}$.
The inverse power method applied to a shifted matrix, $\mathbf{A} - b\,\mathbf{I}$, finds the eigenvalue closest to b.

QR factorization of a matrix $\mathbf{A}$ can be accomplished using Householder matrices, as described in the following algorithm.

Define $\mathbf{R}^{(0)} = \mathbf{A}$.
For $k = 1, \ldots, n-1$
    Find $\mathbf{H}^{(k)}$ to reduce positions $k+1, \ldots n$ in the kth column of $\mathbf{R}^{(k-1)}$ to zero;
    Define $\mathbf{R}^{(k)} = \mathbf{H}^{(k)}\,\mathbf{R}^{(k-1)}$.
End.
Define $\mathbf{Q} = \mathbf{I}$.
For $k = n-1, \ldots, 1$,
    $\mathbf{Q} = \mathbf{H}^{(k)}\,\mathbf{Q}$
End.
Define $\mathbf{R} = \mathbf{R}^{(n-1)}$.

For a matrix in Hessenberg form it is more efficient to perform the factorization using Givens rotations. The preprocessing to Hessenberg form and the use of Givens rotations for QR factorization are discussed in Section 7.3.2.

QR factorization for finding all the eigenvalues of a matrix:
  Define $\mathbf{A}^{(1)} = \mathbf{A}$.
  For $k = 1$ to k_max,
    Factor $\mathbf{A}^{(k)} = \mathbf{Q}^{(k)} \mathbf{R}^{(k)}$
    Define $\mathbf{A}^{(k+1)} = \mathbf{R}^{(k)} \mathbf{Q}^{(k)}$
  End .

SUGGESTIONS FOR FURTHER READING

The following are a few of the many excellent undergraduate texts on linear algebra:

Kolman, B., *Introductory Linear Algebra with Applications*, (6th ed.), Prentice Hall, Upper Saddle River, NJ, 1997.

Leon, S. J., *Linear Algebra with Applications*, (5th ed.), Prentice Hall, Upper Saddle River, NJ, 1998.

Strang, G., *Linear Algebra and Its Applications*, (3rd ed.), Harcourt Brace Jovanovich, San Diego, CA, 1988.

Some references at a somewhat more advanced level include:

Golub, G. H. and C. F. Van Loan, *Matrix Computations*, (3d ed.), Johns Hopkins University Press, Baltimore, 1996.

Fox, L., *An Introduction to Numerical Linear Algebra*, Oxford University Press, New York, 1965. This classic work also contains many bibliographic entries.

Stoer, J. and R. Bulirsch, *Introduction to Numerical Analysis*, Springer-Verlag, New York, 1980.

Wilkinson, J. H. and C. Reinsch, *Linear Algebra*, vol II of *Handbook for Automatic Computation*, Springer-Verlag, New York, 1971.

For further discussion of the application of eigenvalues see:

Thomson, W. T., *Introduction to Space Dynamics*, Dover, New York, 1986. (Originally published by John Wiley & Sons, 1961.)

Greenberg, M. D., *Foundations of Applied Mathematics*, (2d ed.), Prentice Hall, Englewood Cliffs, NJ, 1998.

PRACTICE THE TECHNIQUES

For Problems P7.1 to P7.20 use the power method to find the specified eigenvalue, and a corresponding eigenvector, of the given matrix.

a. Use the basic power method to find the dominant eigenvalue.
b. Use the inverse power method to find the eigenvalue of smallest magnitude.
c. Use the shifted power method to find the eigenvalue that is closest to the average of the values found in parts a and b.

P7.1 $\mathbf{A} = \begin{bmatrix} 1 & 0 & 0 \\ 2 & -1 & 2 \\ 4 & -4 & 5 \end{bmatrix}$

P7.2 $\mathbf{A} = \begin{bmatrix} -2 & 2 & -1 \\ -2 & 2 & 0 \\ 2 & -2 & 3 \end{bmatrix}$

P7.3 $\mathbf{A} = \begin{bmatrix} 5 & -2 & 1 \\ 3 & 0 & 1 \\ 0 & 0 & 2 \end{bmatrix}$

P7.4 $\mathbf{A} = \begin{bmatrix} 2 & 2 & -1 \\ -5 & 9 & -3 \\ -4 & 4 & 1 \end{bmatrix}$

P7.5 $\mathbf{A} = \begin{bmatrix} -19 & 20 & -6 \\ -12 & 13 & -3 \\ 30 & -30 & 12 \end{bmatrix}$

P7.6 $\mathbf{A} = \begin{bmatrix} 25 & -26 & 8 \\ 15 & -16 & 4 \\ -39 & 39 & -15 \end{bmatrix}$

P7.7 $\mathbf{A} = \begin{bmatrix} -36 & 10 & -5 \\ -172 & 47 & -22 \\ -16 & 4 & 1 \end{bmatrix}$

P7.8 $\mathbf{A} = \begin{bmatrix} 2 & 0 & 0 \\ -180 & 47 & -15 \\ -600 & 150 & -48 \end{bmatrix}$

P7.9 $\mathbf{A} = \begin{bmatrix} 104 & -34 & 8 \\ 357 & -117 & 28 \\ 204 & -68 & 18 \end{bmatrix}$

P7.10 $\mathbf{A} = \begin{bmatrix} 66 & -21 & 9 \\ 228 & -73 & 33 \\ 84 & -28 & 16 \end{bmatrix}$

P7.11 $\mathbf{A} = \begin{bmatrix} 4 & 0 & 0 & 0 \\ -4 & 8 & -4 & 2 \\ -7 & 7 & -3 & 3 \\ -2 & 2 & -2 & 4 \end{bmatrix}$

P7.12 $\mathbf{A} = \begin{bmatrix} 11 & -6 & 4 & -2 \\ 4 & 1 & 0 & 0 \\ -9 & 9 & -6 & 5 \\ -6 & 6 & -6 & 7 \end{bmatrix}$

P7.13 $\mathbf{A} = \begin{bmatrix} 20 & -15 & 10 & -5 \\ 26 & -21 & 16 & -8 \\ 11 & -11 & 11 & -5 \\ 4 & -4 & 4 & -1 \end{bmatrix}$

P7.14 $\mathbf{A} = \begin{bmatrix} -6 & 6 & -4 & 2 \\ -8 & 8 & -4 & 2 \\ 0 & 0 & 2 & 0 \\ 0 & 0 & 0 & 2 \end{bmatrix}$

P7.15 $\mathbf{A} = \begin{bmatrix} -4 & 2 & -8 & 0 \\ -21 & 9 & -20 & 0 \\ 0 & 0 & 4 & 0 \\ 0 & 0 & 8 & 0 \end{bmatrix}$

P7.16 $\mathbf{A} = \begin{bmatrix} 11 & -8 & 6 & -4 & 2 \\ 16 & -13 & 12 & -8 & 4 \\ 4 & -4 & 5 & 0 & 0 \\ -9 & 9 & -9 & 12 & -5 \\ -6 & 6 & -6 & 6 & -1 \end{bmatrix}$

P7.17 $\mathbf{A} = \begin{bmatrix} 2 & 0 & 0 & 0 & 0 \\ 3 & -1 & 3 & -2 & 1 \\ 2 & -2 & 4 & 0 & 0 \\ -5 & 5 & -5 & 8 & -3 \\ -4 & 4 & -4 & 4 & 0 \end{bmatrix}$

P7.18 $\mathbf{A} = \begin{bmatrix} -7 & 8 & -6 & 4 & -2 \\ -13 & 14 & -9 & 6 & -3 \\ -2 & 2 & 1 & 0 & 0 \\ 5 & -5 & 5 & -3 & 3 \\ 4 & -4 & 4 & -4 & 5 \end{bmatrix}$

P7.19 $\mathbf{A} = \begin{bmatrix} -8 & 8 & -6 & 4 & -2 \\ -7 & 7 & -3 & 2 & -1 \\ 8 & -8 & 10 & -6 & 3 \\ 8 & -8 & 8 & -5 & 4 \\ 4 & -4 & 4 & -4 & 5 \end{bmatrix}$

P7.20 $\mathbf{A} = \begin{bmatrix} -1 & 4 & -3 & 2 & -1 \\ -11 & 14 & -9 & 6 & -3 \\ -6 & 6 & -2 & 2 & -1 \\ 6 & -6 & 6 & -4 & 4 \\ 6 & -6 & 6 & -6 & 7 \end{bmatrix}$

For the symmetric matrices in Problems P7.21 to P7.40, find the dominant eigenvalue
    a. *using the power method.*
    b. *using the accelerated power method (Rayleigh quotient).*

P7.21 $\mathbf{A} = \begin{bmatrix} 6 & -1 & -2 \\ -1 & 8 & 1 \\ -2 & 1 & 2 \end{bmatrix}$

P7.22 $\mathbf{A} = \begin{bmatrix} 1 & 0 & 1 \\ 0 & 9 & -2 \\ 1 & -2 & 11 \end{bmatrix}$

P7.23 $\mathbf{A} = \begin{bmatrix} 11 & -2 & -4 \\ -2 & 15 & 2 \\ -4 & 2 & 3 \end{bmatrix}$

P7.24 $\mathbf{A} = \begin{bmatrix} 10 & 3 & 6 \\ 3 & 4 & -3 \\ 6 & -3 & 22 \end{bmatrix}$

P7.25 $\mathbf{A} = \begin{bmatrix} 8 & -6 & -9 \\ -6 & 44 & 0 \\ -9 & 0 & 14 \end{bmatrix}$

P7.26 $\mathbf{A} = \begin{bmatrix} -9 & 0 & 1 \\ 0 & -1 & -2 \\ 1 & -2 & 1 \end{bmatrix}$

P7.27 $\mathbf{A} = \begin{bmatrix} 1 & -2 & -4 \\ -2 & 5 & 2 \\ -4 & 2 & -7 \end{bmatrix}$

P7.28 $\mathbf{A} = \begin{bmatrix} 5 & -3 & -9 & -7 \\ -3 & 33 & 3 & 10 \\ -9 & 3 & 9 & -3 \\ -7 & 10 & -3 & 37 \end{bmatrix}$

P7.29 $\mathbf{A} = \begin{bmatrix} 50 & 3 & 2 & -11 \\ 3 & 8 & 10 & 5 \\ 2 & 10 & 18 & -4 \\ -11 & 5 & -4 & 42 \end{bmatrix}$

P7.30 $\mathbf{A} = \begin{bmatrix} 51 & 5 & 11 & 1 \\ 5 & 5 & 3 & -8 \\ 11 & 3 & 37 & 1 \\ 1 & -8 & 1 & 21 \end{bmatrix}$

P7.31 $\mathbf{A} = \begin{bmatrix} 11 & -6 & -8 & 6 \\ -6 & 47 & -4 & 0 \\ -8 & -4 & 19 & 4 \\ 6 & 0 & 4 & 15 \end{bmatrix}$

P7.32 $\mathbf{A} = \begin{bmatrix} 4 & 2/3 & -4/3 & 4/3 \\ 2/3 & 4 & 0 & 0 \\ -4/3 & 0 & 12 & 8 \\ 4/3 & 0 & 8 & 12 \end{bmatrix}$

P7.33 $\mathbf{A} = \begin{bmatrix} 81/2 & 21/2 & -21 & 21 \\ 21/2 & 41/2 & -5 & 5 \\ -21 & -5 & 37 & -1 \\ 21 & 5 & -1 & 37 \end{bmatrix}$

P7.34 $\mathbf{A} = \begin{bmatrix} 32 & -2 & -5 & -8 & 1 \\ -2 & 20 & 8 & 6 & 0 \\ -5 & 8 & 6 & 0 & -3 \\ -8 & 6 & 0 & 32 & -2 \\ 1 & 0 & -3 & -2 & 24 \end{bmatrix}$

P7.35 $\mathbf{A} = \begin{bmatrix} 6 & 2 & -1 & -3 & -5 \\ 2 & 18 & -2 & 5 & 4 \\ -1 & -2 & 20 & -5 & -4 \\ -3 & 5 & -5 & 28 & 1 \\ -5 & 4 & -4 & 1 & 12 \end{bmatrix}$

P7.36 $\mathbf{A} = \begin{bmatrix} 35 & 5 & 9 & -1 & 4 & -3 \\ 5 & 5 & 3 & -8 & 2 & 0 \\ 9 & 3 & 33 & -1 & 4 & -3 \\ -1 & -8 & -1 & 21 & 0 & 1 \\ 4 & 2 & 4 & 0 & 29 & -2 \\ -3 & 0 & -3 & 1 & -2 & 31 \end{bmatrix}$

P7.37 $\mathbf{A} = \begin{bmatrix} 52 & 6 & 15 & 1 & 5 & 3 & 5 \\ 6 & 16 & 0 & -13 & 4 & -9 & -2 \\ 15 & 0 & 58 & -3 & 8 & 1 & 6 \\ 1 & -13 & -3 & 30 & 1 & -7 & -3 \\ 5 & 4 & 8 & 1 & 42 & 3 & 5 \\ 3 & -9 & 1 & -7 & 3 & 28 & -1 \\ 5 & -2 & 6 & -3 & 5 & -1 & 44 \end{bmatrix}$

P7.38

$\mathbf{A} = \begin{bmatrix} 46 & 0 & 5 & 14 & 10 & 2 & 5 & 4 \\ 0 & 46 & -5 & -14 & -10 & -2 & -5 & -4 \\ 5 & -5 & 56 & 9 & 14 & -1 & 7 & 3 \\ 14 & -14 & 9 & 18 & 2 & -11 & 1 & -7 \\ 10 & -10 & 14 & 2 & 64 & -4 & 9 & 2 \\ 2 & -2 & -1 & -11 & -4 & 44 & -3 & -5 \\ 5 & -5 & 7 & 1 & 9 & -3 & 58 & 1 \\ 4 & -4 & 3 & -7 & 2 & -5 & 1 & 42 \end{bmatrix}$

P7.39

$\mathbf{A} = \begin{bmatrix} 84 & -8 & -3 & 25 & -3 & -10 & -23 & -12 & -5 \\ -8 & 132 & -18 & -7 & -12 & 22 & 14 & 18 & 2 \\ -3 & -18 & 110 & 13 & 7 & -18 & -17 & -16 & -3 \\ 25 & -7 & 13 & 62 & 13 & 2 & 29 & 8 & 7 \\ -3 & -12 & 7 & 13 & 100 & -14 & -20 & -14 & -4 \\ -10 & 22 & -18 & 2 & -14 & 126 & 8 & 22 & 0 \\ -23 & 14 & -17 & 29 & -20 & 8 & 60 & 0 & -11 \\ -12 & 18 & -16 & 8 & -14 & 22 & 0 & 122 & -2 \\ -5 & 2 & -3 & 7 & -4 & 0 & -11 & -2 & 94 \end{bmatrix}$

P7.40

$\mathbf{A} = \begin{bmatrix} 84 & 1 & 1 & -9 & 23 & 7 & 1 & -7 & 1 & -3 \\ 1 & 46 & 15 & -5 & -11 & -17 & 7 & 1 & 3 & 1 \\ 1 & 15 & 50 & 5 & 11 & 17 & -7 & -1 & -3 & -1 \\ -9 & -5 & 5 & 68 & -17 & -12 & 3 & 4 & 1 & 2 \\ 23 & -11 & 11 & -17 & 100 & -3 & 9 & -13 & 5 & -5 \\ 7 & -17 & 17 & -12 & -3 & 44 & 11 & -2 & 5 & 0 \\ 1 & 7 & -7 & 3 & 9 & 11 & 58 & -1 & -3 & -1 \\ -7 & 1 & -1 & 4 & -13 & -2 & -1 & 72 & -1 & 3 \\ 1 & 3 & -3 & 1 & 5 & 5 & -3 & -1 & 64 & -1 \\ -3 & 1 & -1 & 2 & -5 & 0 & -1 & 3 & -1 & 70 \end{bmatrix}$

For the matrices given in Problems P7.1 to P7.40:
  a. *Find the QR factorization*
  b. *Find the eigenvalues using the QR method*
  c. *Use a Householder similarity transformation to transform the given matrix to Hessenberg form.*
  d. *Find the eigenvalues of the transformed matrix from Part c.*
  e. *Find the eigenvalues using the Mathcad function* `eigenvals`

EXPLORE SOME APPLICATIONS

A7.1 Show that the quadratic equation in three variables

$$a x^2 + b y^2 + c z^2 + 2d\, xy + 2e\, xz + 2f\, yz = g$$

can be written as $[\mathbf{A}\,\mathbf{v}]\,\mathbf{v} = g$, with

$$\mathbf{A} = \begin{bmatrix} a & d & e \\ d & b & f \\ e & f & c \end{bmatrix}$$

The surfaces can be classified according to the signs of eigenvalues of $\mathbf{A}$, as indicated in the following table. (Degenerate cases may also occur.) (See Fraleigh and Beauregard, 1987, p. 381.)

Eigenvalues	Surface
$+++$ or $---$	Ellipsoid
$++-$ or $+--$	Elliptic cone, hyperboloid (1 or 2 sheets)
$++0$ or $--0$	Elliptic paraboloid or elliptic cylinder
$+-0$	Hyperbolic paraboloid or hyperbolic cylinder
$+00$ or -00	Parabolic cylinder or two parallel planes

For Problems A7.2 to A7.12, write the equation as a quadratic form, find the eigenvalues of $\mathbf{A}$, and classify the surface.

A7.2 $6 x^2 + 8 y^2 + 2 z^2 - 2\, xy - 4\, xz + 2\, yz = 12$
A7.3 $x^2 + 9 y^2 + 11 z^2 + 2\, xz - 4\, yz = 24$
A7.4 $11 x^2 + 15 y^2 + 3 z^2 - 4\, xy - 8\, xz + 4\, yz = 21$
A7.5 $10 x^2 + 4 y^2 + 22 z^2 + 6\, xy + 12\, xz - 6\, yz = 48$
A7.6 $8 x^2 + 44 y^2 + 14 z^2 - 12\, xy - 18\, xz = 36$
A7.7 $-x^2 + 7 y^2 + 9 z^2 + 2\, xz - 4\, yz = 24$
A7.8 $9 x^2 + 13 y^2 + z^2 - 4\, xy - 8\, xz + 4\, yz = 21$
A7.9 $7 x^2 + 10 y^2 + 19 z^2 - 28\, xy - 8\, xz - 20\, yz = 18$
A7.10 $-4 x^2 + 3 y^2 + z^2 - 4\, xy + 12\, xz - 16\, yz = -36$
A7.11 $-16 x^2 - 7 y^2 - 13 z^2 + 4\, xy + 20\, xz - 16\, yz = -28$
A7.12 $16 x^2 - y^2 + 11 z^2 - 20\, xy - 28\, xz + 8\, yz = 82$

For Problems A7.13 to A7.16, find the eigenvalues and eigenvectors of $\mathbf{A}$, the matrix of the quadratic form: $[\,x\;y\,]\,\mathbf{A}\,[\,x\;y\,]^T = g$ (See Problems A6.4 to A6.6.)

If $\mathbf{Q}$ is the matrix whose columns are the eigenvectors of $\mathbf{A}$ (with each eigenvector of unit length), then the change of variables

$$\begin{bmatrix} x \\ y \end{bmatrix} = \mathbf{Q} \begin{bmatrix} t \\ u \end{bmatrix}$$

performs the necessary rotation so that in the new coordinate system (t, u) the cross-product terms do not appear. Use the eigenvalues of $\mathbf{A}$ to find the equation of the quadratic equation in terms of the rotated axes that eliminate the mixed-terms. (Ignore any linear terms, they can be eliminated by a translation of axes after the rotation.)

A7.13 $x^2 - \sqrt{3}\, x y + 2 y^2 = 10$
A7.14 $4 x^2 + 3 \sqrt{3}\, xy + y^2 = 22$
A7.15 $x^2 - \sqrt{3}\, xy = -2$
A7.16 $-x^2 + xy - y + 7 x = 11$

In Problems A7.17 to A7.22, find the principal inertias and principal axes of the three-dimensional object consisting of point masses $m_1, \ldots, m_k$ located at $(x_1, y_1, z_1), \ldots, (x_k, y_k, z_k)$, respectively. The inertial matrix is as given in Example 7-A, but for point masses the moment of inertia and products of inertia are

$$I_{xx} = \sum_{i=1}^{k} (y_i^2 + z_i^2)\, m_i$$

$$I_{yy} = \sum_{i=1}^{k} (x_i^2 + z_i^2)\, m_i$$

$$I_{zz} = \sum_{i=1}^{k} (x_i^2 + y_i^2)\, m_i$$

$$I_{xy} = I_{yx} = \sum_{i=1}^{k} (x_i\, y_i)\, m_i$$

$$I_{xz} = I_{zx} = \sum_{i=1}^{k} (x_i\, z_i)\, m_i$$

$$I_{yz} = I_{zy} = \sum_{i=1}^{k} (y_i\, z_i)\, m_i$$

A7.17 Unit point masses ($m = 1$) at $(1, 0, 0)$, $(1, 1, 0)$, $(1, 1, 1)$.

A7.18 Unit point masses at $(1, 0, 0)$, $(0, 1, 0)$, $(1, 1, 0)$, $(1, 0, 1)$, $(1, 1, 1)$.

A7.19 Unit point masses at $(0, 1, 0)$, $(0, 0, 1)$, $(1, 1, 0)$, $(1, 0, 1)$, $(1, 1, 1)$.

A7.20 Unit point masses at $(1, 0, 0)$, $(0, 2, 0)$, $(0, 0, 1)$, $(1, 1, 0)$, $(1, 0, 1)$, $(0, 1, 1)$, $(1, 1, 2)$.

A7.21 Unit point masses at $(1, 0, 0)$, $(0, 1, 0)$, $(0, 0, 1)$, $(1, 1, 0)$, $(1, 0, 1)$, $(0, 1, 1)$; point mass with $m = 2$ at $(1, 1, 1)$.

A7.22 Unit point masses at $(1, 0, 0)$, $(0, 1, 0)$, $(1, 1, 0)$, $(1, 0, 1)$, point masses with $m = 2$ at $(0, 0, 1)$, $(0, 1, 1)$, $(1, 1, 1)$.

Problems A7.23 to A7.26 explore the eigenvalues and eigenvectors of matrices whose shapes resemble letters of the alphabet. (See Leon et al., 1996.)

A7.23 Find the eigenvalues and eigenvectors of the following matrices:

$$\mathbf{L3} = \begin{bmatrix} 1 & 0 & 0 \\ 1 & 0 & 0 \\ 1 & 1 & 1 \end{bmatrix}$$

$$\mathbf{L4} = \begin{bmatrix} 1 & 0 & 0 & 0 \\ 1 & 0 & 0 & 0 \\ 1 & 0 & 0 & 0 \\ 1 & 1 & 1 & 1 \end{bmatrix}$$

$$\mathbf{L5} = \begin{bmatrix} 1 & 0 & 0 & 0 & 0 \\ 1 & 0 & 0 & 0 & 0 \\ 1 & 0 & 0 & 0 & 0 \\ 1 & 0 & 0 & 0 & 0 \\ 1 & 1 & 1 & 1 & 1 \end{bmatrix}$$

Generate some larger **L** matrices and discuss any patterns you find in the eigenvalues and eigenvectors.

A7.24 Find the eigenvalues and eigenvectors of the following matrices:

$$\mathbf{H3} = \begin{bmatrix} 1 & 0 & 1 \\ 1 & 1 & 1 \\ 1 & 0 & 1 \end{bmatrix}$$

$$\mathbf{H5} = \begin{bmatrix} 1 & 0 & 0 & 0 & 1 \\ 1 & 0 & 0 & 0 & 1 \\ 1 & 1 & 1 & 1 & 1 \\ 1 & 0 & 0 & 0 & 1 \\ 1 & 0 & 0 & 0 & 1 \end{bmatrix}$$

$$\mathbf{H7} = \begin{bmatrix} 1 & 0 & 0 & 0 & 0 & 0 & 1 \\ 1 & 0 & 0 & 0 & 0 & 0 & 1 \\ 1 & 0 & 0 & 0 & 0 & 0 & 1 \\ 1 & 1 & 1 & 1 & 1 & 1 & 1 \\ 1 & 0 & 0 & 0 & 0 & 0 & 1 \\ 1 & 0 & 0 & 0 & 0 & 0 & 1 \\ 1 & 0 & 0 & 0 & 0 & 0 & 1 \end{bmatrix}$$

Generate some larger **H** matrices and discuss any patterns you find in the eigenvalues and eigenvectors.

A 7.25 Find the eigenvalues and eigenvectors of the following matrices:

$$\mathbf{T3} = \begin{bmatrix} 1 & 1 & 1 \\ 0 & 1 & 0 \\ 0 & 1 & 0 \end{bmatrix}$$

$$\mathbf{T5} = \begin{bmatrix} 1 & 1 & 1 & 1 & 1 \\ 0 & 0 & 1 & 0 & 0 \\ 0 & 0 & 1 & 0 & 0 \\ 0 & 0 & 1 & 0 & 0 \\ 0 & 0 & 1 & 0 & 0 \end{bmatrix}$$

$$\mathbf{T7} = \begin{bmatrix} 1 & 1 & 1 & 1 & 1 & 1 & 1 \\ 0 & 0 & 0 & 1 & 0 & 0 & 0 \\ 0 & 0 & 0 & 1 & 0 & 0 & 0 \\ 0 & 0 & 0 & 1 & 0 & 0 & 0 \\ 0 & 0 & 0 & 1 & 0 & 0 & 0 \\ 0 & 0 & 0 & 1 & 0 & 0 & 0 \\ 0 & 0 & 0 & 1 & 0 & 0 & 0 \end{bmatrix}$$

Generate some larger **T** matrices and discuss any patterns you find in the eigenvalues and eigenvectors.

A 7.26 Find the eigenvalues and eigenvectors of the following matrices:

$$\mathbf{N3} = \begin{bmatrix} 1 & 0 & 1 \\ 1 & 1 & 1 \\ 1 & 0 & 1 \end{bmatrix}$$

$$\mathbf{N4} = \begin{bmatrix} 1 & 0 & 0 & 1 \\ 1 & 1 & 0 & 1 \\ 1 & 0 & 1 & 1 \\ 1 & 0 & 0 & 1 \end{bmatrix}$$

$$\mathbf{N5} = \begin{bmatrix} 1 & 0 & 0 & 0 & 1 \\ 1 & 1 & 0 & 0 & 1 \\ 1 & 0 & 1 & 0 & 1 \\ 1 & 0 & 0 & 1 & 1 \\ 1 & 0 & 0 & 0 & 1 \end{bmatrix}$$

$$\mathbf{N6} = \begin{bmatrix} 1 & 0 & 0 & 0 & 0 & 1 \\ 1 & 1 & 0 & 0 & 0 & 1 \\ 1 & 0 & 1 & 0 & 0 & 1 \\ 1 & 0 & 0 & 1 & 0 & 1 \\ 1 & 0 & 0 & 0 & 1 & 1 \\ 1 & 0 & 0 & 0 & 0 & 1 \end{bmatrix}$$

Generate some larger **N** matrices and discuss any patterns you find in the eigenvalues and eigenvectors.

Problems A7.27 to A7.50 introduce the application of eigenvalues and eigenvectors to the solution of linear systems of first order ODE. (These problems are related to problems in Chapter 13.).

*If the real matrix **A** has distinct, real eigenvalues, μ_1 and μ_2, with eigenvectors $\mathbf{v}_1$, $\mathbf{v}_2$ respectively, the general solution of the linear system of ODE, $\mathbf{x}' = \mathbf{A}\mathbf{x}$, is*

$$\mathbf{x} = c_1 e^{\mu_1 t} \mathbf{v}_1 + c_2 e^{\mu_2 t} \mathbf{v}_2.$$

*If the matrix **A** has a complex eigenvalue, $\mu = a + b\,i$, with eigenvector $\mathbf{v}$, the general solution of the linear system of ODE, $\mathbf{x}' = \mathbf{A}\mathbf{x}$, is*

$$\mathbf{x} = c_1 e^{at}[\cos(b\,t)\,Re(\mathbf{v}) - \sin(b\,t)\,Im(\mathbf{v})]$$
$$+ c_2 [\cos(b\,t)\,Im(\mathbf{v}) + \sin(b\,t)\,Re(\mathbf{v})].$$

*Note that the complex conjugate of μ is also an eigenvalue, and its eigenvector is the complex conjugate of $\mathbf{v}$. (See Leon, 1998 for a discussion of these basic ideas.) The extension to n-by-n systems follows in a similar manner as long as **A** has n linearly independent eigenvectors. (for discussion of the more general case, see e.g. Zill, 1986)*

For Problems A7.27 to A7.50, find the general solution of $\mathbf{x}' = \mathbf{A}\,\mathbf{x}$.

$$\text{A7.27} \quad \mathbf{A} = \begin{bmatrix} 3 & -3 & 2 & -1 \\ 12 & -12 & 10 & -5 \\ 15 & -15 & 14 & -7 \\ 6 & -6 & 6 & -3 \end{bmatrix}$$

$$\text{A7.28} \quad \mathbf{A} = \begin{bmatrix} -26 & 21 & -14 & 7 \\ -34 & 29 & -20 & 10 \\ -13 & 13 & -11 & 7 \\ -8 & 8 & -8 & 7 \end{bmatrix}$$

$$\text{A7.29} \quad \mathbf{A} = \begin{bmatrix} 1 & -3 & 2 & -1 \\ 4 & -6 & 2 & -1 \\ -5 & 5 & -8 & 5 \\ -10 & 10 & -10 & 7 \end{bmatrix}$$

$$\text{A7.30} \quad \mathbf{A} = \begin{bmatrix} -6 & 6 & -4 & 2 \\ -16 & 16 & -12 & 6 \\ -17 & 17 & -15 & 9 \\ -10 & 10 & -10 & 8 \end{bmatrix}$$

$$\text{A7.31} \quad \mathbf{A} = \begin{bmatrix} 3 & -3 & 2 & -1 \\ 10 & -10 & 8 & -4 \\ 10 & -10 & 9 & -4 \\ 2 & -2 & 2 & 0 \end{bmatrix}$$

$$\text{A7.32} \quad \mathbf{A} = \begin{bmatrix} 29 & -24 & 16 & -8 \\ 52 & -46 & 34 & -17 \\ 35 & -34 & 30 & -16 \\ 14 & -14 & 14 & -9 \end{bmatrix}$$

$$\text{A7.33} \quad \mathbf{A} = \begin{bmatrix} 3 & -6 & 4 & -2 \\ 30 & -32 & 24 & -12 \\ 38 & -37 & 31 & -17 \\ 14 & -14 & 14 & -10 \end{bmatrix}$$

$$\text{A7.34} \quad \mathbf{A} = \begin{bmatrix} 16 & -15 & 10 & -5 \\ 42 & -40 & 30 & -15 \\ 35 & -34 & 29 & -14 \\ 8 & -8 & 8 & -3 \end{bmatrix}$$

A7.35 $\mathbf{A} = \begin{bmatrix} 9 & -9 & 6 & -3 \\ 24 & -23 & 16 & -8 \\ 14 & -13 & 9 & -3 \\ -4 & 4 & -4 & 5 \end{bmatrix}$

A7.36 $\mathbf{A} = \begin{bmatrix} -9 & 9 & -6 & 3 \\ -10 & 11 & -6 & 3 \\ 3 & -2 & 4 & -2 \\ 4 & -4 & 4 & -2 \end{bmatrix}$

A7.37 $\mathbf{A} = \begin{bmatrix} 20 & -18 & 12 & -6 \\ 34 & -31 & 20 & -10 \\ 14 & -13 & 8 & -5 \\ 2 & -2 & 2 & -3 \end{bmatrix}$

A7.38 $\mathbf{A} = \begin{bmatrix} -19 & 18 & -12 & 6 \\ -26 & 26 & -16 & 8 \\ -2 & 3 & 1 & -1 \\ 6 & -6 & 6 & -4 \end{bmatrix}$

A7.39 $\mathbf{A} = \begin{bmatrix} 14 & -9 & 6 & -3 \\ 3 & 34 & -4 & 2 \\ -23 & 24 & -21 & 12 \\ -15 & 15 & -14 & 10 \end{bmatrix}$

A7.40 $\mathbf{A} = \begin{bmatrix} 11 & -12 & 8 & -4 \\ 25 & -25 & 16 & -8 \\ 7 & -6 & 2 & 0 \\ -9 & 9 & -8 & 6 \end{bmatrix}$

A7.41 $\mathbf{A} = \begin{bmatrix} -2 & 3 & -2 & 1 \\ 1 & 1 & 2 & -1 \\ 10 & -9 & 12 & -7 \\ 9 & -9 & 10 & -7 \end{bmatrix}$

A7.42 $\mathbf{A} = \begin{bmatrix} -13 & 12 & -8 & 4 \\ -24 & 21 & -14 & 7 \\ -14 & 12 & -9 & 6 \\ -6 & 6 & -6 & 6 \end{bmatrix}$

A7.43 $\mathbf{A} = \begin{bmatrix} -16 & 18 & -12 & 6 \\ -39 & 38 & -24 & 12 \\ -19 & 16 & -8 & 4 \\ 4 & -4 & 4 & -2 \end{bmatrix}$

A7.44 $\mathbf{A} = \begin{bmatrix} -11 & 6 & -4 & 2 \\ 3 & -9 & 8 & -4 \\ 18 & -19 & 16 & -7 \\ 2 & -2 & 2 & 1 \end{bmatrix}$

A7.45 $\mathbf{A} = \begin{bmatrix} -16 & 18 & -12 & 6 \\ -39 & 38 & -24 & 12 \\ -18 & 15 & -7 & 3 \\ 6 & -6 & 6 & -4 \end{bmatrix}$

A7.46 $\mathbf{A} = \begin{bmatrix} -24 & 24 & -16 & 8 \\ -50 & 46 & -30 & 15 \\ -22 & 18 & -10 & 4 \\ 6 & -6 & 6 & -5 \end{bmatrix}$

A7.47 $\mathbf{A} = \begin{bmatrix} -36 & 30 & -20 & 10 \\ -61 & 50 & -36 & 18 \\ -34 & 29 & -25 & 13 \\ -10 & 10 & -10 & 6 \end{bmatrix}$

A7.48 $\mathbf{A} = \begin{bmatrix} 30 & -24 & 16 & -8 \\ 38 & -28 & 18 & -9 \\ 6 & -2 & 0 & 0 \\ -2 & 2 & -2 & 1 \end{bmatrix}$

A7.49 $\mathbf{A} = \begin{bmatrix} -9 & 6 & -4 & 2 \\ -7 & 3 & -2 & 1 \\ 6 & -7 & 6 & -5 \\ 8 & -8 & 8 & -8 \end{bmatrix}$

A7.50 $\mathbf{A} = \begin{bmatrix} -28 & 24 & -16 & 8 \\ -42 & 34 & -22 & 11 \\ -10 & 6 & -2 & 0 \\ 6 & -6 & 6 & -5 \end{bmatrix}$

EXTEND YOUR UNDERSTANDING

U7.1 Compare the required number of iterations in using QR to find the eigenvalues of F with the number needed when F is preprocessed to Hessenberg form. Consider a reasonable test for convergence.

$$F = \begin{bmatrix} -631 & 316 & -156 & 144 & -36 \\ -798 & 400 & -195 & 180 & -45 \\ 900 & -450 & 227 & -204 & 51 \\ -28 & 14 & -7 & 12 & -1 \\ 96 & -48 & 24 & -24 & 14 \end{bmatrix}$$

the exact eigenvalues are

$$\text{eig}(F) = 1 \quad 2 \quad 5 \quad 8 \quad 6$$

U7.2 The rate of convergence of the power method is influenced by the separation between the largest eigenvalue and the next largest. Compare the performance of the power method in finding the dominant eigenvalue for the following matrices. Use a reasonable test for convergence. The exact eigenvalues are given for comparison.

$$F = \begin{bmatrix} -631 & 316 & -156 & 144 & -36 \\ -798 & 400 & -195 & 180 & -45 \\ 900 & -450 & 227 & -204 & 51 \\ -28 & 14 & -7 & 12 & -1 \\ 96 & -48 & 24 & -24 & 14 \end{bmatrix}$$

the exact eigenvalues are

$$\text{eig}(F) = 1 \quad 2 \quad 5 \quad 8 \quad 6$$

$$E = \begin{bmatrix} -101 & 51 & -12 & 0 & 0 \\ -174 & 88 & -20 & 0 & 0 \\ 136 & -68 & 19 & 0 & 0 \\ 840 & -420 & 105 & -32 & 18 \\ 2016 & -1008 & 252 & -84 & 46 \end{bmatrix}$$

$$\text{eig}(E) = -10 \quad 1 \quad 2 \quad 3 \quad 4$$

U7.3 Compare QR for matrix **A** and for the preprocessed matrix **T**.

$$A = \begin{bmatrix} 2 & 0 & 0 & 0 \\ -4 & 4 & 0 & 0 \\ 12 & -6 & 6 & 0 \\ -48 & 24 & -8 & 8 \end{bmatrix}$$

$$T = \begin{bmatrix} 2 & 0 & 0 & 0 \\ 49.639 & 11.714 & -19.21 & -16.187 \\ -8.1096\text{e-}15 & 1.1265 & 3.3383 & -0.78038 \\ -1.2199\text{e-}14 & -3.1318\text{e-}16 & -1.425 & 2.9474 \end{bmatrix}$$

Note that because of rounding error, the elements that should be zero have small non-zero values. Consider how this could be remedied.

U7.4 Find the eigenvalues of the following matrices; each is the Jacobi iteration matrix for the indicated exercise.

$$T = \begin{bmatrix} 0.0 & 0.2 & -0.1 \\ 0.2 & 0.0 & 0.2 \\ 0.2 & 0.5 & 0.0 \end{bmatrix} \quad \text{(exercise 4.1)}$$

$$T = \begin{bmatrix} 0.0 & 0.2 & 0.0 \\ 0.2 & 0.0 & 0.2 \\ 0.2 & 0.0 & 0.2 \end{bmatrix} \quad \text{(exercise 4.3)}$$

$$T = \begin{bmatrix} 0 & -1/4 & 0 \\ -1/3 & 0 & 1/3 \\ 0 & 1/4 & 0 \end{bmatrix} \quad \text{(exercise 4.5)}$$

$$T = \begin{bmatrix} 0 & 1/2 & 0 & 0 \\ 1/2 & 0 & 1/2 & 0 \\ 0 & 1/2 & 0 & 1/2 \\ 0 & 0 & 1/2 & 0 \end{bmatrix} \quad \text{(exercise 4.6)}$$

$$T = \begin{bmatrix} 0 & -1/5 & 0 & 0 \\ -1/5 & 0 & -1/5 & 0 \\ 0 & -1/5 & 0 & -1/5 \\ 0 & 0 & -1/5 & 0 \end{bmatrix} \quad \text{(exercise 4.7)}$$

8
Interpolation

8.1 Polynomial Interpolation

8.2 Hermite Interpolation

8.3 Rational Function Interpolation

8.4 Spline Interpolation

8.5 Mathcad's Methods

In this chapter, we study methods for representing a function based on knowledge of its behavior at certain discrete points. From this information, we may wish to obtain estimates of function values at other points, or we may need to use the closed-form representation of the function as the basis for other numerical techniques, such as numerical differentiation or integration, which we consider in Chapter 11.

Interpolation produces a function that matches the given data exactly; we seek a function that also provides a good approximation to the (unknown) data values at intermediate points. The data may come from measured experimental values or computed values from other numerical methods, such as the solution of differential equations (Chapters 12 to 15). Our first interpolation methods provide a polynomial of an appropriate degree to exactly match a given set of data. The two standard forms for the polynomial (Lagrange and Newton) are presented with Mathcad functions and examples. It is also possible to interpolate not only function values, but derivative values. Hermite interpolation is included as the simplest example of "osculatory interpolation" for function and first-derivative values.

For some functions that are not well approximated by polynomial interpolation, rational function interpolation gives better results. In this chapter, we consider the problem of interpolating a set of data values with a rational function. In the next chapter, we present a method for finding a rational function to match the function and derivative values at a single point.

Finally, we investigate piecewise polynomial interpolation. This technique allows us to produce a smooth curve without too many "wiggles" through a large number of data points. After a brief presentation of piecewise linear and piecewise quadratic interpolation, we focus on one of the most important piecewise polynomial interpolation techniques, that of cubic splines.

Interpolation may be needed in any field in which measured data are important; the technique may be used to generate function values at points intermediate between those for which measurements are available. Interpolation may also be used to produce a smooth graph of a function for which values are known only at discrete points, either from measurements or calculations.

Example 8-A Chemical Reaction Product

Consider the observed concentration of the product of a simple chemical reaction, as a function of time:

Time:	0.0	0.5	1.0	1.5	2.0
Product:	0.0	0.19	0.26	0.29	0.31

In this chapter, we discuss several techniques for approximating the concentration of the product at other times. The data are illustrated in Fig. 8.1.

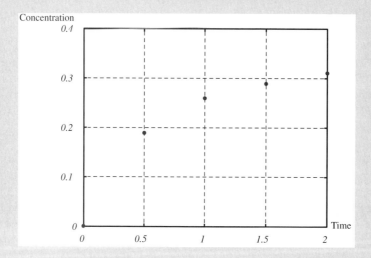

FIGURE 8.1 Chemical reaction product data.

Besides interpolating the data shown above, we also consider the effect of utilizing a more extensive set of data.

Time
0.00 0.10 0.40 0.50 0.60 0.90 1.00 1.10 1.40 1.50 1.60 1.90 2.00

Product:
0.00 0.06 0.17 0.19 0.21 0.25 0.26 0.27 0.29 0.29 0.30 0.31 0.31

One possibility is to use these data directly in the interpolation. Another approach is to use the data to estimate the first derivative of the desired interpolation function by means of the techniques discussed in Chapter 11.

Example 8-B Spline Interpolation

Spline interpolation is used in computer graphics to represent smooth curves (in parametric form). Several points are chosen along the curve and are indexed in terms of a parameter, t. Interpolation is performed on the x and y coordinates separately (each as a function of t). The resulting parametric plot, $(x(t), y(t))$ gives the interpolated curve. For example, we take the points shown in Fig. 8.2 as the data, which gives

$$t = [1 \quad 2 \quad 3 \quad 4 \quad 5 \quad 6 \quad 7 \quad 8 \quad 9 \quad 10 \quad 11 \quad 12]$$

$$x = [0 \quad 1 \quad 2 \quad 2 \quad 1 \quad 1 \quad 2 \quad 3 \quad 3 \quad 3 \quad 4 \quad 5]$$

$$y = [0 \quad 0 \quad 0 \quad 1 \quad 1 \quad 2 \quad 2 \quad 2 \quad 1 \quad 0 \quad 0 \quad 0]$$

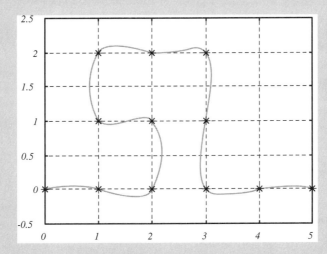

FIGURE 8.2 Spline interpolation for a parametric curve.

Data for several other curves are given in the exercises.

Although the uniform spacing used for the parameter values in this example is probably appropriate, it is not required. For figures in which consecutive pairs of points are not all the same Euclidean distance apart, the parameter values are sometimes adjusted to reflect the distance between the points. It is easy to experiment with the effect of different choices of the parameter t using the Mathcad spline function given in this chapter. For a discussion of the use of splines in computer graphics, see Bartels, Beatty, and Barsky, 1987.

8.1 POLYNOMIAL INTERPOLATION

The two most common forms of polynomial interpolation are Lagrangian interpolation and Newton interpolation. Of course, since the polynomial determined by a set of points is unique, the differences in these forms occur in the process of finding the polynomial (and the form in which it is expressed), not in the resulting function. Each approach has its advantages in different circumstances.

8.1.1 Lagrange Interpolation Polynomials

The Lagrange form of the equation of the straight line passing through two points (x_1, y_1) and (x_2, y_2) is

$$p(x) = \frac{(x - x_2)}{(x_1 - x_2)} y_1 + \frac{(x - x_1)}{(x_2 - x_1)} y_2.$$

It is easy to verify that this equation represents a line and that the given points are on the line.

The Lagrange form of the equation of the parabola passing through three points (x_1, y_1), (x_2, y_2), and (x_3, y_3) is

$$p(x) = \frac{(x - x_2)(x - x_3)}{(x_1 - x_2)(x_1 - x_3)} y_1 + \frac{(x - x_1)(x - x_3)}{(x_2 - x_1)(x_2 - x_3)} y_2 + \frac{(x - x_1)(x - x_2)}{(x_3 - x_1)(x_3 - x_2)} y_3$$

which can also be checked directly.

The general form of the polynomial passing through the n data points (x_1, y_1), ..., (x_n, y_n) has n terms on the right-hand side, one corresponding to each data point:

$$p(x) = L_1 y_1 + L_2 y_2 + \ldots + L_n y_n.$$

The kth term is the product of the kth data value and the $(n-1)^{st}$ degree polynomial

$$L_k(x) = \frac{(x - x_1) \ldots (x - x_{k-1})(x - x_{k+1}) \ldots (x - x_n)}{(x_k - x_1) \ldots (x_k - x_{k-1})(x_k - x_{k+1}) \ldots (x_k - x_n)}.$$

The numerator is the product

$$N_k(x) = (x - x_1) \ldots (x - x_{k-1})(x - x_{k+1}) \ldots (x - x_n),$$

and the denominator is of the same form, but with the variable x replaced by the given value x_k:

$$D_k = N_k(x_k) = (x_k - x_1) \ldots (x_k - x_{k-1})(x_k - x_{k+1}) \ldots (x_k - x_n).$$

Thus $L_k(x_k)$ is 1 and $L_k(x)$ is 0 when x is one of the other specified values of the independent variable, i.e., when $x = x_j$ for $j \neq k$.

Algorithm to Find Coefficients of Lagrange Interpolation Polynomial

Given data vectors **x** and **y**, find **c**, the vector of coefficients for $P(x)$, where
$$P(x) = c_1 (x - x_2) \ldots (x - x_n) + c_2 (x - x_1)(x - x_3) \ldots (x - x_n)$$
$$+ c_3 (x - x_1)(x - x_2)(x - x_4) \ldots (x - x_n)$$
$$+ \ldots + c_n (x - x_1) \ldots (x - x_{n-1})$$

Begin computations

for $k = 1$ to n
    $d_k = 1$
    for $i = 1$ to n
        if $i \neq k$
            $d_k = d_k (x_k - x_i)$
        end
    $c_k = y_k / d_k$
    end
end

Algorithm to Evaluate Lagrange Interpolation Polynomial

Given data vector **x**, and vector of coefficients **c**, evaluate $P(x)$ at $\mathbf{t} = (t_1, \ldots t_m)$.

for $i = 1$ to m
    $P_i = 0$
    for $j = 1$ to n
        $d_j = 1$
        for $k = 1$ to n
            if $(j \neq k)$
                $d_j = d_j (t_i - x_k)$
            end
        end
        $P_i = P_i + c_j d_j$
    end
end

Example 8.1 Lagrange Interpolation Parabola.

We can find a quadratic polynomial using the three given points

$$(x_1, y_1) = (-2, 4), \quad (x_2, y_2) = (0, 2), \quad (x_3, y_3) = (2, 8),$$

by substituting into the general formula, which gives

$$p(x) = \frac{(x-0)(x-2)}{(-2-0)(-2-2)} 4 + \frac{(x-(-2))(x-2)}{(0-(-2))(0-2)} 2 + \frac{(x-(-2))(x-0)}{(2-(-2))(2-0)} 8.$$

This simplifies to

$$p(x) = \frac{x(x-2)}{8} 4 + \frac{(x+2)(x-2)}{-4} 2 + \frac{x(x+2)}{8} 8 = x^2 + x + 2.$$

The data points and the interpolation polynomial are illustrated in Fig. 8.3.

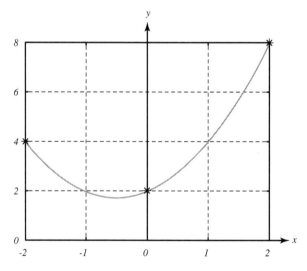

FIGURE 8.3 Parabolic interpolation polynomial.

The coefficients of the Lagrange interpolation polynomial can be found using the following Mathcad function. This function finds the coefficients c_k used in expressing the polynomial in the form

$$p(x) = c_1 N_1 + c_2 N_2 + \ldots + c_n N_n,$$

where

$$c_k = \frac{y_k}{D_k} = \frac{y_k}{(x_k - x_1) \cdots (x_k - x_{k-1})(x_k - x_{k+1}) \cdots (x_k - x_n)}$$

and

$$N_k(x) = (x - x_1) \cdots (x - x_{k-1})(x - x_{k+1}) \cdots (x - x_n).$$

Mathcad Functions for Lagrange Interpolation Polynomial

$\text{Lagrange\_coef}(x, y) := \begin{vmatrix} n \leftarrow \text{length}(x) \\ \text{for } k \in 0..n-1 \\ \quad \begin{vmatrix} d_k \leftarrow 1 \\ \text{for } i \in 0..n-1 \\ \quad \begin{vmatrix} d_k \leftarrow d_k \cdot (x_k - x_i) & \text{if } i \neq k \end{vmatrix} \\ c_k \leftarrow \dfrac{y_k}{d_k} \end{vmatrix} \\ c \end{vmatrix}$

$\text{Lagrange\_Eval}(t, x, c) := \begin{vmatrix} n \leftarrow \text{length}(x) \\ m \leftarrow \text{length}(t) \\ \text{for } i \in 0..m-1 \\ \quad \begin{vmatrix} p_i \leftarrow 0 \\ \text{for } j \in 0..n-1 \\ \quad \begin{vmatrix} N_j \leftarrow 1 \\ \text{for } k \in 0..n-1 \\ \quad \begin{vmatrix} N_j \leftarrow N_j \cdot (t_i - x_k) & \text{if } j \neq k \end{vmatrix} \\ p_i \leftarrow p_i + N_j \cdot c_j \end{vmatrix} \end{vmatrix} \\ p \end{vmatrix}$

Chapter 8 Interpolation

Example 8.2 Additional Data Points

$$x := \begin{pmatrix} -2 \\ 0 \\ -1 \\ 1 \\ 2 \end{pmatrix} \qquad y := \begin{pmatrix} 4 \\ 2 \\ -1 \\ 1 \\ 8 \end{pmatrix} \qquad c := \text{Lagrange\_coef}(x, y) \qquad c = \begin{pmatrix} 0.167 \\ 0.5 \\ 0.167 \\ -0.167 \\ 0.333 \end{pmatrix}$$

Verify interpolated values at data points

$$t := \begin{pmatrix} -2 \\ -1 \\ 0 \\ 1 \\ 2 \end{pmatrix} \qquad \text{Lagrange\_Eval}(t, x, c) = \begin{pmatrix} 4 \\ -1 \\ 2 \\ 1 \\ 8 \end{pmatrix} \qquad yt := \text{Lagrange\_Eval}(t, x, c)$$

Display polynomial

$$p(x) := \frac{1}{6} \cdot (x) \cdot (x+1) \cdot (x-1) \cdot (x-2) + \frac{1}{2} \cdot (x+2) \cdot (x+1) \cdot (x-1) \cdot (x-2) \ldots$$
$$+ \frac{1}{6} \cdot (x+2) \cdot (x) \cdot (x-1) \cdot (x-2) - \frac{1}{6} \cdot (x+2) \cdot (x) \cdot (x+1) \cdot (x-2) \ldots$$
$$+ \frac{1}{3} \cdot (x+2) \cdot (x) \cdot (x+1) \cdot (x-1)$$

Generate points for plotting interpolating polynomial

$x := -2, -1.9 .. 2$

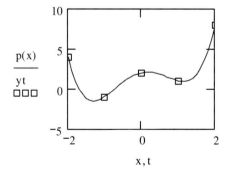

FIGURE 8.4 Quartic interpolation polynomial.

The next example investigates the effect of interpolating using a few data points, as compared to using more data points.

Example 8.3 Chemical Reaction Product Data

The interpolation polynomial and the data points for the observed product concentration data given in Example 8.A are shown in Fig. 8.5.

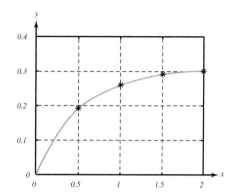

FIGURE 8.5 Quartic interpolation polynomial.

Suppose now that we use the additional data given in Example 8-A to attempt to improve the nice looking production curve. The graph of the new interpolation polynomial, shown in Fig. 8.6, demonstrates the difficulties resulting from using a polynomial of very high degree for interpolation.

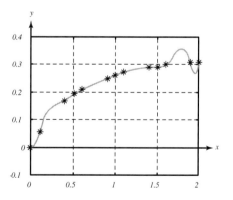

FIGURE 8.6 Polynomial for interpolating additional data.

We note that, in addition to not improving the appearance of the curve, it is necessary to completely rework the problem with the new, expanded data set when one uses the Lagrange form of interpolation.

The next example shows that the (apparent) degree of the Lagrange interpolation polynomial is determined by the number of data points. In fact, the data points all fall on a cubic polynomial, which would be evident if the Lagrange polynomial were simplified.

Example 8.4 Higher Order Interpolation Polynomials

Consider the following data:
$$\mathbf{x} = \begin{bmatrix} -2 & -1 & 0 & 1 & 2 & 3 & 4 \end{bmatrix}$$
$$\mathbf{y} = \begin{bmatrix} -15 & 0 & 3 & 0 & -3 & 0 & 15 \end{bmatrix}$$

The Mathcad function presented earlier produces the coefficients for the polynomial, which appears to be of sixth degree (as is expected for seven data points). The data and the polynomial are shown in Fig. 8.7. This example will be revisited in the next section in Example 8.7. The equation of the polynomial is

$$\begin{aligned} p(x) = &-0.0208(x+1)(x)(x-1)(x-2)(x-3)(x-4) \\ &+0.0625(x+2)(x+1)(x-1)(x-2)(x-3)(x-4) \\ &-0.0625(x+2)(x+1)(x)(x-1)(x-3)(x-4) \\ &+0.0208(x+2)(x+1)(x)(x-1)(x-2)(x-3). \end{aligned}$$

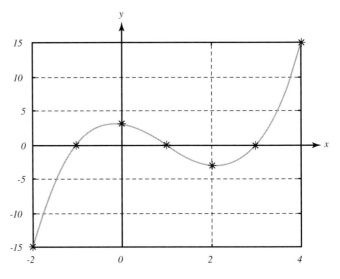

FIGURE 8.7 Higher-order interpolation polynomial.

Discussion

Consider first of all the problem of writing the equation of a line that goes through two points (x_1, y_1) and (x_2, y_2). It is easy to see that

$$p(x) = \frac{(x - x_2)}{(x_1 - x_2)} y_1 + \frac{(x - x_1)}{(x_2 - x_1)} y_2$$

has the desired properties:

1. It is the equation of a straight line. (The highest power of the independent variable x is the first power).
2. If $x = x_1$, then the coefficient of y_1 is 1, and the coefficient of y_2 is 0, so $y = y_1$.
3. If $x = x_2$, then the coefficient of y_1 is 0, and the coefficient of y_2 is 1, so $y = y_2$.

Geometrically this corresponds to determining two straight lines, one of which (call it L_0) is 1 at x_0 and 0 at x_1; the other (call it L_1) is 1 at x_1 and 0 at x_0. These lines are determined by the abscissas of the data points. The final interpolating polynomial is a linear combination of the two lines. Thus, we have

$$L_1: y = \frac{x - x_2}{x_1 - x_2}, \qquad L_2: y = \frac{x - x_1}{x_2 - x_1}; \qquad L(x) = L_1 y_1 + L_2 y_2.$$

The lines L_1 and L_2 are illustrated in Fig. 8.8 for $x_1 = 0$ and $x_2 = 1$. For these particular values of x_1 and x_2, the basis lines are

$$L_1: y = -x + 1, \qquad L_2: y = x.$$

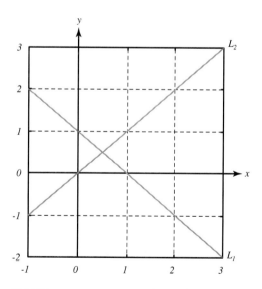

FIGURE 8.8 Lagrange basis functions for linear interpolation polynomial.

Chapter 8 Interpolation

Now let us use the same idea on the problem of writing the equation of a quadratic polynomial through three points: (x_1, y_1), (x_2, y_2), and (x_3, y_3). We see that

$$p(x) = \frac{(x - x_2)(x - x_3)}{(x_1 - x_2)(x_1 - x_3)} y_1 + \frac{(x - x_1)(x - x_3)}{(x_2 - x_1)(x_2 - x_3)} y_2 + \frac{(x - x_1)(x - x_2)}{(x_3 - x_1)(x_3 - x_2)} y_3$$

has the desired properties:

1. It is the equation of a parabola. (The highest power of the independent variable x is the second power.)
2. If $x = x_1$, then $y = y_1$. (The coefficient of y_1 is 1, the coefficient of y_2 is 0, and the coefficient of y_3 is 0.)
3. If $x = x_2$, then $y = y_2$. (The coefficient of y_1 is 0, the coefficient of y_2 is 1, and the coefficient of y_3 is 0.)
4. If $x = x_3$, then $y = y_3$. (The coefficient of y_1 is 0, the coefficient of y_2 is 0, and the coefficient of y_3 is 1).

The basis polynomials for $x_1 = 0$, $x_2 = 1$ and $x_3 = 2$ are $P_1 = (x - 1)(x - 2)/2$, $P_2 = x(x - 20)$, and $P_3 = x(x - 1)/2$; they are illustrated in Fig. 8.9.

In general, there are as many terms on the right-hand side as there are data points; the coefficient of the kth term is a fraction whose numerator is the product $(x - x_1) \ldots (x - x_{k-1})(x - x_{k+1}) \ldots (x - x_n)$ and whose denominator is of the same form with x replaced by x_k. This gives a fraction that is 1 when $x = x_k$ and is 0 when x equals any of the other specified values of the independent variable.

The Lagrange form of polynomial interpolation is particularly convenient when the same abscissas (values of the independent variable x) may occur in different applications (with only the corresponding y values changed). This form is less convenient than the Newton form (discussed in the next section) when

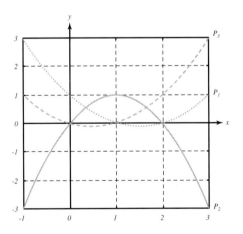

FIGURE 8.9 Lagrange basis polynomials for quadratic interpolation polynomial.

additional data points may be added to the problem or when the appropriate degree of the interpolating polynomial is unknown (i.e., when it might be better to use less than the full set of available data).

Let I be the smallest interval containing $x_1, \ldots, x_n$, and t. Then if $f(x)$ has n continuous derivatives, the error in computing $f(x)$ at $x = t$ using the polynomial that interpolates $f(x)$ at $x_1, \ldots, x_n$, is $\dfrac{(t - x_1) \cdots (t - x_n)}{n!} f^{(n)}(\eta)$, for some η in I. (See Atkinson, 1989, p. 134–135.)

8.1.2 Newton Interpolation Polynomials

The Newton form of the equation of a straight line passing through two points (x_1, y_1) and (x_2, y_2) is

$$p(x) = a_1 + a_2(x - x_1).$$

The Newton form of the equation of a parabola passing through three points (x_1, y_1), (x_2, y_2), and (x_3, y_3) is

$$p(x) = a_1 + a_2(x - x_1) + a_3(x - x_1)(x - x_2),$$

and the general form of the polynomial passing through n points $(x_1, y_1), \ldots, (x_n, y_n)$ is

$$p(x) = a_1 + a_2(x - x_1) + a_3(x - x_1)(x - x_2) + \ldots + a_n(x - x_1) \ldots (x - x_{n-1}).$$

To illustrate the method for finding the coefficients, consider the problem of finding the values of a_1, a_2, and a_3 for the parabola passing through the points (x_1, y_1), (x_2, y_2), and (x_3, y_3).

Substituting (x_1, y_1) into $y = a_1 + a_2(x - x_1) + a_3(x - x_1)(x - x_2)$ gives $a_1 = y_1$.

Substituting (x_2, y_2) into $y = a_1 + a_2(x - x_1) + a_3(x - x_1)(x - x_2)$ gives $y_2 = a_1 + a_2(x_2 - x_1)$, or

$$a_2 = \frac{y_2 - y_1}{x_2 - x_1}.$$

Substituting (x_3, y_3) into $y = a_1 + a_2(x - x_1) + a_3(x - x_1)(x - x_2)$ gives $y_3 = a_1 + a_2(x_3 - x_1) + a_3(x_3 - x_1)(x_3 - x_2)$, or (after a little algebra)

$$a_3 = \frac{\dfrac{y_3 - y_2}{x_3 - x_2} - \dfrac{y_2 - y_1}{x_2 - x_1}}{x_3 - x_1}.$$

The calculations can be performed in a systematic manner by using the "divided differences" of the function values, as illustrated in Examples 8.5 to 8.7.

Example 8.5 Newton Interpolation Parabola

Consider again the data from Example 8.1. We can find a quadratic polynomial passing through the points $(x_1, y_1) = (-2, 4)$, $(x_2, y_2) = (0, 2)$, and $(x_3, y_3) = (2, 8)$. The Newton form of the equation is

$$p(x) = a_1 + a_2(x - (-2)) + a_3(x - (-2))(x - 0),$$

where the coefficients are

$$a_1 = y_1 = 4,$$

$$a_2 = \frac{y_2 - y_1}{x_2 - x_1} = \frac{2 - 4}{0 - (-2)} = -1,$$

and

$$a_3 = \frac{\frac{y_3 - y_2}{x_3 - x_2} - \frac{y_2 - y_1}{x_2 - x_1}}{x_3 - x_0} = \frac{\frac{8 - 2}{2 - 0} - \frac{2 - 4}{0 - (-2)}}{2 - (-2)} = 1.$$

Thus,

$$p(x) = 4 - (x + 2) + x(x + 2) = x^2 + x + 2,$$

as before.

The calculations can be performed in a systematic manner, using a "divided-difference table":

x_i	y_i	$d_i = \frac{y_{i+1} - y_i}{x_{i+1} - x_i}$	$dd_i = \frac{d_{i+1} - d_i}{x_{i+2} - x_i}$
-2	$\boxed{4}$		
		$\frac{y_2 - y_1}{x_2 - x_1} = \frac{2 - 4}{0 - (-2)} = \boxed{-1}$	
0	2		$\frac{d_2 - d_1}{x_3 - x_1} = \frac{3 - (-1)}{2 - (-2)} = \boxed{1}$
		$\frac{y_3 - y_2}{x_3 - x_2} = \frac{8 - 2}{2 - 0} = 3$	
2	8		

The coefficients of the Newton polynomial are the top entries in this table. The graph of the interpolation polynomial is shown in Fig. 8.3.

One of the advantages of the Newton form of polynomial interpolation is that it is easy to add more data points and try a higher order polynomial. This is illustrated in the next example.

Example 8.6 Additional Data Points

If we extend the previous example, adding the points $(x_4, y_4) = (-1, -1)$ and $(x_5, y_5) = (1, 1)$, the polynomial has the form

$$p(x) = a_1 + a_2(x - x_1) + a_3(x - x_1)(x - x_2) + a_4(x - x_1)(x - x_2)(x - x_3) + a_5(x - x_1)(x - x_2)(x - x_3)(x - x_4).$$

The divided-difference table becomes (with new entries shown in bold)

x_i	y_i				
-2	**4**				
		$\dfrac{(2-4)}{(0+2)} = -1$			
0	2		$\dfrac{(3+1)}{(2+2)} = 1$		
		$\dfrac{(8-2)}{(2-0)} = 3$		$\dfrac{(0-1)}{(-1+2)} = \mathbf{-1}$	
2	8		$\dfrac{(3-3)}{(-1-0)} = \mathbf{0}$		$\dfrac{(2+1)}{(1+2)} = \mathbf{1}$
		$\dfrac{(-1-8)}{(-1-2)} = \mathbf{3}$		$\dfrac{(2-0)}{(1-0)} = \mathbf{2}$	
-1	**-1**		$\dfrac{(1-3)}{(1-2)} = \mathbf{2}$		
		$\dfrac{(1+1)}{(1+1)} = \mathbf{1}$			
1	**1**				

The Newton interpolation polynomial is

$$p(x) = 4 - (x + 2) + (x + 2)(x) - (x + 2)(x)(x - 2) + (x + 2)(x)(x - 2)(x + 1).$$

This expression looks quite different from that obtained for the Lagrange interpolation polynomial in Example 8.2; however, the two polynomials are equivalent. The graph of the interpolation polynomial is shown in Fig. 8.4.

Another advantage of the Newton form of the interpolation polynomial is that the degree of the interpolating polynomial may be more evident than from the Lagrange form. This is the case for the next example.

Example 8.7 Higher Order Interpolation Polynomials

Consider again the data from Example 8.4, shown in the first two columns of the following divided-difference table:

x	y						
-2	-15						
		15					
-1	0		-6				
		3		1			
0	3		-3		0		
		-3		1		0	
1	0		0		0		0
		-3		1		0	
2	-3		3		0		
		3		1			
3	0		6				
		15					
4	15						

The fact that the last three columns are zeros means that the degree of the interpolation polynomial is three less than it could have been based on the number of data points. The entry at the top of each column is the coefficient of the corresponding term in the Newton interpolation polynomial:

$$N(x) = -15 + 15(x+2) - 6(x+2)(x+1) + (x+2)(x+1)x.$$

This shows one advantage of the Newton form of the interpolation polynomial: that the polynomial is cubic is clear. (It was not clear from the Lagrange form.) The interpolation polynomial and the data are shown in Fig. 8.7.

If the *y* values are modified slightly, the divided difference table shows the small contribution from the higher degree terms:

−2	−14						
		14.5					
−1	0.5		−5.95				
		2.6		1.0333			
0	3.1		−2.85		−0.0167		
		−3.1		0.9667		0.0042	
1	0		0.05		0.0042		0.0007
		−3.0		0.9833		0.0083	
2	−3		3.00		0.0458		
		3.0		1.1667			
3	0		6.50				
		16.0					
4	16						

The Newton interpolation polynomial is

$$N(x) = -14 + 14.5(x + 2) - 5.95(x + 2)(x + 1) + 1.0333(x + 2)(x + 1)x$$
$$-0.0167(x + 2)(x + 1)x(x - 1)$$
$$+0.0042(x + 2)(x + 1)x(x - 1)(x - 2)$$
$$+0.007(x + 2)(x + 1)x(x - 1)(x - 2)(x - 3).$$

The first four terms are quite similar to those found using the previous data. The last three terms have small coefficients, as should be expected from the fact that the data are only a slight modification to the previous values.

Algorithm to Find Coefficients of Newton Interpolation Polynomial

Given data vectors **x** and **y**, find **a**, the vector of coefficients for $N(x)$, where
$$N(x) = a_1 + a_2(x - x_1) + a_3(x - x_1)(x - x_2)$$
$$+ \ldots + a_n(x - x_1)(x - x_2) \ldots (x - x_{n-1})$$

Begin computations
$a_1 = y_1$

Form 1st divided difference
for $k = 1$ to $n - 1$
$$d(k,1) = \frac{y_{k+1} - y_k}{x_{k+1} - x_k}$$
end

Form jth divided difference
for $j = 2$ to $n - 1$
    for $k = 1$ to $n - j$
$$d(k,j) = \frac{d(k+1, j-1) - d(k, j-1)}{x_{k+j} - x_k}$$
    end
end

Coefficients of interpolating polynomial
for $j = 2$ to n
    $a_j = d(1, j-1)$
end

Algorithm to Evaluate Newton Interpolation Polynomial

Given data vector **x**, and vector of coefficients **a**, evaluate $N(x)$ at $\mathbf{t} = (t_1, \ldots t_m)$.

Begin computation
for $i = 1$ to m
    $d_1 = 1$
    $N_i = a_1$
    for $j = 2$ to n
        $d_j = (t_i - x_{j-1}) d_{j-1}$
        $N_i = N_i + a_j d_j$
    end
end

Mathcad Functions for Newton Interpolation Polynomial

$$\text{Newton\_coef}(x, y) := \begin{vmatrix} n \leftarrow \text{length}(x) \\ a_0 \leftarrow y_0 \\ \text{for } k \in 0..n-2 \\ \quad d_{k,0} \leftarrow \dfrac{y_{k+1} - y_k}{x_{k+1} - x_k} \\ \text{for } j \in 1..n-2 \\ \quad \text{for } k \in 0..n-2-j \\ \quad\quad d_{k,j} \leftarrow \dfrac{d_{k+1,j-1} - d_{k,j-1}}{x_{k+j+1} - x_k} \\ \text{for } j \in 1..n-1 \\ \quad a_j \leftarrow d_{0,j-1} \\ a \end{vmatrix}$$

$$\text{Newton\_Eval}(t, x, a) := \begin{vmatrix} n \leftarrow \text{length}(x) \\ m \leftarrow \text{length}(t) \\ \text{for } i \in 0..m-1 \\ \quad \begin{vmatrix} ddd_0 \leftarrow 1 \\ p_i \leftarrow a_0 \\ \text{for } j \in 1..n-1 \\ \quad \begin{vmatrix} ddd_j \leftarrow (t_i - x_{j-1}) \cdot ddd_{j-1} \\ p_i \leftarrow p_i + a_j \cdot ddd_j \end{vmatrix} \end{vmatrix} \\ p \end{vmatrix}$$

To illustrate the use of the Mathcad functions for the Newton form of polynomial interpolation, we repeat Example 8.6, using these functions.

Chapter 8 Interpolation

Example 8.6 Revisited

$$x := \begin{pmatrix} -2 \\ 0 \\ 2 \\ -1 \\ 1 \end{pmatrix} \quad y := \begin{pmatrix} 4 \\ 2 \\ 8 \\ -1 \\ 1 \end{pmatrix} \quad a := \text{Newton\_coef}(x,y) \quad a = \begin{pmatrix} 4 \\ -1 \\ 1 \\ -1 \\ 1 \end{pmatrix}$$

Evaluate the interpolation polynomial at the data points for confirmation

$$t := \begin{pmatrix} -2 \\ -1 \\ 0 \\ 1 \\ 2 \end{pmatrix} \quad \text{Newton\_Eval}(t,x,a) = \begin{pmatrix} 4 \\ -1 \\ 2 \\ 1 \\ 8 \end{pmatrix} \quad yt := \text{Newton\_Eval}(t,x,a)$$

Define the interpolating polynomial, and graph it, with the data points

$$p(x) := 4 - (x+2) + (x+2) \cdot x \ldots \\ + -1 \cdot (x+2) \cdot x \cdot (x-2) \ldots \\ + (x+2) \cdot x \cdot (x-2) \cdot (x+1)$$

$x := -2, -1.9 .. 2$

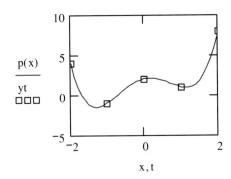

FIGURE 8.10 Quartic interpolation polynomial.

Discussion

The polynomial to interpolate the single point (x_1, y_1) is the constant function

$$N_1(x) = a_1,$$

where $a_1 = y_1$.

The polynomial to interpolate the points (x_1, y_1) and (x_2, y_2) is of the form

$$N_2(x) = a_1 + a_2(x - x_1) = N_1(x) + a_1(x - x_1).$$

We require that $N_2(x_2) = y_2 = a_1 + a_2(x_2 - x_1)$; solving for a_2 yields

$$a_2 = \frac{y_2 - y_1}{x_2 - x_1}.$$

The polynomial to interpolate the points (x_1, y_1), (x_2, y_2), and (x_3, y_3) is of the form

$$N_3(x) = a_1 + a_2(x - x_1) + a_3(x - x_1)(x - x_2) = N_2(x) + a_3(x - x_1)(x - x_2),$$

where we determine a_3 so that

$N_3(x_3) = y_3 = a_1 + a_2(x_3 - x_1) + a_3(x_3 - x_1)(x_3 - x_2)$. After some algebra, we obtain

$$a_3 = \frac{\dfrac{y_3 - y_2}{x_3 - x_2} - \dfrac{y_2 - y_1}{x_2 - x_1}}{x_3 - x_1}.$$

Rather than continuing in this direct computation manner, we wish to find a more general iterative approach. Writing the polynomial to interpolate the points (x_1, y_1), (x_2, y_2), and (x_3, y_3) in the form

$$N_3(x) = a_1 + a_2(x - x_1) + a_3(x - x_1)(x - x_2) = N_2(x) + a_3(x - x_1)(x - x_2)$$

suggests that $N_3(x)$ is formed as an extension of the polynomial (call it $N_2(x)$) that interpolates the points (x_1, y_1) and (x_2, y_2) by adding the point (x_3, y_3). On the other hand, we could just as well have supposed that we already had a polynomial (call it $M_2(x)$) that interpolates the points (x_2, y_2) and (x_3, y_3) that we extended by adding the point (x_1, y_1). From this point of view, we would write

$$M_3(x) = b_1 + b_2(x - x_2) + b_3(x - x_2)(x - x_3) = M_2(x) + b_3(x - x_2)(x - x_3).$$

However, these polynomials are simply different expressions for the same function. Comparing them, we see that the coefficient of the highest power of x must be the same, so $a_3 = b_3$. We now proceed to find an expression for a_3. Setting $N_3(x) = M_3(x)$, we have

$$N_2(x) + a_3(x - x_1)(x - x_2) = M_2(x) + a_3(x - x_2)(x - x_3),$$

so rearranging terms and factoring out the common factor of a_3 yields

$$M_2(x) - N_2(x) = a_3[(x - x_1)(x - x_2) - (x - x_2)(x - x_3)]$$
$$= a_3(x - x_2)(x_3 - x_1).$$

The coefficient of x on the left hand side of the equation is $b_2 - a_2$; the coefficient of x on the right-hand side is $a_3(x_3 - x_1)$. Solving $b_2 - a_2 = a_3(x_3 - x_1)$ for a_3 gives

$$a_3 = \frac{b_2 - a_2}{x_3 - x_1},$$

where $b_2 = \dfrac{y_3 - y_2}{x_3 - x_2}$, and $a_2 = \dfrac{y_2 - y_1}{x_2 - x_1}$.

For the general case, assume that $P_{k-1}(x)$ interpolates $f(x)$ at $x_1, \ldots, x_{k-1}$ and $Q_{k-1}(x)$ interpolates $f(x)$ at $x_2, \ldots, x_k$. Denote the leading coefficient of $P_{k-1}(x)$ as p; this coefficient is the $(k-1)^{\text{st}}$ divided difference formed using the points $(x_1, y_1), \ldots, (x_{k-1}, y_{k-1})$. Similarly, let q be the leading coefficient of $Q_{k-1}(x)$; then q is the $(k-1)^{\text{st}}$ divided difference formed using the points $(x_2, y_2), \ldots, (x_k, y_k)$. We can write the polynomial interpolating $f(x)$ at $x_1, \ldots x_k$ in two ways: Viewed as $P_{k-1}(x)$ with the addition of x_k, the polynomial is written

$$N(x) = P_{k-1}(x) + a(x - x_1) \cdots (x - x_{k-1});$$

Viewed as $Q_{k-1}(x)$ with the addition of x_1, the polynomial is written

$$N(x) = Q_{k-1}(x) + a(x - x_2) \cdots (x - x_k).$$

As before, the coefficient a that we seek to find must be the same in each of these expressions, because it is the coefficient of the highest power of x and the interpolation polynomial is unique.

Equating the two expressions for $N(x)$ and rearranging terms, we have

$$Q_{k-1}(x) + a(x - x_2) \cdots (x - x_k) = P_{k-1}(x) + a(x - x_1) \cdots (x - x_{k-1}),$$

$$\begin{aligned} Q_{k-1}(x) - P_{k-1}(x) &= a(x - x_1) \cdots (x - x_{k-1}) - a(x - x_2) \cdots (x - x_k) \\ &= a(x - x_2) \cdots (x - x_{k-1})[(x - x_1) - (x - x_k)] \\ &= a(x - x_2) \cdots (x - x_{k-1})(x_k - x_1) \end{aligned}$$

Equating the coefficients of x^{k-1} gives $q - p = a(x_k - x_1)$, so

$$a = \frac{q - p}{x_k - x_1}$$

The standard notation for the divided differences is illustrated in the following table, but it is not particularly convenient for computation:

x_1	$f[x_1]$			
		$f[x_1; x_2] = \dfrac{f[x_2] - f[x_1]}{x_2 - x_1}$		
x_2	$f[x_2]$			$f[x_1; x_2; x_3] = \dfrac{f[x_2; x_3] - f[x_1; x_2]}{x_3 - x_1}$
		$f[x_2; x_3] = \dfrac{f[x_3] - f[x_2]}{x_3 - x_2}$		
x_3	$f[x_3]$			

The Newton form of polynomial interpolation is especially convenient when the spacing between the x values of the data is constant. The spacing would have been constant in Example 8.6 if we had taken the points in order. However, instead we placed the additional points at the bottom of the table to emphasize the primary advantage of Newton interpolation, which is that more data points can be incorporated and a higher-degree polynomial generated without repeating the calculations used for the lower-order polynomial. This is in contrast to Lagrange interpolation, where the work to generate a higher-degree polynomial does not make use of calculations to form lower-order functions.

Geometrically, Newton's form of the interpolating polynomial starts with a constant function that has the correct value at $x = x_1$. The next term is a linear function that is 0 at x_1 and has the desired value at x_2, i.e., the value such that the sum of the linear and constant parts is y_2. The three terms in the Newton form of the quadratic passing through the points (0, 1), (1, 3), and (2, 6) are illustrated in Fig. 8.11. These points are

$$P_1(x) = 1,$$

$$P_2(x) = x,$$

and

$$P_3(x) = x(x - 1).$$

The interpolation polynomial is $N(x) = 1 + 2(x - 0) + 0.5\, x\,(x - 1)$

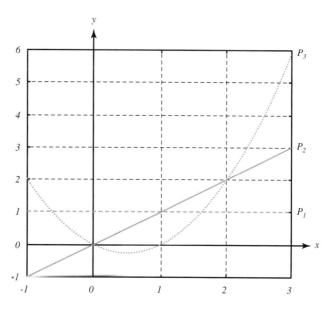

FIGURE 8.11 Newton-basis polynomials to interpolate (0,1), (1,3), and (2,6).

Chapter 8 Interpolation

8.1.3 Difficulties with Polynomial Interpolation

There are many types of problems in which polynomial interpolation through a moderate number of data points works very poorly. We illustrate three well-known cases in Examples 8.8 to 8.10. The first is an example of a function that varies over part of its domain, but is essentially constant over other portions of the domain. The second example is a function that is essentially a straight line. The third is a classic example attributed to Runge.

Example 8.8 Humped and Flat Data

The data

$$x = [-2 \quad -1.5 \quad -1 \quad -0.5 \quad 0 \quad 0.5 \quad 1 \quad 1.5 \quad 2],$$
$$y = [\,0 \quad 0 \quad 0 \quad 0.87 \quad 1 \quad 0.87 \quad 0 \quad 0 \quad 0\,].$$

illustrate the difficulty with using higher-order polynomials to interpolate a moderately large number of points, especially when the curve changes shape significantly over the interval, (being flat in some regions and not in others.) (See Fig. 8.12.)

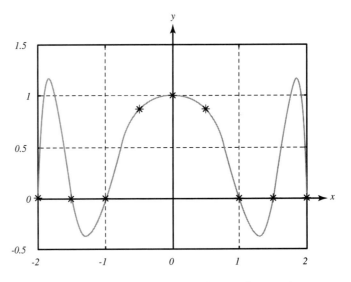

FIGURE 8.12 Difficult data for polynomial interpolation.

The next example illustrates the fact that since polynomial interpolation must go through every data point, noisy data can cause extreme oscillations in the interpolating function.

Example 8.9 Noisy Straight Line

Another example of data to which polynomial interpolation is not well suited is a noisy straight line, with y values given at unevenly spaced x values, like the following data:

$$\mathbf{x} = [0.00 \quad 0.20 \quad 0.80 \quad 1.00 \quad 1.20 \quad 1.90 \quad 2.00 \quad 2.10 \quad 2.95 \quad 3.00],$$

$$\mathbf{y} = [0.01 \quad 0.22 \quad 0.76 \quad 1.03 \quad 1.18 \quad 1.94 \quad 2.01 \quad 2.08 \quad 2.90 \quad 2.95].$$

The following divided-difference table is in the form produced by a computer program for Newton interpolation; the data are not shown. The top row of the array gives the coefficients of the interpolation polynomial, which is shown in Fig. 8.13.

```
1.0500  -0.1875   0.7500  -2.3438   2.3993  -2.4070   2.4061  -1.3332   0.7254
0.9000   0.5625  -2.0625   2.2148  -2.4147   2.6458  -1.5270   0.8429
1.3500  -1.5000   1.7027  -2.1316   2.6123  -1.5535   0.8331
0.7500   0.3730  -0.8552   1.2644  -0.7276   0.2793
1.0857  -0.4821   0.5357  -0.1545  -0.1691
0.7000   0.0000   0.2654  -0.4589
0.7000   0.2786  -0.2394
0.9647   0.0392
1.0000
```

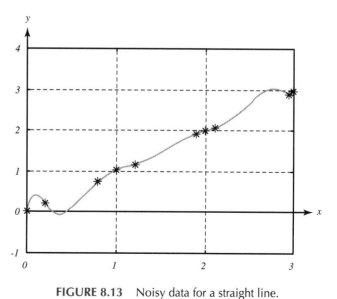

FIGURE 8.13 Noisy data for a straight line.

Example 8.10 Runge Function

The function

$$f(x) = \frac{1}{1 + 25\,x^2}$$

is a famous example of the fact that polynomial interpolation does not produce a good approximation for some functions and that using more function values (at evenly spaced x values) does not necessarily improve the situation. This example is widely known in the literature as *Runge's example*, or the *Runge function*.

First, we interpolate using five equally spaced points in the interval $[-1, 1]$:

$$\mathbf{x} = [\,-1 \quad -0.5 \quad 0.0 \quad 0.5 \quad 1.0\,],$$
$$\mathbf{y} = [\,0.0385 \quad 0.1379 \quad 1.0000 \quad 0.1379 \quad 0.0385\,].$$

The divided-difference table, with data values, is as follows.

x	y				
−1	0.0385				
		0.1989			
−0.5	0.1379		1.5252		
		1.7241		−3.3156	
0.0	1.0000		−3.4483		3.3156
		−1.7241		3.3156	
0.5	0.1379		1.5252		
		−0.1989			
1	0.0385				

The interpolation polynomial (solid line) and the original function (dashed line) are shown in Fig. 8.14.

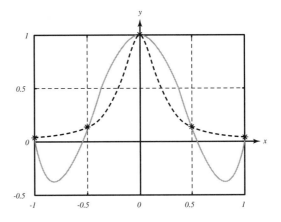

FIGURE 8.14 Runge function and interpolation polynomial.

If we use nine equally spaced data points for the interpolation, e.g.,

$\mathbf{x} = [-1.000 \quad -0.750 \quad -0.500 \quad -0.250 \quad 0.000 \quad 0.250 \quad 0.500 \quad 0.750 \quad 1.000 \ ]$,

$\mathbf{y} = [\ 0.0385 \quad 0.0664 \quad 0.138 \quad 0.3902 \quad 1.000 \quad 0.3902 \quad 0.138 \quad 0.0664 \quad 0.0385]$.

then the divided-difference table (without the data) is as follows:

```
 0.1117
          0.3489
 0.2862              1.4630
          1.4462              0.4215
 1.0093              1.8845             -15.3015
          2.8595            -18.7054              38.1194
 2.4390            -16.8209              41.8777             -53.6893
         -9.7561              33.6417             -55.8369              53.6893
-2.4390             16.8209             -41.8777              53.6893
          2.8595            -18.7054              38.1194
-1.0093             -1.8845              15.3015
          1.4462               0.4215
-0.2862             -1.4630
          0.3489
-0.1117
```

The interpolation polynomial overshoots the true polynomial much more severely than the polynomial formed by using only five points; note the change of vertical scale in Fig. 8.15, compared with Fig. 8.14.

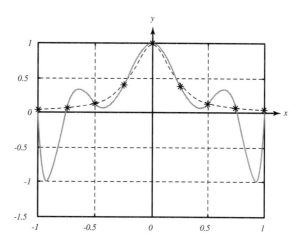

FIGURE 8.15 Interpolating the Runge function with more data points is worse!

8.2 HERMITE INTERPOLATION

Hermite interpolation allows us to find a polynomial that matches both function values and some of the derivative values at specified values of the independent variable; it includes both Taylor polynomials and Lagrange and Newton interpolation as special cases. In this section, we consider the simplest case of Hermite interpolation, that in which function values and first-derivative values are given at each point. A classic example would be data for the position and velocity of a vehicle at several different times; instead, we use the data from two of our previous examples to estimate the desired derivative value at several data points.

Suppose we have data measurements representing the values of a function and its first derivative at several values of the independent variable. The computationally most efficient form of Hermite interpolation is based on the Newton divided-difference tables. However, we require the interpolating polynomial to match each data point twice. For two data points, we obtain a cubic polynomial. The divided-difference table is of the same format as for Newton, except that each data point is entered twice, with values supplied for the first derivative at points where the denominators of the difference quotients would be zero, i.e., between repeated data points:

z_i	x_i	w_i	y_i	$d_i = \dfrac{w_{i+1} - w_i}{z_{i+1} - z_i}$	$dd_i = \dfrac{d_{i+1} - d_i}{z_{i+2} - z_i}$	$ddd_i = \dfrac{dd_{i+1} - dd_i}{z_{i+3} - z_i}$
$z_1 = x_1$		$w_1 = y_1$				
				$d_1 = y'_1$		
$z_2 = x_1$		$w_2 = y_1$			$dd_1 = \dfrac{d_2 - d_1}{z_3 - z_1}$	
				$d_2 = \dfrac{w_3 - w_2}{z_3 - z_2}$		$ddd_1 = \dfrac{dd_2 - dd_1}{z_4 - z_1}$
$z_3 = x_2$		$w_3 = y_2$			$dd_2 = \dfrac{d_3 - d_2}{z_4 - z_2}$	
				$d_3 = y'_2$		
$z_4 = x_2$		$w_4 = y_2$				

The intepolating polynomial is

$$H(x) = a_1 + a_2(x - u_1) + a_3(x - u_1)(x - u_2) + \ldots + a_n(x - u_1)(x - u_2)\ldots(x - u_{n-1})$$

where

$$u_{2i-1} = u_{2i} = x_i,$$

and $a_1 = y_1$, $a_2 = d_1$, $a_3 = dd_1$, etc.

Example 8.11 More Data for Product Concentration

Consider again the product concentration data introduced in Example 8.3. This time, instead of using all the data as separate points, we estimate the values for the derivative (using techniques from Chapter 11). The divided-difference table is as follows, with the data shown in bold:

x_i	y_i	dy						
0	**0**							
		0.6						
0	**0**		−0.44					
		0.38		0.16				
0.5	**0.19**		−0.36		0.08			
		0.2		0.24		−0.24		
0.5	**0.19**		−0.12		−0.16		0.1956	
		0.14		0.08		0.0533		−0.0652
1.00	**0.26**		−0.08		−0.08		0.0978	−0.0519
		0.1		0.00		0.20	−0.1689	0.1259
1.00	**0.26**		−0.08		0.12		−0.24	0.2000
		0.06		0.12		−0.16	0.2311	
1.50	**0.29**		−0.02		−0.12		0.1067	
		0.05		0.00		0.00		
1.50	**0.29**		−0.02		−0.12			
		0.04		−0.12				
2.00	**0.31**		−0.08					
		0.00						
2.00	**0.31**							

The resulting polynomial is illustrated in Fig. 8.16.

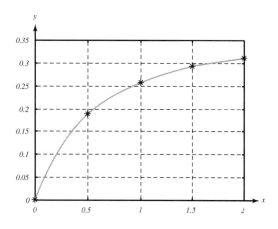

FIGURE 8.16 Hermite interpolation for product concentration data.

Algorithm to Create Coefficients for Hermite Interpolation Polynomial

Given data vectors **x**, **y**, and **dy**, find **a**, the coefficients for $H(x)$;
for $i = 1$ to n (duplicate data entries)
    $u_{2i-1} = x_i$
    $v_{2i-1} = y_i$
    $u_{2i} = x_i$
    $v_{2i} = y_i$
end
for $k = 1$ to $n-1$ (form 1st divided difference)
    $d(2k-1,1) = dy_k$
    $d(2k,1) = \dfrac{v_{2k+1} - v_{2k}}{u_{2k+1} - u_{2k}}$
end
$d(2n-1,1) = dy_n$
for $j = 2$ to $2n-2$ (form jth divided difference)
    for $k = 1$ to $2n-j$
        $d(k,j) = \dfrac{d(k+1,j-1) - d(k,j-1)}{u_{k+j} - u_k}$
    end
end
$d(1,2n-1) = \dfrac{d(2,2n-2) - d((1,2n-2))}{u_{2n} - u_1}$ (form last divided difference)

$a_1 = y_1$ (coefficients)
for $j = 2$ to $2n$
    $a_j = d(1, j-1)$
end

Algorithm to Evaluate Hermite Interpolation Polynomial

Given data vector **x**, and vector of coefficients **a**, evaluate $H(x)$ at $\mathbf{t} = (t_1, \ldots t_m)$.
for $i = 1$ to n (duplicate data entries)
    $u_{2i-1} = x_i$
    $u_{2i} = x_i$
end
for $i = 1$ to m
    $d_1 = 1$
    $H_i = a_1$
    for $j = 2$ to $2n$
        $d_j = (t_i - u_{j-1})\, d_{j-1}$
        $H_i = H_i + a_j d_j$
    end
end

The following Mathcad function finds the coefficients of the Hermite interpolation polynomial, for data given in the arrays **x**, **y**, and **dy**. The process is essentially the same as for Newton interpolation.

Mathcad Functions for Hermite Interpolation Polynomial

$$\text{Hermite\_coef}(x, y, dy) := \begin{array}{|l} n \leftarrow \text{length}(x) \\ a_0 \leftarrow y_0 \\ \text{for } i \in 0..n-1 \\ \quad \begin{array}{|l} xx_{2 \cdot i} \leftarrow x_i \\ yy_{2 \cdot i} \leftarrow y_i \\ xx_{2 \cdot i+1} \leftarrow x_i \\ yy_{2 \cdot i+1} \leftarrow y_i \end{array} \\ \text{for } k \in 0..n-2 \\ \quad \begin{array}{|l} d_{2k, 0} \leftarrow dy_k \\ d_{2k+1, 0} \leftarrow \dfrac{yy_{2k+2} - yy_{2k+1}}{xx_{2k+2} - xx_{2k+1}} \end{array} \\ d_{2n-2, 0} \leftarrow dy_{n-1} \\ \text{for } j \in 1..2n-3 \\ \quad \text{for } k \in 0..2n-j-2 \\ \quad \quad d_{k, j} \leftarrow \dfrac{d_{k+1, j-1} - d_{k, j-1}}{xx_{k+j+1} - xx_k} \\ d_{0, 2n-2} \leftarrow \dfrac{d_{1, 2n-3} - d_{0, 2n-3}}{xx_{2n-1} - xx_0} \\ \text{for } j \in 1..2n-1 \\ \quad a_j \leftarrow d_{0, j-1} \\ a \end{array}$$

$$\text{Hermite\_Eval}(t, x, a) := \begin{array}{|l} n \leftarrow \text{length}(x) \\ m \leftarrow \text{length}(t) \\ \text{for } i \in 0..n-1 \\ \quad \begin{array}{|l} xx_{2i} \leftarrow x_i \\ xx_{2 \cdot i+1} \leftarrow x_i \end{array} \\ \text{for } i \in 0..m-1 \\ \quad \begin{array}{|l} ddd_0 \leftarrow 1 \\ p_i \leftarrow a_0 \\ \text{for } j \in 1..2 \cdot n - 1 \\ \quad \begin{array}{|l} ddd_j \leftarrow (t_i - xx_{j-1}) \cdot ddd_{j-1} \\ p_i \leftarrow p_i + a_j \cdot ddd_j \end{array} \end{array} \\ p \end{array}$$

Example 8.12 Difficult Data

If we use the data from Example 8.8 to estimate both the function values and first derivative values at seven points, we have

$$\mathbf{x} = [-2 \quad -1 \quad -0.5 \quad 0 \quad 0.5 \quad 1 \quad 2],$$
$$\mathbf{y} = [\;0 \quad\; 0 \quad\; 0.87 \quad 1 \quad 0.87 \quad 0 \quad 0],$$
$$\mathbf{dy} = [\;0 \quad\; 0 \quad\; 0.5 \quad 0 \quad 0.5 \quad 0 \quad 0].$$

As with lower-order polynomial interpolation, trying to interpolate in humped and flat regions (see Fig. 8.17) causes overshoots.

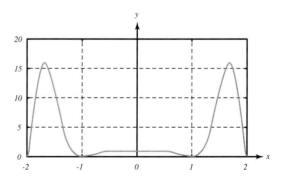

FIGURE 8.18 Hermite interpolation of humped and flat function.

Example 8.11 Revisited

Input data arrays:
$$x := \begin{pmatrix} 0 \\ 0.5 \\ 1 \\ 1.5 \\ 2 \end{pmatrix} \quad y := \begin{pmatrix} 0 \\ 0.19 \\ 0.26 \\ 0.29 \\ 0.31 \end{pmatrix} \quad dy := \begin{pmatrix} 0.6 \\ 0.2 \\ 0.1 \\ 0.05 \\ 0 \end{pmatrix}$$

Evaluation of Hermite coefficients: $\quad a := \text{Hermite\_coef}(x, y, dy)$

$a^T = (0 \quad 0.6 \quad -0.44 \quad 0.16 \quad 0.08 \quad -0.24 \quad 0.196 \quad -0.065 \quad -0.052 \quad 0.126)$

Evaluation of Hermite Interpolation Polynomial at specified t values:

$$t := \begin{pmatrix} 0 \\ 0.5 \\ 1 \\ 1.5 \\ 2 \end{pmatrix} \quad yt := \text{Hermite\_Eval}(t, x, a) \quad yt = \begin{pmatrix} 0 \\ 0.19 \\ 0.26 \\ 0.29 \\ 0.31 \end{pmatrix}$$

Explicit generation of the Hermite polynomial, using the coefficients found with Hermite_coef and the input x-coordinates, and simplified using the Expand command:

$p(x) := 0 + 0.6 \cdot (x - 0) - 0.44 \cdot (x - 0)^2 + 0.16 \cdot (x - 0)^2 \cdot (x - 0.5) \ldots$
$\quad + 0.08 \cdot (x - 0)^2 \cdot (x - 0.5)^2 - 0.24 \cdot (x - 0)^2 \cdot (x - 0.5)^2 \cdot (x - 1) \ldots$
$\quad + 0.196 \cdot (x - 0)^2 \cdot (x - 0.5)^2 \cdot (x - 1)^2 \ldots$
$\quad + (-0.065) \cdot (x - 0)^2 \cdot (x - 0.5)^2 \cdot (x - 1)^2 \cdot (x - 1.5) \ldots$
$\quad + (-0.052) \cdot (x - 0)^2 \cdot (x - 0.5)^2 \cdot (x - 1)^2 \cdot (x - 1.5)^2 \ldots$
$\quad + 0.126 \cdot (x - 0)^2 \cdot (x - 0.5)^2 \cdot (x - 1)^2 \cdot (x - 1.5)^2 \cdot (x - 2)$

$p(x) := .6 \cdot x - .5376250 \cdot x^2 + .126 \cdot x^9 - 6.187500 \cdot x^6 \ldots$
$\quad + 5.6601250 \cdot x^5 - 2.5753750 \cdot x^4 + .6483750 \cdot x^3 - 1.0600 \cdot x^8 + 3.58600 \cdot x^7$

Graph of data points (t,yt) with explicit Hermite polynomial p(x) for x in [0,2]:

$x := 0, 0.01 .. 2$

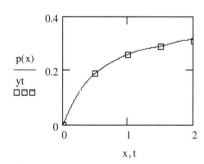

FIGURE 8.17 Data and Hermite interpolation polynomial.

8.3 RATIONAL FUNCTION INTERPOLATION

Polynomials are not always the most effective form of representation of a function or a set of data. A *rational function* (a ratio of two polynomials) may be a better choice, especially if the function to be approximated has a pole (zero of the denominator) in the region of interest. Difficulties occur with polynomial interpolation even if the pole occurs in the complex plane, unless it is far removed from the data being interpolated. The Runge function introduced in Example 8.10, when viewed as a function of the complex variable $z = x + y\, i$, becomes $f(z) = 1/(1 + 25z^2)$; its poles at $z = \pm 0.2\, i$ are too close to the region of interest, namely, the real interval $[-1, 1]$.

In this section, we present a brief description of rational-function interpolation. The Bulirsch-Stoer algorithm produces a "diagonal" rational function, i.e., a rational function in which the degree of the numerator is either the same as, or one less than, the degree of the denominator. The approach is recursive, based on tabulated data (in a manner similar to that for the Newton form of polynomial interpolation). In the next chapter, we consider Pade approximation, which seeks to find a rational function to fit the function value and derivative values at a given point x_0, in a manner more reminiscent of Taylor polynomials.

Given a set of k data points $(x_1, y_1), \ldots, (x_k, y_k)$, we seek an interpolation function of the form

$$r(x) = \frac{p_m(x)}{q_n(x)} = \frac{a_m x^m + \ldots + a_0}{b_n x^n + \ldots + b}.$$

In general, we would need to specify the degree of the numerator and the degree of the denominator. However, for the Bulirsch-Stoer method, $r(x)$ will have either $m = n$ or $m = n - 1$ (depending on whether the number of data points, k, is even or odd). The algorithm can be described recursively; we begin by showing the steps for $k = 3$ in some detail.

The following table shows the Bulirsch-Stoer method for three data points:

Data	First stage	Second stage	Third stage
$x_i\ y_i$	$R_1 = y_1$		
		$R_{12} = R_2 + \dfrac{R_2 - R_1}{\dfrac{x - x_1}{x - x_2}\left[1 - \dfrac{R_2 - R_1}{R_2}\right] - 1}$	
$x_2\ y_2$	$R_2 = y_2$		$R_{123} = R_{23} + \dfrac{R_{23} - R_{12}}{\dfrac{x - x_1}{x - x_3}\left[1 - \dfrac{R_{23} - R_{12}}{R_{23} - R_2}\right] - 1}$
		$R_{23} = R_3 + \dfrac{R_3 - R_2}{\dfrac{x - x_2}{x - x_3}\left[1 - \dfrac{R_3 - R_2}{R_3}\right] - 1}$	
$x_3\ y_3$	$R_3 = y_3$		

The general pattern is established by the third stage. The rational function $R_{123\ldots k}$ to interpolate k points, $(x_1, y_1), \ldots, (x_k, y_k)$ is formed from the function $R_{23\ldots k}$ to interpolate the $(k-1)$ points $(x_2, y_2), \ldots, (x_k, y_k)$, the function $R_{123\ldots(k-1)}$ to interpolate the $(k-1)$ points $(x_1, y_1), \ldots, (x_{k-1}, y_{k-1})$, and the function $R_{23\ldots(k-1)}$ to interpolate the $(k-2)$ points $(x_2, y_2), \ldots, (x_{k-1}, y_{k-1})$, as follows:

$$R_{123\ldots k} = R_{23\ldots k-1} + \frac{R_{23\ldots k} - R_{123\ldots(k-1)}}{\dfrac{x - x_1}{x - x_k}\left[1 - \dfrac{R_{23\ldots k} - R_{12\ldots(k-1)}}{R_{23\ldots k} - R_{23\ldots(k-1)}}\right] - 1}.$$

The computation at the second stage follows this form also, with the understanding that the rational function to interpolate the $k-2$ points is zero at that stage.

It is fairly easy to verify that the expressions at the second stage, i.e., the result of interpolating two data points, have the expected form. However, the algorithm does not yield a simple algebraic expression for the rational function to interpolate several points; rather, it provides a systematic method of computing the interpolated value at specified points. (See Press et al., 1986; and Stoer and Bulirsch, 1980 for further discussion.)

We now write the process as a formal algorithm, and following that, as a Mathcad function. The function treats interpolation at the data points separately, to save computational effort and to avoid many possible divisions by zero. The general expression for the entries in the jth column (the computation of R at the jth stage) has been rewritten to reduce the number of divisions that could lead to the indeterminant form of 0/0. Note that unlike the previous interpolation functions, rat_interp interpolates the data at specified values of x (which are denoted as xx). The results of rational function interpolation are illustrated in Example 8.13.

Algorithm for Bulirsch-Stoer Rational Function Interpolation

Given data vectors $\mathbf{x}$ and $\mathbf{y}$, evaluate $R(x)$ at $\mathbf{t} = (t_1, \ldots t_m)$. $\mathbf{z} = R(\mathbf{t})$.

Begin computation

for $h = 1$ to m
   if $t_h = x_i$ for some i
     set $z_h = y_i$
     proceed to t_{h+1}
   else
     continue with interpolation for t_h
   endif

   $R(:,1) = y(:)$ (form first column of $R = y$)
   for $i = 1$ to $n-1$ (form second column of R)
     $D = R(i+1,1) - R(i,1)$
     $r = \dfrac{t_h - x_i}{t_h - x_{i+1}}$
     $d = r\left(1 - \dfrac{D}{R(i+1,1)}\right) - 1$
     $R(i,2) = R(i+1,1) + \dfrac{D}{d}$
   end

   for $j = 3$ to n (form remaining columns of R)
     for $i = 1$ to $n-j+1$
       $D_1 = R(i+1,j-1) - R(i,j-1)$
       $D_2 = R(i+1,j-1) - R(i+1,i-2)$
       $r = \dfrac{t_h - x_i}{t_h - x_{i+j-1}}$
       if $D_1 = 0$
         $R(i,j) = R(i+1,j-1)$
       else if $D_2 = 0$
         $R(i,j) = R(i+1,j-1)$
       else
         $d = r\left(1 - \dfrac{D_1}{D_2}\right) - 1$
         $R(i,j) = R(i+1,j-1) + \dfrac{D_1}{d}$
       endif
     end
   end
   $z(h) = R(1,n)$ (interpolated value)
end

Mathcad Function for Bulirsch-Stoer Rational-Function Interpolation

$$\text{rat\_interp}(x, y, xx) := \begin{array}{|l} k \leftarrow \text{length}(x) \\ kk \leftarrow \text{length}(xx) \\ \text{test} \leftarrow 0 \\ \text{for } h \in 0..kk-1 \\ \quad \begin{array}{|l} \text{for } i \in 0..k-1 \\ \quad \begin{array}{|l} dd \leftarrow xx_h - x_i \\ \text{if } dd = 0 \\ \quad \begin{array}{|l} yy_h \leftarrow y_i \\ \text{test} \leftarrow 1 \end{array} \end{array} \\ \text{if test} = 0 \\ \quad \begin{array}{|l} R^{\langle 0 \rangle} \leftarrow y \\ \text{for } i \in 0..k-2 \\ \quad \begin{array}{|l} D \leftarrow R_{i+1,0} - R_{i,0} \\ \text{denom} \leftarrow \left(\dfrac{xx_h - x_i}{xx_h - x_{i+1}} \right) \cdot \left(1 - \dfrac{D}{R_{i+1,0}} \right) - 1 \\ R_{i,1} \leftarrow R_{i+1,0} + \dfrac{D}{\text{denom}} \end{array} \\ \text{for } j \in 2..k-1 \\ \quad \begin{array}{|l} \text{for } i \in 0..k-j \\ \quad \begin{array}{|l} D \leftarrow R_{i+1,j-1} - R_{i,j-1} \\ R_{i,j} \leftarrow R_{i+1,j-1} \quad \text{if } D = 0 \\ \text{otherwise} \\ \quad \begin{array}{|l} DD \leftarrow R_{i+1,j-1} - R_{i+1,j-2} \\ \text{denom} \leftarrow \left(\dfrac{xx_h - x_i}{xx_h - x_{i+j-1}} \right) \cdot (DD - D) - DD \\ R_{i,j} \leftarrow R_{i+1,j-1} + \dfrac{D \cdot DD}{\text{denom}} \end{array} \end{array} \end{array} \\ yy_h \leftarrow R_{0,k-1} \end{array} \\ \text{test} \leftarrow 0 \end{array} \\ yy \end{array}$$

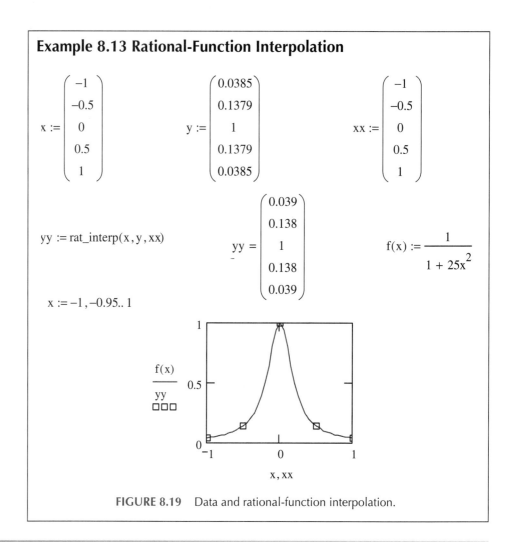

Example 8.13 Rational-Function Interpolation

$$x := \begin{pmatrix} -1 \\ -0.5 \\ 0 \\ 0.5 \\ 1 \end{pmatrix} \quad y := \begin{pmatrix} 0.0385 \\ 0.1379 \\ 1 \\ 0.1379 \\ 0.0385 \end{pmatrix} \quad xx := \begin{pmatrix} -1 \\ -0.5 \\ 0 \\ 0.5 \\ 1 \end{pmatrix}$$

$$yy := \text{rat\_interp}(x, y, xx) \quad yy = \begin{pmatrix} 0.039 \\ 0.138 \\ 1 \\ 0.138 \\ 0.039 \end{pmatrix} \quad f(x) := \frac{1}{1 + 25x^2}$$

$$x := -1, -0.95 .. 1$$

FIGURE 8.19 Data and rational-function interpolation.

8.4 SPLINE INTERPOLATION

The disadvantage of using a single polynomial (of high degree) to interpolate a large number of data points is illustrated in Example 8.10. To avoid these problems, we can use piecewise polynomials. Historically, the design and construction of a ship or aircraft involved the use of full-sized models. One method of forming a smooth curve passing through specified points was to take a thin, flexible metal or wooden lath and bend it around pegs set at the required points. The resulting curve was traced out and used in the design process. Curves generated in this manner were known as *splines* (as were the thin laths used in their construction). Spline curves represent the curve of minimum strain energy; they are also aesthetically pleasing. These physical splines assume the shape of a piecewise cubic polynomial. After a brief discussion of piecewise linear and quadratic interpolation, we consider cubic spline interpolation.

8.4.1 Piecewise Linear Interpolation

To illustrate the simplest form of piecewise polynomial interpolation, namely, piecewise linear interpolation, consider a set of four data points

$$(x_1, y_1), (x_2, y_2), (x_3, y_3), (x_4, y_4),$$

with $x_1 < x_2 < x_3 < x_4$. These points define three subintervals of the x axis:

$$I_1 = [x_1, x_2], \quad I_2 = [x_2, x_3], \quad I_3 = [x_3, x_4].$$

If we use a straight line on each subinterval, we can interpolate the data with the piecewise linear function

$$P(x) = \begin{cases} \dfrac{(x - x_2)}{(x_1 - x_2)} y_1 + \dfrac{(x - x_1)}{(x_2 - x_1)} y_2, & x_1 \leq x \leq x_2; \\ \dfrac{(x - x_3)}{(x_2 - x_3)} y_2 + \dfrac{(x - x_2)}{(x_3 - x_2)} y_3, & x_2 \leq x \leq x_3; \\ \dfrac{(x - x_4)}{(x_3 - x_4)} y_3 + \dfrac{(x - x_3)}{(x_4 - x_3)} y_4, & x_3 \leq x \leq x_4; \end{cases}$$

Example 8.14 Piecewise Linear Interpolation

Using **x** = [0 1 2 3] and **y** = [0 1 4 3], we find the piecewise linear interpolation function illustrated in Fig. 8.20. The function is continuous, but not smooth.

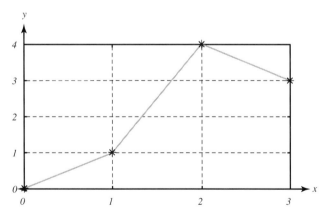

FIGURE 8.20   Piecewise linear interpolation.

8.4.2 Piecewise Quadratic Interpolation

We could use quadratic functions on each subinterval and try to make the first derivatives, as well as the function values agree at the data points. For $n + 1$ data points, there are n intervals. Also, there are three unknowns to determine for each quadratic polynomial, so we have $3n$ unknowns. Furthermore, there are two equations for each interval, corresponding to specified values for the quadratic function at the interval endpoints. That is, (x_1, y_1) and (x_2, y_2) must satisfy the quadratic equation on the first interval, etc. In addition, there are $n - 1$ points at which the intervals meet; we require that the first derivatives of the parabolas on the adjacent intervals be continuous. This gives $2n + n - 1$ equations for the $3n$ unknowns. We have one free parameter; it is not clear how to best define one additional condition.

In order to have a more useful interpolation scheme based on piecewise quadratics, we define the "knots" where the intervals meet to be the midpoints between the data points where the function values are given. We illustrate the process for four data points (x_1, y_1), (x_2, y_2), (x_3, y_3), and (x_4, y_4), with $x_1 < x_2 < x_3 < x_4$. (See Fig. 8.21)

We define the node points (or knots)

$$z_1 = x_1; \quad z_2 = (x_1 + x_2)/2, \quad z_3 = (x_2 + x_3)/2, \quad z_4 = (x_3 + x_4)/2, \quad z_5 = x_4$$

and the spacings between consecutive data points by

$$h_1 = x_2 - x_1, \quad h_2 = x_3 - x_2, \quad \text{and } h_3 = x_4 - x_3.$$

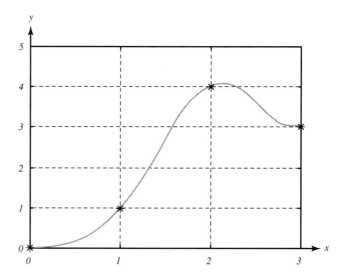

FIGURE 8.21 Piecewise quadratic interpolation of 4 data points.

Then $z_2 - x_1 = h_1/2$, $z_3 - x_2 = h_2/2$, $z_4 - x_3 = h_3/2$,
and $z_2 - x_2 = -h_1/2$, $z_3 - x_3 = -h_2/2$, $z_4 - x_4 = -h_3/2$.

Now we define P_1 on $[z_1, z_2]$: $\quad P_1(x) = a_1(x - x_1)^2 + b_1(x - x_1) + c_1;$

P_2 on $[z_2, z_3]$: $\quad P_2(x) = a_2(x - x_2)^2 + b_2(x - x_2) + c_2;$

P_3 on $[z_3, z_4]$: $\quad P_3(x) = a_3(x - x_3)^2 + b_3(x - x_3) + c_3;$

P_4 on $[z_4, z_5]$: $\quad P_4(x) = a_4(x - x_4)^2 + b_4(x - x_4) + c_4.$

Since $P_k(x_k) = c_k$, imposing the interpolation condition that $P_k(x_k) = y_k$ immediately yields $c_k = y_k$ for $k = 1, 2, 3, 4$. Now if we impose continuity conditions on the polynomials at the interior nodes, we obtain the following:

$P_1(z_2) = P_2(z_2)$: $\quad h_1^2 a_1 - h_1^2 a_2 + 2 h_1 b_1 + 2 h_1 b_2 = 4(y_2 - y_1);$

$P_2(z_3) = P_3(z_3)$: $\quad h_2^2 a_2 - h_2^2 a_3 + 2 h_2 b_2 + 2 h_2 b_3 = 4(y_3 - y_2);$

$P_3(z_4) = P_4(z_4)$: $\quad h_3^2 a_3 - h_3^2 a_4 + 2 h_3 b_3 + 2 h_3 b_4 = 4(y_4 - y_3).$

Similarly, if we impose continuity conditions on the first derivatives of the polynomials at the interior nodes, we obtain another three equations:

$P_1'(z_1) = P_2'(z_2)$: $\quad h_1 a_1 + h_1 a_2 + b_1 - b_2 = 0;$

$P_2'(z_3) = P_3'(z_3)$: $\quad h_2 a_2 + h_2 a_3 + b_2 - b_3 = 0;$

$P_3'(z_4) = P_4'(z_4)$: $\quad h_3 a_3 + h_2 a_4 + b_3 - b_4 = 0.$

At this stage, we have a system of six equations for the eight unknown coefficients ($a_1, a_2, a_3, a_4, b_1, b_2, b_3$, and b_4). Since $P_k'(x) = 2 a_k(x - x_k) + b_k$, we can determine b_1 and b_4 by imposing conditions on the derivative values at the interval endpoints, x_1 and x_4. Setting $P_1'(x_1) = 0$ gives $b_1 = 0$, and setting $P_4'(x_4) = 0$ gives $b_4 = 0$. With these zero-slope conditions at the endpoints of the interval, the equations for the coefficients become

$$\begin{aligned}
a_1 h_1^2 - a_2 h_1^2 \quad\quad\quad\quad\quad\quad + 2 b_2 h_1 \quad\quad\quad\quad\quad &= 4(y_2 - y_1), \\
+ a_2 h_2^2 - a_3 h_2^2 \quad\quad\quad\quad + 2 b_2 h_2 + 2 b_3 h_2 &= 4(y_3 - y_2), \\
+ a_3 h_3^2 - a_4 h_3^2 \quad\quad\quad\quad\quad\quad + 2 b_3 h_3 &= 4(y_4 - y_3), \\
a_1 h_1 + a_2 h_1 \quad\quad\quad\quad\quad\quad\quad - b_2 \quad\quad\quad &= 0, \\
+ a_2 h_2 + a_3 h_2 \quad\quad\quad\quad + b_2 - b_3 &= 0, \\
+ a_3 h_3 + a_4 h_3 \quad\quad\quad\quad\quad\quad + b_3 &= 0.
\end{aligned}$$

Although there are circumstances in which piecewise quadratic interpolation has some theoretical advantages, compared with cubic spline interpolation, we illustrate the former in a simple example and then proceed to the more popular cubic spline interpolation in the next section.

Example 8.15 Piecewise Quadratic Interpolation

Consider the data points (0, 0), (1, 1), (2, 4), and (3, 3). The linear system of equations for the coefficients can be solved using the Mathcad function for Gaussian elimination from Chapter 3. The matrix $\mathbf{A}$ and right-hand side $\mathbf{r}$, are, respectively,

$$\mathbf{A} = \begin{bmatrix} 1 & -1 & 0 & 0 & 2 & 0 \\ 0 & 1 & -1 & 0 & 2 & 2 \\ 0 & 0 & 1 & -1 & 0 & 2 \\ 1 & 1 & 0 & 0 & -1 & 0 \\ 0 & 1 & 1 & 0 & 1 & -1 \\ 0 & 0 & 1 & 1 & 0 & 1 \end{bmatrix}$$

$$\mathbf{r} = \begin{bmatrix} 4 & 12 & -4 & 0 & 0 & 0 \end{bmatrix}^T.$$

The solution vector gives the coefficients a_1, a_2, a_3, a_4, b_2, and b_3:

$$\mathbf{x} = \begin{bmatrix} 0.7429 & 1.7714 & -3.3714 & 2.4571 & 2.5143 & 0.9143 \end{bmatrix}^T$$

The piecewise interpolating polynomial is illustrated in Fig. 8.21 and is given by

$P_1(x) = 0.7429(x - 0)^2$ on $[0.0, 0.5]$,
$P_2(x) = 1.7714(x - 1)^2 + 2.5143(x - 1) + 1$ on $[0.5, 1.5]$,
$P_3(x) = -3.3714(x - 2)^2 + 0.9143(x - 2) + 4$ on $[1.5, 2.5]$,
$P_4(x) = 2.4571(x - 3)^2 + 3$ on $[2.5, 3.0]$,

8.4.3 Piecewise Cubic Interpolation

We can do better, without much more work, if we use cubic splines (i.e., a piecewise cubic polynomial). A simple calculation shows that we have enough information to require continuity of the function and its first and second derivatives at each of the "node points," i.e., the boundaries of the subintervals.

The calculation of the coefficients of the cubic polynomials on each subinterval is simplified by a suitable choice of the algebraic representation of the equations. The n given points $(x_1, y_1), (x_2, y_2), \ldots, (x_i, y_i), \ldots, (x_n, y_n)$ define $n - 1$ subintervals. Since the spacing between the x values is not required to be uniform, let $h_i = x_{i+1} - x_i$. We are looking for a spline function

$$S(x) = \begin{cases} P_1(x), & x_1 \leq x \leq x_2, \\ P_i(x), & x_i \leq x \leq x_{i+1}, \\ P_{n-1}(x), & x_{n-1} \leq x \leq x_n, \end{cases}$$

that is a piecewise cubic with continuous derivatives up to order 2.

For $i = 1, \ldots, n - 1$, we write

$$P_i(x) = a_{i-1} \frac{(x_{i+1} - x)^3}{6h_i} + a_i \frac{(x - x_i)^3}{6h_i} + b_i(x_{i+1} - x) + c_i(x - x_i).$$

The motivation for this form of the cubic and the derivation of the equations for the coefficients a_i, b_i, and c_i are given in the discussion at the end of this section. First we illustrate the method with several examples and a Mathcad function for generating cubic splines. The $n - 2$ equations ($i = 1, \ldots n - 2$) for the n unknowns $a_0, \ldots, a_{n-1}$ have the form

$$\frac{h_i}{6} a_{i-1} + \frac{h_i + h_{i+1}}{3} a_i + \frac{h_{i+1}}{6} a_{i+1} = \frac{y_{i+2} - y_{i+1}}{h_{i+1}} - \frac{y_{i+1} - y_i}{h_i}.$$

There are several possible choices for the conditions on the second derivatives at the endpoints, which provide the additional conditions to determine all of the unknowns. The simplest choice, the natural cubic spline, assigns values of zero to the second derivatives at x_1 and x_n. The values of b_i and c_i ($i = 1, \ldots, n - 1$) are expressed in terms of the a_i coefficients as follows:

$$b_i = \frac{y_i}{h_i} - \frac{a_{i-1} h_i}{6},$$

$$c_i = \frac{y_{i+1}}{h_i} - \frac{a_i h_i}{6}.$$

Example 8.16 Natural Cubic Spline Interpolation

Consider the data points $(-2, 4)$, $(-1, -1)$, $(0, 2)$, $(1, 1)$, and $(2, 8)$. We have $h_i = 1$ for all intervals, and $a_0 = a_4 = 0$ for a natural cubic spline. The equations for a_1, a_2, and a_3 are as follows

$$i = 1 \quad \frac{1}{6}a_0 + \frac{2}{3}a_1 + \frac{1}{6}a_2 \quad = (y_3 - y_2) - (y_2 - y_1);$$

$$i = 2 \quad \frac{1}{6}a_1 + \frac{2}{3}a_2 + \frac{1}{6}a_3 \quad = (y_4 - y_3) - (y_3 - y_2);$$

$$i = 3 \quad \frac{1}{6}a_2 + \frac{2}{3}a_3 + \frac{1}{6}a_4 = (y_5 - y_4) - (y_4 - y_3).$$

Substituting in the values for a_0, a_4, and y_i ($i = 1, \ldots, 5$) and simplifying, we obtain

$$4a_1 + a_2 \quad\quad = 48,$$
$$a_1 + 4a_2 + a_3 = -24,$$
$$a_2 + 4a_3 = 48.$$

This gives a tridiagonal system that can be solved as in Chapter 3. We find that

$$a_1 = 15.4286,$$
$$a_2 = -13.7143,$$
$$a_3 = 15.4286.$$

Solving for the b_i gives

$$b_1 = y_1 - a_0/6 = 4,$$
$$b_2 = y_2 - a_1/6 = -3.5714,$$
$$b_3 = y_3 - a_2/6 = 4.2857,$$
$$b_4 = y_4 - a_3/6 = -1.5714,$$

and for the c_i gives

$$c_1 = y_2 - a_1/6 = -3.5714,$$
$$c_2 = y_3 - a_2/6 = 4.2857,$$
$$c_3 = y_4 - a_3/6 = -1.5714,$$
$$c_4 = y_5 - a_4/6 = 8.$$

The cubic spline is illustrated in Fig. 8.22; it simplifies to

$$S(x) = \begin{cases} 2.57(x+2)^3 - 4(x+1) - 3.57(x+2), & -2 \leq x \leq -1 \\ -2.57x^3 - 2.29(x+1)^3 + 3.57x + 4.29(x+1), & -1 \leq x \leq 0 \\ -2.29(1-x)^3 + 2.57x^3 + 4.29(1-x) - 1.57x, & 0 \leq x \leq 1 \\ 2.57(2-x)^3 - 1.57(2-x) + 8(x-1), & 1 \leq x \leq 2 \end{cases}$$

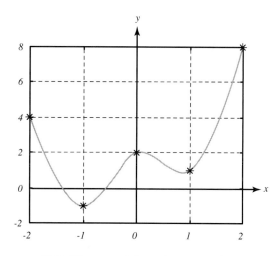

FIGURE 8.22 Cubic spline interpolant.

Algorithm for Natural Cubic Spline Interpolation

Given data vectors **x** and **y**, find the coefficients for a natural cubic spline function ($a_0 = a_{n-1} = 0$)

$$S(x) = \begin{cases} \dfrac{a_1(x-x_1)^3}{6h_1} + b_1(x_2 - x) + c_1(x - x_1) & x_1 < x < x_2 \\ \cdots \\ \dfrac{a_{i-1}(x_{i+1}-x)^3}{6h_i} + \dfrac{a_i(x-x_i)^3}{6h_i} + b_i(x_{i+1} - x) + c_i(x - x_i) & x_i < x < x_{i+1} \\ \cdots \\ \dfrac{a_{n-2}(x_n-x)^3}{6h_{n-1}} + b_{n-1}(x_n - x) + c_{n-1}(x - x_{n-1}) & x_{n-1} < x < x_n \end{cases}$$

Begin computation
Form tridiagonal system
for $k = 1$ to $n-1$
$\quad h_k = x_{(k+1)} - x_k \qquad T_k = \dfrac{y_{k+1} - y_k}{h_k}$
end
for $k = 1$ to $n-2$
$\quad R_k = T_{k+1} - T_k$ $\qquad$ (right-hand side of tridiagonal system)
$\quad D_k = \dfrac{h_k + h_{k+1}}{3}$ $\qquad$ (diagonal of tridiagonal system)
for $k = 1$ to $n-3$
$\quad U_k = \dfrac{h_{k+1}}{6} \qquad L_{k+1} = U_k$
end
$U_{n-2} = 0$
$L_1 = 0$
Solve system, **M a** = **R**, where **M** is tridiagonal, given by vectors **U**, **D**, and **L**
Define coefficients of spline function
$b_1 = \dfrac{y_1}{h_1} \qquad c_1 = \dfrac{y_2}{h_1} - \dfrac{a_1 h_1}{6}$
for $k = 2$ to $n-2$
$\quad b_k = \dfrac{y_k}{h_k} - \dfrac{a_{k-1} h_k}{6} \qquad c_k = \dfrac{y_{k+1}}{h_k} - \dfrac{a_k h_k}{6}$
end
$b_{n-1} = \dfrac{y_{n-1}}{h_{n-1}} - \dfrac{a_{n-2} h_{n-1}}{6} \qquad c_{n-1} = \dfrac{y_n}{h_{n-1}}$

Mathcad Function for Generating the Natural Cubic Spline

The following function for spline interpolation uses the Thomas method to solve the tridiagonal system. The function Thomas should either be given before the function spline, in the same worksheet, or a reference to the function

Thomas should be given, using Insert Reference from the main Mathcad menu to specify the location of the function.

$$
\begin{aligned}
\text{Spline}(xx, yy) := \ & n \leftarrow \text{length}(xx) \\
& \text{for } i \in 0..n-2 \\
& \quad \begin{vmatrix} h_i \leftarrow xx_{i+1} - xx_i \\ T_i \leftarrow \dfrac{yy_{i+1} - yy_i}{h_i} \end{vmatrix} \\
& \text{for } i \in 0..n-3 \\
& \quad \begin{vmatrix} R_i \leftarrow T_{i+1} - T_i \\ D_i \leftarrow \dfrac{h_i + h_{i+1}}{3} \end{vmatrix} \\
& \text{for } i \in 0..n-4 \\
& \quad \begin{vmatrix} U_i \leftarrow \dfrac{h_{i+1}}{6} \\ \\ L_{i+1} \leftarrow \dfrac{h_{i+1}}{6} \end{vmatrix} \\
& U_{n-3} \leftarrow 0 \\
& L_0 \leftarrow 0 \\
& a \leftarrow \text{Thomas}(U, D, L, R) \\
& b_0 \leftarrow \dfrac{yy_0}{h_0} \\
& c_0 \leftarrow \dfrac{yy_1}{h_0} - \dfrac{a_0 \cdot h_0}{6} \\
& \text{for } i \in 1..n-3 \\
& \quad \begin{vmatrix} b_i \leftarrow \dfrac{yy_i}{h_i} - \dfrac{a_{i-1} \cdot h_i}{6} \\ \\ c_i \leftarrow \dfrac{yy_{i+1}}{h_i} - \dfrac{a_i \cdot h_i}{6} \end{vmatrix} \\
& b_{n-2} \leftarrow \dfrac{yy_{n-2}}{h_{n-2}} - \dfrac{a_{n-3} \cdot h_{n-2}}{6} \\
& c_{n-2} \leftarrow \dfrac{yy_{n-1}}{h_{n-2}} \\
& a_{n-2} \leftarrow 0 \\
& s \leftarrow \text{augment}(a, b, c, h) \\
& s
\end{aligned}
$$

Example 8.17 Runge Function

The data values for the Runge function introduced in Example 8.10 are

$$xx := \begin{pmatrix} -1 \\ -0.5 \\ 0 \\ 0.5 \\ 1 \end{pmatrix} \qquad yy := \begin{pmatrix} 0.0385 \\ 0.1379 \\ 1 \\ 0.1379 \\ 0.0385 \end{pmatrix}$$

$s := \text{Spline}(xx, yy) \qquad a := s^{\langle 0 \rangle} \qquad b := s^{\langle 1 \rangle} \qquad c := s^{\langle 2 \rangle} \qquad h := s^{\langle 3 \rangle}$

$$P1(x) := \frac{a_0 \cdot (x - xx_0)^3}{6 \cdot h_0} + b_0 \cdot (xx_1 - x) + c_0 \cdot (x - xx_0)$$

$$P2(x) := \frac{a_0 \cdot (xx_2 - x)^3}{6 \cdot h_1} + \frac{a_1 \cdot (x - xx_1)^3}{6 \cdot h_1} + b_1 \cdot (xx_2 - x) + c_1 \cdot (x - xx_1)$$

$$P3(x) := \frac{a_1 \cdot (xx_3 - x)^3}{6 \cdot h_2} + \left[\frac{a_2 \cdot (x - xx_2)^3}{6 \cdot h_2} + b_2 \cdot (xx_3 - x) + c_2 \cdot (x - xx_2) \right]$$

$$P4(x) := \frac{a_2 \cdot (xx_4 - x)^3}{6 \cdot h_3} + \left[\frac{a_3 \cdot (x - xx_3)^3}{6 \cdot h_3} + b_3 \cdot (xx_4 - x) + c_3 \cdot (x - xx_3) \right]$$

$x := -1, -0.9 .. 1 \qquad S(x) := \begin{vmatrix} P1(x) & \text{if } xx_0 \le x \le xx_1 \\ P2(x) & \text{if } xx_1 < x \le xx_2 \\ P3(x) & \text{if } xx_2 < x \le xx_3 \\ P4(x) & \text{if } xx_3 < x \le xx_4 \end{vmatrix} \qquad f(x) := \frac{1}{1 + 25 \cdot x^2}$

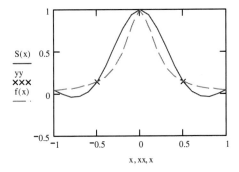

FIGURE 8.23 Runge function and spline interpolation curve.

Example 8.18 Difficult Data

Consider again the data presented in Example 8.8 for the humped and flat function:

$$\mathbf{x} = \begin{bmatrix} -2 & -1.5 & -1 & -0.5 & 0 & 0.5 & 1 & 1.5 & 2 \end{bmatrix}$$
$$\mathbf{y} = \begin{bmatrix} 0 & 0 & 0 & 0.87 & 1 & 0.87 & 0 & 0 & 0 \end{bmatrix}$$

The resulting piecewise function is made up of the following polynomials given over each subinterval:

$P_1 = -0.61 (x + 2)^3 + 0.15 (x + 2)$, $\quad -2 \leq x \leq -1.5$,

$P_2 = -0.61(-1 - x)^3 + 2.45 (x + 1.5)^3 + 0.15 (-1 - x) - 0.61 (x + 1.5)$,
$\quad -1.5 \leq x \leq -1$,

$P_3 = 2.45 (-0.5 - x)^3 - 2.24 (x + 1)^3 - 0.61 (-0.5 - x) + 2.30 (x + 1)$,
$\quad -1 \leq x \leq -0.5$,

$P_4 = -2.24 (-x)^3 + 0.60 (x + 0.5)^3 + 2.30 (-x) + 1.85 (x + 0.5)$, $\quad -0.5 \leq x \leq 0$,

$P_5 = 0.60 (0.5 - x)^3 - 2.24 (x)^3 + (1.85) (0.5 - x) + 2.30 x$, $\quad 0 \leq x \leq 0.5$,

$P_6 = -2.24 (1 - x)^3 + 7.36 (x - 0.5)^3/3 + 2.30 (1 - x) - 0.61 (x - 0.5)$,
$\quad 0.5 \leq x \leq 1$,

$P_7 = 7.36 (1.5 - x)^3/3 - 1.84 (x - 1)^3/3 - 0.61 (1.5 - x) + 0.15 (x - 1)$,
$\quad 1 \leq x \leq 1.5$,

$P_8 = -1.84 (2 - x)^3/3 + 0.15 (2 - x) + 0 (x - 1.5)$, $\quad 1.5 \leq x \leq 2$.

Figure 8.24 shows that the curve is still oscillating in the flat region, as well as near the peak of the hump. It is likely that setting the first derivative, instead of the second, equal to zero at the endpoints of the interval would improve the results obtained.

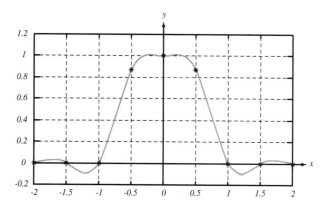

FIGURE 8.24 Cubic spline interpolation of humped and flat data.

Example 8.19 Chemical Reaction Product Data

Consider again the (additional) data presented in Example 8.3 for product concentration as a function of time in a chemical reaction:

$$\mathbf{x} = [0.00\ 0.10\ 0.40\ 0.50\ 0.60\ 0.90\ 1.00\ 1.10\ 1.40\ 1.50\ 1.60\ 1.90\ 2.00]$$

$$\mathbf{y} = [0.00\ 0.06\ 0.17\ 0.19\ 0.21\ 0.25\ 0.26\ 0.27\ 0.29\ 0.29\ 0.30\ 0.31\ 0.31]$$

Figure 8.25 shows that the curve is smoother than the high-degree polynomial (Fig. 8.6) necessary to interpolate all of the points with a single polynomial. However, the limited precision of the data means that especially around $x = 1.5$ and $x = 2.0$, the additional data points do not really improve the appearance of the original curve (Fig. 8.5) based on five data points.

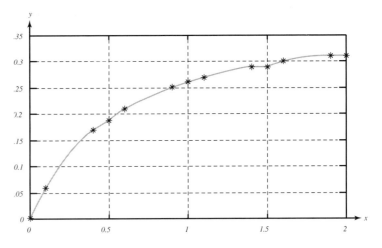

FIGURE 8.25 Cubic spline interpolant of chemical reaction data.

Discussion

To see that we can require continuity of the cubic polynomials and their first and second derivatives at the node points, assume that we have n data points $(x_1, y_1), \ldots, (x_n, y_n)$. In other words, the node points coincide with the data points in this case. We have $n - 1$ intervals and four unknowns to determine a cubic polynomial on each interval; thus, we have $4(n - 1)$ unknowns. Now, two data points must lie on the curve of each cubic polynomial, so there are $2(n - 1)$ equations for the given data. In addition, there are $n - 2$ values of x at which we must have continuity of the first and second derivatives. This gives $2(n - 2)$ more equations, so altogether we have $4n - 6$ equations for $4n - 4$ unknowns. We can obtain two more equations by specifying the value of either the first or the second derivative at each of the two endpoints of the interval (i.e., at x_1 and x_n).

The actual derivation of the spline interpolation formulas is simplified by a suitable choice of the algebraic representation of the equations. Since the spacing between the x values of the data is not required to be uniform, we let $h_i = x_{i+1} - x_i$. We are looking for a spline function of the form

$$S(x) = \begin{cases} P_1(x), & x_1 \leq x \leq x_2, \\ P_i(x), & x_i \leq x \leq x_{i+1}, \\ P_{n-1}(x), & x_{n-1} \leq x \leq x_n. \end{cases}$$

We write

$$P_i(x) = a_{i-1} \frac{(x_{i+1} - x)^3}{6h_i} + a_i \frac{(x - x_i)^3}{6h_i} + b_i(x_{i+1} - x) + c_i(x - x_i).$$

This form of the cubic is motivated by the fact that the second derivatives of the polynomials on adjacent subintervals must be equal at the common node between the two subintervals. Using the definition of h_i, we easily see that $P_i''(x_{i+1}) = a_i$ and that $P_{i+1}''(x_{i+1}) = a_i$ also. Thus, the form of the expression for $P_i(x)$ ensures that the second derivatives are continuous at the interior nodes. By integrating

$$P_i''(x) = a_{i-1} \frac{(x_{i+1} - x)}{h_i} + a_i \frac{(x - x_i)}{h_i}$$ twice and imposing the requirements that

$P_i(x_i) = y_i$ and $P_i(x_{i+1}) = y_{i+1}$, we can determine values for the coefficients b_i and c_i, to obtain

$$P_i(x) = a_{i-1} \frac{(x_{i+1} - x)^3}{6h_i} + \frac{(x - x_i)^3}{6h_i}$$

$$+ \left[y_i - \frac{a_{i-1} h_i^2}{6} \right] \frac{(x_{i+1} - x)}{h_i} + \left[y_{i+1} - \frac{a_i h_i^2}{6} \right] \frac{(x - x_i)}{h_i}.$$

We now apply the requirement that the first derivatives of P_i and P_{i+1} must agree at x_{i+1}. After some algebra, we find that (for $i = 1, 2, \ldots n - 2$)

$$\frac{h_i}{6} a_{i-1} + \frac{h_i + h_{i+1}}{3} a_i + \frac{h_{i+1}}{6} a_{i+1} = \frac{y_{i+2} - y_{i+1}}{h_{i+1}} - \frac{y_{i+1} - y_i}{h_i}.$$

We thus have $n - 2$ equations, but n unknowns ($a_0, \ldots a_{n-1}$). There are several possible choices for specifying two additional conditions. The natural cubic spline assigns values of zero for both a_0 and a_{n-1}; thus, the second derivative is zero at the endpoints. A clamped spline specifies the value of the first derivative at the endpoints.

The error in using cubic spline interpolation is $|S(x) - g(x)|$, where $S(x)$ is the spline interpolation function and $g(x)$ is the function that generated the data. Let h be the maximum spacing between node points, and $G = \max |g^{(4)}(x)|$. Then as long as reasonable choices are made for the two additional conditions,

$$|S(x) - g(x)| < k h^4 G = O(h^4).$$

(See Kahaner et al., 1989)

8.5 MATHCAD'S METHODS

Mathcad has several built-in functions for data interpolation. These are used in various combinations, as described in more detail in the following discussion.

8.5.1 Using the Built-in Functions

Many of Mathcad's built-in functions may be used for either one-dimensional or two-dimensional interpolation. For convenience, we treat the two situations separately.

Interpolation in One Dimension

In the description of the functions, we assume that the data to be interpolated are given in two vectors, **vx** and **vy**. These vectors are real, and of the same size. The elements of **vx** must be in increasing order.

    vx   *x*-values of data to be interpolated
    vy   *y*-values of data to be interpolated

Linear Interpolation

    `linterp(vx, vy, x)`   returns linearly interpolated value.

Spline interpolation

The basic approach to one-dimensional spline interpolation using Mathcad's built-in functions, is to first call the desired function (`cspline`, `lspline`, or `pspline`) which will produce a vector of coefficients, **vs**, and then to call the function `interp` to find the interpolated value at a specified value of the independent variable.

    vs   vector of coefficients

The function `cspline`(**vx**, **vy**) returns a vector **vs** of the coefficients for cubic spline interpolation; the interpolation function is cubic at the ends of the interval.

The function `lspline`(**vx**, **vy**) returns the vector of coefficients for cubic spline intepolation, with the interpolating polynomials at the ends of the interval having linear behavior.

The function `pspline`(**vx**, **vy**) returns the vector of coefficients for cubic spline intepolation, with the interpolating polynomials at the ends of the interval having parabolic behavior.

The function `interp`(**vs**, **vx**, **vy**, **x**) uses the data vectors, **vx**, **vy**, and the vector of coefficients produced by a call to one of the spline or regression functions, to find the interpolated value at the point *x*.

Interpolation in Two Dimensions

The basic approach to two-dimensional interpolation using Mathcad's built-in functions, is to first call the desired function (`cspline`, `lspline`, or `pspline`) which will produce a vector of coefficients, **vs**, and then to call the function `interp` to find the interpolated value at a specified value of the independent variable. The data to be interpolated are given in two arrays, **Mxy** and **Mz**, as described below.

Mxy	n-by-2 matrix, elements are xy coordinates of diagonal of rectangular grid,
Mz	n-by-n matrix, elements are data values at coordinate points of grid
vs	vector of coefficients from `cspline`, etc.

As for the one-dimensional case, the function `cspline`(**Mxy**, **Mz**) returns the vector of coefficients for cubic spline interpolation with cubic ends; the return vector is the first input to the function `interp`. Similarly, `lspline`(**Mxy**, **Mz**) returns the vector of coefficients for cubic spline interpolation with linear ends, and `pspline`(**Mxy**, **Mz**) returns the vector of coefficients for cubic spline interpolation with parabolic ends.

The function `interp`(**vs**, **Mxy**, **Mz**, **v**) interpolates the value from spline coefficients or regression coefficients, where **v** = (x,y) is the point where the interpolated value is desired.

8.5.2 Understanding the Algorithms

The assumption that the behavior of the interpolating function at the ends of the interval is linear corresponds to the assumption that the second derivative of the polynomial P_1 at x_1 is 0, and that the second derivative of P_{n-1} at x_n is also 0. This is the condition generally known as the natural cubic spline. It is achieved by setting a_0 and a_{n-1} to 0.

The assumption that the behavior of the interpolating function at the ends of the interval is parabolic corresponds to the assumption that the third derivative of the polynomial P_1 at x_1 is 0, and that the third derivative of P_{n-1} at x_n is also 0. This condition is achieved by setting $a_0 = a_1$ and $a_{n-1} = a_{n-2}$.

For fully cubic behavior at x_1, we have $a_0 \neq a_1$ and $a_0 \neq 0$. Similar conditions apply to a_{n-1} for cubic behavior at x_n.

SUMMARY

The Lagrange form of the equation of the parabola passing through three points (x_1, y_1), (x_2, y_2), and (x_3, y_3) is

$$p(x) = \frac{(x-x_2)(x-x_3)}{(x_1-x_2)(x_1-x_3)} y_1 + \frac{(x-x_1)(x-x_3)}{(x_2-x_1)(x_2-x_3)} y_2 + \frac{(x-x_1)(x-x_2)}{(x_3-x_1)(x_3-x_2)} y_3$$

The Newton form of the equation is

$$p(x) = a_1 + a_2(x-x_1) + a_3(x-x_1)(x-x_2),$$

where the coefficients are

$$a_1 = y_1, \quad a_2 = \frac{y_2 - y_1}{x_2 - x_1}, \quad a_3 = \frac{\dfrac{y_3 - y_2}{x_3 - x_2} - \dfrac{y_2 - y_1}{x_2 - x_1}}{x_3 - x_0}.$$

The calculations can be performed using a "divided-difference table":

x_i	y_i	$d_i = \dfrac{y_{i+1} - y_i}{x_{i+1} - x_i}$	$dd_i = \dfrac{d_{i+1} - d_i}{x_{i+2} - x_i}$
x_1	y_1		
		$\dfrac{y_2 - y_1}{x_2 - x_1}$	
x_2	y_2		$\dfrac{d_2 - d_1}{x_3 - x_1}$
		$\dfrac{y_3 - y_2}{x_3 - x_2}$	
x_3	y_3		

Hermite interpolation finds a polynomial that agrees with function values and first derivative values at the node points. The polynomial is found in a similar manner to that for Newton interpolation. Data points for which derivative values are given are repeated in the divided difference table.

Rational function interpolation is discussed in Section 8.3. The equations are not repeated here.

Piecewise polynomial interpolation is discussed in Section 8.4. The equations for pieceswise linear and piecewise quadratic interpolation are not repeated here.

Cubic spline interpolation
Let $h_i = x_{i+1} - x_i$. We are looking for a spline function

$$S(x) = \begin{cases} P_1(x), & x_1 \leq x \leq x_2, \\ P_i(x), & x_i \leq x \leq x_{i+1}, \\ P_{n-1}(x), & x_{n-1} \leq x \leq x_n, \end{cases}$$

that is a piecewise cubic with continuous derivatives up to order 2. For $i = 1, \ldots, n-1$, we write

$$P_i(x) = a_{i-1} \frac{(x_{i+1} - x)^3}{6h_i} + a_i \frac{(x - x_i)^3}{6h_i} + b_i(x_{i+1} - x) + c_i(x - x_i).$$

The $n - 2$ equations for the n unknowns $a_0, \ldots, a_{n-1}$ have the form ($i = 1, \ldots n - 2$)

$$\frac{h_i}{6} a_{i-1} + \frac{h_i + h_{i+1}}{3} a_i + \frac{h_{i+1}}{6} a_{i+1} = \frac{y_{i+2} - y_{i+1}}{h_{i+1}} - \frac{y_{i+1} - y_i}{h_i}.$$

The natural cubic spline sets $a_0 = a_{n-1} = 0$.

$$b_i = \frac{y_i}{h_i} - \frac{a_{i-1} h_i}{6},$$

$$c_i = \frac{y_{i+1}}{h_i} - \frac{a_i h_i}{6}.$$

Chapter 8 Interpolation

SUGGESTIONS FOR FURTHER READING

For a more extensive discussion of interpolation, see

Kahaner, D., C. Moler, and S. Nash, *Numerical Methods and Software*, Prentice Hall, Englewood Cliffs, NJ, 1989. Chapter 4.

Press, W. H., S. A. Teukolsky, W. T. Vetterling, and B. P. Flannery, *Numerical Recipes in C, The Art of Scientific Computing* (2d ed.), Cambridge Unversity Press, 1992. See especially sections 3.2 and 3.3.

Stoer, J. and R. Bulirsch, *Introduction to Numerical Analysis*, Springer Verlag, New York, 1980. (Especially for rational function interpolation.)

An excellent description of spline interpolation is given in

deBoor, C., *A Practical Guide to Splines*, Springer-Verlag, New York, 1978.

Applications of spline interpolation for computer graphics are presented in

Bartels, R. H., J. C. Beatty, and B. A. Barsky, *An Introduction to Splines for use in Computer Graphics and Geometric Modeling*, Morgan Kaufmann, Los Altos, CA, 1987.

Farin, G. *Curves and Surfaces for Computer Aided Geometric Design: A Practical Guide*, (2d ed.), Academic Press, Boston, 1990.

Quadratic spline interpolation is discussed in:

Kammer, W. J., G. W. Reddien, and R. S. Varga, "Quadratic Splines," *Numerische Mathematik,* vol. 22, pp. 241–259, 1974.

For further discussion of interpolation in two dimensions, see

Lancaster, P. and K. Salkauskas, *Curve and Surface Fitting: An Introduction*, Academic Press, Boston, 1986.

For a more mathematically advanced treatment of interpolation and approximation, see:

Davis, P. J., *Interpolation and Approximation*, Dover, New York, 1975. (Originally published by Blaisdell Publishing in 1963.)

PRACTICE THE TECHNIQUES

For Problems P8.1 to P8.10 find the
   a. interpolating polynomial in Lagrange form;
   b. interpolating polynomial in Newton form;
   c. piecewise linear interpolating function.

P8.1 $x = [1 \ \ 2 \ \ 3]$,
        $y = [1 \ \ 4 \ \ 8]$.

P8.2 $x = [1 \ \ 4 \ \ 9]$,
        $y = [1 \ \ 2 \ \ 3]$.

P8.3 $x = [4 \ \ 9 \ \ 16]$,
        $y = [2 \ \ 3 \ \ 4]$.

P8.4 $x = [-1 \ \ 0 \ \ 1]$,
        $y = [-2 \ \ 3 \ \ -2]$.

P8.5 $x = [0 \ \ 1 \ \ 2]$,
        $y = [1 \ \ 2 \ \ 4]$.

P8.6 $x = [0 \ \ 1 \ \ 2 \ \ 4]$,
        $y = [1 \ \ 1 \ \ 2 \ \ 5]$.

P8.7 $x = [-1 \ \ 0 \ \ 1 \ \ 2]$,
        $y = [1/3 \ \ 1 \ \ 3 \ \ 9]$.

P8.8 $x = [0 \ \ 1 \ \ 2 \ \ 3]$,
        $y = [1 \ \ 2 \ \ 4 \ \ 8]$.

P8.9 $x = [0 \ \ 1 \ \ 2 \ \ 3]$,
        $y = [0 \ \ 1 \ \ 0 \ \ -1]$.

P8.10 $x = [-1 \ \ 0 \ \ 1 \ \ 2]$,
         $y = [0 \ \ 1 \ \ 2 \ \ 9]$.

For Problems P8.11 to P8.15 find the:
   a. interpolating polynomial in Lagrange form;
   b. interpolating polynomial in Newton form;
   c. piecewise linear interpolating function;
   d. piecewise quadratic interpolating function (knots midway between data points);
   e. cubic spline interpolating function.

P8.11 $x = [0 \ \ 2/3 \ \ 1 \ \ 2]$,
         $y = [2 \ \ -2 \ \ -1 \ \ -1/2]$.

P8.12 $x = [0 \ \ 2/3 \ \ 1 \ \ 2]$,
         $y = [4 \ \ -4 \ \ -2 \ \ -1/2]$.

P8.13 $x = [0 \ \ 2/3 \ \ 1 \ \ 2]$,
         $y = [4 \ \ -4 \ \ -7/2 \ \ -1/2]$.

P8.14 $x = [1 \ \ 2 \ \ 3 \ \ 4]$,
         $y = [2 \ \ 4 \ \ 8 \ \ 16]$.

P8.15 $x = [0 \ \ 1/2 \ \ 1 \ \ 3/2]$,
         $y = [1 \ \ 2 \ \ 1 \ \ 0]$.

For Problems P8.16 to P8.20:
   a. Find the interpolation polynomial (either form).
   b. Find the cubic spline interpolation function.
   c. Compare the interpolated values to the given function.

P8.16 $x = [0 \ \ 1 \ \ 8 \ \ 27]$,
         $y = [0 \ \ 1 \ \ 2 \ \ 3]$.

Compare interpolated values at $x = 0.5, 3.5,$ and 18 to $f(x) = \sqrt[3]{x}$.

P8.17 $x = [0 \ \ 1 \ \ 4 \ \ 9]$,
         $y = [0 \ \ 1 \ \ 2 \ \ 3]$.

Compare interpolated values at $x = 0.5, 2.5,$ and 6.5 to $f(x) = \sqrt{x}$.

P8.18 $x = [-1 \ \ -0.75 \ \ -0.25 \ \ 0.25 \ \ 0.75 \ \ 1]$
         $y = [0 \ \ -0.7 \ \ -0.7 \ \ 0.7 \ \ 0.7 \ \ 0]$

Compare interpolated values $x = -0.5, 0,$ and 0.5 to $f(x) = \sin(\pi x)$.

P8.19 $x = [-2 \ \ -1 \ \ 0 \ \ 1 \ \ 2 \ \ 3 \ \ 4]$
         $y = [-14 \ \ 0.5 \ \ 3.1 \ \ 0 \ \ -3 \ \ 0 \ \ 16]$

Compare to Example 8.4 and 8.7.

P8.20 $x = [0 \ \ 0.5 \ \ 1.0 \ \ 1.5 \ \ 2.0 \ \ 2.5 \ \ 3.0]$
         $y = [4.0 \ \ 0 \ \ -2.0 \ \ 0 \ \ 1.0 \ \ 0 \ \ -0.5]$

Compare the interpolated values to those of the function $y = 2^{(2-x)} \cos(\pi x)$.

For Problems P8.21 to P8.30 find the
  a. interpolation polynomial (either form);
  b. cubic spline interpolation function;
  c. Bulirsch-Stoer rational interpolation function.

P8.21 $x = [0 \quad 1 \quad 2 \quad 3 \quad 4 \quad 5 \quad 6 \quad 7 \quad 8 \quad 9 \quad 10 \quad]$
$y = [2.0 \quad 0 \quad 0.6667 \quad 0 \quad 0.40 \quad 0 \quad 0.2857 \quad 0 \quad 0.2222 \quad 0 \quad 0.1818]$

Compare the interpolated values to $y = \dfrac{1 + \cos(\pi x)}{1 + x}$.

P8.22 $x = [0 \quad 1 \quad 2 \quad 3 \quad 4 \quad 5 \quad 6 \quad 7 \quad 8 \quad 9 \quad 10 \quad]$
$z = [2.0 \quad 0.7702 \quad 0.1946 \quad 0.0025 \quad 0.0693 \quad 0.2139 \quad 0.280 \quad 0.2192 \quad 0.0949 \quad 0.0089 \quad 0.0146]$

Compare the interpolated values to $z = \dfrac{1 + \cos(x)}{1 + x}$.

P8.23 $x = [0 \quad 1 \quad 2 \quad 3 \quad 4 \quad 5 \quad 6 \quad 7 \quad 8 \quad 9 \quad 10 \quad]$
$w = [0 \quad 0.5 \quad 0.5 \quad 0.375 \quad 0.25 \quad 0.1562 \quad 0.0938 \quad 0.0547 \quad 0.0312 \quad 0.0176 \quad 0.0098]$

Compare the interpolated values to $w = x \, 2^{-x}$.

P8.24 $x = [1 \quad 2 \quad 3 \quad 4 \quad 5 \quad 6 \quad 7 \quad 8 \quad 9 \quad 10 \quad]$
$y = [0 \quad 0.3466 \quad 0.3662 \quad 0.3466 \quad 0.3219 \quad 0.2986 \quad 0.2780 \quad 0.2599 \quad 0.2441 \quad 0.2303]$

Compare the interpolated values to $y = \dfrac{\ln(x)}{x}$.

P8.25 $x = [0 \quad 0.50 \quad 1.00 \quad 1.50 \quad 2.00 \quad 2.50 \quad 3.00]$
$y = [-0.3333 \quad -0.2703 \quad -0.2000 \quad -0.1333 \quad -0.0769 \quad -0.0328 \quad 0 \quad]$

Compare the interpolated values to $y = \dfrac{x - 3}{x^2 + 9}$.

P8.26 $x = [0.0 \quad 1.00 \quad 2.00 \quad 3.00 \quad 4.00 \quad 5.00 \quad 6.00]$
$y = [-0.33 \quad -0.20 \quad -0.08 \quad 0.00 \quad 0.04 \quad 0.06 \quad 0.07]$

Compare the interpolated values to $y = \dfrac{x - 3}{x^2 + 9}$.

P8.27 $x = [0 \quad 5 \quad 10 \quad 15 \quad 20 \quad 25 \quad 30 \quad 35 \quad 40 \quad]$
$y = [-0.3333 \quad 0.0588 \quad 0.0642 \quad 0.0513 \quad 0.0416 \quad 0.0347 \quad 0.0297 \quad 0.0259 \quad 0.0230]$

Compare the interpolated values to $y = \dfrac{x - 3}{x^2 + 9}$.

P8.28 Interpolate
$x = [-5 \quad -4 \quad -3 \quad -2 \quad -1 \quad 0 \quad 1 \quad 2 \quad 3 \quad 4 \quad 5 \quad]$
$y = [-0.1923 \quad -0.2353 \quad -0.30 \quad -0.40 \quad -0.50 \quad 0.00 \quad 0.50 \quad 0.40 \quad 0.30 \quad 0.2353 \quad 0.1923]$

Compare the interpolated values at $x = -4.5, -3.5, \ldots, 3.5, 4.5$ to $y = \dfrac{x}{x^2 + 1}$.

P8.29 $x = [0 \quad 1 \quad 2 \quad 3 \quad 4 \quad 5 \quad 6 \quad 7 \quad 8 \quad 9 \quad 10 \quad]$
$y = [1.0 \quad 1.0 \quad 0.5556 \quad 0.3571 \quad 0.2615 \quad 0.2063 \quad 0.1705 \quad 0.1453 \quad 0.1267 \quad 0.1123 \quad 0.1009]$

Compare the interpolated values to $y = \dfrac{x^2 + 1}{x^3 + 1}$.

P8.30 $x = [0 \quad 1 \quad 2 \quad 3 \quad 4 \quad 5 \quad 6 \quad 7 \quad 8 \quad 9 \quad 10 \quad]$
$y = [0.5 \quad 1.0 \quad 1.7 \quad 1.8966 \quad 1.9545 \quad 1.9764 \quad 1.9862 \quad 1.9913 \quad 1.9942 \quad 1.9959 \quad 1.9970]$

Compare the interpolated values to $y = \dfrac{2x^2 + 1}{x^3 + 2}$.

Problems P8.31 to P8.35 provide practice in the use of Hermite interpolation.
   a. *Find the interpolation polynomial using the given data.*
   b. *Find the Hermite interpolation polynomial using the given function and derivative values.*
   c. *Compare the interpolated values (at some intermediate values of x) from Parts a and b to those of the specified function*

P8.31 Interpolate

$x = [-5 \quad -4 \quad -3 \quad -2 \quad -1 \quad 0 \quad 1 \quad 2 \quad 3 \quad 4 \quad 5]$
$w = [0.0385 \quad 0.0588 \quad 0.10 \quad 0.20 \quad 0.50 \quad 1.00 \quad 0.50 \quad 0.20 \quad 0.10 \quad 0.0588 \quad 0.0385]$

for Hermite interpolation

$x = [-5 \quad -3 \quad -1 \quad 1 \quad 3 \quad 5 \quad]$
$w = [0.0385 \quad 0.10 \quad 0.50 \quad 0.50 \quad 0.10 \quad 0.0385 \,]$
$dw = [0.0148 \quad 0.06 \quad 0.50 \quad -0.50 \quad -0.06 \quad -0.0148]$

or

$x = [-4 \quad -2 \quad 0 \quad 2 \quad 4 \quad]$
$w = [0.0588 \quad 0.20 \quad 1.00 \quad 0.20 \quad 0.0588 \,]$
$dw = [0.0277 \quad 0.1600 \quad 0 \quad -0.1600 \quad -0.0277]$

or

$x = [0 \quad 1 \quad 2 \quad 3 \quad 4 \quad 5 \quad]$
$w = [1.00 \quad 0.50 \quad 0.20 \quad 0.10 \quad 0.0588 \quad 0.0385 \,]$
$dw = [0 \quad -0.5000 \quad -0.1600 \quad -0.0600 \quad -0.0277 \quad -0.0148]$

Compare the interpolated values to those from the function $w = \dfrac{1}{x^2 + 1}$.

P8.32 Interpolate

$x = [1 \quad 2 \quad 3 \quad 4 \quad 5 \quad 6 \quad 7 \quad 8 \quad 9 \quad 10 \quad]$
$y = [0 \quad 0.6931 \quad 1.0986 \quad 1.3863 \quad 1.6094 \quad 1.7918 \quad 1.9459 \quad 2.0794 \quad 2.1972 \quad 2.3026]$

For Hermite interpolation

$x - [1 \quad 3 \quad 5 \quad 7 \quad 9 \quad]$
$y = [0 \quad 1.0986 \quad 1.6094 \quad 1.9459 \quad 2.1972]$
$dy = [1.0 \quad 0.3333 \quad 0.20 \quad 0.1429 \quad 0.1111]$

Compare the interpolated values to those of $\ln(x)$ (natural logarithm)

P8.33 $x = [0 \quad 1 \quad 2 \quad 3 \quad 4 \quad 5 \quad 6 \quad 7 \quad 8 \quad]$
$y = [0.50 \quad 0.731 \quad 0.881 \quad 0.953 \quad 0.982 \quad 0.993 \quad 0.997 \quad 0.999 \quad 0.9997]$

For Hermite interpolation, use half of the data above, with the corresponding derivative values.

$dy = [0.25 \quad 0.197 \quad 0.105 \quad 0.045 \quad 0.018 \quad 0.0066 \quad 0.0025 \quad 0.0009 \quad 0.0003]$

Compare the interpolated values to $y = \dfrac{1}{1 + \exp(-x)}$.

P8.34 $x = [0 \quad 1 \quad 2 \quad 3 \quad 4 \quad 5 \quad 6 \quad 7 \quad 8 \quad]$
$y = [0 \quad 0.5000 \quad 0.6667 \quad 0.7500 \quad 0.8000 \quad 0.8333 \quad 0.8571 \quad 0.8750 \quad 0.8889]$

For Hermite interpolation, use half of the data above, with the corresponding derivative values.

$dy = [1 \quad 0.2500 \quad 0.1111 \quad 0.0625 \quad 0.0400 \quad 0.0278 \quad 0.0204 \quad 0.0156 \quad 0.0123]$

Compare the interpolated values to $y = \dfrac{x}{1 + x}$.

P8.35 $x = [0.0100 \quad 1.0000 \quad 4.0000 \quad 9.0000 \quad 16.0000 \quad 25.0000 \quad 36.0000]$
$y = [0.1000 \quad 1.0000 \quad 2.0000 \quad 3.0000 \quad 4.0000 \quad 5.0000 \quad 6.0000 \quad]$
$dy = [5.0000 \quad 0.5000 \quad 0.2500 \quad 0.1667 \quad 0.1250 \quad 0.1000 \quad 0.0833 \quad]$

Compare the interpolated values to $y = \sqrt{x}$.

Problems P8.36 to P8.40 give practice using the Mathcad interpolation functions for data in two dimensions.

P8.36

$Mxy = \begin{bmatrix} 0.0 & 0.0 \\ 0.2 & 0.2 \\ 0.4 & 0.4 \\ 0.6 & 0.6 \\ 0.8 & 0.8 \\ 1.0 & 1.0 \end{bmatrix} \quad Mz = \begin{bmatrix} 0 & 0 & 0 & 0 & 0 & 0 \\ 0 & 0.0047 & 0.0374 & 0.1263 & 0.2994 & 0.5848 \\ 0 & 0.0059 & 0.0472 & 0.1592 & 0.3772 & 0.7368 \\ 0 & 0.0067 & 0.0540 & 0.1822 & 0.4318 & 0.8434 \\ 0 & 0.0074 & 0.0594 & 0.2005 & 0.4753 & 0.9283 \\ 0 & 0.0080 & 0.0640 & 0.2160 & 0.5120 & 1.0000 \end{bmatrix}$

Use $vs = \text{cspline}(Mxy, Mz)$ and $\text{interp}(vs, Mxy, Mz, v)$ to find the interpolated values at the points $v = [0.1 \ 0.1], [0.3 \ 0.3]$, etc. Compare the interpolated results to those found from the function that generated the data, namely $z = x^{1/3} y^3$.

P8.37

$Mxy = \begin{bmatrix} 0.0 & 0.0 \\ 0.1 & 0.1 \\ 0.2 & 0.2 \\ 0.4 & 0.4 \\ 0.7 & 0.7 \\ 1.0 & 1.0 \end{bmatrix} \quad Mz = \begin{bmatrix} 0 & 0 & 0 & 0 & 0 & 0 \\ 0 & 0.0005 & 0.0037 & 0.0297 & 0.1592 & 0.4642 \\ 0 & 0.0006 & 0.0047 & 0.0374 & 0.2006 & 0.5848 \\ 0 & 0.0007 & 0.0059 & 0.0472 & 0.2527 & 0.7368 \\ 0 & 0.0009 & 0.0071 & 0.0568 & 0.3046 & 0.8879 \\ 0 & 0.0010 & 0.0080 & 0.0640 & 0.3430 & 1.0000 \end{bmatrix}$

Use $vs = \text{cspline}(Mxy, Mz)$ and $\text{interp}(vs, Mxy, Mz, v)$ to find the interpolated values at the points $v = [0.3 \ 0.3], [0.5 \ 0.5]$, etc. Compare interpolated results to those found from the function that generated the data, namely $z = x^{1/3} y^3$.

P8.38

$$Mxy = \begin{bmatrix} 0.0 & 0.0 \\ 0.2 & 0.2 \\ 0.4 & 0.4 \\ 0.6 & 0.6 \\ 0.8 & 0.8 \\ 1.0 & 1.0 \end{bmatrix} \qquad Mz = \begin{bmatrix} 0 & 0.0400 & 0.1600 & 0.3600 & 0.6400 & 1.0000 \\ 0.4472 & 0.4872 & 0.6072 & 0.8072 & 1.0872 & 1.4472 \\ 0.6325 & 0.6725 & 0.7925 & 0.9925 & 1.2725 & 1.6325 \\ 0.7746 & 0.8146 & 0.9346 & 1.1346 & 1.4146 & 1.7746 \\ 0.8944 & 0.9344 & 1.0544 & 1.2544 & 1.5344 & 1.8944 \\ 1.0000 & 1.0400 & 1.1600 & 1.3600 & 1.6400 & 2.0000 \end{bmatrix}$$

P 8.39

$$Mxy = \begin{bmatrix} 0.0 & 0.0 \\ 0.2 & 0.2 \\ 0.4 & 0.4 \\ 0.6 & 0.6 \\ 0.8 & 0.8 \\ 1.0 & 1.0 \end{bmatrix} \qquad Mz = \begin{bmatrix} 1.0000 & 1.2214 & 1.4918 & 1.8221 & 2.2255 & 2.7183 \\ 1.7878 & 2.0092 & 2.2796 & 2.6099 & 3.0133 & 3.5061 \\ 2.3511 & 2.5725 & 2.8429 & 3.1732 & 3.5766 & 4.0693 \\ 2.5511 & 2.7725 & 3.0429 & 3.3732 & 3.7766 & 4.2693 \\ 2.3878 & 2.6092 & 2.8796 & 3.2099 & 3.6133 & 4.1061 \\ 2.0000 & 2.2214 & 2.4918 & 2.8221 & 3.2255 & 3.7183 \end{bmatrix}$$

P8.40

$$Mxy = \begin{bmatrix} 0.0 & 0.0 \\ 0.2 & 0.2 \\ 0.4 & 0.4 \\ 0.6 & 0.6 \\ 0.8 & 0.8 \\ 1.0 & 1.0 \end{bmatrix} \qquad Mz = \begin{bmatrix} 0 & 0 & 0 & 0 & 0 & 0 \\ 0.7878 & 0.9622 & 1.1752 & 1.4354 & 1.7532 & 2.1414 \\ 1.3511 & 1.6502 & 2.0155 & 2.4618 & 3.0068 & 3.6726 \\ 1.5511 & 1.8945 & 2.3139 & 2.8262 & 3.4519 & 4.2162 \\ 1.3878 & 1.6950 & 2.0703 & 2.5287 & 3.0886 & 3.7724 \\ 1.0000 & 1.2214 & 1.4918 & 1.8221 & 2.2255 & 2.7183 \end{bmatrix}$$

Compare interpolated values to $z = (x + \sin(\pi x)) \exp(y)$.

Problems P8.41 to P8.50 give practice using spline interpolation for parametric curves. Use the Mathcad function Spline to find xs(t) (that interpolates the data t and x) and also ys(t) (that interpolates t and y). Generate values for these functions for a suitable range of values of t, and then plot y s versus x s. Also plot the data points (x, y).
  a. *Use spline interpolation to find* `xx = spline(t, x)` *and* `yy = spline(t, y)`.
  b. *Plot the curve using* `plot (xx, yy)` *and the data using* `plot (x, y, '*')`.

P8.41 $t = [1\ 2\ 3\ 4\ 5\ 6\ 7\ 8\ 9\ 10\ 11\ 12\ 13]$
$x = [7\ 4\ 3\ 0\ -3\ -4\ -7\ -4\ -3\ 0\ 3\ 4\ 7]$
$y = [0\ 2\ 5\ 8\ 5\ 2\ 0\ -2\ -5\ -8\ -5\ -2\ 0]$

P8.42 $t = [1\ 2\ 3\ 4\ 5\ 6\ 7\ 8\ 9\ 10\ 11\ 12\ 13\ 14\ 15\ 16\ 17\ 18\ 19\ 20\ 21]$
$x = [3\ 4\ 3\ 2\ 1\ 0\ -1\ -2\ -3\ -4\ -3\ -4\ -3\ -2\ -1\ 0\ 1\ 2\ 3\ 4\ 3]$
$y = [0\ 1\ 2\ 3\ 4\ 4\ 3\ 2\ 1\ 0\ -1\ -2\ -3\ -4\ -4\ -4\ -3\ -2\ -1\ 0]$

P8.43 $t = [1\ \ 2\ \ 3\ \ 4\ \ 5\ \ 6\ \ 7]$
$x = [0\ \ 1.5\ \ 3\ \ 0\ \ -3\ \ -1.5\ \ 0]$
$y = [0\ \ 1.5\ \ 6\ \ 4\ \ 6\ \ 1.5\ \ 0]$

Investigate the effect of taking the same points, but starting at a different point; e.g., use

$t = [1\ \ 2\ \ 3\ \ 4\ \ 5\ \ 6\ \ 7]$
$x = [0\ \ -3\ \ -1.5\ \ 0\ \ 1.5\ \ 3\ \ 0]$
$y = [4\ \ 6\ \ 1.5\ \ 0\ \ 1.5\ \ 6\ \ 4]$

P8.44 $t = [1\ \ 2\ \ 3\ \ 4\ \ 5\ \ 6\ \ 7\ \ 8\ \ 9\ \ 10\ \ 11]$
$x = [0\ \ 2\ \ 5\ \ 2\ \ 4\ \ 0\ \ -4\ \ -2\ \ -5\ \ -2\ \ 0]$
$y = [5\ \ 3\ \ 3\ \ 1\ \ -2\ \ -1\ \ -2\ \ 1\ \ 3\ \ 3\ \ 5]$

P8.45 $t = [1\ \ 2\ \ 3\ \ 4\ \ 5\ \ 6\ \ 7\ \ 8\ \ 9]$
$x = [5\ \ 1\ \ 0\ \ -1\ \ -5\ \ -1\ \ 0\ \ 1\ \ 5]$
$y = [0\ \ 2\ \ 5\ \ 2\ \ 0\ \ -1\ \ -5\ \ -1\ \ 0]$

P8.46 $t = [1\ \ 2\ \ 3\ \ 4\ \ 5\ \ 6\ \ 7\ \ 8\ \ 9\ \ 10\ \ 11\ \ 12]$
$x = [-2\ \ -1\ \ 0\ \ 1\ \ 1\ \ 0\ \ -1\ \ -1\ \ 0\ \ 1\ \ 2\ \ 3]$
$y = [0\ \ 0\ \ 0\ \ 1\ \ 2\ \ 3\ \ 2\ \ 1\ \ 0\ \ 0\ \ 0\ \ 0]$

P8.47 $t = [1\ \ 2\ \ 3\ \ 4\ \ 5\ \ 6\ \ 7\ \ 8\ \ 9]$
$x = [0.9\ \ 0.25\ \ 0\ \ -0.25\ \ -0.9\ \ -0.25\ \ 0\ \ 0.25\ \ 0.9]$
$y = [0\ \ 0.25\ \ 0.9\ \ 0.25\ \ 0\ \ -0.25\ \ -0.9\ \ -0.25\ \ 0]$

P8.48 $t = [1\ \ 2\ \ 3\ \ 4\ \ 5\ \ 6\ \ 7\ \ 8\ \ 9]$
$x = [0.9\ \ 0.25\ \ 0\ \ -0.25\ \ -0.9\ \ -0.25\ \ 0\ \ 0.25\ \ 0.9]$
$y = [0\ \ 0.25\ \ 0.9\ \ -0.25\ \ 0\ \ -0.25\ \ -0.9\ \ 0.25\ \ 0]$

P8.49 $t = [1\ \ 2\ \ 3\ \ 4\ \ 5\ \ 6\ \ 7\ \ 8\ \ 9\ \ 10\ \ 11\ \ 12]$
$x = [1\ \ 1\ \ 1/4\ \ 1/4\ \ -1/4\ \ -1/4\ \ -1\ \ -1\ \ -1/4\ \ 0\ \ 1/4\ \ 1]$
$y = [-1/4\ \ 1/4\ \ 1/4\ \ 1\ \ 1\ \ 1/4\ \ 1/4\ \ -1/4\ \ -1/4\ \ -1\ \ -1/4\ \ -1/4]$

Investigate the effect of taking the same points, in a different order.

P8.50 Many of the previous problems on spline interpolation for parametric curves were designed by the author's students. Draw an interesting figure, find the coordinates of some points on it, and use spline interpolation to find the interpolated parametric curve. Compare it to your original curve.

EXPLORE SOME APPLICATIONS

A8.1 Using the following data for the heat capacity C_p (kJ/kg °K) of methylcyclohexane C_7H_{14} as a function of temperature (°K), interpolate to estimate the heat capacity at $T = 175, 225,$ and 275.

T	150	200	250	300
C_p	1.43	1.54	1.70	1.89

(These data are adapted from Vargaftik, 1975.)

A8.2 The drag coefficient C_d for a baseball is a function of velocity (mph)

v	0	50	75	100	125
C_d	0.5	0.5	0.4	0.28	0.23

Approximate the drag coefficient for a baseball at 90 mph.
(These data are adapted from Garcia, 1994, p. 42.)

A8.3 Use interpolation on data for Bessel functions J_0, and J_1.
Data for Bessel function (J_0)

x = [0   1   2   3   4   5   6   7   8   9   10]
y = [1.00 0.77 0.22 −0.26 −0.40 −0.18 0.15 0.30 0.17 −0.09 −0.25]

   a. Compare your interpolated values at $x = 0.5, 1.5, \ldots 9.5$ to those found from the Mathcad function `J0(x)`;
   b. Compare your interpolation polynomial to the function $P(x) = 1 - x^2/4 + x^4/64 - x^6/2304$, the first four terms of the series expansion for J_o.

Data for Bessel function (J_1)

x = [0 1   2   3   4   5   6   7   8   9   10   ]
y = [0 0.44 0.58 0.34 −0.07 −0.33 −0.28 −0.0047 0.23 0.25 0.0435]

Compare your interpolated values at $x = 0.5, 1.5, \ldots 9.5$ to those found from the Mathcad function `J1(x)`;

A8.4 The following data give viscosity at several different temperatures:

$T(°C)$	5	20	30	50	55
μ(N-sec/m^2)	0.08	0.015	0.009	0.006	0.0055

Use interpolation for find an estimate for the viscosity at $T = 25$ and $T = 40$.
(These data are adapted from Ayyub and McCuen, 1996, p. 174)

A8.5 Using the tabulated data for the specific enthalpy (h) of superheated steam as a function of temperature (at constant pressure of 2500 lb/in.^{2}), find an interpolating polynomial and estimate the enthalpy at 1100°F. The dimensions on h are (Btu/lb)

T	800	1000	1200	1400	1600
h	1303.6	1458.4	1585.3	1706.1	1826.2

(These data are adapted from Ayyub and McCuen, 1996, p. 176)

A8.6 Using the tabulated data for the short-wave radiation flux (in gram-calories per cm² per day for September) at the outer limit of the atmosphere, estimate the flux at a latitutde of 35° (°N)

Latitude	0	20	40	60	80
flux	891	856	719	494	219

(These data are adapted from Ayyub and McCuen, 1996, p. 176)

A8.7 Using the tabulated data for the vapor pressure (mm Hg) of water as a function of temperature (°C), find an interpolating polynomial and estimate the pressure at T = 50.

T	40	48	56	64	72
P	55.3	83.7	123.8	179.2	254.5

(These data are adapted from Ayyub and McCuen, p. 151)

A8.8 Using the tabulated data for the saturation values of dissolved oxygen concentration (mg/L) as a function of temperature (°C), find an interpolating polynomial.

$T = [\ 0 \quad 5 \quad 10 \quad 15 \quad 20 \quad 25]$
$D = [14.6 \quad 12.8 \quad 11.3 \quad 10.2 \quad 9.2 \quad 8.4]$

(These data are adapted from Ayyub and McCuen, p. 159)

A8.9 Given the following tabulated values (truncated to 2 decimal places) for the elliptic integrals of the first and second kinds, use interpolation to find values for m = 0.1, 0.3, 0.5, 0.7, 0.9.

$$K(m) = \int_0^{\pi/2} \frac{dt}{\sqrt{1 - m \sin^2 t}} \qquad E(m) = \int_0^{\pi/2} \sqrt{1 - m \sin^2 t}\, dt$$

$m = [0.00 \quad 0.20 \quad 0.40 \quad 0.60 \quad 0.80 \quad 1.00]$
$K = [1.57 \quad 1.66 \quad 1.78 \quad 1.95 \quad 2.26 \quad \infty\ ]$
$E = [1.57 \quad 1.49 \quad 1.40 \quad 1.30 \quad 1.18 \quad 1.00]$

(These data are adapted from Abramowitz and Stegun, pp. 608-609.)

A8.10 Given the following tabulated values (truncated to 3 decimal places) for the Fresnel integrals, use interpolation to find the values for m = 0.5, 1.5, 2.5, 3.5, and 4.5.

$$C(x) = \int_0^x \cos\left(\frac{\pi}{2} t^2\right) dt \qquad S(x) = \int_0^x \sin\left(\frac{\pi}{2} t^2\right) dt$$

$x = [0.00 \quad 1.00 \quad 2.00 \quad 3.00 \quad 4.00 \quad 5.00]$
$C = [0.000 \quad 0.780 \quad 0.488 \quad 0.606 \quad 0.498 \quad 0.564]$
$S = [0.000 \quad 0.438 \quad 0.343 \quad 0.496 \quad 0.421 \quad 0.499]$

(These data are adapted from Abramowitz and Stegun, pp. 321–322.)

A8.11 In modeling a combustion process it is required to find enthalpy as a function of temperature. Find an interpolation polynomial, spline, or rational function interpolation for the following data, and compare the interpolated values to the tabulated values given below.

$T = [60 \quad 80 \quad 100 \quad 120 \quad 140 \quad 160 \quad 180 \quad 200\ ]$
$E = [0.0 \quad 17.2 \quad 45.2 \quad 92.9 \quad 178.8 \quad 349.4 \quad 764.3 \quad 2648.4]$

Data adapted from the *Handbook of Hazardous Waste Incineration*, Tab Professional and Reference books, 1989.

A8.12 Use the data on annual building permits issued (in millions of permits), to estimate the number of permits issued in 1982, 1988, 1993, and 1996.
$Y = [1980 \quad 1985 \quad 1990 \quad 1995]$
$P = [1.19 \quad 1.73 \quad 1.11 \quad 1.33]$
(These data are adapted from the U.S. Census web page.)

A8.13 Use the following data to estimate the average (mean) annual earnings for workers in 1993, 1988, 1983, and 1978.
$y = [1975 \quad 1980 \quad 1985 \quad 1990 \quad 1995]$
salary for high school graduates
$S = [7{,}843 \quad 11{,}314 \quad 14{,}457 \quad 17{,}820 \quad 21{,}431]$
salary for workers with AA degree
$S = [8{,}388 \quad 12{,}409 \quad 16{,}349 \quad 20{,}694 \quad 23{,}862]$
salary for workers with BA/BS degree
$S = [12{,}332 \quad 18{,}075 \quad 24{,}877 \quad 31{,}112 \quad 36{,}980]$
salary for workers with advanced degree
$S = 16{,}725 \quad 23{,}308 \quad 32{,}909 \quad 41{,}458 \quad 56{,}667]$
(These data are adapted from the U. S. Census web page.)

A8.14 Use the following data (with enrollments given in millions) to estimate the school enrollments in 1993, 1988, 1983, and 1978.
$y = [1955 \quad 1960 \quad 1965 \quad 1970 \quad 1975 \quad 1980 \quad 1985 \quad 1990 \quad 1995]$
kindergarten
$K = [1.6 \quad 2.1 \quad 3.1 \quad 3.2 \quad 3.4 \quad 3.2 \quad 3.8 \quad 4.0 \quad .9]$
elementary school
$E = [25.5 \quad 30.3 \quad 32.5 \quad 33.9 \quad 30.5 \quad 27.5 \quad 26.9 \quad 29.2 \quad 31.8]$
high school
$H = [8.0 \quad 10.2 \quad 13.0 \quad 14.7 \quad 15.7 \quad 14.6 \quad 14.0 \quad 12.7 \quad 14.8]$
college
$C = [2.4 \quad 3.6 \quad 5.7 \quad 7.4 \quad 9.7 \quad 10.2 \quad 10.9 \quad 11.3 \quad 12.0]$
These data are adapted from the U. S. Census Web page, which cites the source as the U.S. Bureau of the Census, Current Population Survey.

EXTEND YOUR UNDERSTANDING

U8.2 Investigate the difficulties in interpolating a "noisy line" by generating some data:
$x = [0 \ 1 \ 2 \ 3 \ 4 \ 5 \ 6]$
$y = x + 0.1*\text{rand}(1,7) - 0.05$
Show that even if the interpolating polynomial appears to fit the data well, extrapolating beyond the data is not a good idea.

U8.3 Show that the difficulties in interpolating a "noisy line" are much more severe when the data is not evenly spaced; e.g., generate some data of the form:
$x = [0 \ 0.2 \ 0.4 \ 2 \ 4 \ 4.2 \ 4.4]$
$y = x + 0.1*\text{rand}(1,7) - 0.05$
or
$x = [0 \ 0.1 \ 0.2 \ 2 \ 4.1 \ 4.2 \ 4.3]$
$y = x + 0.1*\text{rand}(1,7) - 0.05$

U8.4 Show that for a cubic spline, specifying the value of $P_1'(x_1)$ gives the following equation relating a_0 and a_1:

$$\frac{a_0 h_1}{3} + \frac{a_1 h_1}{6} = \frac{y_2 - y_1}{h_1} - P_1'(x_1).$$

U8.5 Modify the Mathcad function for the natural cubic spline by adding the equation from U8.4 to the system of equations for $a_0, \ldots, a_{n-1}$ given in Section 8.4.3, to form a function for a spline that is clamped at x_1.

U8.6 Use the function from U8.5 to repeat some of the exercises P8.16 to P8.25.

U8.7 Show that the equation describing a clamped boundary at x_n (see U8.4) is

$$\frac{a_{n-2} h_{n-1}}{6} + \frac{a_{n-1} h_{n-1}}{3} = P_{n-1}'(x_n) - \frac{y_n - y_{n-1}}{h_{n-1}}.$$

U8.8 Use Hermite cubic polynomials for piecewise cubic interpolation by specifying the function value and derivative value at each node point. Use the data from Example 8.12.

U8.9 Use Hermite cubic polynomials for piecewise cubic interpolation by specifying the function value and derivative value at each node point. Compare your results using the following dat to the results of Example 8.10.
$x = [-1 \quad -0.5 \quad 0 \quad 0.5 \quad 1 \quad]$,
$y = [0.0385 \quad 0.1379 \quad 1.00 \quad 0.1379 \quad 0.0385]$,
$dy = [0.074 \quad 0.4756 \quad 0.00 \quad -0.4756 \quad -0.074]$.

U8.10 The error bound formula for cubic spline interpolation given in the discussion at the end of Section 8.4.3 does not state the proportionality constant, since it depends on the choice of endpoint conditions. With clamped boundary conditions, it is

$$|S(x) - g(x)| < (5/384) \, h^4 \, G.$$

(See deBoor, 1978.)
Use $g(x) = \cos(x)$ to generate evenly spaced data on $[0, \pi]$ for different values of h. Find the cubic spline interpolation function $S(x)$, the actual error $|S(x) - g(x)|$, and error bound $(5/384) \, h^4 \, G$.

9
Function Approximation

9.1 Least Squares Approximation

9.2 Continuous Least-Squares Approximation

9.3 Function Approximation at a Point

9.4 Mathcad's Methods

Function approximation is closely related to the idea of function interpolation, discussed in the previous chapter. In function approximation, we do not require the approximating function to match the given data exactly. This avoids some of the difficulties demonstrated previously in regard to trying to match a moderate-to-large amount of data, especially if noise (such as errors in measurement) is present. There are also many applications in which a theoretical functional form is known and the "best" function of that form is required.

The most common approach to "best fit" approximation is to minimize the sum of the squares of the differences between the data values and the values of the approximating function; this is the *method of least squares*. We first investigate *linear* least squares approximation, also known as linear regression (especially in statistics). The same approach is used to find the *quadratic* least squares function or a polynomial of some other (specified) degree. For data that appear to follow an exponential function, the standard approach is to find a linear function that fits the natural logarithm of the data. This gives a close approximation to the "best fit" exponential (but with much less effort).

It may also be desirable in some applications to approximate a given function by the best function of a specified form, on a given interval. This leads to the topic of *continuous* least squares approximation.

In certain situations, a function may be better represented by a rational function than by a polynomial. *Padé approximation* is the rational-function analog of Taylor polynomial approximation. In these cases, knowledge of the function and its derivatives at a single point is used to construct the best local representation of the function.

In the next chapter, we consider fitting periodic data with trigonometric polynomials. In that case, the appropriate function for approximation or interpolation of a given set of data can be found by the least-squares approach.

Example 9-A Oil Reservoir Modeling

In modeling an oil reservoir, it may be necessary to find a relationship between the equilibrium constant of a reaction and the pressure, at constant temperature. The data shown below relate equilibrium constants (K-values) to pressure (expressed in terms of 1000 PSIA) and were obtained from an experimental PVT analysis. Fig. 9.1 is a plot of these data.

Oil reservoir data.

Pressure	K-value
0.635	7.5
1.035	5.58
1.435	4.35
1.835	3.55
2.235	2.97
2.635	2.53
3.035	2.2
3.435	1.93
3.835	1.7
4.235	1.46
4.635	1.28
5.035	1.11
5.435	1.0

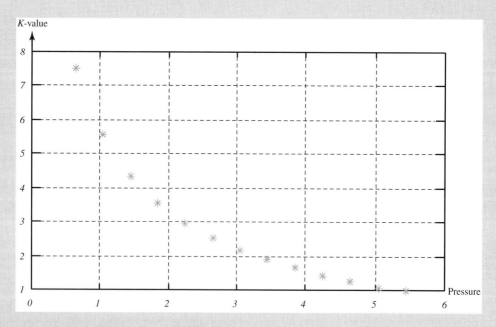

FIGURE 9.1 Equilibrium constant as a function of pressure.

Example 9-B Logistic Population Growth

The data in Fig. 9.2 describe the growth of a population following a logistic model. The plot would also represent a uniform sampling from a cumulative distribution. Several kinds of functions can be used to fit data that display this "s-shaped" form; the appropriate choice depends on the particular application.

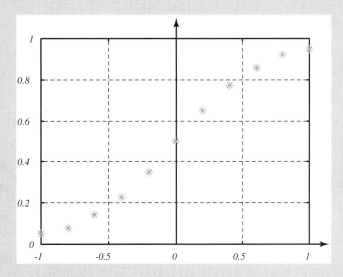

FIGURE 9.2   Logistic population growth.

We see in Example 9.7 that a cubic polynomial provides a reasonable fit to the given data.

On the other hand, if we wish to obtain an approximating function of the appropriate form for the solution of a logistic differential equation, we may look for a function of the form $y = 1/(1 + \exp(ax + b))$.

We transform the given data

$x =$	−1.0	−0.8	−0.6	−0.4	−0.2	0.0	0.2	0.4	0.6	0.8	1.0
$y =$	0.05	0.08	0.14	0.23	0.35	0.50	0.65	0.77	0.86	0.92	0.95

to form $z = 1/y - 1$

| $z=$ | 19.00 | 11.50 | 6.143 | 3.348 | 1.857 | 1.00 | 0.539 | 0.299 | 0.163 | 0.087 | 0.053 |

We then find the coefficients a and b for the exponential function to fit z.

9.1 LEAST SQUARES APPROXIMATION

Some of the most common methods of approximating data are based on the desire to minimize some measure of the difference between the approximating function and the given data points. The method of least squares seeks to minimize the sum (over all data points) of the squares of the differences between the function value and the data value. The method is based on results from calculus demonstrating that a function, in this case the total squared error, attains a minimum value when its partial derivatives are zero.

There are several advantages to using the square of the differences at each point, rather than the difference, or the absolute value of the difference, or some other measure of the error. By squaring the difference,

1. positive differences do not cancel negative differences;
2. differentiation is not difficult; and
3. small differences become smaller, and large differences are magnified.

We begin with an example in which the data can be approximated nicely by a straight line. We then consider several variations, in which the appropriate choice of approximating function is a higher degree polynomial, an exponential function, or the reciprocal of a polynomial. The choice depends to a large extent on the general characteristics of the data.

For data to be approximated by a straight line, we wish to find the best function $f(x) = a x + b$ that approximates the data. In order to find the coefficients a and b, we must define what we mean by the "best fit" of a function to some data. Of course, we want to minimize the difference between the data points and the points on our approximating function, in some sense. The most common method is to minimize the sum of the squares of the differences between the given data values y_i and the computed function values $f_i = a x_i + b$. Another approach, known as total least squares, minimizes the distance from each data point to the straight line, where distance is measured perpendicular to the line, not in the vertical direction. This leads to a much more difficult mathematical problem, which is beyond the scope of our discussion.

9.1.1 Linear Least-Squares Approximation

To introduce the ideas of linear least-square approximation, we consider the error involved in approximating four data points by a straight line, determined by visual inspection of the graph of the points.

Example 9.1 Linear Approximation to Four Points

Consider the data (1, 2.1), (2, 2.9), (5, 6.1), and (7, 8.3), as shown in Fig. 9.3.

If we approximate the data by the straight line $f(x) = 0.9\,x + 1.4$, as shown in Fig. 9.4, the squared errors are as follows:

at $x_1 = 1$ $f(1) = 2.3$ $y_1 = 2.1$ $e_1 = (2.3 - 2.1)^2 = 0.04$;

at $x_2 = 2$ $f(2) = 3.2$ $y_2 = 2.9$ $e_2 = (3.2 - 2.9)^2 = 0.09$;

at $x_3 = 5$ $f(5) = 5.9$ $y_3 = 6.1$ $e_3 = (5.9 - 6.1)^2 = 0.04$;

at $x_4 = 7$ $f(7) = 7.7$ $y_4 = 8.3$ $e_4 = (7.7 - 8.3)^2 = 0.36$.

The total squared error is $0.04 + 0.09 + 0.04 + 0.36 = 0.53$. By finding better values for the coefficients of the straight line, we can make this number smaller.

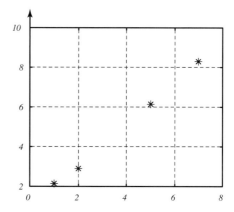

FIGURE 9.3 Data for linear least squares example.

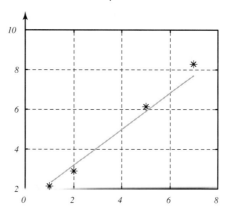

FIGURE 9.4 Approximating data by a straight line.

Chapter 9 Function Approximation

Let us consider a systematic way of finding the coefficients of the "best fit" straight line to approximate data; we illustrate the process for four data points.

The best coefficients for the straight line are those that minimize the total-squared-error function:

$$E = [f(x_1) - y_1]^2 + [f(x_2) - y_2]^2 + [f(x_3) - y_3]^2 + [f(x_4) - y_4]^2$$
$$= [ax_1 + b - y_1]^2 + [ax_2 + b - y_2]^2 + [ax_3 + b - y_3]^2 + [ax_4 + b - y_4]^2$$

The minimum of a function of several variables (two in this case, the coefficients a and b) occurs when the partial derivatives of the function with respect to each of the variables are equal to zero. Here, we have the system of equations $\frac{\partial E}{\partial a} = 0$ and $\frac{\partial E}{\partial b} = 0$. The solution depends on several quantities that can be computed from the data. The required four summations can be denoted as

$$S_{xx} = \sum_{i=1}^{4} x_i^2, \quad S_x = \sum_{i=1}^{4} x_i, \quad S_{xy} = \sum_{i=1}^{4} x_i y_i, \quad S_y = \sum_{i=1}^{4} y_i.$$

The unknowns (a and b) can be found from the following system of equations, known as the normal equations:

$$a S_{xx} + b S_x = S_{xy}$$
$$a S_x + b \, 4 = S_y$$

The solution to the system of equations is

$$a = \frac{4 S_{xy} - S_x S_y}{4 S_{xx} - S_x S_x} \qquad b = \frac{S_{xx} S_y - S_{xy} S_x}{4 S_{xx} - S_x S_x}.$$

Algorithm for Linear Least-Squares Approximation

Given data vectors **x** and **y** (each with n components), find coefficients for approximation function, $y = a x + b$

Define

$$\mathbf{x} = (x_0, x_1, x_2, \ldots x_{n-1})$$
$$\mathbf{y} = (y_0, y_1, y_2, \ldots y_{n-1})$$

Find required sums

Sx = sum of components of vector **x**. Use vector sum (Matrix Toolbar) or [ctrl]4 x.

Sy = sum of components of vector **y**. Use vector sum (Matrix Toolbar) or [ctrl]4 y.

Chapter 9 Function Approximation

Sxx = sum of squares of components of vector **x**.
　　　　This is the same as **x*****x**, so use vector multiplication(Matrix Toolbar), or $x*x$.

Sxy = sum of products of components of vectors **x** and **y**.
　　　　This is the same as **x*****y**, so use vector multiplication(Matrix Toolbar), or $x*y$.

Solve linear system (results from Cramer's rule)

$$a = \frac{n\,S_{xy} - S_x\,S_y}{n\,S_{xx} - S_x\,S_x} \qquad b = \frac{S_{xx}\,S_y - S_{xy}\,S_x}{n\,S_{xx} - S_x\,S_x},$$

Define vector of approximated values

z = *a* **x** + *b*.

Define vector of difference between data and approximated values

d = **y** − **z**

Compute total-squared error

d***d**

The use of algorithm is illustrated in Example 9.2.

Mathcad Function for Linear Least Squares Approximation

$$\text{Lin\_LS}(x, y) := \left| \begin{array}{l} n \leftarrow \text{length}(x) \\ sx \leftarrow \sum x \\ sy \leftarrow \sum y \\ sxx \leftarrow x \cdot x \\ sxy \leftarrow x \cdot y \\ \text{denom} \leftarrow n \cdot sxx - sx \cdot sx \\ s_0 \leftarrow \dfrac{n \cdot sxy - sx \cdot sy}{\text{denom}} \\ s_1 \leftarrow \dfrac{sxx \cdot sy - sxy \cdot sx}{\text{denom}} \\ s \end{array} \right.$$

The use of the Mathcad function Lin_LS is illustrated in Example 9.3.

Example 9.2 Least Squares Straight Line to Fit Four Data Points

For our previous example with the data points (1, 2.1), (2, 2.9), (5, 6.1), and (7, 8.3), the system of equations for a and b is

$$79a + 15b = 96.5,$$

$$15a + 4b = 19.4.$$

The solution is $a = 1.0440$, $b = 0.9352$.

The differences between the data and the computed values are shown in the table below. The total squared error is

$$d_1^2 + d_2^2 + d_3^2 + d_4^2 = 0.0360.$$

Figure 9.5 plots the linear least-squares line with the data.

Linear least-squares straight line.

x_i	y_i	$a x_i + b$	$d_i = y_i - a x_i - b$
1	2.1	1.9791	0.1209
2	2.9	3.0231	−0.1231
5	6.1	6.1549	−0.0549
7	8.3	8.2429	0.0571

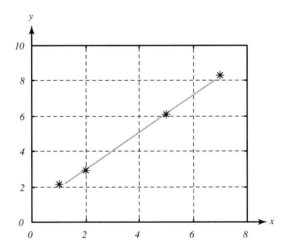

FIGURE 9.5   Linear least-squares line
$f(x) = 1.044 x + 0.9352$.

Example 9.3 Noisy Straight-Line Data

We find the linear least squares approximation for the data introduced in Example 8.9.

$x := (0.00 \ 0.20 \ 0.80 \ 1.00 \ 1.20 \ 1.90 \ 2.00 \ 2.10 \ 2.95 \ 3.00)$

$y := (0.01 \ 0.22 \ 0.76 \ 1.03 \ 1.18 \ 1.94 \ 2.01 \ 2.08 \ 2.90 \ 2.95)$

$a := \text{Lin\_LS}(x^T, y^T)$

$a = \begin{pmatrix} 0.984 \\ 0.017 \end{pmatrix}$

The linear function that best fits the data is $f(x) = 0.984 \, x + 0.017$

The data values (x_i and y_i), computed function values $y(x_i)$, and the actual difference at each x_i, are shown in the table below. The line is plotted in Fig. 9.6.

Data and linear fit for noisy straight line.

x_i	y_i	$a\,x_i + b$	$d_i = y_i - a\,x_i - b$
0.00	0.01	0.0174	−0.0074
0.20	0.22	0.2142	0.0058
0.80	0.76	0.8045	−0.0445
1.00	1.03	1.0013	0.0287
1.20	1.18	1.1981	−0.0181
1.90	1.94	1.8868	0.0532
2.00	2.01	1.9852	0.0248
2.10	2.08	2.0836	−0.0036
2.95	2.90	2.9199	−0.0199
3.00	2.95	2.9691	−0.0191

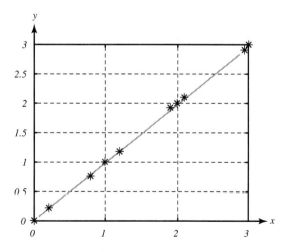

FIGURE 9.6 Linear least-squares line and data for noisy straight line.

Chapter 9 Function Approximation

Discussion

We now consider in a little more detail how the normal equations for linear least-squares approximation arise, illustrating the derivation for four arbitrary data points, namely, $(x_1, y_1), (x_2, y_2), (x_3, y_3), (x_4, y_4)$, that we wish to approximate by the linear function $f(x) = a x + b$. We want to find those values of a and b that minimize the total squared error E over the data points; that is, we have

$$E = [f(x_1) - y_1]^2 + [f(x_2) - y_2]^2 + [f(x_3) - y_3]^2 + [f(x_4) - y_4]^2$$
$$= [ax_1 + b - y_1]^2 + [ax_2 + b - y_2]^2 + [ax_3 + b - y_3]^2 + [ax_4 + b - y_4]^2.$$

Setting $\dfrac{\partial E}{\partial a} = 0$ and $\dfrac{\partial E}{\partial b} = 0$ gives

$$[ax_1 + b - y_1]x_1 + [ax_2 + b - y_2]x_2 + [ax_3 + b - y_3]x_3 + [ax_4 - b - y_4]x_4 = 0,$$

and

$$[ax_1 + b - y_1] + [ax_2 + b - y_2] + [ax_3 + b - y_3] + [ax_4 + b - y_4] = 0.$$

Simplifying gives

$$a[x_1^2 + x_2^2 + x_3^2 + x_4^2] + b[x_1 + x_2 + x_3 + x_4] = x_1y_1 + x_2y_2 + x_3y_3 + x_4y_4,$$

and

$$a[x_1 + x_2 + x_3 + x_4] + b[1 + 1 + 1 + 1] = y_1 + y_2 + y_3 + y_4.$$

We see that no matter how many data points we have, if we want a straight line to minimize the error in the y-coordinate, we get two equations in two unknowns. In general form, the equations are

$$a \sum_{i=1}^{n} x_i^2 + b \sum_{i=1}^{n} x_i = \sum_{i=1}^{n} x_i y_i,$$

$$a \sum_{i=1}^{n} x_i + b \sum_{i=1}^{n} 1 = \sum_{i=1}^{n} y_i.$$

Of course, $\sum_{i=1}^{n} 1 = n$, so there are only four summations that must be calculated. For simplicity, they can be denoted as

$$S_{xx} = \sum_{i=1}^{n} x_i^2, \quad S_x = \sum_{i=1}^{n} x_i, \quad S_{xy} = \sum_{i=1}^{n} x_i y_i, \quad S_y = \sum_{i=1}^{n} y_i.$$

The solution to the system of equations (the normal equations) is

$$a = \frac{n S_{xy} - S_x S_y}{n S_{xx} - S_x S_x}, \quad b = \frac{S_{xx} S_y - S_{xy} S_x}{n S_{xx} - S_x S_x}.$$

9.1.2 Quadratic Least-Squares Approximation

Using the same approach as before, let us now approximate our data with a quadratic function $f(x) = a x^2 + b x + c$. The error function is

$$E = \sum_{i=1}^{n} [f(x_i) - y_i]^2.$$

By equating the partial derivatives of E with respect to a, b, and c to zero, we obtain the normal equations for a, b, and c:

$$a \sum_{i=1}^{n} x_i^4 + b \sum_{i=1}^{n} x_i^3 + c \sum_{i=1}^{n} x_i^2 = \sum_{i=1}^{n} x_i^2 y_i,$$

$$a \sum_{i=1}^{n} x_i^3 + b \sum_{i=1}^{n} x_i^2 + c \sum_{i=1}^{n} x_i = \sum_{i=1}^{n} x_i y_i,$$

$$a \sum_{i=1}^{n} x_i^2 + b \sum_{i=1}^{n} x_i + c[n] = \sum_{i=1}^{n} y_i.$$

Algorithm for Quadratic Least-Squares Approximation

Given data vectors **x** and **y** (each with n components), find the coefficients for the approximation function, $y = a x^2 + b x + c$
Define

$x = (x_0, x_1, x_2, \ldots x_{n-1})$

$y = (y_0, y_1, y_2, \ldots y_{n-1})$

Find required sums:
  Sx, Sy, Sxx, and Sxy are found as for linear least squares.
  $Sx3$, $Sx4$, and $Syx2$ are found as follows:
    Apply the Vectorize operator from Matrix Toolbar (or [ctrl]-)
      to the expression x^2 to form the vector **x2** = $(x_0^2, x_1^2, x_2^2, \ldots x_{n-1}^2)$,
      similarly, vectorize x^3, and x^4 to form vectors **x3** and **x4**,
    then use the dot product (matrix multiplication or *) to find the following sums:

$$Sx3 = \mathbf{x3*x3}, \quad Sx4 = \mathbf{x4*x4} \quad Syx2 = \mathbf{y*x2}$$

Define the linear system

$$\mathbf{A} = \begin{bmatrix} Sx4 & Sx3 & Sx2 \\ Sx3 & Sx2 & Sx \\ Sx2 & Sx & n \end{bmatrix} \quad r = \begin{bmatrix} Syx2 \\ Syx \\ Sy \end{bmatrix}$$

Solve linear system $\mathbf{A}\,z = r$
Results: $a = z_1 \quad b = z_2 \quad c = z_3$
Compute the total-squared error in the corresponding manner as for linear least squares.

Mathcad Function for Quadratic Least-Squares Approximation

$$\text{Quad\_LS}(x,y) := \begin{vmatrix} n \leftarrow \text{length}(x) \\ sx \leftarrow \sum x \\ sy \leftarrow \sum y \\ sx2 \leftarrow \sum x^2 \\ sxy \leftarrow \sum_{i=0}^{n-1} x_i \cdot y_i \\ sx3 \leftarrow \sum x^3 \\ sx4 \leftarrow \sum x^4 \\ sx2y \leftarrow \sum_{i=0}^{n-1} (x_i)^2 \cdot y_i \\ A \leftarrow \begin{pmatrix} sx4 & sx3 & sx2 \\ sx3 & sx2 & sx \\ sx2 & sx & n \end{pmatrix} \\ r \leftarrow \begin{pmatrix} sx2y \\ sxy \\ sy \end{pmatrix} \\ z \leftarrow A^{-1} \cdot r \\ z \end{vmatrix}$$

Example 9.4 Chemical Reaction Data

Consider again the product data from a simple chemical reaction (see Chapter 8):

Find quadratic least squares approximation for the following data

$$xx := \begin{pmatrix} 0 \\ 0.5 \\ 1 \\ 1.5 \\ 2 \end{pmatrix} \qquad y := \begin{pmatrix} 0 \\ 0.19 \\ 0.26 \\ 0.29 \\ 0.31 \end{pmatrix} \qquad a := \text{Quad\_LS}(xx, y) \qquad a = \begin{pmatrix} -0.109 \\ 0.361 \\ 0.012 \end{pmatrix}$$

Define approximating polynomial, and range variable for plotting

$$f(x) := a_0 \cdot x^2 + a_1 \cdot x + a_2 \qquad x := 0, .01 .. 2$$

Plot data and quadratic approximating polynomial

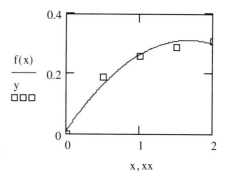

FIGURE 9.7 Data and least-squares quadratic fit for chemical reaction.

Example 9.5 Gompertz Growth Curve

Consider the problem of fitting a quadratic function to data generated from the Gompertz growth curve, $y = \exp(-2\exp(-x))$. The vectors

$$x = \begin{bmatrix} 0 & 1 & 2 & 3 & 4 & 5 & 6 \end{bmatrix},$$
$$y = \begin{bmatrix} 0.135 & 0.479 & 0.763 & 0.905 & 0.964 & 0.987 & 0.995 \end{bmatrix},$$

are input to the Mathcad function Quad LS. The linear system for the coefficients is

$$\begin{bmatrix} 2275 & 441 & 91 \\ 441 & 91 & 21 \\ 91 & 21 & 7 \end{bmatrix} \begin{bmatrix} z_1 \\ z_2 \\ z_3 \end{bmatrix} = \begin{bmatrix} 87.5895 \\ 19.4801 \\ 5.2283 \end{bmatrix}$$

The data, approximations, and errors are listed below. The total squared error is 0.0047.

The least square parabola

$$p(x) = -0.0375 x^2 + 0.3605 x + 0.1528$$

is shown together with the data in Fig. 9.8.

Least-squares parabola and data for Gompertz growth curve.

x_i	y_i	p_i	d_i
0.0000	0.1353	0.1528	−0.0174
1.0000	0.4791	0.4758	0.0033
2.0000	0.7629	0.7238	0.0390
3.0000	0.9052	0.8969	0.0083
4.0000	0.9640	0.9949	−0.0309
5.0000	0.9866	1.0180	−0.0314
6.0000	0.9951	0.9661	0.0290

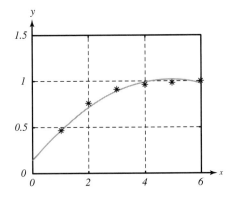

FIGURE 9.8 Least-squares parabola and data for Gompertz growth curve.

Example 9.6 Oil Reservoir

Let us consider the problem of finding the least-squares quadratic function to approximate a subset of the experimental data introduced in Example 9-A. The data represent the relationship between pressure x and reaction rate y for an oil reservoir modeling problem:

$$\mathbf{x} = [5.435 \quad 4.635 \quad 3.835 \quad 3.035 \quad 2.325 \quad 1.435 \quad 0.635]$$
$$\mathbf{y} = [1.00 \quad 1.28 \quad 1.70 \quad 2.20 \quad 2.97 \quad 4.35 \quad 7.50 \;].$$

The normal equations are $\mathbf{A\,z} = \mathbf{b}$, with

$$\mathbf{A} = \begin{bmatrix} 1668.9 & 360.3 & 82.8 \\ 360.3 & 82.8 & 21.3 \\ 82.8 & 21.3 & 7.0 \end{bmatrix} \quad \mathbf{b} = \begin{bmatrix} 130.3413 \\ 42.4743 \\ 21.0000 \end{bmatrix}$$

The desired quadratic function is $p(x) = 0.3582\, x^2 - 3.3833\, x + 9.0748$.
The data, approximations, and errors for this problem are shown below. The data and the least squares parabola are plotted in Fig. 9.9.

Data and least squares parabola for oil reservoir example.

x_i	y_i	p_i	d_i
5.4350	1.0000	1.2664	−0.2664
4.6350	1.2800	1.0877	0.1923
3.8350	1.7000	1.3674	0.3326
3.0350	2.2000	2.1056	0.0944
2.3250	2.9700	3.1447	−0.1747
1.4350	4.3500	4.9573	−0.6073
0.6350	7.5000	7.0708	0.4292

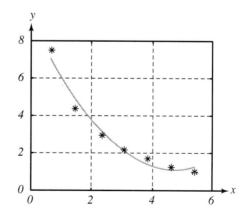

FIGURE 9.9 Data and least-squares parabola.

Discussion

For n data points, $(x_1, y_1) \ldots (x_n, y_n)$ we wish to minimize

$$E = [a(x_1)^2 + b(x_1) + c - y_1]^2 + \ldots + [a(x_n)^2 + b(x_n) + c - y_n]^2.$$

Setting $\dfrac{\partial E}{\partial a} = 0$ gives

$$[a(x_1)^2 + b(x_1) + c - y_1](x_1)^2 + \ldots + [a(x_n)^2 + b(x_n) + c - y_n](x_n)^2 = 0,$$

and similarly setting $\dfrac{\partial E}{\partial b} = 0$ yields

$$[a(x_1)^2 + b(x_1) + c - y_1](x_1) + \ldots + [a(x_n)^2 + b(x_n) + c - y_n](x_n) = 0,$$

while setting $\dfrac{\partial E}{\partial c} = 0$ gives

$$[a(x_1)^2 + b(x_1) + c - y_1] + \ldots + [a(x_n)^2 + b(x_n) + c - y_n] = 0.$$

These equations simplify to

$$a \sum_{i=1}^{n} x_i^4 + b \sum_{i=1}^{n} x_i^3 + c \sum_{i=1}^{n} x_i^2 = \sum_{i=1}^{n} x_i^2 y_i,$$

$$a \sum_{i=1}^{n} x_i^3 + b \sum_{i=1}^{n} x_i^2 + c \sum_{i=1}^{n} x_i = \sum_{i=1}^{n} x_i y_i,$$

$$a \sum_{i=1}^{n} x_i^2 + b \sum_{i=1}^{n} x_i + c[n] = \sum_{i=1}^{n} y_i.$$

9.1.3 Cubic Least-Squares Approximation

Proceeding in a manner analogous to that of quadratic least squares approximation, we find that the system of linear equations to determine the coefficients for the "best-fit" cubic function

$$f(x) = a x^3 + b x^2 + cx + d,$$

is

$$a \sum_{i=1}^{n} x_i^6 + b \sum_{i=1}^{n} x_i^5 + c \sum_{i=1}^{n} x_i^4 + d \sum_{i=1}^{n} x_i^3 = \sum_{i=1}^{n} x_i^3 y_i,$$

$$a \sum_{i=1}^{n} x_i^5 + b \sum_{i=1}^{n} x_i^4 + c \sum_{i=1}^{n} x_i^3 + d \sum_{i=1}^{n} x_i^2 = \sum_{i=1}^{n} x_i^2 y_i,$$

$$a \sum_{i=1}^{n} x_i^4 + b \sum_{i=1}^{n} x_i^3 + c \sum_{i=1}^{n} x_i^2 + d \sum_{i=1}^{n} x_i = \sum_{i=1}^{n} x_i y_i,$$

$$a \sum_{i=1}^{n} x_i^3 + b \sum_{i=1}^{n} x_i^2 + c \sum_{i=1}^{n} x_i + d \sum_{i=1}^{n} 1 = \sum_{i=1}^{n} y_i.$$

Algorithm for Cubic Least-Squares Approximation

Given data vectors **x** and **y** (each with n components), find coefficients for approximation function, $y = a x^3 + b x^2 + c x + d$

Find required sums

$$S_x = \sum_{i=1}^{n} x_i \quad S_{x2} = \sum_{i=1}^{n} x_i^2 \quad S_{x3} = \sum_{i=1}^{n} x_i^3$$

$$S_{x4} = \sum_{i=1}^{n} x_i^4 \quad S_{x5} = \sum_{i=1}^{n} x_i^5 \quad S_{x6} = \sum_{i=1}^{n} x_i^6$$

$$S_y = \sum_{i=1}^{n} y_i \quad S_{yx} = \sum_{i=1}^{n} y_i x_i$$

$$S_{yx2} = \sum_{i=1}^{n} y_i x_i^2 \quad S_{yx3} = \sum_{i=1}^{n} y_i x_i^3$$

Define linear system

$$\mathbf{A} = \begin{bmatrix} S_{x6} & S_{x5} & S_{x4} & S_{x3} \\ S_{x5} & S_{x4} & S_{x3} & S_{x2} \\ S_{x4} & S_{x3} & S_{x2} & S_x \\ S_{x3} & S_{x2} & S_x & n \end{bmatrix} \quad \mathbf{r} = \begin{bmatrix} S_{yx3} \\ S_{yx2} \\ S_{yx} \\ S_y \end{bmatrix}$$

Solve linear system $\mathbf{A} \mathbf{z} = \mathbf{r}$
Results: $a = z_1 \quad b = z_2 \quad c = z_3 \quad c = z_4$

Chapter 9 Function Approximation

Mathcad Function for Cubic Least-Squares Approximation

$$\text{Cubic\_LS}(x,y) := \begin{array}{|l} n \leftarrow \text{length}(x) \\ sx \leftarrow \sum x \\ sy \leftarrow \sum y \\ sxy \leftarrow \sum_{i=0}^{n-1} x_i \cdot y_i \\ sx2 \leftarrow \sum_{i=0}^{n-1} x_i \cdot x_i \\ sx3 \leftarrow \sum_{i=0}^{n-1} (x_i)^3 \\ sx4 \leftarrow \sum_{i=0}^{n-1} (x_i)^4 \\ sx5 \leftarrow \sum_{i=0}^{n-1} (x_i)^5 \\ sx6 \leftarrow \sum_{i=0}^{n-1} (x_i)^6 \\ sx2y \leftarrow \sum_{i=0}^{n-1} (x_i)^2 \cdot y_i \\ sx3y \leftarrow \sum_{i=0}^{n-1} (x_i)^3 \cdot y_i \\ A \leftarrow \begin{pmatrix} sx6 & sx5 & sx4 & sx3 \\ sx5 & sx4 & sx3 & sx2 \\ sx4 & sx3 & sx2 & sx \\ sx3 & sx2 & sx & n \end{pmatrix} \\ r \leftarrow (sx3y \; sx2y \; sxy \; sy)^T \\ z \leftarrow A^{-1} \cdot r \\ z \end{array}$$

Example 9.7 Cubic Least-Squares

The data introduced in Example 9-B looks like it might be approximated reasonably by a cubic equation. The linear system for the coefficients is $\mathbf{A}\,\mathbf{z} = \mathbf{r}$, with

$$\mathbf{A} = \begin{bmatrix} 3.9568 & 0.0000 & 5.0666 & 0.0000 \\ 0.0000 & 5.0666 & 0.0000 & 7.7 \\ 5.0666 & 0.0000 & 7.7 & 0 \\ 0.0000 & 7.7 & 0 & 21 \end{bmatrix}$$

and

$$\mathbf{r} = \begin{bmatrix} 2.5231 & 3.85 & 4.078 & 10.5 \end{bmatrix}^T.$$

Using the preceding Mathcad function for cubic least squares, we find that the least-squares cubic function is

$$p(x) = a\,x^3 + b\,x^2 + c\,x + d,$$

with the computed coefficients

$$a = -0.25719, \quad b = 0.0000, \quad c = 0.69884, \quad d = 0.5.$$

Define the data

$$xx := (-1 \ -0.8 \ -0.6 \ -0.4 \ -0.2 \ 0.0 \ 0.2 \ 0.4 \ 0.6 \ 0.8 \ 1.0)$$

$$y := (0.05 \ 0.08 \ 0.14 \ 0.23 \ 0.35 \ 0.50 \ 0.65 \ 0.77 \ 0.86 \ 0.92 \ 0.95)$$

Compute the coefficients

$$a := \text{Cubic\_LS}(xx^T, y^T)$$

Display the coefficients

$$a = \begin{pmatrix} -0.255 \\ 0 \\ 0.7 \\ 0.5 \end{pmatrix}$$

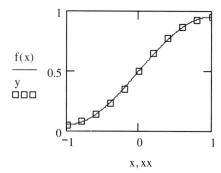

FIGURE 9.10 Data and cubic least-squares function.

Define the approximating polynomial, and the range variable for plotting

$$f(x) := a_0 \cdot x^3 + a_1 \cdot x^2 + a_2 x + a_3 \qquad x := -1, -0.99 .. 1$$

We continue our investigation of this data in the next example.

Example 9.8 Cubic Least-Squares, Continued

We now consider the cubic approximation function found by using only a portion of the data in Example 9-B—specifically, the data on the interval [0, 1]. The linear system of equations for the coefficients is $\mathbf{A}\,\mathbf{z} = \mathbf{r}$, with

$$\mathbf{A} = \begin{bmatrix} 14.791 & 17.171 & 20.503 & 25.502 \\ 17.171 & 20.503 & 25.502 & 33.835 \\ 20.503 & 25.502 & 33.835 & 50.5 \\ 25.502 & 33.835 & 50.5 & 101 \end{bmatrix}$$

and

$$\mathbf{r} = [23.1902, \quad 30.2101, \quad 43.4963, \quad 79.2100]^T$$

The coefficient matrix for this example has a condition number of 14825, which indicates that the problem is very ill conditioned. This is at least partly a result of using a large number of data points, so that we are requiring almost a continuous fit between the data and the cubic approximating function.

Define the data

$xx := (0.00 \quad 0.20 \quad 0.40 \quad 0.60 \quad 0.80 \quad 1.00)$

$y := (0.50 \quad 0.65 \quad 0.77 \quad 0.86 \quad 0.92 \quad 0.95)$

Compute the coefficients

$a := \text{Cubic\_LS}(xx^T, y^T)$

Display the coefficients

$a = \begin{pmatrix} 1.112 \times 10^{-13} \\ -0.375 \\ 0.825 \\ 0.5 \end{pmatrix}$

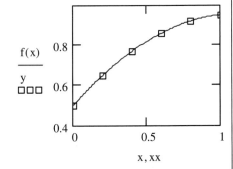

FIGURE 9.11 Cubic least-squares approximation.

Define the approximating polynomial, and the range variable for plotting

$f(x) := a_0 \cdot x^3 + a_1 \cdot x^2 + a_2 \cdot x + a_3$ $\qquad x := 0, 0.01 .. 1$

9.1.4 Least-Squares Approximation for Other Functional Forms

If the data are best fit by an exponential function, it is convenient instead to fit the logarithm of the data by a straight line. This gives a very close approximation to the best fit exponential. (The natural logarithm is denoted `ln` in Mathcad). The following Mathcad function performs the transformation on the data and returns the coefficients for the linear least square fit to the logarithmic data.

Algorithm for Exponential Approximation (Linear Fit to ln x)

> Given data vectors **x** and **y**
> find coefficients for approximation function, $y = \exp(a\,x + b)$.
>
> Transform data: $\mathbf{z} = \ln(\mathbf{y})$.
>
> Apply linear least-squares approximation algorithm to data vectors **x** and **z**.
> The approximating function is $y = \exp(a\,x + b) = e^{ax+b}$.

Mathcad Function for Linear Least-Squares Fit to Log y

We illustrate the use of the function `Ln_Linear_LS` in the next example.

$$
\begin{aligned}
\text{Ln\_Linear\_LS}(x, y) := \ \ & n \leftarrow \text{length}(x) \\
& yy \leftarrow \ln(y) \\
& sx \leftarrow \sum x \\
& sy \leftarrow \sum yy \\
& sxx \leftarrow \sum_{i=0}^{n-1} x_i \cdot x_i \\
& sxy \leftarrow \sum_{i=0}^{n-1} x_i \cdot yy_i \\
& \text{denom} \leftarrow n \cdot sxx - sx \cdot sx \\
& s_0 \leftarrow \frac{n \cdot sxy - sx \cdot sy}{\text{denom}} \\
& s_1 \leftarrow \frac{sxx \cdot sy - sxy \cdot sx}{\text{denom}} \\
& s
\end{aligned}
$$

Chapter 9 Function Approximation

Example 9.9 Oil Reservoir Data

Consider the data introduced in Example 9-A. We find an approximating function of the form $y = \exp(ax + b)$ by finding the linear least-squares fit to $\ln(y)$. The results are summarized in the table below and illustrated in Figures 9.12 and 9.13. The total squared error for the linear approximation and the logarithm of the data is 0.0538; the total squared error for the exponential function and the original data is 71.4017.

Data and best fit exponential for oil reservoir.

x_i	$\ln y_i$	$a x_i + b$	d_i	y_i	$\exp(a x_i + b)$	e_i
5.435	0.0000	−0.0922	0.0922	1.00	0.9120	0.0880
4.635	0.2469	0.2327	0.0142	1.28	1.2619	0.0181
3.835	0.5306	0.5575	0.0269	1.70	1.7463	−0.0463
3.035	0.7885	0.8823	−0.0938	2.20	2.4165	−0.2165
2.325	1.0886	1.1706	−0.0820	2.97	3.2239	−0.2539
1.435	1.4702	1.5320	−0.0618	4.35	4.6272	−0.2772
0.635	2.0149	1.8568	0.1581	7.50	6.4030	1.0970

$x := (5.435 \ 4.635 \ 3.835 \ 3.035 \ 2.325 \ 1.435 \ 0.635)$

$y := (1.0 \ 1.28 \ 1.70 \ 2.20 \ 2.97 \ 4.35 \ 7.50)$

$a := \text{Ln\_Linear\_LS}(x^T, y^T) \qquad a = \begin{pmatrix} -0.406 \\ 2.115 \end{pmatrix}$

$n := \text{length}(y^T) \qquad i := 0..n-1 \qquad yy_i := \ln(y_{0,i}) \qquad xx := x^T$

$x := 0, 0.1 .. 6 \qquad f(x) := a_0 \cdot x + a_1 \qquad g(x) := \exp(f(x))$

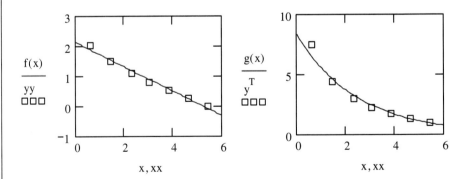

FIGURE 9.12 Natural logarithm of data and straight line fit.

FIGURE 9.13 Data and exponential fit from linear fit to $\ln(y_i)$.

Transformations of the original data may be helpful in other settings also. For example, if the data seem to describe an inverse relation, it may be appropriate to perform a least squares approximation to the reciprocal of the original data.

> **Example 9.10 Least-Squares Approximation of a Reciprocal Relation**
>
> The plot (see Fig. 9.14) of the following data suggests that they could be fit by a function of the form $y = 1/(ax + b)$
>
> $$\mathbf{x} = \begin{bmatrix} 0 & 0.5 & 1 & 1.5 & 2 \end{bmatrix},$$
>
> $$\mathbf{y} = \begin{bmatrix} 1.00 & 0.50 & 0.30 & 0.20 & 0.20 \end{bmatrix}.$$
>
> To find such a function, we consider the reciprocal of the original data and perform a linear least squares approximation to the data given by $\mathbf{x}$ and
>
> $$\mathbf{z} = 1/\mathbf{y} = \begin{bmatrix} 1.00 & 2.00 & 3.3333 & 5.00 & 5.00 \end{bmatrix}.$$
>
> The computed coefficients are $a = 2.2$, and $b = 1.0667$. Figure 9.14 shows the data and the linear least squares fit to the reciprocal of the data.
>
>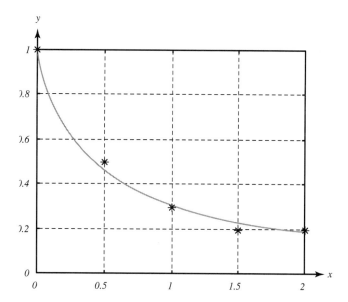
>
> FIGURE 9.14   Linear least-squares fit to reciprocal of data.

Example 9.11 Light Intensity

The following data on the intensity of light as a function of the distance from the light source were measured in a classroom experiment.

d = [30   35   40   45   50   55   60   65   70   75   ];
i = [0.85   0.67   0.52   0.42   0.34   0.28   0.24   0.21   0.18   0.15];

The inverse nature of the relationship is evident from a plot of the data (see Fig. 9.15.) We consider several least-squares approximations to the reciprocal of the data. Figure 9.15 illustrates the linear least-squares fit (dashed line) and the quadratic least squares fit (solid line) to the reciprocal of the data. The agreement between the data and the computed function,

$$z = \frac{1}{0.0013\, d^2 - 0.0208\, d + 0.6161},$$

is quite good.

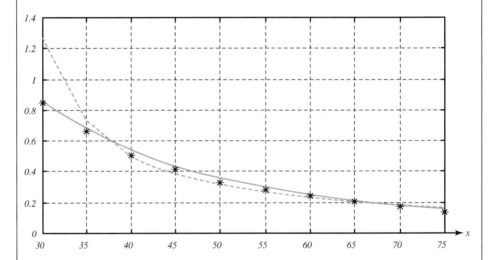

FIGURE 9.15   Linear and quadratic least-squares fit to reciprocal of light-intensity data.

9.2 CONTINUOUS LEAST-SQUARES APPROXIMATION

Suppose that instead of just knowing the value of the function we wish to approximate at some specific points, we know the exact value of the function for all points in an interval—say, for all x in $[0, 1]$. We can then find the best-fit linear, quadratic, or other given form of function, by minimizing the error over the entire interval. The summations are replaced by the corresponding integrals. To approximate a given function $s(x)$ with a quadratic function $p(x) = a x^2 + b x + c$ on the interval $[0, 1]$, we wish to minimize

$$E = \int_0^1 [ax^2 + bx + c - s(x)]^2 \, dx.$$

The resulting equations for a, b, and c are

$$\frac{a}{5} + \frac{b}{4} + \frac{c}{3} = \int_0^1 x^2 s(x) \, dx,$$

$$\frac{a}{4} + \frac{b}{3} + \frac{c}{2} = \int_0^1 x s(x) \, dx,$$

$$\frac{a}{3} + \frac{b}{2} + c = \int_0^1 s(x) \, dx.$$

To approximate a given function $s(x)$ with a quadratic function $p(x) = a x^2 + b x + c$, on the interval $[-1, 1]$, we wish to minimize

$$E = \int_{-1}^1 [ax^2 + bx + c - s(x)]^2 \, dx.$$

The resulting equations for a, b, and c are

$$\frac{2}{5} a + 0 + \frac{2}{3} c = \int_{-1}^1 x^2 s(x) \, dx,$$

$$0 + \frac{2}{3} b + 0 = \int_{-1}^{-1} x s(x) \, dx,$$

$$\frac{2}{3} a + 0 + 2c = \int_{-1}^1 s(x) \, dx.$$

Example 9.12 Continuous Least-Squares

To find the continuous least-squares quadratic approximation to the exponential function on $[-1, 1]$, we have the coefficient matrix

$$\mathbf{A} = \begin{bmatrix} 2/5 & 0 & 2/3 \\ 0 & 2/3 & 0 \\ 2/3 & 0 & 2 \end{bmatrix}.$$

To find the right-hand side of the linear system, we use the following results:

$$\int x^2 e^x \, dx = x^2 e^x - 2 \int x e^x \, dx = x^2 e^x - 2(x e^x - e^x) = e^x(x^2 - 2x + 2);$$

$$\int x e^x \, dx = x^2 e^x - e^x; \text{ and } \int e^x \, dx = e^x.$$

Then

$$\int_{-1}^{1} x^2 e^x \, dx = e - \frac{5}{e} \approx 0.8789,$$

$$\int_{-1}^{1} x e^x \, dx = \frac{2}{e} \approx 0.7358, \qquad \int_{-1}^{1} e^x \, dx = e - \frac{1}{e} \approx 2.3504.$$

Thus, the right-hand side of the system is $\mathbf{r} = [0.8789 \; 0.7358 \; 2.3504]^T$, and solving the linear system gives $\mathbf{z} = [0.5367 \; 1.1036 \; 1.1036]^T$. Figure 9.16 shows the exponential function (solid), the quadratic approximation $p = 0.5367x^2 + 1.1036x + 1.1036$ (dashed), and the Taylor polynomial for the exponential function, $t(x) = 0.5\,x^2 + x + 1$ (dotted).

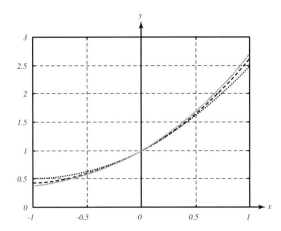

FIGURE 9.16 Exponential function, least-squares approximation, and Taylor polynomial.

Discussion

We wish to minimize $E = \int_0^1 [a x^2 + b x + c - s(x)]^2 \, dx$.

Setting $\dfrac{\partial E}{\partial a} = 0$ gives $\int_0^1 x^2 [a x^2 + b x + c - s(x)] dx = 0$,

$$a \int_0^1 x^4 \, dx + b \int_0^1 x^3 \, dx + c \int_0^1 x^2 \, dx = \int_0^1 x^2 s(x) \, dx,$$

$$\frac{1}{5} a + \frac{1}{4} b + \frac{1}{3} c = \int_0^1 x^2 s(x) \, dx.$$

Setting $\dfrac{\partial E}{\partial b} = 0$ gives $\int_0^1 x [a x^2 + b x + c - s(x)] dx = 0$,

$$a \int_0^1 x^3 \, dx + b \int_0^1 x^2 \, dx + c \int_0^1 x \, dx = \int_0^1 x s(x) \, dx,$$

$$\frac{1}{4} a + \frac{1}{3} b + \frac{1}{2} = \int_0^1 x s(x) \, dx.$$

And setting $\dfrac{\partial E}{\partial c} = 0$ gives $\int_0^1 [a x^2 + b x + c - s(x)] \, dx = 0$,

$$a \int_0^1 x^2 \, dx + b \int_0^1 x \, dx + c \int_0^1 dx = \int_0^1 s(x) \, dx,$$

$$\frac{1}{3} a + \frac{1}{2} b + c = \int_0^1 s(x) \, dx.$$

The coefficient matrix is the three-by-three Hilbert matrix, which is notoriously ill conditioned. If we continue this approach with higher degree polynomial approximations, we find that the coefficient matrix is the correspondingly larger Hilbert matrix. The larger the Hilbert matrix, the worse is the conditioning. The computations for continuous least squares approximation on $[-1, 1]$ are completely analogous. Here, the coefficient matrix is not as ill conditioned as the Hilbert matrix, but its conditioning does become increasingly worse for higher order polynomials.

To utilize continuous least-squares for higher-degree polynomials, it is better to use a basis of polynomials that are orthogonal on the interval over which the approximation is desired. In addition to reducing the computational difficulties encountered in solving a linear system with an ill-conditioned coefficient matrix, the use of orthogonal polynomials allows us to progress easily from a lower-degree approximation to the next higher-degree polynomial.

9.2.1 Continuous Least-Squares with Orthogonal Polynomials

We say that a set of functions $\{f_0, f_1, f_2, \ldots, f_n\}$ is linearly independent on the interval $[a, b]$ if a linear combination of the functions is the zero function only if all of the coefficients are zero. In other words, if $c_0 f_0(x) + c_1 f_1(x) + c_2 f_2(x) + \ldots + c_n f_n(x) = 0$ for all x in $[a, b]$, then $c_0 = c_1 = c_2 = \ldots = c_n = 0$. The functions $f_0 = 1, f_1 = x, f_2 = x^2, \ldots, f_n = x^n$, are linearly independent on any interval $[a, b]$, but in fact, any set of functions $\{p_0, p_1, p_2, \ldots p_n\}$ where p_j is a polynomial of degree j, is linearly independent. Another important set of linearly independent functions, which we will use in the next chapter, is the set of trigonometric functions $\{1, \sin(x), \cos(x), \sin(2x), \cos(2x), \ldots \sin(nx), \cos(nx)\}$.

The set of functions $\{f_0, f_1, f_2, \ldots f_n\}$ is said to be orthogonal on $[a, b]$ if

$$\int_a^b f_i(x) f_j(x)\, dx = \begin{cases} 0 & \text{if } i \neq j \\ d_j > 0 & \text{if } i = j \end{cases}$$

A more general concept of orthogonality includes a weighting function w in the integral, but we restrict our investigations to the case where $w = 1$. If the functions are orthogonal and $d_j = 1$ for all j, the functions are called *orthonormal*.

9.2.2 Gram-Schmidt Process

We now show how we can construct a sequence of polynomials that are orthogonal on the interval $[a, b]$. For $n = 0, 1, 2, \ldots$, we require that

1. p_n be a polynomial of degree n, with the coefficient of x^n positive.
2. $\int_a^b p_n(x) p_m(x)\, dx = 0$ if $n \neq m$.
3. $\int_a^b p_n(x) p_n(x)\, dx = 1$.

We start by taking $p_0(x) = c > 0$.

To satisfy condition 3, we must have:

$$\int_a^b c^2\, dx = 1 \Rightarrow (b-a)c^2 = 1, \text{ so } c = \frac{1}{\sqrt{b-a}}.$$

To construct $p_1(x)$, we begin by letting $q_1(x) = x + c_{1,0} p_0$.

We first require that $q_1(x)$ be orthogonal to p_0:

$$\int_a^b p_0(x + c_{1,0} p_0)dx = 0 \Rightarrow \int_a^b x p_0\, dx + c_{1,0} \int_a^b p_0 p_0\, dx = 0$$

Since $\int_a^b p_0 p_0\, dx = 1$, we have, by construction, $c_{1,0} = -\int_a^b x p_0\, dx$.

With this choice of $c_{1,0}$, $q_1(x) = x + c_{1,0} p_0$ is orthogonal to p_0.

To satisfy condition 3, we normalize $q_1(x)$ to get

$$p_1(x) = \frac{q_1(x)}{\left[\int_a^b q_1 q_1 \, dx\right]^{1/2}}.$$

To construct $p_2(x)$, we begin by letting $q_2(x) = x^2 + c_{2,1} p_1(x) + c_{2,0} p_0(x)$.
We first require that $q_2(x)$ be orthogonal to $p_1(x)$:

$$\int_a^b p_1(x)\,(x^2 + c_{2,1} p_1(x) + c_{2,0} p_0(x))\, dx = 0 \Rightarrow$$

$$\int_a^b x^2 p_1(x)\,dx + c_{2,1} \int_a^b p_1(x) p_1(x)\,dx + c_{2,0} \int_a^b p_1(x) p_0(x)\,dx = 0.$$

Because p_1 is orthogonal to p_0, the third integral is zero; the second integral evaluates to one, so we have

$$c_{2,1} = -\int_a^b x^2 p_1(x)\, dx.$$

Similarly, we require that $q_2(x)$ be orthogonal to $p_0(x)$:

$$\int_a^b p_0(x)\,(x^2 + c_{2,1} p_1(x) + c_{2,0} p_0(x))\, dx = 0 \Rightarrow$$

$$\int_a^b x^2 p_0(x)\,dx + c_{2,1} \int_a^b p_0(x) p_1(x)\,dx + c_{2,0} \int_a^b p_0(x) p_0(x)\,dx = 0.$$

As before, since p_1 is orthogonal to p_0, the second integral is zero and the third integral is one, so we have

$$c_{2,0} = -\int_a^b x^2 p_0(x)\,dx.$$

We complete this step by normalizing $q_2(x)$ to form $p_2(x)$:

$$p_2(x) = \frac{q_2(x)}{\left[\int_a^b q_2(x)\, q_2(x)\, dx\right]^{1/2}}.$$

The process continues in the same manner, as we construct each higher degree polynomial in turn. The polynomial $p_n(x)$ is formed from

$$q_n(x) = x^n + c_{n,0} p_0(x) + c_{n,1} p_1(x) + \cdots + c_{n,n-1} p_{n-1}(x),$$

by normalization. (The coefficients $c_{n,0} \ldots c_{n,n-1}$ are found so that $q_n(x)$ is orthogonal to each of the previously generated polynomials.)

9.2.3 Legendre Polynomials

Although the difficulties with forming and solving a linear system of equations for the coefficients of the continuous least-squares approximation problem are most severe on $[0, 1]$, the situation is not good on any other finite interval. The most convenient interval for finding a continuous least-squares approximating function using orthogonal polynomials is $[-1, 1]$. A function of x defined on any finite interval $a \leq x \leq b$ can be transformed to a function of t defined for $-1 \leq t \leq 1$ by the change of variables $x = \dfrac{b-a}{2} t + \dfrac{b+a}{2}$.

The functions known as *Legendre polynomials* form an orthogonal set on $[-1, 1]$; they are useful for continuous least-squares function approximation on a finite interval (transformed, if necessary, to $[-1, 1]$). We will encounter them again in Chapter 11 when we discuss Gaussian quadrature. We list the first few Legendre polynomials here, normalized so that the coefficient of the leading term is unity.

$$P_0(x) = 1, \qquad P_1(x) = x,$$

$$P_2(x) = x^2 - \frac{1}{3}, \qquad P_3(x) = x^3 - \frac{3}{5}x,$$

$$P_4(x) = x^4 - \frac{6}{7}x^2 - \frac{3}{35}, \quad P_5(x) = x^5 - \frac{10}{9}x^3 - \frac{5}{21}.$$

For this normalization, we have

$$\int_{-1}^{1} P_0(x) P_0(x)\, dx = 2,$$

$$\int_{-1}^{1} P_1(x) P_1(x)\, dx = \int_{-1}^{1} x^2\, dx = \frac{2}{3},$$

$$\int_{-1}^{1} P_2(x) P_2(x)\, dx = \int_{-1}^{1} x^4 - \frac{2}{3}x^2 + \frac{1}{9}\, dx = \frac{8}{45}.$$

An alternative definition of the Legendre polynomials gives

$$p_0(x) = 1 \text{ and } p_n(x) = \frac{(-1)^n}{2^n\, n!} \frac{d^n}{dx^n}[(1-x^2)^n], \text{ for } n \geq 1.$$

The polynomials defined in this way are normalized so that $p_n(1) = 1$ and

$$\int_{-1}^{1} p_n(x) p_n(x)\, dx = \frac{2}{2n+1}.$$

We turn our attention next to illustrating how the property of orthogonality can be used for finding a continuous least-squares approximation function.

9.2.4 Least-Squares Approximation with Legendre Polynomials

To find the quadratic least squares approximation to $f(x)$ on the interval $[-1, 1]$ in terms of the Legendre polynomials, we need to determine the coefficients c_0, c_1, c_2 that minimize

$$E = \int_{-1}^{1} [c_0 P_0(x) + c_1 P_1(x) + c_2 P_2(x) - f(x)]^2 \, dx.$$

The normal equations are found as before, by setting the partial derivatives of E equal to zero. Setting $\dfrac{\partial E}{\partial c_0} = 0$ yields

$$\int_{-1}^{1} 2[c_0 P_0(x) + c_1 P_1(x) + c_2 P_2(x) - f(x)] P_0(x) \, dx = 0,$$

which, after dividing out the common factor of 2, expands to give

$$\int_{-1}^{1} c_0 P_0(x) P_0(x) \, dx + \int_{-1}^{1} c_1 P_1(x) P_0(x) \, dx + \int_{-1}^{1} c_2 P_2(x) P_0(x) \, dx = \int_{-1}^{1} f(x) P_0(x) \, dx.$$

Now, because the Legendre polynomials are orthogonal, the second and third integrals on the left side of the equations are zero, so we have

$$c_0 \int_{-1}^{1} P_0(x) P_0(x) \, dx = \int_{-1}^{1} f(x) P_0(x) \, dx.$$

In a similar manner, the equations formed by setting $\dfrac{\partial E}{\partial c_1} = 0$ and $\dfrac{\partial E}{\partial c_2} = 0$ give

$$c_1 \int_{-1}^{1} P_1(x) P_1(x) \, dx = \int_{-1}^{1} f(x) P_1(x) \, dx,$$

and

$$c_2 \int_{-1}^{1} P_2(x) P_2(x) \, dx = \int_{-1}^{1} f(x) P_2(x) \, dx.$$

The integrations on the left were performed earlier; the result is

$$c_0 = \frac{1}{2} \int_{-1}^{1} f(x) P_0(x) \, dx; \quad c_1 = \frac{3}{2} \int_{-1}^{1} f(x) P_1(x) \, dx; \quad c_2 = \frac{45}{8} \int_{-1}^{1} f(x) P_2(x) \, dx.$$

If we wish a higher-order approximation, we have only to compute the additional integrals; this is in contrast to the system of linear equations approach, in which the entire problem must be reworked if the order of the approximating polynomial is to be increased.

Example 9.13 Least-Squares Approximation Using Legendre Polynomials

To find the quadratic least-squares approximation to $f(x) = e^x$ on the interval $[-1, 1]$ in terms of the Legendre polynomials, we write

$$g(x) = c_0 P_0(x) + c_1 P_1(x) + c_2 P_2(x),$$

where $P_0(x) = 1$, $P_1(x) = x$, and $P_2(x) = x^2 - \frac{1}{3}$. We determine the coefficients c_0, c_1, and c_2 as described previously. We will need the values of the following integrals, found in Example 9.12.

$$\int_{-1}^{1} x^2 f^x \, dx = e - \frac{5}{e} \approx 0.8789, \int_{-1}^{1} x f^x \, dx = \frac{2}{e} \approx 0.7358, \int_{-1}^{1} f^x \, dx - e - \frac{1}{e} \approx 2.3504;$$

$$c_0 = \frac{1}{2} \int_{-1}^{1} f(x) P_0(x) \, dx = \frac{1}{2} \int_{-1}^{1} e^x \, dx \approx \frac{1}{2}(2.3504) \approx 1.1752;$$

$$c_1 = \frac{3}{2} \int_{-1}^{1} f(x) P_1(x) \, dx = \frac{3}{2} \int_{-1}^{1} x e^x \, dx \approx \frac{3}{2}(0.7358) \approx 1.1037;$$

$$c_2 = \frac{45}{8} \int_{-1}^{1} f(x) P_2(x) \, dx = \frac{45}{8} \int_{-1}^{1} e^x \left(x^2 - \frac{1}{3}\right) dx \approx 0.5368.$$

We thus have $g(x) = c_0 P_0(x) + c_1 P_1(x) + c_2 P_2(x) = c_0 + c_1 x + c_2 (x_2 - \frac{1}{3})$; the plot of $g(x)$ and $f(x)$ shown in Fig. 9.17 strongly suggests what we could verify algebraically if we desire, namely, that this is the same quadratic approximation function as that found in Example 9.12.

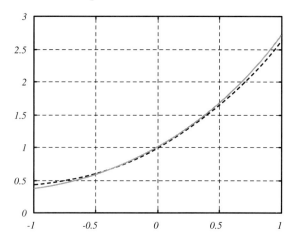

FIGURE 9.17 Exponential and least squares approximation function.

9.3 FUNCTION APPROXIMATION AT A POINT

In this section we consider two important methods of approximating a function $f(x)$ at a specific value of x: Taylor polynomial approximation and Padé rational function approximation.

9.3.1 Taylor Approximation

For a polynomial approximation, the Taylor polynomial (of degree n) gives the highest possible order of contact between the function and the polynomial. That is, the Taylor polynomial agrees with the function and its first n derivatives at $x = a$; the formula is

$$p(x) = f(a) + f'(a)(x-a) + \frac{f''(a)}{2!}(x-a)^2 + \frac{f'''(a)}{3!}(x-a)^3 + \ldots + \frac{f^{(n)}(a)}{n!}(x-a)^n.$$

The Taylor polynomial was included in Example 9.12 for comparison with the exponential function and its least squares approximation.

The Taylor polynomial is widely used in the analysis of numerical techniques. (It was introduced in Chapter 1 in the discussion of truncation error.) The error incurred in using $p(x)$ to approximate the actual value of the function $f(x)$ is given by

$$\frac{f^{(n+1)}(\eta)}{(n+1)!}(x-a)^{n+1},$$

for some η between x and a. Taylor's formula is derived in most standard calculus texts.

We next present a method for finding a rational function that agrees with the values of a given function and its derivatives at a specific point.

9.3.2 Padé Function Approximation

Padé approximation seeks to approximate a function $f(x)$ by finding a rational function that fits the values of the function and its derivatives at a given point x_0. The desired rational function has the form

$$r(x) = \frac{p_m(x)}{q_n(x)} = \frac{a_m x^m + \ldots + a_0}{b_n x^n + \ldots + b_0}.$$

We assume that $f(x_0), f'(x_0), \ldots, f^{(k)}(x_0)$ are given for $k = m + n$, and, in the discussion that follows, we assume, for simplicity, that $x_0 = 0$.

Let $t(x)$ be the Taylor polynomial for the given function $f(x)$. We write

$$t(x) = c_k x^k + \ldots + c_2 x^2 + c_1 x + c_0.$$

In terms of information about the derivatives of $f(x)$, we have $c_k = \dfrac{f^{(k)}(0)}{k!}$. We want $r(x)$ to be the same as $t(x)$—to have the same value at $x = 0$ and the same derivatives of all orders up to, and including, $k = m + n$. Our basic approach is to use the equivalence of the expressions

$$t(x) = r(x) = \frac{p_m(x)}{q_n(x)} \text{ and } q_n(x)\, t(x) = p_m(x).$$

Thus, we work with the relationship

$$(b_n x^n + \ldots + b_0)(c_k x^k + \ldots + c_2 x^2 + c_1 x + c_0) = a_m x^m + \ldots + a_0 \quad (9.1)$$

We begin by considering the requirement that $q_n(0)\, t(0) = p_m(0)$; this gives

$$b_0 c_0 = a_0.$$

The preceding rational function $r(x)$ is determined only up to a scale factor. We choose $b_0 = 1$. With this convention, we have an equation for determining a_0.

Taking the first derivative of Eq. (9.1), we find that

$$(b_n x^n + \ldots + b_0)(c_1 + 2c_2 x + \ldots + k c_k x^{k-1})$$
$$+ (n b_n x^{n-1} + \ldots + b_1)(c_0 + c_1 x + c_2 x^2 + \ldots + c_k x^k)$$
$$= m a_m x^{m-1} + \ldots + a_1.$$

We require that this equation be satisfied when $x = 0$, so we have

$$b_0 c_1 + b_1 c_0 = a_1.$$

This gives us an equation relating the unknown coefficients a_1 and b_1 to the known quantities c_0, c_1, and b_0. We rewrite the equation as

$$a_1 - b_1 c_0 = c_1.$$

To continue the process of finding higher derivatives of the product $g(x) = q(x)t(x)$ efficiently, we note the following formulae, which can be found by repeated applications of the product rule:

$$g'(x) = q(x)t'(x) + q'(x)t(x),$$
$$g''(x) = q(x)t''(x) + 2q'(x)t'(x) + q''(x)t(x),$$

$$\vdots$$

$$g^{(n)}(x) = \sum_{j=0}^{n} \frac{n!}{j!(n-j)!} q^{(j)}(x) t^{(n-j)}(x).$$

We are interested in the values of the derivative at $x = 0$, so we have

$$g^{(n)}(0) = \sum_{j=0}^{n} \frac{n!}{j!(n-j)!} q^{(j)}(0) t^{(n-j)}(0).$$

Taking the second derivative of Eq. (9.1) at $x = 0$, we find that

$$q(0)t''(0) + 2 q'(0)t'(0) + q''(0)t(0) = p''(0).$$

Substituting in the values of these derivatives (in terms of the coefficients), we have

$$b_0 \, 2 \, c_2 + 2 \, b_1 \, c_1 + 2 \, b_2 \, c_0 = 2 \, a_2,$$

or, after simplifying,

$$a_2 - b_1 \, c_1 - b_2 \, c_0 = c_2.$$

We continue in this manner until all required equations are specified (depending on the degrees of the polynomials p_m and q_n).

If we define $a_j = 0$ for $j > m$ and $b_j = 0$ for $j > n$, then the unknowns a_k and b_k can be found from the general relationship

$$a_k - \sum_{i=0}^{k-1} c_i \, b_{k-i} = c_k \qquad k = 0, 1, \ldots, m+n.$$

In general, errors for Padé approximates are less when the degree of the numerator and the degree of the denominator are the same or when the degree of the numerator is one larger than the denominator. (See Ralston & Rabinowitz, p. 295.) It is desired to have the polynomial in the denominator of sufficiently high degree to account for any poles that the original function may have in the complex plane (since such poles are often the cause of the difficulties that arise with polynomial interpolation, even when our interest is restricted to real variables only.)

Example 9.14 Padé Approximation of the Runge Function

The values of the Runge function
$$f(x) = \frac{1}{1 + 25\,x^2}$$
and its first three derivatives at $x = 0$ are
$$f(0) = 1, \quad f'(0) = 0, \quad f''(0) = -50, \quad f'''(0) = 0.$$
The Taylor polynomial of the function (at $x = 0$) is
$$t(x) = 1 + 0\,x - 25\,x^2 + 0\,x^3.$$
We seek a rational-function representation, using $k = 3$, $m = 1$, and $n = 2$. That is, we want
$$r(x) = \frac{a_1 x + a_0}{b_2 x^2 + b_1 x + 1}.$$
For $k = 3$, the linear system of equations is
$$\begin{aligned}
a_0 &= c_0 \\
a_1 - c_0 b_1 &= c_1 \\
a_2 - c_0 b_2 - c_1 b_1 &= c_2 \\
a_3 - c_0 b_3 - c_1 b_2 - c_2 b_1 &= c_3
\end{aligned}$$
In this particular problem, we have
$$c_0 = 1, \quad c_1 = 0, \quad c_2 = -25, \quad \text{and } c_3 = 1.$$
The solution of the linear system is found to be
$$a_0 = 1, \quad a_1 = b_1 = 0, \quad b_2 = 25,$$
so
$$r(x) = \frac{1}{1 + 25\,x^2}.$$
Thus, the Padé approximation has given us the exact representation of the Runge function in this example.

9.4 MATHCAD'S METHODS

Mathcad's built-in functions for regression (least-squares approximation) follow the same basic form as the functions for interpolation discussed in Section 8.5. However, the returned vector gives the parameters of the regression function, rather than input to another function (such as the function `interp` in Ch 8). The functions for linear regression, general polynomial regression, and general least squares approximation using a user specified functional form, are discussed in the next section.

9.4.1 Using the Built-in Functions

The input to each of the following functions includes the vectors **vx**, and **vy**, which contain the data to be approximated.

Linear Regression

Mathcad has five functions for linear regression. They are

> `intercept`(**vx,vy**): This function returns the y-intercept of the least-squares regression line.
>
> `slope`(**vx,vy**): This function returns the slope of the least-squares regression line.
>
> `line`(**vx,vy**): This function returns a vector containing the y-intercept and the slope of the least-squares regression line.
>
> `stderr`(**vx,vy**): This function returns the standard error of the linear regression.
>
> `medfit`(**vx,vy**): This function returns the y-intercept and slope of the median-median regression line. This linear fit is less sensitive to outliers than the standard regression line produced by the function `line`(**vx,vy**). To generate the median-median line, the data is divided into three sets, the median of first and last subsets is calculated, and the result is the line connecting these medians.

Polynomial Regression

> `regress`(**vx, vy**, n): This function returns the parameters for the nth degree polynomial least-squares fit to the data.

General Approximation

Mathcad has a function `genfit` for general least-squares approximation, using a user-defined functional form, as well as several special cases for fitting the data with specified functional forms. Most of these functions requires as input, in addition to the data vectors described previously, a vector **vg** which contains the initial guess of the parameters that are to be determined.

> `genfit`(**vx**, **vy**, **vg**, F): This function returns the parameters that make $f(x)$ approximate the data The user supplied function, F, returns an $n + 1$ element vector, containing the definition of the approximating function f and its partial derivatives with respect to the parameters that are to be determined by `genfit`.

Special cases of `genfit`

> `expfit`(**vx**, **vy**, **vg**): This function returns a vector containing the parameters (a, b, c) so that $f(x) = a \exp^\wedge(b\, x) + c$ fits (approximates in the least squares sense) the data.
>
> `lgsfit`(**vx**, **vy**, **vg**): This function returns a vector containing the parameters (a, b, c) so that $f(x) = a\, (1 + b \exp^\wedge(-c\, x))^{-1}$ fits the data.
>
> `linfit`(**vx**, **vy**, F): The input function F is a function of a single variable, which returns a vector of functions. The function `linfit` returns a vector containing the coefficients so that a linear combination of the functions in F approximates the data. This function is not as general as `genfit`, but does not require an initial guess of the parameters, since it is doing a linear fit.
>
> `logfit`(**vx**, **vy**, **vg**): This function returns a vector containing the parameters (a, b, c) so that $f(x) = a \ln(x + b) + c$ fits the data.
>
> `pwrfit`(**vx**, **vy**, **vg**): This function returns a vector containing the parameters (a, b, c) so that $f(x) = a\, x^b + c$ fits the data
>
> `sinfit`(**vx**, **vy**, **vg**): This function returns a vector containing the parameters (a, b, c) so that $f(x) = a \sin(x + b) + c$ fits the data.

9.4.2 Understanding the Algorithms

The regression function `regress` solves the normal equations by Gauss-Jordan elimination.

The approximation function `linfit` finds the required parameters by solving the linear system using singular value decomposition (SVD). SVD is the method of choice for any linear system that is numerically close to singular. Since the coefficient matrix for the normal equations is ill-conditioned, SVD is also the method of choice for linear least-squares problems. See [Press et al., 1992, pp. 59–70, 416] for a discussion of the method.

The general approximation function `genfit` finds the required parameters using the Levenberg-Marquardt method, described in Section 5.4.2.

SUMMARY

Linear least squares approximation: The coefficients a and b for the straight line $y = a x + b$, to approximate the data

$$\mathbf{x} = [x_1, x_2, \ldots, x_n], \qquad \mathbf{y} = [y_1, y_2, \ldots, y_n]$$

can be found from the following system of equations:

$$a S_{xx} + b S_x = S_{xy}$$
$$a S_x + b\, 4 = S_y$$

where $S_{xx} = \sum_{i=1}^{n} x_i^2,\ S_x = \sum_{i=1}^{n} x_i,\ S_{xy} = \sum_{i=1}^{n} x_i y_i,\ S_y = \sum_{i=1}^{n} y_i.$

The solution to the system of equations is

$$a = \frac{n S_{xy} - S_x S_y}{n S_{xx} - S_x S_x}, \qquad b = \frac{S_{xx} S_y - S_{xy} S_x}{S_{xx} - S_x S_x}.$$

Quadratic least squares approximation: The coefficients for the quadratic function $f(x) = a x^2 + b x + c$ that best fits the data are found by solving the system of equations:

$$a \sum_{i=1}^{n} x_i^4 + b \sum_{i=1}^{n} x_i^3 + c \sum_{i=1}^{n} x_i^2 = \sum_{i=1}^{n} x_i^2 y_i,$$

$$a \sum_{i=1}^{n} x_i^3 + b \sum_{i=1}^{n} x_i^2 + c \sum_{i=1}^{n} x_i = \sum_{i=1}^{n} x_i y_i.$$

$$a \sum_{i=1}^{n} x_i^2 + b \sum_{i=1}^{n} x_i + c[n] = \sum_{i=1}^{n} y_i$$

To find an exponential function of the form $y = \exp(a x + b)$ to fit a set of data, we first find a linear fit to the logarithm of the data.

Continuous least squares approximation may be obtained in an analogous manner, with integrals replacing summations. To approximate a given function $s(x)$ with a quadratic function $p(x) = a x^2 + b x + c$ on the interval $[-1, 1]$, the resulting equations for a, b, and c are

$$\frac{2}{5} a + 0 + \frac{2}{3} c = \int_{-1}^{1} x^2 s(x)\, dx,$$

$$0 + \frac{2}{3} b + 0 = \int_{-1}^{1} x s(x)\, dx,$$

$$\frac{2}{3} a + 0 + 2c = \int_{-1}^{1} s(x)\, dx.$$

Since the coefficient matrix of the linear system is ill conditioned, it is better to use orthogonal polynomials, rather than powers of x, as the basis functions for the approximation. The quadratic least squares approximation to $f(x)$ in terms of the Legendre polynomials (on $[-1, 1]$) is $c_0 P_0(x) + c_1 P_1(x) + c_2 P_2(x)$; the coefficients are

$$c_0 = \frac{1}{2}\int_{-1}^{1} f(x)P_0(x)\,dx; \quad c_1 = \frac{3}{2}\int_{-1}^{1} f(x)P_1(x)\,dx; \quad c_2 = \frac{45}{8}\int_{-1}^{1} f(x)P_2(x)\,dx.$$

Padé approximation: The desired rational function has the form

$$r(x) = \frac{p_m(x)}{q_n(x)} = \frac{a_m x^m + \ldots + a_0}{b_n x^n + \ldots + b_0}.$$

Given $f(0), f'(0), \ldots, f^{(k)}(0)$ for $k = m + n$; the Taylor polynomial for $f(x)$ is

$$t(x) = c_k x^k + \ldots + c_2 x^2 + c_1 x + c_0, \text{ where } c_k = \frac{f^{(k)}(0)}{k!}.$$

We choose $b_0 = 1$, then the coefficients are determined as:

$$b_0 c_0 = a_0.$$
$$a_1 - b_1 c_0 = c_1.$$
$$a_2 - b_1 c_1 - b_2 c_0 = c_2.$$

If we define $a_j = 0$ for $j > m$ and $b_j = 0$ for $j > n$, then the unknowns a_k and b_k can be found from the general relationship

$$a_k - \sum_{i=0}^{k-1} c_i b_{k-i} = c_k \quad k = 0, 1, \ldots, m + n.$$

SUGGESTIONS FOR FURTHER READING

The standard texts on approximation of functions include:

Achieser, N. I., *Theory of Approximation*, Dover, New York, 1993.
Cheney, E. W., *Introduction to Approximation Theory*, McGraw-Hill, 1966.
Rivlin, T. J., *An Introduction to the Approximation of Functions*, Dover, 1981. (Originally published by Blaisdell Publishing, 1969.)
Timan, A. F., C. J. Hyman, N. I. Achieser, *Theory of Approximation*, Dover, New York, 1993.

For further discussion of Padé approximation, see

Brezinski, C., *History of Continued Fractions and Padé Approximants*, Springer-Verlag, Berlin, 1991.
Jensen, J. A., and J. H. Rowland, *Methods of Computation*, Scott, Foresman and Company, Glenview, IL, 1975.

PRACTICE THE TECHNIQUES

For Problems P9.1 to P 9.5, find the linear least-squares approximation to the data.

P9.1 $x = [1\ 2\ 3]$,
$y = [1\ 4\ 8]$.

P9.2 $x = [1\ 4\ 9]$,
$y = [1\ 2\ 3]$.

P9.3 $x = [4\ 9\ 16]$,
$y = [2\ 3\ 4]$.

P9.4 $x = [-1\ 0\ 1]$,
$y = [-2\ 3\ -2]$.

P9.5 $x = [0\ 1\ 2]$,
$y = [1\ 2\ 4]$.

For Problems P9.6 to P 9.15:
  a. Find the linear least-squares approximation to the data.
  b. Find the quadratic least-squares approximation to the data.

P9.6 $x = [0\ 1\ 2\ 4]$,
$y = [1\ 1\ 2\ 5]$.

P9.7 $x = [-1\ 0\ 1\ 2]$,
$y = [1/3\ 1\ 3\ 9]$.

P9.8 $x = [0\ 1\ 2\ 3]$,
$y = [1\ 2\ 4\ 8]$.

P9.9 $x = [0\ 1\ 2\ 3]$,
$y = [1\ 2\ 4\ 8]$.

P9.10 $x = [-1\ 0\ 1\ 2]$,
$y = [0\ 1\ 2\ 9]$.

P9.11 $x = [0\ 2/3\ 1\ 2]$,
$y = [2\ -2\ -1\ -1/2]$.

P9.12 $x = [0\ 2/3\ 1\ 2]$,
$y = [4\ -4\ -2\ -1/2]$.

P9.13 $x = [0\ 2/3\ 1\ 2]$,
$y = [4\ -4\ -3.5\ -0.5]$.

P9.14 $x = [1\ 2\ 3\ 4]$,
$y = [2\ 4\ 8\ 16]$.

P9.15 $x = [0\ 1/2\ 1\ 3/2]$,
$y = [1\ 2\ 1\ 0]$.

For Problems P9.16 to P9.20:
  a. Find the linear least-squares approximation to the data.
  b. Find the quadratic least-squares approximation to the data.
  b. Find the cubic least-squares approximation to the data.

P9.16 $x = [0\ 1\ 2\ 3\ 4]$,
$y = [0\ 1\ 4\ 8\ 16]$.

P9.17 $x = [0\ 1\ 4\ 9\ 16]$,
$y = [0\ 1\ 2\ 3\ 4]$.

P9.18 $x = [-1\ -0.75\ -0.25\ 0.25\ 0.75\ 1]$,
$y = [0\ -0.7\ -0.7\ 0.7\ 0.7\ 0]$.

P9.19 $x = [-1\ -0.5\ 0.0\ 0.5\ 0.75\ 1]$,
$y = [0.4\ 0.6\ 1.0\ 1.6\ 2.1\ 2.7]$.

P9.20 $x = [0.5\ 0.75\ 1.0\ 1.25\ 1.5\ 1.75]$,
$y = [-0.7\ -0.3\ 0.0\ 0.2\ 0.4\ 0.6]$.

For Problems P9.21 to P9.25, find the linear least-squares approximation to the data.

P9.21 $x = [1\ 2\ 3\ 4\ 5\ 6\ 7\ 8\ 9\ 10]$
$y = [6.0\ 9.2\ 13.0\ 14.7\ 19.7\ 21.8\ 22.8\ 29.1\ 30.2\ 32.2]$

P9.22 $x = [1\ 2\ 3\ 4\ 5\ 6\ 7\ 8\ 9\ 10]$
$y = [5.7\ 8.9\ 15.2\ 16.6\ 20.9\ 26.7\ 28.6\ 34.0\ 34.1\ 47.0]$

P9.23 $x = [1\ 2\ 3\ 4\ 5\ 6\ 7\ 8\ 9\ 10]$
$y = [-3.3\ -7.1\ -9.8\ -12.3\ -14.0\ -21.4\ -21.3\ -28.6\ -29.0\ -32.8]$

P9.24 $x = [1\ 2\ 3\ 4\ 5\ 6\ 7\ 8\ 9\ 10]$
$y = [2.9\ 0.5\ -0.2\ -3.8\ -5.4\ -4.3\ -7.8\ -13.8\ -10.4\ -13.9]$

P9.25 $x = [\ 4.6\quad 8.7\quad 9.3\quad 2.6\quad 1.6\quad 8.7\quad 2.4\quad 6.5\quad 9.7\quad 6.6]$
$y = [16.5\quad 35.8\quad 31.6\quad 6.6\quad 2.5\quad 28.0\quad 6.8\quad 21.1\quad 38.8\quad 20.8]$

For Problems P9.26 to P9.30, find the quadratic least-squares approximation to the data.

P9.26 $x = [1\quad 2\quad 3\quad 4\quad 5\quad 6\quad 7\quad 8\quad 9\quad 10\ ]$
$y = [2.9\quad 4.8\quad 6.0\quad 18.9\quad 8.7\quad 30.7\quad 25.7\quad 77.7\quad 55.8\quad 104.8]$

P9.27 $x = [1\quad 2\quad 3\quad 4\quad 5\quad 6\quad 7\quad 8\quad 9\quad 10\ ]$
$y = [-0.4\quad -4.1\quad -7.4\quad -21.5\quad -21.3\quad -45.2\quad -44.7\quad -62.8\quad -80.6\quad -96.5]$

P9.28 $x = [-4\quad -3\quad -2\quad -1\quad 0\quad 1\quad 2\quad 3\quad 4\quad 5\ ]$
$y = [\ 8.0\quad 9.9\quad 1.8\quad 1.4\quad 1.7\quad 3.6\quad 7.5\quad 16.8\quad 14.1\quad 27.2]$

P9.29 $x = [-3.2\quad 0.0\quad -0.8\quad 1.6\quad 1.7\quad 4.6\quad -3.1\quad -3.9\quad 0.7\quad 4.7\ ]$
$y = [57.7\quad 4.4\quad 6.4\quad 17.6\quad 19.6\quad 101.6\quad 49.8\quad 86.2\quad 6.7\quad 117.1]$

P9.30 $x = [\ -3.8\quad -4.3\quad 3.5\quad -3.2\quad -4.7\quad 2.3\quad 0.4\quad -2.2\quad -1.3\quad -4.9]$
$y = [-91.8\quad -120.5\quad -50.4\quad -70.2\quad -125.1\quad -23.5\quad -4.1\quad -36.1\quad -18.1\quad -150.7]$

For Problems P9.31 to P9.45, plot the data, choose an appropriate form for the least-squares approximation function (linear, quadratic, cubic, exponential, reciprocal of linear, or reciprocal of quadratic) and then determine the best fit function of the chosen form.

P9.31 $x = [3.0\quad 0.5\quad 6.9\quad 6.5\quad 9.8\quad 5.5\quad 4.0\quad 2.0\quad 6.3\quad 7.3]$
$y = [5.8\quad -2.5\quad 22.1\quad 23.9\quad 36.5\quad 18.0\quad 10.3\quad 2.2\quad 18.4\quad 21.5]$

P9.32 $x = [\ 5.80\quad 1.40\quad 8.70\quad 3.30\quad 4.90\quad 4.30\quad 2.60\quad 9.80\quad 9.70\quad 9.50]$
$y = [10.90\quad 1.40\quad 15.30\quad 5.90\quad 11.10\quad 9.20\quad 3.30\quad 14.50\quad 17.00\quad 14.00]$

P9.33 $x = [-3.5\quad -1.2\quad -1.9\quad -3.3\quad 4.0\quad -1.8\quad 2.3\quad -0.9\quad -1.0\quad 0.1]$
$y = [32.2\quad -0.9\quad 4.9\quad 25.7\quad 81.2\quad 4.8\quad 28.5\quad -2.0\quad -2.3\quad -1.2]$

P9.34 $x = [\ -2.5\quad 3.0\quad 1.7\quad -4.9\quad 0.6\quad -0.5\quad 4.0\quad -2.2\quad -4.3\quad -0.2]$
$y = [-20.1\quad -21.8\quad -6.0\quad -65.4\quad 0.2\quad 0.6\quad -41.3\quad -15.4\quad -56.1\quad 0.5]$

P9.35 $x = [3.00\quad 1.80\quad 6.90\quad 2.60\quad 4.60\quad 8.40\quad 8.80\quad 7.00\quad 7.60\quad 9.70]$
$y = [0.08\quad 0.14\quad 0.03\quad 0.10\quad 0.05\quad 0.03\quad 0.02\quad 0.03\quad 0.03\quad 0.02]$

P9.36 $x = [4.70\quad 2.30\quad 3.20\quad 7.90\quad 6.30\quad 6.60\quad 5.40\quad 9.20\quad 7.80\quad 3.30]$
$y = [0.13\quad 0.33\quad 0.18\quad 0.10\quad 0.13\quad 0.07\quad 0.13\quad 0.08\quad 0.08\quad 0.17]$

P9.37 $x = [0.70\quad 3.10\quad 9.40\quad 9.80\quad 5.60\quad 9.90\quad 6.90\quad 2.40\quad 8.10\quad 9.30]$
$y = [0.36\quad 0.07\quad 0.02\quad 0.02\quad 0.04\quad 0.02\quad 0.03\quad 0.09\quad 0.03\quad 0.02]$

P9.38 $x = [\ 8.80\quad 4.90\quad 8.90\quad 7.60\quad 6.60\quad 9.70\quad 1.70\quad 1.40\quad 7.60\quad 3.10]$
$y = [-0.02\quad -0.04\quad -0.02\quad -0.02\quad -0.03\quad -0.02\quad -0.11\quad -0.13\quad -0.03\quad -0.06]$

P9.39 $x = [-0.20\quad 0.00\quad -2.10\quad -4.40\quad -2.40\quad -3.10\quad 4.20\quad -3.80\quad -4.90\quad -1.30]$
$y = [\ 0.42\quad 0.63\quad 0.05\quad 0.01\quad 0.04\quad 0.02\quad 0.02\quad 0.01\quad 0.01\quad 0.10]$

P9.40 $x = [4.40\quad -0.10\quad -4.10\quad 1.70\quad 0.10\quad -2.80\quad 2.30\quad -4.30\quad 4.60\quad -2.90]$
$y = [0.02\quad 1.23\quad 0.07\quad 0.08\quad 0.70\quad 0.18\quad 0.05\quad 0.04\quad 0.01\quad 0.14]$

P9.41 $x = [4.20\ 0.20\ -4.10\ 2.40\ -5.00\ 1.00\ 4.60\ -1.00\ 2.30\ 1.80]$
$y = [0.01\ 0.20\ \ 0.01\ \ 0.04\ \ 0.01\ 0.11\ 0.01\ \ 0.11\ 0.04\ 0.05]$

P9.42 $x = [-0.90\ 0.30\ 4.20\ -1.20\ 1.40\ -3.30\ 2.90\ -1.30\ 2.80\ -1.60]$
$y = [\ 1.10\ 1.70\ 6.40\ \ 0.90\ 2.60\ \ 0.50\ 4.80\ \ 0.90\ 4.80\ \ 0.80]$

P9.43 $x = [0.30\ 0.00\ -0.20\ -1.20\ 0.20\ 3.20\ -2.90\ -1.10\ -4.50\ -2.00]$
$y = [0.80\ 0.70\ \ 0.70\ \ 0.50\ 0.70\ 1.60\ \ 0.30\ \ 0.50\ \ 0.30\ \ 0.40]$

P9.44 $x = [-4.90\ 1.40\ -3.90\ -2.80\ 4.10\ -3.60\ -2.90\ 3.50\ 2.10\ -4.20]$
$y = [\ 0.20\ 1.70\ \ 0.30\ \ 0.50\ 3.90\ \ 0.30\ \ 0.50\ 3.40\ 2.60\ \ 0.40]$

P9.45 $x = [-2.00\ 1.00\ -3.20\ 2.30\ 0.50\ -3.40\ 4.30\ -3.00\ -4.90\ 1.90]$
$y = [\ 0.50\ 1.20\ \ 0.30\ 1.80\ 1.00\ \ 0.30\ 2.80\ \ 0.30\ \ 0.10\ 1.40]$

EXPLORE SOME APPLICATIONS

A9.1
The following data describe the product of a chemical reaction as a function of time:
$t = [0.00\ \ 0.10\ \ 0.40\ \ 0.50\ \ 0.60\ \ 0.90\ \ 1.00\ \ 1.10\ \ 1.40\ \ 1.50\ \ 1.60\ \ 1.90\ \ 2.00]$
$y = [0.00\ \ 0.06\ \ 0.17\ \ 0.19\ \ 0.21\ \ 0.25\ \ 0.26\ \ 0.27\ \ 0.29\ \ 0.29\ \ 0.30\ \ 0.31\ \ 0.31]$
Find a least squares approximating function for the data. Compare your function with the functions found in Examples 8.3 and 8.11, in which interpolation was used on a subset of these data.

A9.2
Use the following rounded data to find the linear relationship between C and F:
$C = [\ 50\ \ \ 45\ \ \ \ 40\ \ \ 35\ \ \ 30\ \ \ 25\ \ \ 20\ \ \ 15\ \ \ 10\ \ \ \ 5\ \ \ \ \ 0\ \ \ -5\ \ -10]$
$F = [122\ \ \ 113\ \ \ 104\ \ \ 95\ \ \ 86\ \ \ 77\ \ \ 68\ \ \ 59\ \ \ 50\ \ \ 41\ \ \ 32\ \ \ 23\ \ \ 14]$

A9.3
A laboratory experiment measured the height of a bouncing ball; the following data give the time and height of the peak of each bounce:
$t = [0.43\ \ \ 0.989\ \ 1.462\ \ 1.892\ \ 2.279\ \ 2.623\ \ 2.924\ \ 3.182]$
$h = [0.428\ \ 0.308\ \ 0.239\ \ 0.19\ \ \ \ 0.153\ \ 0.121\ \ 0.097\ \ 0.079]$
Find the best fit function of the form suggested by a plot of the data.

A9.4
According to Newton's law of cooling, the rate of change of the temperature of a cup of hot liquid (such as coffee) is proportional to the difference between the temperature and the surrounding air. The following data were obtained from a simple classroom laboratory experiment.

$$t = \begin{bmatrix} 0 & 5 & 10 & 15 & 20 & 25 & 30 & \ldots \\ 35 & 40 & 45 & 50 & 55 & 60 & \ldots \\ 65 & 70 & 75 & 80 & 85 & 90 & 95 \end{bmatrix};$$

$d = T - A$ (A = room temperature = 74 F)

$$d = \begin{bmatrix} 88.266 & 75.072 & 66.144 & 60.204 & 54.912 & 50.43 & 46.398 & \ldots \\ 42.51 & 38.748 & 35.562 & 32.646 & 29.802 & 27.228 & \ldots \\ 24.906 & 22.602 & 20.514 & 18.678 & 16.626 & 15.006 & 13.584 \end{bmatrix};$$

Using `Ln_Lin_LS`, find the coefficients in the "best fit" exponential function.

A9.5
An experiment was conducted to measure the weight of the water in a cylindrical can. The water drained out through a small hole in the bottom of the can. Find the quadratic best fit to the following data obtained from the experiment:

time (in seconds)
1.00 6.06 11.12 16.18 21.23 26.29 31.35 36.41 41.47 46.53 51.59 56.64
weight of water (in Newtons).
2.85 2.74 2.63 2.46 2.35 2.24 2.14 2.03 1.97 1.92 1.81 1.81

A9.6
An experiment that involved rolling a can up an inclined plane produced the following data for the distance of the can from the measuring device as a function of time:

t = [0.00 0.27 0.54 0.81 1.08 1.34 1.61 1.88 2.15 2.42 2.69 ... 2.96 3.23 3.49 3.76 4.03 4.30 4.57 4.84]
d = [1.91 1.82 1.61 1.43 1.28 1.18 1.10 1.04 1.00 0.99 1.00 ... 1.03 1.08 1.15 1.24 1.34 1.48 1.66 1.86]

Find the best fit quadratic function.

A9.7
An experiment to measure the intensity of light as a function of the distance from source of the light produced the following data:
d = [30 35 40 45 50 55 60 65 70 75];
i = [0.85 0.67 0.52 0.42 0.34 0.28 0.24 0.21 0.18 0.15];
Find the best fit exponential function and the best–fit quadratic function.

EXTEND YOUR UNDERSTANDING

U9.1 Show that the functions $P_2 = x^2 - 1/3$ and $P_3 = x^3 - 3/5\, x$ are orthogonal on $[-1, 1]$ with respect to the weight function $w(x) = 1$. Are the two functions orthonormal? Why or why not?

U9.2 Use the Gram–Schmidt process to construct the first four Legendre polynomials.

U9.3 Use the Gram–Schmidt process to construct the first four polynomials that are orthonomal on the interval $[0, 4]$.

For Problems U9.4 to U9.8, find the Padé approximation function with $k = 6$.

a. Take $m = 2$, and $n = 4$; that is, find the best approximation function of the form

$$\frac{a_2 x^2 + a_1 x + a_0}{b_4 x^4 + b_3 x^3 + b_2 x^2 + b_1 x + 1}$$

b. Take $m = 3$, and $n = 3$; that is, find the best approximation function of the form

$$\frac{a_3 x^3 + a_2 x^2 + a_1 x + a_0}{b_3 x^3 + b_2 x^2 + b_1 x + 1}$$

c. Take $m = 4$, $n = 2$; i.e., find the best approximation function of the form

$$\frac{a_4 x^4 + a_3 x^3 + a_2 x^2 + a_1 x + a_0}{b_2 x^2 + b_1 x + 1}$$

d. Compare the errors incurred in using the approximations from Parts a, b, and c on the interval $[-1, 1]$.

U9.4 $f(x) = \sin(x)$.
U9.5 $f(x) = \cos(x)$.
U9.6 $f(x) = \tan(x)$.
U9.7 $f(x) = e^x$.
U9.8 $f(x) = \ln(x + 1)$.

10
Fourier Methods

10.1 Fourier Approximation and Interpolation

10.2 Fast Fourier Transforms for n=2^r

10.3 Fast Fourier Transofrms for General n

10.4 Mathcad's Methods

In the previous two chapters, we interpolated data by polynomials and approximated data by polynomials, exponential functions, or rational functions, depending on the general characteristics of the data. However, for periodic data, it is more appropriate to use sine and cosine functions for the approximation or interpolation.

We begin this chapter with an investigation into data approximation and interpolation using trigonometric polynomials, also known as (finite) Fourier series. The approach is the same as that discussed previously, namely, we seek to minimize the total squared error by setting the partial derivatives of the error equal to zero. The formulas for the coefficients are found by using the appropriate orthogonality results for the sine and cosine functions. The coefficients are given by sums of the form

$$a_j = \frac{2}{n} \sum_{k=0}^{n-1} x_k \cos(j\, t_k)$$

or

$$b_j = \frac{2}{n} \sum_{k=0}^{n-1} x_k \sin(j\, t_k).$$

If we form a single complex quantity, $c_j = a_j + i\, b_j$, these summations can be combined to give one example of the discrete Fourier transform, which maps the data x_k to the transformed data c_j.

The *fast Fourier transform* (FFT), a computationally efficient method of computing the discrete Fourier transform, has greatly increased the feasibility of using this transform for large sets of data. In Section 10.2, we introduce the FFT in its most common setting, the case in which n is a power of 2; this is called a *radix*-2 FFT. The method is presented in both matrix and algebraic forms for $n = 4$, and a Mathcad function is also given. For higher powers of 2, we use Mathcad's built-in functions fft or FFT.

In Section 10.3, we describe the general FFT for the case when n is not a power of 2. The process is a generalization of the algebraic approach for the radix-2 transform. If the number of data points is not a power of 2, then the Mathcad functions cfft or CFFT are appropriate.

Example 10-A Waveforms for Different Instruments

The characteristic sound of different instruments playing the same pitch can be shown by the vibrational waveform of the sound. (See Fig. 10.1.) Such periodic vibrations are often analyzed by Fourier transform methods.

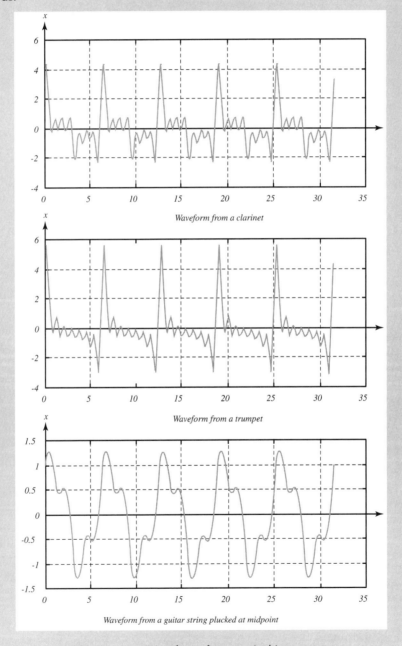

FIGURE 10.1   Waveforms from musical instruments.

Example 10-B Geometric Figures

Fourier transforms may be used to remove angular dependence in data. Consider, for example, describing various geometric figures, centered at the origin, by measuring the distance from the origin to the boundary of the figure at certain (evenly spaced) angles. In different rotations, a square is distinguishable from a triangle or a cross. (See Fig. 10.2.)

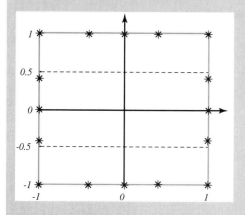

FIGURE 10.2A   Square

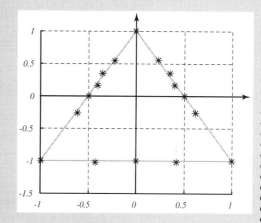

FIGURE 10.2B   Triangle

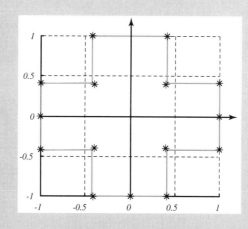

FIGURE 10.2C   Cross

The distances from the origin, measured at angular intervals of $\pi/8$, of these figures are as follows:

Square
[1.00 1.08 1.40 1.08 1.00 1.08 1.40 1.08 1.00 1.08 1.40 1.08 1.00 1.08 1.40 1.08]

Triangle
[0.50 0.43 0.50 0.60 1.00 0.60 0.50 0.43 0.50 0.67 1.40 1.10 1.00 1.10 1.40 0.67]

Cross
[1.00 1.08 0.56 1.08 1.00 1.08 0.56 1.08 1.00 1.08 0.56 1.08 1.00 1.08 0.56 1.08]

10.1 FOURIER APPROXIMATION AND INTERPOLATION

In order to approximate or interpolate a set of data using a trigonometric polynomial, i.e., a function of the form

$$f(t) = \frac{a_0}{2} + a_1\cos t + a_2\cos 2t + \ldots + a_m\cos mt$$
$$+ b_1\sin t + b_2\sin 2t + \ldots + b_m\sin mt,$$

we must find the coefficients $a_0, a_1, \ldots, a_m$, and $b_1, \ldots, b_m$. The function $f(t)$ is a trigonometric polynomial of degree m if a_m and b_m are not both zero.

We assume that the interval $[0, 2\pi)$ is divided into n equal subintervals and that we have the data values given at the points

$$t_0 = 0, \ t_1 = \frac{2\pi}{n}, \ t_2 = 2\frac{2\pi}{n}, \cdots, t_k = k\frac{2\pi}{n}, \cdots, t_{n-1} = (n-1)\frac{2\pi}{n}.$$

We denote the corresponding data values as $x_0, x_1, \ldots, x_{n-1}$.

The functional form for $f(t)$ is appropriate for a least-squares approximation problem when $2m + 1 < n$, i.e., when we have more data points than there are coefficients to be determined.

For exact interpolation, the appropriate form of the trigonometric polynomial depends on whether the number of data points is even or odd. If n is odd, the polynomial for exact interpolation has the same form as that for approximation (except that $n = 2m + 1$). However, if n is even ($n = 2m$), the polynomial for interpolation is

$$f(t) = \frac{a_0}{2} + a_1 \cos t + a_2 \cos 2t + \ldots + \frac{a_m}{2} \cos mt$$
$$+ b_1 \sin t + b_2 \sin 2t + \ldots + b_{m-1} \sin(m-1)t.$$

The reason for the difference in functional form will be made clear when we consider the derivation of the formulas for the coefficients.

The formulas for the coefficients ($j = 0, 1, \ldots, m$) are

$$a_j = \frac{2}{n} \sum_{k=0}^{n-1} x_k \cos(j\,t_k), \qquad b_j = \frac{2}{n} \sum_{k=0}^{n-1} x_k \sin(j\,t_k).$$

We do not have to solve a linear system, since the set of functions $\{1, \cos(t), \cos(2t), \ldots \cos(mt), \sin(t), \ldots \sin(mt)\}$ are orthogonal as long as the data are evenly spaced on an interval of length 2π; we assume that the data are given on $[0, 2\pi)$. The formulas are found in a manner analogous to that for continuous least squares with orthogonal polynomials in Chapter 9.

Algorithm for Fourier Approximation or Interpolation

Given **x**, the data values at $\mathbf{t} = \left[0, \dfrac{2\pi}{n}, \ldots, \dfrac{2k\pi}{n}, \ldots, \dfrac{2(n-1)\pi}{n}\right]$ find the coefficients for the trigonometric approximation function (if $n > 2m + 1$) or interpolation function (if $n = 2m$ or $n = 2m + 1$):

$$x = a_0 + a_1 \cos(t) + a_2 \cos(2t) + \cdots + a_m \cos(m\,t)$$
$$+ b_1 \sin(t) + \cdots + b_m \sin(mt)$$

Find coefficients

$$a_0 = \frac{1}{n} \sum_{k=1}^{n} x_k$$

for $j = 1$ to m

$$a_j = \frac{2}{n} \sum_{k=1}^{n} x_k \cos(j\,t_k)$$

$$b_j = \frac{2}{n} \sum_{k=1}^{n} x_k \sin(j\,t_k)$$

end

if $n = m$

$$a_m = \frac{a_m}{2}$$

end

The next examples show that for more data points, we have more terms in each summation. For exact-fit trigonometric polynomials, we have as many coefficients to be determined as there are data points.

Example 10.1 Trigonometric Interpolation

We consider the trigonometric polynomial that fits three data points exactly. The maximum degree we can use is $m = 1$, so we must find a_0, a_1, and b_1. The data for k, t, and x, are as shown in the following table:

k	0	1	2
t	0	$\frac{2\pi}{3}$	$\frac{4\pi}{3}$
x	0	$\frac{8\pi^3}{27}$	$\frac{-8\pi^3}{27}$

Also,

$$a_0 = \frac{1}{3}[x_0 \cos(0\, t_0) + x_1 \cos(0\, t_1) + x_2 \cos(0\, t_2)] = \frac{1}{3}[x_0 + x_1 + x_2] = 0,$$

$$a_1 = \frac{2}{3}[x_0 \cos(t_0) + x_1 \cos(t_1) + x_2 \cos(t_2)]$$

$$= \frac{2}{3}\left[0 \cos(0) + \frac{8\pi^3}{27} \cos\left(\frac{2\pi}{3}\right) + \frac{-8\pi^3}{27} \cos\left(\frac{4\pi}{3}\right)\right] = 0,$$

$$b_1 = \frac{2}{3}[x_0 \sin(t_0) + x_1 \sin(t_1) + x_2 \sin(t_2)]$$

$$= \frac{2}{3}\left[0 \sin(0) + \frac{8\pi^3}{27} \sin\left(\frac{2\pi}{3}\right) + \frac{-8\pi^3}{27} \sin\left(\frac{4\pi}{3}\right)\right] = \frac{16\pi^3 \sqrt{3}}{81}.$$

The interpolating polynomial is $x(t) = \dfrac{16\pi^3 \sqrt{3}}{81} \sin t$, shown in Fig. 10.3.

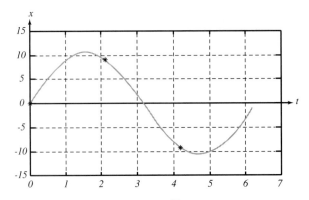

FIGURE 10.3 $x(t) = \dfrac{16\pi^3 \sqrt{3}}{81} \sin t$ and the data points.

Example 10.2 Trigonometric Approximation

We seek the least squares trigonometric polynomial, with $m = 1$, for the five data points shown in the following table (together with the sine and cosine values):

$$k = [0 \quad 1 \quad 2 \quad 3 \quad 4]$$
$$t_k = [0 \quad 2\pi/5 \quad 4\pi/5 \quad 6\pi/5 \quad 8\pi/5]$$
$$x_k = [0 \quad 48\pi^3/125 \quad 24\pi^3/125 \quad -24\pi^3/125 \quad -48\pi^3/125]$$
$$\cos(t_k) = [1 \quad 0.3090 \quad -0.8090 \quad -0.8090 \quad 0.3090]$$
$$\sin(t_k) = [0 \quad 0.9511 \quad 0.5878 \quad -0.5878 \quad -0.9511]$$

The formulas for the coefficients,

$$a_0 = \frac{1}{5}\sum_{k=0}^{4} x_k, \quad a_1 = \frac{2}{5}\sum_{k=0}^{4} x_k \cos(t_k), \quad b_1 = \frac{2}{5}\sum_{k=0}^{4} x_k \sin(t_k),$$

yield

$$a_0 = \frac{1}{5}[x_0 + x_1 + x_2 + x_3 + x_4] = 0$$

$$a_1 = \frac{2}{5}\left[0\cos(0) + \frac{48\pi^3}{125}\cos\left(\frac{2\pi}{5}\right) + \frac{24\pi^3}{125}\cos\left(\frac{4\pi}{5}\right)\right.$$
$$\left. + \frac{-24\pi^3}{125}\cos\left(\frac{6\pi}{5}\right) + \frac{-48\pi^3}{125}\cos\left(\frac{8\pi}{5}\right)\right] = 0$$

$$b_1 = \frac{2}{5}\left[0\sin(0) + \frac{48\pi^3}{125}\sin\left(\frac{2\pi}{5}\right) + \frac{24\pi^3}{125}\sin\left(\frac{4\pi}{5}\right)\right.$$
$$\left. + \frac{-24\pi^3}{125}\sin\left(\frac{6\pi}{5}\right) + \frac{-48\pi^3}{125}\sin\left(\frac{8\pi}{5}\right)\right] = 11.8584.$$

Figure 10.4 shows the approximating function, $f(t) = a_0 + a_1 \cos t + b_1 \sin t$.

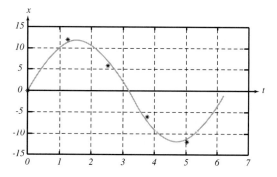

FIGURE 10.4 Approximating trigonometric polynomial.

In the following Mathcad function, the coefficients a_0 and (if necessary) a_m have been divided by 2.

Mathcad Function for Fourier Interpolation or Approximation

The use of the function Trig_poly is illustrated in the next example.

$$\text{Trig\_poly}(x, m) := \begin{array}{l} n \leftarrow \text{length}(x) \\ w \leftarrow \dfrac{2 \cdot \pi}{n} \\ \text{for } k \in 0 \mathinner{\ldotp\ldotp} n - 1 \\ \quad t_k \leftarrow \dfrac{2 \cdot \pi \cdot k}{n} \\ \text{for } k \in 0 \mathinner{\ldotp\ldotp} m - 1 \\ \quad a_k \leftarrow 0 \\ b \leftarrow a \\ \text{for } j \in 1 \mathinner{\ldotp\ldotp} m \\ \quad \begin{array}{|l} a_j \leftarrow x \cdot \cos(j \cdot t) \\ b_j \leftarrow x \cdot \sin(j \cdot t) \end{array} \\ a \leftarrow \dfrac{2 \cdot a}{n} \\ b \leftarrow \dfrac{2 \cdot b}{n} \\ a_0 \leftarrow \dfrac{\sum x}{n} \\ a_m \leftarrow \dfrac{a_m}{2} \quad \text{if } n = 2 \cdot m \\ c \leftarrow \text{stack}(a, b) \\ c \end{array}$$

Chapter 10 Fourier Methods

Example 10.3 A Step Function

To illustrate the use of the Mathcad function `Trig_poly`, consider the problem of interpolating a step function with eight data points.

Define data for a step function with 8 data points; m = 4 for interpolation

$$z := (1\ 1\ 1\ 1\ 0\ 0\ 0\ 0) \qquad m := 4 \qquad x := z^T \qquad c := \text{Trig\_poly}(x, m)$$

$$t := \left(0\ \ \frac{\pi}{4}\ \ \frac{\pi}{2}\ \ 3 \cdot \frac{\pi}{4}\ \ \pi\ \ 5 \cdot \frac{\pi}{4}\ \ 3 \cdot \frac{\pi}{2}\ \ 7 \cdot \frac{\pi}{4}\right)$$

$c^T =$

	0	1	2	3	4	5	6	7	8	9
0	0	0.5	0.25	0	0.25	0	0	0.604	0	0.104	0

Extract the coefficients a and b; display transpose to conserve space.

$$a := \text{submatrix}(c, 0, m, 0, 0) \qquad b := \text{submatrix}(c, m+1, 2 \cdot m + 1, 0, 0)$$

$$a^T = (0.5\ \ 0.25\ \ 0\ \ 0.25\ \ 0) \qquad b^T = (0\ \ 0.604\ \ 0\ \ 0.104\ \ 0)$$

Define the interpolating polynomial, and range variable for ploting

$$tt := 0, 0.1 .. 2 \cdot \pi$$

$$f(tt) := a_0 + a_1 \cdot \cos(tt) + a_2 \cdot \cos(2 \cdot tt) + a_3 \cdot \cos(3 \cdot tt) + b_1 \cdot \sin(tt) + b_2 \cdot \sin(2 \cdot tt) + b_3 \cdot \sin(3 \cdot tt)$$

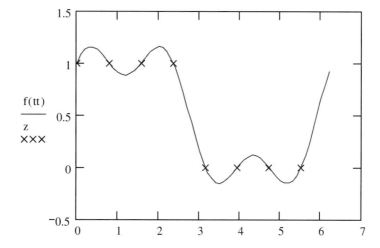

FIGURE 10.5 Data points and interpolating trigonometric polynomial.

Discussion

Derivation of Fourier Approximation and Interpolation Formulas

Fourier approximation and interpolation formulas can be derived in the same manner as polynomial and other least squares approximation formulas were in the previous chapter. However, the special properties of the sine and cosine functions at evenly spaced data points lead to simplifications that make the calculation of the coefficients more manageable.

Given n data at evenly spaced points in $[0, 2\pi)$, viz.,

$$\mathbf{t} = \left[0, \frac{2\pi}{n}, 2\frac{2\pi}{n}, \ldots, k\frac{2\pi}{n}, \ldots, (n-1)\frac{2\pi}{n} \right],$$

$$\mathbf{x} = [x_0, x_1, x_2, \ldots, x_k, \ldots, x_{n-1}],$$

we consider the trigonometric polynomial

$$p(t) = \frac{a_0}{2} + a_1 \cos t + a_2 \cos 2t + \ldots + a_m \cos mt$$

$$+ b_1 \sin t + b_2 \sin 2t + \cdots + b_m \sin mt$$

$$= \frac{a_0}{2} + \sum_{j=1}^{m} a_j \cos(j\, t_k) + \sum_{j=1}^{m} b_j \sin(j\, t_k).$$

where $2m + 1 \le n$.

We wish to minimize

$$E = \sum_{k=0}^{n-1} \left[\frac{a_0}{2} + \sum_{j=1}^{m} a_j \cos(j\, t_k) + \sum_{j=1}^{m} b_j \sin(j\, t_k) - x_k \right]^2.$$

We proceed by setting the partial derivatives of E with respect to each of the unknown coefficients to zero. We use the index J to denote the generic coefficient a_J or b_J during this derivation, to distinguish it from the index of summation. We will return to the more common lowercase later. Because of the restriction that $2m + 1 \le n$, we only need to consider $J \le n/2$.

Setting $\partial E / \partial a_0 = 0$ results in

$$\sum_{k=0}^{n-1} \left[\frac{a_0}{2} + \sum_{j=1}^{m} a_j \cos(j\, t_k) + \sum_{j=1}^{m} b_j \sin(j\, t_k) - x_k \right] = 0. \quad (10.1)$$

Setting $\partial E / \partial a_J = 0$ yields

$$\sum_{k=0}^{n-1} \left[\frac{a_0}{2} + \sum_{j=1}^{m} a_j \cos(j\, t_k) + \sum_{j=1}^{m} b_j \sin(j\, t_k) - x_k \right] \cos(J\, t_k) = 0. \quad (10.2)$$

And setting $\partial E / \partial b_J = 0$ gives

$$\sum_{k=0}^{n-1} \left[\frac{a_0}{2} + \sum_{j=1}^{m} a_j \cos(j\, t_k) + \sum_{j=1}^{m} b_j \sin(j\, t_k) - x_k \right] \sin(J\, t_k) = 0. \quad (10.3)$$

Equation (10.1) can be written as

$$\frac{n}{2}a_0 + \sum_{k=0}^{n-1}\sum_{j=1}^{m} a_j \cos(j\,t_k) + \sum_{k=0}^{n-1}\sum_{j=1}^{m} b_j \sin(j\,t_k) = \sum_{k=0}^{n-1} x_k,$$

which, after interchanging the order of the summations, becomes

$$\frac{n}{2}a_0 + \sum_{j=1}^{m} a_j \sum_{k=0}^{n-1} \cos(j\,t_k) + \sum_{j=1}^{m} b_j \sum_{k=0}^{n-1} \sin(j\,t_k) = \sum_{k=0}^{n-1} x_k.$$

However, since the data are evenly spaced, we have, for each j

$$\sum_{k=0}^{n-1} \cos(j\,t_k) = 0$$

and

$$\sum_{k=0}^{n-1} \sin(j\,t_k) = 0,$$

so that

$$\frac{n}{2}a_0 = \sum_{k=0}^{n-1} x_k$$

which gives

$$a_0 = \frac{2}{n} \sum_{k=0}^{n-1} x_k$$

The analysis of Eqs. (10.2) and (10.3) makes use of the fact that the functions:

$$1,\ \cos t,\ \cos 2t,\ \ldots\ \cos(n-1)t,\ \sin t,\ \ldots\ \sin(n-1)t$$

are orthogonal with respect to n evenly spaced data points in $[\,0, 2\pi\,)$, which gives the following results:

$$\sum_{k=0}^{n-1} \cos(J\,t_k) \cos(j\,t_k) = \begin{cases} 0, & \text{for } j \neq J, \\ n/2 & \text{for } j = J < n/2, \\ n, & \text{for } j = J = n/2, \end{cases}$$

$$\sum_{k=0}^{n-1} \cos(J\,t_k) \sin(j\,t_k) = 0,$$

$$\sum_{k=0}^{n-1} \sin(J\,t_k) \sin(j\,t_k) = \begin{cases} 0, & \text{for } j \neq J, \\ n/2, & \text{for } j = J < n/2, \\ 0, & \text{for } j = J = n/2, \end{cases}$$

The summations that are equal to zero for $j \neq J$ may be verified using the following basic trigonometric identities:

$$\cos j \cos J = (\cos(j - J) + \cos(j + J))/2;$$
$$\cos j \sin J = (\sin(j + J) - \sin(j - J))/2;$$
$$\sin j \sin J = (\cos(j - J) - \cos(j + J))/2.$$

When $j = J < n/2$, we use the half angle formulas

$$(\sin j)^2 = (1 - \cos 2j)/2$$

and

$$(\cos j)^2 = (1 + \cos 2j)/2$$

Finally, in the case of $j = J = n/2$ (which occurs only for the exact interpolation of an even number of data points),

$$\sum_{k=0}^{n-1} \cos^2(j\, t_k) = \sum_{k=0}^{n-1} \cos^2\left(\frac{n}{2} k \frac{2\pi}{n}\right) = \sum_{k=0}^{n-1} \cos^2(k\pi) = n.$$

On the other hand,

$$\sum_{k=0}^{n-1} \sin(J\, t_k) \sin(j\, t_k) = 0,$$

since all terms in the summation involve $\sin(k\pi)$, which is zero.

We now consider Eq. (10.2) in some detail; the simplification of (10.3) follows in a similar manner. Rearranging the terms in Eq. (10.2) yeilds

$$\sum_{k=0}^{n-1} \frac{a_0}{2} \cos(J\, t_k) + \sum_{k=0}^{n-1} \cos(J\, t_k) \sum_{j=0}^{m} a_j \cos(j\, t_k)$$

$$+ \sum_{k=0}^{n-1} \cos(J\, t_k) \sum_{j=0}^{m} b_j \sin(j\, t_k) = \sum_{k=0}^{n-1} \cos(J\, t_k)\, x_k.$$

We simplify each summation on the left hand side of the equation separately:

$$\sum_{k=0}^{n-1} \frac{a_0}{2} \cos(J\, t_k) = \frac{a_0}{2} \sum_{k=0}^{n-1} \cos(J\, t_k) = 0,$$

$$\sum_{k=0}^{n-1} \cos(J\, t_k) \sum_{j=0}^{m} a_j \cos(j\, t_k) = \sum_{j=0}^{m} a_j \sum_{k=0}^{n-1} \cos(j t_k) \cos(J\, t_k))$$

$$= \begin{cases} 0 & \text{for } j \neq J, \\ \dfrac{a_J\, n}{2} & \text{for } j = J < 2, \\ a_J\, n & \text{for } j = J = n/2, \end{cases}$$

$$\sum_{k=0}^{n-1} \cos(J\, t_k) \sum_{j=0}^{m} b_j \sin(j\, t_k) = \sum_{j=0}^{m} b_j \sum_{k=0}^{n-1} \sin(j\, t_k) \cos(J\, t_k)) = 0.$$

Thus, for $J < n/2$, the second equation gives us

$$\frac{n}{2} a_J = \sum_{k=0}^{n-1} \cos(J\, t_k)\, x_k,$$

or

$$a_J = \frac{2}{n} \sum_{k=0}^{n-1} \cos(J\, t_k)\, x_k.$$

For exact interpolation with an even number of data points, we have $n = 2m$. Rather than using a different formula for a_m in this case (which would be required by the difference in the value of the summation of $\cos^2(J\, t_k)$ when $J = n/2$), we choose, for $n = 2m$, to define the interpolating polynomial as

$$p(t) = \frac{a_0}{2} + a_1 \cos t + a_2 \cos 2t + \cdots + \frac{a_m}{2} \cos mt$$

$$+ b_1 \sin t + b_2 \sin 2t + \cdots + b_{m-1} \sin(m-1)t$$

$$= \frac{a_0}{2} + \sum_{j=1}^{m-1} a_j \cos(j\, t_k) + \frac{a_m}{2} \cos m t + \sum_{j=1}^{m-1} b_j \sin(j\, t_k)$$

Using this form, we find that all of the coefficients follow the same formulas:

$$a_j = \frac{2}{n} \sum_{k=0}^{n-1} \cos(j\, t_k)\, x_k \qquad b_j = \frac{2}{n} \sum_{k=0}^{n-1} \sin(j\, t_k)\, x_k \qquad j = 0, \ldots, m.$$

Note that the formula for b_j can be included for $j = 0$; however, b_0 is always zero, so it is not necessary to compute it. Similarly, for $n = 2m$ (interpolation of an even number of data points), we always have $b_m = 0$, since all terms in the summation are of the form $\sin(k\,\pi)$.

Data on Other Intervals

If the given data are evenly spaced on $[-\pi, \pi]$, the formulas for a_j and b_j in terms of t_k remain the same; however, the t_k change. Although the form of the polynomial and the formulas for the coefficients are the same, the values of the coefficients will be different because the t_k are different.

Data on any other interval may be transformed to $[0, 2\pi]$ or $[-\pi, \pi]$ by a linear transformation; the process is the same as for Gaussian integration (discussed in Chapter 11).

The Mathcad function for Fourier interpolation or approximation may be modified to use data that are evenly distributed on any interval of length 2π (say n data points on the interval $[t_0, t_0 + 2\pi]$) by using a range variable, with starting value t_0, second value $t_0 + 2\pi/n$, and final value $t_0 + 2(n-1)\pi/n$.

Example 10.4 Geometric Figures

The following data represent the distance from the origin to the perimeter of a square (shown in Figure 10.2a), measured at evenly spaced angular intervals t_k:

$$\mathbf{t} = [\,0 \quad \pi/4 \quad \pi/2 \quad 3\pi/4 \quad \pi \quad 5\pi/4 \quad 3\pi/2 \quad 7\pi/4\,],$$

$$\mathbf{x} = [\,1.0 \quad 1.4 \quad 1.0 \quad 1.4 \quad 1.0 \quad 1.4 \quad 1.0 \quad 1.4\,].$$

With the use of the Mathcad function `Trig_poly(x,m)`, the coefficients are found to be

$$\mathbf{a} = [\,1.2 \quad 0 \quad 0 \quad 0 \quad -0.2\,],$$

$$\mathbf{b} = [\,0 \quad 0 \quad 0 \quad 0 \quad 0\,].$$

We denote the interpolation function as $r(t)$ because it represents radial distance from the origin:

$$r(t) = 1.2 - 0.2 \cos(4\,t).$$

We can reconstruct a rough approximation of the original figure by plotting $x = r\cos(t)$, $y = r\sin(t)$, as shown in Fig. 10.6.

For the triangle shown in Figure 10.2b, (sampled at angular intervals of $\pi/4$), the data are $\mathbf{d} = [0.5 \quad 0.5 \quad 1.0 \quad 0.5 \quad 0.5 \quad 1.4 \quad 1.0 \quad 1.4]$.

The coefficients are

$$\mathbf{a} = [\,0.85 \quad 0.00 \quad -0.25 \quad 0.00 \quad -0.1\,],$$

$$\mathbf{b} = [\,0.00 \quad -0.3182 \quad 0.00 \quad -0.3182 \quad 0.0\,],$$

so the interpolation function for the radial distance is

$$r(t) = 0.85 - 0.25 \cos(2\,t) - 0.1 \cos(4\,t) - 0.3182 \sin(t)$$
$$- 3.182 \sin(3\,t).$$

The reconstructed triangle is shown in Fig. 10.7. Using more data points improves the figures, as we shall see later.

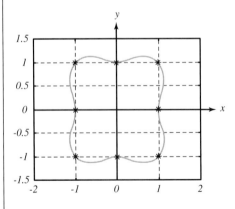

FIGURE 10.6 Reconstructed square.

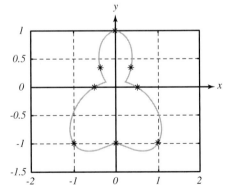

FIGURE 10.7 Reconstructed triangle.

10.2 FAST FOURIER TRANSFORMS FOR $n = 2^r$

Finding the coefficients of a Fourier polynomial leads to the problem of computing, for $j = 0, \ldots, m$,

$$a_j = \frac{2}{n} \sum_{k=0}^{n-1} x_k \cos(j\, t_k),$$

$$b_j = \frac{2}{n} \sum_{k=0}^{n-1} x_k \sin(j\, t_k).$$

The a_j and b_j can be combined into a single complex quantity

$$c_j = a_j + i\, b_j = \frac{2}{n} \sum_{k=0}^{n-1} x_k [\cos(j\, t_k) + i \sin(j\, t_k)], \quad \text{where } i = \sqrt{-1}.$$

These computations can be carried out more conveniently by working in the complex domain. Using Euler's formula relating the trigonometric and exponential functions for complex variables, we may write

$$c_j = \frac{2}{n} \sum_{k=0}^{n-1} x_k \exp(i\, j\, t_k).$$

Recalling that the points t_k are evenly spaced on $[0, 2\pi)$, we define $\omega = \dfrac{2\pi}{n}$ and write t_k as $k\, \omega$ to obtain

$$c_j = \frac{2}{n} \sum_{k=0}^{n-1} x_k \exp(i\, j\, k\, \omega).$$

The coefficients of the Fourier polynomial are the scaled real and imaginary parts of the discrete Fourier transform of the data, which we consider in the remainder of this section.

10.2.1 Discrete Fourier Transform

The discrete Fourier transform of a set of n complex data values z_k evenly spaced on $[0, 2\pi)$ is the set of complex numbers

$$g_j = \sum_{k=0}^{n-1} z_k \exp(i\, j\, k\, \omega), \quad \text{for } j = 0, \ldots, n-1.$$

This transformation has a wide range of applications beyond just the computation of interpolating coefficients discussed in the previous section. Since $e^{i\theta} = \cos \theta + i \sin \theta$, the periodicity of $\cos \theta$ and $\sin \theta$ cause the complex exponential to be periodic also. Direct computation of the g_j requires $O(n^2)$ operations. However, by exploiting the efficiencies of the fast Fourier transform, this can be reduced to $O(n \log n)$ operations.

10.2.2 Fast Fourier Transform

We begin by considering the FFT when n is a power of 2, i.e., $n = 2^r$; this is the case to which the FFT is most often applied. The basic idea in the FFT is to make use of the periodic nature of the complex exponential function and clever reordering of the computations, in order to reduce the total effort required to find the transform.

We define $w = \exp(i\,\omega)$ (with $\omega = \dfrac{2\pi}{n}$, as before). With this notation, we write

$$g_j = \sum_{k=0}^{n-1} z_k w^{jk}, \quad \text{for } j = 0, \ldots, n-1. \tag{10.4}$$

Note that $w^n = 1$, so that some simplifications can be achieved in some of the equations the g_j.

The cleverness in the computations results from the fact that the index on the components of the transform and the index on the summation both run from 0 to $n-1$. Each value of j ($0 \le j \le n-1$) can be written in binary form as $j = 2^{r-1} j_r + \ldots + 2^2 j_3 + 2 j_2 + j_1$, where each of the numbers $j_1, j_2, \ldots, j_r$ is either 0 or 1.

For example, if $n = 4$, we express the numbers $0, 1, \ldots, 3$ as $j = 2j_2 + j_1$:

j	j_2	j_1
0	0	0
1	0	1
2	1	0
3	1	1

We also write the numbers $k = 0, 1, \ldots, 3$ in binary form, but as $k = 2k_1 + k_2$. If we keep the values of k_1 and k_2 in the same order as for j_1 and j_2, the values of k appear in a scrambled form, which is useful for the FFT. Combining the representations of j and k in a single table, clarifies the relationship between the binary coefficients. This reordering of the k values relative to the j values is called *bit reversal*.

Binary coefficients in natural and scrambled order

j	j_2	j_1	k_2	k_1	k
0	0	0	0	0	0
1	0	1	0	1	2
2	1	0	1	0	1
3	1	1	1	1	3

There are several equivalent methods of presenting the FFT method. We first consider the matrix form, which we illustrate for $n = 4$. The resulting computations are shown schematically. Finally, the FFT method for $n = 4$ is developed from a more algebraic point of view, which allows generalization to the case when n has prime factors other than 2.

Algorithm for FFT with $n = 4$

Given $\mathbf{z} = [z_0, z_1, z_2, z_3]$, the data values at $\mathbf{t} = \left[0, \dfrac{\pi}{2}, \pi, \dfrac{3\pi}{2}\right]$ find FFT of the data.

Begin computation
for $h = 0$ to 3
    $w(h) = \exp(i\,\pi\,h/2)$
end

for $k_2 = 0$ to 1   (perform bit reversal)
    for $k_1 = 0$ to 1
        $zz(2\,k_2 + k_1) = z(2\,k_1 + k_2)$
    end
end

for $k_2 = 0$ to 1
    for $j_1 = 0$ to 1
        $s(2\,k_2 + j_1) = zz(2\,k_2)\,w(0) + zz(2\,k_2 + 1)\,w(2\,j_1)$
    end
end

for $j_2 = 0$ to 1
    for $j_1 = 0$ to 1
        $g(2\,j_2 + j_1) = s(j_1)\,w(0) + s(2 + j_1)\,w(j_1 + 2\,j_2)$
    end
end

To form trigonometric polynomial from transform of z, compute

$a_0 = \operatorname{Re}(g_1)/4$ $a_1 = 2\operatorname{Re}(g_2)/4$ $a_2 = \operatorname{Re}(g_3)/4$
$b_1 = 2\operatorname{Im}(g_2)/4$
$y = a_0 + a_1 \cos(t) + b_1 \sin(t) + a_2 \cos(2t)$

Matrix Form of FFT

We begin by writing out the linear system of equations for the Fourier transform components, for the case $n = 4$:

$$w^0 z_0 + w^0 z_1 + w^0 z_2 + w^0 z_3 = g_0,$$
$$w^0 z_0 + w^1 z_1 + w^2 z_2 + w^3 z_3 = g_1,$$
$$w^0 z_0 + w^2 z_1 + w^4 z_2 + w^6 z_3 = g_2,$$
$$w^0 z_0 + w^3 z_1 + w^6 z_2 + w^9 z_3 = g_3.$$

Making use of the fact that $w^4 = w^0 = 1$, to simplify the equations, and interchanging the order of the second and third equations (which corresponds to the previous interchange in the order of the j and k indices) gives the system

$$1z_0 + 1z_1 + 1z_2 + 1z_3 = g_0,$$
$$1z_0 + w^2z_1 + 1z_2 + w^2z_3 = g_2,$$
$$1z_0 + w^1z_1 + w^2z_2 + w^3z_3 = g_1,$$
$$1z_0 + w^3z_1 + w^2z_2 + w^1z_3 = g_3.$$

Writing these equations in matrix form, we have

$$\begin{bmatrix} 1 & 1 & 1 & 1 \\ 1 & w^2 & 1 & w^2 \\ 1 & w & w^2 & w^3 \\ 1 & w^3 & w^2 & w \end{bmatrix} \begin{bmatrix} z_0 \\ z_1 \\ z_2 \\ z_3 \end{bmatrix} = \begin{bmatrix} g_0 \\ g_2 \\ g_1 \\ g_3 \end{bmatrix} \qquad (10.5)$$

We now factor the coefficient matrix:

$$\begin{bmatrix} 1 & 1 & 0 & 0 \\ 1 & w^2 & 0 & 0 \\ 0 & 0 & 1 & w \\ 0 & 0 & 1 & w^3 \end{bmatrix} \begin{bmatrix} 1 & 0 & 1 & 0 \\ 0 & 1 & 0 & 1 \\ 1 & 0 & w^2 & 0 \\ 0 & 1 & 0 & w^2 \end{bmatrix} = \begin{bmatrix} 1 & 1 & 1 & 1 \\ 1 & w^2 & 1 & w^2 \\ 1 & w & w^2 & w^3 \\ 1 & w^3 & w^2 & w \end{bmatrix}.$$

Next, we carry out the computation of the g's in two steps. Substituting the factored form of the coefficient matrix into Eq. (10.5), we obtain

$$\begin{bmatrix} 1 & 1 & 0 & 0 \\ 1 & w^2 & 0 & 0 \\ 0 & 0 & 1 & w \\ 0 & 0 & 1 & w^3 \end{bmatrix} \begin{bmatrix} 1 & 0 & 1 & 0 \\ 0 & 1 & 0 & 1 \\ 1 & 0 & w^2 & 0 \\ 0 & 1 & 0 & w^2 \end{bmatrix} \begin{bmatrix} z_0 \\ z_1 \\ z_2 \\ z_3 \end{bmatrix} = \begin{bmatrix} g_0 \\ g_2 \\ g_1 \\ g_3 \end{bmatrix}. \qquad (10.6)$$

First we find the product

$$\begin{bmatrix} 1 & 0 & 1 & 0 \\ 0 & 1 & 0 & 1 \\ 1 & 0 & w^2 & 0 \\ 0 & 1 & 0 & w^2 \end{bmatrix} \begin{bmatrix} z_0 \\ z_1 \\ z_2 \\ z_3 \end{bmatrix} = \begin{bmatrix} z_0 + z_2 \\ z_1 + z_3 \\ z_0 + w^2 z_2 \\ z_1 + w^2 z_2 \end{bmatrix} = \begin{bmatrix} s_0 \\ s_1 \\ s_2 \\ s_3 \end{bmatrix}.$$

Then we form the second product

$$\begin{bmatrix} 1 & 1 & 0 & 0 \\ 1 & w^2 & 0 & 0 \\ 0 & 0 & 1 & w \\ 0 & 0 & 1 & w^3 \end{bmatrix} \begin{bmatrix} s_0 \\ s_1 \\ s_2 \\ s_3 \end{bmatrix} = \begin{bmatrix} s_0 + s_1 \\ s_0 + w^2 s_1 \\ s_2 + w s_3 \\ s_2 + w^3 s_3 \end{bmatrix} = \begin{bmatrix} g_0 \\ g_2 \\ g_1 \\ g_3 \end{bmatrix}.$$

The efficiencies in this computation result from the structure of the matrices that are employed in the two stages. The method is implemented not by using general matrix multiplication, but by performing only the necessary mutiplications and additions.

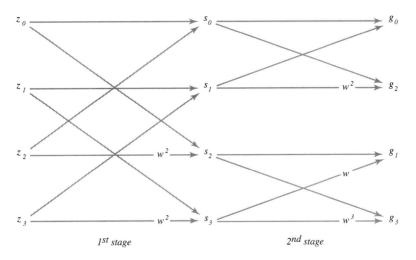

FIGURE 10.8 Two-stage computational procedure.

The computations are shown schematically in Figure 10.8. Pathways with powers of w on them indicate that the quantity on the left is multiplied by that amount.

Algebraic Form of FFT

We now consider an alternative approach to the FFT for $n = 2^r$. As before, each value of j ($0 \le j \le n-1$) is written in binary form, and each value of k is written in binary form with the bits reversed. For example, if $n = 4$, the binary coefficients are given in Table 10.1.

To calculate the discrete Fourier transform of the data z_k, i.e.,

$$g_j = \sum_{k=0}^{3} z_k \exp(i j k \omega) \quad \text{for } j = 0, \ldots 3;$$

we write $w = \exp(i \omega) = \exp\left(i \dfrac{\pi}{2}\right)$. Then the sum is given (as in Eq. 10.4) by

$$g_j = \sum_{k=0}^{3} z_k w^{j k}.$$

Note that $w^4 = 1$, and therefore, $w^{4a} = 1$ for any integer a. Using the binary factorizations of j and k, we have

$$g_j = g(j_1 + 2j_2) = \sum_{k=0}^{3} z(k_2 + 2 k_1) w^{(j_1 + 2 j_2)(k_2 + 2 k_1)}$$

$$= \sum_{k=0}^{3} z(k_2 + 2 k_1) w^{(j_1 + 2j_2)k_2} w^{(j_1 + 2j_2) k_1}$$

Chapter 10 Fourier Methods

$$= \sum_{k_2=0}^{1} \sum_{k_1=0}^{1} z(k_2 + 2k_1) \, w^{(j_1+2j_2)2k_1} \, w^{(j_1+2j_2)k_2}$$

$$= \sum_{k=0}^{1} \left[\sum_{k=0}^{1} z(k_2 + 2k_1) \, w^{(j_1)(2)(k_1)} \right] w^{(j_1+2j_2)k_2}.$$

Since $w = \exp\left(i\frac{2\pi}{4}\right) = \exp\left(i\frac{\pi}{2}\right) = \cos\frac{\pi}{2} + i\sin\frac{\pi}{2}$, we have

$$w^0 = 1; \quad w = 0 + i(1/2); \quad w^2 = -1; \quad w^3 = 0 - i(1/2).$$

We can use these results to simplify our final calculations, but first we observe that the way in which the original summation over all data points has been decomposed into two nested summations forms the basis for the FFT.

We first compute the inner summation, $\left[\sum_{k_1=0}^{1} z(k_2 + 2k_1) \, w^{(j_1)(2)(k_1)}\right]$, for each value of j, but we do so by using each possible value of j_1 (0 or 1) and each possible value of j_2 (0 or 1).

Writing the digits so that j is in natural order, we have $k = k_2 + 2k_1$ and $j = j_1 + 2j_2$; the first stage produces the values of $s(j_1 + 2\, k_2)$:

$$s(j_1 + 2\, k_2) = \sum_{k_1=0}^{1} z(k_2 + 2k_1) \, w^{(j_1)(2)(k_1)}$$

$$= z(k_2 + 0) \, w^{(j_1)(2)(0)} + z(k_2 + 2) \, w^{(j_1)(2)(1)}.$$

These values are summarized in the following table:

$z(k_2+2k_1)$	k_1	k_2	j_1	j_2	$s(j_1 + 2\, k_2)$
z_0	0	0	0	0	$z_0 w^{(0)(2)(0)} + z_2 w^{(0)(2)(1)} = s_0$
z_3	1	0	1	0	$z_0 w^{(1)(2)(0)} + z_2 w^{(1)(2)(1)} = s_1$
z_1	0	1	0	1	$z_1 w^{(0)(2)(0)} + z_3 w^{(0)(2)(1)} = s_2$
z_4	1	1	1	1	$z_1 w^{(1)(2)(0)} + z_3 w^{(1)(2)(1)} = s_3$

The results of this first summation are expressed in terms of the digits j_1 and k_2, since the summation over the possible values of k_1 has been performed, but the second summation, over k_2, remains to be done.

We now compute the outer summation,

$$g(j_1 + 2j_2) = \sum_{k_2=0}^{1} s(j_1 + 2\, k_2) \, w^{(j_1+2j_2)k_2}$$

$$= s(j_1 + 2(0)) w^{(j_1+2j_2)(0)} + s(j_1 + 2(1)) w^{(j_1+2j_2)(1)},$$

and its corresponding table of values:

s	k_1	k_2	j_1	j_2	g
s_0	0	0	0	0	$s_0 w^{(0+2(0))0} + s_2 w^{(0+2(0))1} = s_0 w^0 + s_2 w^0$
s_1	1	0	1	0	$s_1 w^{(1+2(0))0} + s_3 w^{(1+2(0))1} = s_1 w^0 + s_3 w^1$
s_2	0	1	0	1	$s_0 w^{(0+2(1))0} + s_2 w^{(0+2(1))1} = s_0 w^0 + s_2 w^2$
s_3	1	1	1	1	$s_1 w^{(1+2(1))0} + s_3 w^{(1+2(1))1} = s_1 w^0 + s_3 w^3$

The foregoing computations above are implemented in the Mathcad function FFT_4 for the fast Fourier transform on four data points.

Mathcad Function for FFT with $n = 4$

The following Mathcad function requires four data points.

$$\text{FFT\_4}(z) := \begin{array}{|l} n \leftarrow \text{length}(z) \\ \text{for } h \in 0..8 \\ \quad w_h \leftarrow \exp\left(i \cdot \pi \cdot \dfrac{h}{2}\right) \\ \text{for } k2 \in 0..1 \\ \quad \text{for } k1 \in 0..1 \\ \quad\quad zz_{2 \cdot k2 + k1} \leftarrow z_{2 \cdot k1 + k2} \\ \text{for } i \in 0..n-1 \\ \quad s_i \leftarrow 0 \\ \quad g_i \leftarrow 0 \\ \text{for } k2 \in 0..1 \\ \quad \text{for } j1 \in 0..1 \\ \quad\quad \text{for } k1 \in 0..1 \\ \quad\quad\quad s_{2 \cdot k2 + j1} \leftarrow s_{2 \cdot k2 + j1} + zz_{2 \cdot k2 + k1} \cdot w_{j1 \cdot 2 \cdot k1} \\ \text{for } j2 \in 0..1 \\ \quad \text{for } j1 \in 0..1 \\ \quad\quad \text{for } k2 \in 0..1 \\ \quad\quad\quad g_{2 \cdot j2 + j1} \leftarrow g_{2 \cdot j2 + j1} + s_{2 \cdot k2 + j1} \cdot w_{(j1 + 2 \cdot j2) \cdot k2} \\ g \end{array}$$

The use of the function FFT_4 for finding the trig interpolating polynomial for a step function with 4 data points is illustrated in the next example.

Example 10.5 FFT with four points.

Find trigonometric polynomial to interpolate the following data, which represents function values at equally spaced points in the interval $[0, 2\pi)$

$$z := \begin{pmatrix} 1 \\ 1 \\ 0 \\ 0 \end{pmatrix} \qquad g := FFT\_4(z) \qquad n := \text{length}(z)$$

Define coefficients of trig polynomial from the Fourier transform

$$a := \begin{pmatrix} \dfrac{\text{Re}(g_0)}{n} \\ \dfrac{2 \cdot \text{Re}(g_1)}{n} \\ \dfrac{\text{Re}(g_2)}{n} \end{pmatrix} \qquad b_1 := \dfrac{2 \cdot \text{Im}(g_1)}{n}$$

Define the trig polynomial, range variable tt for plotting, and the locations of data points

$$f(tt) := a_0 + a_1 \cdot \cos(tt) + b_1 \cdot \sin(tt) + a_2 \cdot \cos(2 \cdot tt)$$

$$tt := 0, 0.1 .. 2 \cdot \pi \qquad i := 0 .. 3 \qquad t_i := i \cdot \dfrac{\pi}{2}$$

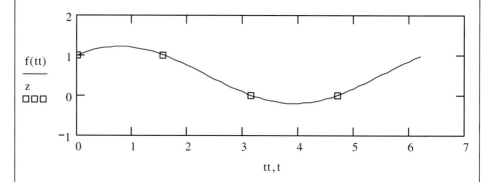

FIGURE 10.9 Fourier interpolation function found from FFT.

10.3 FAST FOURIER TRANSFORMS FOR GENERAL n

Although the FFT is most often applied for $n = 2^r$, the general FFT does not require the factorization of n to have any specific form. We designate the prime factorization of n as $n = r_1 r_2 \ldots r_t$, where we assume that $r_1 \leq r_2 \leq \ldots \leq r_t$. In a generalization of the binary representation of a number, each value of j ($0 \leq j \leq n-1$) can be written in terms of the prime factors of n as $j = r_1 \ldots r_{t-1} j_t + \ldots + r_1 r_2 j_3 + r_1 j_2 + j_1$. The possible values for j_1 are $0, 1, \ldots, r_1 - 1$; for j_2 are $0, 1, \ldots, r_2 - 1$; and for j_t are $0, 1, \ldots, r_t - 1$.

We also write the numbers $k = 0, 1, \ldots n-1$ in terms of the factors of n, but as $k = r_t \ldots r_2 k_1 + \ldots + r_t r_{t-1} k_{t-2} + r_t k_{t-1} + k_t$. The coefficient k_s can take on values $0, 1, \ldots, r_s - 1$.

For example, if $n = 6$ we have $r_1 = 2$ and $r_2 = 3$. We express the numbers $0, 1, \ldots, 5$ as $j = r_1 j_2 + j_1 = 2 j_2 + j_1$; j_1 is 0 or 1, and j_2 is 0, 1, or 2. The following table gives j in terms of j_2 and j_1:

j	j_2	j_1
0	0	0
1	0	1
2	1	0
3	1	1
4	2	0
5	2	1

For this example, $k = r_2 k_1 + k_2 = 3 k_1 + k_2$; k_1 is 0 or 1, k_2 is 0, 1, or 2, and the table is as follows:

k	k_1	k_2
0	0	0
1	0	1
2	0	2
3	1	0
4	1	1
5	1	2

Note that the possible values for k_1 and j_1 are the same, for k_2 and j_2 are the same, etc. However, k_2 is the lowest-order digit in the representation of k, whereas j_2 is the highest-order digit in the representation of j. In what follows, we perform the operations in each computation by letting j_1 and j_2 run through all possible values. We also let k_1 and k_2 run through their range of values, in the same order as j_1 and j_2; this generates all possible values of k also, but in a scrambled order, as shown in the following table.

Chapter 10 Fourier Methods

The Integers 0, . . . 5 Expressed in Terms of Prime Factors.

$j = 2j_2 + j_1$	j_2	j_1	k_1	k_2	$k = 3k_1 + k_2$
0	0	0	0	0	0
1	0	1	1	0	3
2	1	0	0	1	1
3	1	1	1	1	4
4	2	0	0	2	2
5	2	1	1	2	5

Now let us consider the calculation of the discrete Fourier transform of the data z_k, where $w = \exp\left(i\frac{2\pi}{n}\right)$, in which $n = r_1 r_2$. We have

$$g_j = \sum_{k=0}^{n-1} z_k \exp(ijk\omega) = \sum_{k=0}^{n-1} z_k w^{jk} \quad \text{for } j = 0, \ldots, n-1.$$

Since $w^n = 1$, any term of the form $w^{r_1 r_2 a}$ is equal to unity, and can be dropped from the calculations in the formulas that follow. Using the preceding factorization of j and k, we have

$$g_j = g(r_1 j_2 + j_1) = \sum_{k=0}^{n-1} z(r_2 k_1 + k_2) w^{(j_1 + r_1 j_2)(k_2 + r_2 k_1)}$$

$$= \sum_{k=0}^{n-1} z(r_2 k_1 + k_2) w^{(j_1 + r_1 j_2)k_2} w^{(j_1 + r_1 j_2)r_2 k_1}$$

$$= \sum_{k_2=0}^{r_2-1} \sum_{k_1=0}^{r_1-1} z(r_2 k_1 + k_2) w^{(j_1 + r_1 j_2) r_2 k_1} w^{(j_1 + r_1 j_2)k_2}$$

$$= \sum_{k_2=0}^{r_2-1} \left[\sum_{k_1=0}^{r_1-1} z(r_2 k_1 + k_2) w^{j_1 r_2 k_1} \right] w^{(j_1 + r_1 j_2)k_2}$$

The result of first stage is

$$s(2k_2 + j_1) = \sum_{k=0}^{1} z(3k_1 + k_2) w^{3j_1 k_1} = z(3(0) + k_2)w^{3j_1(0)} + z(3(1) + k_2)w^{3j_1(1)},$$

so that we have the following table:

$z(k_2+3k_1)$	k_1	k_2	j_1	j_2	$s(2k_2 + j_1)$
z_0	0	0	0	0	$z_0 w^{(0)(3)(0)} + z_3 w^{(0)(3)(1)} = s_0$
z_3	1	0	1	0	$z_0 w^{(1)(3)(0)} + z_3 w^{(1)(3)(1)} = s_1$
z_1	0	1	0	1	$z_1 w^{(0)(3)(0)} + z_4 w^{(0)(3)(1)} = s_2$
z_4	1	1	1	1	$z_1 w^{(1)(3)(0)} + z_4 w^{(1)(3)(1)} = s_3$
z_2	0	2	0	2	$z_2 w^{(0)(3)(0)} + z_5 w^{(0)(3)(1)} = s_4$
z_5	1	2	1	2	$z_2 w^{(1)(3)(0)} + z_5 w^{(1)(3)(1)} = s_5$

Chapter 10 Fourier Methods

We next compute the outer summation,

$$g(2j_2 + j_1) = \sum_{k_2=0}^{3-1} s(2k_2 + j_1)w^{(j_1+2j_2)k_2}$$

$$= s(2(0) + j_1)w^{(j_1+2j_2)0} + s(2(1) + j_1)w^{(j_1+2j_2)1} + s(2(2) + j_1)w^{(j_1+2j_2)2},$$

yielding the following table:

j	k_1	k_2	j_1	j_2	s	g
0	0	0	0	0	s_0	$s_0 w^{(0+2(0))0} + s_2 w^{(0+2(0))1} + s_4 w^{(0+2(0))2}$
1	1	0	1	0	s_3	$s_1 w^{(1+2(0))0} + s_3 w^{(1+2(0))1} + s_5 w^{(1+2(0))2}$
2	0	1	0	1	s_1	$s_0 w^{(0+2(1))0} + s_2 w^{(0+2(1))1} + s_4 w^{(0+2(1))2}$
3	1	1	1	1	s_4	$s_1 w^{(1+2(1))0} + s_3 w^{(1+2(1))1} + s_5 w^{(1+2(1))2}$
4	0	2	0	2	s_2	$s_0 w^{(0+2(2))0} + s_2 w^{(0+2(2))1} + s_4 w^{(0+2(2))2}$
5	1	2	1	2	s_5	$s_1 w^{(1+2(2))0} + s_3 w^{(1+2(2))1} + s_5 w^{(1+2(2))2}$

Simplifying, we obtain the final table:

j	k_1	k_2	j_1	j_2	s	g
0	0	0	0	0	s_0	$s_0 w^0 + s_2 w^0 + s_4 w^0$
1	1	0	1	0	s_3	$s_1 w^0 + s_3 w^1 + s_5 w^2$
2	0	1	0	1	s_1	$s_0 w^0 + s_2 w^2 + s_4 w^4$
3	1	1	1	1	s_4	$s_1 w^0 + s_3 w^3 + s_5 w^6$
4	0	2	0	2	s_2	$s_0 w^0 + s_2 w^4 + s_4 w^8$
5	1	2	1	2	s_5	$s_1 w^0 + s_3 w^5 + s_5 w^{10}$

Since

$$w = \exp\left(i\frac{2\pi}{n}\right) = \exp\left(i\frac{\pi}{3}\right),$$

we have

$$w^0 = 1 = w^6 = \exp\left(i\frac{6\pi}{3}\right) = 1$$

$$w = \exp\left(i\frac{\pi}{3}\right) = \cos\frac{\pi}{3} + i\sin\frac{\pi}{3} = 1/2 + i\sqrt{3}/2$$

$$w^2 = \exp\left(i\frac{2\pi}{3}\right) = \cos\frac{2\pi}{3} + i\sin\frac{2\pi}{3} = -1/2 + i\sqrt{3}/2 = w^8$$

$$w^3 = \exp(i\pi) = -1$$

$$w^4 = \exp\left(i\frac{4\pi}{3}\right) = \cos\frac{4\pi}{3} + i\sin\frac{4\pi}{3} = -1/2 - i\sqrt{3}/2 = w^{10}$$

$$w^5 = \exp\left(i\frac{5\pi}{3}\right) = \cos\frac{5\pi}{3} + i\sin\frac{5\pi}{3} = 1/2 - i\sqrt{3}/2.$$

Algorithm for FFT with $n = rs$ (data evenly spaced on $[0, 2\pi)$)

Given $\mathbf{z} = [z_0, z_1, z_2, z_3, \ldots z_{n-1}]$, where $n = rs$, r and s prime, $r \leq s$. The data values $\mathbf{z}$ are given at $\mathbf{t} = \left[0, \dfrac{2\pi}{n}, \cdots \dfrac{2(n-1)\pi}{n}\right]$.

Find FFT for $z = [z_1, z_2, z_3, \ldots z_n]$ where n has 2 prime factors

Begin computation
for $h = 0$ to $2n$
$\quad w(h) = \exp(2\pi i h/n)$
end
for $k_2 = 0$ to $s - 1$ (perform bit reversal)
$\quad$for $k_1 = 0$ to $r - 1$
$\quad\quad zz(r k_2 + k_1) = z(s k_1 + k_2)$
$\quad$end
end

for $k_2 = 0$ to $s - 1$
$\quad$for $j_1 = 0$ to $r - 1$
$\quad\quad ss(r k_2 + j_1) = \displaystyle\sum_{k_1=0}^{r-1} zz(r k_2 + k_1) w(s j_1 k_1)$
$\quad$end
end

for $j_2 = 0$ to $s - 1$
$\quad$for $j_1 = 0$ to $r - 1$
$\quad\quad g(r j_2 + j_1) = \displaystyle\sum_{k_2=0}^{s-1} s(r k_2 + j_1) w(k_2 (j_1 + r j_2))$
$\quad$end
end

Example 10.6 FFT for Six Data Points

We illustrate the use of the FFT for six data points by finding the interpolation function for the data $\mathbf{z} = [0\ 1\ 2\ 3\ 2\ 1]$. From before, for $n = 6$, $k = 3 k_1 + k_2$ and $j = 3 j_2 + j_1$. The data are listed in scrambled order so that the final results (the g's) are in natural order.

First, we compute the inner sum for each pair of values of j_1 and k_2:

$$s(2k_2 + j_1) = \sum_{k_1=0}^{2-1} z(3k_1 + k_2) w^{j_1(3)k_1}$$

$$= z(0 + k_2) w^{j_1(3)(0)} + z(3 + k_2) w^{j_1(3)(1)}.$$

This yields the following table:

z	k	j	k_1	k_2	j_1	j_2	$s(2k_2 + j_1)$
0	0	0	0	**0**	**0**	0	$0w^0 + 3w^0 = 3 = s_0$
3	3	1	1	**0**	**1**	0	$0w^0 + 3w^3 = -3 = s_1$
1	1	2	0	**1**	**0**	1	$1w^0 + 2w^0 = 3 = s_2$
2	4	3	1	**1**	**1**	1	$1w^0 + 2w^3 = -1 = s_3$
2	2	4	0	**2**	**0**	2	$2w^0 + 1w^0 = 3 = s_4$
1	5	5	1	**2**	**1**	2	$2w^0 + 1w^3 = 1 = s_5$

Next, we compute the outer sum

$$g_j = \sum_{k_2=0}^{3-1} s(2k_2 + j_1) w^{(j_1 + 2j_2)(k_2)}$$

$$= s(0 + j_1)(w^{(j_1 + 2j_2)(0)}) + s(2 + j_1)w^{(j_1 + 2j_2)(1)} + s(4 + j_1)w^{(j_1 + 2j_2)(2)}.$$

The associated table is as follows:

j	k_1	k_2	j_1	j_2	$g(2j_2 + j_1)$	
0	0	0	0	0	$s_0 w^0 + s_2 w^{0+} s_4 w^0$	$g_0 = 9$
1	1	0	1	0	$s_1 w^0 + s_3 w^{1+} s_5 w^2$	$g_1 = -4$
2	0	1	0	1	$s_0 w^0 + s_2 w^{2+} s_4 w^4$	$g_2 = 0$
3	1	1	1	1	$s_1 w^0 + s_3 w^{3+} s_5 w^6$	$g_3 = -1$
4	0	2	0	2	$s_0 w^0 + s_2 w^{4+} s_4 w^8$	$g_4 = 0$
5	1	2	1	2	$s_1 w^0 + s_3 w^{5+} s_5 w^{10}$	$g_5 = -4$

To use these results to find the interpolating trigonometric polynomial, we note that, in general, $a = 2\mathrm{Re}(g)/n$; in addition, a_0 (and a_m when $n = 2m$) is divided by 2. In this example, all g's are real, so all b's are zero. We have

$$a_0 = g_0/6 = 3/2; \quad a_1 = g_1/3 = -4/3; \quad a_2 = g_2/3 = 0; \quad a_3 = g_3/6 = -1/6.$$

The data and the interpolation function are shown in Figure 10.10.

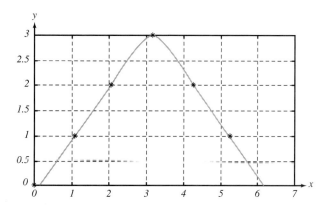

FIGURE 10.10 Data and interpolation function for $n = 6$.

Mathcad Function for FFT with $n = rs$ (data evenly spaced on $[0, 2\pi)$)

The number of data points for the following Mathcad function must be the product of two primes; the prime factors are supplied as input to the function.

$$
\begin{aligned}
\text{FFT\_rs}(z,r,s) := \quad & n \leftarrow \text{length}(z) \\
& \text{for } h \in 0..\, 2 \cdot r \cdot s \\
& \quad w_h \leftarrow \exp\left(i \cdot \pi \cdot 2 \cdot \frac{h}{n}\right) \\
& \text{for } k2 \in 0..\, s-1 \\
& \quad \text{for } k1 \in 0..\, r-1 \\
& \quad\quad zz_{r \cdot k2 + k1} \leftarrow z_{s \cdot k1 + k2} \\
& \text{for } i \in 0..\, n-1 \\
& \quad \left| \begin{array}{l} ss_i \leftarrow 0 \\ g_i \leftarrow 0 \end{array} \right. \\
& \text{for } k2 \in 0..\, s-1 \\
& \quad \text{for } j1 \in 0..\, r-1 \\
& \quad\quad \text{for } k1 \in 0..\, r-1 \\
& \quad\quad\quad ss_{r \cdot k2 + j1} \leftarrow ss_{r \cdot k2 + j1} + zz_{r \cdot k2 + k1} \cdot w_{j1 \cdot s \cdot k1} \\
& \text{for } j2 \in 0..\, s-1 \\
& \quad \text{for } j1 \in 0..\, r-1 \\
& \quad\quad \text{for } k2 \in 0..\, s-1 \\
& \quad\quad\quad g_{r \cdot j2 + j1} \leftarrow g_{r \cdot j2 + j1} + ss_{r \cdot k2 + j1} \cdot w_{(j1 + r \cdot j2) \cdot k2} \\
& g
\end{aligned}
$$

The use of the function FFT_rs is illustrated in the next example, which uses the same data as the previous example.

Example 10.7 FFT for Six Data Points

Find the trigonometric polynomial to interpolate the following data, which represent function values at equally spaced points ti in the interval $[0, 2\pi)$.

$$x := (0 \quad 1 \quad 2 \quad 3 \quad 2 \quad 1) \qquad z := x^T \qquad s := 3 \qquad r := 2$$

$$g := \text{FFT\_rs}(z, r, s) \qquad n := \text{length}(z) \qquad i := 0, 1..5 \qquad t_i := i \cdot \frac{2\pi}{n}$$

Define the coefficients of the interpolating polynomial

$$a_0 := \frac{\text{Re}(g_0)}{n} \qquad a_1 := \frac{2 \cdot \text{Re}(g_1)}{n} \qquad a_2 := \frac{2\,\text{Re}(g_2)}{n} \qquad a_3 := \frac{\text{Re}(g_3)}{n}$$

$$b_1 := \frac{2 \cdot \text{Im}(g_1)}{n} \qquad b_2 := \frac{2 \cdot \text{Im}(g_2)}{n}$$

$$g^T = (9 \quad -4 \quad 0 \quad -1 \quad 2.22i \times 10^{-15} \quad -4)$$

$$a^T = (1.5 \quad -1.333 \quad 0 \quad -0.167) \qquad b^T = (0 \quad 0 \quad 0)$$

$$tt := 0, 0.1.. 2\cdot\pi$$

$$f(tt) := a_0 + a_1 \cdot \cos(tt) + b_1 \cdot \sin(tt) + a_2 \cdot \cos(2\cdot tt) + a_3 \cdot \cos(3\cdot tt) + b_2 \cdot \sin(2\cdot tt)$$

plot function and data points

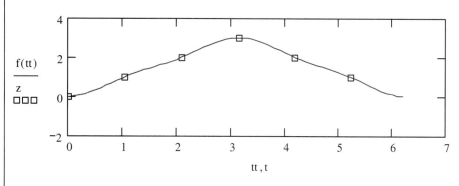

FIGURE 10.11 Trigonometric interpolating polynomial.

We also demonstrate the use of FFT_rs for a step function with 6 data points in the next example.

Example 10.8 FFT for Six Data Points

Find the trigonometric polynomial to interpolate the following data, which represent function values at equally spaced points ti in the interval $[0, 2\pi)$.

$x := (1\ 1\ 1\ 0\ 0\ 0)$ $z := x^T$ $s := 3$ $r := 2$

$g := \text{FFT\_rs}(z, r, s)$ $n := \text{length}(z)$ $i := 0, 1 .. 5$ $t_i := i \cdot \dfrac{2\pi}{n}$

Define the coefficients of the interpolating polynomial

$a_0 := \dfrac{\text{Re}(g_0)}{n}$ $a_1 := \dfrac{2 \cdot \text{Re}(g_1)}{n}$ $a_2 := \dfrac{2\,\text{Re}(g_2)}{n}$ $a_3 := \dfrac{\text{Re}(g_3)}{n}$

$b_1 := \dfrac{2 \cdot \text{Im}(g_1)}{n}$ $b_2 := \dfrac{2 \cdot \text{Im}(g_2)}{n}$

$g^T = (3\ \ 1 + 1.732i\ \ 0\ \ 1\ \ 0\ \ 1 - 1.732i)$

$a^T = (0.5\ \ 0.333\ \ 0\ \ 0.167)$ $b^T = (0\ \ 0.577\ \ 0)$

$tt := 0, 0.1 .. 2 \cdot \pi$

$f(tt) := a_0 + a_1 \cdot \cos(tt) + b_1 \cdot \sin(tt) + a_2 \cdot \cos(2 \cdot tt) + a_3 \cdot \cos(3 \cdot tt) + b_2 \cdot \sin(2 \cdot tt)$

plot function and data points

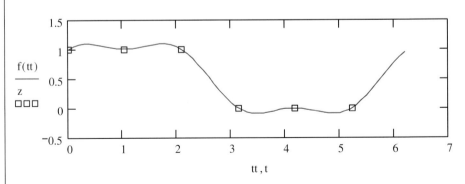

FIGURE 10.12 Trigonometric interpolating polynomial for step function.

Chapter 10 Fourier Methods

10.4 MATHCAD'S METHODS

Mathcad has several functions for performing Fast Fourier Transforms and the corresponding Inverse Fast Fourier Transform. The appropriate form of FFT depends on whether or not the number of data points is a power of 2, whether the data is real or complex valued, and on the desired scaling of the transform. We describe the use of the functions first, and then consider briefly the algorithms on which they are based.

10.4.1 Using the Built-In Functions

The FFT functions fft(**v**) and FFT(**v**), and the corresponding inverse FFT functions, ifft(**v**) and IFFT(**v**) are the basic FFT functions in Mathcad. The functions fft(**v**) and FFT(**v**) return the fast discrete Fourier transform of real data. The input vector has 2^n elements, representing measurements at equally spaced points in the time domain. The function returns a vector with $2^{n-1} + 1$ elements, giving the Fourier transform of the input data. The functions fft(**v**) and FFT(**v**) are the same except for the convention used for the initial factor of the transform, and whether the results of the transform are conjugated.

The function fft(**v**) uses a positive exponent in going from the time domain to the frequency domain. (The inverse transform therefore has a negative exponent.)

The function fft(**v**) returns the vector **c**, where i is the imaginary unit, m is the number of elements in **v**, and

$$c_j = \frac{1}{\sqrt{m}} \sum_{k=0}^{m-1} v_k \, e^{2\pi i (j/m) k}$$

This is the same as the FFT formulas given in Section 10.2, except that in 10.2 we use no normalization on the transform (i.e., we do not multiply the summation by $1/\sqrt{m}$.)

We illustrate the use of fft for a two dimensional interpolation problem in the next example. FFT are useful in pattern recognition because the spectral radius of the transform **z** (the spectral radius is a scaled form of **z***conj(**z**)) remains the same when a figure is rotated. Note that in using the function fft to construct an interpolating polynomial, the result must be scaled by $2/\sqrt{n}$ before forming the coefficients of the polynomial, to compensate for the difference in the normalizations used in Section 10.2 and in fft.

The function FFT(**v**) differs from fft(**v**) in the normalization of the transform, and the choice of sign on the exponent. FFT(**v**) uses a negative exponent in going from the time domain to the frequency domain (and a positive exponent for the inverse transform).

The function FFT(**v**) returns the vector **c**, where i is the imaginary unit, m is the number of elements in **v**, and

$$c_j = \frac{1}{m} \sum_{k=1}^{m} v_k e^{-2\pi i(j/m)k}$$

The inverse transform is given by

$$v_k = \sum_{j=1}^{m} c_j e^{2\pi i(k/m)k}$$

The formulas for these functions are based on [Bracewell, 1986]. The functions fft(**v**) and FFT(**v**) are based on the Cooley-Tukey algorithm, in which the data is stored in bit-reversed order before the transform is computed.

If the sampling frequency is f_s, the frequency corresponding to c_k is $f_k = \frac{k}{m} f_s$.

For further discussion of issues of sampling frequency, see a text on digital signal processing.

More General Fast Fourier Transform Functions

The FFT functions cfft(A) and CFFT(A), and the corresponding inverse FFT functions, icfft(B) and ICFFT(B) are more general than the functions fft(**v**), FFT(**v**), ifft(**v**) and IFFT(**v**). In particular, the functions cfft(A) and CFFT(A) do not require that the data to be transformed are real, and they do not require that the number of data points is a power of 2. These methods work more efficiently when the number of data points can be factored into many, small factors. These functions are based on [Singleton, 1986]. The normalization and sign of exponent conventions for cfft(A) are the same as for fft(**v**). The conventions for CFFT(A) are the same as for FFT(**v**).

10.4.2 Understanding the Algorithms

The standard presentation of the Cooley-Tukey FFT (also called decimation in time) is based on the Danielson-Lanczos Lemma, which shows that a discrete Fourier transform of length N can be expressed as the sum of two transforms of length $N/2$. Applied recursively, this provides an efficicient method of computing the transform when the number of data points is a power of 2. The Fourier transform of a single element is just the identity operation.

The basic recursion is given in the following, in which the number of data points is $n = 2^r$; we write (for $j = 0, \ldots, n-1$)

$$g_j = \sum_{k=0}^{n-1} z_k \exp(2\pi i\, j\, k/n)$$

$$= \sum_{k=0}^{n/2-1} z_{2k} \exp(2\pi i\, j2k/n) + \sum_{k=0}^{n/2-1} z_{2k+1} \exp(2\pi i\, j(2k+1)/n)$$

$$= \sum_{k=0}^{n/2-1} z_{2k} \exp(2\pi i\, jk/(n/2)) + w^j \sum_{k=0}^{n/2-1} z_{2k+1} \exp(2\pi i\, jk/(n/2))$$

$$= g_j^e + W^j g_j^o$$

where $W^j = \exp(2\pi i j/n)$, and g_j^e and g_j^o denote the jth component of the transform of the vectors of length $n/2$ formed by the even components of z_k and the odd components of z_k, respectively.

Example 10.9 Waveform of a Clarinet

We sample the waveform of a clarinet over one period, at 16 evenly spaced points (from 0 to $2\pi - \pi/8$), and use the data together with the Mathcad function fft to find the interpolation function as shown in Figure 10.13.

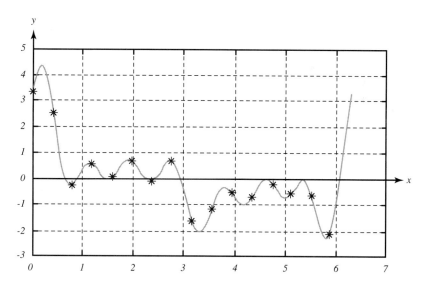

FIGURE 10.13   Waveform of a clarinet.

We apply the function fft to the data representing the amplitude of the wave; i.e.,

```
yy =
  3.304   2.5535 -0.22396   0.56711   0.067979 0.69711 -0.14572   0.70309
 -1.635  -1.1464 -0.55064  -0.70925  -0.18781 -0.55497 -0.62887  -2.1101
```

Chapter 10 Fourier Methods

Example 10.10 Interpolation of a Triangle in the Plane

Define data: $y := (0.5 \ 0.43 \ 0.5 \ 0.6 \ 1.0 \ 0.6 \ 0.5 \ 0.43 \ 0.5 \ 0.67 \ 1.4 \ 1.1 \ 1.0 \ 1.1 \ 1.4 \ 0.67)$

Perform Fast Fourier Transform: $n := \text{length}(y^T) \quad z := \text{fft}(y^T) \quad zz := \dfrac{z \cdot 2}{\sqrt{n}}$

Determine Fourier Coefficients: $nn := \dfrac{n}{2} \quad i := 0..nn$

$a_i := \text{Re}(zz_i) \quad b_i := \text{Im}(zz_i) \quad a_0 := \dfrac{a_0}{2} \quad a_{nn} := \dfrac{a_{nn}}{2}$

Generate trigonometric polynomial: $k := 0..100 \quad t_k := \dfrac{k \cdot 2 \cdot \pi}{100}$

$$r_k := a_0 + \sum_{j=1}^{nn} \left(a_j \cos(j \cdot t_k) + b_j \sin(j \cdot t_k) \right)$$

Find x and y coordinates from trignometric polynomial:

$xx_k := r_k \cdot \cos(t_k) \qquad yy_k := r_k \cdot \sin(t_k) \qquad |z| = 3.242$

Generate x and y coordinates of the original data:

$j := 0..2 \cdot nn - 1 \quad tt_j := j \cdot \dfrac{\pi}{nn} \quad xd_j := (y^T)_j \cdot \cos(tt_j) \quad yd_j := (y^T)_j \cdot \sin(tt_j)$

Plot the data points and trig interpolating function:

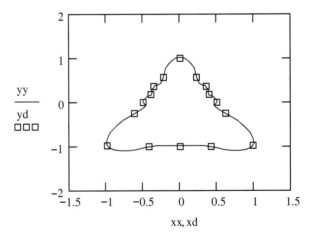

FIGURE 10.14 Data and interpolating curve.

Example 10.11 Rotation-Invariant Figures

We conclude our investigation of the FFT by illustrating its use in pattern recognition, where the figures we wish to identify may appear in different orientations. Using the data from Example 10-B, we can reconstruct a reasonable approximation to the original figures. (See Fig. 10.15.)

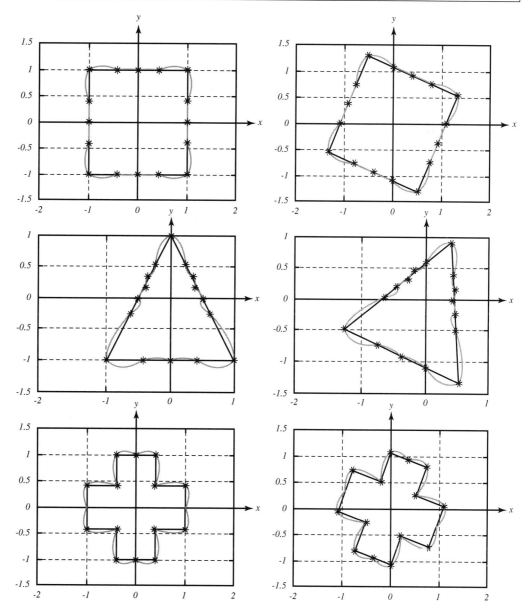

FIGURE 10.15 Square, triangle, and cross reconstructed from measured data.

Chapter 10 Fourier Methods

SUMMARY

Consider data values $x_0, x_1, \ldots, x_{n-1}$ given at points evenly spaced on the interval $[0, 2\pi)$; i.e., at $t_0 = 0$, $t_1 = \dfrac{2\pi}{n}, \ldots, t_k = k\dfrac{2\pi}{n}, \ldots, t_{n-1} = (n-1)\dfrac{2\pi}{n}$.

If $n > 2m + 1$, the trigonometric polynomial

$$f(t) = \frac{a_0}{2} + a_1 \cos t + a_2 \cos 2t + \ldots + a_m \cos mt$$
$$+ b_1 \sin t + b_2 \sin 2t + \ldots + b_m \sin mt,$$

with coefficients

$$a_j = \frac{2}{n} \sum_{k=0}^{n-1} x_k \cos(j\, t_k) \quad \text{and} \quad b_j = \frac{2}{n} \sum_{k=0}^{n-1} x_k \sin(j\, t_k).$$

gives the least squares approximation to the data.

If $n = 2m + 1$, the polynomial for exact interpolation has the same form as that for approximation.

If $n = 2m$, the trigonometric polynomial for interpolation is

$$f(t) = \frac{a_0}{2} + a_1 \cos t + a_2 \cos 2t + \ldots + \frac{a_m}{2} \cos mt$$
$$+ b_1 \sin t + b_2 \sin 2t + \ldots + b_{m-1} \sin(m-1)t.$$

The formulas for the coefficients are the same as for approximation.

SUGGESTIONS FOR FURTHER READING

The following books are recommended for additional discussion of trigonometric interpolation, approximation, and FFT:

Bloomfield, P., *Fourier Analysis of Time Series—An Introduction*, Wiley, New York, 1976.

Bracewell, R. *The Fourier Transform and Its Applications*, Mc-Graw Hill, 1986.

Briggs, W. L. and V. E. Henson, *The DFT: An Owner's Manual for the Discrete Fourier Transform*, SIAM, Philadelphia, 1995.

Bringham, E. O., *The Fast Fourier Transform*, Prentice-Hall, Englewood Cliffs, 1974.

Elliott, D. F. and K. R. Rao, *Fast Transforms: Algorithms, Analyses, Applications*, Academic Press, New York, 1982.

Golub, G. H., and C. F. Van Loan, *Matrix Computations* (3d ed.), Johns Hopkins University Press, Baltimore, 1996.

Nussbaumer, H. J., *Fast Fourier Transform and Convolution Algorithms*, Springer-Verlag, New York, 1982.

Ralston, A., and P. Rabinowitz, *A First Course in Numerical Analysis* (2d ed.), McGraw-Hill, New York, 1978.

Singleton, R.C., *Communications of ACM,* vol. 11, no. 11, 1986.

Singleton, R.C., *IEEE Transactions on Audio and Electroacoustics,* vol. AU-15, pp. 91–97, 1967.

Van Loan, C., *Computational Frameworks for the Fast Fourier Transform,* SIAM, New York, 1992.

PRACTICE THE TECHNIQUES

In Problems P10.1 to P10.5, assume that the data are evenly distributed on $[0, 2\pi)$.

    a. Find the best-fit trigonometric polynomial of degree $m = 1$.

    b. Find the trigonometric interpolating polymonial.

P10.1 $x = [1\ 1\ 0\ 0]$
P10.2 $x = [0\ 1\ 0\ -1]$
P10.3 $x = [0\ 1\ 1\ 0]$
P10.4 $x = [0\ 1/2\ 1\ 1/2]$
P10.5 $x = [0\ 1/3\ 2/3\ 1]$

In Problems P10.6 to P10.10, assume that the data are evenly distributed on $[0, 2\pi)$.

    a. Find the best fit trigonometric polynomial of degrees $m = 1$, and $m = 2$.

    b. Find the trigonometric interpolating polymonial.

P10.6 $x = [1\ 1\ 1\ 0\ 0\ 0]$
P10.7 $x = [0\ 1/3\ 2/3\ 1\ 2/3\ 1/3]$
P10.8 $x = [0\ 1/2\ 1\ 0\ -1\ -1/2]$
P10.9 $x = [1\ 1\ 1\ 1\ 0\ 0\ 0\ 0]$
P10.10 $x = [0\ 1/2\ 1\ 1/2\ 0\ -1/2\ -1\ -1/2]$

Problems P10.11 to P10.20: Use the data values from Problems P10.1 to P10.10, but assume that the data are evenly distributed on $[-\pi, \pi)$.

Problems P10.21 to P10.30: Find the FFT of the data in Problems P10.1 to P10.10, evenly distributed on the interval $[0, 2\pi)$.

EXPLORE SOME APPLICATIONS

A10.1 Find the waveform and power spectrum for the following sounds:

    a. A guitar string plucked at the midpoint;

y = [2.2118 2.1734 1.1321 1.2247 0.8702 0.9085 0.6849 0.5923 2.2118 2.1734 1.1321 1.2247 0.8702 0.9085 0.6849 0.5923]

    b. A guitar string plucked one-fourth of the way from one end of the string;

y = [3.0572 3.2011 2.0739 2.3860 1.7887 0.9738 0.7410 0.5787 1.1081 0.9605 0.4487 0.1454 −0.1605 0.6580 0.7410 0.8945]

    c. A guitar string plucked fairly close to one end;

y = [4.6101 5.3621 2.6796 1.3951 1.4639 0.9843 0.9115 1.0058 0.4658 0.4313 0.6075 0.1599 0.1480 −0.0899 −1.0885 0.5493]

    d. A trumpet;

y = [3.9408 3.8613 1.1210 1.0305 0.5567 0.7601 0.4008 0.6906 0.0800 0.2362 0.2165 0.1217 0.0398 −0.2403 −0.6988 −0.8033]

A 10.1 Find the best fit trigonometric polynomial of degree $m = 1$ for the following data, which give the monthly (average daily) high temperature at various locations (such data are available from the Internet, travel brochures, and a variety of other sources):

	Jan	Feb	Mar	Apr	May	Jun	Jul	Aug	Sep	Oct	Nov	Dec
Paris	43	45	54	60	68	73	76	75	70	60	50	44
Amsterdam	40	42	49	56	64	70	72	71	67	57	48	42
Prague	31	34	44	54	64	70	73	72	65	53	42	34
St. Petersburg, Russia	19	22	32	46	59	68	70	69	60	48	35	26
Rome	52	55	59	66	74	82	87	86	79	71	61	55
Madrid	47	52	59	65	70	80	87	85	77	65	55	48
Marrakesh, Morocco	65	68	74	79	84	92	101	100	92	83	73	66
Palermo, Sicily	60	62	63	68	74	81	85	86	83	77	71	64
Ankara, Turkey	39	42	51	63	73	78	86	87	78	69	57	43
Shanghai, China	46	47	55	66	77	82	90	90	82	74	63	53
Reykjavik, Iceland	35	37	39	43	50	54	57	56	52	45	39	36
Edinburgh, Scotland	42	43	46	51	56	62	65	64	60	54	48	44
London, England	43	44	50	56	62	69	71	71	65	58	50	45
Toronto, Canada	30	30	37	50	63	73	79	77	69	56	43	33
Santa Fe, New Mexico	40	43	51	59	68	78	80	79	73	62	50	40
Melbourne, Florida	73	73	75	79	83	86	89	90	89	86	80	74
Aiken, South Carolina	58	61	65	74	82	88	92	91	89	81	72	62

The data in Problems A10.3 to A10.20 give the radial distance from the origin to the perimeter of a geometric figure. For each problem

a. Find the Fourier transform of the data.
b. Find the power spectrum of the data ($z*conj(z)$).
c. Reconstruct the figure represented by each set of data.

A10.3

$\mathbf{y} = [1.40\ 1.08\ 1.00\ 1.08\ 1.40\ 1.08\ 1.00\ 1.08\ 1.40$
$1.08\ 1.00\ 1.08\ 1.40\ 1.08\ 1.00\ 1.08]$

A10.4

$\mathbf{y} = [1.00\ 1.08\ 1.40\ 1.08\ 1.00\ 1.08\ 1.40\ 1.08\ 1.00$
$1.08\ 1.40\ 1.08\ 1.00\ 1.08\ 1.40\ 1.08]$

A10.5

$\mathbf{y} = [0.5\ 0.6\ 1.0\ 0.6\ 0.5\ 0.43\ 0.5\ 0.67\ 1.4\ 1.1\ 1.0$
$1.1\ 1.4\ 0.67\ 0.5\ 0.43]$

A10.6

$\mathbf{y} = [1.0\ 0.6\ 0.5\ 0.43\ 0.5\ 0.67\ 1.4\ 1.1\ 1.0\ 1.1\ 1.4$
$0.67\ 0.5\ 0.43\ 0.5\ 0.6]$

A10.7

$\mathbf{y} = [0.56\ 1.08\ 1.00\ 1.08\ 0.56\ 1.08\ 1.00\ 1.08\ 0.56$
$1.08\ 1.00\ 1.08\ 0.56\ 1.08\ 1.00\ 1.08]$

A10.8

$\mathbf{y} = [1.00\ 1.08\ 0.56\ 1.08\ 1.00\ 1.08\ 0.56\ 1.08\ 1.00$
$1.08\ 0.56\ 1.08\ 1.00\ 1.08\ 0.56\ 1.08]$

A10.9

$\mathbf{y} = [2.83\ 2.16\ 2.00\ 1.18\ 0.94\ 0.90\ 1.00\ 1.36\ 2.83$
$2.16\ 2.00\ 1.18\ 0.94\ 0.90\ 1.00\ 1.36]$

A10.10
y = [1.26 1.06 1.00 1.06 1.26 1.67 2.00 1.67 1.26 1.06 1.00 1.06 1.26 1.67 2.00 1.67]

A10.11
y = [2.00 0.81 0.57 0.81 2.00 0.81 0.57 0.81 2.00 0.81 0.57 0.81 2.00 0.81 0.57 0.81]

A10.12
y = [0.57 0.81 2.00 0.81 0.57 0.81 2.00 0.81 0.57 0.81 2.00 0.81 0.57 0.81 2.00 0.81]

A10.13
y = [1.08 1.00 1.08 1.40 1.08 1.00 1.08 1.40 1.08 1.00 1.08 1.40 1.08 1.00 1.08 1.40]

A10.14
y = [1.08 1.00 1.08 0.56 1.08 1.00 1.08 0.56 1.08 1.00 1.08 0.56 1.08 1.00 1.08 0.56]

A10.15
y = [1.67 1.26 1.06 1.00 1.06 1.26 1.67 2.00 1.67 1.26 1.06 1.00 1.06 1.26 1.67 2.00]

A10.16
y = [1.36 2.83 2.16 2.00 1.18 0.94 0.90 1.00 1.36 2.83 2.16 2.00 1.18 0.94 0.90 1.00]

A10.17
y = [0.81 0.57 0.81 2.00 0.81 0.57 0.81 2.00 0.81 0.57 0.81 2.00 0.81 0.57 0.81 2.00]

A10.18
y = [2.00 1.67 1.26 1.06 1.00 1.06 1.26 1.67 2.00 1.67 1.26 1.06 1.00 1.06 1.26 1.67]

A10.19
y = [1.00 1.36 2.83 2.16 2.00 1.18 0.94 0.90 1.00 1.36 2.83 2.16 2.00 1.18 0.94 0.90]

A10.20
y = [0.6 1.0 0.6 0.5 0.43 0.5 0.67 1.4 1.1 1.0 1.1 1.4 0.67 0.5 0.43 0.5]

A10.21 Using your results from Part a of Problems A10.3 to A10.20, determine which data represent the same figure (in different rotations).

EXTEND YOUR UNDERSTANDING

U10.1 Modify the Mathcad function `trig_poly` to use n data points evenly spaced on an arbitrary interval of length 2π (i.e., on $[t_0, t_0 + 2\pi)$).

U10.2 Use Mathcad to transform (evenly spaced) data on $[a, b]$ to the interval $[0, 2\pi)$; use the Mathcad function `trig_poly` to find the coefficients for the best fit trigonometric polynomial, and display the plot in terms of the original independent variable.

U10.3 Write a Mathcad function `FFT8` to implement the FFT for $n = 8$.

U10.4 Write an algorithmic description of the process for Fourier least-squares approximation and interpolation.

U10.5 Construct the table showing the bit-reversal scheme for $n = 8$.

U10.6 Write a Mathcad function `FFT_R3` for the case in which $n = r_1 r_2 r_3$.

In Problems U10.7 to U10.10, use your Mathcad function `FFT8` *or Mathcad's built-in function* `fft` *to find the interpolative trigonometric polynomial for the given data.*

U10.7 z = [0 0.5 1 0.5 0 −0.5 −1 −0.5]

U10.8 $z = \left[1 \quad \frac{\sqrt{2}}{2} \quad 1 \quad \frac{\sqrt{2}}{2} \quad 1 \quad \sqrt{2} \quad 1 \quad \sqrt{2}\right]$

U10.9 $z = \left[0 \quad \frac{1}{4} \quad 1/2 \quad 3/4 \quad 1 \quad 3/4 \quad 1/2 \quad \frac{1}{4}\right]$

U10.10 z = [0 0 1 1 1 1 0 0]

U10.11 Show that the matrix form of the FFT can be applied with the order of the data shuffled, so that the transform components appear in natural order. Use the factorization of the coefficient matrix to create a schematic diagram for the two-stage computations.

11
Numerical Differentiation and Integration

11.1 Differentiation
11.2 Basic Numerical Integration
11.3 Better Numerical Integration
11.4 Gaussian Quadrature
11.5 Mathcad's Methods

We now turn our attention to the use of numerical methods for solving problems from calculus and differential equations. In this chapter, we investigate numerical techniques for finding derivatives and definite integrals. Several formulas for approximating a first or second derivative by a difference quotient are given. These formulas can be found with the use of Taylor polynomials or Lagrange interpolation polynomials.

Numerical methods approximate the definite integral of a given function by a weighted sum of function values at specified points. We first consider several methods, known as Newton-Cotes formulas, that use evenly spaced data points. These methods are based on the integral of a simple interpolating polynomial. The trapezoid rule uses the function values at the ends of the interval of integration; Simpson's rule is based on a parabola through the ends of the interval and the midpoint of the interval. Improved accuracy can be obtained by subdividing the interval of integration and applying one of these simple techniques on each subinterval. Finally, we present a powerful integration technique, Gaussian quadrature, in which the points used in evaluting a function are chosen to provide the best possible result for a certain class of functions.

Applications of numerical differentiation are especially common in converting differential equations into difference equations for numerical solution. One must be very careful when using numerical techniques to estimate the rate of change of measured data, since small errors are exaggerated by differentiation. Integration, on the other hand, tends to smooth out errors. Numerical integration is widely used in applications, because some simple functions are difficult or impossible to integrate exactly. We present a few representative problems and use them, together with other examples, to illustrate the techniques of the chapter. We consider techniques for ordinary differential equations in Chapters 12 to 14 and for partial differential equations in Chapter 15.

Example 11-A Simple Functions That Do Not Have Simple Antiderivatives

The normal distribution is a very important function in statistics. Gaussian noise is one of many ways in which this function is used in engineering and science. The normal distribution function (Fig. 11.1) is a scaled form of the function

$$f(x) = e^{-x^2}.$$

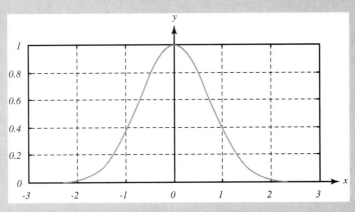

FIGURE 11.1 $f(x) = e^{-x^2}$

The indefinite integral of this function cannot be represented as a simple function. Instead, we find numerical approximations to the area under the graph of the function between any two finite values of x ($x = a$ and $x = b$), that is,

$$A = \int_a^b e^{-x^2}\, dx.$$

Another function that is important in optics and other applications, but does not have a simple antiderivative, is $f(x) = \sin(x)/x$; its graph is illustrated in Fig. 11.2.

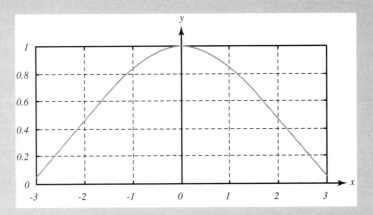

FIGURE 11.2 $f(x) = \sin(x)/x$

Example 11-B Length of an Elliptical Orbit

The simplest form of planetary orbit is an ellipse. Example 2-B in Chapter 2 describes the motion of a planet in an elliptical orbit with an eccentricity of 0.5 according to Kepler's law. If, instead of wishing to find the position of the planet at certain times, we desire to find the length of the orbit or the distance traveled between certain positions (measured by the central angle), we are faced with an example of the difficulty of calculating arc length, even for fairly simple functions. The well known formula for the arc length of a curve described parametrically as $x(r), y(r)$ is

$$L = \int_a^b \sqrt{(x')^2 + (y')^2}\, dr.$$

If $x(r) = \cos(r)$ and $y(r) = \dfrac{3}{4} \sin(r)$, the function to be integrated is

$$f(r) = 0.25 \sqrt{16 \sin^2(r) + 9 \cos^2(r)}.$$

We will find approximate values for the length of this ellipse using several different techniques.

For example, we may wish to compare the length of the arc traversed from $t = 0$ to $t = 10$ with the arc from $t = 60$ to $t = 70$. (See Fig. 11.3.) The central angles at 10 day intervals were found in Chapter 2 to be

$$r = [0.00 \quad 1.07 \quad 1.75 \quad 2.27 \quad 2.72 \quad 3.14 \quad 3.56 \quad 4.01 \quad 4.53 \quad 5.22 \quad 6.28],$$

so we are interested in the arc length from $r = 0.00$ to $r = 1.07$ and the length from $r = 3.56$ to $r = 4.01$.

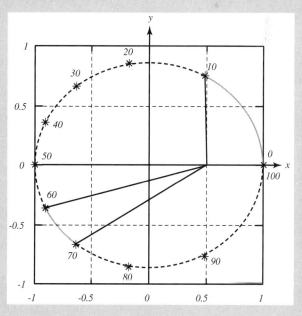

FIGURE 11.3 Length of arc from day 0 to day 10 and day 60 to day 70.

11.1 DIFFERENTIATION

Numerical differentiation requires us to find estimates for the derivative or slope of a function by using the function values at only a set of discrete points. We begin by considering methods of approximating a first derivative. We then present formulas for second and higher derivatives. The final topic in our treatment of numerical differentiation is the use of acceleration (introduced in Chapter 1) to improve an approximate derivative value.

11.1.1 First Derivatives

The simplest difference formulas are based on using a straight line to interpolate the given data; that is, they use two data points to estimate the derivative. We assume that we have function values at x_{i-1}, x_i, and x_{i+1}; we let $f(x_{i-1}) = y_{i-1}$, $f(x_i) = y_i$, and $f(x_{i+1}) = y_{i+1}$. The spacing between the values of x is constant, so that $h = x_{i+1} - x_i = x_i - x_{i-1}$. Then we have the standard two-point formulas:

Forward difference formula

$$f'(x_i) \approx \frac{f(x_{i+1}) - f(x_i)}{x_{i+1} - x_i} = \frac{y_{i+1} - y_i}{x_{i+1} - x_i} = \frac{y_{i+1} - y_i}{h}$$

Backward difference formula

$$f'(x_i) \approx \frac{f(x_{i-1}) - f(x_i)}{x_{i-1} - x_i} = \frac{y_i - y_{i-1}}{x_i - x_{i-1}} = \frac{y_i - y_{i-1}}{h}$$

A more balanced approach gives an approximation to the derivative at x_i using function values $f(x_{i-1})$ and $f(x_{i+1})$. Taking the average of the approximations from the forward and backward difference formulas gives the central difference formula.

Central difference formula

$$f'(x_i) \approx \frac{f(x_{i+1}) - f(x_{i-1})}{x_{i+1} - x_{i-1}} = \frac{y_{i+1} - y_{i-1}}{x_{i+1} - x_{i-1}} = \frac{y_{i+1} - y_{i-1}}{2h}$$

The three kinds of difference formula are shown in Fig. 11.4.

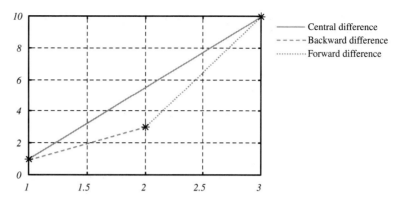

FIGURE 11.4 Three difference approximations to $f'(x_i)$.

Example 11.1 Forward, backward, and central differences

To illustrate the three kinds of difference formula, consider the data points $(x_0, y_0) = (1, 2)$, $(x_1, y_1) = (2, 4)$, $(x_2, y_2) = (3, 8)$, $(x_3, y_3) = (4, 16)$, and $(x_4, y_4) = (5, 32)$. Using the forward difference formula, we estimate $f'(x_2) = f'(3)$, with $h = x_3 - x_2 = 1$, as

$$f'(x_2) \approx \frac{f(x_3) - f(x_2)}{x_3 - x_2} = \frac{y_3 - y_2}{1} = 16 - 8 = 8.$$

Using the backward difference formula, we find that

$$f'(x_2) \approx \frac{f(x_2) - f(x_1)}{x_2 - x_1} = \frac{y_2 - y_1}{1} = 8 - 4 = 4.$$

With the central difference formula, the estimate for $f'(x_2)$, with $h = 1$, is

$$f'(x_2) \approx \frac{f(x_3) - f(x_1)}{x_3 - x_1} = \frac{y_3 - y_1}{2} = \frac{16 - 4}{2} = 6.$$

We also observe that we can use any of these formulas with the given data and $h = 2$. For example, the central difference formula estimate for $f'(x_2)$ with $h = 2$ is

$$f'(x_2) \approx \frac{f(x_4) - f(x_0)}{x_4 - x_0} = \frac{y_4 - y_0}{4} = \frac{32 - 2}{4} = 7.5.$$

Although it may seem surprising that we would want to use a larger step size (like $h = 2$), we will use this result in Example 11.4.

The data are taken from the function $y = f(x) = 2^x$, so we can compare estimates of the derivative with the true value, found by evaluating $f'(x) = 2^x (\ln 2)$ at $x = 3$. The result is $f'(3) \approx 2^3 (0.693) = 5.544$.

Interpolating the data by a polynomial rather than a straight line gives a difference formula that makes use of more than two data points. The forward and backward three-point formulas are given next, for evenly spaced data.

Three-point forward difference formula

$$f'(x_i) \approx \frac{-f(x_{i+2}) + 4f(x_{i+1}) - 3f(x_i)}{x_{i+2} - x_i} = \frac{-y_{i+2} + 4y_{i+1} - 3y_i}{2h}$$

Three-point backward difference formula

$$f'(x_i) \approx \frac{3f(x_i) - 4f(x_{i-1}) + f(x_{i-2})}{x_i - x_{i-2}} = \frac{3y_i - 4y_{i-1} + y_{i-2}}{2h}$$

Example 11.2 Three-point Difference Formulas

We illustrate these difference formulas by using the data from Example 11.1. Using the three-point forward difference formula, we find that

$$f'(x_2) \approx \frac{-f(x_4) + 4f(x_3) - 3f(x_2)}{x_4 - x_2} = \frac{-y_4 + 4y_3 - 3y_2}{2} = \frac{-3(8) + 4(16) - 32}{2} = 4.$$

Using the three-point backward difference formula, we find that

$$f'(x_2) \approx \frac{3f(x_2) - 4f(x_1) + f(x_0)}{x_2 - x_0} = \frac{3y_2 - 4y_1 + y_0}{2} = \frac{3(8) - 4(4) + 2}{2} = 5.$$

Discussion

The forward difference formula can be found from the Taylor polynomial with remainder:

$$f(x + h) = f(x) + hf'(x) + \frac{h^2}{2}f''(\eta). \tag{11.1}$$

For $h = x_{i+1} - x_i$, this gives $f'(x_i) = \dfrac{f(x_{i+1}) - f(x_i)}{h} - \dfrac{h}{2}f''(\eta)$, for some $x_i \leq \eta \leq x_{i+1}$.

Thus, the truncation error for the forward difference formula is $O(h)$. The formula can also be obtained by considering the Lagrange interpolating polynomial for the points (x_i, y_i) and (x_{i+1}, y_{i+1}).

Similarly, the backward difference formula can be found from Eq. (11.1) by letting $h = x_{i-1} - x_i$. This gives $f(x_{i-1}) = f(x_i) + hf'(x_i) + \dfrac{h^2}{2}f''(\eta)$, or

$$f'(x_i) = \frac{f(x_{i-1}) - f(x_i)}{h} - \frac{h}{2} f''(\eta), \quad \text{for some } x_{i-1} \leq \eta \leq x_i.$$

The central difference formula for the first derivative of f at the point x_i can be found from the next higher order Taylor polynomial, with $h = x_{i+1} - x_i = x_i - x_{i-1}$.

$$f(x_{i+1}) = f(x_i + h) = f(x_i) + h f'(x_i) + \frac{h^2}{2!} f''(x_i) + \frac{h^3}{3!} f'''(\eta_1),$$

$$f(x_{i-1}) = f(x_i - h) = f(x_i) - h f'(x_i) + \frac{h^2}{2!} f''(x_i) - \frac{h^3}{3!} f'''(\eta_2),$$

where $x_i \leq \eta_1 \leq x_i + h$ and $x_i - h \leq \eta_2 \leq x_i$. Although the error term involves the third derivative at two unknown points in two different intervals, if we assume that the third derivative is continuous on $[x_i - h, x_i + h]$ we can write the central difference formula with the error term as

$$f'(x_i) = \frac{f(x_{i+1}) - f(x_{i-1})}{2h} + \frac{h^2}{6} f'''(\eta) \quad \text{for some point } x_{i-1} \leq \eta \leq x_{i+1}.$$

The central difference formula can also be found from the three-point Lagrange interpolating polynomial and is therefore known as a three-point formula (although $f(x_i)$ does not appear in it).

General Three-Point Formulas

Three-point approximation formulas for the first derivative, based on the Lagrange interpolation polynomial, do not require that the data points be equally spaced; given the three points, (x_1, y_1), (x_2, y_2), and (x_3, y_3), with $x_1 < x_2 < x_3$, the formula that follows can be used to approximate the derivative at any point in the interval $[x_1, x_3]$. The first derivative at each of the data points is given by

$$f'(x_1) \approx \frac{2x_1 - x_2 - x_3}{(x_1 - x_2)(x_1 - x_3)} y_1 + \frac{x_i - x_3}{(x_2 - x_1)(x_2 - x_3)} y_2 + \frac{x_1 - x_2}{(x_3 - x_1)(x_3 - x_2)} y_3,$$

$$f'(x_2) \approx \frac{x_2 - x_3}{(x_1 - x_2)(x_1 - x_3)} y_1 + \frac{2x_2 - x_1 - x_3}{(x_2 - x_1)(x_2 - x_3)} y_2 + \frac{x_2 - x_1}{(x_3 - x_1)(x_3 - x_2)} y_3,$$

$$f'(x_3) \approx \frac{x_3 - x_2}{(x_1 - x_2)(x_1 - x_3)} y_1 + \frac{x_3 - x_1}{(x_2 - x_1)(x_2 - x_3)} y_2 + \frac{2x_3 - x_1 - x_2}{(x_3 - x_1)(x_3 - x_2)} y_3.$$

For evenly spaced data, the formula for $f'(x_2)$ reduces to the central difference formula presented earlier.

As discussed in Chapter 8, the Lagrange interpolation polynomial for the points (x_1, y_1), (x_2, y_2), and (x_3, y_3) can be written as

$$L(x) = L_1(x) y_1 + L_2(x) y_2 + L_3(x) y_3,$$

where

$$L_1(x) = \frac{(x-x_2)(x-x_3)}{(x_1-x_2)(x_1-x_3)}, \quad L_2(x) = \frac{(x-x_1)(x-x_3)}{(x_2-x_1)(x_2-x_3)}, \quad L_3(x) = \frac{(x-x_1)(x-x_2)}{(x_3-x_1)(x_3-x_2)}.$$

The approximation to the first derivative of f comes from $f'(x) \approx L'(x)$, which can be written as

$$L'(x) = L_1'(x)y_1 + L_2'(x)y_2 + L_3'(x)y_3,$$

where

$$L_1'(x) = \frac{2x - x_2 - x_3}{(x_1 - x_2)(x_1 - x_3)},$$

$$L_2'(x) = \frac{2x - x_1 - x_3}{(x_2 - x_1)(x_2 - x_3)},$$

$$L_3'(x) = \frac{2x - x_1 - x_2}{(x_3 - x_1)(x_3 - x_2)}.$$

Thus,

$$f'(x) \approx \frac{2x - x_2 - x_3}{(x_1 - x_2)(x_1 - x_3)} y_1 + \frac{2x - x_1 - x_3}{(x_2 - x_1)(x_2 - x_3)} y_2 + \frac{2x - x_1 - x_2}{(x_3 - x_1)(x_3 - x_2)} y_3.$$

11.1.2 Higher Derivatives

Formulas for higher derivatives can be found by differentiating the interpolating polynomial repeatedly or by using Taylor expansions. For example, given data at three equally spaced abscissas x_{i-1}, x_i, and x_{i+1}, the formula for the second derivative is

$$f''(x_i) \approx \frac{1}{h^2}[f(x_{i+1}) - 2f(x_i) + f(x_{i-1})], \quad \text{with truncation error } O(h^2).$$

Example 11.3 Second Derivative

Using the data given in Example 11.1, we estimate the second derivative at $x_2 = 3$, using the points $(x_1, y_1) = (2, 4)$, $(x_2, y_2) = (3, 8)$, and $(x_3, y_3) = (4, 16)$; for this example, $h = 1$, so we have

$$f''(3) \approx [f(4) - 2f(3) + f(2)] = [16 - 2(8) + 4] = 4.$$

Derivation of Formula for the Second Derivative

From the Taylor polynomial with remainder, we find that

$$f(x+h) = f(x) + hf'(x) + \frac{h^2}{2!}f''(x) + \frac{h^3}{3!}f'''(x) + \frac{h^4}{4!}f^{(4)}(\eta_1),$$

$$f(x-h) = f(x) - hf'(x) + \frac{h^2}{2!}f''(x) - \frac{h^3}{3!}f'''(x) + \frac{h^4}{4!}f^{(4)}(\eta_2),$$

where $x \leq \eta_1 \leq x+h$ and $x-h \leq \eta_2 \leq x$. Adding gives

$$f(x+h) + f(x-h) = 2f(x) + f''(x)h^2 + \frac{h^4}{4!}[f^{(4)}(\eta_1) + f^{(4)}(\eta_2)],$$

or $f''(x) \approx \dfrac{1}{h^2}[f(x+h) - 2f(x) + f(x-h)]$,

with truncation error $O((h^4))$. The error depends on even powers of h. If we assume that the fourth derivative is continuous on $[x-h, x+h]$, we can write the error term as $-\dfrac{h^2}{12}f^{(4)}(\eta)$ for some point $x-h \leq \eta \leq x+h$.

To find formulas for the third and fourth derivatives, we seek a linear combination of the Taylor expansions for $f(x+2h)$, $f(x+h)$, $f(x-h)$ and $f(x-2h)$ so that all derivatives below the desired derivative cancel.

Centered Difference Formulas, all $O(h^2)$

$$f'(x_i) \approx \frac{1}{2h}[f(x_{i+1}) - f(x_{i-1})]$$

$$f''(x_i) \approx \frac{1}{h^2}[f(x_{i+1}) - 2f(x_i) + f(x_{i-1})]$$

$$f'''(x_i) \approx \frac{1}{2h^3}[f(x_{i+2}) - 2f(x_{i+1}) + 2f(x_{i-1}) - f(x_{i-2})]$$

$$f''''(x_i) \approx \frac{1}{h^4}[f(x_{i+2}) - 4f(x_{i+1}) + 6f(x_i) - 4f(x_{i-1}) + f(x_{i-2})]$$

11.1.3 Partial Derivatives

Finite-difference approximations for partial derivatives of a function of two variables are based on a discrete mesh of points for both variables. We denote a general point as (x_i, y_j), and the value of the function $u(x,y)$ at that point as $u_{i,j}$; the spacing in the x- and y-directions is the same, h. The simplest partial-derivative formulas are direct analogs of the preceding ordinary-derivative formulas; we use subscripts to indicate partial differentiation. Also, each formula is given in a schematic form, indicating only the coefficients on each function value. All of the following formulas are $O(h^2)$; the approximation to each partial derivative is given at (x_i, y_j):

$$u_x(x_i, y_j) \approx \frac{1}{2h}[-u_{i-1,j} + u_{i+1,j}] \qquad u_x \approx \frac{1}{2h}\{\,\boxed{-1}\!\!-\!\!\boxed{0}\!\!-\!\!\boxed{1}\,\}\ j;$$
$$\phantom{u_x(x_i, y_j) \approx \frac{1}{2h}[-u_{i-1,j} + u_{i+1,j}]} \qquad \phantom{u_x \approx \frac{1}{2h}\{} i-1\quad i\quad i+1$$

$$u_{xx}(x_i, y_j) \approx \frac{1}{h^2}[u_{i-1,j} - 2u_{i,j} + u_{i+1,j}] \qquad u_{xx} \approx \frac{1}{h^2}\{\,\boxed{1}\!\!-\!\!\boxed{-2}\!\!-\!\!\boxed{1}\,\}\ j;$$
$$\phantom{u_{xx}(x_i, y_j) \approx \frac{1}{h^2}[u_{i-1,j} - 2u_{i,j} + u_{i+1,j}]} \qquad \phantom{u_{xx} \approx \frac{1}{h^2}\{} i-1\quad i\quad i+1$$

For the mixed second partial derivative and higher derivatives, the schematic form is especially convenient. The Laplacian operator is $\nabla^2 u = u_{xx} + u_{yy}$, and the biharmonic operator is $\nabla^4 u = u_{xxxx} + u_{xxyy} + u_{yyyy}$. We thus have

$$u_{xy} \approx \frac{1}{4h^2}\left\{\begin{array}{ccc}\boxed{-1} & \boxed{0} & \boxed{1}\\ \boxed{0} & \boxed{0} & \boxed{0}\\ \boxed{1} & \boxed{0} & \boxed{-1}\end{array}\right\}\begin{array}{l}j+1\\ j,\\ j-1\end{array}$$
$$\phantom{u_{xy} \approx \frac{1}{4h^2}\left\{\right.} i-1\quad i\quad i+1$$

$$\nabla^2 u \approx \frac{1}{h^2}\left\{\begin{array}{ccc} & \boxed{1} & \\ \boxed{1} & \boxed{-4} & \boxed{1}\\ & \boxed{1} & \end{array}\right\}\begin{array}{l}j+1\\ j,\\ j-1\end{array}$$
$$\phantom{\nabla^2 u \approx \frac{1}{h^2}\left\{\right.} i-1\quad i\quad i+1$$

$$\nabla^4 u \approx \frac{1}{h^4}\left\{\begin{array}{ccccc} & & \boxed{1} & & \\ & \boxed{2} & \boxed{-8} & \boxed{2} & \\ \boxed{1} & \boxed{-8} & \boxed{20} & \boxed{-8} & \boxed{1}\\ & \boxed{2} & \boxed{-8} & \boxed{2} & \\ & & \boxed{1} & & \end{array}\right\} + O(h^2).$$

(See Ames, 1992, for a discussion of these and other formulas.)

11.1.4 Richardson Extrapolation

The technique known as Richardson extrapolation, introduced in Chapter 1, provides a method of improving the accuracy of a low-order approximation formula $A(h)$ whose error can be expressed as

$$A - A(h) = a_2 h^2 + a_4 h^4 + \ldots,$$

where A is the true (unknown) value of the quantity being approximated by $A(h)$ and the coefficients of the error terms do not depend on the step size h. To apply Richardson extrapolation, we form approximations to A separately using the step sizes h and $h/2$. These are combined to give an $O(h^4)$ approximation to A by means of two applications of an $O(h^2)$ formula.

$$A = \frac{4\,A(h/2) - A(h)}{3}.$$

To continue the extrapolation process, consider

$$A = B(h) + b_4 h^4 + b_6 h^6 + b_8 h^8 + \ldots,$$

where $B(h)$ is simply the extrapolated approximation to A, using step sizes h and $h/2$. If we can also find an approximation to A using step sizes $h/2$ and $h/4$, this would correspond to $B(h/2)$. If we extrapolate using $B(h)$ and $B(h/2)$, we get

$$C(h) \approx \frac{16 B(h/2) - B(h)}{15},$$

which has error $O(h^6)$.

The central difference formula can be written as

$$D(h) = f'(x) = \frac{1}{2h}[f(x+h) - f(x-h)] - \frac{h^2}{6} f'''(x) + O(h^4).$$

We can also find $f'(x_i)$ using one-half the previous value of h (whatever it may have been):

$$D(h/2) = f'(x) = \frac{1}{h}[f(x+h/2) - f(x-h/2)] - \frac{h^2}{24} f'''(x) + O(h^4).$$

Since the coefficient $(f''(x))$ of the h^2 term does not change (although we do not, in general, know its value), the two estimates can be combined to give

$$D = \frac{4\,D(h/2) - D(h)}{3}.$$

Example 11.4 Improved Estimate of the Derivative

We illustrate the use of Richardson extrapolation with the values from Example 11.1. The value of h is 2, and the approximation to $f'(x_2)$ is based on $D(h) = 7.5$ and $D\left(\dfrac{h}{2}\right) = 6$. We have

$$D = \frac{4(6) - 7.5}{3} = \frac{16.5}{3} \approx 5.5.$$

The data in the example are points on the curve $f(x) = 2^x$. The actual value of $f'(x)$ is $(\ln 2) 2^x$, which gives $f'(3) \approx 5.54$.

Discussion

Richardson extrapolation forms a linear combination of approximations $A(h)$ and $A(h/2)$, the first using a step size h, the second based on half the original step size; the combination is chosen so that the dominant error term, which depends on h^2, cancels. Representing A in terms of the approximation and the error terms, we have

$$A = A(h) + a_2 h^2 + a_4 h^4 + \ldots . \tag{11.2}$$

If the same approximation formula is used with step size $h/2$ in place of h, the true value can be expressed as

$$A = A\left(\frac{h}{2}\right) + a_2 \frac{h^2}{4} + a_4 \frac{h^4}{16} + \ldots ,$$

or

$$4A = 4A\left(\frac{h}{2}\right) + a_2 h^2 + a_4 \frac{h^4}{4} + \ldots \tag{11.3}$$

Subtracting (11.2) from (11.3) gives

$$3A = 4A(h/2) - A(h) + O(h^4),$$

or

$$A = \frac{1}{3}\left[4A\left(\frac{h}{2}\right) - A(h)\right] + O(h^4)$$

The h^2 error terms cancel, although the higher-order terms do not. However, we now have an $O(h^4)$ approximation to A derived by using two applications of an $O(h^2)$ formula.

To continue the extrapolation, we write

$$A = B(h) + b_4 h^4 + b_6 h^6 + b_8 h^8 + \ldots , \tag{11.4}$$

where $B(h)$ is simply the extrapolated approximation to A, using step sizes h and $h/2$. If we can also find an approximation to A using step sizes $h/2$ and $h/4$, this would correspond to $B(h/2)$. We begin with

$$A = B\left(\frac{h}{2}\right) + b_4 \frac{h^4}{16} + b_6 \left(\frac{h^6}{64}\right) + b_8 \left(\frac{h^8}{2^8}\right) + \ldots \tag{11.5}$$

Multiplying Eq. (11.5) by 16 and subtracting Eq. (11.4) from the result, so that the h^4 terms cancel, yields

$$15A = 16B(h/2) - B(h) + c_6 h^6 + c_8 h^8 + \ldots$$

Now we define the second-level extrapolated approximation to A as

$$C(h) \approx \frac{16}{15} B\left(\frac{h}{2}\right) - \frac{1}{15} B(h).$$

11.2 BASIC NUMERICAL INTEGRATION

Numerical integration (quadrature) rules are very important because even simple functions may not have exact formulas for their antiderivatives (indefinite integrals). Even when an exact formula for the antiderivative does exist, it may be difficult to find.

In general, a numerical integration formula approximates a definite integral by a weighted sum of function values at points within the interval of integration. A numerical integration rule has the form

$$\int_a^b f(x)\, dx \approx \sum_{i=0}^n c_i f(x_i)$$

where the coefficients c_i depend on the particular method. In this section, we consider the most common numerical integration formulas that are based on equally spaced data points; these are known as Newton-Cotes formulas. We first present the most basic formulas, namely, the trapezoid, Simpson, and midpoint rules. We then consider two methods of improving these simple formulas: composite quadrature and extrapolation. In Section 11.3, we present Gaussian quadrature, which is an effective method of obtaining more accurate results (for a given number of function evaluations) if the function to be integrated may be evaluated at any desired points.

We start by investigating several basic quadrature formulas that use function values at equally spaced points. There are two basic types of Newton-Cotes formulas, depending on whether or not the function values at the ends of the interval of integration are used. In this section, we consider the trapezoid and Simpson rules, "closed" formulas in which the endpoint values are used, and the midpoint rule, an "open" formula in which the endpoints are not used. Each of these formulas can be derived by approximating the function to be integrated by its Lagrange interpolating polynomial (different methods use polynomials of different degree) and then integrating the polynomial exactly. Since the interpolation polynomials also have an explicit formula for the error bound, error bounds can be obtained for the numerical integration formulas.

One way to achieve greater accuracy in numerical integration might be to use a method based on a higher-order interpolating polynomial; however, interpolation with higher-degree polynomials is not generally a good idea. We next consider integration formulas derived from linear and quadratic interpolation of $f(x)$. Later, more accurate methods based on subdivision of the original interval of integration are presented.

11.2.1 Trapezoid Rule

One of the simplest ways to approximate the area under a curve is to approximate the curve by a straight line. The trapezoid rule approximates the curve by the straight line that passes through the points $(a, f(a))$ and $(b, f(b))$, the two ends of the interval of interest. We have $x_0 = a$; $x_1 = b$, and $h = b - a$, and then

$$\int_a^b f(x)\, dx \approx \frac{h}{2}[f(x_0) + f(x_1)].$$

Example 11.5 Integral of e^{-x^2} Using the Trapezoid Rule

Consider now a very important function for which the exact value of the integral is not known:

$$f(x) = \exp(-x^2) \quad x_0 = a = 0, \quad x_1 = b = 2.$$

Using the trapezoid rule, we find (since $(b - a)/2 = 1$ for this example) that

$$\int_0^2 \exp(-x^2)\, dx \approx [\exp(-0^2) + \exp(-2^2)] = 1 + \exp(-4) = 1.0183.$$

The function and the straight-line approximation are shown in Fig. 11.5.

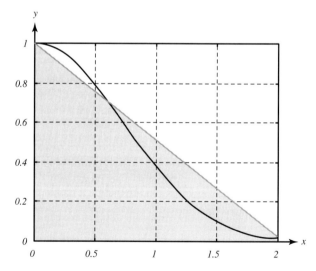

FIGURE 11.5 $f(x) = e^{-x^2}$ and straight line for integral obtained by means of the trapezoid rule.

Discussion

The trapezoid rule can be derived from the foregoing geometric reasoning, or, more formally, from the Lagrange form of linear interpolation of $f(x)$ using the endpoints of the interval of integration. To demonstrate the latter method, define $h = b - a = x_1 - x_0$ and let

$$L(x) = L_0(x)\, y_0 + L_1(x)\, y_1 = \frac{(x - x_1)}{(x_0 - x_1)} f(x_0) + \frac{(x - x_0)}{(x_1 - x_0)} f(x_1).$$

Then

$$\int_a^b f(x)\, dx \approx \int_a^b \frac{(x - x_1)}{(x_0 - x_1)} f(x_0) + \frac{(x - x_0)}{(x_1 - x_0)} f(x_1)\, dx$$

$$= \int_a^b \frac{(x_1 - x)}{h} f(x_0) + \frac{(x - x_0)}{h} f(x_1)\, dx$$

$$= \frac{1}{h} \int_a^b x_1 f(x_0) - x_0 f(x_1) + x[f(x_1) - f(x_0)]\, dx$$

$$= x_1 f(x_0) - x_0 f(x_1) + \frac{1}{h} \int_a^b x\, [f(x_1) - f(x_0)]\, dx$$

$$= x_1 f(x_0) - x_0 f(x_1) + \frac{b^2 - a^2}{2h} [f(x_1) - f(x_0)]$$

$$= bf(x_0) - a f(x_1) + \frac{b + a}{2} [f(x_1) - f(x_0)]$$

$$= \frac{b - a}{2} f(x_0) + \frac{b - a}{2} f(x_1)$$

$$= \frac{h}{2} [f(x_0) + f(x_1)].$$

The degree of precision r of an integration formula is the (highest) degree of polynomial for which the method gives an exact result. Error analysis (see, e.g., Atkinson, 1989) shows that the trapezoid rule gives exact results for polynomials of degree ≤ 1, i.e., linear functions for which the interpolating polynomial is exact. Thus, the trapezoid rule has a degree of precision, $r = 1$. In general, if we assume that $f(x)$ is twice continuously differentiable on $[a, b]$, then

$$\int_a^b f(x)\, dx = \frac{b - a}{2} [f(a) + f(b)] - \frac{(b - a)^3}{12} f''(\eta), \quad \text{for some } \eta \in [a, b] \quad (11.6)$$

11.2.2 Simpson Rule

Approximating the function to be integrated by a quadratic polynomial leads to the basic Simpson rule ($n = 2$):

$$h = \frac{b-a}{2}, \quad x_0 = a, \quad x_1 = x_0 + h = \frac{b+a}{2}, \quad x_2 = b.$$

The approximate integral is given by

$$\int_a^b f(x)\,dx \approx \frac{h}{3}[f(x_0) + 4f(x_1) + f(x_2)] = \frac{b-a}{6}\left[f(a) + 4f\left(\frac{b+a}{2}\right) + f(b)\right].$$

Example 11.6 Integral Using Simpson's Rule

Consider the problem of finding $\int_0^1 \frac{1}{1+x^2}\,dx$ numerically. Using Simpson's rule with $f(x) = 1/(1+x^2)$, $a = 0$, and $b = 1$ gives $h = 1/2$, and

$$\int_0^1 \frac{1}{1+x^2}\,dx \approx \frac{1}{6}[f(0) + 4f(1/2) + f(1)] = \frac{1}{6}\left[\frac{1}{1} + (4)\frac{4}{5} + \frac{1}{2}\right] = \frac{47}{60} \approx 0.78333.$$

The exact value of the integral is $\arctan(1) = \pi/4 \approx 0.7853\ldots$. The graphs of $f(x) = 1/(1+x^2)$ and the quadratic polynomial that passes through the points $(0,1)$, $(1/2, 4/5)$, and $(1, 1/2)$ are shown in Fig. 11.6. It is not surprising that the approximate value of the integral from Simpson's rule is quite good, because the two functions are very similar.

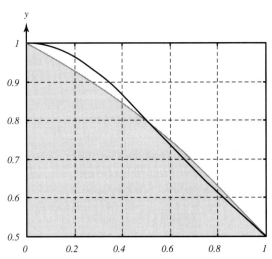

FIGURE 11.6 $f(x) = \dfrac{1}{1+x^2}$ and parabolic approximation for Simpson's rule.

Example 11.7 Integral of e^{-x^2} Using Simpson's Rule

Consider the problem of finding the integral of $f(x) = \exp(-x^2)$ on $[0, 2]$. The required values for applying Simpson's rule are

$$h = \frac{b-a}{2} = 1, \quad x_0 = a = 0, \quad x_1 = (2+0)/2 = 1, \quad x_2 = b = 2,$$

which gives

$$\int_0^2 \exp(-x^2)\, dx \approx \frac{1}{3}[\exp(-0^2) + 4\exp(-1^2) + \exp(-2^2)] = .8299.$$

The graphs of $f(x) = \exp(-x^2)$ and the quadratic polynomial passing through the points $(0, 1)$, $(1, \exp(-1))$, and $(2, \exp(-4))$ are shown in Fig. 11.7.

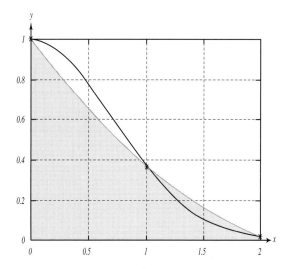

FIGURE 11.7 $f(x) = e^{-x^2}$ and parabola for finding the integral using Simpson's rule.

Discussion

To simplify the derivation of Simpson's Rule, we assume that the function values are given at $x_0 = -h$, $x_1 = 0$, and $x_2 = h$; there is no loss of generality in this assumption, since it is a simple matter to shift any other interval of length $2h$ to $[-h, h]$. We write the polynomial that interpolates these points as $p(x) = a x^2 + b x + c$, so that

$$p(-h) = a(-h)^2 + b(-h) + c \quad \text{or} \quad y_0 = a h^2 - b h + c$$
$$p(0) = a(0)^2 + b(0) + c \quad \text{or} \quad y_1 = + c$$
$$p(h) = a(h)^2 + b(h) + c \quad \text{or} \quad y_2 = a h^2 + b h + c$$

The second equation gives

$$c = y_1$$

Adding the first and third equations yields

$$2 a h^2 + 2 c = y_0 + y_2$$

or

$$2a = (y_0 + y_2 - 2 y_1)/h^2$$

Integrating $p(x)$, we find

$$\int_{-h}^{h} a x^2 + b x + c \, dx = \frac{2}{3} a h^3 + 2 c h$$

Substituting for a and c (in terms of y_0, y_1, and y_2), and simplifying gives

$$\int_{-h}^{h} a x^2 + b x + c \, dx = \frac{h}{3}(y_0 + 4 y_1 + y_2)$$

Error analysis shows that Simpson's rule gives the exact value of the integral for polynomials of degree ≤ 3, even though quadratic interpolation is exact only if $f(x)$ is a polynomial of degree ≤ 2. This surprising result indicates that Simpson's rule is significantly more accurate than the trapezoid rule. The degree of precision for Simpson's rule is $r = 3$. If $f(x)$ is four times continuously differentiable on $[a, b]$, then

$$\int_a^b f(x) \, dx = \frac{h}{3}\left[f(a) + 4 f\left(\frac{a+b}{2}\right) + f(b)\right] - \frac{h^5}{90} f^{(4)}(\eta) \tag{11.7}$$

for some $\eta \in [a, b]$.

The derivation of the error term for Simpson's Rule makes use of the Lagrange interpolating polynomial, and its error term. (See Atkinson, 1989 for details.)

11.2.3 The Midpoint Formula

The trapezoid and Simpson rules are the simplest examples of Newton-Cotes closed formulas; that is, they use function evaluations at the endpoints of the interval of integration. If we use only function evaluations at points within the interval, the simplest formula (a Newton-Cotes *open* formula) is the midpoint rule. This formula uses only one function evaluation (so $n = 1$), at the midpoint of the interval, $x_m = (a + b)/2$. Interpolating the function by the constant value $f(x_m)$, we get the midpoint rule:

$$\int_a^b f(x)\,dx \approx (b - a) f\left(\frac{a + b}{2}\right).$$

Assuming that f is twice continuously differentiable yields

$$\int_a^b f(x)\,dx = (b - a) f\left(\frac{a + b}{2}\right) + \frac{(b - a)^3}{24} f''(\eta), \quad \text{for some } \eta \in [a, b].$$

Example 11.8 The Midpoint Rule

Using the midpoint rule to approximate the integral of the sinc function, $f(x) = \sin(x)/x$, on $[0, \pi]$ we find that

$$\int_0^\pi \frac{\sin(x)}{x}\,dx \approx \pi \frac{\sin(\pi/2)}{\pi/2} = \pi \frac{1}{\pi/2} = 2$$

Figure 11.8 compares the actual value of the area with that found by using the midpoint rule.

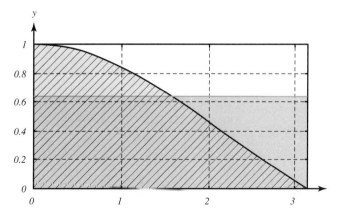

FIGURE 11.8 The area given by the integral S (hatched), and the approximation using the midpoint rule (shaded).

11.2.4 Other Newton-Cotes Open Formulas

The Newton-Cotes open formula that uses two function evaluations is given by

$$x_0 = a, \quad x_1 = \frac{2a + b}{3}, \quad x_2 = \frac{a + 2b}{3}, \quad f_1 = f(x_1), f_2 = f(x_2).$$

$$\int_a^b f(x)\,dx = \frac{b - a}{2}[f_1 + f_2] + \frac{(b - a)^3}{108} f''(\eta), \text{ for some } \eta \in [a, b]. \quad (11.8)$$

The coefficient of the error term is smaller than that in the trapezoid rule (Eq. (11.6)), which uses the same number of function evaluations. (See Isaacson and Keller, 1994, p. 316.)

For the Newton-Cotes open formula using three functions evaluations, we define $f_1 = f(x_1), f_2 = f(x_2)$, and $f_3 = f(x_3)$, where

$$x_0 = a, \quad x_1 = \frac{3a + b}{4}, \quad x_2 = \frac{2a + 2b}{4}, \quad x_3 = \frac{a + 3b}{4}.$$

Then, taking $h = \frac{b - a}{4}$, we have

$$\int_a^b f(x)\,dx = \frac{4h}{3}[2f_1 - f_2 + 2f_3] + \frac{14 h^5}{45} f^{(4)}(\eta), \text{ for some } \eta \in [a, b]. \quad (11.9)$$

The coefficient of the error term is smaller than that in Simpson's rule [Eq. (11.7)], which uses the same number of function evaluations. [Note the difference in definition of h in Eqs. (11.7) and (11.9)]; (see Isaacson and Keller, 1994, p. 316.)

Care must be taken in comparing open and closed formulas, since a comparison can be made on the number of nodes or on the number of function evaluations. Issacson and Keller define the number of nodes to be the number of subintervals used, so that closed formulas use $n + 1$ function evaluations and open formulas use $n - 1$ evaluations, for a given number of subintervals.

Rather than continuing to use quadrature formulas based on interpolating polynomials of ever higher order, we now consider three more effective methods of improving the accuracy of integration. The first two of these improvements are based (as are all our formulas so far) on function evaluations at evenly spaced points. Composite integration formulas, which we examine in the next section, are based on subdividing the interval of integration into subintervals and applying the basic integration rule in each subinterval. The formal statement of the algorithm and Mathcad function for each rule is presented for the more general, composite, form of the method.

11.3 BETTER NUMERICAL INTEGRATION

The easiest method of improving the accuracy of numerical integration is to apply one of the lower order methods presented in the previous section repeatedly on several subintervals. This is known as *composite integration*.

11.3.1 Composite Trapezoid Rule

If we divide the interval of integration, $[a,b]$, into two or more subintervals and use the trapezoid rule on each subinterval, we obtain the composite trapezoid rule. For the simplest case, which uses the same number of function evaluations (at the same points) as the Simpson rule, consider two subintervals $[a, x_1]$ and $[x_1, b]$, where $x_1 = (b + a)/2$. The value of h is the same for each subinterval, namely $h = (b - a)/2$. Then

$$\int_a^b f(x)\,dx = \int_a^{x_1} f(x)\,dx + \int_{x_1}^b f(x)\,dx \approx \frac{h}{2}[f(a) + f(x_1)] + \frac{h}{2}[f(x_1) + f(b)],$$

or

$$\int_a^b f(x)\,dx \approx h/2\,[f(a) + 2f(x_1) + f(b)].$$

If we divide the interval into n subintervals, we get $h = \dfrac{b-a}{n}$, and

$$\int_a^b f(x)\,dx = \int_a^{x_1} f(x)\,dx + \ldots + \int_{x_{n-1}}^b f(x)\,dx$$

$$\approx \frac{h}{2}[f(a) + f(x_1)] + \ldots + \frac{h}{2}[f(x_{n-1}) + f(b)],$$

so that

$$\int_a^b f(x)\,dx \approx \frac{h}{2}[f(a) + 2f(x_1) + \ldots + 2f(x_{n-1}) + f(b)].$$

Algorithm for Trapezoid Rule

Approximate the integral of $f(x)$ on $[a,b]$ using trapezoid rule, with n subdivisions.

Begin computations
$h = (b - a)/n$
for $i = 1$ to $n - 1$
$\quad x_i = a + h\,i$
end

$$S = \frac{h}{2}\left[f(a) + f(b) + 2\sum_{i=1}^{n-1} f(x_i)\right]$$

Mathcad Function for Trapezoid Rule

$$\text{Trap}(f, a, b, n) := \begin{array}{|l} h \leftarrow \dfrac{(b-a)}{n} \\ S \leftarrow f(a) \\ \text{for } i \in 1..n-1 \quad \text{if } n > 1 \\ \quad \begin{array}{|l} x_i \leftarrow a + h \cdot i \\ S \leftarrow S + 2 \cdot f(x_i) \end{array} \\ S \leftarrow S + f(b) \\ I \leftarrow \dfrac{h \cdot S}{2} \\ I \end{array}$$

Example 11.9 Integral of 1/x Using the Composite Trapezoid Rule

Consider the problem of finding

$$\int_1^2 \frac{dx}{x} \approx \frac{h}{2}[f(a) + 2f(x_1) + \cdots 2f(x_{n-1}) + f(b)].$$

For $n = 2$ subintervals, $h = (2-1)/2 = 1/2$, and we have

$$I_1 = \frac{1}{4}[f(1) + 2f(1.5) + f(2)] = \frac{1}{4}\left[\frac{1}{1} + \frac{2}{1.5} + \frac{1}{2}\right] = \frac{17}{24} \approx 0.7083.$$

The function and the straight-line approximations are shown in Fig. 11.9. For $n = 2^2 = 4$ subintervals, $h = 1/4$, and the composite trapezoid rule gives

$$I_2 = \frac{1}{8}[f(1) + 2f(5/4) + 2f(3/2) + 2f(7/4) + f(2)] = \frac{1}{8}\left[1 + \frac{8}{5} + \frac{4}{3} + \frac{8}{7} + \frac{1}{2}\right]$$

$$= 0.6970\ldots.$$

For $n = 2^3 = 8$ subintervals, $h = 1/8$, and the composite trapezoid rule yields

$$I_3 = \frac{1}{16}[f(1) + 2f(9/8) + 2f(5/4) + 2f(11/8) + 2f(3/2)$$

$$+ 2f(13/8) + 2f(7/4) + 2f(15/8) + f(2)]$$

$$= 0.6941\ldots.$$

The exact value of the integral is $ln(2) \approx 0.693147\ldots$.

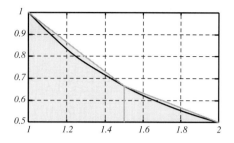

FIGURE 11.9 $y = 1/x$ and trapezoid rule approximations on $[1,1.5]$ and $[1.5,2]$.

11.3.2 Composite Simpson's Rule

Applying the same idea of subdivision of intervals to Simpson's rule and requiring that n be even gives the composite Simpson rule. If we divide the interval of integration $[a, b]$ into two subintervals, we have $n = 4$, and we can apply Simpson's rule twice. Accordingly, consider the two subintervals $[a, x_2]$ and $[x_2, b]$, where $x_2 = (b + a)/2$ and $h = (b - a)/4$. Then

$$\int_a^b f(x)\, dx = \int_a^{x_2} f(x)\, dx + \int_{x_2}^b f(x)\, dx$$

$$\approx \frac{h}{3}[f(a) + 4f(x_1) + f(x_2)] + \frac{h}{3}[f(x_2) + 4f(x_3) + f(b)],$$

or

$$\int_a^b f(x)\, dx \approx \frac{h}{3}[f(a) + 4f(x_1) + 2f(x_2) + 4f(x_3) + f(b)].$$

In general, for n even, we have $h = (b - a)/n$ and Simpson's rule is

$$\int_a^b f(x)\, dx \approx \frac{h}{3}[f(a) + 4f(x_1) + 2f(x_2) + 4f(x_3) \\ + 2f(x_4) + \ldots + 2f(x_{n-2}) + 4f(x_{n-1}) + f(b)].$$

Algorithm for Simpson's Rule

Approximate the integral of $f(x)$ on $[a,b]$ using Simpson's rule, with $n = 2m$ subdivisions.

Begin computations
$h = (b - a)/n$
$m = n/2$
for $i = 1$ to $n - 1$
$\quad x_i = a + h\, i$
end

$$S = \frac{h}{3}\left[f(a) + f(b) + 2\sum_{i=0}^{m-1} f(x_{2i+1}) + 4\sum_{i=1}^{m-1} f(x_{2i})\right]$$

Mathcad Function for Simpson's Rule

$$\text{Simp}(f, a, b, n) := \begin{array}{|l} h \leftarrow \dfrac{(b-a)}{n} \\ S \leftarrow f(a) \\ \text{for } i \in 1, 3 .. n - 1 \\ \quad \begin{array}{|l} x_i \leftarrow a + h \cdot i \\ S \leftarrow S + 4 \cdot f(x_i) \end{array} \\ \text{for } i \in 2, 4 .. n - 2 \\ \quad \begin{array}{|l} x_i \leftarrow a + h \cdot i \\ S \leftarrow S + 2 \cdot f(x_i) \end{array} \\ S \leftarrow S + f(b) \\ I \leftarrow \dfrac{h \cdot S}{3} \\ I \end{array}$$

Example 11.10 Integral of $\exp(-x^2)$ Using Composite Simpson's Rule

For $n = 4$ and $f(x) = \exp(-x^2)$, the integral of f is approximately 0.84232. The function and the two quadratic approximations are shown in Fig. 11.10.

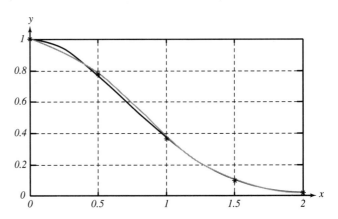

FIGURE 11.10 $y = \exp(-x^2)$ and Simpson's rule approximations for $n = 4$.

Example 11.11 Length of Elliptical Orbit

Consider the problem of finding the length of an elliptical orbit, where the ellipse is described parametrically as $x(r) = \cos(r)$ and $y(r) = \frac{3}{4}\sin(r)$. The length of the orbit is

$$L = \int_a^b \sqrt{(x')^2 + (y')^2}\, dr = 0.25 \int_a^b \sqrt{16\sin^2(r) + 9\cos^2(r)}\, dr.$$

We can approximate the entire length of the orbit by using the preceding Simpson integration function, with $a = 0$ and $b = 2\pi$.

$$f(x) := 0.25 \cdot \sqrt{16 \cdot (\sin(x))^2 + 9 \cdot (\cos(x))^2}$$

$n := 4$ $\text{Simp}(f, a, b, n) = 5.759587$

$n := 20$ $\text{Simp}(f, a, b, n) = 5.525879$

As described in Example 11-B, we can estimate the difference in the speed of the orbiting object when it is close to the planet (say, between day 0 and day 10) and the speed when it is further away (say, between day 60 and day 70). The arc length covered from day 0 to day 10 is approximately $I = 0.8556$; the arc length covered from day 60 to day 70 is approximately $I = 0.3702$. Thus, during the 10 days that the orbiting object is closest to the object it is revolving around, it is traveling more than twice as fast as it travels during days 60 to 70 (not quite the time it is farthest away).

11.3.3 Extrapolation Methods for Quadrature

As we saw earlier, an approximation formula whose error can be expressed as even powers of the step size may be extrapolated by using results from two step sizes (h and $h/2$) to obtain an estimate that is more accurate than either of the original results. The extrapolated form of the trapezoid rule is known as *Romberg integration*. The midpoint rule is also suitable for extrapolation.

The composite trapezoid rule can be expressed as

$$\int_a^b f(x)\,dx = \frac{h}{2}[f(a) + 2f(x_1) + \ldots + 2f(x_{n-1}) + f(b)] + \sum_{j=1}^{\infty} c_j h^{2j},$$

which indicates that we can apply Richardson extrapolation. We start with a simple example.

Algorithm for Romberg integration

Approximate the integral of $f(x)$ on $[a,b]$ using Romberg integration (extrapolated trapezoid rule), with kmax steps of extrapolation.

Use calls to a routine called trap(f,a,b,n) which implements trapezoid rule

Begin computations
$h = (b - a)/n$
$Q(1, 1) = \text{trap}(f, a, b, 1)$
for $k = 1$ to kmax
    $n = 2^k$
    $Q(k + 1, 1) = \text{trap}(f, a, b, n)$ (one more entry in col 1)
    for $j = 2$ to $k + 1$
        $c = 4^{j-1}$

$$Q(k - j + 2, j) = \frac{c\,Q(k - j + 3, j - 1) - Q(k - j + 2, j - 1)}{c - 1}$$

        (another column of extrapolated values)
    end
    (test for convergence if desired)
end

The full algorithm for Romberg integration makes use of the fact that the same function evaluations that were used in the previous approximation are needed (along with some additional ones) in the current approximation, to achieve further efficiencies that are not included in the Mathcad function that follows. We choose instead to rely on repeated calls to the function for the trapezoid rule given earlier.

Example 11.12 Integral of 1/x Using Romberg Integration

Consider the problem of finding $\int_1^2 \frac{dx}{x}$ using the trapezoid rule.

We start with one subinterval, so $h = 1$, and the trapezoid rule gives

$$\int_1^2 \frac{dx}{x} \approx I_0 = \frac{1}{2}[f(1) + f(2)] = \frac{1}{2}\left[\frac{1}{1} + \frac{1}{2}\right] = \frac{3}{4} = 0.75.$$

For two subintervals, $h = 1/2$, and the composite trapezoid rule gives

$$I_1 = \frac{1}{4}[f(1) + 2f(1.5) + f(2)] = \frac{1}{4}\left[\frac{1}{2} + \frac{2}{1.5} + \frac{1}{2}\right] = \frac{17}{24} \approx 0.7083.$$

To apply Richardson extrapolation, we use the formula

$$A \approx \frac{1}{3}\left[4A\left(\frac{h}{2}\right) - A(h)\right];$$

we let I_0 be $A(h)$ and I_1 be $A\left(\frac{h}{2}\right)$ to give the extrapolated value

$$I \approx \frac{1}{3}[4I_1 - I_0] = [4(0.7083) - 0.7500]/3 = 0.6944.$$

The exact value of the integral is $ln(2) \approx 0.693147\ldots$.

In the following table of approximations, Column I gives the result from the composite trapezoid rule, and column II gives the extrapolated value:

	I	II
$h = 1$	0.7500	
		0.6944
$h = 1/2$	0.7083	

We extend this process by applying the composite trapezoid rule several times, using twice as many subintervals each time. This produces a corresponding succession of values of h. We denote the approximation formed using $n = 2^k$ subintervals as I_k. Each time we increase k by 1, we double the number of points at which the function is evaluated and halve the value of h (as is required for Richardson extrapolation). Furthermore, the function values that were used for a previous estimate appear again on each refinement. Before giving the general formulas, we continue the problem of finding the integral of $f(x) = 1/x$.

For $k = 0$, we found that $I_0 = 0.75$. For $k = 1$, we found that $I_1 = 0.7083$.
For $k = 2$, we have $n = 2^2 = 4$ subintervals; so $h = 1/4$ and

$$I_2 = \frac{1}{8}\left[1 + \frac{8}{5} + \frac{4}{3} + \frac{8}{7} + \frac{1}{2}\right] \approx 0.6970.$$

For $k = 3$, we have $n = 2^3 = 8$ subintervals; so $h = 1/8$ and $I_3 \approx 0.6941$

To apply Richardson extrapolation to the sequence we have generated, we have several approximations that can play the role of $A(h/2)$ and $A(h)$ in the basic formula;

$$A \approx \frac{1}{3}\left[4 A\left(\frac{h}{2}\right) - A(h)\right].$$

First, we could let I_0 be $A(h)$ and I_1 be $A(h/2)$ as we did previously. Then

$$I \approx \frac{1}{3}[4 I_1 - I_0] = [4(0.7083) - 0.7500]/3 = 0.6944.$$

Or we could use I_1 as $A(h)$ and I_2 as $A(h/2)$, yielding $I \approx \frac{1}{3}[4 I_2 - I_1]$

Extending the previous table of the approximations, with the first column giving the values computed by the composite trapezoid rule and the second column giving the extrapolated approximation, results in the following new table:

	I	II
$h = 1$	0.7500	
		0.6944
$h = 1/2$	0.7083	
		0.6932
$h = 1/4$	0.6970	
		0.6931
$h = 1/8$	0.6941	

If we use the fact that the error term in the composite trapezoid rule can be represented as a series with only even powers of h (see, e.g., Ralston and Rabinowitz, 1978), the extrapolation can be extended for as many levels as is desired. For the second level of extrapolation, we have

$$C(h) \approx \frac{16}{15} B\left(\frac{h}{2}\right) - \frac{1}{15} B(h)$$

and the following table:

	I	II	III
$h = 1$	0.7500		
		0.6944	
$h = 1/2$	0.7083		0.6932
		0.6932	
$h = 1/4$	0.6970		

Mathcad Function for Romberg Integration

The following Mathcad function uses the function Trap defined previously. The function Trap must be included in the worksheet prior to the function Romberg, or a reference to Trap must be given (using Insert/Ref from the menu).

$$\text{Romberg}(f,a,b,\text{kmax}) := \begin{vmatrix} Q_{0,0} \leftarrow \text{Trap}(f,a,b,1) \\ Q_{1,0} \leftarrow \text{Trap}(f,a,b,2) \\ Q_{0,1} \leftarrow \dfrac{4 \cdot Q_{1,0} - Q_{0,0}}{3} \\ \text{for } k \in 2..\text{kmax} \\ \quad \begin{vmatrix} n \leftarrow 2^k \\ Q_{k,0} \leftarrow \text{Trap}(f,a,b,n) \\ \text{for } j \in 1..k \\ \quad \begin{vmatrix} c \leftarrow 4^j \\ Q_{k-j,j} \leftarrow \dfrac{c \cdot Q_{k-j+1,j-1} - Q_{k-j,j-1}}{c - 1} \end{vmatrix} \end{vmatrix} \\ Q \end{vmatrix}$$

Example 11.12 (Continued) Integral of 1/x

To use the Mathcad function Romberg, we define the function to be integrated, the interval [a,b] of integration, and the number of levels of extrapolation to be performed.

$$f(x) := \dfrac{1}{x} \qquad a := 1 \qquad b := 2 \qquad \text{kmax} := 4$$

$$\text{Romberg}(f,a,b,\text{kmax}) = \begin{pmatrix} 0.75 & 0.694444 & 0.693175 & 0.693147 & 0.693147 \\ 0.708333 & 0.693254 & 0.693148 & 0.693147 & 0 \\ 0.697024 & 0.693155 & 0.693147 & 0 & 0 \\ 0.694122 & 0.693148 & 0 & 0 & 0 \\ 0.693391 & 0 & 0 & 0 & 0 \end{pmatrix}$$

11.4 GAUSSIAN QUADRATURE

The Newton-Cotes formulas are based on evaluations of a function at equally spaced values of the independent variable. Gaussian integration formulas evaluate functions at points which are chosen so that the formula is exact for polynomials of as high a degree as possible. Gaussian integration formulas are usually expressed in terms of the interval of integration $[-1, 1]$. For other intervals, a change of variable is used to transform the problem so that it utilizes the interval $[-1, 1]$.

11.4.1 Gaussian Quadrature on [−1,1]

The basic form of a Gaussian quadrature formula is

$$\int_{-1}^{1} f(t)\, dt \approx \sum_{i=1}^{n} c_i f(t_i),$$

where the appropriate values of the points t_i and the coefficients c_i depend on the choice of n.

By choosing the quadrature points $t_1, \ldots, t_n$ as the n zeros of the nth-degree Legendre polynomial, and by using the appropriate coefficients, the integration formula is exact for polynomials of degree up to $2n - 1$.

For example, the Gauss-Legendre quadrature rule for two evaluation points, which is exact for polynomials up to and including degree 3, has the form

$$\int_{-1}^{1} f(t)\, dt \approx c_1 f(t_1) + c_2 f(t_2) = f(-1/\sqrt{3}) + f(1/\sqrt{3})$$

$$\approx f(-0.577) + f(0.577).$$

Example 11.13 Integral of exp(−t^2) on [−1, 1] Using Gaussian Quadrature

To find the integral of $\exp(-t^2)$ on $[-1, 1]$ using Gaussian quadrature with two quadrature points, we compute

$$\int_{-1}^{1} \exp(-t^2)\, dt \approx c_1 f(t_1) + c_2 f(t_2) = f(-1/\sqrt{3}) + f(1/\sqrt{3})$$

$$= \exp[-(-1/\sqrt{3})^2] + \exp[-(1/\sqrt{3})^2]$$

$$= \exp[-(1/3)] + \exp[-(1/3)] \approx 2(0.7165) \approx 1.433.$$

Similarly, the Gauss-Legendre quadrature rule for $n = 3$ evaluation points, which is exact for polynomials up to and including degree 5, has the form

$$\int_{-1}^{1} f(t)\, dt \approx c_1 f(t_1) + c_2 f(t_2) + c_3 f(t_3)$$

$$= \frac{5}{9} f(-\sqrt{3/5}) + \frac{8}{9} f(0) + \frac{5}{9} f(\sqrt{3/5})$$

$$\approx \frac{5}{9} f(-0.775) + \frac{8}{9} f(0) + \frac{5}{9} f(0.775).$$

The values of the Gaussian quadrature parameters x_i and c_i for values of $n = 2, \ldots, 4$ are given in the following table:

Parameters for Gaussian Quadrature

n	t_i	c_i
2	$\pm 1/\sqrt{3} \approx \pm 0.57735$	1
3	0	8/9
	$\pm 1/\sqrt{3/5} \approx \pm 0.77459$	5/9
4	$\approx \pm 0.861136$	0.34785
	$\approx \pm 0.339981$	0.652145

To illustrate the use of the Mathcad function `Gauss_quad` (given in the next section) we compute the integral of $\exp(-x^2)$ on $[-1, 1]$ using 3 and 4 quadrature points.

Example 11.14 Gaussian Quadrature Using More Quadrature Points

To illustrate the use of Gaussian quadrature, we approximate the integral of f(t) on the interval from –1 to 1, using 3 and 4 quadrature points.

$f(t) := \exp(-t^2)$ $a := -1$ $b := 1$

$k := 3$ $\text{Gauss\_quad}(f, a, b, k) = 1.49868$

$k := 4$ $\text{Gauss\_quad}(f, a, b, k) = 1.493334$

11.4.2 Gaussian Quadrature on [a,b]

If we have an integral on an interval $[a,b]$ that is not $[-1, 1]$, we must make a change of variable to transform the integral to the required interval. We start by writing the desired integral in terms of some variable other than t, say, x:

$$\int_a^b f(x)\, dx.$$

The change of variable that is required to convert an integral on the interval $x \in [a, b]$ to the required interval $t \in [-1, 1]$ for Gaussian quadrature is simply the linear transformation

$$x = \frac{(b-a)t + b + a}{2}, \quad \text{or} \quad \frac{2x - b - a}{b - a} = t.$$

Thus, the integral from a to b of the function $f(x)$ is changed into the integral from -1 to 1 of the function

$$f\left[\frac{(b-a)t + b + a}{2}\right]\frac{b-a}{2},$$

where the factor $\dfrac{b-a}{2}$ comes from the conversion from dx to dt. Accordingly, we now apply Gaussian quadrature to the integral

$$\int_{-1}^{1} f\left[\frac{(b-a)t + b + a}{2}\right]\frac{b-a}{2}\, dt.$$

This change of variable is included in the following algorithm and Mathcad function for Gaussian quadrature. Since it adds very little extra computation, it is not important to distinguish the cases where the interval of integration of the original problem is $[-1, 1]$.

Algorithm for Gaussian Quadrature

Approximate the integral of $f(x)$ on $[a,b]$ using Gaussian quadrature at $k = 2, \ldots 5$ points

Set k, the desired number of quadrature points.

Find points x_i in $[a, b]$ that corresponds to the quadrature points $t(i, k-1)$.

```
for i = 1 to k
    x_i = 0.5 ((b - a) t(i,k - 1) + b + a)
end
```

Define parameters

$$t = \begin{bmatrix} -0.5773502692 & -0.7745966692 & -0.8611363116 & -0.9061798459 \\ 0.5773502692 & 0.0000000000 & -0.3399810436 & -0.5384693101 \\ 0.0 & 0.7745966692 & 0.3399810436 & 0.0000000000 \\ 0.0 & 0.0 & 0.8611363116 & 0.5384693101 \\ 0.0 & 0.0 & 0.0 & 0.9061798459 \end{bmatrix}$$

$$c = \begin{bmatrix} 1.0 & 0.5555555556 & 0.3478548451 & -0.2369268850 \\ 1.0 & 0.8888888889 & 0.6521451549 & -0.4786286705 \\ 0.0 & 0.5555555556 & 0.6521451549 & 0.5688888889 \\ 0.0 & 0.0 & 0.3478548451 & 0.4786286705 \\ 0.0 & 0.0 & 0.0 & 0.2369268850 \end{bmatrix}$$

Approximate the integral.

$$S = \frac{b-a}{2} \sum_{i=1}^{K} c(i, k-1) f(x_i)$$

Mathcad Function for Gaussian Quadrature

$$\text{Gauss\_quad}(f, a, b, k) := \begin{vmatrix} t \leftarrow \begin{pmatrix} -0.577350269 & -0.77459667 & -0.86113632 & -0.906179846 \\ 0.577350269 & 0 & -0.33998144 & -0.53846931 \\ 0 & 0.77459667 & 0.33998144 & 0 \\ 0 & 0 & 0.86113632 & 0.53846931 \\ 0 & 0 & 0 & 0.906179846 \end{pmatrix} \\ c \leftarrow \begin{pmatrix} 1 & 0.555555556 & 0.347854845 & 0.236926885 \\ 1 & 0.888888889 & 0.652145155 & 0.4788628671 \\ 0 & 0.555555556 & 0.652145155 & 0.568888889 \\ 0 & 0 & 0.347854845 & 0.4788628671 \\ 0 & 0 & 0 & 0.236926885 \end{pmatrix} \\ S \leftarrow 0 \\ \text{for } i \in 0..k-1 \\ \quad \begin{vmatrix} x_i \leftarrow 0.5 \cdot [(b-a) \cdot t_{i, k-2} + b + a] \\ S \leftarrow S + c_{i, k-2} \cdot f(x_i) \end{vmatrix} \\ I \leftarrow \frac{S \cdot (b-a)}{2} \\ I \end{vmatrix}$$

Chapter 11 Numerical Differentiation and Integration

We now illustrate the use of Gaussian quadrature on an interval other than $[-1, 1]$.

> **Example 11.15 Integral of $\exp(-x^2)$ on $[0,2]$ Using Gaussian Quadrature.**
>
> Consider again the integral
>
> $$f(x) = \exp(-x^2), \quad x_0 = a = 0, \quad x_1 = b = 2.$$
>
> The required change of variable gives
>
> $$x = \frac{(b-a)t + b + a}{2} = \frac{(2-0)t + 2 + 0}{2} = t + 1,$$
>
> or
>
> $$t = \frac{2x - b - a}{b - a} = \frac{2x - 2 - 0}{2 - 0} = x - 1.$$
>
> Thus, the integral from a to b of the function $f(x)$ is changed into the integral from -1 to 1 of the function $\exp[-(t+1)^2]$.
>
> We now apply Gaussian quadrature to the integral $\int_{-1}^{1} \exp[-(t+1)^2]\, dt$:
>
> $$\int_{-1}^{1} f(t)\, dt \approx c_1 f(t_1) + c_2 f(t_2) = \exp[-(1.5774)^2] + \exp[-(0.4226)^2] = 0.9195.$$
>
> $$f(t) := \exp\left[-(t+1)^2\right] \qquad a := -1 \quad b := 1$$
>
> $k := 2$ Gauss_quad$(f, a, b, k) = 0.919486$
>
> $k := 3$ Gauss_quad$(f, a, b, k) = 0.878865$
>
> $k := 4$ Gauss_quad$(f, a, b, k) = 0.882229$
>
> $k := 5$ Gauss_quad$(f, a, b, k) = 0.882289$

Discussion

It is fairly simple to directly derive the coefficients for the case of $n = 2$ by requiring that the integration formula give the exact result for the polynomials $f_0 = 1$, $f_1 = x$, $f_2 = x^2$ and $f_3 = x^3$. For $f_0 = 1$, we require that

$$\int_{-1}^{1} 1 \, dx = 2 = c_1 + c_1;$$

for $f_1 = x$, we require that

$$\int_{-1}^{1} x \, dx = 0 = c_1 x_1 + c_1 x_2;$$

for $f_2 = x^2$, we require that

$$\int_{-1}^{1} x^2 \, dx = \frac{2}{3} = c_1 x_1^2 + c_1 x_2^2;$$

and for $f_3 = x^3$, we require that

$$\int_{-1}^{1} x^3 \, dx = 0 = c_1 x_1^3 + c_1 x_2^3.$$

These four equations must be solved for the points x_1 and x_2 and the coefficients c_1 and c_2. First, we observe that none of the unknowns can be zero. Then, solving the second equation for c_1 and substituting into the fourth equation gives $x_1^2 = x_2^2$. Since we assume that $x_1 \neq x_2$, we must have $x_1 = -x_2$. Substituting the expressions for c_1 and x_1 into the first equation gives $c_2 = c_1 = 1$. Finally, using the third equation gives $x_1 = -1/\sqrt{3}$ and $x_2 = +1/\sqrt{3}$.

Using a direct algebraic approach to derive higher-degree Gaussian quadrature is not practical. Instead, the analysis utilizes the orthogonality of the Legendre polynomials (introduced in Chapter 9). The best approximation to the integral is obtained when the function is evaluated at the zeros of the appropriate Legendre polynomial. (See Atkinson, 1989, for details.)

The n-point Gauss-Legendre quadrature rule evaluates a function at the n zeros of the nth-degree Legendre polynomial; the quadrature rule has a degree of precision of at least $2n - 1$ (i.e., it is exact for polynomials of degree at least $2n - 1$).

The coefficients are

$$c_i = \frac{-2}{(n+1) P_n'(x_i) P_{n+1}(x_i)}.$$

(See Atkinson, 1989, p. 276; Ralston and Rabinowitz, 1978, p. 105.)

The error in Gaussian quadrature goes to zero more rapidly for integrands that are smoother, whereas the composite trapezoid rule converges as h^2, regardless of the smoothness of $f(x)$. (See Atkinson, 1989.)

11.5 MATHCAD'S METHODS

Mathcad has operators on the Calculus Toolbar for finding numerical approximations to derivatives and definite integrals.

11.5.1 Using the Operators

Derivative Operator

To use the derivative operator on the Calculus Toolbar, one must first define the function whose derivative is desired. The function is a scalar (not vector) valued function of a scalar variable. The result of applying the operator is a number, not a function. It is, however, possible to define a function in terms of the derivative of another function. For example, if the function $df(x)$ is defined as the $[d/dx]\ f$, (where $f(x)$ is a previously defined function of x), then whenever the function df is called, Mathcad will compute the numerical value of the derivative of $f(x)$ at the appropriate value of x.

The value of the first derivative computed by Mathcad is usually accurate to 7 or 8 significant digits, unless the point where the derivative is being approximated is close to a singularity of the function. The accuracy of higher derivatives decreases by one significant digit for each higher order of derivative.

Definite Integral

To use the definite integral operator on the Calculus Toolbar, first define the function $f(t)$ to be integrated. All variables in the expression $f(t)$, except the variable of integration t, must be defined. One must also define the limits of integration, which must be real scalars, or positive or negative infiinity.

The numerical integration algorithm makes successive estimates of the integral and returns a value when the two most recent estimates differ by less than the built-in tolerance. The tolerance can be reset by defining the desired value of TOL in your worksheet before using the integration routine.

To evaluate a double, or multiple, integral, press the ampersand key [&] twice, and fill in the integrand, and limits on the integrals. In general, multiple integrals take much longer to compute than single integrals.

Mathcad has several methods of numerical integration, and selects the appropriate method according to the characteristices of the integrand. In Mathcad 2000Pro, the user may specify the desired method.

11.5.2 Understanding the Algorithms

Numerical Derivatives

Although the intuitive idea of a finite difference approximation for computing a numerical derivative is attractive, there are severe difficulties in choosing an appropriate step size to minimize both truncation and roundoff error. See [Press, et al., 1992, pp. 186–189] for a discussion of the difficulties involved in using a simple finite difference approximation for computing a numerical derivative, as well as a presentation of Ridder's method for derivatives (which is the method implemented by Mathcad). The basic idea of Ridder's method is to apply extrapolation to a sequence of finite difference approximations to the derivative.

Integration

Mathcad's AutoSelect chooses the appropriate method of integration depending on the characteristice of the integrand, and the interval of integration.

The most basic method is Romberg integration with Richarson extrapolation. The method is essentially as described in Section 11.2.3.

If the integrand varies rapidly on the interval, an adaptive integration routine is used, with unequal subintervals. This corresponds to rewriting the problem

$$I = \int_a^b f(x)\, dx$$

as

$$\frac{dy}{dx} = f(x); \quad y(a) = 0$$

and solving for $I = y(b)$.

See [Press et al., 1992, Chapter 16] for discussion of adaptive step-size ODE solution techniques.

If one or both of the limits of the integration are infinite, we need an integration formula that does not require a function evaluation at the end of the interval. Such formulas are known as open formulas, and were discussed briefly in Section 11.2. The prefered open method [Press et al., 1992, p. 142] is the Second Euler-Maclaurin summation formula, namely

$$\int_{x_1}^{x_N} f(x)\, dx = h[f_{3/2} + f_{5/2} + \cdots f_{N-3/2} + f_{N-1/2}]$$

where the function evaluations are at the midpoints of the subintervals.

The first Euler-Maclaurin summation formula is the extended or composite trapezoid rule

$$\int_{x_1}^{x_N} f(x)\,dx = h\left[\frac{1}{2}f_1 + f_2 + \cdots f_{N-1} + \frac{1}{2}f_N\right]$$

The importance of these formulas comes from the fact (not at all trivial) that the error for each of these formulas can be expressed as a power series in h, with only even powers of h. The second Euler-Maclaurin formula can be found by taking the first with step-size h, and again with step-size $h/2$, and subtracting the first from twice the second.

In order to extrapolate the second Euler-Maclaurin formula (and reuse the function evaluations from one stage to the next), we triple the number of subintervals in the second stage. If the result of the first stage is A and the result of the second stage is B, then the extrapolated integral is

$$I = (9B - A)/8$$

since tripling the number of subintevals reduces the error by a factor of 1/9 rather than the 1/4 that occurs when the number of subintervals is doubled.

To use an open integration formula, such as this, for an integral on an infinite interval, the integration routine includes a change of variable, $t = e^{-x}$ or $x = -\ln t$, so that an integral from $x = a$ to $x = \infty$ becomes an integral from $t = 0$ to $t = e^{-a}$. See [Press et al., 1992, pp. 141–147].

Mathcad also implements the Gauss-Kronod method, which is an extension of Gaussian quadrature in which each higher order method uses the abscissas of the previous lower order method, thus allowing the reuse of previous function evaluations. The original paper by Kronod was published in 1964, in Russian. An automatic integration routine in the popular subroutine package QUADPACK [Piessens et al., 1983] uses a sequence of Gauss-Kronod integrations with $N = 10$, 21, 43, 87, ... points.

Other integration methods are based on ideas from function approximation; in particular, Mathcad incorporates Clenshaw-Curtis quadrature. The basic idea is that for a smooth function, the coefficients c_j which represent the function in terms of Chebyshev polynomials, decrease very rapidly as j increases. Furthermore, if the coefficients are known, the coefficients of the antiderivative function (or the derivative function) can be found directly from the c_j. For an adaptive method, one can compute the coefficients as a sum of products of cosine functions; the sum can be computed efficiently using fast cosine transforms. See [Press et al., 1992, p. 196] for further discussion.

SUMMARY

First derivatives:

Forward difference formula

$$f'(x_i) \approx \frac{y_{i+1} - y_i}{x_{i+1} - x_i}.$$

Backward difference formula

$$f'(x_i) \approx \frac{y_i - y_{i-1}}{x_i - x_{i-1}}.$$

Central difference formula

$$f'(x_i) \approx \frac{f(x_{i+1}) - f(x_{i-1})}{x_{i+1} - x_{i-1}}.$$

Three-point forward formula

$$h = x_{i+1} - x_i = x_{i+2} - x_{i+1},$$

$$f'(x_i) \approx \frac{1}{2h}[-3f(x_i) + 4f(x_{i+1}) - f(x_{i+2})].$$

Second derivative:

$$f''(x_i) \approx \frac{1}{h^2}[f(x_{i+1}) - 2f(x_i) + f(x_{i-1})].$$

Integration: $h = (b-a)/n$

Trapezoid rule

$$\int_a^b f(x)\, dx \approx (h/2)\,[f(a) + 2f(x_1) + \cdots + 2f(x_{n-1}) + f(b)]$$

Simpson's rule (n must be even)

$$\int_a^b f(x)\, dx \approx (h/3)\,[f(a) + 4f(x_1) + 2f(x_2) + 4f(x_3)$$
$$+ 2f(x_4) + \cdots + 2f(x_{n-2}) + 4f(x_{n-1}) + f(b)]$$

Midpoint rule

$$\int_a^b f(x)\, dx \approx h \sum_{j=1}^n f(x_j); \quad x_j = a + \left(j - \frac{1}{2}\right)h$$

Gaussian quadrature

$$\int_{-1}^{1} f(x)\, dx \approx \sum_{i=1}^{n} c_i f(x_i)$$

n	x_i	c_i
2	$\pm 1/\sqrt{3} \approx \pm 0.57735$	1
3	0	8/9
	$\pm\sqrt{3/5} \approx \pm 0.77459$	5/9
4	$\approx \pm 0.861136$	0.34785
	$\approx \pm 0.339981$	0.652145

SUGGESTIONS FOR FURTHER READING

Basic formulas
- Ames, W. F., *Numerical Methods for Partial Differential Equations*, 3rd ed., Academic Press, Boston, 1992.
- Atkinson, K. E., *An Introduction to Numerical Analysis*, 2d ed., John Wiley & Sons, New York, 1989.
- Isaacson, E., and H. B. Keller, *Analysis of Numerical Methods*, Dover, New York, 1994 (originally published by John Wiley & Sons, 1966).
- Ralston, A., and P. Rabinowitz, *A First Course in Numerical Analysis*, 2d ed., McGraw-Hill, New York, 1978.

Applications
- Ayyub, B. M., and R. H. McCuen, *Numerical Methods for Engineers*, Prentice Hall, Upper Saddle River, NJ, 1996.
- Abramowitz, M., and I. A. Stegun (eds.), *Handbook of Mathematical Functions, with Formulas, Graphs, and Mathematical Tables*, Dover, New York, 1965.
- Ritger, P. D., and N. J. Rose, *Differential Equations with Applications*, McGraw-Hill, New York, 1968.
- Jensen, J. A., and J. H. Rowland, *Methods of Computation*, Scott, Foresman and Company, Glenview, IL, 1975.

More advanced methods:
- Piessens, R., E. de Doncker, C. W. Uberhuber, and D. K. Kahaner, *QUADPACK: A Subroutine Package for Automatic Integration*, Springer-Verlag, New York, 1983.
- Clenshaw, C. W. and A. R. Curtis, *Numerische Mathematik*, vol. 2, pp. 197–205, 1960.
- Ridders, C. J. F., *Advances in Engineering Software*, vol. 4, no. 2, 1982, pp. 75–76.

PRACTICE THE TECHNIQUES

For problems P11.1 to P11.5, approximate the specified derivative

a. Using the forward-difference formula.
b. Using the backward-difference formula.
c. Using the central-difference formula.
d. Using Richardson extrapolation to improve your answer to Part c.

P11.1 Approximate $y'(1.0)$ if

$\mathbf{x} = [0.8 \quad 0.9 \quad 1.0 \quad 1.1 \quad 1.2]$

$\mathbf{y} = [0.992 \quad 0.999 \quad 1.000 \quad 1.001 \quad 1.008]$

P11.2 Approximate $y'(2)$ if

$\mathbf{x} = [0 \quad 1 \quad 2 \quad 3 \quad 4]$

$\mathbf{y} = [0 \quad 1 \quad 4 \quad 9 \quad 16]$

P11.3 Approximate $y'(1)$ if

$\mathbf{x} = [-1 \quad 0 \quad 1 \quad 2 \quad 3]$

$\mathbf{y} = [1/3 \quad 1 \quad 3 \quad 9 \quad 27]$

P11.4 Approximate $y'(1)$ if

$\mathbf{x} = [-1 \quad 0 \quad 1 \quad 2 \quad 3]$

$\mathbf{y} = [1/2 \quad 1 \quad 2 \quad 4 \quad 8]$

P11.5 Approximate $y'(4)$ if

$\mathbf{x} = [0 \quad 1 \quad 4 \quad 9 \quad 16]$

$\mathbf{y} = [0 \quad 1 \quad 2 \quad 3 \quad 4]$

For Problems P11.6 to P11.15, approximate the specified integral

a. Using the composite trapezoid method with 2 subintervals.
b. Using the composite trapezoid method with 10 subintervals.
c. Using Simpson's rule with 2 subintervals.
d. Using the composite Simpson's rule with 10 subintervals.
e. Using Gaussian quadrature with $n = 2$.
f. Using Romberg integration.

P11.6 $\int_0^1 x \sin(\pi x)\, dx$.

P11.7 $\int_0^4 2^x\, dx$.

P11.8 $\int_0^2 \sqrt{x}\, dx$.

P11.9 $\int_1^2 \dfrac{dx}{1+x}$.

P11.10 $\int_1^2 \dfrac{dx}{x}$.

P11.11 $\int_{-1}^1 \dfrac{dx}{1+x^2}$.

P11.12 $\int_0^2 e^x\, dx$.

P11.13 $\int_0^1 \dfrac{1+x}{1+x^3}\, dx$.

P11.14 $\int_1^2 \sqrt{x^3 - 1}\, dx$.

P11.15 $\int_2^3 \sqrt{x^2 - 4}\, dx$.

P11.16 $\int_0^\pi x^2 \sin(2x)\, dx$.

P11.17 $\int_0^3 \dfrac{1}{\sqrt{x^3 + 1}}\, dx$.

P11.18 $\int_0^\pi x^3 \sin(x^2)\, dx$.

P11.19 $\int_0^3 \sqrt[3]{x^3 + 1}\, dx$.

P11.20 $\int_0^2 \ln(x^3 + 1)\, dx$.

EXPLORE SOME APPLICATIONS

A11.1 The flow rate of an incompressible fluid in a pipe of radius 1 is given by

$$Q = \int_0^1 2\pi r V \, dr$$

where r is the distance from the center of the pipe and V is the velocity of the fluid. Find Q if only the following tabulated velocity measurements V are available:

$r = [0.0 \;\; 0.1 \;\; 0.2 \;\; 0.3 \;\; 0.4 \;\; 0.5 \;\; 0.6 \;\; 0.7 \;\; 0.8 \;\; 0.9 \;\; 1.0]$

$V = [1.0 \;\; 0.99 \;\; 0.96 \;\; 0.91 \;\; 0.84 \;\; 0.75 \;\; 0.64 \;\; 0.51 \;\; 0.36 \;\; 0.19 \;\; 0.0]$

Compare your result with the value obtained by using $V = 1 - r^2$. (See Ayyub and McCuen, 1996, for a discussion of similar problems.)

Problems A11.2 to A11.7 investigate some important nonelementary functions that are defined by integrals. Tabulated values are available for these functions in many reference books.

A11.2 The error function is defined as

$$\text{Erf}(x) = \frac{2}{\sqrt{\pi}} \int_0^x e^{-t^2} \, dt.$$

Find Erf(2).

A11.3 The sine-integral is defined as

$$Si(x) = \int_0^x \frac{\sin t}{t} \, dt.$$

Find $Si(2)$.

A11.4 The cosine-integral is defined as

$$Ci(x) = \int_1^x \frac{\cos t}{t} \, dt.$$

Find $Ci(2)$.

A11.5 The exponential-integral is defined as

$$Ei(x) = \int_1^x \frac{e^{-t}}{t} \, dt.$$

Find $Ei(2)$.

A11.6 The Fresnel integrals are defined as

$$C(x) = \int_0^x \cos\left(\frac{\pi}{2} t^2\right) dt$$

$$S(x) = \int_0^x \sin\left(\frac{\pi}{2} t^2\right) dt.$$

Find $C(2)$ and $S(2)$.

A11.7 There are several forms of elliptic integrals.

a. The complete elliptic integral of the first kind is defined as

$$K(m) = \int_0^{\pi/2} \frac{dt}{\sqrt{1 - m \sin^2 t}}.$$

Find $K(1)$ and $K(4)$.

b. The complete elliptic integral of the second kind is defined as

$$E(m) = \int_0^{\pi/2} \sqrt{1 - m \sin^2 t} \, dt.$$

Find $E(1)$ and $E(4)$.

c. The elliptic integral of the first kind is

$$K(k, x) = \int_0^x \frac{dt}{\sqrt{1 - k^2 \sin^2 t}}.$$

Find $K(1,1)$ and $K(2,1)$.

d. The elliptic integral of the second kind is

$$E(k, x) = \int_0^x \sqrt{1 - k^2 \sin^2 t} \, dt.$$

Find $E(1,1)$ and $E(2,1)$. (See Ritger and Rose, 1968, or Abramowitz and Stegun, 1965, for tabulated values.)

Problems A11.8 to A11.15 illustrate some integrals that arise in the measurement of arc length. For many functions, the integral that measures its arc length cannot be evaluated exactly.

A11.8 Find the arc length of the curve described by the function $y = x^2$, $0 < x < 2$; the length is given by the integral

$$\int_0^2 \sqrt{1 + 4x^2}\, dx.$$

A11.9 Find the arc length of the curve described by the function $y = x^3$, $0 < x < 2$; the length is given by the integral

$$\int_0^2 \sqrt{1 + 9x^4}\, dx.$$

A11.10 Find the arc length of the curve described by the function $y = x^{-1}$, $1 < x < 2$; the length is given by the integral

$$\int_1^2 \sqrt{1 + x^{-4}}\, dx.$$

A11.11 Find the arc length of the curve described by the function $y = 1/x$, $1 < x < 2$; the length is given by the integral

$$\int_1^2 \sqrt{1 + (1/4)x^{-1}}\, dx.$$

A11.12 Find the arc length of the curve described by the function $y = \sin(x)$, $0 < x < \pi$; the length is given by the integral

$$\int_0^\pi \sqrt{1 + \cos^2(x)}\, dx.$$

A11.13 Find the arc length of the curve described by the function $y = \tan(x)$, $0 < x < \pi/4$; the length is given by the integral

$$\int_0^{\pi/4} \sqrt{1 + \sec^4(x)}\, dx.$$

A11.14 Find the arc length of the curve described by the function $y = e^x$, $0 < x < 2$; the length is given by the integral

$$\int_0^2 \sqrt{1 + (e^x)^2}\, dx = \int_0^2 \sqrt{1 + e^{2x}}\, dx.$$

A11.15 Find the arc length of the curve described by the function $y = \ln(x)$, $1 < x < 2$; the length is given by the integral

$$\int_1^2 \sqrt{1 + x^{-2}}\, dx.$$

EXTEND YOUR UNDERSTANDING

U11.1 Use the definition of the Legendre polynomial

$$P_n(x) = \frac{(-1)^n}{2^n n!} \frac{d^n}{dx^n}[(1 - x^2)^n], \quad n \geq 1,$$

to find a relationship between $P_n'(x)$ and $P_{n+1}(x)$, and then show the equivalence of the following expressions for the coefficients for Gaussian quadrature:

$$c_i = \frac{-2}{(n + 1)\, P_n'(x_i) P_{n+1}(x_i)} \quad \text{and}$$

$$c_i = \frac{2(1 - x_i^2)}{(n + 1)^2 P_{n+1}^2(x_i)}.$$

U11.2 The coefficients for Gauss-Legendre quadrature can be given by the integral of the Lagrange interpolating polynomial (Jensen and Rowland, 1975, p. 225) as follows:

$$c_j = \int_{-1}^{1} \frac{(x - x_1)\ldots(x - x_{j-1})(x - x_{j+1})\ldots(x - x_n)}{(x_j - x_1)\ldots(x_j - x_{j-1})(x_j - x_{j+1})\ldots(x_j - x_n)}\, dx.$$

Use this form to compute the c_j for $n = 1, 2$, and 3.

U11.3 Use Gaussian quadrature with $n = 3$ and exact arithmetic to approximate $\int_{-1}^{1} x^4\, dx$.

Compare your results to the exact value of the integral and discuss the two.

U11.4 Consider again the computations for I_0, I_1, and I_2 in Romberg integration, but now pay special attention to the function evaluations that can be reused at each stage. Since the value of h changes at each stage, it may be helpful to denote the value at stage i as h_i.

$$\int_a^b f(x)\, dx \approx \frac{h}{2}[f(a) + 2f(x_1) + \ldots + 2f(x_{n-1}) + f(b)]$$

$$h = \frac{b-a}{n}, n \text{ subinternvals}$$

First take 1 subinterval, i.e., $n = 1, h_0 = b - a$:

$$I_0 \approx \frac{b-a}{2}[f(a) + f(b)]$$

Next, 2 subintervals, $n = 2, h_1 = \frac{b-a}{2}$

$$I_1 \approx \frac{b-a}{4}\left[f(a) + 2f\left(\frac{a+b}{2}\right) + f(b)\right]$$

$$= \frac{b-a}{4}[f(a) + f(b)] + \frac{b-a}{4} 2f\left(\frac{a+b}{2}\right)$$

$$= \frac{1}{2}I_0 + h_1 f\left(\frac{a+b}{2}\right)$$

Subdividing again, 4 subintervals, $n = 4, h_2 = \frac{b-a}{4}$

$$I_2 \approx \frac{b-a}{8}\left[f(a) + 2f\left(\frac{3a+b}{4}\right) + 2f\left(\frac{a+b}{2}\right)\right.$$

$$\left. + 2f\left(\frac{a+3b}{4}\right) + f(b)\right]$$

$$= \frac{b-a}{8}\left[f(a) + 2f\left(\frac{a+b}{2}\right) + f(b)\right]$$

$$+ \frac{b-a}{8}\left[2f\left(\frac{3a+b}{4}\right) + 2f\left(\frac{a+3b}{4}\right)\right]$$

$$= \frac{1}{2}I_1 + h_2\left[f\left(\frac{3a+b}{4}\right) + f\left(\frac{a+3b}{4}\right)\right]$$

Expand the algorithm and Mathcad functions for Romberg integration to minimize the number of function evaluations required by replacing the calls to the trapezoid function by the computation shown here. Compare the compuational effort and the clarity of the process for the two forms.

U11.5 Show that the first extrapolated value obtained in Romberg integration is identical to that found by Simpson's rule.

U11.6 The error incurred in using the composite trapezoid rule to integrate $f(x)$ for $a \le x \le b$, with n subdivisions and $h = (b - a)/n$ is

$$E_T = \frac{-1}{12}(b-a) h^2 f''(\eta) \quad \text{for some } \eta \in [a, b].$$

Use this formula to find a bound on the error in the results obtained for Problems P11.6, P11.8, P11.10, P11.12, P11.15
Now, find the actual error for each of the preceding integrals—i.e., the difference between the exact value and the approximate value of the given integral. Compare the actual error with the error bound.

U11.7 The error incurred in using the composite Simpson rule to integrate $f(x)$ for $a \le x \le b$, with n subdivisions and $h = (b - a)/n$ is

$$E_S = \frac{-1}{180}(b-a) h^4 f^{(4)}(\eta), \quad \text{for some } \eta \in [a, b].$$

Use this formula to find a bound on the error in the results obtained for Problems P11.6, P11.8, P11.10, P11.12, P11.15
Now, find the actual error for each of the preceding integrals—i.e., the difference between the exact value and the approximate value of the given integral. Compare the actual error with the error bound.

U11.8 The error incurred in using the composite midpoint rule to integrate $f(x)$ for $a \le x \le b$, with n subdivisions and $h = (b - a)/n$ is

$$E_M = \frac{1}{24}(b-a) h^2 f''(\eta), \quad \text{for some } \eta \in [a, b].$$

Use this formula to find a bound on the error in the results obtained for Problems P11.6, P11.8, P11.10, P11.12, P11.15
Now, find the actual error for each of the preceding integrals—i.e., the difference between the exact value and the approximate value of the given integral. Compare the actual error with the error bound.

12

Ordinary Differential Equations: Initial-Value Problems

12.1 Taylor Methods
12.2 Runge-Kutta Methods
12.3 Multistep Methods
12.4 Stability
12.5 Mathcad's Methods

The numerical differentiation formulas presented in the previous chapter are used extensively in the numerical solution of ordinary and partial differential equations; techniques for solving first-order ordinary differential equations are the subject of this chapter. We assume that the differential equation is written in the form $y' = f(x, y)$ with the value of the function $y(x)$ given at x_0; i.e., $y(x_0) = y_0$. The basic idea is to divide the interval of interest into discrete steps (of fixed width h) and find approximations to the function y at those values of x. In other words, we find solutions at $x_1, x_2, x_3, \ldots, x_n$.

The first methods we consider are based on the Taylor polynomial representation of the unknown function $y(x)$. The simplest method, Euler's, retains only the first-derivative term in the Taylor expansion. In order to use higher order Taylor polynomials, it is necessary to find higher derivatives of the function $f(x, y)$ that defines the slope of the unknown function $y(x)$. The Runge-Kutta method achieves a more accurate solution than Euler's method does, without computing higher derivatives of f. Each step of a Runge-Kutta method involves evaluating $f(x, y)$ at several different values of x and y and combining the results to form the approximation to y at the next x.

The third group of techniques that we study are the "multistep methods," which include both explicit and implicit forms. The term "multistep" refers to the fact that these methods make use of the computed value of the solution at several previous points. Implicit methods have superior stability characteristics, but are more difficult to solve than the explicit methods. An implicit and an explicit method are often combined to form a predictor-corrector formula.

In the next chapter, we investigate techniques for solving higher order ordinary differential equation intial-value problems (ODE-IVPs) and systems of first order ODE-IVPs. In Chapter 14, we consider techniques for solving ODE boundary-value problems (BVPs). Finally, Chapter 15 presents an introduction to methods for numerically solving partial differential equations.

Example 12-A Motion of a Falling Body

The motion of a falling body is described by Newton's second law, $F = m\,a$. The acceleration a is the rate of change of the velocity of the body with respect to time. The forces acting on the body may include, in addition to gravity, air resistance that is proportional to a power of the velocity. Empirical studies suggest that air resistance can be modeled as

$$F = k\,v^p$$

where $1 \leq p \leq 2$ and the value of the proportionality constant k depends on the size and shape of the body, as well as the density and viscosity of the air. Typically, $p = 1$ for relatively slow velocities and $p = 2$ for high velocities. For intermediate velocities (with $0 < p < 1$), numerical methods may be especially appropriate.

For $p = 1$, the differential equation takes the form

$$m\frac{dv}{dt} = -k\,v - mg,$$

with the positive y-direction upward and $y = 0$ at ground level. Thus, when the body is falling ($v < 0$), the resistance force is positive (upward); when the body is rising ($v > 0$), the resistance acts in the downward direction.

The ratio k/m is known as the drag coefficient; for a parachutist, a typical value is $k/m \approx 1.5$; the terminal velocity is on the order of 21 ft/s. (See Edwards and Penney, 1996.)

For resistance proportional to the square of the velocity, we must distinguish between upward and downward motion, to be sure that the resistance force is acting opposite to the motion of the object. If we leave our coordinate axis as before (with up positive), we have

$$m\frac{dv}{dt} = -k\,v^2 - m\,g \quad \text{(for a body moving upward)}$$

or

$$m\frac{dv}{dt} = k\,v^2 - m\,g \quad \text{(for a body moving downward).}$$

The velocity for downward motion with air resistance proportional to velocity squared is

$$v(t) = b\,\frac{1 + C\,e^{pt}}{1 - C\,e^{pt}},$$

where $p = 2\sqrt{g\,k/m}$ and $b = \sqrt{g\,m/k}$. The terminal velocity is $v_f = -b$.

12.1 TAYLOR METHODS

In this section, we consider two methods that are based on the Taylor series representation of the unknown function $y(x)$. We first discuss Euler's method, which uses only the first term in the Taylor series; we then present a higher order Taylor method.

12.1.1 Euler's Method

The simplest method of approximating the solution of the differential equation
$$y' = f(x, y)$$
is to treat the function as a constant, $f(x_0, y_0)$ and replace the derivative y' by the forward difference quotient. This gives
$$y_1 - y_0 = f(x_0, y_0)(x_1 - x_0),$$
or
$$y_1 = y_0 + h f(x_0, y_0),$$
where $h = (b - a) / n$, and n is the number of values of the independent variable where we wish to calculate the approximate solution. Geometrically, this corresponds to using the line tangent to the true solution curve $y(x)$ to find the value of y_1, the approximation to the value of y at x_1.

In general, Euler's method gives
$$y_i = y_{i-1} + h f(x_{i-1}, y_{i-1}) \quad \text{for } i = 1, \ldots, n.$$

If we take $f(x, y) = x + y$, $x_0 = 0$, $y_0 = 2$, and $h = 1/2$, the first step of Euler's method proceeds along the straight line shown in Fig. 12.1.

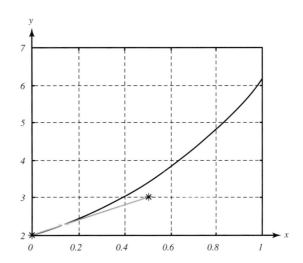

FIGURE 12.1 Exact solution and first step of Euler's method for $y' = x + y$.

A step-by-step procedure for carrying out the computations is given in the following algorithm.

Algorithm for Euler's Method

Approximate the solution of the ODE $y' = f(x, y)$ with initial condition $y(a) = y0$; using n steps of Euler's method; step size $h = (b - a)/n$.

Begin computations

$$h = \frac{b - a}{n}$$

Define grid points
for $i = 0$ to n
$\quad x_i = a + h\,i$
end

for $i = 1$ to n
$\quad y_i = y_{i-1} + h\,f(x_{i-1}, y_{i-1})$
end

Display results

Mathcad Function for Euler's Method

A simple Mathcad function to implement Euler's method is given here. It's use is illustrated in the next examples.

$\text{Euler}(f, \text{tspan}, y0, n) :=$
$\quad a \leftarrow \text{tspan}_0$
$\quad b \leftarrow \text{tspan}_1$
$\quad h \leftarrow \dfrac{b - a}{n}$
$\quad x_0 \leftarrow a$
$\quad y_0 \leftarrow y0$
$\quad \text{for } i \in 1..n$
$\quad\quad x_i \leftarrow a + h \cdot i$
$\quad \text{for } i \in 1..n$
$\quad\quad y_i \leftarrow y_{i-1} + h \cdot f(x_{i-1}, y_{i-1})$
$\quad s \leftarrow \text{stack}(x^T, y^T)$
$\quad s$

Example 12.1 Solving a simple ODE with Euler's Method

Consider the differential equation $y' = f(x, y)$ on $a \leq x \leq b$. Let

$$y' = x + y; \quad 0 \leq x \leq 1 \quad (\text{i.e.}, a = 0, b = 1), y(0) = 2.$$

First, we find the approximate solution for $h = 0.5$ ($n = 2$), a very large step size. The approximation at $x_1 = 0.5$ is

$$y_1 = y_0 + h(x_0 + y_0) = 2.0 + 5.0\,(0.0 + 2.0) = 3.0$$

Next, we find the approximate solution y_2 at $x_2 = 0.0 + 2h = 1.0$

$$y_2 = y_1 + h(x_1 + y_1) = 3.0 + 0.5\,(0.5 + 3.0) = 4.75.$$

To find a better approximate solution, we use $n = 20$ intervals, so that $h = 0.05$. We find the approximate solution using the Mathcad function for Euler's method.

$f(x,y) := x + y \qquad y0 := 2$

$\text{tspan} := \begin{pmatrix} 0 \\ 1 \end{pmatrix} \qquad n := 20$

$B := \text{Euler}(f, \text{tspan}, y0, n)$

Display results as the transpose of B; graph approximate and exact solutions.

$x := \left(B^T\right)^{\langle 0 \rangle} \qquad y := \left(B^T\right)^{\langle 1 \rangle}$

$t := 0, 0.1 .. 1 \qquad g(t) := 3 \cdot \exp(t) - t - 1$

$B^T = \begin{pmatrix} 0 & 2 \\ 0.05 & 2.1 \\ 0.1 & 2.208 \\ 0.15 & 2.323 \\ 0.2 & 2.447 \\ 0.25 & 2.579 \\ 0.3 & 2.72 \\ 0.35 & 2.871 \\ 0.4 & 3.032 \\ 0.45 & 3.204 \\ 0.5 & 3.387 \\ 0.55 & 3.581 \\ 0.6 & 3.788 \\ 0.65 & 4.007 \\ 0.7 & 4.24 \\ 0.75 & 4.487 \\ 0.8 & 4.749 \\ 0.85 & 5.026 \\ 0.9 & 5.32 \\ 0.95 & 5.631 \\ 1 & 5.96 \end{pmatrix}$

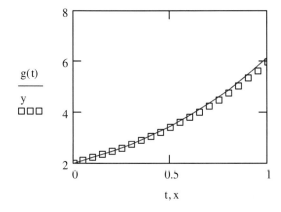

FIGURE 12.2   Euler solution and exact solution.

Example 12.2 Another Example of the Use of Euler's Method

Consider the differential equation

$$y' = f(x, y) = \begin{cases} y\left(-2x + \dfrac{1}{x}\right) & x \neq 0 \\ 1 & x = 0 \end{cases}$$

on the interval $0 \leq x \leq 2$ with initial value $y(0) = 0.0$. With $n = 10$ intervals, the step size $h = (b - a)/n = 0.2$.

We solve the ODE y' = f(x,y), with initial value y0, using Euler's method.

$$f(x, y) := \text{if}\left[x = 0, 1, y \cdot \left(-2 \cdot x + \dfrac{1}{x}\right)\right] \qquad y0 := 0 \qquad \text{tspan} := \begin{pmatrix} 0 \\ 2 \end{pmatrix} \qquad n := 10$$

$$B := \text{Euler}(f, \text{tspan}, y0, n) \qquad x := \left(B^T\right)^{\langle 0 \rangle} \qquad y := \left(B^T\right)^{\langle 1 \rangle}$$

$x^T =$

	0	1	2	3	4	5	6	7	8	9
0	0	0.2	0.4	0.6	0.8	1	1.2	1.4	1.6	1.8

$y^T =$

	0	1	2	3	4	5	6	7	8	9
0	0	0.2	0.384	0.515	0.563	0.523	0.419	0.287	0.168	0.081

The approximate solution, and the exact solution g(t).

$t := 0, 0.1 .. 2 \qquad g(t) := t \cdot \exp\left(-t^2\right)$

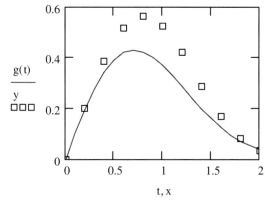

FIGURE 12.3   Euler solution ($n=10$) and exact solution.

Discussion

The question of how well a numerical technique for solving an initial-value ordinary differential equation works is closely related to the truncation error of the method. This is a measure of the error introduced by the approximation to the first derivative of y that is used in deriving the method. Euler's method employs only the first term in the Taylor expansion of the unknown function $y(x)$; that is,

$$y(x + h) = y(x) + h\, y'(x) + \frac{h^2}{2} y''(\eta)$$

with $y(x + h) = y_{i+1}$, $y(x) = y_i$, $y'(x) = f(x_i, y_i)$, $x < \eta < x + h$. Thus the *local truncation error* is $O(h^2)$. Of course, the actual error also depends on the higher derivatives of y, so if, in fact, y is linear, there will be no error. However, since the actual form of y is, in general, unknown, the dependence of the error on the step size is the most direct way to compare methods.

The total truncation error ε_n in going from x_0 to $x_0 + nh$ is bounded by an expression that depends on the Lipschitz constant for f, the bound for f', and the step size h. If f satisfies the Lipschitz condition $| f(x, y_2) - f(x, y_1)| < L\, |y_2 - y_1|$ and the second derivative of y is bounded ($| y''(\eta)| \leq N$), then

$$|\varepsilon_n| \leq \frac{h}{2} N\, \frac{\exp(L(x - x_0)) - 1}{L}.$$

Thus, the total truncation error for Euler's method is $O(h)$ and Euler's method is a first order method.

The usefulness of Euler's method is primarily a result of its simplicity, which makes it convenient for hand calculations. It may be used to provide a very few starting values for a multistep method (Section 12.3), although the Runge-Kutta methods presented in Section 12.2 give more accurate results for only slightly more computational effort.

The results of the basic Euler's method can be improved by using it with Richardson extrapolation (Froberg, 1985, p. 323). We sketch the process here, following the discussion in Jain (1979, pp. 57–58). For a given value of h, the value of y at x_{i+1} is

$$y_{i+1} = y_i + \sum_{j=1}^{\infty} c_i h^i.$$

If we denote the computed value as $Y(h)$, we can also find y_{i+1} by taking two steps with $h/2$ or by taking four steps with $h/4$, etc. Because the error expansion includes all powers of h, the extrapolation formula is

$$Y_m(k) = \frac{2^m Y_{m-1}(k + 1) - Y_{m-1}(k)}{2^m - 1},$$

where k designates the step size. ($k = 0, 1, \ldots$ correspond to step sizes of $h/2^0$, $h/2$, $h/2^2, \ldots, h/2^k$.) The parameter m gives the extrapolation level, with $Y_0\,(m = 1)$ being the values computed directly from Euler's method.

12.1.2 Higher-Order Taylor Methods

The discussion in the previous section suggests that one way to obtain a better solution technique is to use more terms in the Taylor series for y, in order to obtain higher-order truncation error. For example, a second-order Taylor method uses

$$y(x+h) = y(x) + h\,y'(x) + \frac{h^2}{2} y''(x) + O(h^3).$$

The local trucation error is $O(h^3)$ and the total (or global) trucation error is $O(h^2)$. However, we do not have a formula for $y''(x)$. If $y'(x) = f(x, y)$ is not too complicated, it may be practical to differentiate f with respect to x (using the chain rule, since y is a function of x) to find a representation for $y''(x)$.

Example 12.3 Solving a Simple ODE with Taylor's Method

Consider again the differential equation

$$y' = x + y;\ 0 \le x \le 1, \text{ with initial condition } y(0) = 2.$$

To apply the second-order Taylor method to the equation, we find

$$y'' = \frac{d}{dx}(x + y) = 1 + y' = 1 + x + y$$

This gives the approximation formula

$$y(x+h) = y(x) + h\,y'(x) + \frac{h^2}{2} y''(x),$$

or

$$y_{i+1} = y_i + h(x_i + y_i) + \frac{h^2}{2}(1 + x_i + y_i)$$

For $n = 2$ ($h = 0.5$), we find

$$y_1 = y_0 + h(x_0 + y_0) + \frac{h^2}{2}(1 + x_0 + y_0) = 2 + \frac{0+2}{2} + \frac{1+0+2}{8} = \frac{27}{8},$$

$$y_2 = y_1 + h(x_1 + y_1) + \frac{h^2}{2}(1 + x_1 + y_1)$$

$$= \frac{27}{8} + \frac{1}{2}\left(\frac{1}{2} + \frac{27}{8}\right) + \frac{1}{8}\left(1 + \frac{1}{2} + \frac{27}{8}\right) = 5.9219.$$

Chapter 12 Ordinary Differential Equations: Initial-Value Problems

Algorithm for Second-order Taylor's Method

Approximate the solution of the ODE $y' = f(x,y)$.

Define
  $f(x,y)$ right-hand side of ODE $y' = f(x,y)$
  $g(x, y) = f_x + f_y\, y' = f_x + f_y\, f$
  $h = (b - a)/n;$ $hh = h^2/2$
  $x_i = a + h\, i$ (for $i = 0$ to n)

for $i = 0$ to $n - 1$
  $y_{i+1} = y_i + h\, f(x_i, y_i) + hh\, g(x_i, y_i)$
end

A minor modification to the Mathcad function given for Euler's method produces the second-order Taylor approximation; in addition to the function $y' = f(x, y)$, the function for $y'' = g(x,y)$ must be provided.

Mathcad Function for Second-Order Taylor Method

$\text{Taylor}(f, g, \text{tspan}, y0, n) :=$
$\quad a \leftarrow \text{tspan}_0$
$\quad b \leftarrow \text{tspan}_1$
$\quad h \leftarrow \dfrac{b - a}{n}$
$\quad hh \leftarrow \dfrac{h \cdot h}{2}$
$\quad x_0 \leftarrow a$
$\quad y_0 \leftarrow y0$
$\quad \text{for } i \in 1..n$
$\quad\quad x_i \leftarrow a + h \cdot i$
$\quad \text{for } i \in 0..n-1$
$\quad\quad y_{i+1} \leftarrow y_i + h \cdot f(x_i, y_i) + hh \cdot g(x_i, y_i)$
$\quad s \leftarrow \text{stack}\left(x^T, y^T\right)$
$\quad s$

To illustrate the use of the Mathcad function Taylor, we return to Example 12.3.

Example 12.3 Solving a Simple ODE with Taylor's Method, continued

$f(x,y) := x + y$ $\qquad y0 := 2 \qquad$ tspan $:= \begin{pmatrix} 0 \\ 1 \end{pmatrix} \qquad n := 10$

$g(x,y) := 1 + x + y$

$B := \text{Taylor}(f, g, \text{tspan}, y0, n)$

$x := (B^T)^{\langle 0 \rangle}$

$x^T =$

	0	1	2	3	4	5	6	7	8	9
0	0	0.1	0.2	0.3	0.4	0.5	0.6	0.7	0.8	0.9

$y := (B^T)^{\langle 1 \rangle}$

$y^T =$

	0	1	2	3	4	5	6	7	8	9
0	2	2.215	2.463	2.748	3.073	3.442	3.861	4.335	4.868	5.469

$t := 0, 0.1 .. 1 \qquad g(t) := 3 \cdot \exp(t) - t - 1$

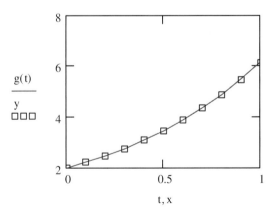

FIGURE 12.4 Simple example of Taylor's Method for ODEs.

12.2 RUNGE-KUTTA METHODS

In the previous section, we saw that using more terms in the Taylor series representation for the unknown function y gives more accurate results, but the necessity of computing derivatives of $f(x,y)$ is often too difficult to make the higher order Taylor methods attractive.

In this section, we consider several forms of a method of obtaining estimates for the slope of our unknown function y that do not require differentiating $f(x, y)$ in order to use higher order Taylor series expansions, but that still improve on the accuracy we can obtain from Euler's method. For example, if we could use the slope of y (i.e., the value of f) at the midpoint of the interval, it would seem to be a more balanced approximation than relying only on information at the left end of the interval. We do not know the value of y at the midpoint, but we can estimate it, as we shall see.

12.2.1 Midpoint Method

The simplest Runge-Kutta method is based on approximating the value of y at $x_i + h/2$ by taking one-half of the change in y that is given by Euler's method and adding that on to the current value y_i. This method is known as the *midpoint method*. The formulas are

$$k_1 = h f(x_i, y_i) \quad \text{(change in } y \text{ given by Euler's method)},$$

$$k_2 = h f\left(x_i + \frac{1}{2}h, y_i + \frac{1}{2}k_1\right) \quad \text{(change in } y \text{ using slope estimate at midpoint)},$$

$$y_{i+1} = y_i + k_2.$$

Although the geometric interpretation helps the method seem more intuitively plausible, the derivation of a Runge-Kutta method depends on finding ways of approximating the slope of y, using the function f evaluated at various points in the interval, so that the accuracy agrees with that obtained from Taylor series approximations. Not too surprisingly, using more involved approximations gives higher-order (more accurate) methods.

Algorithm for the Midpoint Method

Approximate the solution of the ODE $y' = f(x,y)$.

Define right-hand side of ODE, initial value, and parameters
$f(x, y) := \qquad y_0 :=$
$a := \qquad b := \qquad n := \qquad h := \dfrac{b - a}{n}$

for $i = 0$ to $n - 1$

$\quad x_i = a + h\,i \qquad k_1 = h f(x_i, y_i) \qquad k_2 = h f\left(x_i + \dfrac{1}{2}h, y_i + \dfrac{1}{2}k_1\right)$

$\quad y_{i+1} = y_i + k_2$

end

Example 12.4 Solving a Simple ODE with the Midpoint Method

Consider the differential equation $y' = f(x,y)$ on $a \leq x \leq b$.

$$y' = x + y, \quad 0 \leq x \leq 1 \text{ (i.e., } a = 0, b = 1), \quad y(0) = 2.$$

First, we find the approximate solution for $h = 0.5$ ($n = 2$), a very large step size. The approximation at $x_1 = 0.5$ is

$$k_1 = h\,f(x_0, y_0) \qquad\qquad = 0.5\,(x_0 + y_0) \qquad\qquad = 1.0,$$

$$k_2 = h\,f\!\left(x_0 + \frac{h}{2}, y_0 + \frac{k_1}{2}\right) = 0.5\,(x_0 + 0.25 + y_0 + 0.5) = 1.375,$$

$$y_1 = y_0 + k_2 \qquad\qquad\qquad\qquad\qquad\qquad\qquad\qquad\qquad = 3.375.$$

Next, we find the approximate solution y_2 at point $x_2 = 0.0 + 2h = 1.0$:

$$k_1 = h\,f(x_1, y_1) \qquad\qquad = 0.5\,(0.5 + 3.375) \qquad\qquad = 1.9375,$$

$$k_2 = h\,f\!\left(x_1 + \frac{h}{2}, y_1 + \frac{k_1}{2}\right) = 0.5\,(0.5 + 0.25 + 3.375 + 0.97) = 2.547,$$

$$y_2 = y_1 + k_2 \qquad\qquad\qquad = 3.375 + 2.5469 \qquad\qquad = 5.922.$$

Midpoint Method to solve y'=f(x,y)

Define function and initial condition: $\quad f(x,y) := x + y \quad yy_0 := 2$

Define interval and subintervals:

$$a := 0 \quad b := 1 \quad n := 5 \quad h := \frac{(b-a)}{n} \quad i := 0, 1 .. n \quad xx_i := a + i \cdot h$$

Begin midpoint method: $\quad i := 0$

$$z_i := f(xx_i, yy_i) \quad k1_i := h \cdot z_i \quad zz_i := f\!\left(xx_i + \frac{h}{2}, yy_i + \frac{k1_i}{2}\right) \quad k2_i := h \cdot zz_i \quad yy_{i+1} := yy_i + k2_i$$

Step 2: $\quad i := 1$

$$z_i := f(xx_i, yy_i) \quad k1_i := h \cdot z_i \quad zz_i := f\!\left(xx_i + \frac{h}{2}, yy_i + \frac{k1_i}{2}\right) \quad k2_i := h \cdot zz_i \quad yy_{i+1} := yy_i + k2_i$$

Step 3: $\quad i := 2$

$$z_i := f(xx_i, yy_i) \quad k1_i := h \cdot z_i \quad zz_i := f\!\left(xx_i + \frac{h}{2}, yy_i + \frac{k1_i}{2}\right) \quad k2_i := h \cdot zz_i \quad yy_{i+1} := yy_i + k2_i$$

Step 4: $\quad i := 3$

$$z_i := f(xx_i, yy_i) \quad k1_i := h \cdot z_i \quad zz_i := f\!\left(xx_i + \frac{h}{2}, yy_i + \frac{k1_i}{2}\right) \quad k2_i := h \cdot zz_i \quad yy_{i+1} := yy_i + k2_i$$

Step 5: $\quad i := 4$

$$z_i := f(xx_i, yy_i) \quad k1_i := h \cdot z_i \quad zz_i := f\!\left(xx_i + \frac{h}{2}, yy_i + \frac{k1_i}{2}\right) \quad k2_i := h \cdot zz_i \quad yy_{i+1} := yy_i + k2_i$$

Display approximate results, yy $yy^T = (2 \; 2.46 \; 3.065 \; 3.848 \; 4.846 \; 6.108)$

Display exact soution

$g(x) := 3 \cdot \exp(x) - x - 1 \quad gg := g(xx) \quad gg^T = (2 \; 2.464 \; 3.075 \; 3.866 \; 4.877 \; 6.155)$

Graph approximate and exact solutions

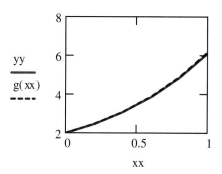

FIGURE 12.5 Approximate solution using midpoint method, and exact solution.

With $n = 10$, the computed values appear to fall directly on the graph of the exact solution. For comparison, the midpoint method with $n = 10$ requires approximately the same number of function evaluations as does Euler's method with $n = 20$ (Fig. 12.2) and the fourth-order Runge-Kutta method presented in the next section, with $n = 5$.

Note that it is necessary to repeat the computations for each step of the method, which is easy to do by copying the appropriate lines of code and changing the value of the index i. A range variable can be used to compute all of the values of xx_i at the same time, because the value of each xx_i is independent of the others, However, a range variable is not suitable for computing the other quantities, such as z_i or yy_{i+1}, beause each of these depends on the value of other variables computed in a previous step.

Mathcad Function for the Midpoint Method

$$\text{RK2}(f, \text{tspan}, y0, n) := \begin{array}{l} a \leftarrow \text{tspan}_0 \\ b \leftarrow \text{tspan}_1 \\ h \leftarrow \dfrac{b-a}{n} \\ x_0 \leftarrow a \\ y_0 \leftarrow y0 \\ \text{for } i \in 1..n \\ \quad x_i \leftarrow a + h \cdot i \\ \text{for } i \in 0..n-1 \\ \quad \left| \begin{array}{l} k1 \leftarrow h \cdot f(x_i, y_i) \\ k2 \leftarrow h \cdot f\left(x_i + \dfrac{h}{2}, y_i + \dfrac{k1}{2}\right) \\ y_{i+1} \leftarrow y_i + k2 \end{array} \right. \\ s \leftarrow \text{stack}\left((x)^T, (y)^T\right) \\ s \end{array}$$

Example 12.5 ODE for Dawson's Integral

$$f(x,y) := 1 - 2 \cdot x \cdot y \qquad y0 := 0 \qquad tspan := \begin{pmatrix} 0 \\ 1 \end{pmatrix} \qquad n := 10$$

$$B := RK2(f, tspan, y0, n)$$

$$x := \left(B^T\right)^{\langle 0 \rangle}$$

$x^T = $

	0	1	2	3	4	5	6	7	8	9
0	0	0.1	0.2	0.3	0.4	0.5	0.6	0.7	0.8	0.9

$$y := \left(B^T\right)^{\langle 1 \rangle}$$

$y^T = $

	0	1	2	3	4	5	6	7	8	9
0	0	0.1	0.195	0.283	0.36	0.425	0.475	0.51	0.532	0.54

Graph approximate and exact solutions

$$t := 0, 0.1 .. 1 \qquad g(t) := \exp\left(-t^2\right) \cdot \int_0^t \exp\left(s^2\right) ds$$

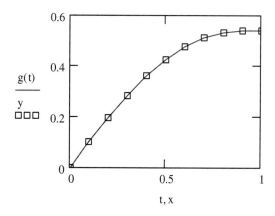

FIGURE 12.6 Approximate solution and Dawson's integral function.

12.2.2 Other Second-Order Runge-Kutta Methods

The general form for a second-order Runge-Kutta method is

$$k_1 = hf(x_n, y_n), \qquad k_2 = hf(x_n + c_2 h, y_n + a_{21} k_1),$$

$$y_{n+1} = y_n + w_1 k_1 + w_2 k_2.$$

We can summarize any such method by listing its parameters in an array whose standard form is as follows:

c_2	a_{21}
w_1	w_2

The modified Euler's (or Euler-Cauchy) method is given by the formulas

$$k_1 = hf(x_n, y_n), \qquad k_2 = hf(x_n + h, y_n + k_1),$$

$$y_{n+1} = y_n + \frac{1}{2} k_1 + \frac{1}{2} k_2.$$

The parameter array for the modified Euler method is as follows:

1	1
1/2	1/2

Heun's method, with

$$k_1 = hf(x_n, y_n), \qquad k_2 = hf\left(x_n + \frac{2}{3} h, y_n + \frac{2}{3} k_1\right),$$

$$y_{n+1} = y_n + \frac{1}{4} k_1 + \frac{3}{4} k_2,$$

has the following parameter array:

2/3	2/3
1/4	3/4

For comparison, the formulas for the midpoint method are:

$$k_1 = hf(x_n, y_i), \qquad k_2 = hf\left(x_n + \frac{1}{2} h, y_n + \frac{1}{2} k_1\right),$$

$$y_{n+1} = y_n + k_2.$$

The parameters for the midpoint method are as follows:

1/2	1/2
0	1

Discussion

To derive the second-order Runge-Kutta formulas, we write them as

$$y(x + h) = y(x) + h\,[w_1\, f(x, y) + w_2\, f(x + c_2 h, y + a_{21} k_1)],$$

form the Taylor expansion of $f(x + c_2 h, y + a_{21} k_1)$, and find values of the parameters $w_1, w_2, c_2,$ and a_{21}, so that the method agrees with the second-order Taylor method. The Taylor expansion for $f(x + c_2 h, y + a_{21} k_1)$ is

$$f(x + c_2 h, y + a_{21} k_1) = f(x, y) + c_2 h\, f_x(x, y) + a_{21}\, k_1\, f_y(x, y).$$

Thus, we want the Runge-Kutta formula

$$y(x + h) = y(x) + h\, w_1\, f + h\, w_2\, f + w_2\, c_2 h^2\, f_x + w_2\, a_{21}\, k_1\, h\, f_y) \qquad (12.1)$$

to agree with the second-order Taylor formula

$$y(x + h) = y(x) + h f + \frac{h^2}{2}[f_x + f_y f]; \qquad (12.2)$$

where $f, f_x,$ and f_y are all evaluated at (x, y).

Matching the terms involving f in eqs. (12.1) and (12.2) gives

$$w_1 + w_2 = 1.$$

Matching the terms with f_x gives $w_2\, c_2 h = \dfrac{h}{2}$, so

$$w_2\, c_2 = \frac{1}{2}.$$

Matching the terms with f_y gives $w_2\, a_{21}\, k_1 = \dfrac{h}{2} f$; since $k_1 = h f$, we have

$$w_2\, a_{21} = \frac{1}{2}.$$

The midpoint method is obtained by taking $w_1 = 0$ and $w_2 = 1$; this then requires that $c_2 = \dfrac{1}{2}$ and $a_{21} = \dfrac{1}{2}$. The modified Euler method is obtained by taking $w_1 = \dfrac{1}{2}$ and $w_2 = \dfrac{1}{2}$; this then requires that $c_2 = 1$ and $a_{21} = 1$. The Heun method is obtained by taking $w_1 = \dfrac{1}{4}$ and $w_2 = \dfrac{3}{4}$; this then requires that $c_2 = \dfrac{2}{3}$ and $a_{21} = \dfrac{2}{3}$.

12.2.3 Third-Order Runge-Kutta Methods

The general form for a third-order Runge-Kutta method is

$$k_1 = h f(x_n, y_n),$$

$$k_2 = h f(x_n + c_2 h, y_n + a_{21} k_1),$$

$$k_3 = h f(x_n + c_3 h, y_n + a_{31} k_1 + a_{32} k_2),$$

$$y_{n+1} + y_n + w_1 k_1 + w_2 k_2 + w_3 k_3.$$

The array of parameters has the following form:

c_2	a_{21}		
c_3	a_{31}	a_{32}	
	w_1	w_2	w_3

The parameters for four well-known third-order methods are as follows (see Jain, 1979 for further discussion):

2/3	2/3		
2/3	0	2/3	
	2/8	3/8	3/8

Nystrom

1/2	1/2		
1	−1	2	
	1/6	4/6	1/6

Classical

1/2	1/2		
3/4	0	3/4	
	2/9	3/9	4/9

Nearly Optimal

1/3	1/3		
2/3	0	2/3	
	1/4	0	3/4

Heun

12.2.4 Classic Runge-Kutta Method

Probably the most common form of the Runge-Kutta method is the classic fourth-order method. It uses a linear combination of four function evaluations:

$$k_1 = h f(x_i, y_i),$$

$$k_2 = h f\left(x_i + \frac{1}{2}h, y_i + \frac{1}{2}k_1\right),$$

$$k_3 = h f\left(x_i + \frac{1}{2}h, y_i + \frac{1}{2}k_2\right),$$

$$k_4 = h f(x_i + h, y_i + k_3).$$

These four equations, together with the recursion equation

$$y_{i+1} = y_i + \frac{1}{6}k_1 + \frac{1}{3}k_2 + \frac{1}{3}k_3 + \frac{1}{6}k_4,$$

make up the method. The classic fourth-order Runge-Kutta method is implemented in the following Mathcad function.

Algorithm for Classic Runge-Kutta Method (Fourth-Order)

Approximate the solution of the ODE $y' = f(x,y)$.

Begin computations

$$h = \frac{b-a}{n}$$

Define grid points
for $i = 0$ to n
    $x_i = a + h\,i$
end

Find solution
for $i = 0$ to $n - 1$
    $k_1 = h f(x_i, y_i)$
    $k_2 = h f\left(x_i + \frac{1}{2}h, y_i + \frac{1}{2}k_1\right)$
    $k_3 = h f\left(x_i + \frac{1}{2}h, y_i + \frac{1}{2}k_2\right)$
    $k_4 = h f(x_i + h, y_i + k_3)$
    $y_{i+1} = y_i + \frac{1}{6}k_1 + \frac{1}{3}k_2 + \frac{1}{3}k_3 + \frac{1}{6}k_4$
end

Mathcad Function for Classic Runge-Kutta Method

$\text{RK4}(f, \text{tspan}, y0, n) := \begin{vmatrix} a \leftarrow \text{tspan}_0 \\ b \leftarrow \text{tspan}_1 \\ h \leftarrow \dfrac{b-a}{n} \\ x_0 \leftarrow a \\ y_0 \leftarrow y0 \\ \text{for } i \in 1..n \\ \quad x_i \leftarrow a + h \cdot i \\ \text{for } i \in 0..n-1 \\ \quad \begin{vmatrix} k1 \leftarrow h \cdot f(x_i, y_i) \\ k2 \leftarrow h \cdot f\left(x_i + \dfrac{h}{2}, y_i + \dfrac{k1}{2}\right) \\ k3 \leftarrow h \cdot f\left(x_i + \dfrac{h}{2}, y_i + \dfrac{k2}{2}\right) \\ k4 \leftarrow h \cdot f(x_i + h, y_i + k3) \\ y_{i+1} \leftarrow y_i + \dfrac{k1}{6} + \dfrac{k2}{3} + \dfrac{k3}{3} + \dfrac{k4}{6} \end{vmatrix} \\ s \leftarrow \text{stack}\left(((x))^T, ((y))^T\right) \\ s \end{vmatrix}$

Example 12.6 Solving a Simple ODE with the Classic Runge-Kutta Method

Consider again the differential equation

$$y' = f(x, y) = x + y, \quad y(0) = 2, \quad a = 0, b = 1, n = 5, h = 0.2.$$

For comparison with the results of Euler's method, observe that this choice of n will require approximately the same number of evaluations of $f(x, y)$ as are used in Euler's method with $n = 20$.

Solve the ODE y'=f(x,y) with initial value y0, on the interval tspan, with n subintervals,

$$f(x, y) := x + y \qquad y0 := 2 \qquad \text{tspan} := \begin{pmatrix} 0 \\ 1 \end{pmatrix} \qquad n := 5$$

using the classic fourth-order Runge-Kutta method.

$$B := RK4(f, \text{tspan}, y0, n)$$

Display the results

$$x := (B^T)^{\langle 0 \rangle} \qquad\qquad x^T = (0 \;\; 0.2 \;\; 0.4 \;\; 0.6 \;\; 0.8 \;\; 1)$$

$$y := (B^T)^{\langle 1 \rangle} \qquad\qquad y^T = (2 \;\; 2.464 \;\; 3.075 \;\; 3.866 \;\; 4.877 \;\; 6.155)$$

Graph the exact and approximate solutions.

$$t := 0, 0.1 .. 1 \qquad\qquad g(t) := 3 \cdot \exp(t) - t - 1$$

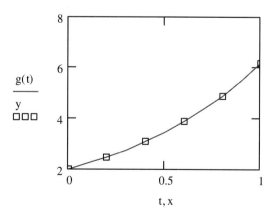

FIGURE 12.7 Approximate solution using classic Runge-Kutta, together with exact solution.

Example 12.7 Another Example of the Use of the Classic Runge-Kutta Method

Consider once more the differential equation

$$y' = f(x, y) = \begin{cases} y\left(-2x + \dfrac{1}{x}\right) & x \neq 0 \\ 1 & x = 0 \end{cases}$$

$$f(x,y) := \text{if}\left[x = 0, 1, y \cdot \left(-2 \cdot x + \dfrac{1}{x}\right)\right] \qquad y0 := 0 \qquad \text{tspan} := \begin{pmatrix} 0 \\ 2 \end{pmatrix} \qquad n := 10$$

$$B := \text{RK4}(f, \text{tspan}, y0, n)$$

$$x := \left(B^T\right)^{\langle 0 \rangle} \qquad y := \left(B^T\right)^{\langle 1 \rangle}$$

$y^T =$	0	1	2	3	4	5	6	7	8	9
0	0	0.192	0.341	0.419	0.422	0.368	0.284	0.197	0.124	0.071

Graph the approximate and exact solutions.

$$t := 0, 0.1 .. 2 \qquad\qquad g(t) := t \cdot \exp\left(-t^2\right)$$

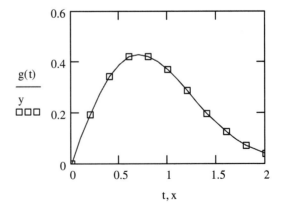

FIGURE 12.8 Approximate solution using classic Runge-Kutta, and exact solution.

12.2.5 Other Runge-Kutta Methods

The Runge-Kutta methods described in the previous sections are only the most common, simple forms of a very extensive field of study. We write a general fourth-order Runge-Kutta method as

$$k_1 = h f(x_n, y_n),$$
$$k_2 = h f(x_n + c_2 h, y_n + a_{21} k_1),$$
$$k_3 = h f(x_n + c_3 h, y_n + a_{31} k_1 + a_{32} k_2),$$
$$k_4 = h f(x_n + c_4 h, y_n + a_{41} k_1 + a_{42} k_2 + a_{43} k_3),$$
$$y_{n+1} = y_n + w_1 k_1 + w_2 k_2 + w_3 k_3 + w_4 k_4.$$

and the array of parameters has the form

c_2	a_{21}		
c_3	a_{31}	a_{32}	
c_4	a_{41}	a_{42}	a_{43}
	w_1	w_2	w_3	w_4

The parameter arrays for two fourth-order methods are as follows:

1/2	1/2		
1/2	0	1/2	
1	0	0	1
	1/6	2/6	2/6	1/6

Classic fourth-order Runge-Kutta method

1/3	1/3		
2/3	−1/3	1	
1	1	−1	1
	1/8	3/8	3/8	1/8

Kutta's method

Higher-order Runge-Kutta methods are described in a similar manner;

$$k_1 = h f(x_n, y_n),$$
$$k_2 = h f(x_n + c_2 h, y_n + a_{21} k_1),$$
$$k_3 = h f(x_n + c_3 h, y_n + a_{31} k_1 + a_{32} k_2),$$
$$k_4 = h f(x_n + c_4 h, y_n + a_{41} k_1 + a_{42} k_2 + a_{43} k_3),$$

$$\cdot$$
$$\cdot$$
$$\cdot$$

$$k_m = h f(x_n + c_m h, y_n + a_{m1} k_1 + a_{m2} k_2 + \ldots + a_{m,m-1} k_{m-1}),$$
$$y_{n+1} = y_n + w_1 k_1 + w_2 k_2 + \ldots + w_m k_m.$$

Thus, to describe a Runge-Kutta method, we need to specify the parameters $c_2, \ldots, c_m, a_{21}, \ldots, a_{m,m-1}$, and $w_1, \ldots, w_m$. The parameters for a fifth-order method and a sixth-order method are given in the following tables (see Jain, 1979, for further discussion):

1/2	1/2				
1/4	3/16	1/16			
1/2	0	0	1/2		
3/4	0	−3/16	6/16	9/16	
1	1/7	4/7	6/7	−12/7	8/7
	7/90	0	32/90	12/90	32/90	7/90

Larson's fifth-order Runge-Kutta method

1/3	1/3					
2/3	0	2/3				
1/3	1/12	1/3	−1/12			
1/2	−1/16	9/8	−3/16	−3/8		
1/2	0	9/8	−3/8	−3/4	1/2	
1	9/44	−9/11	63/44	18/11	0	−16/11
	11/120	0	27/40	27/40	−4/15	−4/15	11/120

Butcher's sixth-order Runge-Kutta method

12.2.6 Runge-Kutta-Fehlberg Methods

The Runge-Kutta-Fehlberg methods use a pair of Runge-Kutta methods to obtain both the computed solution and an estimate of the truncation error. The estimate of the error can be used in programs of variable step size to decide when to adjust the step size. We outline here the most well-known Runge-Kutta-Fehlberg formulas, which combine Runge-Kutta formulas of orders 4 and 5. In general, six function evaluations are required for a method of order 5; Fehlberg developed a fourth-order method which uses five of the function evaluations that are used in the higher order method, so the extra computational burden is slight. The parameters for the fourth- and fifth-order methods are given in the following tables (See Atkinson, 1989, for further discussion.):

$\frac{1}{4}$	$\frac{1}{4}$				
$\frac{3}{8}$	$\frac{3}{32}$	$\frac{9}{32}$			
$\frac{12}{13}$	$\frac{1932}{2197}$	$\frac{-7200}{2197}$	$\frac{7296}{2197}$		
1	$\frac{439}{216}$	-8	$\frac{3680}{513}$	$\frac{-845}{4104}$	
	$\frac{25}{216}$	0	$\frac{1408}{2565}$	$\frac{2197}{4104}$	$\frac{-1}{5}$

Fourth-order Runge-Kutta method

$\frac{1}{4}$	$\frac{1}{4}$					
$\frac{3}{8}$	$\frac{3}{32}$	$\frac{9}{32}$				
$\frac{12}{13}$	$\frac{1932}{2197}$	$\frac{-7200}{2197}$	$\frac{7296}{2197}$			
1	$\frac{439}{216}$	-8	$\frac{3680}{513}$	$\frac{-845}{4104}$		
$\frac{1}{2}$	$\frac{-8}{27}$	2	$\frac{-3544}{2565}$	$\frac{1859}{4104}$	$\frac{-11}{40}$	
	$\frac{16}{135}$	0	$\frac{6656}{12825}$	$\frac{28561}{56430}$	$\frac{-9}{50}$	$\frac{2}{55}$

Fifth-order Runge-Kutta method

The estimated error is the value computed from the fifth-order method minus the value computed from the fourth-order method; it can be expressed directly as

$$\text{error} = \frac{1}{360} k_1 + 0\, k_2 + \frac{-128}{4275} k_3 + \frac{-2197}{75240} k_4 + \frac{1}{50} k_5 + \frac{2}{55} k_6.$$

12.3 MULTISTEP METHODS

Many approximation methods use more than one previous approximate solution value or function evaluation (of the right-hand side of the differential equation) involving approximate solution values at several previous points; such methods are known as *multistep methods*. The methods we have discussed so far use only one previous approximation and are therefore known as *one-step methods*. The general form of a two-step method is

$$y_{i+1} = a_1 y_i + a_2 y_{i-1} + h[b_0 f(x_{i+1}, y_{i+1}) + b_1 f(x_i, y_i) + b_2 f(x_{i-1}, y_{i-1})],$$

where the coefficients $a_1, a_2, b_0, b_1,$ and b_2 depend on the particular method. We use the notation

$$f_{i+1} = f(x_{i+1}, y_{i+1}); \quad f_i = f(x_i, y_i); \quad f_{i-1} = f(x_{i-1}, y_{i-1}); \quad h = \frac{b-a}{n}.$$

Multistep methods are further distinguished according to whether the coefficient of the $f(x_{i+1}, y_{i+1})$ term is zero. If the coefficient is not zero, then the unknown y_{i+1} appears on the right hand side of the equation, necessitating an iterative solution procedure, in general. Such methods are called *implicit*. Not too surprisingly, they have some nice properties that make them important techniques and that compensate for the apparent disadvantages of the difficulty of their solution. Multistep methods in which the coefficient of the $f(x_{i+1}, y_{i+1})$ term is zero are known as *explicit methods*.

Multistep methods require starting values, in addition to the initial condition specified for the differential equation. For a two-step method, y_1 must be found by some other method, such as a Runge-Kutta solution; for an n-step method, the first $n-1$ values must be computed by another method.

12.3.1 Adams-Bashforth Methods

Among the most popular *explicit* multistep methods are the Adams-Bashforth methods.

The formulas for the Adams-Bashforth two-step method are as follows:
y_0 is given by the initial condition for the differential equation.
y_1 is found from a one-step method, such as a Runge-Kutta technique. Then

$$y_{i+1} = y_i + \frac{h}{2}[3f_i - f_{i-1}], \quad \text{for } i = 1, \ldots, n-1;$$

Comparing this formula with the general form of a two-step method, we see that

$$a_1 = 1, \quad a_2 = 0, \quad b_0 = 0, \quad b_1 = 3/2, \quad \text{and} \quad b_2 = -1/2.$$

It is quite common to have $a_2 = 0$. Because the method is explicit, $b_0 = 0$. This is a second-order method with local truncation error $O(h^3)$.

Following are the formulas for higher order Adams-Bashforth methods:

Adams-Bashforth three-step method:

y_0 is given,

y_1 and y_2 are found from a one-step method, such as a Runge-Kutta technique.

Then

$$y_{i+1} = y_i + \frac{h}{12}[23 f_i - 16 f_{i-1} + 5 f_{i-2}], \quad i = 2, \ldots n - 1$$

This a third-order method with local truncation error $O(h^4)$.

Adams-Bashforth four-step method:

y_0 is given,

y_1, y_2 and y_3 are found from a one-step method, such as Runge-Kutta technique.

Then

$$y_{i+1} = y_i + \frac{h}{24}[55 f_i - 59 f_{i-1} + 37 f_{i-2} - 9 f_{i-3}], \quad i = 3, \ldots n - 1$$

This is a fourth-order method, with local truncation error $O(h^5)$.

Adams-Bashforth five-step method:

y_0 is given,

y_1 through y_4 are found from a one-step method, such as Runge-Kutta technique.

Then

$$y_{i+1} = y_i + \frac{h}{720}[1901 f_i - 2774 f_{i-1} + 2616 f_{i-2} - 1274 f_{i-3} + 251 f_{i-4}]$$

This is a fifth-order method, with local truncation error $O(h^6)$.

In general, it is desirable to use a kth order Runge-Kutta method for finding $y_1, y_2 \ldots y_{k-1}$ in a k-step (k^{th} order) Adams-Bashforth method.

The primary use of the explicit Adams-Bashforth methods presented in this section is in conjunction with the implicit Adams-Moulton methods that we consider next. Especially for the higher order Adams-Bashforth methods, stability requirements severely limit the step size for which the method gives reasonable results. (The large negative coefficients appearing in the formula for the fifth-order Adams-Bashforth method are one indication that the method may have difficulty.)

Algorithm for Third-Order Adams-Bashforth Method

Approximate the solution of the ODE $y' = f(x,y)$ with initial condition $y(a) = y0$ using n steps of the third order Adams-Bashforth method; step-size $h = (b - a)/n$.
Begin computations
$$h = \frac{b - a}{n}$$
$$hh = h/12$$
Define grid points
    for $i = 0$ to n
        $x_i = a + h\,i$
    end
Use midpoint method to start
    for $i = 0$ to 1
        $z_i = f(x_i, y_i)$
        $k_1 = h\,z_i$
        $k_2 = h\,f\left(x_i + \frac{1}{2}h, y_i + \frac{1}{2}k_1\right)$
        $y_{i+1} = y_i + k_2$
    end
Use 3rd order AB method to continue
    for $i = 2$ to $n - 1$
        $z_i = f(x_i, y_i)$
        $y_{i+1} = y_i + hh\,(23\,z_i - 16\,z_{i-1} + 5\,z_{i-2})$
    end
Display results

The use of the algorithm for the Adams-Bashforth method is illustrated in the next example. Note that, as for the step-by-step solution of the same problem using the midpoint method, a range variable may be used to compute all of the points where the solution is desired (xxi), but the other quantities must be computed sequentially, since each depends on values computed in the previous step.

Example 12.8 Solving a Simple ODE with an Adams-Bashforth Method Worksheet

Midpoint Method to solve y'=f(x,y)

Define function and initial condition: $f(x,y) := x + y \quad yy_0 := 2$

Define interval and subintervals:

$$a := 0 \quad b := 1 \quad n := 5 \quad h := \frac{(b-a)}{n} \quad i := 0, 1 .. n \quad xx_i := a + i \cdot h$$

Begin midpoint method: $\quad i := 0$

$z_i := f(xx_i, yy_i) \quad k1_i := h \cdot z_i \quad zz_i := f\left(xx_i + \frac{h}{2}, yy_i + \frac{k1_i}{2}\right) \quad k2_i := h \cdot zz_i \quad yy_{i+1} := yy_i + k2_i$

Step 2: $\quad i := 1$

$z_i := f(xx_i, yy_i) \quad k1_i := h \cdot z_i \quad zz_i := f\left(xx_i + \frac{h}{2}, yy_i + \frac{k1_i}{2}\right) \quad k2_i := h \cdot zz_i \quad yy_{i+1} := yy_i + k2_i$

Begin AB3: $\quad i := 2 \quad hh := \frac{h}{12}$

$z_i := f(xx_i, yy_i) \quad\quad yy_{i+1} := yy_i + hh \cdot (23 \cdot z_i - 16 \cdot z_{i-1} + 5 \cdot z_{i-2})$

Step 4: $\quad i := 3$

$z_i := f(xx_i, yy_i) \quad\quad yy_{i+1} := yy_i + hh \cdot (23 \cdot z_i - 16 \cdot z_{i-1} + 5 \cdot z_{i-2})$

Step 5: $\quad i := 4$

$z_i := f(xx_i, yy_i) \quad\quad yy_{i+1} := yy_i + hh \cdot (23 \cdot z_i - 16 \cdot z_{i-1} + 5 \cdot z_{i-2})$

Display approximate results, yy $\quad\quad yy^T = (2 \;\; 2.46 \;\; 3.065 \;\; 3.851 \;\; 4.855 \;\; 6.124)$

Display exact soution

$g(x) := 3 \cdot \exp(x) - x - 1 \quad gg := g(xx) \quad\quad gg^T = (2 \;\; 2.464 \;\; 3.075 \;\; 3.866 \;\; 4.877 \;\; 6.155)$

Graph approximate and exact solutions

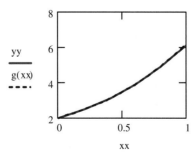

FIGURE 12.9 Adams-Bashforth and exact solutions.

Mathcad Function for Adams-Bashforth Third-Order Method

$AB3(f, tspan, y0, n) :=$
$\quad a \leftarrow tspan_0$
$\quad b \leftarrow tspan_1$
$\quad h \leftarrow \dfrac{b - a}{n}$
$\quad hh \leftarrow \dfrac{h}{12}$
$\quad x_0 \leftarrow a$
$\quad y_0 \leftarrow y0$
$\quad \text{for } i \in 1..n$
$\quad\quad x_i \leftarrow a + h \cdot i$
$\quad \text{for } i \in 0..1$
$\quad\quad z_i \leftarrow f(x_i, y_i)$
$\quad\quad k1 \leftarrow h \cdot z_i$
$\quad\quad k2 \leftarrow h \cdot f\left(x_i + \dfrac{h}{2}, y_i + \dfrac{k1}{2}\right)$
$\quad\quad y_{i+1} \leftarrow y_i + k2$
$\quad \text{for } i \in 2..n-1$
$\quad\quad z_i \leftarrow f(x_i, y_i)$
$\quad\quad y_{i+1} \leftarrow y_i + hh \cdot (23 \cdot z_i - 16 \cdot z_{i-1} + 5 \cdot z_{i-2})$
$\quad s \leftarrow \text{stack}(x^T, y^T)$
$\quad s$

Example 12.9 Solving a Simple ODE with an Adams-Bashforth Method Function

We illustrate the use of the third-order AB method to solve the ODE $y' = f(x,y)$ with initial value given by y0, on the interval defined by tspan, using n subintervals.

$$f(x,y) := x + y \qquad y0 := 2 \qquad tspan := \begin{pmatrix} 0 \\ 1 \end{pmatrix} \qquad n := 5$$

$$B := AB3(f, tspan, y0, n)$$

The solution is given at the points in the first row of matrix B; the solution values are given in the second row.
The rows are most conveninetly accessed by means of the columns of the transpose of B.

$$x := (B^T)^{\langle 0 \rangle} \qquad x^T = (0 \ \ 0.2 \ \ 0.4 \ \ 0.6 \ \ 0.8 \ \ 1)$$

$$y := (B^T)^{\langle 1 \rangle} \qquad y^T = (2 \ \ 2.46 \ \ 3.065 \ \ 3.851 \ \ 4.855 \ \ 6.124)$$

The approximate solution, and the exact solution, which is g(t) are graphed below.

$$t := 0, 0.1 .. 1 \qquad g(t) := 3 \cdot \exp(t) - t - 1$$

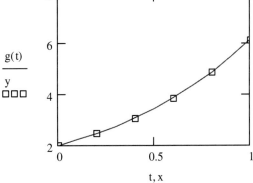

FIGURE 12.10   Adams-Bashforth and exact solutions.

12.3.2 Adams-Moulton Methods

Among the most popular *implicit* multistep methods are the Adams-Moulton methods. We first give the formulas for the Adams-Moulton two-, three- and four-step methods. We then show how these are used together with the Adams-Bashforth methods in practice.

Adams-Moulton two-step method:
  y_0 is given;
  y_1 is found from a one-step method, such as a Runge-Kutta technique. Then

$$y_{i+1} = y_i + \frac{h}{12}[5f_{i+1} + 8f_i - f_{i-1}], \quad \text{for } i = 1, \ldots, n-1.$$

This is a third-order method with local truncation error $O(h^4)$.

Comparing this formula with the general form of a two-step method, we see that

$$a_1 = 1, \quad a_2 = 0, \quad b_0 = 5/12, \quad b_1 = 8/12, \quad \text{and} \quad b_2 = -1/12.$$

As with the Adams-Bashforth two-step method, $a_2 = 0$. Also, note that $b_0 \neq 0$, because this method is implicit. The coefficients of the error term are, in general, smaller for implicit methods than for the corresponding explicit method of the same order. (See summary at end of chapter.) Thus, implicit methods have less round-off error than do explicit methods.

Adams-Moulton three-step method:
  y_0 is given by the initial condition;
  y_1 and y_2 are found from a one-step method. Then

$$y_{i+1} = y_i + \frac{h}{24}[9f_{i+1} + 19f_i - 5f_{i-1} + f_{i-2}], \quad \text{for } i = 2, \ldots, n-1.$$

This is a fourth-order method with local truncation error $O(h^5)$.

Adams-Moulton four-step method:
  y_0 is given by the initial condition;
  $y_1, y_2,$ and y_3 are found from a one-step method. Then

$$y_{i+1} = y_i + \frac{h}{720}[251f_{i+1} + 646f_i - 264f_{i-1} + 106f_{i-2} - 19f_{i-3}].$$

This is a fifth-order method with local truncation error $O(h^6)$.

12.3.3 Predictor-Corrector Methods

In order to take advantage of the beneficial properties of the implicit multistep methods while avoiding the difficulties inherent in solving the implicit equation, an explicit and implicit method can be combined. The explicit method is used to predict a value of y_{i+1}, which we denote y^*_{i+1}. This value is then used in the right hand side of the implicit method, which produces an improved, or corrected, value of y_{i+1}. We let $f_i = f(x_i, y_i)$ and $f^*_{i+1} = f(x_{i+1}, y^*_{i+1})$

Adams Second-order Predictor-Corrector Method

The simplest predictor-corrector method combines the second-order Adams-Bashforth method (presented in Section 12.3.1) with a one-step second-order Adams-Moulton method.

y_0 is given by the initial condition
y_i is found from a one-step method, such as a second-order Runge-Kutta method.
Then, for $i = 1, \ldots n-1$

$$y^*_{i+1} = y_i + \frac{h}{2}[3f_i - f_{i-1}]$$

$$y_{i+1} = y_i + \frac{h}{2}[f^*_{i+1} + f_i]$$

Since the local truncation error is $O(h^3)$ for each of these methods, the combination is also $O(h^3)$.

Adams Third-order Predictor-Corrector Method

As another example, we use the third-order Adams-Bashforth three-step method as a predictor with the third-order Adams-Moulton two-step method as a corrector:

y_0 is given by the initial condition;
y_1 and y_2 are found from a one-step method.
Then for $i = 2, \ldots, n - 1$

$$y^*_{i+1} = y_i + \frac{h}{12}[23f_i - 16f_{i-1} + 5f_{i-2}]$$

$$y_{i+1} = y_i + \frac{h}{12}[5f^*_{i+1} + 8f_i - f_{i-1}]$$

Since the local truncation error is $O(h^4)$ for each of these methods, the combination is also $O(h^4)$.

Algorithm for Third-Order Adams-Bashforth-Moulton Method

Approximate the solution of the ODE $y' = f(x,y)$ with initial condition $y(a) = y_0$ using the third-order Adams-Bashforth-Moulton predictor corrector methdod.

Begin computations
$$h = \frac{b-a}{n}$$
Define grid points (range variable is suitable)
 (for $i = 0$ to n) $x_i = a + h\,i$

Use midpoint method to start (loop as indicated)
 for $i = 0$ to 1
  $z_i = f(x_i, y_i)$
  $k_1 = h\,z_i$
  $k_2 = h\,f\left(x_i + \frac{1}{2}h,\, y_i + \frac{1}{2}k_1\right)$
  $y_{i+1} = y_i + k_2$
 end

Use Adams-Bashforth-Moulton method (loop as indicated)
 for $i = 2$ to $n-1$
  Use 3rd order AB method to predict
  $z_i = f(x_i, y_i)$
  $y^* = y_i + \frac{h}{12}(23\,z_i - 16\,z_{i-1} + 5\,z_{i-2})$
  Use 3rd order AM method to correct
  $z^* = f(x_{i+1}, y^*)$
  $y_{i+1} = y_i + \frac{h}{12}(5\,z^* + 8\,z_i - z_{i-1})$
 end
Display results

Mathcad Function for Adams-Bashforth-Moulton Third-Order Method

$\text{ABM3}(f, \text{tspan}, y0, n) :=$
$\quad a \leftarrow \text{tspan}_0$
$\quad b \leftarrow \text{tspan}_1$
$\quad h \leftarrow \dfrac{b - a}{n}$
$\quad hh \leftarrow \dfrac{h}{12}$
$\quad y_0 \leftarrow y0$
$\quad \text{for } i \in 0..n$
$\quad\quad x_i \leftarrow a + h \cdot i$
$\quad \text{for } i \in 0..1$
$\quad\quad z_i \leftarrow f(x_i, y_i)$
$\quad\quad k1 \leftarrow h \cdot z_i$
$\quad\quad k2 \leftarrow h \cdot f\left(x_i + \dfrac{h}{2}, y_i + \dfrac{k1}{2}\right)$
$\quad\quad y_{i+1} \leftarrow y_i + k2$
$\quad \text{for } i \in 2..n-1$
$\quad\quad z_i \leftarrow f(x_i, y_i)$
$\quad\quad yy_{i+1} \leftarrow y_i + hh \cdot (23 \cdot z_i - 16 \cdot z_{i-1} + 5 \cdot z_{i-2})$
$\quad\quad zz \leftarrow f(x_{i+1}, yy_{i+1})$
$\quad\quad y_{i+1} \leftarrow y_i + hh \cdot (5 \cdot zz + 8 \cdot z_i - z_{i-1})$
$\quad s \leftarrow \text{stack}(x^T, y^T)$
$\quad s$

Example 12.10 Using the Adams Predictor Corrector Method

We illustrate the use of the third-order ABM method to solve the ODE
y' = f(x,y) with initial value given by y0, on the interval defined by tspan,
using n sub-intervals

$$f(x,y) := x + y \qquad y0 := 2 \qquad tspan := \begin{pmatrix} 0 \\ 1 \end{pmatrix} \qquad n := 5$$

$$B := ABM3(f, tspan, y0, n)$$

The solution is given at the points given in the first row of matrix B,
which is accessed as the first column of the transpose of B.
The solution values are given in the second row of B (second column of the transpose.)

$$x := \left(B^T\right)^{\langle 0 \rangle} \qquad\qquad x^T = (0 \ \ 0.2 \ \ 0.4 \ \ 0.6 \ \ 0.8 \ \ 1)$$

$$y := \left(B^T\right)^{\langle 1 \rangle} \qquad\qquad y^T = (2 \ \ 2.46 \ \ 3.065 \ \ 3.854 \ \ 4.861 \ \ 6.136)$$

The exact solution to this problem is g(t);
the solution and the approximate solution are graphed below.

$$t := 0, 0.1 .. 1 \qquad\qquad g(t) := 3 \cdot \exp(t) - t - 1$$

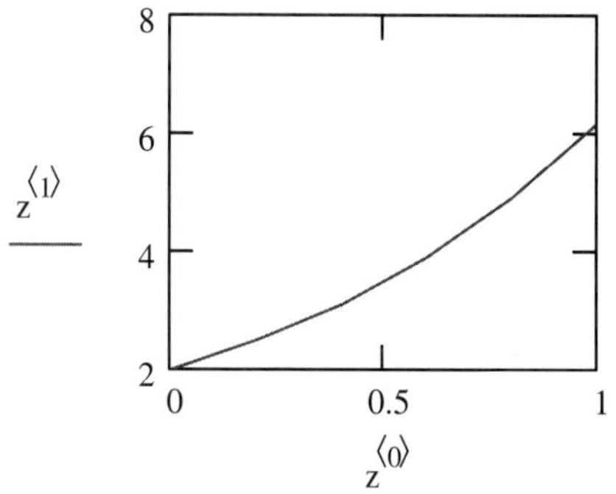

FIGURE 12.11 Predictor-Corrector and exact solutions.

Example 12.11 Velocity of Falling Parachutist

Consider the situation described in Example 12-A, in which the velocity of a parachutist with drag coefficient $k/m = 1.5$, $g = 32$, and $v_0 = 0$ is given by

$$m\frac{dv}{dt} = k(-v) - mg.$$

The parachutist reaches a terminal velocity of approximately 21 ft/sec after only about 3 sec.

If we modify the model slightly, so that

$$m\frac{dv}{dt} = k(-v)^{1.1} - mg$$

we find that the terminal velocity is somewhat slower. Figure 12.12 shows the velocity of the parachutist under both model of air resistance.

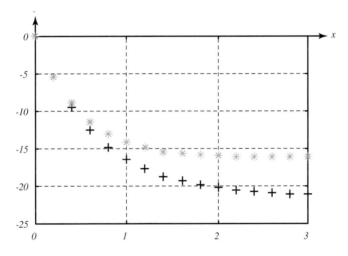

FIGURE 12.12 Velocity of parachutist with air resistance: + denotes resistance proportional to v; * denotes resistance proportional to $v^{1.1}$.

12.4 STABILITY

The term stability is used in a variety of ways in the description of differential equations, and in particular, numerical methods for solving differential equations. For some differential equations, any errors that occur in computation will be magnified regardless of the numerical method. Such problems are called *ill conditioned*. Other differential equations require extremely small step sizes to achieve accurate results; these problems are called *stiff*. Since these types of difficulties occur more often with higher order ODE, or systems of ODE, we postpone our our discussion of stiff and ill conditioned problems until Chapter 13.

We now consider the stability of the numerical methods presented in the previous sections. We call a numerical method *stable* if errors incurred at one stage of the process do not tend to be magnified at later stages. The analysis of the stability of a method often involves the investigation of the error for a simple problem, such as $y' = \lambda y$. If the method is unstable for the model equation, it is likely to behave badly for other problems as well, and the method is unstable. If $\lambda > 0$, the true solution grows exponentially, and it is not reasonable to expect the error to remain small as x increases. The most we could hope for is that the error remains small relative to the solution. On the other hand, for $\lambda < 0$ the exact solution is a decaying exponential, and we would like the error to also go to zero as $x \to \infty$.

If we apply Euler's method to the model equation, with initial condition $y(0) = y_0$, and to the same equation with error introduced in the form of a perturbation of the initial condition to $y(0) = y_0 + \varepsilon$, we find that the difference of the two solutions $z(x)$ satisfies the differential equation $z' = \lambda z$ with $z(0) = \varepsilon$. Applying Euler's method (with step-size h) to this equation leads to the stability requirement $-2 < h\lambda < 0$. This gives the *region of absolute stability*.

In general, a method with a larger region of absolute stability will impose less restriction on the step-size h. Similar analysis for the Adams-Bashforth second-order method shows that the region of absolute stability is $-1 < h\lambda < 0$. Note that although this is a smaller region than for Euler's method, the fact that the Adams-Bashforth method is higher order than Euler's method gives it some advantage. The second-order Adams-Moulton method, an implicit method, is absolutely stable for $-\infty < h\lambda < 0$. (See Atkinson, 1993 for further details.)

The stability results summarized in the previous paragraphs give some indication of the restrictions on the step size that may be necessary to achieve a stable numerical solution. We now consider more directly the stability of the difference equation that defines a numerical method. We generalize the notation introduced in Section 12.3 for a two-step method to represent an m-step method as

$$y_{i+1} = a_1 y_i + a_2 y_{i-1} \ldots + a_m y_{i+1-m} + h(b_0 f_{i+1} + b_1 f_i + \ldots + b_m f_{i+1-m}).$$

The method is stable if all roots of the characteristic polynomial

$$p(\lambda) = \lambda^m - a_1 \lambda^{m-1} + a_2 \lambda^{m-2} + \ldots + a_m$$

satisfy $|\lambda_k| \le 1$, and any root with $|\lambda_k| = 1$ is simple. It can be shown that for any method that is at least first-order accurate we must have $a_1 + a_2 + \ldots + a_m = 1$, so $\lambda_k = 1$ is a root. If the other $m - 1$ roots satisfy $|\lambda_k| < 1$, the method is called *strongly stable*. If the method is stable, but not strongly stable, it is called *weakly stable*. A strongly stable method is stable for $y' = \lambda y$ regardless of the sign of λ; a method that is only weakly stable can yield unstable numerical solutions when $\lambda < 0$, as the following example illustrates. The stability analysis based on the roots of the characteristic polynomial reveals the behavior of the method in the limit as the step size becomes arbitrarily small. Even a stable method can exhibit unstable behavior if the step size is too large.

Example 12.12 A Weakly Stable Method

Consider the simple two-step method

$$y_{i+1} = y_{i-1} + 2 h f(x_i, y_i)$$

and the differential equation

$$y' = -4y, \quad y(0) = 1$$

for which the exact solution is $y = e^{-4x}$. If we perturb the initial condition slightly, to $y(0) = 1 + \varepsilon$, the solution becomes $y = (1 + \varepsilon)e^{-4x}$; the solution is stable since the change in the solution is only $y = \varepsilon e^{-4x}$.

However, the numerical method is only weakly stable; the characteristic polynomial is $\lambda^2 - 1 = 0$, which has roots $\lambda = 1$ and $\lambda = -1$. The numerical results for $h = 0.1$ are illustrated together with the exact solution in Fig. 12.13.

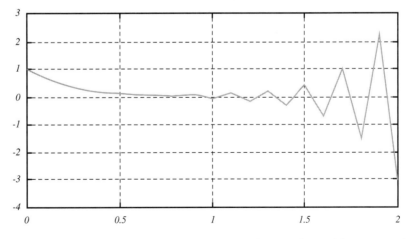

FIGURE 12.13 The solution of $y' = -4y$ with a weakly stable numerical method, $h = 0.1$.

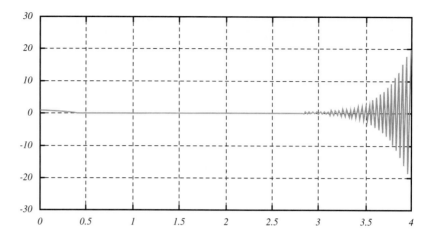

FIGURE 12.14 Instability occurs even with a much smaller step size, $h = 0.02$.

Figure 12.14 illustrates the fact that a smaller step size ($h = 0.02$) delays the onset of the instability, but does not prevent it from occurring.

On the other hand, for the differential equation

$$y' = 4y, \quad y(0) = 1$$

the weakly stable method yields acceptable results, even for $h = 0.1$, as illustrated in Fig. 12.15.

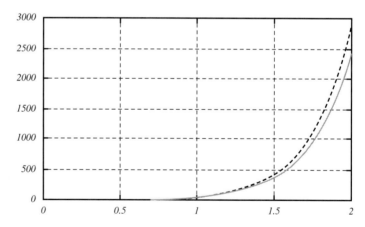

FIGURE 12.15 The solution of $y' = 4y$ with a weakly stable numerical method, $h = 0.1$.

12.5 MATHCAD'S METHODS

Mathcad2000 implements the fourth-order Runge-Kutta method as its method for solving ODE. Mathcad2000Pro also includes functions for solving ODE with special characteristics. If the functions that define the ODE are known to be smooth, the Bulirsch-Stoer method is recommended. For ODEs that require an adaptive step size method (for example, if the solution functions for the ODE vary much more rapidly in some regions than in others), an adaptive Runge-Kutta method is provided. In addition, functions are also included for stiff differential equations; since stiffness is usually acssociated with higher-order ODE or systems of ODE, we delay their consideration until the next chapter. The Mathcad2000Pro function `odesolve`, which can be used in a Solve Block structure to solve either initial value or boundary value problems, is discussed in Chapter 14.

12.5.1 Using the Built-In Functions

The function `rkfixed` provides Mathcad's basic approach for solving ODEs, either a single first-order equation, a single higher-order equation, or a system of first-order equations. We consider the use of `rkfixed` for higher-order ODE and systems of ODE in the next chapter.

The function `rkfixed` returns the values of the solution to the ODE at equally spaced points across the specified interval. The call to this function has the form

$$\text{rkfixed}\,(y, x1, x2, \text{npts}, D)$$

where the input parameters are

- y the initial value for the ODE (y is a vector for higher-order ODE or systems of ODE)
- $x1$ left end of interval on which solution is desired
- $x2$ right end of interval on which solution is desired
- $npts$ number of points beyond the initial point where solution will be approximated vector of values returned by `rkfixed` has npts + 1 rows
- D derivative of the unknown function (D is a vector for higher order ODE or systems of ODE)

For a first-order ODE, the function `rkfixed` returns a vector (npts + 1 by 2). The first column contains the values of the independent variable at which the solution is given; the second column contains the corresponding solution values.

Because Mathcad distinguishes between a scalar and a vector with one component, it is necessary in using `rkfixed` for a single first order ODE to specify the first component of the vector y and the vector D, as y_0 and $D(x,y) := f(x, y_0)$. Note that the notation y_0 signifies the first component of y, not the initial value.

Example 12.13 Using the Mathcad function `rkfixed`

We illustrate the use of the function `rkfixed` to solve the differential equation

$$y' = f(x, y) = x + y; \quad y(0) = 2,$$

investigated in several previous examples. We find the solution at 5 intermediate points on the interval [0, 1]. To plot the result, use the matrix toolbar (or [ctrl] 6) to get the superscripts which access the columns of matrix **z**. The first column (column 0) gives the independent variable values; the second column (column 1) gives the corresponding values of the solution.

Define initial condition, and right hand side of differential equation

$$y_0 := 2 \qquad D(x, y) := x + y$$

Define interval on which solution is desired, and number of intermediate points

$$x1 := 0 \qquad x2 := 1 \qquad npts := 5$$

Compute solution, and display result

$$z := \text{rkfixed}(y, x1, x2, npts, D)$$

$$z = \begin{pmatrix} 0 & 2 \\ 0.2 & 2.464 \\ 0.4 & 3.075 \\ 0.6 & 3.866 \\ 0.8 & 4.877 \\ 1 & 6.155 \end{pmatrix}$$

Graph the result

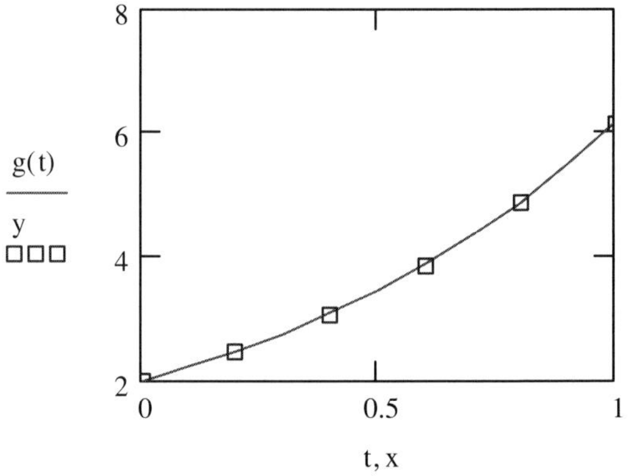

FIGURE 12.16 Approximate solution from Mathcad's function `rkfixed`.

Mathcad2000Pro includes several other functions for solving ODE. As with `rkfixed`, the functions can be used for a single first-order ODE (which we discuss here), or for a single higher-order ODE or a system of ODE (considered in Chapter 13). These functions are an adaptive step-size Runga-Kutta method, the Bulirsch-Stoer method for smooth ODE, and Bulirsch-Stoer and Rosenbrock methods for stiff ODE. Stiff ODE are considered in Chapter 13.

Adaptive Step-Size

For ODE which need an adaptive step-size solution method, the functions `rkadapt` and `Rkadapt` are provided. The function `rkadapt` gives the solution of the ODE at a single point. The function `Rkadapt` gives the solution at equally spaced points on the specified interval, by making repeated calls to `rkadapt`.

The usage of `Rkadapt` is the same as for `rkfixed`. In other words, the call to this function has the form

$$\text{Rkadapt}\ (y, x1, x2, \text{npts}, D)$$

where the parameters are as above. Although the function uses variable step sizes for its internal computations, the returned values of the solution are at equally spaced points between $x1$ and $x2$.

The function `rkadapt`, used internally by `Rkadapt`, may also be called by the user to obtain the solution of the ODE at the point $x2$, given initial the value at $x1$. The call has the form

$$\text{rkadapt}\ (y, x1, x2, \text{acc}, D, \text{kmax}, \text{save})$$

The parameters y, $x1$, $x2$, and D are as defined above. The additional parameters are as follows:

acc	controls the size of the steps (and accuracy of the solution); smaller values of acc require smaller step-size; a value of 0.001 is indicated as generally giving good solutions
kmax	maximum number of intermediate points where solution will be approximated
save	smallest spacing between values of x where solution will be found

The adaptation of the step size is based on a combination of fifth-order and fourth-order Runge-Kutta methods.

Smooth ODE

Although Runge-Kutta methods are generally accepted as the method of choice for an ODE that is not known to have special characteristics (either particularly nice, or particularly troublesome), other methods may be more efficient in special cases. If the ODE is known to have a smooth solution, the Bulirsch-Stoer method is recommended for highly accurate results with minimal computational effort. It is not suitable if the right-hand side of the ODE involves a function evaluated from a look-up table, or if the solution function has a singularity inside the interval of interest.

For smooth ODE, the functions `bulstoer` and `Bulstoer` are provided. As for the functions `rkadapt` and `Rkadapt`, the function `bulstoer` gives the solution of the ODE at a single point and the function `Bulstoer` gives the solution at equally spaced points on the specified interval, by making repeated calls to `bulstoer`. The calling parameters for `Bulstoer` are the same as for `Rkadapt` and `rkfixed`. The call to `bulstoer` follows the same form as for `rkadapt`.

12.5.2 Understanding the Algorithms

The documentation of the Mathcad functions given in [Mathcad2000 Reference Manual] indicates that the functions are based on discussions in [Press et al., 1992]. In this section we summarize the basic ideas for the adaptive step size Runga-Kutta method, and the Bulirsch-Stoer method. The basic Runge-Kutta method is the standard algorithm, presented in Section 12.2.4.

Adaptive Runge-Kutta Methods

The related problems of monitoring the truncation error and selecting an appropriate step-size (and adjusting it if necessary) occur whenever we use numerical methods to approximate the solution of an ODE. One straightforward approach is to compute the solution from x to $x + 2h$, first by taking a single step of length $2h$, and again by taking two steps of length h. By taking the difference of these approximate solutions, one can obtain an estimate of the truncation error. The cost of the extra computations is primarily related to the extra evaluations of the function which defines the right hand side of the ODE.

For Runge-Kutta there is another approach to stepsize adjustment, known as the embedded Runge-Kutta formulas, or Runge-Kutta-Fehlberg methods. One reason that the classical fourth-order Runge-Kutta method is so popular is that for Runge-Kutta methods of order 5 and higher, more function evaluations are required than the order of the method (for order M, $M > 4$, either $M + 1$ or $M + 2$ evaluations are needed). Fehlberg found fourth-order and fifth-order Runge-Kutta methods that use the same six function evaluations (or five of the same evaluations for the fourth-order method). The coefficients for the Runge-Kutta-Fehlberg method are given in Section 12.2.6. An alternative set of coefficients [Cash and Karp, 1990] are recommended as giving a more efficient method with better error properties.

The parameters for the fourth- and fifth-order methods are given in the following tables, which follow the same form as in Section 12.2.:

$\frac{1}{5}$	$\frac{1}{5}$			
$\frac{3}{10}$	$\frac{3}{40}$	$\frac{9}{40}$		
$\frac{3}{5}$	$\frac{3}{10}$	$\frac{-9}{10}$	$\frac{6}{5}$	
1	$\frac{-11}{54}$	$\frac{5}{2}$	$\frac{-70}{27}$	$\frac{35}{27}$
	$\frac{37}{378}$	0	$\frac{250}{621}$	$\frac{125}{594}$	0

Fourth-order Runge-Kutta Fehlberg method

$$
\begin{array}{c|cccccc}
\frac{1}{5} & \frac{1}{5} \\
\frac{3}{10} & \frac{3}{40} & \frac{9}{40} \\
\frac{3}{5} & \frac{3}{10} & \frac{-9}{10} & \frac{6}{5} \\
1 & \frac{-11}{54} & \frac{5}{2} & \frac{-70}{27} & \frac{35}{27} \\
\frac{7}{8} & \frac{1631}{55296} & \frac{175}{512} & \frac{575}{13824} & \frac{44275}{110592} & \frac{253}{4096} \\
\hline
 & \frac{2825}{27648} & 0 & \frac{18575}{48384} & \frac{13525}{55296} & \frac{277}{14336} & \frac{1}{4}
\end{array}
$$

Fifth-order Runge-Kutta-Cash-Karp method

The estimated error is the value computed from the fifth-order method minus the value computed from the fourth-order method. See [Press et al., 1992, p. 717].

Bulirsch-Stoer method

The Bulirsch-Stoer method uses the idea of acceleration to combine the results of many applications of the midpoint method (with modified first and last steps) to obtain a highly accurate solution with minimal computational effort. This approach is closely related to the Richardson extrapolation for integration discussed in Chapter 11, in which the results from several applications of the trapezoid rule are combined to give a more accurate result than could be obtained just by reducing the step size for the integration. Bulirsch and Stoer used rational function interpolation to model the dependence of the true solution on the step size; this can give good approximations for some functions for which polynomial interpolation is not satisfactory. However, for smooth functions, polynomial interpolation is somewhat more efficient, and is used as the default approach in the presentation of the method in [Press et al., 1992].

In using acceleration, it is highly advantageous to use a method for which the error is strictly an even function of h, i.e., the error expressed in powers of the step size h contains only even powers. In this case, the order of the accelerated method increases by 2 at each stage. Gragg (1965) has shown that the error expansion for the modified midpoint method has this form.

Modified Midpoint Method

To solve $y' = f(x,y)$, $y(a) = y_a$, on $[a,b]$, using $n - 1$ intermediate points, set $x_0 = a$, and $x_k = a + k\,h$ (for $k = 1, \ldots n$). Then $y_0 = y_a$, $y_1 = y_0 + h\,f(x_0, y_0)$ and (for $k = 1, \ldots n - 1$) $y_{k+1} = y_{k-1} + 2\,h\,f(x_k, y_k)$. Finally $y_n = \frac{1}{2}(y_n + y_{n-1}) + h\,f(x_n, y_n)$.

Bulirsch-Stoer-Deuflhard Extrapolation

Apply the modified midpoint method using a sequence of n steps to go from a to b, taking $n = 2, 4, 6, 8, \ldots$. Each application of the acceleration gives both an improved estimate of the solution (the extrapolated value) and an error estimate. We begin by computing the solution at x_1, starting with the modified midpoint method. Using $n = 2$, we find the solution we denote as z_2, using $n = 4$ we find z_4, and the extrapolated value, $(4 z_4 - z_2)/3$. If the error estimate indicates that the extrapolated value is not a sufficiently accurate approximation to the solution at x_1, we take $n = 6$, and extrapolate again, $(4 z_6 - z_4)/3$; the extrapolation continues until the error estimate indicates that the solution is satisfactory (or we reach an upper bound on the number of extrapolation steps allowed). When we have a satifactory value for y_1, we go to the next subinterval (beginning the solution process with $n = 2$ and $n = 4$).

For a more detailed discussion, including the use of the error estimate to adjust step-size, see [Press et al., 1992, pp. 724–732].

SUMMARY

Explicit One-Step Methods: *Euler* (first-order)

$$y_{i+1} = y_i + h f_i + \frac{h^2}{2} y''(\eta_i) + O(h^2)$$

Taylor method (second-order)

$$y_{i+1} = y_i + h f_i + \frac{h^2}{2} g_i + O(h^3) \quad \text{where } g_i = \frac{d}{dx} f(x_i, y_i) = y''(x)$$

Runge-Kutta
   Midpoint (second-order)

$$y_{i+1} = y_i + h f\left(x_i + \frac{h}{2}, y_i + \frac{h}{2} f(x_i, y_i)\right) + O(h^3)$$

(this can also be viewed as an explicit two-step method with step size $H = 2h$)

Modified Euler's method (second-order)

$$y_{i+1} = y_i + \frac{h}{2} [f(x_i, y_i) + f(x_i + h, y_i + h f(x_i, y_i))] + O(h^3)$$

Heun's method (second-order)

$$y_{i+1} = y_i + \frac{h}{4} \left[f(x_i, y_i) + 3 f\left(x_i + \frac{2h}{3}, y_i + \frac{2h}{3} f(x_i, y_i)\right) \right] + O(h^3)$$

Classic Runge-Kutta (fourth-order)

$$k_1 = h f(x_i, y_i) \qquad k_2 = h f\left(x_i + \frac{1}{2}h, y_i + \frac{1}{2}k_1\right)$$

$$k_3 = h f\left(x_i + \frac{1}{2}h, y_i + \frac{1}{2}k_2\right) \quad k_4 = h f(x_i + h, y_i + k_3)$$

$$y_{i+1} = y_i + \frac{1}{6}k_1 + \frac{1}{3}k_2 + \frac{1}{3}k_3 + \frac{1}{6}k_4.$$

Explicit Multi-Step Methods: *Midpoint* (two-step, second-order)

$$y_{i+1} = y_{i-1} + 2h f_i + \frac{h^3}{3} y^{(3)}(\eta_i)$$

General form of two-step, second-order methods. (See Atkinson, 1989, p. 382.)

$$y_{i+1} = a_1 y_i + a_2 y_{-1} + h \left[b_0 f_{i+1} + b_1 f_i + b_2 f_{i-1}\right]$$

with $a_1 = 1 - a_2 \quad b_0 = 1 - \frac{1}{4}a_2 - \frac{1}{2}b_1 \quad b_2 = 1 - \frac{3}{4}a_2 - \frac{1}{2}b_1$

Adams-Bashforth (See Atkinson, 1989, p. 387.)
Two-Step (second-order)

$$y_{i+1} = y_i + \frac{h}{2}[3 f_i - f_{i-1}] + \frac{5}{12} h^3 y^{(3)}(\eta_i)$$

Three-Step (third-order)

$$y_{i+1} = y_i + \frac{h}{12}[23 f_i - 16 f_{i-1} + 5 f_{i-2}] + \frac{3}{8} h^4 y^{(4)}(\eta_i)$$

Four-Step (fourth-order)

$$y_{i+1} = y_i + \frac{h}{24}[55 f_i - 59 f_{i-1} + 37 f_{i-2} - 9 f_{i-3}] + \frac{251}{720} h^5 y^{(5)}(\eta_i)$$

Five-Step (fifth order)

$$y_{i+1} = y_i + \frac{h}{720}[1901 f_i - 2774 f_{i-1} + 2616 f_{i-2} - 1274 f_{i-3} + 251 f_{i-4}]$$

Implicit Multistep Methods *Adams-Moulton* methods. (See Atkinson, 1989.)

Trapezoid (one-step, second-order)

$$y_{i+1} = y_i + \frac{h}{2}[f_{i+1} + f_i] - \frac{h^3}{12} y^{(3)}(\eta_i)$$

Two-Step method (third-order)

$$y_{i+1} = y_i + \frac{h}{12}[5 f_{i+1} + 8 f_i - f_{i-1}] - \frac{1}{24} h^4 y^{(4)}(\eta_i)$$

Three-Step (fourth-order)

$$y_{i+1} = y_i + \frac{h}{24}[9 f_{i+1} + 19 f_i - 5 f_{i-1} + f_{i-2}] + \frac{19}{720} h^5 y^{(5)}(\eta_i)$$

Four-Step (fifth-order)

$$y_{i+1} = y_i + \frac{h}{720}[251 f_{i+1} + 646 f_i - 264 f_{i-1} + 106 f_{i-2} - 19 f_{i-3}]$$

Adams Predictor-Corrector Methods: Let $f^*_{i+1} = f(x_{i+1}, y^*_{i+1})$

Second-order

$$y^*_{i+1} = y_i + \frac{h}{2}[3 f_i - f_{i-1}] + \frac{5}{12} h^3 y^{(3)}(\eta_1)$$

$$y_{i+1} = y_i + \frac{h}{2}[f^*_{i+1} + f_i] - \frac{1}{12} h^3 y^{(3)}(\eta_2)$$

Third-order

$$y^*_{i+1} = y_i + \frac{h}{12}[23 f_i - 16 f_{i-1} + 5 f_{i-2}] + \frac{3}{8} h^4 y^{(4)}(\eta_1)$$

$$y_{i+1} = y_i + \frac{h}{12}[5 f^*_{i+1} + 8 f_i - f_{i-1}] - \frac{1}{24} h^4 y^{(4)}(\eta_2)$$

Fourth-order

$$y^*_{i+1} = y_i + \frac{h}{24}[55 f_i - 59 f_{i-1} + 37 f_{i-2} - 9 f_{i-3}] + \frac{251}{720} h^5 y^{(5)}(\eta_1)$$

$$y_{i+1} = y_i + \frac{h}{24}[9 f^*_{i+1} + 19 f_i - 5 f_{i-1} + f_{i-2}] + \frac{19}{720} h^5 y^{(5)}(\eta_2)$$

SUGGESTIONS FOR FURTHER READING

Edwards, C. H. Jr. and D. E. Penney, *Differential Equations and Boundary Value Problems: Computing and Modeling*, Prentice Hall, Englewood Cliffs, NJ, 1996.

Finizio, N., and G. Ladas, *An Introduction to Differential Equations, with Difference Equations, Fourier Series, and Partial Differential Equations*, Wadsworth Publishing, Belmont, CA, 1982.

Froberg, C. E., *Numerical Mathematics: Theory and Computer Applications*, Benjamin/Cummings, Menlo Park, CA, 1985.

Garcia, A. L., *Numerical Methods for Physics*, Prentice Hall, Englewood Cliffs, NJ, 1994.

Gear, C. W., *Numerical Initial Value Problems in Ordinary Differential Equations*, Prentice-Hall, Englewood Cliffs, NJ, 1971.

Golub, G. H., and J. M. Ortega, *Scientific Computing and Differential Equations: An Introduction to Numerical Methods*, Academic Press, Boston, 1992.

Hanna, O. T., and O. C. Sandall, *Computational Methods in Chemical Engineering*, Prentice Hall, Upper Saddle River, NJ, 1995.

Jain, M. K., *Numerical Solution of Differential Equations*, John Wiley & Sons, New York, 1979.

Ortega, J. M., and W. G. Poole, *An Introduction to Numerical Methods for Differential Equations*, Pitman Publishing, Marshfield, MA, 1981.

Ritger, P. D., and N. J. Rose, *Differential Equations with Applications*, McGraw-Hill, New York, 1968.

Roberts, C. E., *Ordinary Differential Equations: A Computational Approach*, Prentice-Hall, Englewood Cliffs, NJ, 1979.

Zill, D. G. *Differential Equations with Boundary-Value Problems*, Prindle, Weber, & Schmidt, Boston, 1986.

Rice has excellent discussion of Runge-Kutta methods, and the difficulties in using simple methods for various sample problems.

Rice, J. R., *Numerical Methods, Software, and Analysis*, McGraw-Hill, New York, 1983.

Discussion of the methods underlying the Mathcad functions:

Cash, J. R. and A. H. Karp, *ACM Transactions on Mathematical Software*, vol. 16, pp. 201–222, 1990.

Deuflhard, P, *Numerische Mathematik*, vol. 41, pp. 399–422, 1983.

Deuflhard, P, *SIAM Review*, vol. 27, pp. 505–535, 1985.

Press, W. H., S. A. Teukolsky, W. T. Vetterling, and B. P. Flannery, *Numerical Recipes in C, The Art of Scientific Computing*, (2d ed.), Cambridge Unversity Press, 1992.

Stoer, J. and R. Bulirsch, *Introduction to Numerical Analysis*, Springer-Verlag, New York, 1980, Sect. 7.2.14.

PRACTICE THE TECHNIQUES

For Problems P12.1 to P12.10 solve the initial value problem and compare your results to the exact solution given. Use $h = 0.5$ (suggested for hand calculation) or $h = 0.1$ (for using Mathcad functions).
   a. Use Euler's method.
   b. Use the second order Taylor method.
   c. Use the midpoint method.
   d. Use the classic Runge-Kutta method.

P12.1 Solve $y' = y$, $y(0) = 2$; on $[0, 1]$.
The exact solution is $y = 2 e^x$.

P12.2 Solve $y' = x + y$, $y(0) = 2$; on $[0, 1]$.
The exact solution is $y = 3 e^x - x - 1$.

P12.3 Solve $y' = -y^2$, $y(0) = 1$, on $[0, 1]$.
The exact solution is $y = 1/(x + 1)$.

P12.4 Solve $y' = 1 + x - y - xy$, $y(0) = 2$, on $[0, 1]$.
The exact solution is
$$y = 1 + \exp\left(-x - \frac{x^2}{2}\right).$$

P12.5 Solve $y' = y x^{-2}$, $y(1) = 2$, on $[1, 2]$.
The exact solution is $y = 2 \exp((x-1)/x)$.

P12.6 Solve $y' = xy + x$, $y(0) = 0$, on $[0, 1]$.
The exact solution is $y = -1 + \exp\left(\dfrac{x^2}{2}\right)$.

P12.7 Solve $y' = -2 xy$, $y(0) = 2$, on $[0, 1]$.
The exact solution is $y = 2 \exp(-x^2)$.

P12.8 Solve $y' = x + 4 y x^{-1}$, $y(1) = 1/2$, on $[1, 2]$.
The exact solution is $y = \dfrac{1}{2} x^2 + x^4$.

P12.9 Solve $y' = 3 x^2 y$, $y(0) = 1$, on $[0, 1]$.
The exact solution is $y = \exp(x^3)$.

P12.10 Solve $y' = y \cos(x)$, $y(0) = 1$, on $[0, 1]$.
The exact solution is $y = e^{\sin x}$.

For Problems P12.11 to P12.20 solve the initial value problem and compare your results to the exact solution given. Investigate the effect of using different step sizes.
   a. Use Euler's method.
   c. Use the midpoint method.
   d. Use the classic Runge-Kutta method.
   e. Use the Adams-Bashforth-Moulton predictor-corrector method.

P12.11 Solve $y' = -y + \sin(x)$, $y(0) = 1$, on $[0, \pi]$.
(compare the results using $n = 10, 20, 40$)
The exact solution is $y = 1.5 e^{-x} + 0.5 \sin(x) - 0.5 \cos(x)$.

P12.12 Solve $y' = y \tan(x) + x$, $y(0) = 3$, on $[0, \pi/4]$.
Exact solution is $y = x \tan(x) + 2 \sec(x) + 1$.

P12.13 Solve $y' = \dfrac{x^2 + y^2}{2xy}$, $y(1) = 2$, on $[1,2]$.
The exact solution is $y^2 = x(x + 3)$.

P12.14 Solve $y' = - y \tan(x) + \sec(x)$, $y(0) = 2$, on $[0, \pi/4]$.
The exact solution is $y = \sin(x) + 2 \cos(x)$.

P12.15 Solve $y' = 2\sqrt{y - 1}$, $y(0) = 5$, on $[0, 2]$.
The exact solution is $y = 1 + (x - 2)^2$.

P12.16 Solve $y' = x + 2 y x^{-1}$, $y(1) = 1$, on $[1, 2]$.
The exact solution is $y = x^2 \ln(x) + x^2$.

P12.17 Solve $y' = 4 x y^{-1} - x y$; $y(0) = 3$ on $[0, 2]$
The exact solution is $y = \sqrt{4 + 5 \exp(-x^2)}$.

P12.18 Solve $y' = x y^{-1} - x y$; $y(0) = 2$ on $[0, 2]$
The exact solution is $y = \sqrt{1 + 3 \exp(-x^2)}$.

P12.19 $y' = (y + x)^2$ $y(0) = -1$
The exact solution is $y = -x + \tan(x - \pi/4)$.

P12.20 Solve $y' = \dfrac{3x}{y} - x y$; $y(0) = 2$ on $[0, 2]$
The exact solution is $y = \sqrt{3 + \exp(-x^2)}$.

For Problems P12.21 to P12.25 solve the initial value problem. Investigate the effect of using different step sizes, and intervals of different lengths.

P12.21 $y' = y + x^2$, $y(0) = 1$
P12.22 $y' = y + \cos(x)$, $y(0) = 1$
P12.23 $y' = y + \ln(x + 1)$, $y(0) = 1$
P12.24 $y' = y + x^{-1}$, $y(1) = 1$
P12.25 $y' = y \tan(x)$, $y(0) = 1$

EXPLORE SOME APPLICATIONS

A12.1 The concentration of a chemical in a batch reactor can be modeled by the differential equation

$$\frac{dC}{dt} = \frac{-k_1 C}{1 + k_2 C}.$$

Find a numerical solution for $0 \leq t \leq 1$.
  a. Use $k_1 = 2$, $k_2 = 0.1$, and $C(0) = 1$.
  b. Use $k_1 = 1$, $k_2 = 0.3$, and $C(0) = 0.8$.

(see Hanna and Sandall, 1995 for a discussion of similar problems.)

A 12.2 Solve the Ginzburg-Landau equation on the interval $[0, 3]$ for the given values of the parameter k, and initial condition.

$$\frac{dx}{dt} = k^2 x^3 - x$$

  a. $k = 1$; $x(0) = 0.7$.
  b. $k = -0.1$; $x(0) = 0.9$.
  c. $k = -0.8$; $x(0) = 0.9$.
  d. $k = 0.5$; $x(0) = 1.2$.

(Adapted from Garcia, 1994.)

A 12.3 The velocity of a body subject to the force of gravity and air resistance proportional to v is given by the differential equation:

$$\frac{dv}{dt} = g - pv$$

where g represents the gravitational acceleration (32 ft/sec) and p is the drag coefficient.
  a. Find the velocity of an arrow with initial velocity $v_0 = 300$ ft/sec, and drag coefficient $p = 0.05$.
  b. Find the velocity of a parachutist with $v_0 = 0$ ft/sec, and drag coefficient $p = 1.5$.

(See Edwards and Penney, 1993 for discussion of similar problems.)

A12.4 According to Torricelli's law, the depth y of the water in a tank with a hole in the bottom changes according to the differential equation

$$\frac{dy}{dt} = -k \sqrt{y} A(y).$$

$A(y)$ is the cross-sectional area of the tank at depth y. The parameter $k = a \sqrt{2g}$, where g is the gravitational constant (32 ft/sec) and a is the area of the hole. (See Edwards and Penney, 1996 for a derivation of this equation.)

  a. Find the water depth in a tank with cross sectional area $A(y) = \pi y$; i.e. formed by rotating the curve $y = x^2$ around the y axis. Let the initial water depth be 2 ft, and the area of the hole be 0.01. When is the tank empty?
  b. Find the water depth in a tank with cross-sectional area $A(y) = \pi y^{2/3}$; i.e. formed by rotating the curve $y = x^3$ around the y axis. Take the initial water depth to be 2 ft, and the area of the hole to be 0.01. When is the tank empty?

A12.5 A simple model of the spread of disease gives $P' = k P (C - P)$, where $P(t)$ represents the number of individuals in the population who are infected, and C is the constant size of the total population. The solution of this differential equation is the logistic function. Suppose now that the parameter k fluctuates (e.g., perhaps because the population is more susceptible during certain seasons). Solve the modified problem and compare the results to those of the original model.

$$P' = (k + 0.1 \sin(t)) P (C\text{-}P);$$
$$k = 2, C = 2000, P(0) = 10.$$

A12.6 The balance in a bank account in which interest is being earned at the rate of 5%, compounded continuously and reinvested, obeys the differential equation

$$B' = 0.05 B; B(0) = B_0.$$

Suppose now that additional deposits are made on a regular basis, but with larger deposits made during certain months. The change in the bank balance could then be modeled by the ODE

$B' = 0.05 B + C (\sin(2\pi t))^4$; $B(0) = B_0$.

Take an initial deposit of $B_0 = 1000$, and a deposit schedule of $2(\sin(2\pi t))^4$. Compare the balance in the account after 2 years to that given by the initial deposit and reinvestment only.

Problems A12.7 to A12.13 explore some Riccati differential equations; i.e., differential equations of the form $y' = P(x) y + Q(x) y^2 + R(x)$.

A12.7 Solve $y' = x y + y^2 + x^2$, $y(0) = 1$, on $[0, 0.5]$.

A12.8 Solve $y' = \dfrac{-1}{2} y - y^2 + \dfrac{1}{x^2}$, $y(1) = -1/3$, on $[1, 2]$.

A12.9 Solve $y' = \dfrac{1}{x} y + \dfrac{1}{x} y^2 + \dfrac{-2}{x^2}$, $y(0.1) = 1$, on $[0.1, 1]$.

A12.10 Solve $y' = \dfrac{1}{x} y + \dfrac{1}{x} y^2 + \dfrac{-2}{x}$, $y(2) = -3$, on $[2, 3]$.

A12.11 Solve $y' = y^2 + x^{-4}$, $y(1) = 0.1$, on $[1, 2]$.

A12.12 Solve $y' = y^2 + x^{-8/5}$, $y(1) = 0.1$, on $[1, 2]$.

A12.13 Solve $y' = y^2 + x^{-8/3}$, $y(1) = 0.1$, on $[1, 3]$.

Problems A12.14 to A12.16 explore some Bernoulli differential equations; i.e., differential equations of the form $y' = -P(x) y + Q(x) y^n$.

A12.14 Solve $y' = 1.5 x^{-1} y + 2 x y^{-1}$, $y(1) = 0$, on $[1, 30]$.

A12.15 Solve $y' = 6 x^{-1} y + 3 y^{4/3}$, $y(1) = 1/8$, on $[1, 3]$.

A12.16 Solve $y' = x y + y^2$, $y(1) = 0.1$, on $[0, 2]$.

EXTEND YOUR UNDERSTANDING

U12.1 Solve $y' = x^2 + y^2$, $y(0) = 1$, on $[0, 0.9]$. Discuss what happens if you try to extend the interval to $[0, 1]$.

U12.2 Solve $y' = -x^2 + y^2$, $y(0) = 1$, on $[0, 1]$.

U12.3 Solve $y' = x^2 - y^2$, $y(0) = 1$, on $[0, 1]$.

U12.4 Solve $y' = 1 - y^2$,
  a. $y(0) = 0$ on $[0, 5]$.
  b. $y(1) = 4$ on $[1, 5]$.

U12.5 Solve $y' = 4 y - 2 x^2$; $y(0) = 1/16$ on $[0, 3]$. Compare your results to the exact solution, $y = 1/2 x^2 + 1/4 x + 1/16$.

Problems U12.6 to U12.9 explore the appication of numerical methods to differential equations for which the solution may not be unique. See a standard differential equaitons text for discussion of the conditions that guarantee the existence and uniqueness of the solution of a first order ODE initial value problem; e.g. Edwards and Penney, 1996.

U12.6 Solve $y' = y - \sin x$; $y(-\pi) = -0.5$ on $[-\pi, \pi]$

U12.7 Solve $y' = 2 y x^{-1}$; $y(-1) = 1$ on $[-1, 1]$

U12.8 Solve $y' = 2 \sqrt{y}$; $y(0) = 0$ on $[0, 1]$ compare to $y(0) = 0.001$ on $[0, 1]$

U12.9 Solve $y' = y^2$; $y(0) = 1$ on $[0, 0.5]$ compare to $y(2) = -0.5$ on $[2, 3]$

U12.10 Investigate the stability of the predictor and corrector formulas for Milne's method; this is a fourth-order method. The predictor is given by

$$y^*_{i+1} = y_{i-3} + \frac{4h}{3} [2 f_i - f_{i-1} + 2 f_{i-2}]$$

$$+ \frac{14}{45} h^5 y^{(5)}(\eta_1),$$

and the corrector is

$$y_{i+1} = y_{i-1} + \frac{h}{3} [f^*_{i+1} + 4 f_i + f_{i-1}]$$

$$- \frac{1}{90} h^5 y^{(5)}(\eta_2).$$

(See Froberg, 1985, p. 338 for further discussion.)

13

Systems of Ordinary Differential Equations

13.1 Higher-Order ODEs

13.2 Systems of Two First-Order ODE

13.3 Systems of First-Order ODE-IVP

13.4 Stiff ODE and Ill-Conditioned Problems

13.5 Mathcad's Methods

In this chapter, we extend the techniques presented in the previous chapter to higher-order ODEs and systems of first-order ODEs. We begin by showing how a higher-order ODE can be converted into a system of first-order ODEs. In the second section, we treat the Euler and midpoint methods for second-order ODEs and systems of two first-order ODEs in some detail to emphasize the direct relationship between each of the methods for a single ODE and the corresponding method for a system of ODEs.

For larger first-order systems, we can update all components of the solution very easily by utilizing Mathcad's vector capabilities. The function for each of the methods presented in the previous chapter can be applied to systems of arbitrary size with only minor modifications. Although the methods presented in this chapter are direct extensions of those seen in the last chapter, the variety of applications that can be solved is greatly expanded. We illustrate the methods using simple examples and problems, including the motion of a nonlinear pendulum, a spring-mass system, and a two-link robot arm. Sample problems describing chemical reactions are also solved.

Ordinary differential equations can be used to describe a wide variety of processes. Population growth models, predator-prey models, radioactive carbon dating, combat models, traffic flow models, and mechanical and electrical vibrations are a few of the most common applications; the list of possibilities is almost endless.

The solution of ODE initial-value problems forms the basis for the shooting method, one of the approaches to solving boundary-value problems for ordinary differential equations. The study of numerical methods for ODE-BVPs is the subject of the next chapter.

Example 13-A Motion of a Nonlinear Pendulum

The motion of a pendulum of length L subject to damping can be described by the angular displacement of the pendulum from vertical, θ, as a function of time. (See Fig. 13.1.) If we let m be the mass of the pendulum, g the gravitational constant, and c the damping coefficient (i.e., the damping force is $F = -c\theta'$), then the ODE initial-value problem describing this motion is

$$\theta'' + \frac{c}{mL}\theta' + \frac{g}{L}\sin\theta = 0.$$

The initial conditions give the angular displacement and velocity at time zero; for example, if $\theta(0) = a$ and $\theta'(0) = 0$, the pendulum has an initial displacement, but is released with 0 initial velocity.

Analytic (closed-form) solutions rely on approximating $\sin\theta$; the exact solutions to this approximated system do not have the characteristics of the physical pendulum, namely, a decreasing amplitude and a decreasing period. (See Greenspan, 1974 for further discussion.)

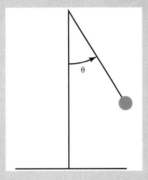

FIGURE 13.1A Simple pendulum.

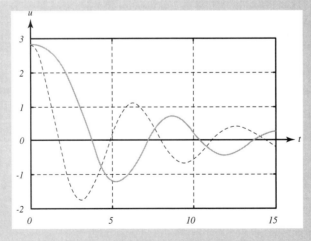

FIGURE 13.1B Motion of nonlinear (solid line) and linear (dashed line) pendulum;

Example 13-B Chemical Flow

A *circular reaction* involving three chemical reactions can be described as

$$A + A' \xrightarrow{k_1} B,$$

$$B + B' \xrightarrow{k_2} C,$$

$$C + C' \xrightarrow{k_3} A.$$

We assume that compounds A', B' and C' are present in excess, so that changes in their quantities can be neglected, and we simplify the notation by defining $r_1 = k_1 A'$, $r_2 = k_2 B'$, and $r_3 = k_3 C'$; the differential equations can be written as

$$\frac{dA}{dt} = r_3 C - r_1 A,$$

$$\frac{dB}{dt} = r_1 A - r_2 B,$$

$$\frac{dC}{dt} = r_2 B - r_3 C.$$

If the reaction rates are constants, the solution can be found from the eigenvalues and eigenvectors of the coefficient matrix R when the differential equation is written in matrix-vector form, $\mathbf{x}' = \mathbf{R}\mathbf{x}$, with $\mathbf{x} = [A, B, C]'$ and

$$\mathbf{R} = \begin{bmatrix} -r_1 & 0 & r_3 \\ r_1 & -r_2 & 0 \\ 0 & r_2 & -r_3 \end{bmatrix}.$$

This is discussed briefly in problems A7.27–A7.50. On the other hand, if the reactoin rates are not constant, numerical methods may be especially useful. For example, we can take $r_2 = 2$, $r_3 = 1$, and r_1 changing from 0.1 to 10. (The units for these parameters are $\sec^{-1}$.) The appropriate initial values of the unknown functions A, B, and C depend on the total amount of the three chemicals that is present ($Q = A + B + C$) and the rate constants; the initial values are chosen to be consistent with the equilibrium values, which are

$$A = \frac{Q}{1 + r_1/r_2 + r_1/r_3}, \quad B = \frac{r_1}{r_2} A, \quad C = \frac{r_1}{r_3} A.$$

(For further discussion, see Simon, 1986, p. 118.)

13.1 HIGHER-ORDER ODEs

A second-order ODE of the form

$$y'' = g(x, y, y')$$

can be converted to a system of two first-order ODEs by a simple change of variables:

$$u = y,$$
$$v = y'.$$

The differential equations relating these variables (functions) are

$$u' = v = f(x, u, v),$$
$$v' = g(x, u, v).$$

The initial conditions for the original ODE,

$$y(0) = \alpha_0, \quad y'(0) = \alpha_1,$$

become the initial conditions for the system, i. e.,

$$u(0) = \alpha_0, \quad v(0) = \alpha_1.$$

Example 13.1 Nonlinear Pendulum

Consider the nonlinear pendulum described at the beginning of the chapter, with angular displacement $y(x)$ given by

$$y'' + \frac{c}{mL} y' + \frac{g}{L} \sin y = 0; \quad y(0) = a, y'(0) = b,$$

Choosing $g/L = 1$ and $c/(mL) = 0.3$, $a = \pi/2$, and $b = 0$, we get the second-order ODE-IVP

$$y'' = -0.3\, y' - \sin y$$

which can be converted to a system of first-order ODEs by means of the change of variables

$$u = y$$
$$v = y'$$

The differential equations relating these variables are

$$u' = v = f(x, u, v)$$
$$v' = -0.3\, v - \sin u = g(x, u, v)$$

with initial conditions $u(0) = \pi/2$, $v(0) = 0$.

We investigate the application of Euler's method and the midpoint method to this and other systems of two first-order ODEs in the next section.

A higher-order ODE may be converted to a system of first-order ODEs by a similar change of variables. The nth-order ODE

$$y^{(n)} = f(x, y, y', y'', \ldots y^{(n-1)}),$$

$$y(0) = \alpha_0, \quad y'(0) = \alpha_1, \quad y''(0) = \alpha_2, \ldots, y^{(n-1)}(0) = \alpha_{n-1},$$

becomes a system of first order ODEs by the following change of variables:

$$u_1 = y,$$
$$u_2 = y',$$
$$u_3 = y'',$$
$$\vdots$$
$$u_n = y^{(n-1)}.$$

The differential equations relating these variables are

$$u_1' = u_2,$$
$$u_2' = u_3,$$
$$u_3' = u_4,$$
$$\vdots$$
$$u_n' = f(x, u_1, u_2, u_3, \ldots u_n),$$

with the initial conditions

$$u_1(0) = \alpha_0, \quad u_2(0) = \alpha_1, \quad u_3(0) = \alpha_2, \ldots, u_n(0) = \alpha_{n-1}.$$

We investigate the application of several of the methods from Chapter 12 to general systems of ODE in Section 13.3. By utilizing Mathcad's vector capabilities, only minor changes are required to the functions presented in Chapter 12.

13.2 SYSTEMS OF TWO FIRST-ORDER ODE

Any of the methods for solving ODE-IVPs discussed in Chapter 12 can be generalized to apply to systems of equations. In this section we consider systems of two first-order ODEs in detail, using the Euler and Runge-Kutta methods. In Section 13.3, we treat systems of arbitrary size, using Mathcad's vector capabilities extensively.

13.2.1 Euler's Method for Solving Two ODE-IVPs

To apply the basic Euler's method

$$y_{i+1} = y_i + h f(x_i, y_i)$$

to the system of ODEs

$$u' = f(x, u, v), \quad v' = g(x, u, v),$$

we update the function u using $f(x, u, v)$ and update v using $g(x, u, v)$. The same step size h is used for each function (since that refers to the spacing of the independent variable x):

$$u(i+1) = u(i) + h f(x(i), u(i), v(i)),$$
$$v(i+1) = v(i) + h g(x(i), u(i), v(i)).$$

The process is implemented in the following Mathcad function.

Mathcad Function for Euler's Method for System of Two First-Order ODEs

$$\text{Euler\_sys2}(f, g, a, b, u0, v0, n) := \begin{vmatrix} h \leftarrow \dfrac{b-a}{n} \\ u_0 \leftarrow u0 \\ v_0 \leftarrow v0 \\ \text{for } i \in 0..n \\ \quad x_i \leftarrow a + h \cdot i \\ \text{for } i \in 0..n-1 \\ \quad \begin{vmatrix} u_{i+1} \leftarrow u_i + h \cdot f(x_i, u_i, v_i) \\ v_{i+1} \leftarrow v_i + h \cdot g(x_i, u_i, v_i) \end{vmatrix} \\ s \leftarrow \text{augment}(x, u) \\ s \leftarrow \text{augment}(s, v) \\ s \end{vmatrix}$$

Example 13.2 Nonlinear Pendulum Using Euler's Method

To illustrate the effect of the number of subdivisions, we solve the following system of ODE with 50, 100 and 200 subdivisions of the interval.

$$f(x,u,v) := v \qquad g(x,u,v) := -0.3 \cdot v - \sin(u) \qquad u0 := \frac{\pi}{2} \qquad v0 := 0$$

$$a := 0 \qquad b := 15$$

$$B := \text{Euler\_sys2}(f, g, a, b, u0, v0, 50)$$

$$C := \text{Euler\_sys2}(f, g, a, b, u0, v0, 100)$$

$$D := \text{Euler\_sys2}(f, g, a, b, u0, v0, 200)$$

$$x50 := B^{\langle 0 \rangle} \qquad x100 := C^{\langle 0 \rangle} \qquad x200 := D^{\langle 0 \rangle}$$

$$u50 := B^{\langle 1 \rangle} \qquad u100 := C^{\langle 1 \rangle} \qquad u200 := D^{\langle 1 \rangle}$$

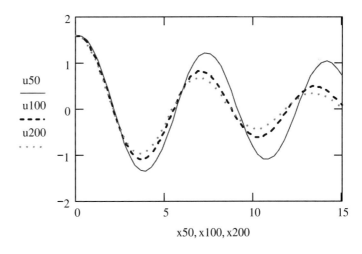

FIGURE 13.2 Euler solution for nonlinear pendulum.

Example 13.3 Series Dilution Problem Using Euler's Method

To illustrate the use of Euler's method for a system of two ODEs, consider the concentration of a dye in a two-compartment dilution process. A pure substance flows into the first tank at the same rate that a mixture leaves the first tank and flows into the second tank; the dye leaves the first tank at a rate that is proportional to the concentration. The loss from the first tank becomes the influx to the second tank, which in turn loses fluid at the same rate; thus, the volume of fluid in each tank is constant. The differential equations describing the concentration of dye in the two tanks are

$$\frac{dC_1}{dt} = -\frac{L}{V_1} C_1, \quad \frac{dC_2}{dt} = -\frac{L}{V_2}[C_2 - C_1].$$

Taking $C_1(0) = 0.3$ moles/liter, $C_2(0) = 0$, $L = 2$ liters/min, $V_1 = 10$ liters, and $V_2 = 5$ liters, we find the concentration in the two tanks for the first 10 minutes of the process. The computed values of C_1 and C_2 are plotted in Fig. 13.3. The exact solutions are also shown for comparison; the equations are given in Example 13.5.

$L := 2 \quad V_1 := 10 \quad V_2 := 5 \quad C10 := 0.3 \quad C20 := 0$

$f(t, C1, C2) := \frac{-L}{V_1} \cdot C1 \qquad g(t, C1, C2) := \frac{-L}{V_2} \cdot (C2 - C1)$

$a := 0 \qquad b := 10 \qquad n := 20$

$B := \text{Euler\_sys2}(f, g, a, b, C10, C20, n) \quad t := B^{\langle 0 \rangle} \quad c1 := B^{\langle 1 \rangle} \quad c2 := B^{\langle 2 \rangle}$

$C1(t) := 0.3 \cdot \exp(-0.2 \cdot t) \qquad C2(t) := 0.6 \cdot (\exp(-0.2 \cdot t) - \exp(-0.4 \cdot t))$

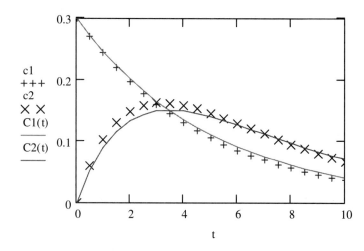

FIGURE 13.3 Concentration of dye.

13.2.2 Midpoint Method for Solving Two ODE-IVPs

The idea in generalizing Runge-Kutta methods to systems of two equations is the same as for Euler's method; that is, we update each unknown function u and v, using the basic Runge-Kutta formulas and the appropriate right-hand-side function, f or g, from the differential equation for the unknown.

$$u' = f(x, u, v) \qquad v' = g(x, u, v)$$

We rewrite the basic second-order Runge-Kutta formulas (the midpoint method),

$$k_1 = h f(x_i, y_i),$$

$$k_2 = h f\left(x_i + \frac{1}{2}h, y_i + \frac{1}{2}k_1\right),$$

$$y_{i+1} = y_i + k_2,$$

using k_1 and k_2 to represent the update quantities for the unknown function u and calling m_1 and m_2 the corresponding quantities for the function v. We must update function u by the appropriate multiple of k_1 or k_2 and function v by the corresponding amount of m_1 or m_2. This means that k_1 and m_1 must be computed before k_2 and m_2 can be found. The method is stated in the following algorithm.

Algorithm for the Midpoint Method for a System of Two First-Order ODE

Approximate the solution $u(x)$, $v(x)$ of the ODE initial value problem
$u' = f(x, u, v)$,
$v' = g(x, u, v)$,
$u(a) = u_0$,
$v(a) = v_0$,
$h = \dfrac{b - a}{n}$
$x_i = a + h\,i$ \qquad (for $i = 0$ to n)
Find solution (loop as indicated)
  for $i = 0$ to $n - 1$
    $k_1 = h f(x_i, u_i, v_i)$
    $m_1 = h g(x_i, u_i, v_i)$
    $k_2 = h f\left(x_i + \dfrac{1}{2}h, u_i + \dfrac{1}{2}k_1, v_i + \dfrac{1}{2}m_1\right)$
    $m_2 = h g\left(x_i + \dfrac{1}{2}h, u_i + \dfrac{1}{2}k_1, v_i + \dfrac{1}{2}m_1\right)$
    $u_{i+1} = u_i + k_2$
    $v_{i+1} = v_i + m_2$
  end

Mathcad Function for Runge-Kutta Two-Step Method for Solving Two ODEs

$$\text{RK2\_sys2}(f, g, a, b, u0, v0, n) := \begin{vmatrix} h \leftarrow \dfrac{b-a}{n} \\ hh \leftarrow \dfrac{h}{2} \\ u_0 \leftarrow u0 \\ v_0 \leftarrow v0 \\ \text{for } i \in 0..n \\ \quad x_i \leftarrow a + h \cdot i \\ \text{for } i \in 0..n-1 \\ \quad \begin{vmatrix} k1 \leftarrow h \cdot f(x_i, u_i, v_i) \\ m1 \leftarrow h \cdot g(x_i, u_i, v_i) \\ k2 \leftarrow h \cdot f(x_i + hh, u_i + 0.5 \cdot k1, v_i + 0.5 \cdot m1) \\ m2 \leftarrow h \cdot g(x_i + hh, u_i + 0.5 \cdot k1, v_i + 0.5 \cdot m1) \\ u_{i+1} \leftarrow u_i + k2 \\ v_{i+1} \leftarrow v_i + m2 \end{vmatrix} \\ s \leftarrow \text{augment}(x, u) \\ s \leftarrow \text{augment}(s, v) \\ s \end{vmatrix}$$

Example 13.4 Nonlinear Pendulum Using Runge-Kutta Method

Consider again the system of ODEs obtained from the second-order ODE for the motion of the nonlinear pendulum described in Examples 13-A, 13.1, and 13.2, i.e.,

$$u' = v = f(x, u, v),$$
$$v' = -0.3\, v - \sin u = g(x, u, v),$$

with initial conditions

$$u_0 = \pi/2; \quad v_0 = 0.$$

$f(x,u,v) := v \qquad\qquad g(x,u,v) := -0.3 \cdot v - \sin(u) \qquad\qquad u0 := \dfrac{\pi}{2}$

$a := 0 \qquad b := 15$

$B := RK2\_sys2(f, g, a, b, u0, v0, 50)$

$C := RK2\_sys2(f, g, a, b, u0, v0, 100)$

$D := RK2\_sys2(f, g, a, b, u0, v0, 200)$

$x50 := B^{\langle 0 \rangle} \qquad\qquad x100 := C^{\langle 0 \rangle} \qquad\qquad x200 := D^{\langle 0 \rangle}$

$u50 := B^{\langle 1 \rangle} \qquad\qquad u100 := C^{\langle 1 \rangle} \qquad\qquad u200 := D^{\langle 1 \rangle}$

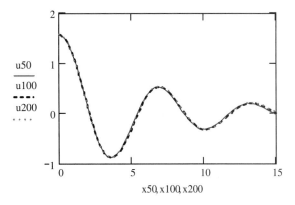

FIGURE 13.4 Runge-Kutta solution of nonlinear pendulum.

Comparing the graphs of the solutions with 50, 100, and 200 subintervals shows that there is virtually no change in the solutions obtained by using more subintervals. This is in marked contrast to the results found with Euler's method in Example 13.2.

Example 13.5 Series Dilution Using the Midpoint Method

We consider again a two-tank dilution process introduced in Example 13.3. The concentrations of the dye in the two tanks satisfy the following differential equations:

$$\frac{dC_1}{dt} = -\frac{L}{V_1} C_1, \qquad \frac{dC_2}{dt} = -\frac{L}{V_2}[C_1 - C_2].$$

Taking $C_1(0) = 0.3$, $C_2(0) = 0$, $L = 2$, $V_1 = 10$, $V_2 = 5$, and $n = 20$, the computed results are shown in Fig. 13.6, together with the exact solutions, viz.,

$$C_1(t) = C_1(0) \exp[(-L/V_1)t] = 0.3 \exp[(-0.2)t]$$

$$C_2(t) = \frac{V_1 C_1(0)}{V_1 - V_2}[\exp[-(L/V_1)t] - \exp[-(L/V_2)t]]$$

$$= 0.6[\exp(-0.2t) - \exp(-0.4t)].$$

$L := 2 \qquad V_1 := 10 \qquad V_2 := 5 \qquad C10 := 0.3 \qquad C20 := 0$

$f(t, C1, C2) := \frac{-L}{V_1} \cdot C1 \qquad g(t, C1, C2) := \frac{-L}{V_2} \cdot (C2 - C1)$

$a := 0 \qquad b := 10 \qquad n := 20$

$B := \text{RK2\_sys2}(f, g, a, b, C10, C20, n) \qquad t := B^{\langle 0 \rangle} \qquad c1 := B^{\langle 1 \rangle} \qquad c2 := B^{\langle 2 \rangle}$

$C1(t) := 0.3 \cdot \exp(-0.2 \cdot t) \qquad\qquad C2(t) := 0.6 \cdot (\exp(-0.2 \cdot t) - \exp(-0.4 \cdot t))$

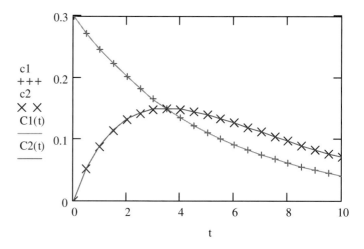

FIGURE 13.5 Concentration of dye in two tanks.

13.3 SYSTEMS OF FIRST-ORDER ODE-IVP

Systems of ODEs may arise directly from applications such as chemical reactions, predator-prey models, and many others. They also come from the conversion of higher order ODEs into system form. The process of making this conversion is illustrated in the next example.

Example 13.6 A Higher-Order System of ODEs

Consider the equation
$$y''' = f(x, y, y', y'') = x + 2y - 3y' + 4y''$$
with initial conditions
$$y(0) = 4, \quad y'(0) = 3, \quad y''(0) = 2.$$
The system of ODEs is
$$u_1' = u_2,$$
$$u_2' = u_3,$$
$$u_3' = x + 2u_1 - 3u_2 + 4u_3.$$
We write this system as
$$u_1' = f_1(x, u_1, u_2, u_3) = u_2,$$
$$u_2' = f_2(x, u_1, u_2, u_3) = u_3,$$
$$u_3' = f_3(x, u_1, u_2, u_3) = x + 2u_1 - 3u_2 + 4u_3.$$

For systems that come from a single higher-order ODE, the structure of the right-hand side in the previous example is a direct result of the definitions of the transformed functions. For systems of ODEs in general, each of the right-hand-side functions $f_1, f_2, \ldots$ may contain any or all of the indicated variables.

A system of ODE can be expressed compactly in vector notation as
$$\mathbf{u}' = \mathbf{f}(x, \mathbf{u}).$$

Since the components of the vectors, $\mathbf{u}$ and $\mathbf{f}$, are denoted by subscripts, we indicate the approximate solutions at the grid points as $u_1(i)$, etc.

13.3.1 Euler's Method for Solving Systems of ODEs

To apply the basic Euler method, $y_{i+1} = y_i + h\, f(x_i, y_i)$, to the system of ODEs

$$u_1' = f_1(x, u_1, u_2, u_3), \quad u_2' = f_2(x, u_1, u_2, u_3), \quad u_3' = f_3(x, u_1, u_2, u_3),$$

we update the function u_1 using f_1, u_2 using f_2, and u_3 using f_3. The same step size h is used for each function. We have

$$u_1(i+1) = u_1(i) + h\, f_1(x(i), u_1(i), u_2(i), u_3(i)),$$
$$u_2(i+1) = u_2(i) + h\, f_2(x(i), u_1(i), u_2(i), u_3(i)),$$
$$u_3(i+1) = u_3(i) + h\, f_3(x(i), u_1(i), u_2(i), u_3(i)).$$

Example 13.7 Solving a Higher-Order System Using Euler's Method

We apply Euler's method with $n = 2$ to find an approximate solution of the system of ODEs

$$u_1' = u_2,$$
$$u_2' = u_3,$$
$$u_3' = x + 2u_1 - 3u_2 + 4u_3,$$

with initial conditions $u_1(0) = 4$, $u_2(0) = 3$, and $u_3(0) = 2$ on $[0, 1]$. The solution at $i = 1$ corresponds to $x(i=1) = 0.5$:

$$u_1(1) = u_1(0) + 0.5\, u_2(0) = 4 + 0.5\,(3) = 5.5,$$

$$u_2(1) = u_2(0) + 0.5\, u_3(0) = 3 + 0.5\,(2) = 4.$$

$$u_3(1) = u_3(0) + 0.5(x(0) + 2u_1(0) - 3u_2(0) + 4u_3(0))$$
$$= 2 + 0.5(0 + 2(4) - 3(3) + 4(2)) = 5.5.$$

The solution at $i = 2$ corresponds to $x(i=2) = 1.0$:

$$u_1(2) = u_1(1) + 0.5\, u_2(1) = 5.5 + 0.5\,(4) = 7.5,$$

$$u_2(2) = u_2(1) + 0.5\, u_3(1) = 4 + 0.5\,(5.5) = 6.75,$$

$$u_3(2) = u_3(1) + 0.5(x(1) + 2u_1(1) - 3u_2(1) + 4u_3(1))$$
$$= 5.5 + 0.5(0.5 + 2(5.5) - 3(4) + 4(5.5)) = 11.25.$$

The next example illustrates the use of Euler's method to solve a system of ODEs that occur as part of the shooting method for solving boundary value problems, that is discussed in Chapter 14. This particular problem, which we revisit in Example 14.1 in the next chapter, deals with finding the electrostatic potential between two concentric spheres.

Example 13.8 Solving Another Higher-Order System Using Euler's Method

Consider the system

$$u_1' = u_2, \quad u_2' = \frac{-2}{x} u_2, \quad u_3' = u_4, \quad u_4' = \frac{-2}{x} u_4,$$

with initial conditions

$$u_1(1) = 10, \quad u_2(1) = 0, \quad u_3(1) = 0, \quad u_4(1) = 1,$$

on [1, 2] with $n = 2$ ($h = 0.5$). The following calculations show the values of each component of the solution as a function of x (not the mesh index):

$$u_1(3/2) = 10 + 1/2(0) = 10,$$
$$u_2(3/2) = 0 + 1/2(0) = 0.$$

$$u_3(3/2) = 0 + 1/2(1) = 1/2,$$
$$u_4(3/2) = 1 + 1/2(-2/1) = 0;$$

second step:

$$u_1(2) = 10 + 1/2(0) = 10,$$
$$u_2(2) = 0 + 1/2(0) = 0,$$
$$u_3(2) = 1/2 + 1/2(0) = 1/2,$$
$$u_4(2) = 0 + 1/2(0) = 0.$$

13.3.2 Runge-Kutta Methods for Solving Systems of ODEs

The idea in generalizing Runge-Kutta methods for use on systems of equations is the same as for Euler's method; that is, we update each unknown function $u_1, u_2, ...,$ using the basic Runge-Kutta formulas and the appropriate right-hand-side function $f_1, f_2, ...,$ from the differential equation for the unknown. We now consider the midpoint method, and the classic fourth-order Runge-Kutta method for systems.

Midpoint Method

For the midpoint method, if we denote the two update parameters as k and m, then the basic second-order Runge-Kutta formulas (the midpoint method) are

$$k = h f(x_i, y_i),$$

$$m = h f\left(x_i + \frac{1}{2}h, y_i + \frac{1}{2}k\right),$$

$$y_{i+1} = y_i + m.$$

To apply these formulas to a system, we must compute k and m for each unknown function (i.e., for each component of the unknown vector **u**). Note that k must be computed for each unknown before m can be found.

We illustrate the process for a system of three ODEs:

$$u_1' = f_1(x, u_1, u_2, u_3),$$
$$u_2' = f_2(x, u_1, u_2, u_3),$$
$$u_3' = f_3(x, u_1, u_2, u_3).$$

The values of the parameter k for the unknown functions $u_1, u_2,$ and u_3 are

$$k_1 = h f_1(x(i), u_1(i), u_2(i), u_3(i)),$$
$$k_2 = h f_2(x(i), u_1(i), u_2(i), u_3(i)),$$
$$k_3 = h f_3(x(i), u_1(i), u_2(i), u_3(i)).$$

Similarly, the values of m are $m_1, m_2,$ and m_3. Of course, to find the value of m for the first ODE, we use f_1; however, we must evaluate f_1 at the appropriate values of $x, u_1, u_2,$ and u_3. Remembering that we are approximating the value of the unknown function employed in evaluating f makes it clear that we approximate each u using its value of k:

$$m_1 = h f_1\left(x(i) + \frac{1}{2}h, u_1(i) + \frac{1}{2}k_1, u_2(i) + \frac{1}{2}k_2, u_3(i) + \frac{1}{2}k_3\right),$$

$$m_2 = h f_2\left(x(i) + \frac{1}{2}h, u_1(i) + \frac{1}{2}k_1, u_2(i) + \frac{1}{2}k_2, u_3(i) + \frac{1}{2}k_3\right),$$

$$m_3 = h f_3\left(x(i) + \frac{1}{2}h, u_1(i) + \frac{1}{2}k_1, u_2(i) + \frac{1}{2}k_2, u_3(i) + \frac{1}{2}k_3\right).$$

Finally, the values of the unknown functions at the next grid point are found:
$$u_1(i + 1) = u_1(i) + m_1,$$
$$u_2(i + 1) = u_2(i) + m_2,$$
$$u_3(i + 1) = u_3(i) + m_3.$$

We summarize the steps of the midpoint method in the following algorithm, and then give a Mathcad function to implement the method. In the algorithm we denote the solution vector at each step as $\mathbf{u}^{(i)}$. The Mathcad function RK2_sys stores the solution at step i as a column in matrix $\mathbf{u}$ (denoted as the superscript <i>) since it is easier to access a column of a matrix than a row. The final solution is displayed as the transpose of the matrix (with the values of the independent variable in the first column.)

Algorithm for the Midpoint Method for a System of First-Order ODE

Approximate the solution $\mathbf{u}(x)$, of the system of ODE $\mathbf{u}' = \mathbf{f}(x, \mathbf{u})$, with initial conditions $\mathbf{u}(a) = \mathbf{u}^{(0)}$, using n steps of a second-order Runge-Kutta method; step size $h = (b - a)/n$.

Begin computations
$$h = \frac{b - a}{n}$$
Define grid points (range variable is suitable)
$\quad (i = 0$ to $n) \quad x_i = a + h\,i$
Find solution (loop is required)
$\quad$ for $i = 0$ to $n - 1$
$$\mathbf{k} = h\,\mathbf{f}(x_i, \mathbf{u}^{(i)})$$
$$\mathbf{m} = h\,\mathbf{f}\left(x_i + \frac{1}{2}h, \mathbf{u}^{(i)} + \frac{1}{2}\mathbf{k}\right)$$
$$\mathbf{u}^{(i+1)} = \mathbf{u}^{(i)} + \mathbf{m}$$
$\quad$ end
Display solution as desired.

We illustrate the use of the algorithm in the next example, which revisits the problem introduced in Example 13.8.

Example 13.9 Solving a Fourth Order System Using the Midpoint Method

Consider again the system

$$u_1' = u_2, \quad u_2' = \frac{-2}{x} u_2, \quad u_3' = u_4, \quad u_4' = \frac{-2}{x} u_4,$$

with initial conditions

$$u_1(1) = 10, \quad u_2(1) = 0, \quad u_3(1) = 0, \quad u_4(1) = 1,$$

on the interval $[1, 2]$. Using $n = 2$ (and $h = 0.5$), we calculate the values of each component of the solution as a function of x (at $x_0 = 1$, $x_1 = 1.5$, and $x_2 = 2.0$). First, find k for each component:

$$k_1 = 0.5\,(u_2(1)) = 0,$$

$$k_2 = 0.5\left(\frac{-2}{x} u_2(1)\right) = 0,$$

$$k_3 = 0.5\,(u_4(1)) = 0.5,$$

$$k_4 = 0.5\left(\frac{-2}{x} u_4(1)\right) = -1.$$

Next, find m for each component:

$$m_1 = 0.5\,(u_2(1) + 0.5\,k_2) = 0,$$

$$m_2 = 0.5\left(\frac{-2}{1.25}\right)(u_2(1) + 0.5\,k_2) = 0,$$

$$m_3 = 0.5\,(u_4(1) + 0.5\,k_4) = 0.25,$$

$$m_4 = 0.5\left(\frac{-2}{1.25}\right)(u_4(1) + 0.5\,k_4) = -0.4.$$

The approximate solution at $x = 1.5$ is

$$u_1(1.5) = 10 + 0 = 10, \qquad u_2(1.5) = 0 + 0 = 0.$$
$$u_3(1.5) = 0 + 0.25 = 0.25, \qquad u_4(1.5) = 1 = 0.4 = 0.6.$$

Now find k for each component:

$$k_1 = 0.5\,(u_2(1.5)) = 0,$$

$$k_2 = 0.5\left(\frac{-2}{1.5} u_2(1.5)\right) = 0,$$

$$k_3 = 0.5\,(u_4(1.5)) = 0.3,$$

$$k_4 = 0.5\left(\frac{-2}{1.5} u_4(1.5)\right) = -0.4.$$

Next, find *m* for each component:

$$m_1 = 0.5 \, (u_2 \, (1.5) + 0.5 \, k_2) = 0,$$

$$m_2 = 0.5 \left(\frac{-2}{1.75}\right)(u_2 \, (1.5) + 0.5 k_2) = 0,$$

$$m_3 = 0.5 \, (u_4 \, (1.5) + 0.5 \, k_4) = 0.2,$$

$$m_4 = 0.5 \left(\frac{-2}{1.75}\right)(u_4 \, (1.5) + 0.5 k_4) = -0.2286.$$

The approximate solution at $x = 2.0$ is

$$u_1(2.0) = 10 + 0 = 10, \qquad u_2(2.0) = 0 + 0 = 0,$$

$$u_3(2.0) = 0.25 + 0.2 = 0.45, \quad u_4(2.0) = 0.6 - 0.2286 = 0.3714.$$

Mathcad Function for Systems Using a Second-Order Runge-Kutta Method

$$\text{RK2\_sys}(f, \text{tspan}, u0, n) := \begin{vmatrix} a \leftarrow \text{tspan}_0 \\ b \leftarrow \text{tspan}_1 \\ h \leftarrow \dfrac{b-a}{n} \\ u^{\langle 0 \rangle} \leftarrow u0 \\ \text{for } i \in 0..n \\ \quad x_i \leftarrow a + h \cdot i \\ \text{for } i \in 0..n-1 \\ \quad \begin{vmatrix} k \leftarrow h \cdot f\left(x_i, u^{\langle i \rangle}\right) \\ m \leftarrow h \cdot f\left(x_i + 0.5 \cdot h, u^{\langle i \rangle} + 0.5 \cdot k\right) \\ u^{\langle i+1 \rangle} \leftarrow u^{\langle i \rangle} + m \end{vmatrix} \\ s \leftarrow \text{augment}\left(x, u^T\right) \\ s \end{vmatrix}$$

Example 13.10 Using Mathcad Function for Midpoint Method

Consider again the problem from Examples 13.8 and 13.9.

$$f(x,u) := \begin{pmatrix} u_1 \\ \dfrac{-2}{x} \cdot u_1 \\ u_3 \\ \dfrac{-2}{x} \cdot u_3 \end{pmatrix} \qquad u0 := \begin{pmatrix} 10 \\ 0 \\ 0 \\ 1 \end{pmatrix} \qquad tspan := \begin{pmatrix} 1 \\ 2 \end{pmatrix}$$

$n := 10$

$RK2_sys(f, tspan, u0, n) = $

	0	1	2	3	4
0	1	10	0	0	1
1	1.1	10	0	0.09	0.828571
2	1.2	10	0	0.165325	0.697572
3	1.3	10	0	0.229269	0.595261
4	1.4	10	0	0.284216	0.513858
5	1.5	10	0	0.331931	0.448044
6	1.6	10	0	0.373749	0.394086
7	1.7	10	0	0.410694	0.349303
8	1.8	10	0	0.44357	0.311731
9	1.9	10	0	0.473011	0.279903
10	2	10	0	0.499528	0.252706

Classic Runge-Kutta for Systems

We now consider the classic fourth-order Runge-Kutta method for systems of ODE. We denote the four update parameters as **k1**, **k2**, **k3**, and **k4** (each of which is a vector). They could easily be stored in a single matrix if desired.

In the following algorithm and Mathcad functions, the solution vector at x_i is denoted using superscripts, and is stored as a column in the solution matrix for easier acces during computation. The final solution is displayed as the transpose of the solution matrix, with the values of the independent variable shown in the first column.

Algorithm for Fourth-Order Runge-Kutta Method for System of ODE

Approximate the solution **u**(x), of the system of ODE
$\mathbf{u}' = \mathbf{f}(x, \mathbf{u})$
with initial conditions $\mathbf{u}(a) = \mathbf{u}^{(0)}$, using n steps of the classic fourth-order Runge-Kutta method; step size $h = (b - a)/n$.

Begin computations
$$h = \frac{b - a}{n}$$
Define grid points (use range variable instead of loop if desired)
  (for $i = 0$ to n) $x_i = a + h\, i$
Begin computations (loop is required)
  for $i = 0$ to $n - 1$
   $\mathbf{k1} = h\, \mathbf{f}(x_i, \mathbf{u}^{(i)})$
   $\mathbf{k2} = h\, \mathbf{f}\left(x_i + \frac{1}{2}h, \mathbf{u}^{(i)} + \frac{1}{2}\mathbf{k1}\right)$
   $\mathbf{k3} = h\, \mathbf{f}\left(x_i + \frac{1}{2}h, \mathbf{u}^{(i)} + \frac{1}{2}\mathbf{k2}\right)$
   $\mathbf{k4} = h\, \mathbf{f}(x_i + h, \mathbf{u}^{(i)} + \mathbf{k3})$
   $\mathbf{u}^{(i+1)} = \mathbf{u}^{(i)} + \frac{1}{6}\mathbf{k1} + \frac{1}{3}\mathbf{k2} + \frac{1}{3}\mathbf{k3} + \frac{1}{6}\mathbf{k4}$
  end

Mathcad Function for Systems Using a Fourth-Order Runge-Kutta Method

$$\text{RK4\_sys}(f, \text{tspan}, u0, n) := \begin{vmatrix} a \leftarrow \text{tspan}_0 \\ b \leftarrow \text{tspan}_1 \\ h \leftarrow \dfrac{b-a}{n} \\ u^{\langle 0 \rangle} \leftarrow u0 \\ \text{for } i \in 0..n \\ \quad x_i \leftarrow a + h \cdot i \\ \text{for } i \in 0..n-1 \\ \quad \begin{vmatrix} k1 \leftarrow h \cdot f\left(x_i, u^{\langle i \rangle}\right) \\ k2 \leftarrow h \cdot f\left(x_i + \dfrac{h}{2}, u^{\langle i \rangle} + \dfrac{k1}{2}\right) \\ k3 \leftarrow h \cdot f\left(x_i + \dfrac{h}{2}, u^{\langle i \rangle} + \dfrac{k2}{2}\right) \\ k4 \leftarrow h \cdot f\left(x_i + h, u^{\langle i \rangle} + k3\right) \\ u^{\langle i+1 \rangle} \leftarrow u^{\langle i \rangle} + \dfrac{k1}{6} + \dfrac{k2}{3} + \dfrac{k3}{3} + \dfrac{k4}{6} \end{vmatrix} \\ s \leftarrow \text{augment}\left(x, (u)^T\right) \\ s \end{vmatrix}$$

Example 13.11 Solving a Circular Chemical Reaction Using a Runge-Kutta Method

Consider the circular reaction involving three chemical reactions described in Example 13-B; the differential equations are

$$\frac{dA}{dt} = r_3 C - r_1 A, \quad \frac{dB}{dt} = r_1 A - r_2 B, \quad \frac{dC}{dt} = r_2 B - r_3 C.$$

We assume that the total quantity of the chemicals is $A + B + C = Q = 1$; the initial values of A, B, and C are chosen to satisfy the relations

$$A = \frac{1}{1 + r_1/r_2 + r_1/r_3}, \quad B = \frac{r_1}{r_2} A, \quad C = \frac{r_1}{r_3} A$$

for the initial rate parameters r_1, r_2, and r_3. We let $r_2 = 2$, let $r_3 = 1$, and allow r_1 to change slowly from an initial value of 0.1 according to the linear equation $r_1 = 0.1(t + 1)$. The initial values of A, B, and C are

$$A(0) = \frac{1}{1.15} = 0.8696, \quad B(0) = \frac{0.05}{1.15} = 0.0435, \quad C(0) = \frac{0.1}{1.15} = 0.0870.$$

$$f(x,u) := \begin{bmatrix} u_2 - 0.1 \cdot (x + 1) \cdot u_0 \\ 0.1 \cdot (x + 1) \cdot u_0 - 2 \cdot u_1 \\ 2 \cdot u_1 - u_2 \end{bmatrix} \qquad u0 := \begin{pmatrix} \dfrac{1}{1.15} \\ \dfrac{0.05}{1.15} \\ \dfrac{0.1}{1.15} \end{pmatrix} \qquad \text{tspan} := \begin{pmatrix} 0 \\ 40 \end{pmatrix}$$

$n := 100 \qquad M := \text{RK4\_sys}(f, \text{tspan}, u0, n)$

$t := M^{\langle 0 \rangle} \qquad A := M^{\langle 1 \rangle} \qquad B := M^{\langle 2 \rangle} \qquad C := M^{\langle 3 \rangle}$

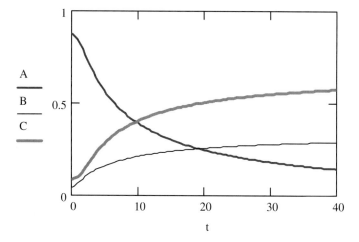

FIGURE 13.6   Concentration of three reactants.

13.3.3 Multistep Methods for Systems

Adams-Bashforth Methods for Systems

The basic two-step Adams-Bashforth method, in which y_0 is given by the initial condition for the differential equation, y_1 is found from a one-step method, such as a Runge-Kutta technique, and for $i = 1, \ldots, n-1$, and $h = \dfrac{b-a}{n}$,

$$y_{i+1} = y_i + \frac{h}{2}[3f(x_i, y_i) - f(x_{i-1}, y_{i-1})],$$

can be extended for use with a system of three ODEs

$$u_1' = f_1(x, u_1, u_2, u_3),$$
$$u_2' = f_2(x, u_1, u_2, u_3),$$
$$u_3' = f_3(x, u_1, u_2, u_3),$$

in a similarly straightforward manner. That is,

$u_1(i=0), u_2(i=0), u_3(i=0)$ are given by the initial conditions,

$u_1(i=1), u_2(i=1), u_3(i=1)$ are found from a 1-step method,

and for $i = 1, \ldots, n-1$:

$$u_1(i+1) = u_1(i) + \frac{h}{2}[3f_1(x(i), u_1(i), u_2(i), u_3(i))$$
$$- f_1(x(i-1), u_1(i-1), u_2(i-1), u_3(i-1))],$$

$$u_2(i+1) = u_2(i) + \frac{h}{2}[3f_2(x(i), u_1(i), u_2(i), u_3(i))$$
$$- f_2(x(i-1), u_1(i-1), u_2(i-1), u_3(i-1))],$$

$$u_3(i+1) = u_3(i) + \frac{h}{2}[3f_3(x(i), u_1(i), u_2(i), u_3(i))$$
$$- f_3(x(i-1), u_1(i-1), u_2(i-1), u_3(i-1))].$$

Predictor-Corrector Methods for Systems

The Adams-Bashforth-Moulton predictor-corrector methods are extended for use with systems of ODEs in a similar manner. The third-order method is implemented in the Mathcad function ABM3_sys that follows. The dimensions of the vectors **u**, **u0**, and **tspan** depend on the ODE system being solved, as illustrated in Examples 13.12 and 13.13.

Algorithm for Third Order Predictor-Corrector Method for a System of ODE

Approximate the solution vector $\mathbf{u}(x)$ of the system of ODE
$$\mathbf{u}' = \mathbf{f}(x, \mathbf{u})$$
with initial conditions $\mathbf{u}(a) = \mathbf{u}^{(0)}$
using the third-order Adams-Bashforth-Moulton predictor corrector method.

Initialize
$$h = \frac{b-a}{n}$$
Define grid points
    for $i = 0$ to n
        $x_i = a + h\,i$
    end
Begin computations
    $\mathbf{z}^{(0)} = \mathbf{f}(x_0, \mathbf{u}^{(0)})$
Use the midpoint method to find the first two points
    for $i = 0$ to 1
        $\mathbf{k}_1 = h\,\mathbf{f}(x_i, \mathbf{u}^{(i)})$
        $\mathbf{k}_2 = h\,\mathbf{f}\!\left(x_i + \tfrac{1}{2}h,\, \mathbf{u}^{(i)} + \tfrac{1}{2}\mathbf{k}_1\right)$
        $\mathbf{u}^{(i+1)} = \mathbf{u}^{(i)} + \mathbf{k}_2$
    end
    $\mathbf{z}^{(1)} = \mathbf{f}(x_1, \mathbf{u}^{(1)})$

for $i = 2$ to $n - 1$
    Use 3rd order AB method to predict
    $\mathbf{z}^{(i)} = \mathbf{f}(x_i, \mathbf{u}^{(i)})$
    $\mathbf{u}^* = \mathbf{u}^{(i)} + \dfrac{h}{12}\left(23\,\mathbf{z}^{(i)} - 16\,\mathbf{z}^{(i-1)} + 5\,\mathbf{z}^{(i-2)}\right)$
    Use 3rd order AM method to correct
    $\mathbf{z}^* = \mathbf{f}(x_{i+1}, \mathbf{u}^*)$
    $\mathbf{u}^{(i+1)} = \mathbf{u}^{(i)} + \dfrac{h}{12}\left(5\,\mathbf{z}^* + 8\,\mathbf{z}^{(i)} - \mathbf{z}^{(i-1)}\right)$
end

Display results

Mathcad Function for Predictor-Corrector Method for System of ODEs

$\text{ABM3\_sys}(f, \text{tspan}, u0, n) := \begin{vmatrix} a \leftarrow \text{tspan}_0 \\ b \leftarrow \text{tspan}_1 \\ h \leftarrow \dfrac{b-a}{n} \\ hh \leftarrow \dfrac{h}{12} \\ u^{\langle 0 \rangle} \leftarrow u0 \\ \text{for } i \in 0..n \\ \quad x_i \leftarrow a + h \cdot i \\ \text{for } i \in 0..1 \\ \quad \begin{vmatrix} k \leftarrow h \cdot f\left(x_i, u^{\langle i \rangle}\right) \\ m \leftarrow h \cdot f\left(x_i + \dfrac{h}{2}, u^{\langle i \rangle} + \dfrac{k}{2}\right) \\ u^{\langle i+1 \rangle} \leftarrow u^{\langle i \rangle} + m \end{vmatrix} \\ \text{for } i \in 2..n-1 \\ \quad \begin{vmatrix} z^{\langle i \rangle} \leftarrow f\left(x_i, u^{\langle i \rangle}\right) \\ uu^{\langle i+1 \rangle} \leftarrow u^{\langle i \rangle} + hh \cdot \left(23 \cdot z^{\langle i \rangle} - 16 \cdot z^{\langle i-1 \rangle} + 5 \cdot z^{\langle i-2 \rangle}\right) \\ zz \leftarrow f\left(x_{i+1}, uu^{\langle i+1 \rangle}\right) \\ u^{\langle i+1 \rangle} \leftarrow u^{\langle i \rangle} + hh \cdot \left(5 \cdot zz + 8 \cdot z^{\langle i \rangle} - z^{\langle i-1 \rangle}\right) \end{vmatrix} \\ s \leftarrow \text{stack}\left(x^T, u\right) \\ s \end{vmatrix}$

Example 13.12 Mass-and-Spring System

The vertical displacements of two masses m_1 and m_2 suspended in series by springs with spring constants s_1 and s_2 are given by Hooke's law as a system of two second-order ODEs. The displacements are x_1 and x_2. (See Fig. 13.7; the displacement of each mass is measured from its equilibrium position, with positive direction downward.) The ODEs are

$$m_1 x_1'' = -s_1 x_1 + s_2 (x_2 - x_1),$$

$$m_2 x_2'' = -s_2 (x_2 - x_1).$$

The differential equations relating these variables (functions) are

$$u_1' = u_2, \quad u_2' = \frac{s_1}{m_1} u_1 + \frac{s_2}{m_1} (u_3 - u_1),$$

$$u_3' = u_4, \quad u_4' = -\frac{s_2}{m_2} (u_3 - x_1),$$

with the initial conditions

$$u_1(0) = \alpha_1, \quad u_2(0) = \alpha_2, \quad u_3(0) = \alpha_3, \quad u_4(0) = \alpha_4$$

The displacement profiles of a 10-kg mass and a 2-kg mass, are illustrated in Fig. 13.8. The spring constants are 100 and 120, respectively.

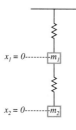

FIGURE 13.7 Two mass system

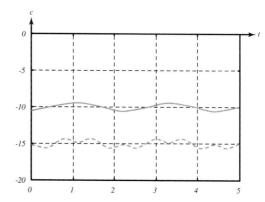

FIGURE 13.8 Displacement profiles of two masses connected by springs.

Example 13.13 Motion of a Baseball

Air resistance is one of the factors influencing how far a fly ball travels. In this example, we illustrate the effect of changing assumptions about the form of the air resistance. If a ball is hit with an initial velocity of [100, 45] and is subject to air resistance proportional to its velocity (acting on the horizontal component only), the motion can be found by the Mathcad script that follows. The trajectories are shown in Figure 13.9 for two choices of the form of the air resistance.

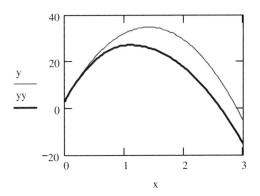

FIGURE 13.9 *y* corresponds to air resistance proportional to velocity in *x* direction; *yy* gives path for air resistance proportional to velocity.

$$f(x,z) := \begin{pmatrix} z_1 \\ -0.1 \cdot z_1 \\ z_3 \\ -32 \end{pmatrix} \qquad g(x,z) := \begin{bmatrix} z_1 \\ -0.1 \cdot \sqrt{(z_1)^2 + (z_3)^2} \\ z_3 \\ -32 - 0.1 \cdot \sqrt{(z_1)^2 + (z_3)^2} \cdot \text{sign}(z_3) \end{bmatrix}$$

$$u0 := \begin{pmatrix} 0 \\ 100 \\ 3 \\ 45 \end{pmatrix} \qquad n := 100 \qquad \text{tspan} := \begin{pmatrix} 0 \\ 3 \end{pmatrix}$$

$B := \text{ABM3\_sys}(f, \text{tspan}, u0, n)$ \qquad $C := \text{ABM3\_sys}(g, \text{tspan}, u0, n)$

$x := (B^T)^{\langle 0 \rangle}$

$y := (B^T)^{\langle 3 \rangle}$ \qquad $yy := (C^T)^{\langle 3 \rangle}$

13.4 STIFF ODE AND ILL-CONDITIONED PROBLEMS

There are ODE for which any error that occurs will increase, regardless of the numerical method employed. Such problems are called *ill-conditioned*. As an illustration, consider the system

$$u_1' = 2 u_2$$
$$u_2' = 2 u_1$$

for which the general solution is

$$u_1 = a e^{2x} + b e^{-2x}$$
$$u_2 = a e^{2x} - b e^{-2x}$$

With the initial conditions

$$u_1(0) = 3$$
$$u_2(0) = -3$$

we have $a = 0$, $b = 3$. However, for any numerical error that occurs, a component of the positive exponential will be introduced and will eventually dominate the true solution.

Ill-conditioning can also occur for a single first-order ODE, as the following problem shows. Consider the ODE

$$y' = 3 y - t^2$$

for which the general solution is

$$y = C e^{3t} + \frac{1}{3} t^2 + \frac{2}{9} t + \frac{2}{27}.$$

If we take the initial condition as $y(0) = \frac{2}{27}$, the exact solution is

$$y = \frac{1}{3} t^2 + \frac{2}{9} t + \frac{2}{27}.$$

However, any error in the numerical solution process will introduce the exponential component which will eventually dominate the true solution. The exponential term is known as a parasitic solution.

An ODE in which there is a rapidly decaying transient solution also causes difficulties for numerical solution, requiring an extremely small step size in order to obtain an accurate solution. One source of such equations is in the description of a spring-mass system with large spring constants, hence these problems are known as *stiff ODE*. Stiff ODEs are very common in chemical kinetic studies, and also occur in many network analysis and simulation problems.

As an illustration, consider the system
$$u' = 98\,u + 198\,v,$$
$$v' = -99\,u - 199\,v,$$
with initial conditions $u(0) = 1$, $v(0) = 0$.
The exact solution is
$$u(t) = 2\,e^{-t} - e^{-100t},$$
$$v(t) = -e^{-t} + e^{-100t}.$$

It is also possible for a single first-order ODE to be stiff, as the following problem shows. Consider the ODE
$$y' = \lambda\,(y - g(t)) + g'(t)$$
with $\lambda \ll 0$ and $g(t)$ a smooth, slowly varying function.
The solution is
$$y = (y_0 - g(0))\,e^{\lambda t} + g(t).$$
The first term in the solution will soon be insignificant compared with $g(t)$, but stability will continue to be governed by $h\lambda$, necessitating a very small step size.

For a system of equations
$$\mathbf{y}' = \mathbf{A}(\mathbf{y} - \mathbf{g}(t)) + \mathbf{g}'(t)$$
the eigenvalues of $\mathbf{A}$ correspond to λ; if all of the eigenvalues have negative real parts, the solution will converge towards $\mathbf{g}(t)$ as $t \to \infty$.

The simplest method for stiff problems is the backward Euler method
$$y_{i+1} = y_i + h\,f(t_{i+1} + y_{i+1}).$$
The error is amplified by $(1 - h\,\lambda)^{-1}$ at each step, which is less than one if $\mathrm{Re}(\lambda) < 0$. Thus, the backward Euler's method is A-stable, according to the following definition.

A method is called A-stable if any solution produced when the method is applied (with fixed step size $h > 0$) to the problem $y' = \lambda\,y$ (with $\lambda = \alpha + \beta\,i$ and $\alpha < 0$) tends to zero as $n \to \infty$.

Dahlquist (1963) showed that a multistep method that is A-stable cannot have order greater than two. The trapezoidal method is the second-order multistep method with the smallest error constant (see summary for Chapter 12).

Since A-stability is difficult to achieve, a somewhat less restrictive stability condition, known as stiff-stability is often sufficient. Methods for stiff ODE are implicit and often require iterative techniques for their solution. Newton's method may be used, with the required Jacobian either supplied by the user, or generated numerically. (See Gear, 1971 for further discussion.)

13.5 MATHCAD'S METHODS

The Mathcad functions for solving the ODE described at the end of the previous chapter are directly applicable to systems of ODEs. We discuss their use for systems and higher order ODE here. In addition, we describe the Mathcad2000Pro functions for stiff ODE, `stiffb`, `Stiffb`, `stiffr`, and `Stiffr`. The Mathcad function `odesolve`, discussed in Section 14.5 can be used to solve a single higher order ODE.

13.5.1 Using the Built-In Functions

The function `rkfixed` provides Mathcad's basic approach for solving ODEs, either a single first order equation, a single higher order equation, or a system of first order equations. We considered the use of `rkfixed` for a single first-order ODE in Section 12.5; we now consider its use for higher order ODE and systems of ODE.

The function `rkfixed` returns the values of the solution to the ODE at equally spaced points across the specified interval. The call to this function has the form

$$\text{rkfixed}\,(y, x1, x2, npts, D)$$

where the input parameters are

y	the vector of initial values for the ODE
$x1$	left end of interval on which solution is desired
$x2$	right end of interval on which solution is desired
$npts$	number of points beyond the initial point where solution will be approximated
D	vector containing the derivatives of the unknown function (the right hand side of the ODE, $\mathbf{y}' = \mathbf{f}(x,y)$)

The function `rkfixed` returns a vector with $npts + 1$ rows and $j + 1$ columns, where j is the order of the ODE or system of ODE. The first column contains the values of the independent variable at which the solution is given; the second column contains the corresponding solution values, the third column contains the values of the second component of the solution (first derivative of y if the ODE was a second order ODE), etc.

Example 13.14 Using the Mathcad function `rkfixed`

We illustrate the use of the function `rkfixed` to solve the system of differential equations

$$u_1' = u_2, \quad u_2' = \frac{-2}{x} u_2, \quad u_3' = u_4, \quad u_4' = \frac{-2}{x} u_4,$$

with initial conditions

$$u_1(1) = 10, \quad u_2(1) = 0, \quad u_3(1) = 0, \quad u_4(1) = 1,$$

investigated in several previous examples. We find the solution at 5 intermediate points on the interval $[1, 2]$. Remember that Mathcad indexing starts with 0.

$$u := \begin{pmatrix} 10 \\ 0 \\ 0 \\ 1 \end{pmatrix} \qquad D(x, u) := \begin{pmatrix} u_1 \\ -2 \cdot \frac{u_1}{x} \\ u_3 \\ -2 \cdot \frac{u_3}{x} \end{pmatrix} \qquad x1 := 1 \quad x2 := 2$$

Solution using 2 subintervals

$$z := \text{rkfixed}(u, x1, x2, 2, D) \qquad z = \begin{pmatrix} 1 & 10 & 0 & 0 & 1 \\ 1.5 & 10 & 0 & 0.33 & 0.447 \\ 2 & 10 & 0 & 0.497 & 0.251 \end{pmatrix}$$

Solution using 5 subintervals

$$z := \text{rkfixed}(u, x1, x2, 5, D) \qquad z = \begin{pmatrix} 1 & 10 & 0 & 0 & 1 \\ 1.2 & 10 & 0 & 0.167 & 0.694 \\ 1.4 & 10 & 0 & 0.286 & 0.51 \\ 1.6 & 10 & 0 & 0.375 & 0.391 \\ 1.8 & 10 & 0 & 0.444 & 0.309 \\ 2 & 10 & 0 & 0.5 & 0.25 \end{pmatrix}$$

Use [ctrl] 6 or matrix toolbar to get the superscripts which access the columns of matrix z. The first column (column 0) gives the independent variable values; the remaining 4 columns gives the corresponding values of $u1$, $u2$, $u3$, and $u4$ (in the notation of the problem statement).

The results using `rkfixed` with 5 subintervals compare well to those found in Example 13.10, which used the Midpoint Method with 10 subintervals.

Mathcad2000Pro includes several other functions for solving ODE. As with `rkfixed`, the functions can be used for a single first order ODE (which we discussed in Section 12.5), or for a single higher-order ODE or a system of ODE. These functions are an adaptive step size Runga-Kutta method, the Bulirsch-Stoer method for smooth ODE, and Bulirsch-Stoer and Rosenbrock methods for stiff ODE. The modifications to the `rkadapt/Rkadapt` and `bulstoer/Bulstoer` functions to go from a single ODE to a system of ODE (or a higher-order ODE) are the same as described for `rkfixed`. We turn our attention now to the routines for stiff ODE, namely `stiffb`, `Stiffb`, `stiffr`, and `Stiffr`. When solving stiff ODE it is important that the variables be scaled properly. See [Press et al. 1992] or other advanced texts on numerical methods for further discussion.

Bulirsch-Stoer Methods for Stiff ODE

The functions `stiffb` and `Stiffb` may be used to solve a stiff ODE using the Bulirsch-Stoer method with adaptive step size. The function `stiffb` gives the solution of the ODE at a single point. The function `Stiffb` gives the solution at equally spaced points on the specified interval, by making repeated calls to `stiffb`.

The usage of `Stiffb` is similar to that for `rkfixed`, with the exception that a matrix must be supplied giving the partial derivatives of the right-hand side of the ODE with respect to the independent variable, and each of the dependent variables. This is the Jacobian matrix augmented with a first column giving the derivatives with respect to x. The call to this function has the form

   `Stiffb (y, x1, x2, npts, D, J)`

where the parameters are as above, and $J(x,y)$ is a vector-valued function whose first column contains the derivative dD/dx and whose remaining columns contain the Jacobian matrix dD/dy for the system of ODEs.

The usage of the function `stiffb` is the same as for `rkadapt`, with the addition of the augmented Jacobian matrix as desribed above.

A system of ODEs expressed in the form $\mathbf{y'} = \mathbf{A}\,\mathbf{y}$ is stiff if the matrix $\mathbf{A}$ is nearly singular.

Rosenbrock Methods for Stiff ODE

The functions `stiffr` and `Stiffr` may be used to solve a stiff ODE using the Rosenbrock method with adaptive step size. The function `stiffr` gives the solution of the ODE at a single point. The function `Stiffr` gives the solution at equally spaced points on the specified interval, by making repeated calls to `stiffr`. The usage of `Stiffr` is the same as for `Stiffb`. The call to this function has the form

   `Stiftr (y, x1, x2, npts, D, J)`

where the parameters are as for `Stiffb`. The usage of `stiffr` is as for `stiffb`.

13.5.2 Understanding the Algorithms

The algorithms used in the built-in Mathcad functions for solving stiff ODEs are either semi-implicit extrapolation methods (extensions of the Bulirsch-Stoer methods) and Rosenbrock methods (extensions of the Runge-Kutta methods). The basic ideas are summarized briefly here.

Semi-Implicit Extrapolations Methods

The Bulirsch-Stoer method with adaptive step size is used for the functions `stiffb` and `Stiffb`. These methods are semi-implict extrapolation methods, attributed to Bader and Dueflhard, which use a semi-implicit midpoint rule for the extrpolation. [Press et al., 1992, pp. 742–747]. The method is based on the implicit midpoint rule

$$y_{n+1} - y_{n-1} = 2h\,f\left(\frac{y_{n+1} + y_{n-1}}{2}\right)$$

The semi-implicit form comes from linearizing the right-hand side. This, together with a special first step (the semi-implicit Euler step), and a smoothing last step, forms the basic algorithm. For computational efficiency, the equations are written in terms of the change in the solution, $\mathbf{D}_k = \mathbf{y}_{k+1} - \mathbf{y}_k$, as follows:

$$\mathbf{D}_0 = [\mathbf{I} - h\,\mathbf{J}]^{-1} h\,\mathbf{f}(\mathbf{y}_0)$$

$$\mathbf{y}_1 = \mathbf{y}_0 + \mathbf{D}_0$$

For $k = 1, \ldots m-1$

$$\mathbf{D}_k = \mathbf{D}_{k-1} + 2[\mathbf{I} - h\,\mathbf{J}]^{-1}[h\,\mathbf{f}(\mathbf{y}_k) - \mathbf{D}_{k-1}]$$

$$\mathbf{y}_{k+1} = \mathbf{y}_k + \mathbf{D}_k$$

Finally, to smooth the last step, compute

$$\mathbf{D}_m = [\mathbf{I} - h\,\mathbf{J}]^{-1}[h\,\mathbf{f}(\mathbf{y}_m) - \mathbf{D}_{m-1}]$$

$$\mathbf{y}_m = \mathbf{y}_m + \mathbf{D}_m$$

The matrix $\mathbf{J}$ is the Jacobian matrix, $\left[\dfrac{\partial \mathbf{f}}{\partial \mathbf{y}}\right]$; m is the number of steps to be used in going across the interval, from x_0 to x_1, and $h = \dfrac{x_1 - x_0}{m}$ is the step size. See [Press et al., 1992, p. 742–747] for further details.

Rosenbrock Method

The Rosenbrock method, used for the functions `stiffr` and `Stiffr`, is a generalization of the Runge-Kutta methods, for stiff equations. The first of the practical implementations of these methods is due to Kaps and Rentrop [Kaps and Rentrop, 1979]. The Rosenbrock methods are competitive with more complicated algorithms for moderate-sized systems (on the order of 10 or fewer equations) with moderate accuracy criterion (relative error of 10^{-4} or 10^{-5}). The general idea is to find a solution of the form $y(x + h) = y(x) +$ corrections, where the corrections are solutions of a set of linear equations involving Runga-Kutta terms and terms depending on the Jacobian of the system. The effectiveness of the method relies on an effective embedded Runge-Kutta method for step-size adjustment. Several different forms of the update equations have been proposed; the form favored by Press et al., and presumably implemented in Mathcad, was originally developed by [Shampine, 1982].

Example 13.15 Robot Motion

We illustrate the use of the function `rkfixed` to solve the problem of simulating the motion of a two-link planar robot arm. This is an example of a forward dynamics problem—i.e., given the applied joint torques, we solve for the resulting motion of the system. Attaining a solution involves integrating the equations of motion, which are two nonlinear coupled ODEs. (See Spong and Vidyasagar, 1989 for derivation of the equations and description of the coordinate system conventions.) Figure 13.10 shows the motion of the arm.

Define global variables:

$L1 \equiv 1.0 \quad Lc1 \equiv 0.5 \quad L2 \equiv 1.0 \quad Lc2 \equiv 0.5$

$m1 \equiv 50.0 \quad m2 \equiv 50.0 \quad g \equiv 9.81$

$K\_p \equiv 200 \cdot identity(2) \quad K\_v \equiv 1000 \cdot identity(2)$

$q1\_goal \equiv -0.0 \quad q2\_goal \equiv \frac{\pi}{2}$

$I1 \equiv \frac{m1 \cdot L1^2}{12} \quad I2 \equiv \frac{m2 \cdot L2^2}{12}$

Establish parameters to be passed to S_robot_motion:

$y0 := \begin{pmatrix} \frac{-\pi}{2} \\ \frac{-\pi}{2} \\ 0 \\ 0 \end{pmatrix} \quad t0 := 0 \quad tf := 25 \quad n := 15000$

Chapter 13 Systems of Ordinary Differential Equations

Execute S_robot_motion:

$$B := S\_robot\_motion(y0, t0, tf, n, robot2)$$

Organize data returned from S_robot_motion:

$$t := B^{\langle 0 \rangle} \qquad q1 := B^{\langle 1 \rangle} \qquad q2 := B^{\langle 2 \rangle}$$

$$q1d := B^{\langle 3 \rangle} \qquad q2d := B^{\langle 4 \rangle} \qquad torq1 := B^{\langle 5 \rangle} \qquad torq2 := B^{\langle 6 \rangle}$$

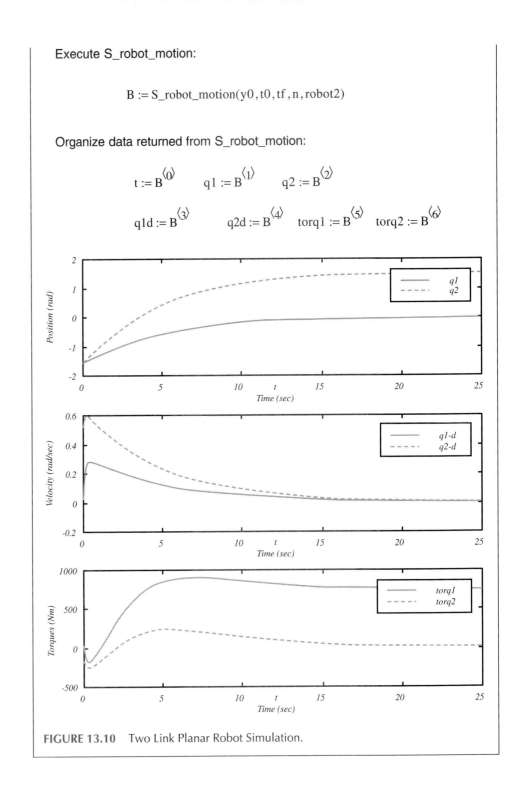

FIGURE 13.10 Two Link Planar Robot Simulation.

Robot Motion

Define the function to calculate the accelerations for a two-link planar robot:

$$\text{robot2}(t, y) := \begin{array}{|l} q1 \leftarrow y_0 \\ q2 \leftarrow y_1 \\ q1d \leftarrow y_2 \\ q2d \leftarrow y_3 \\ M_{0,0} \leftarrow m1 \cdot Lc1^2 + m2 \cdot \left(L1^2 + Lc2^2 + 2 \cdot L1 \cdot Lc2 \cdot \cos(q2)\right) + I1 + I2 \\ M_{0,1} \leftarrow m2 \cdot \left(Lc2^2 + L1 \cdot Lc2 \cdot \cos(q2)\right) + I2 \\ M_{1,0} \leftarrow M_{0,1} \\ M_{1,1} \leftarrow m2 \cdot Lc2^2 + I2 \\ h \leftarrow -m2 \cdot L1 \cdot Lc2 \cdot \sin(q2) \\ C \leftarrow h \cdot \begin{pmatrix} q2d & q2d + q1d \\ -q1d & 0 \end{pmatrix} \\ \text{phi}_0 \leftarrow (m1 \cdot Lc1 + m2 \cdot L1) \cdot g \cdot \cos(q1) + m2 \cdot Lc2 \cdot g \cdot \cos(q1 + q2) \\ \text{phi}_1 \leftarrow m2 \cdot Lc2 \cdot g \cdot \cos(q1 + q2) \\ e \leftarrow \begin{pmatrix} q1 - q1\_\text{goal} \\ q2 - q2\_\text{goal} \end{pmatrix} \\ ed \leftarrow \begin{pmatrix} q1d \\ q2d \end{pmatrix} \\ \text{tau} \leftarrow -K\_p \cdot e - K\_v \cdot ed + \text{phi} \\ qd \leftarrow ed \\ qdd \leftarrow M^{-1}(\text{tau} - C \cdot qd - \text{phi}) \\ ydot \leftarrow \begin{pmatrix} q1d \\ q2d \\ qdd_0 \\ qdd_1 \end{pmatrix} \\ ydot \end{array}$$

Define the function that calculates the robot motion. Note: rkfixed (Runge-Kutta fixed step size) will not evaluate the solution accurately with n less than approximately 10000. If you use Rkadapt (Runge-Kutta adaptive step size) or Bulstoer (Bulirsch-Stoer method) you can reduce n down to approximately 200. Both however, are only available with the professional edition.

$$\text{S\_robot\_motion}(y0, t0, tf, n, f) := \begin{vmatrix} A \leftarrow \text{rkfixed}(y0, t0, tf, n, f) \\ t \leftarrow A^{\langle 0 \rangle} \\ q1 \leftarrow A^{\langle 1 \rangle} \\ q2 \leftarrow A^{\langle 2 \rangle} \\ q1d \leftarrow A^{\langle 3 \rangle} \\ q2d \leftarrow A^{\langle 4 \rangle} \\ \text{imax} \leftarrow \text{rows}(t) \\ \text{for } i \in 0..\text{imax} - 1 \\ \quad \begin{vmatrix} \text{phi}_0 \leftarrow (m1 \cdot Lc1 + m2 \cdot L1) \cdot g \cdot \cos(q1_i) + m2 \cdot Lc2 \cdot g \cdot \cos(q1_i + q2_i) \\ \text{phi}_1 \leftarrow m2 \cdot Lc2 \cdot g \cdot \cos(q1_i + q2_i) \\ ed \leftarrow \begin{pmatrix} q1d_i \\ q2d_i \end{pmatrix} \\ e \leftarrow \begin{pmatrix} q1_i - q1\_\text{goal} \\ q2_i - q2\_\text{goal} \end{pmatrix} \\ \text{torq} \leftarrow -K\_p \cdot e - K\_v \cdot ed + \text{phi} \\ \text{torq1}_i \leftarrow \text{torq}_0 \\ \text{torq2}_i \leftarrow \text{torq}_1 \end{vmatrix} \\ B \leftarrow \text{augment}(A, \text{torq1}, \text{torq2}) \\ B \end{vmatrix}$$

SUMMARY

Convert higher-order ODE to system of first-order ODE: The nth-order ODE

$$y^{(n)} = f(x, y, y', y'', \ldots, y^{(n-1)}),$$

$$y(0) = \alpha_0, \quad y'(0) = \alpha_1, \quad y''(0) = \alpha_2, \ldots, y^{(n-1)}(0) = \alpha_{n-1},$$

becomes a system of first-order ODEs by the following change of variables:

$$u_1 = y, \quad u_2 = y', \quad u_3 = y'', \ldots, \quad u_n = y^{(n-1)}.$$

The differential equations relating these variables are

$$u_1' = u_2, \quad u_2' = u_3, \quad u_3' = u_4, \ldots, \quad u_n' = f(x, u_1, u_2, u_3, \ldots, u_n),$$

with the initial conditions

$$u_1(0) = \alpha_0, \quad u_2(0) = \alpha_1, \quad u_3(0) = \alpha_2, \ldots, \quad u_n(0) = \alpha_{n-1}.$$

Solve the System of Two First-Order ODEs $u' = f(x, u, v)$, $v' = g(x, u, v)$.
Euler's method updates u using $f(x, u, v)$ and updates v using $g(x, u, v)$:

$$u_{i+1} = u_i + h f(x_i, u_i, v_i),$$

$$v_{i+1} = v_i + h g(x_i, u_i, v_i).$$

The midpoint method can be written using k_1 and k_2 to represent the update quantities for the unknown function u; m_1 and m_2 give the corresponding quantities for function v. We have

$$k_1 = h f(x_i, u_i, v_i),$$

$$m_1 = h g(x_i, u_i, v_i),$$

$$k_2 = h f\left(x_i + \frac{h}{2}, u_i + \frac{1}{2}k_1, v_i + \frac{1}{2}m_1\right),$$

$$m_2 = h g\left(x_i + \frac{h}{2}, u_i + \frac{1}{2}k_1, v_i + \frac{1}{2}m_1\right),$$

$$u_{i+1} = u_i + k_2,$$

$$v_{i+1} = v_i + m_2.$$

The basic two-step Adams-Bashforth method for a system of three ODEs

$$u_1' = f_1(x, u_1, u_2, u_3),$$
$$u_2' = f_2(x, u_1, u_2, u_3),$$
$$u_3' = f_3(x, u_1, u_2, u_3),$$

is described as follows:

$u_1(i=0), u_2(i=0), u_3(i=0)$ are given by the initial condition,
$u_1(i=1), u_2(i=1), u_3(i=1)$ are found from a one-step method,
and for $i = 1, \ldots, n-1$, and $h = \dfrac{b-a}{n}$:

$$u_1(i+1) = u_1(i) + \frac{h}{2}[3f_1(x(i), u_1(i), u_2(i), u_3(i)) \\ - f_1(x(i-1), u_1(i-1), u_2(i-1), u_3(i-1))],$$

$$u_2(i+1) = u_2(i) + \frac{h}{2}[3f_2(x(i), u_1(i), u_2(i), u_3(i)) \\ - f_2(x(i-1), u_1(i-1), u_2(i-1), u_3(i-1))],$$

$$u_3(i+1) = u_3(i) + \frac{h}{2}[3f_3(x(i), u_1(i), u_2(i), u_3(i)) \\ - f_3(x(i-1), u_1(i-1), u_2(i-1), u_3(i-1))].$$

SUGGESTIONS FOR FURTHER READING

Bader, G. and P. Deuflhard, *Numerische Mathematik,* vol. 41, pp. 373–398, 1983.

Kaps, P. and P. Rentrop, *Numerische Mathematik*, vol. 33, pp. 55–68, 1979.

Press, W. H., S. A. Teukolsky, W. T. Vetterling, and B. P. Flannery, *Numerical Recipes in C, The Art of Scientific Computing*, (2d ed.), Cambridge Unversity Press, 1992.

Shampine, L. F., *ACM Transactions on Mathematical Software*, vol. 8, pp. 93–113, 1982.

The suggested readings for Chapter 12 are also excellent references for the topics in this chapter. In addition, the following texts include discussion of applications of ODEs:

Ayyub, B. M. and R. H. McCuen, *Numerical Methods for Engineers*, Prentice Hall, Upper Saddle River, NJ, 1996

Greenberg, M. D., *Advanced Engineering Mathematics*, (2d ed.), Prentice Hall, Upper Saddle River, NJ, 1998.

Greenberg, M. D., *Foundations of Applied Mathematics*, Prentice-Hall, Englewood Cliffs, NJ, 1978.

Grossman, S. I., and W. R. Derrick, *Advanced Engineering Mathematics*, Harper & Row, New York, 1988.

Hanna, O. T., and O. C. Sandall, *Computational Methods In Chemical Engineering*, Prentice Hall, Upper Saddle River, NJ, 1995.

Hildebrand, F. B., *Advanced Calculus for Applications,* (2d ed.), Prentice Hall, Englewood Cliffs, NJ, 1976.

Inman, D. J., *Engineering Vibration*, Prentice Hall, Englewood Cliffs, NJ, 1996.

Simon, W., *Mathematical Techniques for Biology and Medicine*, Dover, New York, 1986.

Spong, M. W. and M. Vidyasagar, *Robot Dynamics and Control*, John Wiley & Sons, New York, 1989. (See p. 145, ex. 6.4.2.)

Thomson, W. T., *Theory of Vibrations with Applications*, Prentice Hall, Englewood Cliff, NJ, 1993.

PRACTICE THE TECHNIQUES

For Problems P13.1 to P13.10 solve the initial value problem by first converting the problem to a system of first order ODE. For each solution method, investigate the effect of increasing n.

   a. Solve the system using Euler's method.
   b. Solve the system using the midpoint method.
   c. Solve the system using the classic Runge-Kutta method.
   d. Solve the system using the Adams-Bashforth-Moulton method.
   e. Solve the system using the built-in Mathcad function `rkfixed`.

P13.1 $y'' = y + x$, $y(0) = 2$, $y'(0) = 0$, on $[0, 2]$.
P13.2 $y'' = y' + y + x$, $y(0) = 1$, $y'(0) = 0.5$, on $[0, 2]$.
P13.3 $y'' = -2y' - y + x^2$, $y(0) = 7$, $y'(0) = -4$, on $[0, 2]$.
P13.4 $y'' = -4y' - 4y$, $y(0) = 1$, $y'(0) = 8$, on $[0, 4]$.
P13.5 $y'' = y + x^2 - 4x$, $y(0) = -2$, $y'(0) = 2$, on $[0, 4]$.
P13.6 $y'' = 5y' - 6y$, $y(0) = 2$, $y'(0) = 5$, on $[0, 1]$.
P13.7 $y'' = -y' + 6y$, $y(0) = 1$, $y'(0) = -2$, on $[0, 4]$.
P13.8 $y'' = -9y$, $y(0) = 1$, $y'(0) = 6$, on $[0, 2\pi]$.
P13.9 $y'' + \dfrac{-1}{1+x} y' + \dfrac{-3}{1+x} y = 0$, $y(0) = 1$, $y'(0) = -1$, on $[0, 2]$.
P13.10 $y'' + \dfrac{4}{1+x^2} y' + \dfrac{2}{1+x^2} y = 0$, $y(0) = 1$, $y'(0) = -1$, on $[0, 2]$.

For Problems P13.11 to P13.20 solve the initial value problem by the numerical method of your choice. Investigate the effect of modifying the initial conditions (for either y or y').

P13.11 $y'' + \dfrac{-6x^2}{1+x^3} y' + \dfrac{-6x}{1+x^3} y = 0$, $y(0) = 1$, $y'(0) = 0$, on $[0, 1]$.

P13.12 $y'' + y' - y^2 = 0$, $y(0) = 1$, $y'(0) = 0$, on $[0, 2]$.
P13.13 $y'' + y + y^3 = 0$, $y(0) = 2$, $y'(0) = 0$, on $[0, 10]$.
P13.14 $y'' = -2yy'$, $y(1) = 1$, $y'(1) = -1$, on $[1, 5]$.
P13.15 $y'' = -2(y+x)(y'+1)$, $y(1) = 0$, $y'(1) = -2$, on $[1, 5]$.
P13.16 $y'' = \dfrac{6x^4}{y}$, $y(1) = 1$, $y'(1) = 3$, on $[1, 2]$.
P13.17 $yy'' + (y')^2 = 0$, $y(1) = 2$, $y'(1) = 1/2$, on $[1, 2]$.
P13.18 $y'' + 4x^{-1} y' + 2x^{-2} y = 0$, $y(1) = 2$; $y'(1) = -3$, on $[1, 2]$.
P13.19 $y'' - 4x^{-1} y' + 6x^{-2} y = 0$, $y(1) = 5$, $y'(1) = 13$, on $[1, 2]$.
P13.20 $x^2 y'' + 4xy' + 2y = x$, $y(1) = 1/6$, $y'(1) = -5/6$, on $[1, 2]$.
P13.21 $y'' - 2x^{-2} y = 0$, $y(1) = 3$, $y'(1) = 3$, on $[1, 2]$.
P13.22 $x^2 y'' - xy' + y = 0$, $y(1) = 1$, $y'(1) = 2$, on $[1, 2]$.
P13.23 $x^2 y'' + xy' + y = 0$, $y(1) = 0$, $y'(1) = 3$, on $[1, 3]$.
P13.24 Solve the Bessel equation of order zero

$$x^2 y'' + xy' + x^2 y = 0, \; y(1) = 1, \; y'(1) = 0, \text{ on } [1, 4].$$

P13.25 Solve the Bessel equation of order one

$$x^2 y'' + xy' + (x^2 - 1)y = 0, \; y(1) = 1, \; y'(1) = 0, \text{ on } [1, 4].$$

P13.26 Solve Legendre's equation for $\alpha = 1, 2,$ or 3.

$$(1-x^2) y'' - 2xy' + \alpha(\alpha+1)y = 0, \; y(0) = 1,$$
$$y'(0) = 1, \text{ on } [0, 0.9].$$

P13.27 Solve the Chebyshev equation for $\alpha = 1, 2,$ or 3.

$$(1-x^2) y'' - xy' + \alpha^2 y = 0, \; y(0) = 1, \; y'(0) = 1,$$
$$\text{on } [0, 0.9].$$

P13.28 Solve Airy's equation

$$y'' - xy = 0, \; y(0) = 1, \; y'(0) = 1, \text{ on } [0, 5].$$

P13.29 Solve the Hermite equation for $\lambda = 1, 2, 3,$ or 4.

$$y'' - 2xy' + \lambda y = 0, \; y(1) = 1, \; y'(1) = 0, \text{ on } [1, 4].$$

P13.30 Solve the Laguerre equation for $\lambda = 1, 2, 3,$ or 4.

$$xy'' + (1-x) y' + \lambda y = 0, \; y(1) = 1, \; y'(1) = 0, \text{ on } [1, 4].$$

For problems P13.31 to P13.40, solve the initial value problem $\mathbf{x}' = A\mathbf{x}$.

P13.31

$$A = \begin{bmatrix} 3 & -3 & 2 & -1 \\ 12 & -12 & 10 & -5 \\ 15 & -15 & 14 & -7 \\ 6 & -6 & 6 & -3 \end{bmatrix}$$

a. $x0 = [1\ 0\ 0\ 0]^T$ b. $x0 = [0\ 1\ 0\ 0]^T$
c. $x0 = [0\ 0\ 1\ 0]^T$ d. $x0 = [0\ 0\ 0\ 1]^T$

P13.32

$$A = \begin{bmatrix} 1 & -3 & 2 & -1 \\ 4 & -6 & 2 & -1 \\ -5 & 5 & -8 & 5 \\ -10 & 10 & -10 & 7 \end{bmatrix}$$

a. $x0 = [1\ 0\ 0\ 0]^T$ b. $x0 = [0\ 1\ 0\ 0]^T$
c. $x0 = [0\ 0\ 1\ 0]^T$ d. $x0 = [0\ 0\ 0\ 1]^T$

P13.33

$$A = \begin{bmatrix} 3 & -3 & 2 & -1 \\ 10 & -10 & 8 & -4 \\ 10 & -10 & 9 & -4 \\ 2 & -2 & 2 & 0 \end{bmatrix}$$

a. $x0 = [1\ 0\ 0\ 0]^T$ b. $x0 = [0\ 1\ 0\ 0]^T$
c. $x0 = [0\ 0\ 1\ 0]^T$ d. $x0 = [0\ 0\ 0\ 1]^T$

P13.34

$$A = \begin{bmatrix} -9 & 9 & -6 & 3 \\ -10 & 11 & -6 & 3 \\ 3 & -2 & 4 & -2 \\ 4 & -4 & 4 & -2 \end{bmatrix}$$

a. $x0 = [1\ 0\ 0\ 0]^T$ b. $x0 = [0\ 1\ 0\ 0]^T$
c. $x0 = [0\ 0\ 1\ 0]^T$ d. $x0 = [0\ 0\ 0\ 1]^T$

P13.35

$$A = \begin{bmatrix} -2 & 3 & -2 & 1 \\ 1 & 1 & 2 & -1 \\ 10 & -9 & 12 & -7 \\ 9 & -9 & 10 & -7 \end{bmatrix}$$

a. $x0 = [1\ 0\ 0\ 0]^T$ b. $x0 = [0\ 1\ 0\ 0]^T$
c. $x0 = [0\ 0\ 1\ 0]^T$ d. $x0 = [0\ 0\ 0\ 1]^T$

P13.36

$$A = \begin{bmatrix} -13 & 12 & -8 & 4 \\ -24 & 21 & -14 & 7 \\ -14 & 12 & -9 & 6 \\ -6 & 6 & -6 & 6 \end{bmatrix}$$

a. $x0 = [1\ 0\ 0\ 0]^T$ b. $x0 = [0\ 1\ 0\ 0]^T$
c. $x0 = [0\ 0\ 1\ 0]^T$ d. $x0 = [0\ 0\ 0\ 1]^T$

P13.37

$$A = \begin{bmatrix} -11 & 6 & -4 & 2 \\ 3 & -9 & 8 & -4 \\ 18 & -19 & 16 & -7 \\ 2 & -2 & 2 & 1 \end{bmatrix}$$

a. $x0 = [1\ 0\ 0\ 0]^T$ b. $x0 = [0\ 1\ 0\ 0]^T$
c. $x0 = [0\ 0\ 1\ 0]^T$ d. $x0 = [0\ 0\ 0\ 1]^T$

P13.38

$$A = \begin{bmatrix} -36 & 30 & -20 & 10 \\ -61 & 50 & -36 & 18 \\ -34 & 20 & -25 & 13 \\ -10 & 10 & -10 & 6 \end{bmatrix}$$

a. $x0 = [1\ 0\ 0\ 0]^T$ b. $x0 = [0\ 1\ 0\ 0]^T$
c. $x0 = [0\ 0\ 1\ 0]^T$ d. $x0 = [0\ 0\ 0\ 1]^T$

P13.39

$$A = \begin{bmatrix} 30 & -24 & 16 & -8 \\ 38 & -28 & 18 & -9 \\ 6 & -2 & 0 & 0 \\ -2 & 2 & -2 & 1 \end{bmatrix}$$

a. $x0 = [1\ 0\ 0\ 0]^T$ b. $x0 = [0\ 1\ 0\ 0]^T$
c. $x0 = [0\ 0\ 1\ 0]^T$ d. $x0 = [0\ 0\ 0\ 1]^T$

P13.40

$$A = \begin{bmatrix} -28 & 24 & -16 & 8 \\ -42 & 34 & -22 & 11 \\ -10 & 6 & -2 & 0 \\ 6 & -6 & 6 & -5 \end{bmatrix}$$

a. $x0 = [1\ 0\ 0\ 0]^T$ b. $x0 = [0\ 1\ 0\ 0]^T$
c. $x0 = [0\ 0\ 1\ 0]^T$ d. $x0 = [0\ 0\ 0\ 1]^T$

EXPLORE SOME APPLICATIONS

A13.1 The motion $x(t)$, $y(t)$ of an object (such as a baseball), subject to the forces of gravity and air resistance proportional to velcoity, can be described by the system of second-order ODEs

$$x'' = -c\,v\,x', \qquad y'' = -c\,v\,y' - g,$$

where the speed of the object is $v = \sqrt{(x')^2 + (y')^2}$, $g = 32$ ft/s^2, and $c = 0.002$ is a typical value for a baseball. (In mks units, $g = 9.81$ m/s^2, and $c = 0.006$.) Solve with initial conditions $x(0) = 0$; $y(0) = 0$; $x'(0) = 100$ ft/s ; $y'(0) = 100$ ft/s. Does the ball clear a fence which is 400 ft from home plate and 10 ft tall? Investigate other initial conditions for the velocity of the ball.

A13.2 The motion of one body around another, such as a comet orbiting around the sun, can be described by the system of ODEs

$$x'' = -K\,x/r^3, \qquad y'' = -K\,y/r^3,$$

where $r = \sqrt{x^2 + y^2}$. With distance measured in AU (1 $AU = 1.496 \times 10^{11}$ m) and time measured in years, we have $K \approx 40$ (for an object rotating around the sun). Take as initial conditions, $x(0) = 1$, $x'(0) = 0$, $y(0) = 0$, $y'(0) = 2$, and solve for $0 \le t \le 4$; investigate the effect of different values of $y'(0)$. (For further discussion, see Garcia, 1994, or Greenberg, 1978.)

A13.3 The motion of a spherical pendulum of length L can be described, in terms of its angular displacement from the vertical ϕ and its angular dispacement from the positive x-axis θ, by the ODEs

$$\phi'' = -2\,\phi'\,\theta'\,\cot(\theta),$$

$$\theta'' = (\phi')^2 \sin(\theta) \cos(\theta) - (g/L) \sin(\theta).$$

Find the motion for the following sets of initial conditions

    a. $\phi = 0$, $\theta = 0.2$, $\phi' = 0$, $\theta' = 0.2$.
    b. $\phi = 0$, $\theta = 0$, $\phi' = 0.2$, $\theta' = 0$.
    c. $\phi = 0$, $\theta = 0.2$, $\phi' = 0.2$, $\theta' = 0.2$.
    d. $\phi = 0$, $\theta = 0.2$, $\phi' = 0.01$, $\theta' = 0.2$.
    e. $\phi = 0$, $\theta = 0.2$, $\phi' = 0.2$, $\theta' = 0$.

(See Thomson, 1986, p. 275.)

A13.4 Consider a system of four blocks coupled by springs, between two walls a distance L_w apart, as introduced in Chapter 3 (A3.6). Let the unstretched lengths of the springs be $L_1, \ldots, L_5$, the spring constants be $k_1, \ldots, k_5$, and the masses of the blocks be $m_1, \ldots m_4$. The equations of motion for each block ($i = 1, \ldots, 4$) are

$$\frac{dx_i}{dt} = v_i \qquad \frac{dv_i}{dt} = \frac{F_i}{m_i}$$

where
$$F_1 = -k_1(x_1 - L_1) + k_2(x_2 - x_1 - L_2)$$
$$F_2 = -k_2(x_2 - x_1 - L_2) + k_3(x_3 - x_2 - L_3)$$
$$F_3 = -k_3(x_3 - x_2 - L_3) + k_4(x_4 - x_3 - L_4)$$
$$F_4 = -k_4(x_4 - x_3 - L_4) + k_5(L_w - x_4 - L_5)$$

Solve the system for

$L_1 = 2$; $L_2 = 2$; $L_3 = 2$; $L_4 = 2$; $L_5 = 2$; $L_w = 8$.

$k_1 = 1$; $k_2 = 1$; $k_3 = 1$; $k_4 = 1$; $k_5 = 5$, $m_1 = \ldots = m_4 = 4$.

Investigate the effect of changing various parameter values, including making the blocks of different masses. (See Garcia, 1994, p. 103 for further discussion.)

A13.5 The time-evolution of the concentrations of two components (A and C) in a nonlinear chemical reaction of the form $A + B \leftrightarrow C \to D + E$, occuring in a constant volume batch reactor, can be modeled by the equations

$$\frac{dy_1}{dt} = -r_1\,y_1\,(y_1 - K) + r_2 y_2 \qquad y_1(0) = 1$$

$$\frac{dy_2}{dt} = -(r_2 + r_3)\,y_2 + r_1\,y_1\,(y_1 - K) \qquad y_2(0) = 0$$

where y_1 is the concentration of A, y_2 is the concentration of C, r_1, r_2, and r_3 are rate constants, and the parameter K depends on the initial composition of the mixture. Solve using $r_1 = r_2 = r_3 = 1$; $K = 0$; investigate the effect of modifying these values. (See Hanna and Sandall, 1995, pp. 285 for further discussion.)

A13.6 The time evolution of the concentration of two chemical species in an oscillatory chemical system such as the Belousov-Zhabotinski reaction can be described by the Brusselator model:

$$\frac{dx}{dt} = A + x^2 y - (B + 1)x \qquad \frac{dy}{dt} = Bx - x^2 y$$

The parameters A and B are positive, as are the initial conditions for x and y; investigate the solutions for various choices of A, B, $x(0)$ and $y(0)$. Consider cases where $B/(1 + A^2) < 1$; $B/(1 + A^2) > 1$; and $B/(1 + A^2) = 1$. (See Garcia, 1994, p. 98 for further discussion; the original reference is Nicolis and Prigogine, 1977.)

A13.7 A fairly general two-compartment chemical flow problem describes the concentration of two chemicals, C_1 and C_2, in two compartments with volumes V_1 and V_2 respectively. The concentrations change with time as a result of concentration-independent input into each compartment, I_1 and I_2, concentration-dependent output ($L_1 C_1$ and $L_2 C_2$) and diffusion from V_1 into V_2, $K(C_1 - C_2)$. Any of the inputs or outputs can be taken to be negative to represent flow in the opposite direction. The system of differential equations is

$$\frac{dC_1}{dt} = \frac{1}{V_1}[I_1 - L_1 C_1 - K(C_1 - C_2)]$$

$$\frac{dC_2}{dt} = \frac{1}{V_2}[I_2 - L_2 C_2 - K(C_2 - C_1)]$$

The behavior of the system depends on the relative values of the various rates and volumes. For example, let $C_1(0) = C_2(0) = 0$, $I_1 = 1$, $I_2 = 0$, $V_1 = 10$, $V_2 = 5$, and investigate the following flows:

  a. $L_1 = 2, L_2 = 2, K = 10$;
  b. $L_1 = 2, L_2 = 3, K = 20$;

find the concentration in the two tanks for the first 10 minutes of the process. (See Simon, 1986, p. 88, for further discussion.)

A13.8 For a mother-daughter radioactive decay process, we assume that for each mother atom that decays, a daughter atom is produced; daughter atoms are in turn lost at a rate proportional to their quantity. This process is described by the equations

$$\frac{dM}{dt} = -L_1 M(t), \qquad \frac{dD}{dt} = -L_2 D(t) + L_1 M(t).$$

Find the amount of each substance as a function of time if $M(0) = 10$, $D(0) = 0$, $L_1 = 2$, $L_2 = 0.1$. Investigate the effect of varying any of these values. (See Simon, 1986, p. 37 for further discussion.)

A13.9 The Lorenz equations

$$\frac{dx}{dt} = p(y - x) \qquad \frac{dy}{dt} = rx - y - xz \qquad \frac{dz}{dt} = xy - qz$$

are a well-known example of a system with chaotic behavior for certain values of the parameters. The system was studied by Lorenz in connection with the problem of finding the effect of heating a horizontal fluid layer from below. Investigate the solutions for the following values of the parameters and initial conditions.

  a. $p = 10, q = 8/3, r = 28, [x\ y\ z] = [1\ 1\ 2]$
  b. $p = 10, q = 8/3, r = 28, [x\ y\ z] = [1\ 2\ 2]$
  c. $p = 10, q = 8/3, r = 28, [x\ y\ z] = [2\ 2\ 2]$
  d. $p = 10, q = 3,\ \ \ r = 18, [x\ y\ z] = [1\ 1\ 2]$
  e. $p = 10, q = 3,\ \ \ r = 18, [x\ y\ z] = [1\ 2\ 2]$
  f. $p = 10, q = 3,\ \ \ r = 18, [x\ y\ z] = [2\ 2\ 2]$
  g. $p = 10, q = 8,\ \ \ r = 18, [x\ y\ z] = [1\ 1\ 2]$
  h. $p = 10, q = 8,\ \ \ r = 18, [x\ y\ z] = [1\ 2\ 2]$
  i. $p = 10, q = 8,\ \ \ r = 18, [x\ y\ z] = [2\ 2\ 2]$

The solutions are often plotted in the x-y plane or the y-z plane.

A13.10 Solve the classical Van der Pol differential equation for an oscillator

$$\frac{d^2 x}{dt^2} - \mu(1 - x^2)\frac{dx}{dt} + x = 0,$$

for a variety of values of the parameter μ and the initial conditions.

  a. $\mu = 0.4,\ x(0) = 0.1,\ x'(0) = 0.$
  b. $\mu = 1,\ \ \ x(0) = 0.5,\ x'(0) = 0.$
  c. $\mu = 2,\ \ \ x(0) = 1,\ \ \ x'(0) = 1.$
  d. $\mu = 5,\ \ \ x(0) = 1,\ \ \ x'(0) = 0.$
  e. $\mu = 5,\ \ \ x(0) = 1,\ \ \ x'(0) = 1.$

A13.11 A simple predator-prey relationship is described by the Lotka-Volterra model, which we write in terms of a fox population $f(t)$, with birth rate b_f and death rate d_f and a geese population $g(t)$ with birth rate b_g and death rate d_g.

$$\frac{df}{dt} = f(t)\,(b_f g(t) - d_f) \qquad \frac{dg}{dt} = g(t)\,(b_g - d_g f(t))$$

Find the populations as a function of time for the following initial conditions and parameter values.

a. $b_f = 1, d_f = 1, b_g = 1, d_g = 1, f(0) = 2, g(0) = 2$.
b. $b_f = 1, d_f = 1, b_g = 1, d_g = 1, f(0) = 10, g(0) = 2$.
c. $b_f = 1, d_f = 1, b_g = 1, d_g = 1, f(0) = 2, g(0) = 10$.
d. $b_f = 1, d_f = 1, b_g = 1, d_g = 1, f(0) = 2, g(0) = 2$.
e. $b_f = 1, d_f = 1, b_g = 1, d_g = 1, f(0) = 10, g(0) = 2$.
f. $b_f = 1, d_f = 0.5, b_g = 1, d_g = 0.5, f(0) = 2, g(0) = 10$.

A13.12 If two species compete for food but do not prey on each other, the populations can be described by the equations

$$\frac{dx}{dt} = x(a_1 - b_1 x - c_1 y) \qquad \frac{dy}{dt} = y(a_2 - b_2 y - c_2 x)$$

where all constants are positive; each population would have logistic growth if the other were not present. Find the solutions for the following combinations of parameter values and initial conditions.

a. $a_1 = 2; b_1 = 1; c_1 = 1; a_2 = 10; b_2 = 10; c_2 = 1$; $x(0) = 4; y(0) = 2$.
b. try other combinations of parameters for which $a_2/c_2 > a_1/b_1$ and $a_1/c_1 > a_2/b_2$
c. $a_1 = 2, b_1 = 1, c_1 = 1, a_2 = 10, b_2 = 1, c_2 = 1$, $x(0) = 4; y(0) = 2$.
d. try other combinations of parameters for which $a_2/c_2 > a_1/b_1$ and $a_2/b_2 > a_1/c_1$
e. $a_1 = 2, b_1 = 1, c_1 = 1, a_2 = 10, b_2 = 1, c_2 = 10$, $x(0) = 4; y(0) = 2$.
f. try other combinations of parameters for which $a_1/b_1 > a_2/c_2$ and $a_2/b_2 > a_1/c_1$

A13.13 The equations for the deflection y and rotation z of a simply supported beam with a uniformly distributed load of intensity 2 kips/ft and bending moment $M(x) = 10x - x^2$, can be expressed as

$$\frac{dz}{dx} = \frac{M}{EI} = \frac{10x - x^2}{EI} \qquad \frac{dy}{dx} = z$$

where E is the modulus of elasticity, and I is the moment of inertia of the cross section of the beam. Taking $EI = 3600$ kip/ft, $y(0) = 0$ and $z(0) = -0.02$, find y and z (for $0 \le x \le 10$). (See Ayyub and McCuen, 1996, p. 239 for further discussion.)

A13.14 The shape of a cantilever beam with a uniformly distributed load of intensity w kips/ft can expressed in terms of the deflection y, the slope of the tangent to deflected shape of the beam s, the bending moment m, and the shear force v by the following system of ODEs:

$$\frac{dy}{dx} = s; \quad \frac{ds}{dx} = \frac{m}{EI}; \quad \frac{dm}{dx} = v; \quad \frac{dv}{dx} = -w.$$

Taking $EI = 3600$ kip/ft, $w = 1.5$ kips/ft, and initial conditions

$$y(0) = 0,\ s(0) = 0,\ v(0) = 20 \text{ kips},\ m(0) = -100 \text{ kip-ft}$$

find y and z (for $0 \le x \le 10$). (See Ayyub and McCuen, 1996, p. 260 for a discussion of a similar problem.)

EXTEND YOUR UNDERSTANDING

U13.1 Investigate the numerical solution of the ODE $y'' = 16y$, $y(0) = 1$, $y'(0) = -4$; using the techniques discussed in this chapter, including Mathcad's built-in functions. Compare each numerical solution to the exact solution. What is the nature of the difficulty displayed by this problem?

U13.2 Compare the computational effort required in using the fourth-order Runge-Kutta method with that for the third-order Adams-Bashforth-Moulton method; use several of the previous problems to investigate these two methods.

14
Ordinary Differential Equations–Boundary Value Problems

14.1 Shooting Method for Linear BVP

14.2 Shooting Method for Nonlinear BVP

14.3 Finite-Difference Method for Linear BVP

14.4 Finite-Difference Method for Nonlinear BVP

14.5 Mathcad's Methods

For the higher-order ordinary differential equations considered in the previous chapter, all of the required information about the solution is specified at the same point, say, $x = a$, and the solution function is sought on an interval $a \leq x \leq b$. In many important applications, the information that is known about the desired solution is given at the endpoints of the interval. Such problems are called *ordinary differential equation-boundary value problems* (ODE–BVPs).

In this chapter, we consider two standard approaches to solving ODE–BVPs, the shooting method and the finite-difference method. The shooting method is motivated by the example of trying to hit a target at a specified distance. In the initial-value problem for the motion of an object, both the initial position and the initial velocity are given. From the solution, the location at which the object lands can be determined. In an experimental setting, one could try different velocities, observe the landing points, and eventually make the necessary adjustments to hit the target. Fortunately, for a linear problem, this basic idea leads to a numerical method that does not rely on repeated trial-and-error corrections. For a nonlinear ODE, we obtain an iterative technique analogous to the secant method or Newton's method from Chapter 2.

The finite-difference method is based on dividing the interval of interest into a number of subintervals and replacing the derivatives by the appropriate finite-difference approximations, as discussed in Chapter 11. If the differential equation is linear, it is transformed into a system of linear algebraic equations, which can be solved by the techniques investigated in previous chapters. For a nonlinear ODE, the system of algebraic equations is nonlinear, and the methods of Chapter 5 may be used. The finite-difference approach is also very important for partial differential equations, which are the subject of the next chapter.

Example 14-A Deflection of a Beam

Several boundary-value problems arise in the study of the deflection of a horizontal beam. For example, we consider a beam (see Fig. 14.1) that is freely hinged at its ends (i.e., at $x = 0$ and $x = L$), with a uniform transverse load w and tension T. In this case, the deflection, $y(x)$, is described by the ODE boundary-value problem

$$y'' - \frac{T}{EI}y = \frac{wx(x-L)}{2EI}, \quad 0 \leq x \leq L.$$

The physical parameters of the beam are the modulus of elasticity, E, and the central moment of inertia, I. We assume that the beam is of uniform thickness, so that the product EI is a constant. For convenience, the downward direction is taken as positive.

The boundary conditions state that the beam is supported, and therefore has no deflection, at $x = 0$ and $x = L$; i.e.,

$$y(0) = y(L) = 0.$$

The exact solution of the ODE is

$$y(x) = A\sinh(\alpha x) + B\sinh(\alpha(L-x)) - \frac{w}{\alpha^2 T} + \frac{wLx}{2T} - \frac{wx^2}{2T},$$

with $\alpha^2 = T/EI$, and $A\sinh(\alpha L) = B\sinh(\alpha L) = w/\alpha^2 T$.

However, the ODE is based on the assumption that y' is small, so that $(y')^2$ is negligible and $1/R \approx y''$. In general, the radius of curvature, R, is related to the deflection by

$$\frac{1}{R} = \frac{y''}{[1 + (y')^2]^{3/2}},$$

which leads to the differential equation

$$\frac{y''}{[1 + (y')^2]^{3/2}} - \frac{T}{EI}y = \frac{wx(x-L)}{2EI}, \quad 0 \leq x \leq L,$$

for which numerical methods are important. (See Jaeger, 1963, for discussion.)

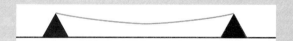

FIGURE 14.1 Bending of a beam.

Example 14-B A Well Hit Ball

Many factors influence the path of a baseball after it is hit. As an example, we consider a ball hit so that it lands 300 feet from home plate after 3 seconds. We assume that air resistance acts only against the horizontal component of the flight and is proportional to the horizontal velocity, so that the motion of the ball is given by

$$\frac{d^2x}{dt^2} = -c\frac{dx}{dt}$$

and

$$\frac{d^2y}{dt^2} = -g.$$

The initial position is $x(0) = 0$, $y(0) = 3$, and the desired landing time is $t_f = 3$, so that $x(3) = 300$, $y(3) = 0$. We take the drag coefficient $c = 0.5$. The vertical component of the motion can be found directly by

$$y(t) = -16\,t^2 + 47\,t + 3.$$

The solution for $x(t)$ can be found by the methods of this chapter. Plotting y versus x gives the path of the ball, as illustrated in Fig. 14.2.

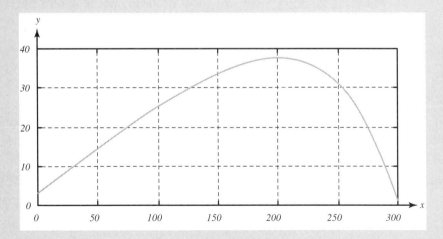

FIGURE 14.2 Flight of a baseball subject to some air resistance.

Higher-order ODEs do not necessarily have all of the information about the solution given at the same (initial) point. Problems in which the value of the unknown function or its derivative is given at two different points are known as *boundary-value problems*. In particular, we investigate the second-order, two-point boundary-value problem of the form

$$y'' = f(x, y, y'), \quad a \le x \le b$$

with Dirichlet boundary conditions

$$y(a) = \alpha, \quad y(b) = \beta,$$

or Neuman boundary conditions

$$y'(a) = \alpha, \quad y'(b) = \beta,$$

or mixed boundary conditions

$$y'(a) + c_1 y(a) = \alpha, \quad y'(b) + c_2 y(b) = \beta.$$

The first method we consider is based on the techniques presented in Chapter 13 for IVPs. The idea is to guess an initial value for $y'(a)$, generate a solution, and then adjust the solution so that it matches the specified value for $y(b)$. This is known as the *shooting method*. If the ODE is linear, we can solve two IVPs and form a linear combination of the solutions that will solve the BVP. If the ODE is nonlinear, an iterative process can be used. The linear case is described in the next section.

14.1 SHOOTING METHOD FOR LINEAR BVP

A linear two-point boundary value problem can be solved by forming a linear combination of the solutions to two initial-value problems. The form of the IVP depends on the form of the boundary conditions. We begin with the simplest case, Dirichlet boundary conditions, in which the value of the function is given at each end of the interval. We then consider some more general boundary conditions.

14.1.1 Simple Boundary Conditions

Suppose the two-point boundary value problem is linear, i.e., of the form $y'' = p(x) y' + q(x) y + r(x)$, $a \le x \le b$, with boundary conditions $y(a) = ya$, $y(b) = yb$. The approach is to solve the two IVPs

$$u'' = p(x) u' + q(x) u + r(x), \quad u(a) = ya, \quad u'(a) = 0,$$

$$v'' = p(x) v' + q(x) v, \qquad\qquad v(a) = 0, \quad v'(a) = 1.$$

If $v(b) \ne 0$, the solution of the original two-point BVP is given by

$$y(x) = u(x) + \frac{yb - u(b)}{v(b)} v(x).$$

This solution is based on the standard technique of solving a linear ODE by finding a general solution of the homogeneous equation (expressed as the ODE for v) and a particular solution of the nonhomogeneous equation (expressed as the

ODE for u). The arbitrary constant A that would appear in the solution $y(x) = u(x) + A\,v(x)$ is found from the requirement that $y(b) = u(b) + A\,v(b) = yb$, which yields

$$A = \frac{yb - u(b)}{v(b)}.$$

Algorithm for Linear Shooting Method to Solve Boundary Value Problem

> Approximate the solution of the linear ODE–BVP
> $$y'' = p(x)\,y' + q(x)\,y + r(x)$$
> with boundary conditions
> $$y(a) = ya,\ y(b) = yb$$
> using the linear shooting method.
>
> Input
> a initial point
> b final point
> ya solution value at $x = a$
> yb solution value at $x = b$
> n number of subdivisions of $[a,b]$
> Define functions that specify the ODE
> $p(x), q(x), r(x)$
> The shooting method is based on two IVP, which we denote as
> $u'' = p(x)\,u' + q(x)\,u + r(x);\quad u(a) = ya,\ u'(a) = 0$
> $v'' = p(x)\,v' + q(x)\,v;\quad v(a) = 0,\ v'(a) = 1$
> Convert ODE-BVP to system of 1st order ODE-IVP, by defining variables $z_1, \ldots z_4$
> $z_1 = u$
> $z_2 = u'$
> $z_3 = v$
> $z_4 = v'$
> The 1st order system is
> $z'_1 = z_2$
> $z'_2 = p(x)\,z_2 + q(x)\,z_1 + r(x)$
> $z'_3 = z_4$
> $z'_4 = p(x)\,z_4 + q(x)\,z_3$
> with initial conditions
> $z0 = [ya\ \ 0\ \ 0\ \ 1]$
>
> Solve this 1st order system (at the grid points $x(i)$ for $i = 1$ to n)
>   Use any of the methods presented in Chapter 13.
>   The solution is denoted $z_1(i), z_2(i), z_3(i)$ and $z_4(i)$.
>
> Construct the solution of the BVP at the grid points (for $i = 1$ to n) as
> $y(i) = z_1(i) + (yb - z_1(n))\,z_3(i)/z_3(n)$

Example 14.1 Electrostatic Potential Between Two Spheres

The electrostatic potential between two concentric spheres can be represented by the second-order ODE

$$y'' = \frac{-2}{x} y'.$$

If we assume that the radius of the inner sphere is one and its potential is 10, while the radius of the outer sphere is two and its potential is zero, then the boundary conditions are $y(1) = 10;\quad y(2) = 0$.

This equation can be converted to a pair of initial value problems, which in turn becomes a system of four first-order initial value problems by defining the variables $w_1, \ldots w_4$ as $w_1 = u, w_2 = u', w_3 = v, w_4 = v'$.

The differential equations become

$$w_1' = w_2, \quad w_2' = \frac{-2}{x} w_2, \quad w_3' = w_4, \quad w_4' = \frac{-2}{x} w_4,$$

with initial conditions $w_1(1) = 10, \quad w_2(1) = 0, \quad w_3(1) = 0, \quad w_4(1) = 1$.

The solution of the original ODE is the linear combination of the solutions u and v, or, in terms of our system,

$$y(x) = w_1(x) + \frac{yb - w_1(b)}{w_3(b)} w_3(x).$$

This system was solved in Example 13.10; using the values computed there, we find (approximately) that $(y_b - w_1(b))/w_3(b) = -20$, and $y(x) = 10 + 20\, w_3(x)$. thus at the grid points

$w_3 = [\,0 \quad 0.09 \quad 0.17 \quad 0.23 \quad 0.28 \quad 0.33 \quad 0.37 \quad 0.41 \quad 0.44 \quad 0.47 \quad 0.50]$

and

$y = [10 \quad 8.2 \quad 6.6 \quad 5.4 \quad 4.4 \quad 3.4 \quad 2.6 \quad 1.9 \quad 1.2 \quad 0.6 \quad 0.0]$

Example 14.2 A Simple Linear Shooting Problem

Consider the ODE

$$y'' = \frac{2x}{x^2 + 1} y' - \frac{2}{x^2 + 1} y + x^2 + 1,$$

with boundary conditions $y(0) = 2, y(1) = 5/3$. For reference, we note that the general solution of the ODE is $y(x) = d_1 x + d_2(x^2 - 1) + x^4/6 + x^2/2$. The exact solution of the BVP is $y = x^4/6 - 3x^2/2 + x + 2$.

Include function for RK4-sys before the computations, or include reference to it from the menu using Insert/Reference

Define function, boundary conditions, and vector of initial values:

$$f\_14\_2(x,z) := \begin{pmatrix} z_1 \\ \dfrac{z_1 \cdot 2 \cdot x}{x^2+1} - \dfrac{z_0 \cdot 2}{x^2+1} + x^2 + 1 \\ z_3 \\ \dfrac{z_3 \cdot 2 \cdot x}{x^2+1} - \dfrac{z_3 \cdot 2}{x^2+1} \end{pmatrix} \qquad ya := 2 \qquad yb := \dfrac{5}{3} \qquad z0 := \begin{pmatrix} ya \\ 0 \\ 0 \\ 1 \end{pmatrix}$$

Define interval: $\quad a := 0 \qquad b := 1 \qquad tspan := \begin{pmatrix} a \\ b \end{pmatrix}$

Solve initial value problem using Runge-Kutta method:

$$M := RK4\_sys(f\_14\_2, tspan, z0, 10)$$

Extract (x,y) values

$$x := M^{\langle 0 \rangle} \qquad n := rows(M) \qquad y := M^{\langle 1 \rangle} + \dfrac{(yb - M_{n-1,1}) \cdot M^{\langle 3 \rangle}}{M_{n-1,3}}$$

The solution is illustrated in Fig. 14.3.

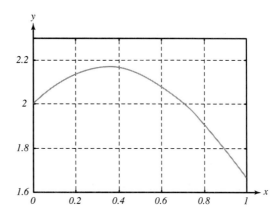

FIGURE 14.3 Electrostatic potential between two spheres.

Example 14.3 Deflection of a Simply Supported Beam

As described in Example 14-A, the deflection of a beam supported at both ends, subject to uniform loading along its length, is described by the ODE-BVP

$$y'' = \frac{T}{EI} y + \frac{w x (x - L)}{2 EI}, \quad 0 \leq x \leq L, \quad y(0) = y(L) = 0.$$

We illustrate the problem for the following parameter values:

$$L = 100, \quad w = 100, \quad E = 10^7, \quad T = 500, \quad I = 500.$$

The problem is linear, of the form

$$y'' = p(x) y' + q(x) y + r(x), \quad a \leq x \leq b,$$

with $a = 0$, $b = L = 100$, $p(x) = 0$, $q(x) = T/EI = 10^{-7}$, and $r(x) = 10^{-8}[x(x - L)]$; the boundary conditions are $y(a) = 0$, and $y(b) = 0$.

The solution can be obtained by solving the following IVPs over $a \leq x \leq b$:

$$u'' = q(x) u + r(x), \quad u(a) = 0, \quad u'(a) = 0,$$

$$v'' = q(x) v, \quad v(a) = 0, \quad v'(a) = 1.$$

The two second-order IVPs are converted to a system of four first-order IVPs by the change of variables

$$u_1 = u, \quad u_2 = u', \quad u_3 = v, \quad u_4 = v'.$$

The differential equations relating these variables are

$$u_1' = u_2, \quad u_2' = 10^{-7} u_1 + 10^{-8} [x(x - L)],$$

$$u_3' = u_4, \quad u_4' = 10^{-7} u_3,$$

with the initial conditions $u_1(0) = 0$, $u_2(0) = 0$, $u_3(0) = 0$, $u_4(0) = 1$. The computed solution is illustrated in Fig. 14.4.

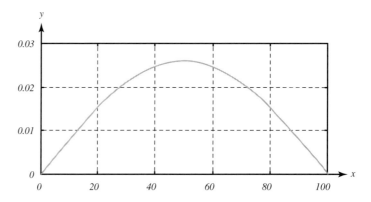

FIGURE 14.4 Deflection of a simply supported beam with uniform loading.

14.1.2 General Boundary Condition at x = b

Suppose that the linear ODE

$$y'' = p(x)\, y' + q(x)\, y + r(x);$$

has boundary conditions consisting of the value of y given at $x = a$, but the condition at $x = b$ involves a linear combination of $y(b)$ and $y'(b)$:

$$y(a) = ya; \qquad y'(b) + c\, y(b) = yb.$$

As in the previous discussion, the approach is to solve the two IVP:

$$u'' = p(x)\, u' + q(x)\, u + r(x) \quad u(a) = ya; \quad u'(a) = 0,$$

$$v'' = p(x)\, v' + q(x)\, v \qquad v(a) = 0; \quad v'(a) = 1.$$

The linear combination, $y = u + d\, v$, satisfies the conditions at $x = a$, since $y(a) = ya$. We now need to find d (if possible) so that y satisfies

$$y'(b) + c\, y(b) = yb$$

If $v'(b) + c\, v(b) \neq 0$, there is a unique solution, given by

$$y(x) = u(x) + \frac{yb - u'(b) - c\, u(b)}{v'(b) + c\, v(b)}\, v(x)$$

Example 14.4 More General Boundary Conditions

Consider the ODE from Example 14.2

$$y'' = \frac{2x}{x^2 + 1}\, y' - \frac{2}{x^2 + 1}\, y + x^2 + 1;$$

but with the boundary conditions $y(0) = 1; \quad y'(1) + y(1) = 0;$
The exact solution, $y = (x^4 - 3x^2 - x + 6)/6$, and the computed solution are indistinguishable in Figure 14.5.

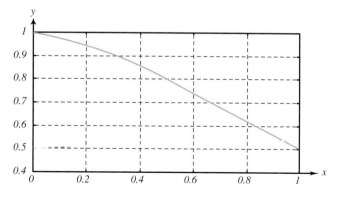

FIGURE 14.5 A BVP with a mixed-boundary condition.

14.1.3 General Boundary Conditions at Both Ends of the Interval

Suppose that the linear ODE
$$y'' = p(x)\, y' + q(x)\, y + r(x)$$
has mixed boundary conditions at both $x = a$ and $x = b$; i.e.,
$$y'(a) + c_1\, y(a) = ya, \qquad y'(b) + c_2\, y(b) = yb.$$
As in the previous discussion, the approach is to solve two IVPs, however, the appropriate forms are now
$$u'' = p(x)\, u' + q(x)\, u + r(x), \quad u(a) = 0, \quad u'(a) = ya,$$
$$v'' = p(x)\, v' + q(x) v, \quad v(a) = 1, \quad v'(a) = -c_1.$$
The linear combination $y = u + d\, v$, satisfies $y'(a) + c_1\, y(a) = ya$; we need to find d (if possible) such that y satisfies $y'(b) + c_2\, y(b) = yb$.

If $v'(b) + c_2\, v(b) \neq 0$, there is a unique solution, given by
$$y(x) = u(x) + \frac{yb - u'(b) - c_2\, u(b)}{v'(b) + c_2\, v(b)}\, v(x).$$

Example 14.5 Linear Shooting with Mixed Boundary Conditions

Consider again the ODE from Examples 14.2 and 14.4, i.e.,
$$y'' = \frac{2x}{x^2 + 1}\, y' - \frac{2}{x^2 + 1}\, y + x^2 + 1,$$
but with the boundary conditions $y'(0) + y(0) = 0, \quad y'(1) - y(1) = 3$. The exact solution of the BVP, $y = x^4/6 + 3x^2/2 + x - 1$, and the computed solution appear as a single curve in Fig. 14.6.

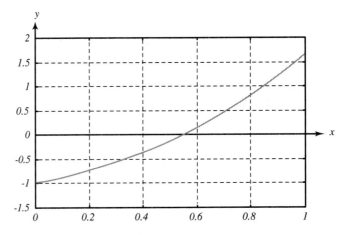

FIGURE 14.6 Solution to linear shooting problem with mixed boundary conditions.

14.2 SHOOTING METHOD FOR NONLINEAR BVP

We now consider the shooting method for nonlinear problems of the form $y'' = f(x, y, y')$ on the interval $[a, b]$. We assume that $y(a)$ is given and that some condition on the solution is also given at $x = b$. The idea is the same as for linear problems, namely, to solve the appropriate initial-value problems and use the results to find a solution to the nonlinear problem. However, for a nonlinear BVP, we have an iterative procedure rather than a simple formula for combining the solutions of two IVPs. In both the linear and the nonlinear case, we need to find a zero of the function representing the error—that is, the amount by which the solution to the IVP fails to satisfy the boundary condition at $x = b$. We assume the continuity of f, f_x and f_y on an appropriate domain, to ensure that the initial-value problems have unique solutions.

We begin by solving the initial value problem

$$u'' = f(x, u, u'), \quad u(a) = ya, \quad u'(a) = t, \quad (14.1)$$

for some particular value of t. We then find the error associated with this solution; that is, we evaluate the boundary condition at $x = b$ using $u(b)$ and $u'(b)$. Unless it happens that $u(x)$ satisfies the boundary condition at $x = b$, we take a different initial value for $u'(a)$ and solve the resulting IVP. Thus, the error (the amount by which our shot misses its mark) is a function of our choice for the initial slope. We denote this function as $m(t)$.

The first approach we consider uses the secant method to find the zero of the error function. This allows us to treat a fairly general boundary condition at $x = b$. The second approach is based on Newton's method.

14.2.1 Nonlinear Shooting Based on the Secant Method

To solve a nonlinear BVP of the form

$$y'' = f(x, y, y'), \quad y(a) = ya, \quad h(y(b), y'(b)) = 0,$$

we may use an iterative process based on the secant method presented in Chapter 2. We need to find a value of t, the initial slope, so that solving Eq. (14.1) gives a solution that is within a specified tolerance of the boundary condition at $x = b$. We begin by solving the equation with $u'(a) = t(1) = 0$; the corresponding error is $m(1)$. Unless the absolute value of $m(1)$ is less than the tolerance, we continue by solving Eq. (14.1) with $u'(a) = t(2) = 1$. If this solution does not happen to satisfy the boundary condition (at $x = b$) either, we continue by updating our initial slopes according to the secant rule (until our stopping condition is satisfied), i. e.,

$$t(i) = t(i-1) - \frac{t(i-1) - t(i-2)}{m(i-1) - m(i-2)} m(i-1).$$

Example 14.6 Nonlinear Shooting Method

To illlustrate the use of the nonlinear shooting method, consider the BVP

$$y'' = -2y\,y', \quad y(0) = 1, \quad y(1) + y'(1) - 0.25 = 0.$$

The Mathcad function S_nonlinear_shoot given below uses the secant method to update the initial slopes, and includes the boundary conditions for this example in the computation of the $m(i)$ values and the test for the stopping condition.

The use of this function to solve the given BVP is illustrated in the following Mathcad worksheet.

Define ODE-BVP $z' = f(x,z)$, on the interval $a < x < b$, where $z = (z0, z1)$, and $z0(a) = za$

$$f(x,z) := \begin{pmatrix} z_1 \\ -2 \cdot z_0 \cdot z_1 \end{pmatrix}$$

$za := 1 \qquad a := 0 \qquad b := 1 \qquad \text{max\_it} := 10 \qquad \text{tol} := 0.00001 \qquad \text{tspan} := \begin{pmatrix} a \\ b \end{pmatrix}$

$A := \text{S\_nonlinear\_shoot}\,(f, za, \text{tspan}, \text{max\_it}, \text{tol})$

$x := A^{\langle 0 \rangle} \qquad z0 := A^{\langle 1 \rangle} \qquad z1 := A^{\langle 2 \rangle} \qquad zz := z0 + z1$

$zz^T = \begin{pmatrix} -8.191 \times 10^{-6} & 0.083 & 0.139 & 0.178 & 0.204 & 0.222 & 0.234 & 0.242 & 0.247 & 0.249 & 0.25 \end{pmatrix}$

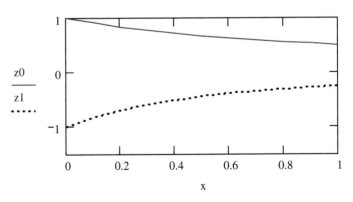

FIGURE 14.7 Solution $y(x)$ and $y'(x)$ for nonlinear shooting problem.

Mathcad Function for Nonlinear Shooting Using Secant Method

The function S_nonlinear_shoot incorporates the boundary conditions for Example 14.6.

$$S\_nonlinear\_shoot\,(f, ya, tspan, max\_it, tol) := \begin{vmatrix} t_0 \leftarrow 0 \\ t_1 \leftarrow 1 \\ test \leftarrow 1 \\ i \leftarrow 0 \\ m \leftarrow t \\ while\ (test > tol) \wedge i \leq max\_it \\ \quad \begin{vmatrix} t_i \leftarrow t_{i-1} - \dfrac{(t_{i-1} - t_{i-2})}{(m_{i-1} - m_{i-2})} \cdot m_{i-1} \quad if\ i > 1 \\ z0 \leftarrow \begin{pmatrix} ya \\ t_i \end{pmatrix} \\ A \leftarrow rkfixed(z0, tspan_0, tspan_1, 10, f) \\ n \leftarrow rows(A) \\ x \leftarrow A^{\langle 0 \rangle} \\ z \leftarrow submatrix(A, 0, n-1, 1, 2) \\ z1 \leftarrow z_{n-1, 0} \\ z2 \leftarrow z_{n-1, 1} \\ m_i \leftarrow z1 + z2 - 0.25 \\ test \leftarrow |m_i| \\ i \leftarrow i + 1 \end{vmatrix} \\ A \end{vmatrix}$$

Example 14.7 Flight of a Baseball

Let us now investigate the effect of air resistance on the flight of a baseball if the resistance is proportional to the square of the velocity. As in Example 14-B, we neglect the effect of the air resistance on the vertical component of the motion. With a small drag coefficient $c = -0.0025$, the path is not dissimilar to that found for the linear BVP, as illustrated in Fig. 14.8.

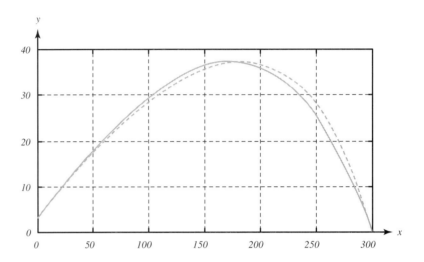

FIGURE 14.8 Air resistance proportional to velocity squared (solid line), and proportional to velocity (dashed line), each acting on x component only.

14.2.2 Nonlinear Shooting Using Newton's Method

We next illustrate how Newton's method can be used to find the value of $y'(a) = t$ in the initial-value problem for nonlinear shooting. We consider the following nonlinear BVP with simple boundary conditions at $x = a$ and $x = b$:

$$y'' = f(x, y, y'), \qquad y(a) = ya, \qquad y(b) = yb.$$

We begin by solving the initial-value problem

$$u'' = f(x, u, u'), \qquad u(a) = ya, \qquad u'(a) = t.$$

The error in this solution is the amount by which $y(b)$ misses the desired value, yb. For different choices of t, we get different errors, so we define

$$m(t) = u(b, t) - yb.$$

We need to find t such that $m(t) = 0$ [or $m(t)$ is as close to zero as we wish to continue the process]. In the previous section, we found a sequence of t using linear interpolation between the two previous solutions; in order to use Newton's method, we need to have the derivative of the function whose zero is required, namely, $m(t)$. Although we do not have an explicit formula for $m(t)$, we can construct an additional differential equation whose solution allows us to update t at each iteration. The solution process is outlined in the algorithm that follows. The derivation of the auxiliary ODE is given in the discussion at the end of the section.

Algorithm for Nonlinear Shooting Using Newton's Method

Given an estimate for t, t_k, solve
$$u'' = f(x, u, u'), \quad u(a) = ya, \quad u'(a) = t_k,$$
$$v'' = v f_u(x, u, u') + v' f_{u'}(x, u, u'), \quad v(a) = 0, \quad v'(a) = 1.$$
Check for convergence
$m = u(b, t_k) - yb;$
if $|m| <$ tol, stop;
otherwise, update t:
$$t_{k+1} = t_k - m/v(b, t_k).$$

Mathcad Function for Nonlinear Shooting Using Newton's Method

Nonlinear_shoot_Newton(f, ya, yb, tspan, max_it, tol) :=
$\begin{vmatrix} t_0 \leftarrow 0 \\ \text{test} \leftarrow 1 \\ i \leftarrow 0 \\ \text{while (test} > \text{tol)} \wedge i \leq \text{max\_it} \\ \quad \begin{vmatrix} z0 \leftarrow \begin{pmatrix} ya \\ t_i \\ 0 \\ 1 \end{pmatrix} \\ A \leftarrow \text{rkfixed}(z0, \text{tspan}_0, \text{tspan}_1, 10, f) \\ x \leftarrow A^{\langle 0 \rangle} \\ z \leftarrow A^{\langle 1 \rangle} \\ n \leftarrow \text{rows}(A) \\ vb \leftarrow A_{n-1, 3} \\ m_i \leftarrow z_{n-1} - yb \\ \text{test} \leftarrow |m_i| \\ t_{i+1} \leftarrow t_i - \dfrac{m_i}{vb} \\ i \leftarrow i + 1 \end{vmatrix} \\ A \end{vmatrix}$

Example 14.8 Nonlinear Shooting with Newton's Method

$$y'' = -\frac{[y']^2}{y}, \quad y(0) = 1, \quad y(1) = 2.$$

The exact solution of this ODE is $y = \sqrt{3x + 1}$.

The initial value problem consists of two second-order ODEs; the first corresponds to the original problem, but with the boundary condition replaced by an initial condition on the first derivative, $u'(0) = t_k$. The second ODE is an auxiliary equation that allows us to update the value of the parameter t_k. The first ODE is

$$u'' = -[u']^2 u^{-1}, \quad u(0) = 1, \quad u'(0) = t_k.$$

To construct the auxiliary ODE, we need the following partial derivatives:

$$f_u(x, u, u') = -[u']^2 (-1) u^{-2}$$

$$f_{u'}(x, u, u') = -2[u'] u^{-1}.$$

Thus, the auxiliary ODE is

$$v'' = v [u']^2 u^{-2} - v' 2[u'] u^{-1}, \quad v(0) = 0, \quad v'(0) = 1.$$

Writing this as a system of four first-order ODEs, we define the components of our unknown function z as follows:

$$z_1 = u,$$
$$z_2 = u',$$
$$z_3 = v,$$
$$z_4 = v',$$

The ODEs are

$$z_1' = z_2,$$
$$z_2' = -\frac{z_2^2}{z_1},$$
$$z_3' = z_4,$$
$$z_4' = z_3 \left(\frac{z_2}{z_1}\right)^2 - \frac{2 z_4 z_2}{z_1}.$$

Define ODE-BVP $z' = f(x,z)$ on the interval (a,b), with $z = (z_0, z_1, z_2, z_3)^T$

$$f(x,z) := \begin{bmatrix} z_1 \\ \dfrac{-(z_1)^2}{z_0} \\ z_3 \\ z_2 \cdot \left(\dfrac{z_1}{z_0}\right)^2 - \dfrac{2 \cdot z_3 \cdot z_1}{z_0} \end{bmatrix}$$

$a := 0 \qquad b := 1$

$\text{tspan} := \begin{pmatrix} a \\ b \end{pmatrix}$

$za := 1 \qquad zb := 2 \qquad \text{max\_it} := 10 \qquad \text{tol} := 0.00001$

$A := \text{Nonlinear\_shoot\_Newton}(f, za, zb, \text{tspan}, \text{max\_it}, \text{tol})$

$x := A^{\langle 0 \rangle} \qquad z0 := A^{\langle 1 \rangle}$

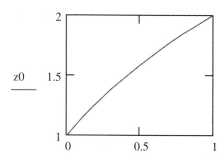

FIGURE 14.9 Solution of nonlinear shooting problem using Newton's method.

Discussion

In order to use Newton's method to find the value of the parameter t so that the amount by which the solution to the initial value problem

$$u'' = f(x, u, u'), \qquad u(a) = ya, \qquad u'(a) = t,$$

misses the solution to the original boundary value problem

$$y'' = f(x, y, y'), \qquad y(a) = ya, \qquad y(b) = yb,$$

we need to know how the error function, $m(t) = u(b, t) - yb$, varies with t. Newton's method updates t as

$$t_i = t_{i-1} - \frac{m(t_{i-1})}{m_t(t_{i-1})}.$$

We continue to use a prime to denote differentiation with respect to x, and we use a subscript to denote a partial derivative. In order to find an expression for m_t, we make use of the fact that, for the given form of boundary condition, we have $m_t(t) = u_t(b, t)$. To find u_t, we use the chain rule for partial derivatives to differentiate $u'' = f(x, u, u')$:

$$(u'')_t = f_t(x, u, u') = f_x x_t + f_u u_t + f_{u'}(u')_t.$$

Since x and t are independent, $x_t = 0$, so we have

$$(u'')_t = f_u u_t + f_{u'}(u')_t.$$

We introduce a new variable $v = u_t$ and assume sufficient continuity that we can interchange the order of differentiation with respect to x and t, so that $(u'')_t = (u_t)''$ and $(u')_t = (u_t)'$. This gives the linear ODE for v:

$$v'' = f_v v + f_u v'.$$

The initial conditions $u(a, t) = ya$ and $u'(a) = t$ yield the initial conditions for v, namely, $v(a) = 0$ and $v'(a) = 1$. Thus, solving the ODE-IVP for v allows us to use the fact that $v = u_t = m_t$ to update t in the formula for Newton's method.

Shooting methods can suffer from instabilities in the IVP; however, for a nonlinear second-order ODE-BVP, the resulting nonlinear zero-finding problem depends on only one variable. Newton's method generalizes to systems more easily than the secant method, and may converge more rapidly, but requires solving twice as many ODEs. In the next sections, we investigate an alternative approach to solving ODE-BVPs.

14.3 FINITE-DIFFERENCE METHOD FOR LINEAR BVP

The second type of solution technique we consider for ODE–BVPs is based on replacing the derivatives in the differential equation by finite-difference approximations (discussed in Chapter 11). The interval of interest, $[a, b]$, is divided into n subintervals by specifying evenly spaced values of the independent variable, $x_0, x_1, x_2, \ldots, x_n$, with $x_0 = a$ and $x_n = b$. The length of each subinterval is $h = x_{i+1} - x_i$. The approximate solution at x_i is denoted y_i. We first illustrate the finite-difference method with a simple example.

Example 14.9 A Finite Difference Problem

Use the finite difference method to solve the problem

$$y'' = y + x(x - 4), \quad 0 \le x \le 4,$$

with $y(0) = y(4) = 0$ and $n = 4$ subintervals. The finite-difference method will find an approximate solution at the points $x_1 = 1$, $x_2 = 2$, and $x_3 = 3$.

Using the central difference formula for the second derviative, we find that the differential equation becomes the system

$$y''(x_i) \approx \frac{y_{i+1} - 2y_i + y_{i-1}}{h^2} = y_i + x_i(x_i - 4), \, i = 1, 2, 3.$$

For this example, $h = 1$. In writing out the system of algebraic equations, we make use of the fact that at $i = 1$, $y_0 = 0$ (from the boundary condition at $x = 0$), and similarly, at $i = 3$, $y_4 = 0$. Substituting in the values of x_1, x_2, and x_3 we obtain

$$y_2 - 2y_1 + 0 = y_1 + 1(1 - 4),$$
$$y_3 - 2y_2 + y_1 = y_2 + 2(2 - 4),$$
$$0 - 2y_3 + y_2 = y_3 + 3(3 - 4).$$

Combining like terms and simplifying gives

$$-3y_1 + y_2 = -3,$$
$$y_1 - 3y_2 + y_3 = -4,$$
$$ y_2 - 3y_3 = -3.$$

Solving, we find that $y_1 = 13/7$, $y_2 = 18/7$, and $y_3 = 13/7$.

We note for comparison, that the exact solution of this problem is

$$y = \frac{2(1 - e^4)}{e^{-4} - e^4} e^{-x} - \frac{2(1 - e^{-4})}{e^{-4} - e^4} e^x - x^2 + 4x - 2.$$

At $x = 1$, the exact solution is 1.8341 (to four decimal places); the finite-difference solution is $y_1 = 13/7 = 1.8571$. (See Fig. 14.10.)

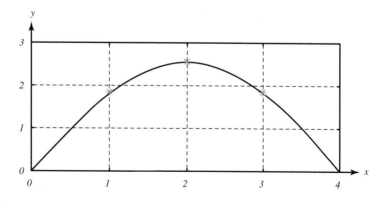

FIGURE 14.10 Small finite difference example.

We now consider the general linear two-point boundary value problem

$$y'' = p(x) y' + q(x) y + r(x), \quad a \leq x \leq b,$$

with boundary conditions

$$y(a) = \alpha, \quad y(b) = \beta.$$

To solve this problem using finite differences, we divide the interval $[a, b]$ into n subintervals, so that $h = (b - a)/n$. To approximate the function $y(x)$ at the points $x_1 = a + h, \ldots, x_{n-1} = a + (n - 1)h$, we use the central difference formulas from Chapter 11:

$$y''(x_i) \approx \frac{y_{i+1} - 2 y_i + y_{i-1}}{h^2}, \quad y'(x_i) \approx \frac{y_{i+1} - y_{i-1}}{2h}.$$

Substituting these expressions into the BVP and writing $p(x_i)$ as p_i, $q(x_i)$ as q_i, and $r(x_i)$ as r_i, gives

$$\frac{y_{i+1} - 2 y_i + y_{i-1}}{h^2} = p_i \frac{y_{i+1} - y_{i-1}}{2h} + q_i y_i + r_i.$$

Further algebraic simplification leads to a tridiagonal system for the unknowns $y_1, \ldots, y_{n-1}$, viz.,

$$\left(1 + p_i \frac{h}{2}\right) y_{i-1} - (2 + h^2 q_i) y_i + \left(1 - p_i \frac{h}{2}\right) y_{i+1} = h^2 r_i, \, i = 1, \ldots, n - 1,$$

where $y_0 = y(a) = \alpha$ and $y_n = y(b) = \beta$.

Expanding this expression into the full system gives

$$-(2 + h^2 q_1) y_1 + \left(1 - p_1 \frac{h}{2}\right) y_2 \quad = h^2 r_1 - \left(1 + p_1 \frac{h}{2}\right)\alpha,$$

$$\left(1 + p_2 \frac{h}{2}\right) y_1 - (2 + h^2 q_2) y_2 + \left(1 - p_2 \frac{h}{2}\right) y_3 \quad = h^2 r_2,$$

$$\vdots$$

$$\left(1 + p_i \frac{h}{2}\right) y_{i-1} - (2 + h^2 q_i) y_i + \left(1 - p_i \frac{h}{2}\right) y_{i+1} \quad = h^2 r_i,$$

$$\vdots$$

$$\left(1 + p_{n-2} \frac{h}{2}\right) y_{n-3} - (2 + h^2 q_{n-2}) y_{n-2} + \left(1 - p_{n-2} \frac{h}{2}\right) y_{n-1} \quad = h^2 r_{n-2},$$

$$\left(1 + p_{n-1} \frac{h}{2}\right) y_{n-2} - (2 + h^2 q_{n-1}) y_{n-1} \quad = h^2 r_{n-1} - \left(1 - p_{n-1} \frac{h}{2}\right)\beta.$$

Algorithm for Linear Finite-Difference Method

Approximate the solution $u(x)$ of the linear ODE - BVP
$$y'' = p(x) y' + q(x) y + r(x)$$
with boundary conditions
$$y(a) = y_a, \ y(b) = y_b,$$
using the linear finite-difference method.

Define
    functions which specify the ODE: $p(x), q(x), r(x)$
    interval of interest: $[a \ b]$
    boundary conditions: $y_a \quad y_b$
    number of steps n
    step size $h = (b - a)/n$

Define grid points, $x_0 = a, x_1, \ldots, x_{n-1}, x_n = b$; (for $k = 0$ to n)
$$x_k = av + h\,k$$
Evaluate $p(x), q(x)$, and $r(x)$ at the grid points; (for $k = 1$ to $n - 1$)
$$p_k = p(x_k), \ q_k = q(x_k), \ r_k = r(x_k).$$
$$r_k = 0$$

Define the tridiagonal system, $T\,y = C$, where T has
    upper diagonal (A), diagonal (D), lower diagonal (B)
for $k = 1$ to $n - 1$
$$A_k = 1 - p_k \frac{h}{2}$$
$$D_k = -(2 + h^2 \, q_k)$$
$$B_k = 1 + p_k \frac{h}{2}$$
end
$B_1 = 0$
$A_{n-1} = 0$

Define right-hand side of equation (C)
$$C_1 = h^2 r_1 - y_a \left(1 + p_1 \frac{h}{2}\right)$$
$$C_k = h^2 r_k \quad \text{(for } k = 2 \text{ to } n - 2\text{)}$$
$$C_{n-1} = h^2 r_{n-1} - y_b \left(1 - p_{n-1} \frac{h}{2}\right)$$

Solve tridiagonal system $T\,y = C$, to find y, which is the desired solution
Display results, as desired.

Example 14.10 Mathcad Function for a Linear Finite-Difference Problem

The Mathcad function that follows solves the BVP:

$$y_{xx} = 2y_x - 2y,$$
$$y(0) = 0.1,$$
$$y(3) = 0.1\, e^3 \cos(3).$$

The function S_linear_FD uses the function Thomas (from Chapter 3). Either include Thomas, or a reference to it, before using S_linear_FD.

The definition of the linear system for this problem is included in S_linear_FD

$A := \text{S\_linear\_FD} \qquad xx := A^{\langle 0 \rangle} \qquad yy := A^{\langle 1 \rangle}$

Compute the exact solution

$f(x) := 0.1 \cdot \exp(x) \cdot \cos(x) \qquad x := 0, 0.1 .. 3$

Plot approximate and exact solutions

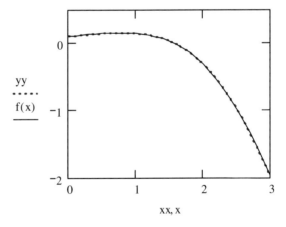

FIGURE 14.11 Linear finite-difference problem.

$$
\begin{aligned}
\text{S\_linear\_FD} := \;\; & aa \leftarrow 0 \\
& bb \leftarrow 3 \\
& n \leftarrow 300 \\
& \text{for } i \in 0..n-2 \\
& \quad\quad p_i \leftarrow 2 \\
& \quad\quad q_i \leftarrow -2 \\
& \quad\quad r_i \leftarrow 0 \\
& ya \leftarrow 0.1 \\
& yb \leftarrow 0.1 \cdot \exp(3) \cdot \cos(3) \\
& h \leftarrow \frac{bb - aa}{n} \\
& h2 \leftarrow \frac{h}{2} \\
& hh \leftarrow h \cdot h \\
& \text{for } i \in 0..n \\
& \quad\quad x_i \leftarrow aa + h \cdot i \\
& \text{for } i \in 0..n-2 \\
& \quad\quad a_i \leftarrow 0 \\
& b \leftarrow a \\
& \text{for } i \in 0..n-3 \\
& \quad\quad a_i \leftarrow 1 - p_i \cdot h2 \\
& \quad\quad b_{i+1} \leftarrow 1 + p_{i+1} \cdot h2 \\
& \quad\quad c_i \leftarrow hh \cdot r_i \\
& d \leftarrow -(2 + hh \cdot q) \\
& c_0 \leftarrow hh \cdot r_0 - \left(1 + p_0 \cdot h2\right) \cdot ya \\
& c_{n-2} \leftarrow hh \cdot r_{n-2} - \left(1 - p_{n-2} \cdot h2\right) \cdot yb \\
& y \leftarrow \text{Thomas}(a, d, b, c) \\
& A^{\langle 0 \rangle} \leftarrow x \\
& A_{0,1} \leftarrow ya \\
& \text{for } i \in 1..n-1 \\
& \quad\quad A_{i,1} \leftarrow y_{i-1} \\
& A_{n,1} \leftarrow yb \\
& A
\end{aligned}
$$

Example 14.11 Deflection of a Beam, Using Finite Differences

Consider again the deflection of a simply supported beam, described by the ODE boundary-value problem

$$y'' - \frac{T}{EI}y = \frac{w\,x\,(x-L)}{2\,EI}, \qquad 0 \le x \le L$$

where we use the specific parameter values given in Example 14.3:

$$L = 100, \quad w = 100, \quad E = 10^7, \quad T = 500, \quad I = 500.$$

Thus, the problem reduces to $y'' = 10^{-7}\,y + 10^{-8}[x(x-L)]$, $0 \le x \le 100$, with

$$y(0) = y(100) = 0,$$

$$p(x) = 0, \quad q(x) = 10^{-7}, \quad r(x) = 10^{-8}\,[x(x-L)].$$

We let $n = 20$ be the number of subintervals.

The only portion of S_linear_FD that must be changed for this example is the section on the definition of the problem and the plot of the exact solution. The solution is shown in Fig. 14.12.

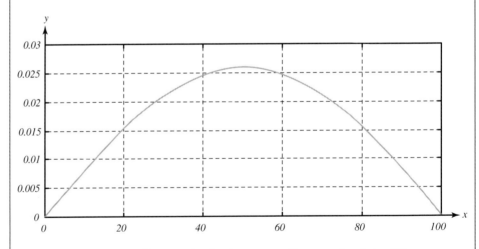

FIGURE 14.12 Deflection of a simply supported beam, using finite differences.

14.4 FINITE DIFFERENCE METHOD FOR NONLINEAR BVP

We next discuss briefly the use of finite differences in nonlinear boundary-value problems. As has been remarked earlier, nonlinear problems are, in general, significantly more difficult than linear problems.

We consider the nonlinear ODE–BVP of the form

$$y'' = f(x, y, y'), \quad y(a) = \alpha, \quad y(b) = \beta.$$

We assume that $f(x, y, y')$ has continuous derivatives that satisfy

$$0 < Q_* \leq f_y(x, y, y') \leq Q^* \quad \text{and} \quad |f_{y'}(x, y, y')| \leq P^*$$

for some constants Q_*, Q^*, and P^*.

We use a finite difference grid with spacing $h \leq 2/P^*$. Let us apply the central difference formula for y' and y'' to obtain a nonlinear system of equations. We denote the result of evaluating f at x_i using $(y_{i+1} - y_{i-1})/2h$ for y' as f_i. The ODE then becomes the system

$$\frac{y_{i+1} - 2y_i + y_{i-1}}{h^2} - f_i = 0$$

An explicit iteration scheme, analogous to the SOR method, is given by

$$y_i = \frac{1}{2(1 + \omega)}[y_{i-1} + 2\omega\, y_i + y_{i+1} - h^2 f_i],$$

where $y_0 = \alpha$, and $y_n = \beta$.

The remarkable result is that, for $\omega \geq h^2 Q^*/2$, the process will converge. (See Keller, 1968, Section 3.2.) We illustrate the process for the nonlinear BVP introduced in Example 14.8.

Example 14.12 Solving a Nonlinear BVP by Using Finite Differences

Consider again the nonlinear BVP

$$y'' = -\frac{[y']^2}{y}, \quad y(0) = 1, \quad y(1) = 2.$$

We illustrate the use of the iterative procedure just outlined, by taking a grid with $h = 1/4$, so that $x_0 = 0$, $x_1 = 0.25$, $x_2 = 0.5$, $x_3 = 0.75$, and $x_4 = 1$.

The general form of the difference equation is

$$y_i = \frac{1}{2(1+\omega)}[y_{i-1} + 2\omega\, y_i + y_{i+1} - h^2 f_i],$$

where

$$f_i = -\frac{[(y_{i+1} - y_{i-1})/2h]^2}{y_i} = -\frac{y_{i+1}^2 - 2y_{i+1}\,y_{i-1} + y_{i-1}^2}{4h^2\, y_i}.$$

Substituting the rightmost expression for f_i into the equation for y_i, we obtain

$$y_i = \frac{1}{2(1+\omega)}\left[y_{i-1} + 2\omega\, y_i + y_{i+1} + \frac{y_{i+1}^2 - 2y_{i+1}\,y_{i-1} + y_{i-1}^2}{4 y_i}\right].$$

We can carry out these computations using the following Mathcad function, which includes the definition of the the right-hand side of the BVP and the number of subdivisions of the interval.

The computed solution after 10 iterations agrees very closely with the exact solution, as is shown in Fig. 14.13. The Mathcad function to solve this problem is presented after the figure; to use a finer grid, a loop can be added to generate the equations for $i = 2, ..., n - 1$, since only the first and last equations have a special form to accommodate the boundary conditions.

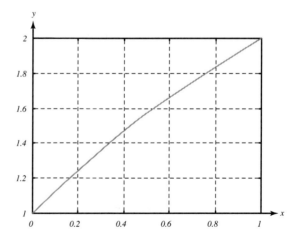

FIGURE 14.13 Solution of nonlinear ODE-BVPs using finite differences.

Mathcad Function for Nonlinear Finite Differences

The following Mathcad function solves the ODE in Example 14.12.

$$S\_nonlinear\_FD := \begin{vmatrix} n \leftarrow 4 \\ y_0 \leftarrow 1 \\ y_n \leftarrow 2 \\ a \leftarrow 0 \\ b \leftarrow 1 \\ max\_it \leftarrow 10 \\ w \leftarrow 0.1 \\ ww \leftarrow \dfrac{1}{2 \cdot (1 + w)} \\ h \leftarrow \dfrac{b - a}{n} \\ x_0 \leftarrow a \\ x_n \leftarrow b \\ \text{for } i \in 1..n - 1 \\ \quad \begin{vmatrix} y_i \leftarrow 1 \\ x_i \leftarrow x_0 + i \cdot h \end{vmatrix} \\ \text{for } k \in 1..max\_it \\ \quad \text{for } i \in 1..n - 1 \\ \quad\quad y_i \leftarrow ww \cdot \left[y_{i-1} + 2 \cdot w \cdot y_i + y_{i+1} + \dfrac{(y_{i-1})^2 - 2 \cdot y_{i-1} \cdot y_{i+1} + (y_{i+1})^2}{4 \cdot y_i} \right] \\ A^{\langle 0 \rangle} \leftarrow x \\ A^{\langle 1 \rangle} \leftarrow y \\ A \end{vmatrix}$$

14.5 MATHCAD'S METHODS

Mathcad2000Pro has three functions for use in solving ODE two-point boundary value problems. The functions `sbval` and `bvalfit` convert a boundary value problem into an initial value problem; `sbval` is used for continuous problems, `bvalfit` if there is a discontinuity on the interval ($x1$, $x2$). The function `odesolve`, which is used in a Solve Block structure, can be used for either initial value or boundary value problems.

14.5.1 Using the Built-In Functions

The function `sbval` converts a two-point boundary value problem into an initial value problem, using the shooting method, with Newton's method to reduce the discrepancy between the computed solution at the end of the interval and the given boundary values at that point. The call to the function is

$$\text{sbval}(\mathbf{v},\ x1,\ x2,\ \mathbf{D},\ \mathbf{load},\ \mathbf{score})$$

The input parameters are

v	vector of initial guesses for the initial values of the unknowns for which initial values are not specified in the problem.
$x1, x2$	endpoints of the interval on which the solution is desired.
D	vector of the derivatives of the unknown functions. For a higher-order ODE, this includes the conversion to a first-order system, as described in Section 13.1.
load($x1$,**v**)	vector of the values of each of the unknown functions at $x1$; use initial values if given, otherwise use initial guesses from vector **v**, that is, give the value as a component of vector **v**.
score($x2$,**y**)	vector of the difference between the computed value and the desired value of the unknown functions at $x2$; there is a component for each function for which a boundary value is given.

This function finds the required initial values for the solution so that the solution will match the specified boundary values. In order to actually obtain the solution to the ODE, one must call `rkfixed`, or any of the other ODE-IVP solution functions.

The function `bvalfit` is used in a manner similar to that for `sbval`; however, `bvalfit` is used when the derivatives (right-hand side of the differential equation) have a discontinuity at a single point xf between $x1$ and $x2$. The method is two-point shooting, in which solutions computed starting at $x1$ (towards $x2$) and starting at $x2$ (going towards $x1$), are constrained to be equal at an interior point, xf. The call to the function is

$$\text{bvalfit}(\mathbf{v1},\ \mathbf{v2},\ x1,\ x2,\ xf,\ \mathbf{D},\ \text{load1},\ \text{load2},\ \text{score})$$

The input parameters are

v1	vector of guesses for the initial values of the unknowns for which initial values are not specified at *x*1.
v2	vector of guesses for the initial values of the unknowns for which initial values are not specified at *x*2.
*x*1, *x*2	endpoints of the interval on which the solution is desired.
xf	point between *x*1 and *x*2 where the solutions
D	vector of the derivatives of the unknown functions. For a higher-order ODE, this includes the conversion to a first-order system, as described in Section 13.1.
load(*x*1,**v1**)	vector of the values of each unknown function at *x*1; use initial values if given, otherwise use corresponding component of vector **v1**.
load(*x*2,**v2**)	as for **load**(*x*1,**v1**), but using values given at *x*2, or corresponding component of *v*2.
score(*xf*,**y**)	specifies how the two solutions are to match at *xf*; setting **score**(*xf*,**y**) := **y** requires all components of the solution to match at *xf*.

As for sbval, bvalfit gives the initial values needed (at *x*1 and *x*2) for the solution to match at *xf*. The solution may be obrtained by calling rkfixed, or any of the other ODE solvers.

Odesolve

The Mathdad2000Pro function odesolve, which is used in a Solve Block structure, can be used for either initial value or boundary value problems. The basic structure of a Solve Block was introduced in Chapter 4, and discussed further in Chapter 5. A Solve Block begins with the word Given (typed in any style, but not in a text region). The ODE and the constraints (initial and/or boundary values) are then typed, followed by the call to the function odesolve. The function odesolve solves a single ODE, with either initial or boundary value constraints. Boundary constraints must be specified at exactly two points. The differential equation must be linear in the highest derivative. The call to the function is

    odesolve(x, b, [nstep])

where the parameters are

x	variable of integration
b	terminal point of the integration interval
nstep	(optional) integer number of steps (the default value is ten times the length of the interval)

In typing the ODE and the initial and boundary constraints, one must observe the following

1. Use [ctrl]= to type the equal signs
   (or use corresponding = from Boolean toolbar)
2. Explicitly indicate the independent variable on all functions $y(x)$, $y'(x)$, etc.
3. Use prime notation for derivatives, or derivative operators from calculus toolbar.
4. Limit constraints to simple forms such as $y(a) = c$ or $y'(a) = d$.
   Mathcad cannot handle constraints such as $y(a) + y'(a) = e$.
5. Do not enter a numerical value for the variable of integration in the call to odesolve.

The return value from odesolve is a function (which Mathcad interpolates from the table of values generated by rkfixed). The independent variable is not specified in naming the solution function; e.g.

```
f:=odesolve(x, 10)
```

14.5.2 Understanding the Algorithms

The function sbval uses the shooting method with the classical fourth-order Runge-Kutta method for the solution of the initial value problems. The approach is based on using Newton's method for a nonlinear problem, as was the discussion in Section 14.2.2. However, the implementation in sbval is more general, allowing for an arbitrary number of unknown functions, and boundary conditions which are not restricted in form. Newton's method should give the exact result in one step if the problem is linear, although a second step may be needed to improve the round-off error. See [Press et al., 1992, 757-759].

The function bvalfit implements a two-point shooting method; see [Press et al., 1992, sect 17.2], or [Keller, 1968] for details.

The function odesolve, used in a Solve Block structure, can be used for either initial value or boundary value problems. For an initial value problem, the solution is based on the Mathcad function rkfixed (or if desired, the user may select rkadapt), with interpolation of function values at points between those computed by the ODE solver. For a boundary value problem, odesolve calls sbval, followed by a call to rkfixed or rkadapt.

SUMMARY

Linear Shooting:

$$y'' = p(x) y' + q(x) y + r(x), \quad y(a) = \alpha, \quad y(b) = \beta.$$

Solve the IVP $u'' = p(x)u' + q(x)u + r(x)$, $u(a) = \alpha$, $u'(a) = 0$, and (assuming $u(b) \neq \beta$)

$$v'' = p(x) v' + q(x) v, \quad v(a) = 0, \quad v'(a) = 1.$$

If $v(b) \neq 0$, the solution of the original two-point BVP is given by

$$y(x) = u(x) + \frac{\beta - u(b)}{v(b)} v(x).$$

Nonlinear Shooting (stop when $|m_i| <$ tol):

$$y'' = f(x, y, y'), \quad y(a) = ya, \quad h(y(b), y'(b)) = yb.$$

Solve the IVP $u'' = f(x, u, u')$, $u(a) = ya$, $u'(a) = 0$.
  Calculate $h_1 = h(u(b), u'(b))$ and $m_1 = yb - h_1$ (the error at $x = b$).
Solve the IVP $v'' = f(x, v, v')$, $v(a) = ya$, $v'(a) = 1$.
  Find $h_2 = h(v(b), v'(b))$, and $m_2 = yb - h_2$ (error at $x = b$).
Form a new initial guess for $y'(a)$:

$$t = u'(a) + \frac{[v'(a) - u'(a)]m_1}{h_2 - h_1}.$$

Solve $w'' = f(x, y, y')$, $w(a) = ya$, $w'(a) = t$.
  Find $h_3 = h(w(b), w'(b))$, and $m_3 = yb - h_3$ (error at $x = b$).

Finite Difference Method for Linear BVP:

$$y'' = p(x) y' + q(x) y + r(x) \quad a \leq x \leq b$$

$$\frac{y_{i+1} - 2 y_i + y_{i-1}}{h^2} = p(x_i) \frac{y_{i+1} - y_{i-1}}{2h} + q(x_i) y_i + r(x_i).$$

$$y_{i+1} - 2 y_i + y_{i-1} = (y_{i+1} - y_{i-1}) p(x_i) \frac{h}{2} + h^2 q(x_i) y_i + h^2 r(x_i).$$

$$\left(1 + p(x_i)\frac{h}{2}\right) y_{i-1} - (2 + h^2 q(x_i)) y_i + \left(1 - p(x_i)\frac{h}{2}\right) y_{i+1} = h^2 r(x_i).$$

Finite Difference Method for Nonlinear BVP:

$$y'' = f(x, y, y'), \quad y(a) = \alpha, \quad y(b) = \beta.$$

Assume that there are constants Q_*, Q^*, and P^* such that

$$0 < Q^* \leq f_y(x, y, y') \leq Q^* \quad \text{and} \quad |f_{y'}(x, y, y')| \leq P^*$$

Use a finite difference grid with spacing $h \leq 2/P^*$, and let f_i denote the result of evaluating f at x_i using $(y_{i+1} - y_{i-1})/2h$ for y'. The ODE then becomes the system

$$\frac{y_{i+1} - 2y_i + y_{i-1}}{h^2} - f_i = 0.$$

An explicit iteration scheme, analogous to the SOR method, is given by

$$y_i = \frac{1}{2(1 + \omega)} [y_{i-1} + 2\omega y_i + y_{i+1} - h^2 f_i],$$

where $y_0 = \alpha$, and $y_n = \beta$. The process will converge for $\omega \geq h^2 Q^*/2$.

SUGGESTIONS FOR FURTHER READING

For the basic theory, see suggestions from Chapter 12. In addition, see

Ascher, U. M., R. M. M. Mattheij, and R. D. Russell, *Numerical Solution of Boundary Value Problems for Ordinary Differential Equations,* SIAM, Philadelphia, 1995.

Fox, L., *Numerical Solution of Two-Point Boundary Value Problems in Ordinary Differential Equations,* Dover, New York, 1990.

Keller, H. B., *Numerical Methods for Two-point Boundary-value Problems,* Blaisdell, Waltham, MA, 1968.

Troutman, J. L., and M. Bautista, *Boundary Value Problems of Applied Mathematics,* Prindle, Weber & Schmidt Publishing, 1994.

For further disucssion of applications of two-point boundary value problems, see

Haberman, R., *Elementary Applied Partial Differential Equations, with Fourier Series and Boundary Value Problems,* Prentice-Hall, Englewood Cliffs, NJ, 1983.

Hanna, O. T., and O. C. Sandall, *Computational Methods In Chemical Engineering,* Prentice Hall, Upper Saddle River, NJ, 1995.

Hornbeck, R. W., *Numerical Methods,* Prentice-Hall, Englewood Cliffs, NJ, 1975.

Inman, D. J. *Engineering Vibration,* Prentice Hall, Upper Saddle River, NJ, 1996.

Jaeger, J. C., *An Introduction to Applied Mathematics,* Clarenden Press, Oxford, U. K., 1951.

Roberts, C. E., *Ordinary Differential Equations: A Computational Approach,* Prentice-Hall, Englewood Cliffs, NJ, 1979.

For discussion of the methods implemented in Mathcad, see

Acton, F. S., *Numerical Methods That (usually) Work,* corrected edition, Mathematical Association of America, Washington, DC, 1990.

Keller, H. B., *Numerical Methods for Two-point Boundary-value Problems,* Blaisdell, Waltham, MA, 1968.

Press, W. H., S. A. Teukolsky, W. T. Vetterling, and B. P. Flannery, *Numerical Recipes in C, The Art of Scientific Computing* (2d ed.), Cambridge Unversity Press, 1992.

Stoer, J., and R. Bulirsch, *Introduction to Numerical Analysis,* Springer Verlag, New York, 1980.

PRACTICE THE TECHNIQUES

For Problems P14.1 to P14.14, solve the boundary-value problem
 a. Using the linear shooting method.
 b. Using the finite-difference method.

P14.1 $y'' = -y$, $\quad y(0) = 1$, $\quad y(\pi) = -1$.

P14.2 $y'' = y + x$, $\quad y(0) = 2$, $\quad y(1) = 2.5$.

P14.3 $y'' = -2y' - y + x^2$, $\quad y(0) = 10$, $\quad y(1) = 2$.

P14.4 $y'' = y/(e^x + 1)$, $\quad y(0) = 1$, $\quad y(1) = 5$.

P14.5 $y'' = -2y' - 4y$, $y(0) = 2$, $y(1) = 2$.
P14.6 $y'' = -9y$, $y(0) = 0$, $y(\pi/6) = 1$.
P14.7 $y'' = y/4 + 8$, $y(0) = 0$, $y(\pi) = 2$.
P14.8 $y'' = -2xy'$, $y(0) = 1$, $y(10) = 0$.
P14.9 $x^2 y'' + x y' + x^2 y = 0$, $y(1) = 1$, $y(8) = 0$.
P14.10 $x^2 y'' - x y' + y = 0$, $y(1) = 1$, $y(3) = 4$.
P14.11 $x^2 y'' + x y' + y = 0$, $y(1) = 0$, $y(10) = 1/2$.
P14.12 $6x^2 y'' + x y' + y = 0$, $y(1) = 2$, $y(64) = 12$.
P14.13 $x y'' - y' - x^5 = 0$, $y(1) = 1/2$, $y(2) = 4$.
P14.14 $x y'' + y' + x = 0$, $y(2) = -1$, $y(4) = 15$.

For Problems P14.15 to P14.25, solve the boundary-value problem
  a. *Using the nonlinear shooting method.*
  b. *Using the finite-difference method.*

P14.15 $y'' + y' - y^2 = 0$ $y(0) = 1$; $y(1) = 2$.
P14.16 $y'' = 2y y'$, $y(1) = 1$, $y(2) = 1/2$.
P14.17 $y'' = -2(y+x)(y'+1)$, $y(1) = 0$, $y(2) = -2$.
P14.18 $y'' = -x(y')^2 - x^2 y$, $y(0) = 1$, $y(1) = -1$.
P14.19 $y'' = e^y$, $y(0) = 1$, $y(1) = 0$.
P14.20 $2y y'' = (y')^2 - 4y^2$, $y(\pi/6) = 1/4$, $y(\pi/2) = 1$.
P14.21 $y'' = 2y^3$, $y(1) = 1$, $y(2) = 1/2$.
P14.22 $y y'' = -(y')^2 - 1$, $y(1) = 1$, $y(1/2) = \sqrt{3/4}$.
P14.23 $(1+x^2) y'' = 4xy' - 6y$, $y(0) = 1$, $y(1) = -4/3$.
P14.24 $x^3 y'' = x^2 y' + 3 - x^2$, $y(1) = 4$, $y(1) = 15/2$.
P14.25 $y'' = x(y')^3$, $y(0) = 0$, $y(1) = \pi/2$.

EXPLORE SOME APPLICATIONS

A14.1 The steady-state temperature distribution for a rod of length L, with source term $Q(x) = 1$, and temperatures given at the two ends of the rod, is described by the BVP:

$$u_{xx} + Q(x) = 0; \quad u(0) = A, \quad u(L) = B.$$

Solve for $A = 0$, $B = 100$.

A14.2 To find the steady state temperature distribution for a rod of length L, with source term $Q(x) = \sin(2\pi x/L)$, and insulated ends, solve the BVP:

$$u_{xx} + Q(x) = 0 \quad u_x(0) = 0, \quad u_x(L) = 0.$$

A14.3 The steady state temperature distribution for a rod of length L, with source term $Q(x) = x^2$, temperature fixed at $x = 0$, and the end of the rod at $x = L$ insulted, is described by the BVP:

$$u_{xx} + Q(x) = 0, \quad u(0) = T, \quad u_x(L) = 0.$$

Solve for $T = 100$, $L = 1$.

A14.4 The steady-state temperature distribution between two concentric spheres, with fixed temperature at $r = a$ and $r = b$, is given by the BVP:

$$r u_{rr} + 2 u_r = 0; \quad u(a) = T_1, \quad u(b) = T_2.$$

Solve for $a = 1$, $u(1) = 0$, $b = 4$, $u(4) = 80$.

A14.5 The steady-state temperature distribution of a rod with heat source $Q > 0$ proportional to temperature, and with the temperature at the ends of the rod fixed at 0, is given by the BVP:

$$u_{xx} + Q u = 0, \quad u(0) = 0, \quad u(L) = 0.$$

Solve for $L = 10$, $Q = (\pi/L)^2$.

A14.6 Modeling a second-order chemical flow reactor with axial dispersion leads to the ODE-BVP:

$$e P_{zz} - P_z - B P^2 = 0; \; e P_z(0) = P(0) - 1; \; P_z(1) = 0.$$

Solve for $B = 7.5$, and $e = 0.1$ to find $P(1)$.
(See Hanna and Sandall, 1995, pp. 16–18.)

A14.7 The Blasius equation describes laminar boundary layer flow on a flat plate

$$f_{xxx} + f f_{xx} = 0; \quad f(0) = 0; \quad f_x(0) = 0; \quad f_x(\infty) = 1.$$

Investigate a numerical solution by approximating ∞ by some large value of x.
(See Hanna and Sandall, 1995, pp. 256 for discussion of a series approach to the solution).

A14.8 A simple model of pseudo-homogeneous, isothermal chemical reaction and diffusion in a cylindrical catalyst pellet with irreversible first-order reaction kinetics can be written (in terms of dimensionless variables) as

$$y_{xx} + \frac{1}{x} y_x = B y \quad y_x(0) = 0 \quad y(1) = 1$$

Investigate the numerical solution of this problem for $B = 1$, 10, and 100.
(see Hanna and Sandall, pp. 257 to 258 for discussion of a series approach to the solution).

A14.9 Equations for the deflection (y) and rotation (z) of a simply supported beam with a uniformly distributed load of intensity 2 kips/ft and bending moment $M(x) = 10x - x^2$ can expressed as

$$\frac{dz}{dx} = \frac{M}{EI} = \frac{10x - x^2}{EI} \qquad \frac{dy}{dx} = z$$

where E is the modulus of elasticity and I is the moment of inertia of the cross section of the beam. Taking $EI = 3600$ kip/ft, $y(0) = 0$ and $y(10) = 0$, find y and z for $0 \le x \le 10$. (See Ayyub and McCuen, 1996, p. 239.)

A14.10 The deflection of a uniform beam of length L, with both ends fixed, subject to a load proportional to the distance from one end, i.e., $w = bx$, is described by the ODE-BVP

$$EI \frac{d^4 y}{dx^4} = w(x) = bx, \qquad y(0) = y'(0) = 0,$$

$$y(L) = y'(L) = 0.$$

Solve using $EI = 3600$ kip/ft, $L = 10$, $b = 2$.

A14.11 To find the deflection of a cantilever beam of unit length, with a distributed load, $w(x) = x$, solve the BVP

$$u_{xxxx} = x$$

with the boundary conditions

$$u(0) = u'(0) = u''(1) = u'''(1) = 0.$$

The conditions at $x = 1$ correspond to no bending moment and no shear there.

A14.12 To find the deflection of a simply supported beam of unit length, with a point load P at the midpoint, $w(x) = \delta(x - 1/2)$, solve the BVP

$$u_{xxxx} = w(x)$$

with the boundary conditions $\quad u(0) = u'(0) = u(1) = u'(1) = 0$.

EXTEND YOUR UNDERSTANDING

U14.1 Consider the numerical solutions to the BVP

$$y'' = -y; \quad y(0) = 0; \quad y(\pi) = 0.$$

The exact solution is $y = A \sin(x)$ for any constant A. How do the shooting method and the finite difference method respond to the nonunique solution?

U14.2 Consider the numerical solutions to the BVP

$$y'' = -y; \quad y(0) = 0; \quad y(\pi) = 1.$$

The general solution is $y = A \sin(x) + B \cos(x)$, but no choice of A and B will satisfy these boundary conditions. How do the shooting method and the finite difference method repsond to this situation?

U14.3 $y'' = \frac{2x}{x^2 + 1} y' - \frac{2}{x^2 + 1} y + x^2 + 1;$

$$y(0) = 2; \quad y'(1) - y(1) = -3$$

This BVP has infinitely many solutions (See Roberts, 1979, pp. 347.)

U14.4 $y'' = \frac{2x}{x^2 + 1} y' - \frac{2}{x^2 + 1} y + x^2 + 1;$

$$y(0) = 2; \quad y'(1) - y(1) = -1$$

U14.5 The finite difference method presented in the text uses the central difference approximation for y'. The linear system that is obtained in using the finite difference method is guaranteed to be diagonally dominant if the step size $\Delta x < 2/M$, where M is an upper bound on $|p(x)|$.

$$y'' = -2xy', \quad y(0) = 1, \quad y(10) = 0.$$

Investigate the solution of this problem using finite differences with different values of Δx, both larger and smaller than 0.1.

U14.6 Forward differences may be preferable for the first derivative approximation in cases where $p(x) \le 0$. Develop a program implementing this method and compare the results to those found in U14.5.

15

Partial Differential Equations

15.1 Classification of PDE

15.2 Heat Equation: Parabolic PDE

15.3 Wave Equation: Hyperbolic PDE

15.4 Poisson Equation: Elliptic PDE

15.5 Finite-Element Method for an Elliptic PDE

15.6 Mathcad's Methods

We conclude our investigation into numerical methods for solving differential equations with an introduction to some numerical techniques for solving linear second-order partial differential equations (PDEs) with constant coefficients. These equations fall into three basic categories: parabolic, hyperbolic, and elliptic. As an example of a parabolic PDE, we consider the heat equation, which describes the temperature distribution in a slender rod. A hyperbolic PDE is illustrated by the wave equation for a vibrating string. To illustrate numerical methods for elliptic PDEs, we investigate the Laplace (potential) and Poisson equations for steady-state temperature distribution in a two-dimensional region.

Numerical techniques for solving partial differential equations are primarily of two types: finite-difference methods and finite-element methods. In the first three sections of the chapter, we investigate finite-difference methods for parabolic, hyperbolic, and elliptic equations, using the heat equation, wave equation, and Poisson equation as examples. These techniques are a direct extension of the ideas presented in the previous two chapters. Replacing the partial derivatives by finite- difference approximations leads to a system of algebraic equations for the values of the unknown function at the grid points. Depending on the type of PDE and the choice of difference approximation (forward, backward, or central), the resulting equations may be solved directly or may require iterative techniques. Stability issues restrict the choice of mesh spacing for some types of problems also.

The final section of the chapter provides an introduction to finite-element methods for dealing with elliptic problems. The finite-element approach seeks to find a solution as a linear combination of relatively simple basis functions. For many elliptic problems, the solution of the PDE is equivalent to minimizing an integral over the relevant domain or, for certain problems with derivative boundary conditions, a combination of an integral over the region and an integral along the boundary. Finite-element methods are especially suitable for regions that are not rectangular. We illustrate the process using triangular subregions and basis functions that are piecewise linear on the subregions.

Example 15-A Heat Equation

The temperature in a thin rod can be described by the one-dimensional heat equation

$$u_t = c\, u_{xx}, \quad \text{for } 0 < x < a, \quad 0 < t.$$

The initial temperature is given for each point in the rod:

$$u(x,0) = f(x), \quad 0 < x < a.$$

In addition, information must be supplied describing what happens at each end of the rod. If the ends of the rod are kept at specified temperatures (by immersing each end in a fluid bath with a temperature that fluctuates with time, for example), then the boundary conditions are

$$u(0,t) = g_1(t) \quad u(a,t) = g_2(t), \quad 0 < t.$$

Other physical situations give different forms for the boundary conditions. For example, keeping an end of the rod insulated corresponds to specifying that the partial derivative u_x is zero at that end of the rod:

$$u_x(0,t) = 0 \quad \text{or} \quad u_x(a,t) = 0.$$

A third possibility is that an end of the rod is subject to convective cooling (a warm rod exposed to cooler air, for example). The corresponding boundary condition involves a combination of u and u_x at the appropriate end.

This combination of PDE, initial condition, and boundary conditions is a standard example of a parabolic PDE. (See Fig. 15.1.)

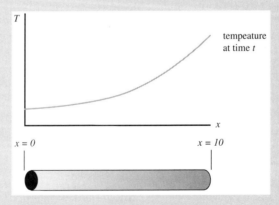

FIGURE 15.1   A rod that is cold at $x = 0$ and hot at $x = 10$.

Example 15-B Wave Equation

The motion of a vibrating string can be described by the one-dimensional wave equation

$$u_{tt} - c^2 u_{xx} = 0, \quad \text{for } 0 < x < a \text{ and } 0 \leq t.$$

The initial displacement $u(x, 0)$ and initial velocity $u_t(x, 0)$ are given for each point in the string:

$$u(x, 0) = f_1(x) \quad u_t(x, 0) = f_2(x), \quad \text{for } 0 < x < a.$$

In addition, information must be supplied about the motion of the ends of the string. The string may be fixed at each end (with zero displacement), which gives the boundary conditions

$$u(0, t) = 0, \quad u(a, t) = 0, \quad 0 < t.$$

If the ends of the string are allowed to move in a prescribed manner, the boundary conditions have the more general form

$$u(0, t) = g_1(t), \quad u(a, t) = g_2(t), \quad 0 < t,$$

where either g_1 or g_2 could be the zero function.

If an end of the string is attached to a frictionless vertical track, the appropriate boundary condition specifies that, at $x = 0$ or $x = a$,

$$u_x = 0;$$

this corresponds to the insulated boundary condition for the one-dimensional heat equation.

More complicated boundary conditions result when an end of the string is attached to a spring-mass system. For example, the so-called elastic boundary condition is analogous to the Newton-cooling boundary condition for the one-dimensional heat equation.

This combination of PDE, initial conditions and boundary conditions, is a standard example of a hyperbolic PDE. (See Fig. 15.2.)

Although the one-dimensional wave equation does not usually require a numerical solution, an example will serve as an introduction to higher dimensional wave equations, wherein numerical methods are more likely to be worthwhile.

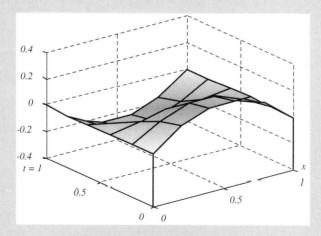

FIGURE 15.2 Vibrating string.

Example 15-C Poisson's Equation

The steady-state temperature distribution in a rectangular plate can be described by the Poisson equation

$$u_{xx} + u_{yy} = f(x, y), \quad 0 \leq x \leq a, \quad 0 \leq y \leq b.$$

If there are no heat sources (i.e., if $f(x,y) = 0$), the equation is known as Laplace's equation. Gravitational and electrostatic potentials also satisfy the Poisson equation (or the Laplace or potential equation if there are no sources).

The temperature may be prescribed along each boundary:

$$u(0, y) = g_1(y), \quad u(a, y) = g_2(y), \quad 0 \leq y \leq b;$$
$$u(x, 0) = g_3(x), \quad u(x, b) = g_4(x), \quad 0 \leq x \leq a.$$

As with the one-dimensional heat equation, other possible boundary conditions include having the boundary or part of the boundary insulated, so that the directional derivative of u in the outward normal direction is zero along that part of the boundary:

$$u_x(0, y) = 0, \text{ or } u_x(a, y) = 0, \text{ or } u_y(x, 0) = 0, \text{ or } u_y(x, b) = 0.$$

The Newton cooling boundary condition corresponds to heat flowing out at a rate proportional to the difference between the temperature of the plate and that of the surrounding medium.

The Poisson equation is a standard example of an elliptic PDE. (See Fig. 15.3.)

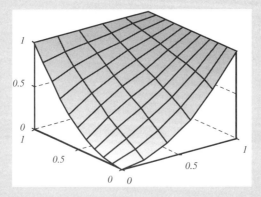

FIGURE 15.3 Steady-state temperature in a plate.

15.1 CLASSIFICATION OF PDE

Although the theory of solutions of PDEs, and many of the techniques for numerical solutions of PDEs, are beyond the scope of this book, we examine several standard techniques for the numerical solution of linear second-order PDEs involving two independent variables—either spatial variables x and y or a single spatial variable and a time variable. The general form of PDE we consider is

$$a\, u_{xx} + b\, u_{xy} + c\, u_{yy} + d\, u_x + e\, u_y + f u + g = 0$$

or

$$a\, u_{xx} + b\, u_{xt} + c\, u_{tt} + d\, u_x + e\, u_t + f u + g = 0.$$

The coefficients $a, b, \ldots, g$ may depend on the independent variables (but not on the unknown function u).

PDEs of the preceding form are normally classified into three types—parabolic, hyperbolic, and elliptic—depending on the sign of $b^2 - 4ac$. The PDE is

- parabolic if $b^2 - 4ac = 0$,
- hyperbolic if $b^2 - 4ac > 0$,
- elliptic if $b^2 - 4ac < 0$.

Of course, if the coefficients are not constants, the PDE may have a different classification in different parts of the solution domain.

The most widespread numerical techniques for solving PDEs are finite-difference methods and finite-element methods. Finite differences are based on subdividing the domain of the problem by introducing a mesh of discrete points for each of the independent variables. Derivatives are replaced by the appropriate difference quotients (see Chapter 11), and the resulting system of algebraic equations is solved by methods presented in previous chapters. Finite-element methods are based on restricting the form of the functions used (rather than the points at which the solution is sought) and finding a linear combination of these simple functions that minimizes the appropriate integral functional (which includes information from both the differential equation and the boundary conditions). For finite elements, the simple functions are required to be zero except on a small subregion of the problem domain. Finite-element methods are especially popular for elliptic problems.

15.2 HEAT EQUATION: PARABOLIC PDE

A finite-difference solution of the one-dimensional heat equation

$$u_t = c\, u_{xx}, \quad \text{for } 0 < x < a, \quad 0 < t \leq T,$$
with initial conditions $\quad u(x, 0) = f(x), \quad 0 < x < a,$
and boundary conditions $\quad u(0, t) = g_1(t), \quad u(a, t) = g_2(t), 0 < t \leq T,$

begins with the definition of a mesh of points at which the solution is sought. We divide the interval $[0, a]$ into n pieces, each of length $h = \Delta x = a/n$. The corresponding points are denoted x_i, for $i = 0, \ldots, n$. The ends of the interval are at $x_0 = 0$ and $x_n = a$; the interior points are $x_i = i\,h$, for $i = 1, \ldots, n - 1$. In a similar manner, we define a mesh for the time interval, with m subdivisions, with $k = \Delta t = T/m$ and $t_j = j\,k, j = 0, 1, \ldots, m$. As with the variable x, the ends of the time interval are $t_0 = 0$ and $t_m = T$. (See Fig. 15.4.)

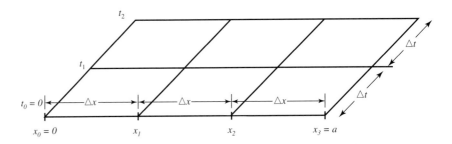

FIGURE 15.4 Spatial mesh, $n = 3$; temporal mesh, $m = 2$.

The solution at a grid point $u(x_i, t_j)$ is denoted u_{ij}. Similarly, values of the initial condition $f(x)$ at grid points are abbreviated as f_i, and values of the boundary conditions are either g_{1j} or g_{2j}. Finite difference techniques replace the partial derivatives in the PDE with difference quotients.

For the heat equation, we use the forward difference formula for u_t:

$$u_t \Rightarrow \frac{1}{k}\left[u_{i,j+1} - u_{i,j}\right].$$

Similarly, we replace the second derivative (with respect to the spatial variable) by the finite difference formula from Chapter 11, using the fact that the spacing between points in the x direction is h. If this difference formula is applied at the jth time step, we have

$$c\, u_{xx} \Rightarrow \frac{c}{h^2}\left[u_{i-1,j} - 2\,u_{i,j} + u_{i+1,j}\right].$$

This expression yields an explicit method that is easy to solve, but imposes restrictions on the relative values of the mesh spacing in the x and t directions (to maintain stability of the solution, i.e., to prevent the inevitable errors in the solution from becoming larger as the calculations proceed from $t = 0$ to $t = T$).

On the other hand, if the difference is used at the $(j+1)^{st}$ time step, we have

$$c\, u_{xx} \Rightarrow \frac{c}{h^2}[u_{i-1,j+1} - 2\, u_{i,j+1} + u_{i+1,j+1}].$$

This expression gives an implicit method that is somewhat more difficult to solve, but that is stable without placing any restriction on the mesh spacing.

Finally, we consider the Crank-Nicolson method which averages the second derivative difference formulas at the jth and $(j + 1)^{st}$ time steps; this gives a stable method with better truncation error than the simple implicit method.

15.2.1 Explicit Method for Solving the Heat Equation

Replacing the space derivative by the difference formula at the jth time step and the time derivative by a forward difference gives a linear system of equations for the temperature u at the grid points:

$$\frac{1}{k}[u_{i,j+1} - u_{i,j}] = \frac{c}{h^2}[u_{i-1,j} - 2\, u_{i,j} + u_{i+1,j}].$$

This equation can be simplified by introducing the parameter $r = \frac{c\, k}{h^2}$; solving for $u_{i,j+1}$, we have

$$u_{i,j+1} = r\, u_{i-1,j} + (1 - 2\, r)\, u_{i,j} + r\, u_{i+1,j}, \quad \text{for } i = 1, \cdots, n - 1.$$

Since the solution is known for $t = 0$, we can solve explicitly for the first time step and proceed from there in a step-by-step manner.

The points that are involved in the calculations at time steps j and $j+1$ with the explicit method are shown schematically in Fig. 15.5.

The x and t meshes must be chosen so that $0 < r \le 0.5$ in order to ensure stability. This requirement is discussed further following an example.

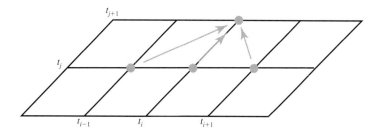

FIGURE 15.5   The temperature at the next time step is determined from information at the previous time step.

Algorithm for Solving the Heat Equation, Explicit Method

To approximate the solution of the parabolic PDE
$$u_t = c\, u_{xx}, \quad \text{for } 0 < x < a, \quad 0 < t \le T,$$
with initial conditions $u(x, 0) = f(x), \quad 0 < x < a,$
and boundary conditions $u(0, t) = g_1(t), \quad u(a, t) = g_2(t), 0 < t \le T,$

Begin by defining the parameters
$$a = \qquad T = \qquad c = \qquad n = \qquad m =$$
Compute
$$h = a/n \qquad k = T/m \qquad r = \frac{c\,k}{h^2} \qquad s = 1 - 2r$$
Define the functions specifying the initial condition, $f(x)$, and boundary conditions, $g1(t)$ and $g2(t)$.
$$f(x) = \qquad g1(t) = \qquad g2(t) =$$
Define vectors of grid points for x and t; and evaluate the initial and boundary conditions.

$i = 0$ to n
    $x_i = i\,h$
    $u(i, 0) = f(x_i)$
$j = 0$ to m
    $t_j = j\,k$
    $u(0, j) = g_1(t_j)$
    $u(n, j) = g_2(t_j)$

Finally, find the solution at time steps $jj = 1, \ldots m$
$jj = 1$ to m
    $ii = 1$ to $n - 1$
        $u(ii, jj) = r\, u(ii - 1, jj - 1) + s\, u(ii, jj - 1) + r\, u(ii+1, jj - 1)$

Example 15.1a Temperature in a Rod, Explicit Method, Stable Solution

Consider the temperature in a rod of unit length given by the PDE
$$u_t - u_{xx} = 0, \quad 0 < x < 1, \quad 0 < t.$$
The initial temperature of the rod is
$$u(x, 0) = x^4, \quad 0 < x < 1,$$
and the temperatures at $x = 0$ and $x = 1$ are, respectively,
$$u(0, t) = 0, \quad u(1, t) = 1, \quad 0 < t.$$
We take a fairly coarse mesh with $h = \Delta x = 0.2$ and the largest time step allowed for stability, so that $\Delta t = 0.02$ and $r = 0.5$. We illustrate the use of the algorithm for the explicit finite difference method using a Mathcad worksheet to approximate the solution of this parabolic PDE

Begin by defining the parameters

$$a := 1 \quad T := 0.1 \quad c := 1 \quad n := 5 \quad m := 3$$

and let Mathcad compute the additional parameters

$$h := a/n \quad k := T/m \quad r := \frac{c \cdot k}{h^2} \quad s := 1 - 2r$$

Display results if desired for verification

$$h = 0.2 \quad k = 0.02 \quad r = 0.5 \quad s = 0$$

Define the functions that give the initial and boundary values

$$f(x) := x^{\wedge}4 \quad g1(t) := 0 \quad g2(t) := 1$$

Define vectors of grid points for x and t using range variables and evaluate the initial and boundary conditions.

$$i := 0 \ldots n \quad x_i := i*h \quad u(i, 0) = f(x_i)$$
$$j := 0 \ldots m \quad t_j := j*k \quad u(0, j) = g1(t_j) \quad u(n, j) = g2(t_j)$$

Finally, find the solution at time steps $jj = 1, \ldots m$

$$jj := 1 \quad ii := 1 \ldots n - 1 \quad u(ii, jj) := r*u(ii - 1, jj - 1) \\ + s*u(ii, jj - 1) + r*u(ii + 1, jj - 1)$$

(copy the preceeding line and change the value of jj)

$$jj := 2 \quad ii := 1 \ldots n - 1 \quad u(ii, jj) := r*u(ii - 1, jj - 1) \\ + s*u(ii, jj - 1) + r*u(ii + 1, jj - 1)$$

$$jj := 3 \quad ii := 1 \ldots n - 1 \quad u(ii, jj) := r*u(ii - 1, jj - 1) \\ + s*u(ii, jj - 1) + r*u(ii + 1, jj - 1)$$

Continuing with $j = 4$ and $j = 5$ gives the following solution, with x increasing from left to right, and t increasing from top to bottom.

$$u^T = \begin{pmatrix} 0 & 1.6 \times 10^{-3} & 0.0256 & 0.1296 & 0.4096 & 1 \\ 0 & 0.0128 & 0.0656 & 0.2176 & 0.5648 & 1 \\ 0 & 0.0328 & 0.1152 & 0.3152 & 0.6088 & 1 \\ 0 & 0.0576 & 0.174 & 0.362 & 0.6576 & 1 \\ 0 & 0.087 & 0.2098 & 0.4158 & 0.681 & 1 \\ 0 & 0.1049 & 0.2514 & 0.4454 & 0.7079 & 1 \end{pmatrix}$$

Results for the problem introduced in the previous example, carried out for more time steps using Mathcad functions to implement the explicit, implicit, and Crank-Nicolson methods are given in the remainder of this section. We begin with the Mathcad function for the explicit method.

Mathcad Function for Solving the Heat Equation, Explicit Method

$$\text{Heat}(f, g1, g2, L, T, n, m, a) := \begin{vmatrix} h \leftarrow \dfrac{L}{n} \\ k \leftarrow \dfrac{T}{m} \\ r \leftarrow \dfrac{a \cdot k}{h^2} \\ rr \leftarrow 1 - 2 \cdot r \\ \text{for } i \in 0..n \\ \quad x_i \leftarrow h \cdot i \\ \text{for } i \in 0..m \\ \quad t_i \leftarrow k \cdot i \\ u^{\langle 0 \rangle} \leftarrow f(x) \\ \text{for } i \in 0..m \\ \quad \begin{vmatrix} u_{0,i} \leftarrow g1(t_i) \\ u_{n,i} \leftarrow g2(t_i) \end{vmatrix} \\ \text{for } j \in 0..m-1 \\ \quad \text{for } i \in 1..n-1 \\ \quad\quad u_{i,j+1} \leftarrow r \cdot u_{i-1,j} + rr \cdot u_{i,j} + r \cdot u_{i+1,j} \\ u^T \end{vmatrix}$$

The next example illustrates the use of this function to solve the heat equation.

Example 15.1b Temperature in a Rod, Explicit Method, Stable Solution

To solve the simple heat equation defined in Example 15.1a using the Mathcad function, Heat, we begin by defining the functions which specify the initial and boundary conditions, and the other required parameters

$$f(x) := x\wedge 4 \qquad g1(t) := 0 \qquad g2(t) := 1$$

$$L := 1 \qquad T := 0.1 \qquad n := 5 \qquad m := 5 \qquad a := 1$$

$$\text{Heat}(f, g1, g2, L, T, n, m, a) = \begin{pmatrix} 0 & 1.6 \times 10^{-3} & 0.026 & 0.13 & 0.41 & 1 \\ 0 & 0.013 & 0.066 & 0.218 & 0.565 & 1 \\ 0 & 0.033 & 0.115 & 0.315 & 0.609 & 1 \\ 0 & 0.058 & 0.174 & 0.362 & 0.658 & 1 \\ 0 & 0.087 & 0.21 & 0.416 & 0.681 & 1 \\ 0 & 0.105 & 0.251 & 0.445 & 0.708 & 1 \end{pmatrix}$$

The solution for $0 < x < 1$ and $0 < T < 1$ is shown in Figure 15.6 (using larger values of T and m); x increases from the origin to the right, and t increases from front to back. The exact steady state solution of this problem is $u(x) = x$.

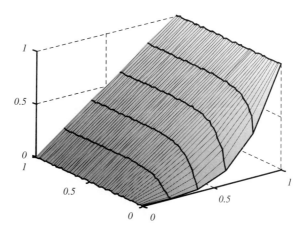

FIGURE 15.6 Temperature in a rod.

Discussion

There are two primary considerations in choosing the step sizes, h and k, for a finite difference solution of the heat equation. One issue is the effect that the step-sizes have on the order of the truncation error for the method. The other important issue is the stability of the method.

Choosing Step-Sizes for Higher-Order Truncation Error

The finite-difference representations of the partial derivatives in the heat equation are based on the Taylor series formulas developed in Chapter 11. The partial derivative of u with respect to t is

$$u_t(x_i, t_j) = \frac{u(x_i, t_{j+1}) - u(x_i, t_j)}{k} + O(k) = \frac{1}{k}[u_{i,j+1} - u_{i,j}] + O(k).$$

Similarly,

$$u_{xx}(x_i, t_j) = \frac{1}{h^2}[u_{i-1,j} - 2u_{i,j} + u_{i+1,j}] + O(h^2).$$

Substituting into the PDE gives

$$\frac{1}{k}[u_{i,j+1} - u_{i,j}] + O(k) = \frac{c}{h^2}[u_{i-1,j} - 2u_{i,j} + u_{i+1,j}] + O(h^2).$$

Combining the two expressions for the truncation error shows that the truncation error for the explicit method is $O(h^2 + k)$:

$$\frac{1}{k}[u_{i,j+1} - u_{i,j}] = \frac{c}{h^2}[u_{i-1,j} - 2u_{i,j} + u_{i+1,j}] + O(h^2) + O(k).$$

If we make use of the actual form of the first term of the error, we find that

$$\frac{1}{k}[u_{i,j+1} - u_{i,j}] = \frac{c}{h^2}[u_{i-1,j} - 2u_{i,j} + u_{i+1,j}] + \frac{ch^2}{12}u_{xxxx} - \frac{k}{2}u_{tt}.$$

Since u satisfies the PDE $u_t = c\, u_{xx}$, assuming sufficient continuity for the derivatives, we find by calculus that $u_{tt} = c\, u_{xxxx}$. Thus, we can obtain a method with truncation error $O(k^2)$ if we choose h and k so that

$$\frac{ch^2}{12}u_{xxxx} - \frac{k}{2}u_{tt} = \frac{h^2}{12}u_{tt} - \frac{k}{2}u_{tt} = \frac{1}{2}\left(\frac{h^2}{6} - k\right)u_{tt} = 0.$$

Hence, for $r = \dfrac{k}{h^2} = \dfrac{1}{6}$ the truncation error is $O(k^2) = O(h^4)$.

Restrictions on Step-Sizes to Ensure Stability

The primary difficulty with the explicit method is the stability condition, which requires that

$$r = \frac{ck}{h^2} \le \frac{1}{2}.$$

A numerical method is stable if errors that may be present at one stage of the computation do not grow as the process proceeds. To consider the stability of the forward difference solution of the heat equation, it is useful to express the computation in matrix-vector form. The solution at time step $j + 1$, which we denote by the column vector $\mathbf{u}(\,:\,,j+1)$, is found by multiplying the tridiagonal matrix $\mathbf{A}$ by the solution at the jth time step:

$$\begin{bmatrix} 1-2r & r & & & \\ r & 1-2r & r & & \\ & \vdots & \cdot & \vdots & \\ & & r & 1-2r & r \\ & & & r & 1-2r \end{bmatrix} \begin{bmatrix} u(1,j) \\ \vdots \\ u(n,j) \end{bmatrix} = \begin{bmatrix} u(1,j+1) \\ \vdots \\ u(n,j+1) \end{bmatrix}$$

Suppose the true solution of the PDE at time step j is $\mathbf{U}(\,:\,,j)$ and the computed solution is $\mathbf{u}(\,:\,,j) = \mathbf{U}(\,:\,,j) + \mathbf{E}$. Then the computed solution at step $j+1$ is

$$\mathbf{A}(:,j) = \mathbf{A}\{\mathbf{U}(:,j) + \mathbf{E}\} = \mathbf{A}\,\mathbf{U}(:,j) + \mathbf{A}\,\mathbf{E}.$$

After m time steps, the effect of the error $\mathbf{E}$ has become $\mathbf{A}^m \mathbf{E}$. From matrix algebra, it follows that

$$\|\mathbf{A}^m\,\mathbf{E}\| \le |\lambda|^m\,\|\mathbf{E}\|$$

where λ is the dominant eigenvalue (eigenvalue of largest magnitude) of $\mathbf{A}$. Stability is assured if $|\lambda| \le 1$. The Gerschgorin theorem (Chapter 1) bounds the eigenvalues of $\mathbf{A}$ inside circles centered at $1 - 2r$; the radius of each of the largest circles is $2r$, so any eigenvalue μ satisfies

$$(1 - 2r) - 2r \le \mu \le (1 - 2r) + 2r;$$

thus, we are guaranteed that $-1 \le \lambda \le 1$ if

$$-1 \le 1 - 4r,$$

or

$$r \le \frac{1}{2}.$$

Example 15.2 Temperature in a Rod, Explicit Method, Unstable Solution

Using $h = 0.2$, but disregarding the stability requirement by taking $k = 0.04$ so that $r = 1.0$, results in the update equation

$$u_{i,j+1} = u_{i-1,j} - u_{i,j} + u_{i+1,j}.$$

The wild oscillations in the computed solution are evidence of the numerical instability of the method, with an inappropriate ratio of step sizes. The temperature is plotted in Fig. 15.7; distance along the rod goes from left to right, and time goes from front to back. The following tabulation shows the computed solution starting with the initial condition at $t = 0$ and continuing until $t = 0.2$.

$x = 0$		0.2	0.4	0.6	0.8	1
$t=$						
0.00	0	0.0016	0.0256	0.1296	0.4096	1
0.04	0	0.024	0.1056	0.3056	0.72	1
0.08	0	0.0816	0.224	0.52	0.5856	1
0.12	0	0.1424	0.3776	0.2896	0.9344	1
0.16	0	0.2352	0.0544	1.0224	0.3552	1
0.20	0	−0.1808	1.2032	−0.6128	1.6672	1

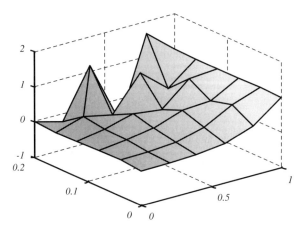

FIGURE 15.7 Unstable solution of heat equation.

15.2.2 Implicit Method for Solving the Heat Equation

Consider again the PDE

$$u_t = c\, u_{xx}, \quad \text{for } 0 < x < a, \quad 0 < t \leq T.$$

Replacing the space derivative by a centered difference at the *forward* time step $j + 1$ and the time derivative by a *forward* difference gives

$$\frac{1}{k}[u_{i,j+1} - u_{i,j}] = \frac{c}{h^2}[u_{i-1,j+1} - 2u_{i,j+1} + u_{i+1,j+1}],$$

or

$$u_{i,j} = (-r)\, u_{i-1,j+1} + (1 + 2r)\, u_{i,j+1} + (-r)\, u_{i+1,j+1},$$

where $r = \dfrac{c\,k}{h^2}$. This method is unconditionally stable.

The points involved in the calculations are illustrated in Fig. 15.8. Unlike calculations in the explicit method, calculations for a point at the $(j + 1)^{\text{st}}$ time level depend both on the results from one point at the jth time level and on other points at the $(j + 1)^{\text{st}}$ level.

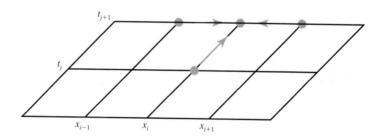

FIGURE 15.8 Values needed for computation at time $j+1$.

The resulting system of equations can be solved by techniques discussed in Chapter 6. The tridiagonal system must be solved at each time step, with a different right-hand side, so LU factorization of the tridiagonal system is an efficient approach, as given in the Mathcad function that follows Example 15.3.

The next example illustrates the solution of the heat equation using the implicit method, with a time step that is twice as large as the maximum vaue that can be used for stable computations with the explicit method.

Example 15.3a Temperature in a Rod, Implicit Method

Let the temperature in a rod (of unit length) be given by the PDE

$$u_t - u_{xx} = 0, \quad \text{for } 0 < x < 1, \quad 0 < t < T.$$

with $u(x, 0) = x^4$, $0 \le x \le 1$; and $u(0, t) = 0$, $u(1, t) = 1$, $0 < t < T$.

Consider a fairly coarse mesh, with $n = 5$, so that $h = \Delta x = 0.2$, and take $m = 5$ and $T = 0.2$, so that $k = \Delta t = 0.04$. With these parameter values, we get $r = \dfrac{ck}{h^2} = 1.0$, and the general equation

$$u_{i,j} = -r\, u_{i-1,j+1} + (1 + 2r)\, u_{i,j+1} - r\, u_{i+1,j+1}$$

simplifies to

$$-u_{i-1,j+1} + 3\, u_{i,j+1} - u_{i+1,j+1} = u_{i,j}.$$

We must find values of u at the node points $i = 1, 2, 3$, and 4 for each time step; u is given by the boundary conditions for $i = 0$ and $i = 5$. To go from the initial conditions ($t = 0$) to the solution at the first time step ($t = 0.04$) requires that the following tridiagonal system be solved:

$$
\begin{aligned}
3u_{1,1} - u_{2,1} &= u_{1,0} + u_{0,1} = 0.0016 + 0.0 \\
-u_{1,1} + 3u_{2,1} - u_{3,1} &= u_{2,0} = 0.0256 \\
-u_{2,1} + 3u_{3,1} - u_{4,1} &= u_{3,0} = 0.1296 \\
-u_{3,1} + 3u_{4,1} &= u_{4,0} + u_{5,1} = 0.4096 + 1.0.
\end{aligned}
$$

The computed values are $u_{1,1} = 0.037033$, $u_{2,1} = 0.1095$, $u_{3,1} = 0.26586$, $u_{4,1} = 0.55849$. The right-hand side for the second time step is

$$(0.037033, \quad 0.1095, \quad 0.26586, \quad 1.55849).$$

Repeating the calculations, using the right-hand side found from the solution at the previous time step, gives the solution with values listed in Table 15.1. The solution has almost reached its steady-state linear temperature distribution; the plot is essentially the same as is shown in Figure 15.6.

Table 15.1 Temperature in a rod; solution for $t = 0.00$ to $t = 0.20$.

	$x = 0$	0.2	0.4	0.6	0.8	1
$t =$						
0.00	0	0.0016	0.0256	0.1296	0.4096	1
0.04	0	0.037033	0.1095	0.26586	0.55849	1
0.08	0	0.072904	0.18168	0.36264	0.64038	1
0.12	0	0.10387	0.2387	0.43055	0.69031	1
0.16	0	0.1286	0.28192	0.47846	0.72292	1
0.20	0	0.14753	0.314	0.51254	0.74515	1

Algorithm for Solving the Heat Equation, Implicit Method

To approximate the solution of the parabolic PDE
$$u_t = c\, u_{xx}, \quad \text{for } 0 < x < a, \quad 0 < t \le T,$$
with initial conditions $\quad u(x, 0) = f(x), \quad 0 < x < a,$
and boundary conditions $\quad u(0, t) = g_1(t), \quad u(a, t) = g_2(t),\ 0 < t \le T,$

Begin by defining the parameters
$$a = \qquad T = \qquad c = \qquad n = \qquad m =$$
Compute
$$h = a/n \qquad k = T/m \qquad r = \frac{c\,k}{h^2} \qquad s = 1 + 2r$$

Define the functions specifying the initial condition, $f(x)$, and boundary conditions, $g1(t)$ and $g2(t)$.
$$f(x) = \qquad g1(t) = \qquad g2(t) =$$

Define vectors of grid points for x and t; and evaluate the initial and boundary conditions.

$i = 0$ to n
  $x_i = i\,h$
  $\mathbf{u}(i, 0) = f(x_i)$
$j = 0$ to m
  $t_j = j\,k$
  $\mathbf{u}(0, j) = g_1(t_j)$
  $\mathbf{u}(n, j) = g_2(t_j)$

Define the linear system $\mathbf{A\,u = b}$

$ii = 1$ to $n - 1$
  $\mathbf{A}(ii, ii) = s$
$jj = 1$ to $n - 2$
  $\mathbf{A}(jj+1, jj) = -r$
  $\mathbf{A}(jj, jj+1) = -r$

(all other entries in $\mathbf{A}$ are 0)

Define right hand side for time step tt

$tt = 1$
$\mathbf{b}(1) = \mathbf{u}(1, tt-1) + \mathbf{u}(0, tt)$
$kk = 2$ to $n - 2$
  $\mathbf{b}(kk) = \mathbf{u}(kk, tt\text{-}1)$
$\mathbf{b}(n-1) = \mathbf{u}(n-1, tt-1) + \mathbf{u}(0, tt)$

Solve the linear system
(it is efficient to use LU decomposition since the same system must be solved several times, with different right hand sides);
the solution becomes the next column in the solution matrix $\mathbf{u}$.

Mathcad Function for Solving the Heat Equation, Implicit Method

The following Mathcad function uses the functions `LU_factor_T` and `LU_solve_T`; a reference giving their location should be inserted in the Mathcad worksheet prior to the definition of `Heat_Imp`, or else the functions themselves must be defined in the same worksheet as `Heat_Imp` (and prior to `Heat_Imp`).

We illustrate the use of this function in the next example, which repeats the same problem as in Example 15.3a. Note that time increases from the top of the array to the bottom of the array; x increases from left to right.

$$
\begin{aligned}
&\text{Heat\_Imp}(f, g1, g2, a, T, n, m, c) := \\
&\quad h \leftarrow \frac{a}{n} \\
&\quad k \leftarrow \frac{T}{m} \\
&\quad r \leftarrow \frac{c \cdot k}{h^2} \\
&\quad \text{for } i \in 0..n \\
&\quad\quad x_i \leftarrow h \cdot i \\
&\quad w \leftarrow f(x) \\
&\quad \text{for } i \in 1..m \\
&\quad\quad t_i \leftarrow k \cdot i \\
&\quad\quad w_{0,i} \leftarrow g1(t_i) \\
&\quad\quad w_{n,i} \leftarrow g2(t_i) \\
&\quad \text{for } i \in 0..n-2 \\
&\quad\quad d_i \leftarrow 1 + 2 \cdot r \\
&\quad\quad aa_i \leftarrow \begin{vmatrix} (-r) & \text{if } i < n-2 \\ 0 & \text{otherwise} \end{vmatrix} \\
&\quad\quad b_i \leftarrow \begin{vmatrix} (-r) & \text{if } i > 0 \\ 0 & \text{otherwise} \end{vmatrix} \\
&\quad B \leftarrow \text{LU\_factor\_T}(aa, d, b) \\
&\quad bb \leftarrow B^{\langle 0 \rangle} \\
&\quad dd \leftarrow B^{\langle 1 \rangle} \\
&\quad \text{for } j \in 0..m-1 \\
&\quad\quad cc_0 \leftarrow r \cdot w_{0,j} + w_{1,j} \\
&\quad\quad \text{for } i \in 1..n-3 \\
&\quad\quad\quad cc_i \leftarrow w_{i+1,j} \\
&\quad\quad cc_{n-2} \leftarrow r \cdot w_{n,j} + w_{n-1,j} \\
&\quad\quad x \leftarrow \text{LU\_solve\_T}(aa, dd, bb, cc) \\
&\quad\quad \text{for } i \in 1..n-1 \\
&\quad\quad\quad w_{i,j+1} \leftarrow x_{i-1} \\
&\quad (w)^T
\end{aligned}
$$

> **Example 15.3b Solving the Heat Equation Using an Implicit Method**
>
> $f(x) := x^4 \quad g1(t) := 0 \quad g2(t) := 1 \quad L := 1 \quad T := 0.2 \quad a := 1 \quad c := 1$
>
> $n := 5 \quad m := 5$
>
> $\text{Heat\_Imp}(f, g1, g2, a, T, n, m, c) = \begin{pmatrix} 0 & 1.6 \times 10^{-3} & 0.026 & 0.13 & 0.41 & 1 \\ 0 & 0.037 & 0.109 & 0.266 & 0.558 & 1 \\ 0 & 0.073 & 0.182 & 0.363 & 0.64 & 1 \\ 0 & 0.104 & 0.239 & 0.431 & 0.69 & 1 \\ 0 & 0.129 & 0.282 & 0.478 & 0.723 & 1 \\ 0 & 0.148 & 0.314 & 0.513 & 0.745 & 1 \end{pmatrix}$

Discussion

The finite-difference representations of the partial derivatives in the heat equation are as given for the explicit method, except that the second derivative is approximated at step $j + 1$, rather than at step j. Thus, we have

$$u_t(x_i, t_j) = \frac{1}{k}[u_{i,j+1} - u_{i,j}] + O(k),$$

and

$$u_{xx}(x_i, t_{j+1}) = \frac{1}{h^2}[u_{i-1,j+1} - 2u_{i,j+1} + u_{i+1,j+1}] + O(h^2).$$

Substituting into the PDE and simplifying shows that the truncation error for the implicit method is the same as for the explicit method, i.e., $O(h^2+k)$.

To show that the implicit method is unconditionally stable, consider the matrix-vector form of the process, viz.,

$$\begin{bmatrix} 1+2r & -r & & & \\ -r & 1+2r & -r & & \\ & \ddots & \ddots & \ddots & \\ & & -r & 1+2r & -r \\ & & & -r & 1+2r \end{bmatrix} \begin{bmatrix} \mathbf{u}(1,j+1) \\ \vdots \\ \mathbf{u}(n,j+1) \end{bmatrix} = \begin{bmatrix} \mathbf{u}(1,j) \\ \vdots \\ \mathbf{u}(n,j) \end{bmatrix}$$

or

$$\mathbf{A}\,\mathbf{u}(:, j+1) = \mathbf{u}(:, j).$$

For analysis (but not for actual computation!), we write the latter equation as

$$\mathbf{A}^{-1}\mathbf{u}(:, j) = \mathbf{u}(:, j+1),$$

so that, by the same reasoning as for the explicit method, stability is assured if λ, the dominant eigenvalue of $\mathbf{A}^{-1}$, satisfies $|\lambda| \leq 1$. In terms of the eigenvalues of $\mathbf{A}$, the condition becomes

$$|\mu| \geq 1,$$

where μ is the eigenvalue of $\mathbf{A}$ with the smallest magnitude. By the Gerschgorin theorem, this condition is true regardless of the value of r.

15.2.3 Crank-Nicolson Method for Solving the Heat Equation

Using the average of the centered difference at the forward time step $j+1$ and the current time step j gives

$$\frac{1}{k}[u_{i,j+1} - u_{i,j}] = \frac{c}{2h^2}[u_{i-1,j} - 2u_{i,j} + u_{i+1,j}] + \frac{c}{2h^2}[u_{i,j+1} - 2u_{i,j+1} + u_{i+1,j+1}].$$

Defining $r = \dfrac{ck}{h^2}$ as before, we can write the equations as

$$-\frac{r}{2}u_{i-1,j+1} + (1+r)u_{i,j+1} - \frac{r}{2}u_{i+1,j+1} = \frac{r}{2}u_{i-1,j} + (1-r)u_{i,j} + \frac{r}{2}u_{I+1,j}.$$

A general two-stage method with weighting factor λ, for $0 \leq \lambda \leq 1$, gives

$$\frac{1}{k}[u_{i,j+1} - u_{i,j}] = \frac{\lambda}{h^2}[u_{i+1,j} - 2u_{i,j} + u_{i-1,j}] + \frac{1-\lambda}{h^2}[u_{i+1,j+1} - 2u_{i,j+1} + u_{i-1,j+1}],$$

or, in terms of r and λ,

$$-r\lambda u_{i-1,j+1} + (1+2r\lambda)u_{i,j+1} - r\lambda u_{i+1,j+1} =$$
$$r(1-\lambda)u_{i-1,j} + (1-2r(1-\lambda))u_{i,j} + r(1-\lambda)u_{i+1,j}.$$

The resulting system of equations can be solved by techniques discussed in Chapter 6.

The Crank-Nicolson method is unconditionally stable and has better truncation error, $O(h^2 + k^2)$, than the basic implicit method. For the appropriate choice of λ, the truncation error for the general two-stage method is $O(h^4)$; this occurs when $2r\lambda = r - 1/6$. The truncation error is reduced to $O(h^6)$ if the step sizes are chosen so that $r = \dfrac{\sqrt{5}}{10}$ and $\lambda = \dfrac{3 - \sqrt{5}}{6}$. (See Ames, 1992, p. 65, for further discussion.)

Algorithm for Solving the Heat Equation, Crank-Nicolson Method

To approximate the solution of the parabolic PDE
$$u_t = c\, u_{xx}, \quad \text{for } 0 < x < a, \quad 0 < t \le T,$$
with initial conditions $\quad u(x, 0) = f(x), \quad 0 < x < a,$
and boundary conditions $\quad u(0, t) = g_1(t), \quad u(a, t) = g_2(t), \; 0 < t \le T,$

Begin by defining the parameters
$\quad a = \quad\quad T = \quad\quad c = \quad\quad n = \quad\quad m =$
Compute
$$h = a/n \quad\quad k = T/m \quad\quad r = \frac{c\,k}{h^2} \quad\quad s = \frac{r}{2}$$
Define the functions specifying the initial condition, $f(x)$, and boundary conditions, $g1(t)$ and $g2(t)$.
$\quad f(x) = \quad\quad g1(t) = \quad\quad g2(t) =$
Define vectors of grid points for x and t; and evaluate the initial and boundary conditions.
$i = 0$ to n
$\quad x_i = i\,h$
$\quad \mathbf{u}(i, 0) = f(x_i)$
$j = 0$ to m
$\quad t_j = j\,k$
$\quad \mathbf{u}(0,j) = g_1(t_j)$
$\quad \mathbf{u}(n, j) = g_2(t_j)$
Define the linear system $\mathbf{A}\,\mathbf{u} = \mathbf{b}$
$ii = 1$ to $n - 1$
$\quad \mathbf{A}(ii, ii) = 1 + r$
$jj = 1$ to $n - 2$
$\quad \mathbf{A}(jj + 1, jj) = -s$
$\quad \mathbf{A}(jj, jj + 1) = -s$
(all other entries in $\mathbf{A}$ are 0)
Define right hand side for time step tt
$tt = 1$
$\mathbf{b}(1) = s\,\mathbf{u}(0,tt - 1) + (1 - r)\,\mathbf{u}(1,tt - 1) + s\,\mathbf{u}(2, tt - 1) + s\,\mathbf{u}(0,tt)$
$kk = 2$ to $n - 2$
$\quad \mathbf{b}(kk) = s\,\mathbf{u}(kk - 1,tt - 1) + (1 - r)\,\mathbf{u}(kk,tt - 1) + s\,\mathbf{u}(kk + 1, tt - 1)$
$\mathbf{b}(n - 1) = s\,\mathbf{u}(n - 2,tt\text{-}1) + (1 - r)\,\mathbf{u}(n - 1,tt - 1) + s\,\mathbf{u}(n,tt - 1) + s\,\mathbf{u}(n,tt)$
Solve the linear system
$\quad\quad$ (it is efficient to use LU decomposition since the same system must be solved several times, with different right hand sides);
The solution becomes the next column in the solution matrix $\mathbf{u}$.

Mathcad Function for Solving the Heat Equation, Crank-Nicolson Method

The following Mathcad function uses the functions `LU_factor_T` and `LU_solve_T`; a reference giving their location should be inserted in the Mathcad worksheet prior to the definition of `Heat_Imp`, or else the functions themselves must be defined in the same worksheet as `Heat_Imp` (and prior to `Heat_Imp`).

$$
\begin{aligned}
\text{Heat\_CN}(f, g1, g2, a, T, n, m, c) := \,& h \leftarrow \frac{a}{n} \\
& k \leftarrow \frac{T}{m} \\
& r \leftarrow \frac{c \cdot k}{h^2} \\
& rr \leftarrow 0.5 \cdot r \\
& \text{for } i \in 0..n \\
& \quad x_i \leftarrow h \cdot i \\
& w \leftarrow f(x) \\
& \text{for } i \in 0..m \\
& \quad t_i \leftarrow k \cdot i \\
& \quad w_{0,i} \leftarrow g1(t_i) \\
& \quad w_{n,i} \leftarrow g2(t_i) \\
& \text{for } i \in 0..n-2 \\
& \quad d_i \leftarrow 1 + r \\
& \quad aa_i \leftarrow -rr \\
& \quad b_i \leftarrow -rr \\
& b_0 \leftarrow 0 \\
& a_{n-2} \leftarrow 0 \\
& B \leftarrow \text{LU\_factor\_T}(aa, d, b) \\
& bb \leftarrow B^{\langle 0 \rangle} \\
& dd \leftarrow B^{\langle 1 \rangle} \\
& \text{for } j \in 0..m-1 \\
& \quad cc_0 \leftarrow rr \cdot w_{0,j} + rr \cdot w_{0,j+1} + (1-r) \cdot w_{1,j} + rr \cdot w_{2,j} \\
& \quad \text{for } i \in 1..n-3 \\
& \quad \quad cc_i \leftarrow rr \cdot w_{i,j} + (1-r) \cdot w_{i+1,j} + rr \cdot w_{i+2,j} \\
& \quad cc_{n-2} \leftarrow rr \cdot w_{n,j} + rr \cdot w_{n-2,j} + (1-r) \cdot w_{n-1,j} + rr \cdot w_{n,j+1} \\
& \quad x \leftarrow \text{LU\_solve\_T}(aa, dd, bb, cc) \\
& \quad \text{for } i \in 1..n-1 \\
& \quad \quad w_{i,j+1} \leftarrow x_{i-1} \\
& ((w))^T
\end{aligned}
$$

Chapter 15 Partial Differential Equations

Example 15.4 Temperature in a Rod, Crank-Nicolson Method

Consider again the temperature of a rod of unit length, given by the PDE

$$u_t - u_{xx} = 0, \quad \text{for } 0 < x < 1, \quad 0 < t,$$

with initial temperature

$$u(x, 0) = x^4, \quad 0 < x < 1,$$

and temperatures of

$$u(0, t) = 0, \quad u(1, t) = 1, \quad 0 < t.$$

at $x = 0$ and $x = 1$, respectively. Take a fairly coarse mesh, with $h = 0.2$ and $k = 0.04$, so that $r = \dfrac{c\,k}{h^2} = 1$. In this case, the general equation

$$-\frac{r}{2} u_{i-1,j+1} + (1 + r)u_{i,j+1} - \frac{r}{2} u_{i+1,j+1} = \frac{r}{2} u_{i-1,j} + (1 - r)u_{i,j} + \frac{r}{2} u_{i+1,j}.$$

simplifies to

$$-0.5\, u_{i-1,j+1} + 2\, u_{i,j+1} - 0.5\, u_{i+1,j+1} = 0.5\, u_{i-1,j} + 0.5\, u_{i+1,j}.$$

To find the solution at the first time step, the equations are

$$\begin{aligned}
+\ 2u_{1,1} - 0.5u_{2,1} &= 0.5(u_{0,0} + u_{2,0} + u_{1,0}) = 0.0128,\\
-0.5u_{1,1} + 2u_{2,1} - 0.5u_{3,1} &= 0.5(u_{1,0} + u_{3,0}) = 0.0656,\\
-0.5u_{2,1} + 2u_{3,1} - 0.5u_{4,1} &= 0.5(u_{2,0} + u_{4,0}) = 0.2176,\\
-0.5u_{3,1} + 2u_{4,1} &= 0.5(u_{3,0} + u_{5,0} + u_{5,1}) = 1.0648.
\end{aligned}$$

The solution from the Mathcad function HEAT_CN is shown in Table 15.2 for the first five time steps. The solution has almost reached its steady-state linear temperature distribution; the plot is essentially the same as is shown in Figure 15.6.

Table 15.2 Temperature in a Rod, Solution by Crank Nicolson Method

$t \setminus x$	0.0	0.2	0.4	0.6	0.8	1.0
0.00	0.00	0.0016	0.0256	0.1296	0.4096	1.00
0.04	0.00	0.034794	0.11358	0.28831	0.60448	1.00
0.08	0.00	0.078313	0.19968	0.39728	0.6714	1.00
0.12	0.00	0.11578	0.26343	0.46235	0.71491	1.00
0.16	0.00	0.14258	0.3069	0.50689	0.74231	1.00
0.20	0.00	0.16092	0.33678	0.53672	0.7609	1.00

15.2.4 Heat Equation with Insulated Boundary

The form of the boundary conditions depends on the physical situation being described. Keeping an end of the rod insulated corresponds to specifying that the partial derivative u_x is zero at that end of the rod; i.e.,

$$u_x(0, t) = 0 \quad \text{or} \quad u_x(a, t) = 0.$$

We illustrate the modification to the explicit method for the case of the temperature of the rod given at $x = 0$, but the end of the rod at $x = a$ insulated, so that $u_x(a, t) = 0$. The recommended approach to a boundary condition specified by a derivative is to add a fictitious point to the grid; for the situation described here, we extend the grid to include the point x_{n+1} at each time step. Using the central difference formula, we find that the boundary condition $u_x(a, t)$ becomes

$$\frac{1}{2k}[u_{n+1,j} - u_{n-1,j}] = 0,$$

or

$$u_{n+1,j} = u_{n-1,j}.$$

Applying the general update equation at $i = n$ gives

$$u_{n,j+1} = r u_{n-1,j} + (1 - 2r) u_{n,j} + r u_{n+1,j},$$

which includes the fictitious point. Substituting the information from the boundary condition, we get

$$u_{n,j+1} = r u_{n-1,j} + (1 - 2r) u_{n,j} + r u_{n-1,j} = 2 r u_{n-1,j} + (1 - 2r) u_{n,j}.$$

The system of equations is

$$u_{1,j+1} = r u_{0,j} + (1 - 2r) u_{1,j} + r u_{2,j}, \quad i = 1,$$

$$u_{i,j+1} = r u_{i-1,j} + (1 - 2r) u_{i,j} + r u_{i+1,j}, \quad i = 2, \cdots, n - 1,$$

$$u_{n,j+1} = 2 r u_{n-1,j} + (1 - 2r) u_{n,j}, \quad i = n.$$

For comparison, the equations for the explicit method (with u_0 and u_n given by boundary conditions) are

$$u_{1,j+1} = r u_{0,j} + (1 - 2r) u_{1,j} + r u_{2,j}, \quad i = 1,$$

$$u_{i,j+1} = r u_{i-1,j} + (1 - 2r) u_{i,j} + r u_{i+1,j}, \quad i = 2, \ldots, n - 2,$$

$$u_{n-1,j+1} = r u_{n-2,j} + (1 - 2r) u_{n-1,j} + r u_{n,j}, \quad i = n - 1.$$

15.3 WAVE EQUATION: HYPERBOLIC PDE

The standard example of a hyperbolic equation is the one-dimensional wave equation

$$u_{tt} - c^2 u_{xx} = 0, \quad \text{for } 0 \leq x \leq a \text{ and } 0 \leq t.$$

Initial conditions are given for $u(x,0)$ and $u_t(x,0)$:

$$u(x, 0) = f_1(x), \quad u_t(x, 0) = f_2(x), \quad \text{for } 0 \leq x \leq a.$$

Boundary conditions are given at $x = 0$ and $x = a$:

$$u(0, t) = g_1(t), \quad u(a, t) = g_2(t), \quad \text{for } 0 < t.$$

Although the one-dimensional wave equation does not usually require a numerical solution, it serves as an introduction to higher dimensional wave equations for which numerical methods are more likely to be worthwhile.

The mesh is given as before:

$$x_i = ih, i = 0, 1, \ldots, n, \quad h = \Delta x = a/n,$$

$$t_j = jk, j = 0, 1, \ldots, m, \quad k = \Delta t = T/m.$$

As with the heat equation, there are several choices for finite-difference approximations for u_{xx} and u_{tt}; each choice yields a method with certain characteristics—explicit or implicit technique, stability requirements, and truncation error. We consider an explicit method and an implicit method, each based on central difference formulas for the second derivatives. A sample solution is shown in Fig. 15.9.

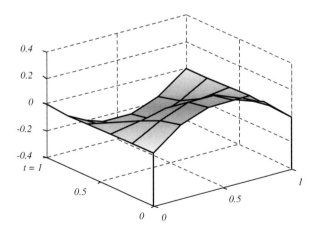

FIGURE 15.9 Vibrating string.

15.3.1 Explicit Method for Solving Wave Equations

Replacing the space derivative in the wave equation by the difference formula at the jth time step, i. e.,

$$c^2 u_{xx} \Rightarrow \frac{c^2}{h^2}[u_{i-1,j} - 2u_{i,j} + u_{i+1,j}].$$

and replacing the time derivative by the difference formula at the ith space step, i.e.,

$$u_{tt} \Rightarrow \frac{1}{k^2}[u_{i,j-1} - 2u_{i,j} + u_{i,j+1}],$$

gives

$$\frac{1}{k^2}[u_{i,j-1} - 2u_{i,j} + u_{i,j+1}] = \frac{c^2}{h^2}[u_{i-1,j} - 2u_{i,j} + u_{i+1,j}].$$

In a similar manner to our approach for the heat equation, we define the parameter

$$p = \frac{c\,k}{h} = c\,\frac{\Delta t}{\Delta x},$$

solve for the unknown $u_{i,j+1}$, and rearrange the order of the terms on the right-hand side to obtain

$$u_{i,j+1} = p^2\,u_{i-1,j} + 2(1 - p^2)\,u_{i,j} + p^2\,u_{i+1,j} - u_{i,j-1}.$$

Since the solution is known for $t = 0$, we can solve for $u_{i,j+1}$, starting with $j = 0$. However, we do not know $u_{i,-1}$. To overcome this difficulty, we use the initial condition for $u_t(x, 0) = f_2(x)$ and replace the time derivative by the centered difference formula to give

$$u_{i,1} - u_{i,-1} = 2\,k\,f_2(x_i).$$

The equation for u at the first time step now becomes

$$u_{i,1} = 0.5\,p^2\,u_{i-1,0} + (1 - p^2)\,u_{i,0} + 0.5\,p^2 u_{i+1,0} + k\,f_2(x_i),$$

where the values of $u_{i-1,0}$, $u_{i,0}$, and $u_{i+1,0}$ are available from the initial condition $u(x, 0) = f_1(x)$. The value of u at each subsequent time step can be found from the general equation

$$u_{i,j+1} = p^2\,u_{i-1,j} + 2(1 - p^2)\,u_{i,j} + p^2 u_{i+1,j} - u_{i,j-1}.$$

There are two stability requirements, determined by the matrix of coefficients of u at the jth time step:

1. The sum of the coefficients of the $u_{\cdot,j}$ terms must be less than or equal to 2; this is satisfied for all choices of p, since $p^2 + 2(1 - p^2) + p^2 = 2$.
2. No coefficient of $u_{\cdot,j}$ is negative. (A negative coefficient on $u_{i,j-1}$ is fine.) This requires that $1 - p^2 \geq 0$, or $p \leq 1$ (i.e., $c\,k \leq h$). (For further discussion, see Ames, 1992, p. 266.)

Algorithm to Solve Wave Equation, Finite Differences

Approximate the solution of the hyperbolic PDE
$u_{tt} = c\, u_{xx}$, for $0 < x < a$, $0 < t \le T$
and boundary conditions $u(0, t) = g_1(t)$, $u(a, t) = g_2(t), 0 < t$
with initial conditions $u(x, 0) = f_1(x)$, $u_t(x, 0) = f_2(x), 0 \le x \le a$
using finite differences, implicit method.

Define parameters
   $h = a/n$ step size for x
   $k = T/m$ step size for t
   $p = (c\, k/h)\wedge 2$

Define vectors of grid points for x and t; evaluate initial and boundary conditions
for $i = 0$ to n
   $x_i = i\, h$
   $u(i,0) = f_1(x_i)$
   $d_i = f_2(x_i)$
end
for $j = 0$ to m
   $t_j = j\, k$
   $u(0,j) = g_1(t_j)$
   $u(n,j) = g_2(t_j)$
end

Solution at first time step
for $i = 1$ to $n - 1$
$$u(i, 1) = \frac{p}{2} u(i - 1,0) + (1 - p)\, u(i,0) + \frac{p}{2} u(i + 1,0) + k\, d_i$$
end

Solution at remaining time steps
for $j = 2$ to m
   for $i = 1$ to $n - 1$
     $u(i,j) = p\, u(i - 1, j - 1) + 2(1 - p)\, u(i, j - 1) + p\, u(i + 1, j - 1) - u(i, j - 2)$
   end
end

Display results

Example 15.5 Vibrating String, Explicit Method, Stable Solution

The motion of a vibrating string of unit length with both ends held fixed and an initial displacement is described by the PDE

$$u_{tt} - u_{xx} = 0, \quad \text{for } 0 < x < 1, \quad 0 < t.$$

with initial conditions
$u(x, 0) = x(1 - x), \quad u_t(x, 0) = 0, \quad 0 < x < 1;$
and boundary conditions
$u(0, t) = 0, \quad u(1, t) = 0, \quad 0 < t.$

With a fairly coarse mesh, $h = \Delta x = 0.2$, stability requires that $p = \dfrac{\Delta t}{\Delta x} \le 1.0$, or $k = \Delta t \le 0.2$. Using the maximum allowed value of Δt (so that $p = 1$) results in a simplification of the general equation for the first time step,

$$u_{i,1} = 0.5\, p^2\, u_{i-1,0} + (1 - p^2)\, u_{i,0} + 0.5\, p^2 u_{i+1,0} + \Delta t\, g(x_i),$$

to the form
$$u_{i,1} = 0.5\, u_{i-1,0} + 0.5\, u_{i+1,0} + 0.2\, g(x_i) = 0.5\, u_{i-1,0} + 0.5\, u_{i+1,0}.$$

The initial and boundary conditions (bold), and the solution at the first time step (italics) are given in the following tabulation:

x =		0.0	0.2	0.4	0.6	0.8	1.0
t	j/i	0	1	2	3	4	5
0.0	0	**0.0**	**0.16**	**0.24**	**0.24**	**0.16**	**0.0**
0.2	1	**0.0**	*0.12*	*0.20*	*0.20*	*0.12*	**0.0**

The value of u at each subsequent time step can be found from the general equation $u_{i,j+1} = u_{i-1,j} + u_{i+1,j} - u_{i,j-1}$. The solution at the second time step (shown in italics) is shown in the following table:

	x =	0.0	0.2	0.4	0.6	0.8	1.0
t	j/i	0	1	2	3	4	5
0.0	0	**0.0**	**0.16**	**0.24**	**0.24**	**0.16**	**0.0**
0.2	1	**0.0**	0.12	0.20	0.20	0.12	**0.0**
0.4	2	**0.0**	*0.04*	*0.08*	*0.08*	*0.04*	**0.0**

The solution for $0 \le t \le 1$ was shown in Fig. 15.9.

Mathcad Function for the Wave Equation

$S\_15\_5(f1, f2, a, T, n, m) :=$
$\quad h \leftarrow \dfrac{a}{n}$
$\quad k \leftarrow \dfrac{T}{m}$
$\quad c \leftarrow 1$
$\quad \text{for } i \in 0..n$
$\quad\quad x_i \leftarrow i \cdot h$
$\quad\quad z_{0,i} \leftarrow f1(x_i)$
$\quad\quad dwx0_i \leftarrow f2(x_i)$
$\quad \text{for } j \in 0..m$
$\quad\quad t_j \leftarrow k \cdot j$
$\quad \text{for } j \in 1..m$
$\quad\quad z_{j,0} \leftarrow 0$
$\quad\quad z_{j,n} \leftarrow 0$
$\quad p \leftarrow \left(\dfrac{c \cdot k}{h}\right)^2$
$\quad p2 \leftarrow \dfrac{p}{2}$
$\quad pm2 \leftarrow 2 \cdot (1 - p)$
$\quad z_{1,1} \leftarrow (1 - p) \cdot z_{0,1} + p2 \cdot z_{0,2} + k \cdot dwx0_1$
$\quad \text{for } i \in 2..n-2$
$\quad\quad z_{1,i} \leftarrow p2 \cdot z_{0,i-1} + (1 - p) \cdot z_{0,1} + p2 \cdot z_{0,i+1} + k \cdot dwx0_i$
$\quad z_{1,n-1} \leftarrow p2 \cdot z_{0,n-2} + (1 - p) \cdot z_{0,n-1} + k \cdot dwx0_{n-1}$
$\quad \text{for } j \in 2..m$
$\quad\quad z_{j,1} \leftarrow pm2 \cdot z_{j-1,1} + p \cdot z_{j-1,2} - z_{j-2,1}$
$\quad\quad \text{for } i \in 2..n-2$
$\quad\quad\quad z_{j,i} \leftarrow p \cdot z_{j-1,i-1} + pm2 \cdot z_{j-1,i} + p \cdot z_{j-1,i+1} - z_{j-2,i}$
$\quad\quad z_{j,n-1} \leftarrow p \cdot z_{j-1,n-2} + pm2 \cdot z_{j-1,n-1} - z_{j-2,n-1}$
$\quad \text{return } z$

15.3.2 Implicit Method for Solving Wave Equation

As is the case with parabolic equations, implicit finite-difference solutions have stability advantages for hyperbolic equations also. A simple implicit scheme (Ames, 1992, p. 284) results from replacing the time derivative by the difference formula at the ith space step:

$$u_{tt} \Rightarrow \frac{1}{k^2}[u_{i,j-1} - 2u_{i,j} + u_{i,j+1}],$$

and replacing the space derivative by an average of the differences at the time steps $(j+1)$ and $(j-1)$:

$$c^2 u_{xx} \Rightarrow \frac{c^2}{2h^2}[u_{i-1,j+1} - 2u_{i,j+1} + u_{i+1,j+1} + u_{i-1,j-1} - 2u_{i,j-1} + u_{i+1,j-1}].$$

The difference equation then becomes
$$u_{i,j-1} - 2u_{i,j} + u_{i,j+1} =$$
$$\frac{c^2 k^2}{2h^2}[u_{i-1,j+1} - 2u_{i,j+1} + u_{i+1,j+1} + u_{i-1,j-1} - 2u_{i,j-1} + u_{i+1,j-1}].$$

In a manner similar to that for the heat equation, we define $p = \dfrac{c\,k}{h}$, so that

$$-p^2 u_{i-1,j+1} + 2(1+p^2)u_{i,j+1} - p^2 u_{i+1,j+1} = 4u_{i,j} + p^2 u_{i-1,j-1} - 2(1+p^2)u_{i,j-1} + p^2 u_{i+1,j-1}.$$

Since the solution is known for $t = 0$, we can solve for $u_{i,j+1}$, starting with $j = 0$. However, we do not know $u_{i,-1}$. To overcome this difficulty, we use the initial condition for $u_t(x, 0) = f_2(x)$ and replace the time derivative by the centered difference formula to give

$$u_{i,-1} = u_{i,1} - 2k f_2(x_i).$$

The equations for $j = 0$ then have the form
$$4(1+p^2)u_{1,1} - 2p^2 u_{2,1} =$$
$$4u_{1,0} - 2p^2 k f_2(x_0) + 4(1+p^2)k f_2(x_1) - 2p^2 k f_2(x_2) + 2p^2 u_{0,1}.$$
$$-2p^2 u_{1,1} + 4(1+p^2)u_{2,1} - 2p^2 u_{3,1} =$$
$$4u_{2,0} - 2p^2 k f_2(x_1) + 4(1+p^2)k f_2(x_2) - 2p^2 k f_2(x_3),$$
$$\vdots$$
$$-2p^2 u_{i-1,1} + 4(1+p^2)u_{i,1} - 2p^2 u_{i+1,1} =$$
$$4u_{i,0} - 2p^2 k f_2(x_{i-1}) + 4(1+p^2)k f_2(x_i) - 2p^2 k f_2(x_{i+1}),$$
$$\vdots$$
$$-2p^2 u_{n-1,1} + 4(1+p^2)u_{n,1} =$$
$$4u_{n,0} - 2p^2 k f_2(x_{n-1}) + 4(1+p^2)k f_2(x_n) - 2p^2 k f_2(x_{n+1}) + 2p^2 u_{n+1,1},$$

The value of u at each subsequent time step can be found from

$$2(1 + p^2)u_{1,j+1} - p^2 u_{2,j+1} = 4u_{1,j} - 2(1 + p^2)u_{1,j-1} + p^2 u_{2,j-1} + p^2 u_{0,j+1} + p^2 u_{0,j-1},$$

$$-p^2 u_{1,j+1} + 2(1 + p^2)u_{2,j+1} - p^2 u_{3,j+1} = 4u_{2,j} + p^2 u_{1,j-1} - 2(1 + p^2)u_{2,j-1} + p^2 u_{3,j-1},$$

$$\vdots$$

$$-p^2 u_{i-1,j+1} + 2(1 + p^2)u_{i,j+1} - p^2 u_{i+1,j+1} =$$
$$4u_{i,j} + p^2 u_{i-1,j-1} - 2(1 + p^2)u_{i,j-1} + p^2 u_{i+1,j-1},$$

$$\vdots$$

$$-p^2 u_{n-1,j+1} + 2(1 + p^2)u_{n,j+1} =$$
$$4u_{n,j} + p^2 u_{n-1,j-1} - 2(1 + p^2)u_{n,j-1} + p^2 u_{n+1,j-1} + p^2 u_{n+1,j+1}.$$

This system of equations is tridiagonal, and thus can be solved by the Thomas algorithm of Chapter 3. Iterative methods can also be used, since the linear system is diagonally dominant. (The diagonal elements are $2(1 + p^2)$, and the off diagonal elements, two in each equation, are $-p^2$.)

This implicit method has unrestricted stability (Ames, 1992, p. 285; see Ames as well for a discussion of the general three-level implicit form. The preceding implicit method corresponds to $\lambda = 1/2$, the explicit method corresponds to $\lambda = 0$, and the general method has unrestricted stability for $\lambda \geq 1/4$.)

15.4 POISSON EQUATION: ELLIPTIC PDE

The standard example of an elliptic equation is the two-dimensional Laplacian or potential equation

$$u_{xx} + u_{yy} = 0, \quad a \leq x \leq b, \quad c \leq y \leq d,$$

or Poisson's equation

$$u_{xx} + u_{yy} = f(x, y), \quad a \leq x \leq b, \quad c \leq y \leq d.$$

The simplest boundary conditions specify the value of the function along each of the four sides of the rectangular domain:

$$u(a, y) = g_1(y), \quad u(b, y) = g_2(y), \quad c < y < d,$$
$$u(x, c) = g_3(y), \quad u(x, d) = g_4(y), \quad a < x < b.$$

We define a mesh in the x–y plane:

$$x_i = a + ih, \quad i = 0, 1, \cdots, n, \quad h = \Delta x = (b - a)/n,$$
$$y_j = c + jk, \quad j = 0, 1, \cdots, m, \quad k = \Delta y = (d - c)/m.$$

We denote the (approximate) value of u(x,y) at the point (x_i, y_j) as u_{ij} and the value of the right-hand side, $f(x_i, y_j)$ as f_{ij}. Replacing the second derivatives by centered differences gives a system of algebraic equations for the function values at the mesh points:

$$\frac{1}{k^2}[u_{i,j+1} - 2u_{i,j} + u_{i,j-1}] + \frac{1}{h^2}[u_{i+1,j} - 2u_{i,j} + u_{i-1,j}] = f_{i,j}.$$

However, unlike the situation with the wave equation, no information is given that allows us to solve these equations in a sequential manner. Due to the somewhat more extensive computations needed to obtain interesting results, we first present a Mathcad function for Poisson's equation and then illustrate the process with examples.

The first example is the potential equation problem solved earlier in Chapter 3. However, using the Mathcad function, one need not form the coefficient matrix explicitly. The second example is a Poisson equation with the same boundary conditions as those for the potential equation. Gauss-Seidel iteration is built directly into the Mathcad function Poisson. This means that changing the order of the updates would change the details of the intermediate results, but not the final answer.

Algorithm to Solve Poisson Equation, Finite Differences

Use finite differences to approximate the solution of the elliptic PDE
$$u_{xx} + u_{yy} = f(x,y) \quad \text{for } a < x < b, c < y < d$$
with boundary conditions
$$u(a,y) = g_1(y) \quad u(b,y) = g_2(y) \quad \text{for } c < y < d$$
$$u(x,c) = g_3(x) \quad u(x,d) = g_4(x) \quad \text{for } a < x < b$$

Define parameters
$\quad h = (b - a)/n \quad$ step size for x
$\quad k = (d - c)/m \quad$ step size for y
$\quad r = \dfrac{h^2}{k^2}$
$\quad c_1 = \dfrac{1}{2r + 2}$
$\quad c_2 = r\, c_1$
$\quad c_3 = \dfrac{h^2}{2r + 2}$

Define grid points
for $i = 0$ to n
$\quad x_i = a + i\,h$
end
for $j = 0$ to m
$\quad y_j = c + j\,k$
end

Evaluate boundary conditions at $x = a$ and $x = b$
for $j = 1$ to $m - 1$
$\quad u(0,j) = g_1(y_j)$
$\quad u(n,j) = g_2(y_j)$
end

Further initializations
for $i = 1$ to $n - 1$
$\quad u(i,0) = g_3(x_i)$ $\qquad$ (evaluate boundary conditions at $y = c$ and $y = d$)
$\quad u(i,m) = g_4(x_i)$
$\quad$ for $j = 1$ to $m - 1$
$\quad\quad ff(i,j) = f(x_i, y_j)$ $\qquad$ (evaluate right-hand side at grid points)
$\quad\quad u(i, j) = 0$ $\qquad$ (initialize solution at interior points)
$\quad$ end
end

Begin iterations
for it $= 1$ to max_it
$\quad$ for $i = 1$ to $n - 1$
$\quad\quad$ for $j = 1$ to $m - 1$
$\quad\quad\quad u(i,j) = c_2\,(u(i,j+1) + u(i,j-1)) + c_1\,(u(i+1,j) + u(i-1,j)) - c_3\,ff(i,j)$
$\quad\quad$ end
$\quad$ end
end $\qquad$ (add test for convergence?)

Display solution

Mathcad Function for the Poisson Equation

$\text{Poisson}(f,g,a,b,c,d,n,m,\text{max\_it}) :=$
$\begin{array}{l}
h \leftarrow \dfrac{b-a}{n} \\[4pt]
k \leftarrow \dfrac{d-c}{m} \\[4pt]
r \leftarrow \left(\dfrac{h}{k}\right)^2 \\[4pt]
c1 \leftarrow \dfrac{1}{2 \cdot r + 2} \\[4pt]
c2 \leftarrow r \cdot c1 \\
c3 \leftarrow c1 \cdot h^2 \\
\text{for } i \in 0..n \\
\quad x_i \leftarrow a + h \cdot i \\
\text{for } i \in 0..m \\
\quad y_i \leftarrow c + k \cdot i \\
\text{for } i \in 1..m-1 \\
\quad \left| \begin{array}{l} w_{0,i} \leftarrow g\big[(y)_i\big]^{\langle 1 \rangle} \\ w_{n,i} \leftarrow g(y_i)^{\langle 2 \rangle} \end{array} \right. \\
\text{for } i \in 0..n \\
\quad \left| \begin{array}{l} w_{i,0} \leftarrow g(x_i)^{\langle 3 \rangle} \\ w_{i,m} \leftarrow g(x_i)^{\langle 4 \rangle} \end{array} \right. \\
\text{for } i \in 1..n-1 \\
\quad \text{for } j \in 1..m-1 \\
\qquad \left| \begin{array}{l} \text{fij}_{i,j} \leftarrow f(x_i, y_j) \\ w_{i,j} \leftarrow 0 \end{array} \right. \\
\text{for } \text{it} \in 1..\text{max_it} \\
\quad \left| \begin{array}{l}
\text{for } i \in 1..n-1 \\
\quad \left| \begin{array}{l} w_{i,1} \leftarrow c2 \cdot (w_{i,2} + w_{i,0}) + c1 \cdot (w_{i+1,1} + w_{i-1,1}) - c3 \cdot \text{fij}_{i,1} \\ w_{i,m-1} \leftarrow c2 \cdot (w_{i,m-2} + w_{i,m}) + c1 \cdot (w_{i+1,m-1} + w_{i-1,m-1}) - c3 \cdot \text{fij}_{i,m-1} \end{array} \right. \\
\text{for } j \in 1..m-1 \\
\quad \left| \begin{array}{l} w_{1,j} \leftarrow c2 \cdot (w_{1,j+1} + w_{1,j-1}) + c1 \cdot (w_{2,j} + w_{0,j}) - c3 \cdot \text{fij}_{1,j} \\ w_{n-1,j} \leftarrow c2 \cdot (w_{n-1,j+1} + w_{n-1,j-1}) + c1 \cdot (w_{n-2,j} + w_{n,j}) - c3 \cdot \text{fij}_{n-1,j} \\ \text{for } i \in 2..n-2 \\ \quad w_{i,j} \leftarrow c2 \cdot (w_{i,j+1} + w_{i,j-1}) + c1 \cdot (w_{i+1,j} + w_{i-1,j}) - c3 \cdot \text{fij}_{i,j} \end{array} \right.
\end{array} \right. \\
w
\end{array}$

Example 15.6 Potential Equation

Consider the equation

$$u_{xx} + u_{yy} = 0, \quad 0 \le x \le 1, \quad 0 \le y \le 1.$$

with boundary conditions

$$u(0, y) = y^2, \quad u(1, y) = 1, \quad 0 < y < 1,$$
$$u(x, 0) = x^2, \quad u(x, 1) = 1, \quad 0 < x < 1.$$

The solutions after 1, 5, 20, and 50 iterations are shown in Fig. 15.10.

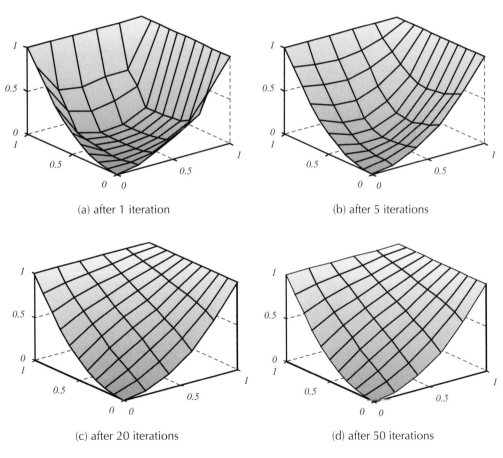

(a) after 1 iteration (b) after 5 iterations

(c) after 20 iterations (d) after 50 iterations

FIGURE 15.10 Solution of potential equation.

Example 15.7 Poisson's Equation

Consider the equation

$$u_{xx} + u_{yy} = x + y, \quad 0 \le x \le 1, \quad 0 \le y \le 1,$$

with boundary conditions

$$u(0, y) = y^2, \quad u(1, y) = 1, \quad 0 < y < 1,$$
$$u(x, 0) = x^2, \quad u(x, 1) = 1, \quad 0 < x < 1.$$

The solutions after 1, 5, 20, and 50 iterations are shown in Fig. 15.11. The solution values change only slightly after the first 20 iterations.

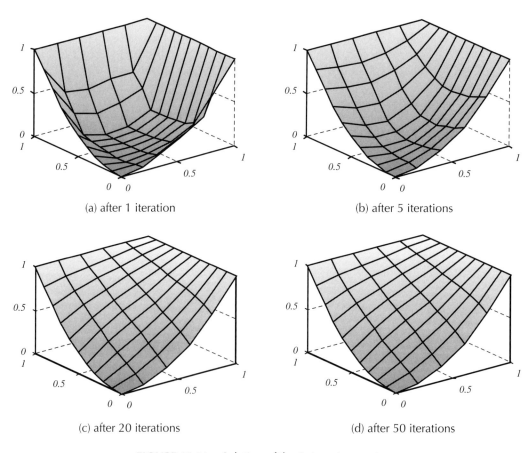

(a) after 1 iteration

(b) after 5 iterations

(c) after 20 iterations

(d) after 50 iterations

FIGURE 15.11 Solution of the Poisson's equation.

15.5 FINITE-ELEMENT METHOD FOR AN ELLIPTIC PDE

Finite-difference methods for solving a two-dimensional Poisson equation

$$u_{xx} + u_{yy} = f(x, y)$$

on a rectangular region are based on dividing the domain of the problem into rectangular subdomains and approximating the solution at the mesh points. An alternative approach, known as the finite-element method, allows the domain to be divided into any convenient set of subregions (often triangular, but not necessarily of the same size). Rather than just finding the solution at the mesh or node points, an approximate solution of suitably simple form is found over the entire region. In the following discussion, we assume that the subregions are triangular.

The problem of solving the differential equation is converted into a corresponding problem of minimizing a functional that consists of an integral over the region (and, for certain types of boundary conditions a line integral along the boundary).

To begin, consider a finite-element solution to an elliptic PDE of the form

$$u_{xx} + u_{yy} + r(x, y)\, u = f(x, y) \qquad \text{on the region } R$$

$$u(x, y) = g(x, y) \qquad \text{on the boundary of } R.$$

The corresponding functional to be minimized is

$$I[u] = \iint_R [u_x^2 + u_y^2 - r(x, y)\, u^2 + 2f(x, y)u]\, dx\, dy.$$

The region R is divided into p triangular subregions $T_1, T_2, \ldots, T_p$; the vertices of these regions are the nodes, $V_1, \ldots, V_n, V_{n+1}, \ldots, V_m$. Nodes $j = 1, \ldots n$ are in the interior of R; the remaining nodes $j = n + 1, \ldots, m$ are on the boundary of R. Figure 15.12 shows a region divided into 14 triangular subregions.

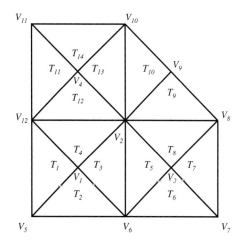

FIGURE 15.12 Subdivided regions.

A finite-element solution of the PDE is a function

$$U = \sum_{j=1}^{m} c_j \Phi_j$$

where the Φ_j are basis functions. There is a basis function corresponding to each node.

The solution process consists of the following steps:

1. Define the subdivision of R by specifying the locations of the nodes.
2. Define each of the basis functions Φ_j, $j = 1, \ldots, m$.
3. Determine coefficients c_j for the basis functions that correspond to boundary nodes ($j = n + 1, \ldots, m$), so that the solution U satisfies the boundary conditions at those nodes.
4. Determine the coefficients c_j for the basis functions that correspond to interior nodes ($j = 1, \ldots, n$), so that U minimizes the integral $I[u]$.

To find the coefficients corresponding to the interior nodes, we must minimize

$$\iint_R [U_x^2 + U_y^2 - r(x,y)U^2 + 2f(x,y)U]\, dx\, dy, \quad \text{with } U = \sum_{j=1}^{m} c_j \Phi_j.$$

The minimum occurs where $\dfrac{\partial U}{\partial c_i} = 0$, for $1 \le i \le n$. This gives a linear system of equations, $\mathbf{Ac} = \mathbf{d}$, where $\mathbf{A} = [a_{ij}]$, $1 \le i, j \le n$,

$$a_{ij} = \iint_R [\Phi_i]_x [\Phi_j]_x + [\Phi_i]_y [\Phi_j]_y - r(x,y)\Phi_i\Phi_j\, dx\, dy, \tag{15.1}$$

and

$$d_i = -\iint_R f(x,y)\Phi_i\, dx\, dy - \sum_{j=n+1}^{m} c_j b_{ij}, \tag{15.2}$$

in which

$$b_{ij} = \iint_R [\Phi_i]_x [\Phi_j]_x + [\Phi_i]_y [\Phi_j]_y - r(x,y)\Phi_i\Phi_j\, dx\, dy, \tag{15.3}$$

$$1 \le i \le n; \quad n + 1 \le j \le m.$$

We now consider the basis functions ϕ_j. We define ϕ_j to be one at node j, zero at all other nodes, and linear on each triangular subdomain. For each triangle that has node j as a vertex, we find a plane that is one at node j, and zero at the other two vertices; ϕ_j is identically zero on any subdomain that does not have node j as a vertex.

Example 15.8 Finding Basis Functions

To illustrate the definition of the basis functions ϕ_j, consider a region $0 \leq x \leq 3$, $0 \leq y \leq 3$, as shown in Fig. 15.13, that is decomposed into four triangular subregions, designated $T_1, \ldots, T_4$. The nodes are the vertices of the triangles, denoted $1, \ldots, 5$. Node 1, located at $(1, 1)$, is the only interior node in this diagram; nodes 2, 3, 4, and 5 are on the boundary.

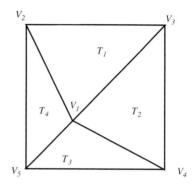

FIGURE 15.13 Triangular decomposition of the square $0 \leq x \leq 3$, $0 \leq y \leq 3$.

We now find the five basis functions for this region. The basis functions corresponding to nodes $2, \ldots, 5$ are each nonzero on only two of the triangular subregions. The second and third basis functions are illustrated in Fig. 15.14.

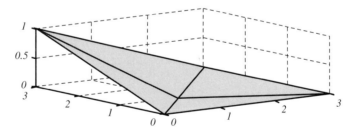

FIGURE 15.14A Second basis function.

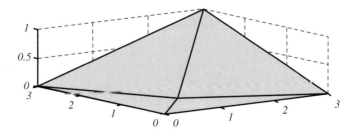

FIGURE 15.14B Third basis function.

The definition of the basis functions involves the determination of a plane that passes through three specified points. We want the jth basis function to be one at node j and zero at the other nodes. The equation of the plane that is equal to one at (x_1, y_1), zero at (x_2, y_2), and zero at (x_3, y_3) has the form

$$z = a + bx + cy,$$

where the constants a, b, and c are found by solving the system

$$1 = a + bx_1 + cy_1,$$
$$0 = a + bx_2 + cy_2,$$
$$0 = a + bx_3 + cy_3.$$

The First Basis Function

Φ_1 is equal to one at node 1, and zero at nodes 2, 3, 4, and 5. Φ_1 is a plane on each of the four triangular subregions.

On T_1, $\Phi_1 = a + bx + cy$, where a, b, and c satisfy

$$1 = a + b + c,$$
$$0 = a + 0 + 3c,$$
$$0 = a + 3b + 3c \Rightarrow a = 3/2, b = 0, c = -1/2.$$

On T_2, $\Phi_1 = a + bx + cy$, where a, b, and c satisfy

$$1 = a + b + c,$$
$$0 = a + 3b + 3c,$$
$$0 = a + 3b + 0 \Rightarrow a = 3/2, b = -1/2, c = 0.$$

On T_3, $\Phi_1 = a + bx + cy$, where a, b, and c satisfy

$$1 = a + b + c,$$
$$0 = a + 3b + 0,$$
$$0 = a + 0 + 0 \Rightarrow a = 0, b = 0, c = 1.$$

On T_4, $\Phi_1 = a + bx + cy$, where a, b, and c satisfy

$$1 = a + b + c,$$
$$0 = a + 0 + 0,$$
$$0 = a + 0 + 3c \Rightarrow a = 0, b = 1, c = 0.$$

Thus,

$$\Phi_1 = \begin{cases} (3-y)/2 & \text{on } T_1 \\ (3-x)/2 & \text{on } T_2 \\ y & \text{on } T_3 \\ x & \text{on } T_4 \end{cases}$$

The Second Basis Function

Φ_2 is equal to zero at node 1, one at node 2, and zero at nodes 3, 4, and 5, so $\Phi_2 = 0$ on T_2 and T_3. On T_1, $\Phi_2 = a + bx + cy$, where a, b, and c satisfy

$$1 = a + 0 + 3c, \qquad 0 = a + b + c,$$
$$0 = a + 3b + 3c \Rightarrow a = 0, b = -1/3, c = 1/3.$$

On T_4, $\Phi_2 = a + bx + cy$, where a, b, and c satisfy

$$1 = a + 0 + 3c, \qquad 0 = a + b + c,$$
$$0 = a + 0 + 0 \Rightarrow a = 0, b = -1/3, c = 1/3.$$

Hence,

$$\Phi_2 = \begin{cases} (-x+y)/3 & \text{on } T_1 \\ 0 & \text{on } T_2 \\ 0 & \text{on } T_3 \\ (-x+y)/3 & \text{on } T_4 \end{cases}$$

The Third Basis Function

Φ_3 is equal to zero at nodes 1 and 2, one at node 3, and zero at nodes 4 and 5; $\phi_3 = 0$ on T_3 and T_4.

As with Φ_1 and Φ_2, we form the linear systems to find the values of a, b, and c that define ϕ_3 on T_1 and on T_2; we find that on T_1, $a = -1/2$, $b = 1/3$, and $c = 1/6$, and on T_2, $a = -1/2$, $b = 1/6$, and $c = 1/3$, so

$$\Phi_3 = \begin{cases} (-3 + 2x + y)/6 & \text{on } T_1 \\ (-3 + x + 2y)/6 & \text{on } T_2 \\ 0 & \text{on } T_3 \\ 0 & \text{on } T_4 \end{cases}$$

Note the symmetric form of Φ_3 on T_1 and T_2.

The Fourth Basis Function

We can exploit the symmetry of the figure to find Φ_4 from Φ_2 by interchanging the roles of x and y and interchanging the triangular regions $T_1 \leftrightarrow T_2$ and $T_3 \leftrightarrow T_4$; doing this gives

$$\Phi_4 = \begin{cases} 0 & \text{on } T_1 \\ (x-y)/3 & \text{on } T_2 \\ (x-y)/3 & \text{on } T_3 \\ 0 & \text{on } T_4 \end{cases}$$

The Fifth Basis Function

Φ_5 is equal to zero at nodes 1, 2, 3, and 4, and one at node 5; $\Phi_5 = 0$ on T_1 and T_2. We find that

$$\Phi_5 = \begin{cases} 0 & \text{on } T_1 \\ 0 & \text{on } T_2 \\ (3 - x - 2y)/3 & \text{on } T_3 \\ (3 - 2x - y)/3 & \text{on } T_4 \end{cases}$$

Example 15.9 A Finite-element Solution

Consider the problem
$$u_{xx} + u_{yy} = 0$$
with boundary conditions

$$\begin{aligned}
u &= x/3 && \text{for } y = 0, 0 \le x \le 3, \\
u &= y/3 && \text{for } x = 0, 0 \le y \le 3, \\
u &= 1 && \text{for } y = 1, 0 \le x \le 3, \\
u &= 1 && \text{for } x = 1, 0 \le y \le 3.
\end{aligned}$$

The basis functions found in Example 15.8 are appropriate for problems defined on $0 \le x \le 3; 0 \le y \le 3$, with the triangular subdivision shown in Fig. 15.16.

To Find the Coefficients for the Nodes on the Boundary

For the given region, $U(\text{node } 2) = 1$, $U(\text{node } 3) = 1$, $U(\text{node } 4) = 1$, and $U(\text{node } 5) = 0$. Since we are looking for $U = c_1 \phi_1 + c_2 \phi_2 + c_3 \phi_3 + c_4 \phi_4 + c_5 \phi_5$ to satisfy the boundary conditions, and because each ϕ_j is zero except at node j, we must have $c_2 = c_3 = c_4 = 1$, and $c_5 = 0$.

To Find the Coefficients for the Interior Node

We have only one interior node, so we have only one coefficient to determine from the given equations. We must solve a $c_1 = d$, where

$$a = \iint_R [\Phi_1]_x [\Phi_1]_x + [\Phi_1]_y [\Phi_1]_y \, dx \, dy \quad \text{and} \quad d = -\sum_{j=2}^{5} c_j b_j,$$

in which

$$b_j = \iint_R [\Phi_1]_x [\Phi_j]_x + [\Phi_1]_y [\Phi_j]_y \, dx, dy, \quad 2 \le j \le 5.$$

Because $c_5 = 0$, the calculation of b_5 is not required, but we do have a number of integrations to perform. We have used the fact that both $r(x, y)$ and $f(x, y)$ are zero in this example to simplify the expressions for a and d also. We note, for use in the following calculations, the areas of the triangular regions:

$$A_1 = \text{area}(T_1) = 3; \quad A_2 = 3; \quad A_3 = 3/2; \quad A_4 = 3/2.$$

To Find a

$$\Phi_1 = \begin{cases} (3-y)/2 \\ (3-x)/2 \\ y \\ x \end{cases} \quad [\Phi_1]_x = \begin{cases} 0 \\ -1/2 \\ 0 \\ 1 \end{cases} \quad [\Phi_1]_y = \begin{cases} -1/2 & \text{on } T_1 \\ 0 & \text{on } T_2 \\ 1 & \text{on } T_3 \\ 0 & \text{on } T_4 \end{cases}$$

$$a = \iint_R [\Phi_1]_x [\Phi_1]_x + [\Phi_1]_y [\Phi_1]_y \, dx \, dy$$

$$= \iint_{T1} 1/4 \, dxdy + \iint_{T2} 1/4 \, dxdy + \iint_{T3} 1 \, dxdy + \iint_{T4} 1 \, dxdy$$

$$= 1/4 \, (\text{area of } T_1 + \text{area of } T_2) + \text{area of } T_3 + \text{area of } T_4$$

$$= (1/4)(3 + 3) + 3/2 + 3/2 = 9/2.$$

To Find b_2

$$\Phi_2 = \begin{cases} (-x+y)/3 \\ 0 \\ 0 \\ (-x+y)/3 \end{cases} \quad [\Phi_2]_x = \begin{cases} -1/3 \\ 0 \\ 0 \\ -1/3 \end{cases} \quad [\Phi_2]_y = \begin{cases} 1/3 & \text{on } T_1 \\ 0 & \text{on } T_2 \\ 1 & \text{on } T_3 \\ 1/3 & \text{on } T_4 \end{cases}$$

$$b_2 = \iint_R [\Phi_1]_x [\Phi_2]_x + [\Phi_1]_y [\Phi_2]_y \, dx \, dy$$

$$= \iint_{T_1} (-1/2)(1/3) dx dy + \iint_{T_2} 0 \, dx \, dy + \iint_{T_3} 0 \, dxdy + \iint_{T_4} (1)(-1/3) \, dx \, dy$$

$$= (-1/6)(\text{area of } T_1) + (-1/3)(\text{area of } T_4) = -1.$$

To Find b_3

$$\Phi_3 = \begin{cases} (-3+2x+y)/6 \\ (-3+x+2y)/6 \\ 0 \\ 0 \end{cases} \quad [\Phi_3]_x = \begin{cases} 1/3 \\ 1/6 \\ 0 \\ 0 \end{cases} \quad [\Phi_3]_y = \begin{cases} 1/6 & \text{on } T_1 \\ 1/3 & \text{on } T_2 \\ 0 & \text{on } T_3 \\ 0 & \text{on } T_4 \end{cases}$$

$$b_3 = \iint_R [\Phi_1]_x [\Phi_3]_x + [\Phi_1]_y [\Phi_3]_y \, dx \, dy$$

$$= \iint_{T_1} (-1/2)(1/6) \, dxdy + \iint_{T_2} (-1/2)(1/3) \, dxdy + \iint_{T_3} 0 \, dxdy + \iint_{T_4} 0 \, dxdy$$

$$= (-1/2)(\text{area of } T_1) + (-1/6)(\text{area of } T_2) = -3/4.$$

To Find b_4

$$\Phi_4 = \begin{cases} 0 \\ (x-y)/3 \\ (x-y)/3 \\ 0 \end{cases} \quad [\Phi_4]_x = \begin{cases} 0 \\ 1/3 \\ 1/3 \\ 0 \end{cases} \quad [\Phi_4]_y = \begin{cases} 0 & \text{on } T_1 \\ -1/3 & \text{on } T_2 \\ -1/3 & \text{on } T_3 \\ 0 & \text{on } T_4 \end{cases}$$

$$b_4 = \iint_R [\Phi_1]_x [\Phi_4]_x + [\Phi_1]_y [\Phi_4]_y \, dxdy$$

$$= \iint_{T_1} 0 \, dxdy + \iint_{T_2} (-1/2)(1/3) \, dxdy + \iint_{T_3} (1)(-1/3) \, dx \, dy + \iint_{T_4} 0 \, dxdy$$

$$= (-1/6)(\text{area of } T_2) + (-1/3)(\text{area of } T_3) = -1.$$

The Right-Hand Side

$d = [c_2 b_2 + c_3 b_3 + c_4 b_4] = -[b_2 + b_3 + b_4] = -[(-1) + (-3/4) + (-1)] = 11/4$

The equation to be solved, $a c_1 = d$ is $9/2 \, c_1 = 11/4 \Rightarrow c_1 = 11/18$.
The solution of the potential equation is $U = (11/8) \phi_1 + \phi_2 + \phi_3 + \phi_4$, which simplifies to

$$U = \begin{cases} 5/12 + 7/36 \, y & \text{on } T_1 \\ 5/12 + 7/36 \, x & \text{on } T_2 \\ 1/3 \, x + 5/18 \, y & \text{on } T_3 \\ 5/18 \, x + 1/3 \, y & \text{on } T_4 \end{cases}$$

Example 15.10 Using Mathcad to Find the Basis Functions

We illustrate the use of a Mathcad function to generate the basis functions corresponding to each node in a finite-element grid with triangular elements. We describe the geometry in two arrays. The first, V, gives the coordinates of each of the nodes (interior nodes first). The second array, T, specifies the triangles by giving the indices of the three vertices that define each triangle. For the region illustrated in Fig. 15.15, there are 12 vertices and 14 triangles; the arrays are

$$V := \begin{pmatrix} \frac{1}{2} & \frac{1}{2} \\ 1 & 1 \\ \frac{3}{2} & \frac{1}{2} \\ \frac{1}{2} & \frac{3}{2} \\ 0 & 0 \\ 1 & 0 \\ 2 & 0 \\ 2 & 1 \\ \frac{3}{2} & \frac{3}{2} \\ 1 & 2 \\ 0 & 2 \\ 0 & 1 \end{pmatrix} \quad T := \begin{pmatrix} 1 & 5 & 12 \\ 1 & 5 & 6 \\ 1 & 2 & 6 \\ 1 & 2 & 12 \\ 3 & 2 & 6 \\ 3 & 6 & 7 \\ 3 & 7 & 8 \\ 3 & 8 & 2 \\ 2 & 8 & 9 \\ 2 & 9 & 10 \\ 4 & 11 & 12 \\ 4 & 12 & 2 \\ 4 & 2 & 10 \\ 4 & 10 & 11 \end{pmatrix}$$

For each node, we must find a basis function, which has the form $A + Bx + Cy$ on each triangle. We store the information in three arrays:

$$A(m, p); \quad B(m, p); \quad C(m, p);$$

The row index gives the node number; the column index gives the triangular element. The computations to set up the three-by-three linear system for each triangular region and solve the system are given in the Mathcad function `basis_func` which follows.

To use the function, we define

$m := 12 \quad n := 4 \quad p := 14 \quad M := \text{basis\_func}(m, p, V, T)$

$A := \text{submatrix}(M, 0, m - 1, 0, p - 1)$

$B := \text{submatrix}(M, 0, m - 1, p, 2p - 1)$

$C := \text{submatrix}(M, 0, m - 1, 2p, 3p - 1)$

Mathcad Function to Form Basis Functions

$$\text{basis\_func}(m, p, V, T) := \begin{array}{|l} \text{for } j \in 0..m-1 \\ \quad \begin{array}{|l} \text{for } k \in 0..p-1 \\ \quad \begin{array}{|l} \text{for } i \in 0..2 \\ \quad \begin{array}{|l} AA_{i,0} \leftarrow 1 \\ q \leftarrow T_{k,i} - 1 \\ AA_{i,1} \leftarrow V_{q,0} \\ AA_{i,2} \leftarrow V_{q,1} \\ bb_i \leftarrow 1 \text{ if } q = j \\ bb_i \leftarrow 0 \text{ otherwise} \end{array} \\ x \leftarrow AA^{-1} \cdot bb \\ A_{j,k} \leftarrow x_0 \\ B_{j,k} \leftarrow x_1 \\ C_{j,k} \leftarrow x_2 \end{array} \end{array} \\ M \leftarrow \text{augment}(A, B, C) \\ M \end{array}$$

Although we do not need to display the definition of each basis function, it is informative to consider briefly what the information in arrays **A, B,** and **C** can tell us for this example. Each row of the matrix corresponds to a basis function, each column to a triangular element. The information about the basis function corresponding to node 1 is in the first row of matrices **A, B,** and **C**. Node 1 is a vertex of triangles 1, 2, 3, and 4, so the only possible nonzero coefficients for the first basis function occur in the first four rows. Specifically, the first row of the three arrays are $A(1, :) = [\ 0\ 0\ 2\ 2\ 0\ \quad ......\ 0\ ]$, $B(1, :) = [\ 2\ 0\ -2\ 0\ 0\ \quad ......\ 0\ ]$, $C(1, :) = [\ 0\ 2\ 0\ -2\ 0\ \quad ......\ 0\ ]$;

This tells us that the first basis function is

$$\begin{aligned} 0 + 2x + 0y & \quad \text{on } T_1, \\ 0 + 0x + 2y & \quad \text{on } T_2, \\ 0 - 2x + 0y & \quad \text{on } T_3, \\ 2 + 0x - 2y & \quad \text{on } T_4, \\ 0 & \quad \text{on all other triangles.} \end{aligned}$$

It is easy to verify that the function defined in this way has the required values at each of the vertices.

We now solve the potential equation on $0 \leq x \leq 2, 0 \leq y \leq 2$, with the upper corner removed, as in Fig. 15.16. The boundary conditions are

$$u = 0 \quad \text{for} \quad 0 \leq x \leq 2, y = 0, \text{ and } 0 \leq y \leq 2, x = 0,$$
$$u = x \quad \text{for} \quad 0 \leq x \leq 1, y = 2,$$
$$u = y \quad \text{for} \quad 0 \leq y \leq 1, x = 2,$$
$$u = 1 \quad \text{along the diagonal boundary.}$$

We continue the computations from Example 15.9; that is, we assume that the geometry has been defined and the basis functions have been computed. We also need the area of each of the triangular elements. In this example, each triangle is of area 1/4. In general, for a triangle with corners at (x_1, y_1), (x_2, y_2) and (x_3, y_3), the area is Area(T) = 0.5 det (TT), where

$$TT = \begin{bmatrix} x_1 & y_1 & 1 \\ x_2 & y_2 & 1 \\ x_3 & y_3 & 1 \end{bmatrix}.$$

To solve the problem, we must find the coefficients of the basis functions so that the linear combination $U = c_1 \phi_1 + \ldots + c_m \phi_m$ is the desired solution. We initialize the vector **c** so that the linear combination will satisfy the boundary conditions. For this problem, the boundary conditions give $U = 1$ at nodes 8, 9, and 10 and $U = 0$ at all other nodes; hence, we begin with

$$\mathbf{c} = [0\ 0\ 0\ 0\ 0\ 0\ 0\ 1\ 1\ 1\ 0\ 0].$$

We now set up the equations to determine the coefficients for the basis functions corresponding to the interior nodes. The Mathcad function `coeff` performs the computations analogous to Eqs. (15.1) to (15.3).

The computations of the coefficients may be carried out in the same worksheet as for Example 15.10, or a reference to that worksheet may be inserted.

The final solution is linear on each triangular element and is of the form $A + Bx + Cy$, where A, B, and C are found by taking the linear combination of the basis elements. For this example, the coefficients for the linear combination are found to be

$$\mathbf{c} = [0.1154, 0.4615, 0.3654, 0.3654, 0, 0, 0, 1, 1, 1, 0, 0].$$

The solution to this problem found using the Mathcad function is illustrated in Figure 15.15

Mathcad Function for Coefficients

$$\text{coeff}(A,B,C,n,m,p,c) := \begin{array}{|l} \text{for } i \in 0..p-1 \\ \quad H_i \leftarrow 0.25 \\ \text{for } i \in 0..n-1 \\ \quad \text{for } j \in 0..m-1 \\ \quad\quad \begin{array}{|l} \text{for } k \in 0..p-1 \\ \quad s_k \leftarrow B_{i,k} \cdot B_{j,k} + C_{i,k} \cdot C_{j,k} \\ G_{i,j} \leftarrow s \cdot H \end{array} \\ \text{for } i \in 0..n-1 \\ \quad \begin{array}{|l} d_i \leftarrow 0 \\ \text{for } j \in n..m-1 \\ \quad d_i \leftarrow d_i - G_{i,j} \cdot c_j \end{array} \\ \text{for } i \in 0..n-1 \\ \quad \text{for } j \in 0..n-1 \\ \quad\quad AAA_{i,j} \leftarrow G_{i,j} \\ Q \leftarrow AAA^{-1} \cdot d \\ \text{for } i \in 0..n-1 \\ \quad c_i \leftarrow Q_i \\ c \end{array}$$

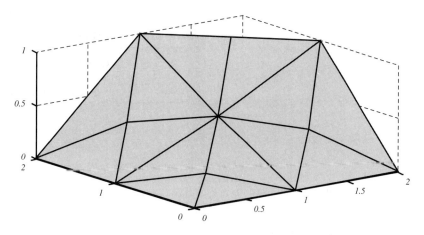

FIGURE 15.15 Solution to Poisson's equation using finite elements.

Example 15.11 Solving the Laplace Equation Using Finite Elements

Use the parameters n, m, p and the matrices A, B, C from Example 15.10:

Initialize coefficients so that the linear combination satisfies boundary conditions:

$$c\_init := \begin{pmatrix} 0 \\ 0 \\ 0 \\ 0 \\ 0 \\ 0 \\ 0 \\ 1 \\ 1 \\ 1 \\ 0 \\ 0 \end{pmatrix}$$

Call function to calculate the coefficients:

$$c := \text{coeff}(A, B, C, n, m, p, c\_init)$$

Combine coefficients with basis functions found in Example 15.10

$$UA := A^T \cdot c \qquad UB := B^T \cdot c \qquad UC := C^T \cdot c \qquad U := \text{augment}(UA, UB, UC)$$

U =

	0	1	2
0	0	0.231	0
1	0	0	0.231
2	-0.231	0.231	0.462
3	-0.231	0.462	0.231
4	-0.269	0.269	0.462
5	0	0	0.731
6	-0.538	0.269	1
7	-0.808	0.538	0.731
8	-0.615	0.538	0.538
9	-0.615	0.538	0.538
10	0	0.731	0
11	-0.269	0.462	0.269
12	-0.808	0.731	0.538
13	-0.538	1	0.269

Thus, the final solution is

$$U = \begin{cases} 0 & + 0.2308\,x + 0\,y & \text{on } T_1 \\ 0 & + 0\,x + 0.2308\,y & \text{on } T_2 \\ -0.2308 + 0.2308\,x + 0.4615\,y & \text{on } T_3 \\ -0.2308 + 0.4615\,x + 0.2308\,y & \text{on } T_4 \\ -0.2692 + 0.2692\,x + 0.4615\,y & \text{on } T_5 \\ 0 & + 0\,x + 0.7308\,y & \text{on } T_6 \\ -0.5385 + 0.2692\,x + 1.0000\,y & \text{on } T_7 \\ -0.8077 + 0.5385\,x + 0.7308\,y & \text{on } T_8 \\ -0.6154 + 0.5385\,x + 0.5385\,y & \text{on } T_9 \\ -0.6154 + 0.5385\,x + 0.5385\,y & \text{on } T_{10} \\ 0 & + 0.7308\,x + 0\,y & \text{on } T_{11} \\ -0.2692 + 0.4615\,x + 0.2692\,y & \text{on } T_{12} \\ -0.8077 + 0.7308\,x + 0.5385\,y & \text{on } T_{13} \\ -0.5385 + 1.0000\,x + 0.2692\,y & \text{on } T_{14} \end{cases}$$

The computed values at the nodes can be found from these formulas, as shown in the following table:

Node	x	y	U	Results from formula for
1	1/2	1/2	0.1154	T_1, T_2, T_3 or T_4
2	1	1	0.4615	$T_3, T_4, T_5, T_8, T_9, T_{10}, T_{12}$, or T_{13}
3	3/2	1/2	0.3654	T_5, T_6, T_7 or T_8
4	1/2	3/2	0.3654	T_{11}, T_{12}, T_{13} or T_{14}
5	0	0	0	T_1, or T_2
6	1	0	0	T_2, T_3, T_5 or T_6
7	2	0	0	T_6 or T_7
8	2	1	1	T_7, T_8 or T_9
9	3/2	3/2	1	T_9 or T_{10}
10	1	2	1	T_{10}, T_{13} or T_{14}
11	0	2	0	T_{11} or T_{14}
12	0	1	0	T_1, T_4, T_{11} or T_{12}

15.6 MATHCAD'S METHODS

Mathcad2000Pro has two built-in functions, relax and multigrid, for solving a Poisson PDE on a square region in the plane. The use of each of these functions is described in the next section.

15.6.1 Using the Built-In Functions

The function relax solves the linear system that arises from using the finite-difference approximations for the partial derivatives in the Poisson PDE on a square region in the plane. The function is called as

relax(**A, B, C, D, E, F, U**, rjac)

The matrices **A, B, C, D, E** give the coefficients of the linear system of equations. The matrix **F** contains the source term for each point inside the region where the solution is sought. The matrix **U** contains the boundary values and initial guesses for the solution inside the region.

The parameter rjac is the spectral radius (largest eigenvalue) of the Jacobi iteration matrix.

The function relax is more general than the function multigrid (discussed next) in that relax allows for arbitrary boundary conditions, and does not restrict the number of mesh subdivisions along the x- or y-axes (as long as the discretization has the same number of divisions in each direction). For simplicity in the following discussion, we take the square region to be the unit square, $0 < x < 1$, $0 < y < 1$.

The PDE to be solved is of the form

$$u_{xx} + u_{yy} = f(x,y) \quad \text{for } 0 < x < 1, 0 < y < 1$$

with boundary conditions

$$u(0,y) = g_1(y) \quad u(0,y) = g_2(y) \quad \text{for } 0 < y < 1$$
$$u(x,1) = g_3(x) \quad u(x,1) = g_4(x) \quad \text{for } 0 < x < 1$$

We discretize the region by taking the same number of subdivisions in the x and y directions; i.e., let $0 \le j \le n$ and $0 \le k \le n$ (and let the boundary conditions apply for $j = 0, j = n, k = 0$ and $k = n$.) If the approximate solution is denoted u, then at each point in the grid, we have the discretized equation

$$a_{j,k} u_{j+1,k} + b_{j,k} u_{j-1,k} + c_{j,k} u_{j,k+1} + d_{j,k} u_{j,k-1} + e_{j,k} u_{j,k} = f_{j,k}$$

For the standard finite difference approximations, $a_{jk} = b_{jk} = c_{jk} = d_{jk} = 1$, and $e_{j,k} = -4$. The value of f_{jk} depends on the right-hand side of the PDE. The information about the boundary conditions is included in matrix **U**.

The function returns a matrix **S**, whose elements give the approximate solution at each grid point of the region.

The function `multigrid`, is generally faster than `relax` for the problems for which `multigrid` is suitable. The method is restricted to problems with 0 boundary conditions, and the number of interior subdivisions in each direction, $n-1$, is a power of 2. The function is called as

`multigrid(`**M**`, ncycle)`

where **M** is an (2^k+1) by (2^k+1) matrix containing the source term at each point of the region, and ncycle is the number of cycles to be performed at each level of the multigrid iteration. In general, ncycle = 2 gives good results.

15.6.2 Understanding the Algorithms

The function `relax` uses Gauss-Seidel with SOR to solve the system of linear equations which give the approximate solution to the Poisson PDE on a discrete mesh of points in a a square planar region.

SUMMARY

Heat equation: Parabolic PDE: A finite-difference solution of the one-dimensional heat equation

$$u_t = c\, u_{xx} \quad \text{for } 0 < x < a, \quad 0 < t \le T,$$

with initial conditions: $\quad u(x, 0) = f(x), \quad 0 < x < a,$
and boundary conditions: $\quad u(0, t) = g_1(t), u(a, t) = g_2(t), 0 < t \le T,$

utilizes the mesh $\quad h = \Delta x = a/n, \quad k = \Delta t = T/m, \quad \text{with } r = \dfrac{c\,k}{h^2}.$

Explicit Method

$$u_{i,j+1} = r\, u_{i-1,j} + (1 - 2r)\, u_{i,j} + r\, u_{i+1,j}. \quad \text{for } i = 1, \ldots, n-1;$$

$$(0 < r \le 0.5 \text{ to ensure stability})$$

Implicit Method

$$u_{i,j} = (-r)\, u_{i-1,j+1} + (1 + 2r)\, u_{i,j+1} + (-r)\, u_{i+1,j+1}; \quad \text{(unconditionally stable)}$$

Crank-Nicolson Method

$$-\frac{r}{2} u_{i-1,j+1} + (1+r)\, u_{i,j+1} - \frac{r}{2} u_{i+1,j+1} = \frac{r}{2} u_{i-1,j} + (1-r)\, u_{i,j} + \frac{r}{2} u_{i+1,j};$$

(unconditionally stable; better truncation error than the basic implicit method.)

Wave equation: Hyperbolic PDE: A finite-difference solution of

$$u_{tt} - c^2 u_{xx} = 0 \quad \text{for } 0 \leq x \leq a, \text{ and } 0 \leq t,$$

with initial conditions: $u(x, 0) = f_1(x)$, $u_t(x, 0) = f_2(x)$, for $0 \leq x \leq a$,

boundary conditions: $u(0, t) = g_1(t)$, $u(a, t) = g_2(t)$, for $0 < t$,

utilizes the mesh $h = \Delta x = a/n$, $k = \Delta t$, with $p = \dfrac{ck}{h} = c\dfrac{\Delta t}{\Delta x}$.

Explicit Method. The general form of the difference equation is

$$u_{i,j+1} = p^2 u_{i-1,j} + 2(1 - p^2) u_{i,j} + p^2 u_{i+1,j} - u_{i,j-1}.$$

The equation for u at the first time step

$$u_{i,1} = 0.5 p^2 u_{i-1,0} + (1 - p^2) u_{i,0} + 0.5 p^2 u_{i+1,0} + k f_2(x_i);$$

(stability requires $p \leq 1$)

Implicit Method (Unrestricted Stability). The general form of the difference equation is

$$-p^2 u_{i-1,j+1} + 2(1 + p^2)u_{i,j+1} - p^2 u_{i+1,j+1} = 4u_{i,j} + p^2 u_{i-1,j-1} - 2(1 + p^2)u_{i,j-1} + p^2 u_{i+1,j-1}$$

The equations for $j = 0$ have the form:

$$4(1 + p^2)u_{1,1} - 2p^2 u_{2,1} = 4u_{1,0} - 2p^2 k f_2(x_0) + 4(1 + p^2)k f_2(x_1) - 2p^2 k f_2(x_2) + 2p^2 u_{0,1}$$

$$-2p^2 u_{1,1} + 4(1 + p^2)u_{2,1} - 2p^2 u_{3,1} = 4u_{2,0} - 2p^2 k f_2(x_1) + 4(1 + p^2)k f_2(x_2) - 2p^2 k f_2(x_3)$$

. . . . . .

$$-2p^2 u_{i-1,1} + 4(1 + p^2)u_{i,1} - 2p^2 u_{i+1,1} =$$
$$4u_{i,0} - 2p^2 k f_2(x_{i-1}) + 4(1 + p^2)k f_2(x_i) - 2p^2 k f_2(x_{i+1})$$

. . . .

$$-2p^2 u_{n-1,1} + 4(1 + p^2)u_{n,1} =$$
$$4u_{n,0} - 2p^2 k f_2(x_{n-1}) + 4(1 + p^2)k f_2(x_n) - 2p^2 k f_2(x_{n+1}) + 2p^2 u_{n+1,1}$$

The value of u at each subsequent time step can be found from:

$$2(1 + p^2)u_{1,j+1} - p^2 u_{2,j+1} = 4u_{1,j} - 2(1 + p^2)u_{1,j-1} + p^2 u_{2,j-1} + p^2 u_{0,j+1} + p^2 u_{0,j-1}$$

$$-p^2 u_{1,j+1} + 2(1 + p^2)u_{2,j+1} - p^2 u_{3,j+1} = 4u_{2,j} + p^2 u_{1,j-1} - 2(1 + p^2)u_{2,j-1} + p^2 u_{3,j-1}$$

. . .

$$-p^2 u_{i-1,j+1} + 2(1 + p^2)u_{i,j+1} - p^2 u_{i+1,j+1} = 4u_{i,j} + p^2 u_{i-1,j-1} - 2(1 + p^2)u_{i,j-1} + p^2 u_{i+1,j-1}$$

. . .

$$-p^2 u_{n-1,j+1} + 2(1 + p^2)u_{n,j+1} = 4u_{n,j} + p^2 u_{n-1,j-1} - 2(1 + p^2)u_{n,j-1} + p^2 u_{n+1,j-1} + p^2 u_{n+1,j+1}$$

Elliptic PDE

A Finite-Difference Solution of Poisson's Equation

$$u_{xx} + u_{yy} = f(x, y) \quad a \leq x \leq b; \quad c \leq y \leq d.$$

with boundary conditions

$$u(a, y) = h_1(y) \quad u(b, y) = h_2(y) \quad c < y < d$$
$$u(x, c) = k_1(x) \quad u(x, d) = k_2(x) \quad a < x < b$$

utilizes a mesh in the x-y plane:

$$x_i = a + ih, \quad i = 0, 1, \cdots n \quad h = \Delta x = (b - a)/n$$
$$y_j = c + jk, \quad j = 0, 1, \cdots m \quad k = \Delta y = (d - c)/m$$

the general form of the difference equation is

$$\frac{1}{k^2}[u_{i,j+1} - 2u_{i,j} + u_{i,j-1}] + \frac{1}{h^2}[u_{i+1,j} - 2u_{i,j} + u_{i-1,j}] = f_{ij}.$$

A Finite-Element Solution of an Elliptic PDE of the Form

$$u_{xx} + u_{yy} + r(x, y)u = f(x, y) \quad \text{on the region } R$$
$$u(x, y) = g(x, y) \quad \text{on boundary of } R$$

seeks to minimize the functional

$$I[u] = \iint_R [u_x^2 + u_y^2 - r(x, y)u^2 + 2f(x, y)u] \, dx \, dy$$

where u is a linear combination of the basis functions. If R is divided into p triangular subregions, $T_1, T_2, \ldots T_p$, there is a piecewise-linear basis function corresponding to each node. The basis function corresponding to node j has the value one at node j and zero at all other nodes.

SUGGESTIONS FOR FURTHER READING

Our discussion of numerical methods for partial differential equations in this chapter provides only a brief introduction to an extensive area of research and application. The following are some of the many excellent sources for further study of these topics.

Ames, W. F., *Numerical Methods for Partial Differential Equations*, (3rd ed.), Academic Press, 1992.

Birkhoff, G., and R. E. Lynch, *Numerical Solution of Elliptic Problems*, SIAM, Philadelphia, 1984.

Boyce, W. E., and R. C. DiPrima, *Elementary Differential Equations and Boundary Value Problems*, (4th ed.) John Wiley & Sons, New York, 1986.

Celia, M. A., and W. G. Gray, *Numerical Methods for Differential Equations*, Prentice Hall, Englewood Cliffs, NJ. 1992.

Colton, D., *Partial Differential Equations*, Random House, New York, 1988. (This text includes a brief history of PDE, pp. 49–53.)

Forsythe, G. E., and W. R. Wasow, *Finite-Difference Methods for Partial Differential Equations*, John Wiley & Sons, New York, 1960.

Garcia, A. L., *Numerical Methods for Physics*, Prentice-Hall, Englewood Cliffs, NJ, 1994.

Golub, G. H., and J. M. Ortega, *Scientific Computing and Differential Equations: An Introduction to Numerical Methods*, Academic Press, Boston, 1992.

Haberman, R., *Elementary Applied Partial Differential Equations, with Fourier Series and Boundary Value Problems*, Prentice-Hall, 1983.

Hall, C. A., and T. A. Porsching, *Numerical Analysis of Partial Differential Equations*, Prentice-Hall, Englewood Cliffs, NJ, 1990.

Meis, T., and U. Marcowitz, *Numerical Solution of Partial Differential Equations*, Springer-Verlag, New York, 1981.

Mitchell, A. R., *Computational Methods in Partial Differential Equations*, John Wiley & Sons, London, 1969.

Svobony, T., *Mathematical Modeling for Industry and Engineering*, Prentice Hall, Upper Saddle River, NJ, 1998.

Troutman, J. L. and M. Bautista, *Boundary Value Problems of Applied Mathematics*, PWS Publishing, 1994.

Zauderer, E., *Partial Differential Equations of Applied Mathematics*, (2^d ed.), John Wiley & Sons, New York, 1989.

The finite-element method is important in mathematics and engineering; the following references include discussion of the method from several different points of view.

Axelsson, O., and V. A. Barker, *Finite Element Solution of Boundary Value Problems*, Academic Press, New York, 1984.

Becker, E. B., G. F. Carey, and J. T. Oden, *Finite Elements: An Introduction*, Vol. 1, Prentice-Hall, Englewood Cliffs, NJ, 1981.

Davies. A. J., *The Finite Element Method: A First Approach*, Clarendon Press, Oxford, U.K., 1980.

Mitchell, A. R., and R. Wait, *The Finite Element Method in Partial Differential Equations*, John Wiley & Sons, London, 1977.

Silvester, P. P., and R. L. Ferrari, *Finite Elements for Electrical Engineers*, (3rd ed.) Cambridge University Press, Cambridge, 1996.

Strang, G. and G. Fix, *An Analysis of the Finite Element Method*, Prentice-Hall, Englewood Cliffs, NJ, 1973.

Zienkiewicz, O. C., and Taylor, R. L., *The Finite Element Method*, (4th ed.) Vol 1, London:McGraw-Hill, 1989.

PRACTICE THE TECHNIQUES

For Problems P15.1 to P15.10, solve the one-dimensional heat equation using the specified finite-difference method.

    a. Use the explicit finite-difference method with $\Delta x = 0.2$ and $\Delta t = 0.01$.
    b. Use the explicit finite-difference method with $\Delta x = 0.2$ and $\Delta t = 0.02$.
    c. Use the explicit finite-difference method with $\Delta x = 0.2$ and $\Delta t = 0.1$.
    d. Use the implicit finite-difference method with $\Delta x = 0.2$ and $\Delta t = 0.1$.
    e. Use the Crank-Nicolson method with $\Delta x = 0.2$ and $\Delta t = 0.1$.

P15.1 Solve $u_t - u_{xx} = 0$, for $0 \leq x \leq 1$, $0 < t < 0.5$, with initial condition $u(x, 0) = 0$, (for $0 \leq x \leq 1$), and boundary conditions $u(0, t) = 0$, $u(1, t) = t$ (for $0 < t < 0.5$).

P15.2 Solve $u_t - u_{xx} = 0$, for $0 \leq x \leq 1$, $0 < t < 0.5$, with initial condition $u(x, 0) = x$, (for $0 \leq x \leq 1$), and boundary conditions $u(0, t) = 0$, $u(1, t) = 1$ (for $0 < t < 0.5$).

P15.3 Solve $u_t - u_{xx} = 0$, for $0 \leq x \leq 1$, $0 < t < 0.5$, with initial condition $u(x, 0) = x(1 - x)$, (for $0 \leq x \leq 1$), and boundary conditions $u(0, t) = 0$, $u(1, t) = 0$ (for $0 < t < 0.5$).

P15.4 Solve $u_t - u_{xx} = 0$, for $0 \leq x \leq 1$, $0 < t < 0.5$, with initial condition $u(x, 0) = 10$, (for $0 \leq x \leq 1$), and boundary conditions $u(0, t) = 0$, $u(1, t) = 0$ (for $0 < t < 0.5$).

P15.5 Solve $u_t - u_{xx} = 0$, for $0 \leq x \leq 1$, $0 < t < 0.5$, with initial condition $u(x, 0) = \sin(4\pi x)$, (for $0 \leq x \leq 1$), and boundary conditions $u(0, t) = 0$, $u(1, t) = 0$ (for $0 < t < 0.5$).

P15.6 Solve $u_t - u_{xx} = 0$, for $0 \leq x \leq 1$, $0 < t < 0.5$, with initial condition $u(x, 0) = \sin(\pi x) + \sin(2\pi x) + \sin(3\pi x)$, ($0 \leq x \leq 1$), and boundary conditions $u(0, t) = 0$, $u(1, t) = 0$ ($0 < t < 0.5$).

P15.7 Solve $u_t - u_{xx} = 0$, for $0 \leq x \leq 1$, $0 < t < 0.5$, with initial condition $u(x, 0) = \sin(\pi x)$, (for $0 \leq x \leq 1$), and boundary conditions $u(0, t) = 0$, $u(1, t) = 1$ (for $0 < t < 0.5$).

P15.8 Solve $u_t - u_{xx} = 0$, for $0 \leq x \leq 1$, $0 < t < 0.5$, with initial condition $u(x, 0) = x$, ($0 \leq x \leq 1$), and boundary conditions $u_x(0, t) = 0$, $u_x(1, t) = 0$ ($0 < t < 0.5$), i.e., ends insulated.

P15.9 Solve $u_t - u_{xx} = 0$, for $0 \leq x \leq 1$, $0 < t < 0.5$, with initial condition

$$u(x, 0) = \begin{cases} 0, & \text{for } 0 \leq x \leq 0.2, \\ 20, & \text{for } 0.2 < x < 0.8, \\ 0, & \text{for } 0.8 \leq x \leq 1, \end{cases}$$

and boundary conditions $u(0, t) = 0$, and $u(1, t) = 0$ ($0 < t < 0.5$).

P15.10 Solve $u_t - u_{xx} = 0$, for $0 \leq x \leq 1$, $0 < t < 0.5$, with initial condition $u(x, 0) = x^2$, (for $0 \leq x \leq 1$), and boundary conditions $u(0, t) = 0$, $u(1, t) = 0$ (for $0 < t < 0.5$).

For Problems P15.11 to P15.15, solve the wave equation using the following finite-difference schemes.

    a. Use the explicit finite-difference method with $\Delta x = 0.1$ and $\Delta t = 0.1$.
    b. Use the explicit finite-difference method with $\Delta x = 0.1$ and $\Delta t = 0.2$.
    c. Use the implicit finite-difference method with $\Delta x = 0.1$ and $\Delta t = 0.2$.

P15.11 Solve $u_{tt} - u_{xx} = 0$, for $0 \leq x \leq 1$, $0 < t < 2$, with initial conditions (string plucked in the middle, zero initial velocity)

$$u(x, 0) = 0.2\,x, \qquad \text{for } 0 \leq x \leq 5.0,$$
$$u(x, 0) = 0.2\,(1 - x), \quad \text{for } 0.5 < x < 1,$$
$$u_x(x, 0) = 0,$$

and boundary conditions $u(0, t) = 0$, and $u(1, t) = 0$ ($0 < t < 2$).

P15.12 Solve $u_{tt} - u_{xx} = 0$, for $0 \leq x \leq 1$, $0 < t < 2$, with initial conditions (string plucked at $x = 0.2$, zero initial velocity)

$$u(x, 0) = \begin{cases} 0.5\,x, & \text{for } 0 \leq x \leq 0.2, \\ 0.1 - (x - 0.2)/8, & \text{for } 0.2 < x < 1, \end{cases}$$

$u_x(x, 0) = 0$, and boundary conditions $u(0, t) = 0$, and $u(1, t) = 0$ ($0 < t < 2$).

P15.13 Solve $u_{tt} - u_{xx} = 0$, for $0 \leq x \leq 1$, $0 < t < 2$, with initial conditions $u(x, 0) = 0$ and $u_x(x, 0) = 0$, and boundary conditions $u(0, t) = 0$, $u(1, t) = 0.2 \sin(t)$ (for $0 < t < 2$).

P15.14 Solve $u_{tt} - u_{xx} = 0$, for $0 \le x \le 1$, $0 < t < 2$, with initial conditions $u(x, 0) = 0$ and $u_x(x, 0) = 0$, and boundary conditions $u(0, t) = 0$, $u(1, t) = 0.2 \sin(\pi t)$ (for $0 < t < 2$).

P15.15 Solve $u_{tt} - u_{xx} = 0$, for $0 \le x \le 1$, $0 < t < 2$, with initial conditions $u(x, 0) = 0$ and $u_x(x, 0) = \sin(\pi x)$, and boundary conditions $u(0, t) = 0$, $u(1, t) = 0$ (for $0 < t < 2$).

For Problems P15.16 to P15.25, solve the potential equation using the finite-difference method with the specified mesh.

    a. Use $\Delta x = 0.25$ and $\Delta y = 0.25$.
    b. Use $\Delta x = 0.2$ and $\Delta y = 0.2$.
    c. Use $\Delta x = 0.1$ and $\Delta y = 0.1$.

P15.16 Solve $u_{xx} + u_{yy} = 0$, for $0 \le x \le 1$, $0 \le y \le 1$, with boundary conditions $u(x, 0) = x$, $u(0, y) = y$, $u(x, 1) = 1$, and $u(1, y) = 1$.

P15.17 Solve $u_{xx} + u_{yy} = 0$, for $0 \le x \le 1$, $0 \le y \le 1$, with boundary conditions $u(x, 0) = 0$, $u(0, y) = 0$, $u(x, 1) = x$, and $u(1, y) = y$.

P15.18 Solve $u_{xx} + u_{yy} = 0$, for $0 \le x \le 1$, $0 \le y \le 1$, with boundary conditions $u(x, 0) = x$, $u(0, y) = 2y$, $u(x, 1) = 2$, and $u(1, y) = 1+y$.

P15.19 Solve $u_{xx} + u_{yy} = 0$, for $0 \le x \le 1$, $0 \le y \le 1$, with boundary conditions $u(x, 0) = x^2$, $u(0, y) = -y^2$, $u(x, 1) = x^2 + x - 1$, and $u(1, y) = 1 + y - y^2$.

P15.20 Solve $u_{xx} + u_{yy} = 0$, for $0 \le x \le 1$, $0 \le y \le 1$, with boundary conditions $u(x, 0) = x^2 + 1$, $u(0, y) = -y^2 + 1$, $u(x, 1) = x^2$, and $u(1, y) = 2 - y^2$.

P15.21 Solve $u_{xx} + u_{yy} = 0$, for $0 \le x \le 1$, $0 \le y \le 1$, with boundary conditions $u(x, 0) = x^2 - 1$, $u(0, y) = -y^2 - 1$, $u(x, 1) = x^2 + x - 2$, and $u(1, y) = y - y^2$.

P15.22 Solve $u_{xx} + u_{yy} = 0$, for $0 \le x \le 1$, $0 \le y \le 1$, with boundary conditions $u(x, 0) = x^3$, $u(0, y) = y^3$, $u(x, 1) = x^3 - 3x^2 - 2x + 1$, and $u(1, y) = y^3 - 3y^2 - 2y + 1$.

P15.23 Solve $u_{xx} + u_{yy} = 0$, for $0 \le x \le 1$, $0 \le y \le 1$, with boundary conditions $u(x, 0) = x^3$, $u(0, y) = y^3$, $u(x, 1) = x^3 - 3x^2 + x + 1$, and $u(1, y) = y^3 - 3y^2 + y + 1$.

P15.24 Solve $u_{xx} + u_{yy} = 0$ for $0 \le x \le 1$, $0 \le y \le 1$, with boundary conditions $u(x, 0) = x^3$, $u(0, y) = y$, $u(x, 1) = x^3 - 3x + 1$, and $u(1, y) = y + 1 - 3y^2$.

P15.25 Solve $u_{xx} + u_{yy} = 0$ for $0 \le x \le 1$, $0 \le y \le 1$, with boundary conditions $u(x, 0) = 0$, $u(0, y) = y^3 + y$, $u(x, 1) = 2 + x - 3x^2$, and $u(1, y) = y^3 - y$.

For Problems P15.26 to P15.35, solve the potential equation (given in P15.16 to P15.25, respectively) using the finite-element method with a mesh of 16 triangles. This mesh is similar to (an extension and rescaling of) that shown in Fig. 15.12; the triangles are defined in the matrices V and T, (compare to those given in Example 15.10.) Note that V_9 is now an interior node, the new node, V_{13} is a boundary node.

$$V = \begin{bmatrix} 1/4 & 1/4 \\ 1/2 & 1/2 \\ 3/4 & 1/4 \\ 1/4 & 3/4 \\ 0 & 0 \\ 1/2 & 0 \\ 1 & 0 \\ 1 & 1/2 \\ 3/4 & 3/4 \\ 1/2 & 1 \\ 0 & 1 \\ 0 & 1/2 \\ 1 & 1 \end{bmatrix}$$

and

$$T = \begin{bmatrix} 1 & 5 & 12 \\ 1 & 5 & 6 \\ 1 & 2 & 6 \\ 1 & 2 & 12 \\ 3 & 2 & 6 \\ 3 & 6 & 7 \\ 3 & 7 & 8 \\ 3 & 8 & 2 \\ 2 & 8 & 9 \\ 2 & 9 & 10 \\ 4 & 11 & 12 \\ 4 & 12 & 2 \\ 4 & 2 & 10 \\ 4 & 10 & 11 \\ 8 & 9 & 13 \\ 9 & 10 & 13 \end{bmatrix}$$

For Problems P15.36 to P15.40, solve the Poisson equation.

a. Use the finite-difference method with $\Delta x = 0.25$ and $\Delta y = 0.25$.
b. Use the finite-difference method with $\Delta x = 0.2$ and $\Delta y = 0.2$.
c. Use the finite-difference method with $\Delta x = 0.1$ and $\Delta y = 0.1$.
d. Use the finite-element method with a mesh of 16 triangles, as described for P15.26-P15.35.

P15.36 Solve $u_{xx} + u_{yy} = 6x$ for $0 \leq x \leq 1$, $0 \leq y \leq 1$, with boundary conditions $u(x, 0) = x^3$, $u(0, y) = y$, $u(x, 1) = x^3 - 3x + 1$, and $u(1, y) = -2y + 1$.

P15.37 Solve $u_{xx} + u_{yy} = 2x + 2 - 2y$ for $0 \leq x \leq 1$, $0 \leq y \leq 1$, with boundary conditions $u(x, 0) = x^2$, $u(0, y) = y$, $u(x, 1) = x^2 + 1$, and $u(1, y) = 1 + y^2$.

P15.38 Solve $u_{xx} + u_{yy} = 2x + 2 - 2y$ for $0 \leq x \leq 1$, $0 \leq y \leq 1$, with boundary conditions $u(x, 0) = x^2$, $u(0, y) = y$, $u(x, 1) = x + 1$, and $u(1, y) = 1 + y^2$.

P15.39 Solve $u_{xx} + u_{yy} = 2(1 + y)$ for $0 \leq x \leq 1$, $0 \leq y \leq 1$, with boundary conditions $u(x, 0) = x^3$, $u(0, y) = y^2$, $u(x, 1) = x^3 + x^2 + 1$, and $u(1, y) = -2y^2 + 4y + 1$.

P15.40 Solve $u_{xx} + u_{yy} = \sin(\pi x)$, for $0 \leq x \leq 1$, $0 \leq y \leq 1$, with boundary conditions $u(x, 0) = x$, $u(0, y) = y$, $u(x, 1) = 1$, and $u(1, y) = 1$.

EXPLORE SOME APPLICATIONS

A15.1 The telegrapher's equation, $v_{tt} - c^2 v_{xx} + 2 a v_t = 0$, which governs propagation of signals on telegraph lines, is an example of a wave equation with damping. Investigate the numerical solution of this equation for a variety of initial and boundary conditions, and parameter values for a and c. (See Zauderer, 1989, or Svobodny, 1998, for discussion.)

A15.2 The standard inviscid Burger's equation,

$$\frac{\partial}{\partial t}\rho = -\frac{\partial}{\partial x}\left(\left(\frac{1}{2}\rho\right)\rho\right) = -\rho\frac{\partial}{\partial x}\rho$$

is a simple nonlinear PDE with wave solutions which describes the evolution of the density of an inviscid fluid. The generalized inviscid Burger's equation is

$$\frac{\partial}{\partial t}\rho = -\frac{\partial}{\partial x}\left(\left(a + \frac{1}{2}b\rho\right)\rho\right).$$

One application of this equation is in the modeling of traffic flow. The density of the traffic ρ depends on both density and velocity according to the equation

$$\frac{\partial}{\partial t}\rho = -\frac{\partial}{\partial x}F(\rho).$$

where $F(\rho) = \rho v(\rho) = \rho v_m(1 - \rho/\rho_m)$; that is, velocity is a linear function of density, with maximum velocity denoted v_m and maximum density denoted ρ_m. Investigate the solution of this PDE using different choices of finite-difference approximations for the partial derivatives. Take the initial density to be ρ_m for $-100 < x < 0$, and 0 for $0 < x < 500$; find the solution for $-100 < x < 500$. (See Garcia, 1994 for further discussion.)

A15.3 In the heat equation $u_t = c u_{xx}$, the parameter c is the thermal diffusivity of the material. Some typical values (cm^2/sec) are given in the following table, (adapted from Boyce and DiPrima, 1986, p. 515.)

Material	Diffusivity
silver	1.71
copper	1.14
aluminium	0.86
cast iron	0.12

Compare the termperature profile for rods of different materials, each with initial temperature of 100C and ends held at 0C.

A15.4 Compare the termperature profile for rods of different materials (see A15.3), each with initial temperature of 100°C, with one end held at 0°C and the other end insulated.

EXTEND YOUR UNDERSTANDING

U15.1 Compare the computational effort required to solve the one-dimensional heat equation using the explicit finite-difference method, the implicit finite-difference method, or the Crank-Nicolson method. The implicit methods (both basic and Crank-Nicolson) allow a larger time step than the explicit method; how much larger does it need to be to have the computational effort be the same as for the explicit method? Investigate these questions both analytically and experimentally for some of the problems P15.1 to P15.10.

U15.2 Compare the computational effort required to solve the one-dimensional wave equation using the explicit finite-difference method and the implicit finite-difference method. The implicit method allows a larger time step than the explicit method; how much larger does it need to be to have the computational effort be the same as for the explicit method? Investigate these questions both analytically and experimentally for some of the problems P15.11 to P15.15.

U15.3 Compare the computational effort required to solve the potential equation using the finite-difference method and the finite-element method. Investigate these questions both analytically and experimentally for some of the problems P15.16 to P15.25.

U15.4 The finite-element method is well suited to domains that are more general than the simple rectangular regions illustrated in the text. Modify the domain for some of the problems given in the practice the techniques section (P15.16 to P15.25) and solve the resulting problem using finite elements.

U15.5 The finite-element method is well suited to problems in which boundary conditions are more general than those illustrated in the text. Modify a portion of the boundary condition for some of the problems given in the practice the techniques section (P15.16 to P15.25) to reflect a no-flux or insulated condition, i.e., take the outward normal to the boundary to be zero; then solve the resulting problem using finite elements.

U15.6 Consider the Poisson equation $u_{xx} + u_{yy} = f(x, y)$, $0 \le x \le 1; 0 \le y \le 1$, with boundary conditions

$$u(x, 0) = x \quad u(x, 1) = 1 \quad 0 < x < 1$$
$$u(0, y) = y \quad u(1, y) = 1 \quad 0 < y < 1$$

Take

$$f(x,x) = \begin{cases} 1 & \text{for } (x, y) = (1/3, 2/3) \\ 0 & \text{otherwise} \end{cases}$$

Choose an appropriate mesh, and solve using the finite-element method.

U15.7 Consider the Poisson equation $u_{xx} + u_{yy} = f(x, y)$, $0 \le x \le 1; 0 \le y \le 1$, with boundary conditions

$$u(x, 0) = x \quad u(x, 1) = 1 \quad 0 < x < 1$$
$$u(0, y) = y \quad u(1, y) = 1 \quad 0 < y < 1$$

Take

$$f(x, y) = \begin{cases} 1 & \text{for } (x, y) = (1/4, 1/2) \\ 0 & \text{otherwise} \end{cases}$$

Choose an appropriate mesh, and solve using the finite-difference method.

U15.8 Consider the Poisson equation $u_{xx} + u_{yy} = f(x, y)$; $0 \le x \le 1; 0 \le y \le 1$. with boundary conditions

$$u(x, 0) = x \quad u(x, 1) = 1 \quad 0 < x < 1$$
$$u(0, y) = y \quad u(1, y) = 1 \quad 0 < y < 1$$

Take

$$f(x, y) = \begin{cases} 1 & \text{for } (x, y) = (1/3, 2/3) \\ 1 & \text{for } (x, y) = (1/4, 1/2) \\ 0 & \text{otherwise} \end{cases}$$

Choose an appropriate mesh, and solve using the finite-element method or the finite-difference method.

Bibliography

Abramowitz, M., and I. A. Stegun (eds.), *Handbook of Mathematical Functions, with Formulas, Graphs, and Mathematical Tables*, Dover, New York, 1965.

Achieser, N. I., *Theory of Approximation*, Dover, New York, 1993.

Acton, F. S., *Numerical Methods That (usually) Work*, Harper and Row, New York, 1970; corrected edition, Mathematical Association of America, Washington, DC, 1990.

Ames, W. F., *Numerical Methods for Partial Differential Equations*, 3rd ed., Academic Press, Boston, 1992.

Ascher, U. M., R. M. M. Mattheij, and R. D. Russell, *Numerical Solution of Boundary Value Problems for Ordinary Differential Equations*, SIAM, Philadelphia, 1995. (Originally published by Prentice Hall, Englewood Cliffs, NJ, 1988.)

Atkinson, K. E., *An Introduction to Numerical Analysis*, 2nd ed., John Wiley, New York, 1989.

Axelsson, O., and V. A. Barker, *Finite Element Solution of Boundary Value Problems*, Academic Press, New York, 1984.

Ayyub, B. M. and R. H. McCuen, *Numerical Methods for Engineers*, Prentice Hall, Upper Saddle River, NJ, 1996.

Bader, G. and P. Deuflhard, "A semi-implicit midpoint rule for stiff systems of ordinary systems of differential equations," *Numerische Mathematik,* vol. 41, pp. 373–398, 1983.

Barrett, R., J. Donato, J. Dongarra, V. Eijkhout, R. Pozo, C. Romine, and H. van der Vorst, *Templates for the Solution of Linear Systems: Building Blocks for Iterative Methods*, SIAM, Philadelphia, 1993.

Bartels, R. H., J. C. Beatty, and B. A. Barsky, *An Introduction to Splines for Use in Computer Graphics and Geometric Modeling*, Morgan Kaufmann, Los Altos, CA, 1987.

Becker, E. B., G. F. Carey, and J. T. Oden, *Finite Elements: An Introduction*, vol. 1, Prentice Hall, Englewood Cliffs, NJ, 1981.

Birkhoff, G., and R. E. Lynch, *Numerical Solution of Elliptic Problems*, SIAM, Philadelphia, 1984.

Bloomfield, P., *Fourier Analysis of Time Series—An Introduction*, Wiley, New York, 1976.

Boyce, W. E. and R. C. DiPrima, *Elementary Differential Equations and Boundary Value Problems*, 4th ed. John Wiley & Sons, New York, 1986.

Boyer, C. B., *The History of the Calculus and Its Conceptual Development*, Dover, New York, 1949.

Bracewell, R. *The Fourier Transform and Its Applications*, Mc-Graw Hill, 1986.

Brent, R., *Algorithms for Minimization Without Derivatives*, Prentice Hall, Englewood Cliffs, NJ, 1973.

Brezinski, C., *History of Continued Fractions and Padé Approximants,* Springer-Verlag, Berlin, 1991.

Briggs, W. L. and V. E. Henson, *The DFT: An Owner's Manual for the Discrete Fourier Transform*, SIAM, Philadelphia, 1995.

Bringham, E. O., *The Fast Fourier Transform*, Prentice Hall, Englewood Cliffs, 1974.

Burden, R. L. and J. D. Faires, *Numerical Analysis*, 6th ed., Prindle, Weber & Schmidt, Boston, 1996.

Cash, J. R. and A. H. Karp, *ACM Transactions on Mathematical Software*, vol. 16, pp. 201–222, 1990.

Celia, M. A. and W. G. Gray, *Numerical Methods for Differential Equations*, Prentice Hall, Englewood Cliffs, NJ, 1992.

Cheney, E. W., *Introduction to Approximation Theory*, McGraw-Hill, New York, 1966.

Clenshaw, C. W. and A. R. Curtis, *Numerische Mathematik*, vol. 2, pp. 197–205, 1960.

Coleman, T. F. and C. Van Loan, *Handbook for Matrix Computations*, SIAM, Philadelphia, 1988.

Colton, D., *Partial Differential Equations*, Random House, New York, 1988.

Conte, S. D. and C. de Boor, *Elementary Numerical Analysis*, 2nd ed. McGraw-Hill, New York, 1972.

Dahlquist, G. "A Special Stability Problem for Linear Multistep Methods," *BIT*, vol. 3, 1963, pp. 27–43.

Dahlquist, G. and A. Bjorck, *Numerical Methods*, (translated by Ned Anderson), Prentice Hall, Englewood Cliffs, NJ, 1974.

Danzig, G. B., *Linear Programming and Extensions*, Princeton University Press, Princeton, NJ, 1963

Datta, B. N., *Numerical Linear Algebra and Applications*, Brooks Cole, Pacific Grove, CA, 1995.

Davies, A. J., *The Finite Element Method: A First Approach*, Clarendon Press, Oxford, U.K., 1980.

Davis, P. J., *Interpolation and Approximation*, Dover, New York, 1975. (Originally published by Blaisdell Publishing in 1963.)

deBoor, C., *A Practical Guide to Splines*, Springer-Verlag, New York, 1978.

Dennis, J. E. and R. B. Schnabel, *Numerical Methods for Unconstrained Optimization and Nonlinear Equations*, Prentice Hall, Englewood Cliffs, NJ, 1983.

Dennis, J. E. Jr. and D. J. Woods, *New Computing Environments: Microcomputers in Large-Scale Computing*, edited by A. Wouk, SIAM, Philadelphia, 1987, pp. 116-122.

Deuflhard, P., *Numerische Mathematik*, vol. 41, pp. 399–422, 1983.

Deuflhard, P., *SIAM Review*, vol. 27, pp. 505-535, 1985.

Dillon, W. R. and M. Goldstein, *Multivariate Analysis: methods and applications*, John Wiley and Sons, New York, 1984.

Dongarra, J. J., et al., *LINPACK User's Guide*, SIAM, Philadelphia, 1979.

Edwards, C. H. Jr. and D. E. Penney, *Calculus and Analytic Geometry*, 5th ed., Prentice Hall, Upper Saddle River, NJ, 1998.

Edwards, C. H. Jr. and D. E. Penney, *Differential Equations and Boundary Value Problems: Computing and Modeling*, Prentice Hall, Englewood Cliffs, NJ, 1996.

Edwards, C. H. Jr. and D. E. Penney, *Elementary Differential Equations with Boundary Value Problems* 3rd ed., Prentice Hall, Englewood Cliffs, NJ, 1993.

Elliott, D. F. and K. R. Rao, *Fast Transforms: Algorithms, Analyses, Applications*, Academic Press, New York, 1982.

Eves, H., *Great Moments in Mathematics* (v. 1, before 1650; v. 2, after 1650), the Mathematical Association of America, Washington, DC, 1983.

Faires, J. D. and R. Burden, *Numerical Methods*, 2nd ed., Brooks/Cole, Pacific Grove, CA, 1998.

Farin, G., *Curves and Surfaces for Computer Aided Geometric Design: A Practical Guide*, 2nd ed., Academic Press, Boston, 1990.

Finizio, N. and G. Ladas, *An Introduction to Differential Equations, with Difference Equations, Fourier Series, and Partial Differential Equations*, Wadsworth Publishing, 1982.

Forsythe, G. E. and W. R. Wasow, *Finite-Difference Methods for Partial Differential Equations*, John Wiley & Sons, New York, 1960.

Forsythe, G. E., M. A. Malcolm, and C. B. Moler, *Computer Methods for Mathematical Computations*, Prentice Hall, Englewood Cliffs, NJ, 1977.

Fox, L., *Numerical Solution of Two-Point Boundary Value Problems in Ordinary Differential Equations*, Dover, New York, 1990. (Originally published by Clarendon Press, Oxford, 1957.)

Fox, L., *An Introduction to Numerical Linear Algebra*, Oxford University Press, New York, 1965.

Fraleigh, J. B. and R. A. Beauregard, *Linear Algebra*, Addison-Wesley, Reading, MA, 1987.

Freund, R. W., G. H. Golub, and N. M. Nachtigal, "Iterative Solution of Linear Systems," *Acta Numerica I*, Cambridge University Press, Cambridge, U.K., pp. 57–100, 1992.

Froberg, C. E., *Numerical Mathematics: Theory and Computer Applications*, Benjamin/Cummings, Menlo Park, CA, 1985.

Garcia, A. L., *Numerical Methods for Physics*, Prentice Hall, Englewood Cliffs, NJ, 1994.

Gear, C. W., *Numerical Initial Value Problems in Ordinary Differential Equations*, Prentice Hall, Englewood Cliffs, NJ, 1971.

Gill, P. E., W. Murray, and M. H. Wright, *Numerical Linear Algebra and Optimization*, Addison-Wesley, Redwood City, CA, 1991.

Golub, G. H. and J. M. Ortega, *Scientific Computing and Differential Equations: An Introduction to Numerical Methods*, Academic Press, Boston, 1992.

Golub, G. H. and C. F. Van Loan, *Matrix Computations*, 3rd ed., Johns Hopkins University Press, Baltimore, 1996.

Gragg, W., "On extrapolation algorithms for ordinary initial value problems," *J. SIAM Numer. Anal. Ser. B*, Vol 2, pp. 384–403, 1965.

Greenbaum, A., *Iterative Methods for Solving Linear Systems*, SIAM, Philadelphia, 1997.

Greenberg, M. D., *Advanced Engineering Mathematics*, 2nd ed., Prentice Hall, Upper Saddle River, NJ, 1998.

Greenberg, M. D., *Foundations of Applied Mathematics*, Prentice Hall, Englewood Cliffs, NJ, 1978.

Greenspan, D., *Discrete Numerical Methods in Physics and Engineering*, Academic Press, New York, 1974.

Grossman, S. I., and W. R. Derrick, *Advanced Engineering Mathematics*, Harper & Row, New York, 1988.

Haberman, R., *Elementary Applied Partial Differential Equations, with Fourier Series and Boundary Value Problems*, Prentice Hall, Englewood Cliffs, NJ, 1983.

Hager, W. W., *Applied Numerical Linear Algebra*, William W. Hager, Dept. of Mathematics, Univ. of Florida, Gainesville, FL. (Originally published by Prentice Hall, Englewood Cliffs, NJ, 1988.)

Hair, J. F., R. E. Anderson, R. L. Tatham, and W. Black, *Multivariate Data Analysis*, 5th ed., Prentice Hall, Englewood Cliffs, NJ, 1998.

Hall, C. A. and T. A. Porsching, *Numerical Analysis of Partial Differential Equations*, Prentice Hall, Englewood Cliffs, NJ, 1990.

Hamming, R. W., *Numerical Methods for Scientists and Engineers*, 2nd ed., McGraw-Hill, New York, 1973.

Hanna, O. T. and O. C. Sandall, *Computational Methods In Chemical Engineering*, Prentice Hall, Upper Saddle River, NJ, 1995.

Hibbeler, R. C., *Engineering Mechanics:Dynamics,* 7th ed., Prentice Hall, Englewood Cliffs, NJ, 1995.

Hibbeler, R. C., *Engineering Mechanics:Statics,* 7th ed., Prentice Hall, Englewood Cliffs, NJ, 1995.

Hildebrand, F. B., *Introduction to Numerical Analysis*, Dover, New York, 1987.

Hildebrand, F. B., *Advanced Calculus for Applications,* 2nd ed., Prentice Hall, Englewood Cliffs, NJ, 1976.

Himmelblau, D. M., *Basic Principles and Calculations in Engineering*, 3rd ed., Prentice Hall, Englewood Cliffs, NJ, 1974.

Hornbeck, R. W., *Numerical Methods*, Prentice Hall, Englewood Cliffs, NJ, 1975.

Inman, D. J., *Engineering Vibration*, Prentice Hall, Englewood Cliffs, NJ, 1996.

Isaacson, E. and H. B. Keller, *Analysis of Numerical Methods*, Dover, New York, 1994. (Originally published by John Wiley & Sons, 1966).

Jacobs, D. A. H. (ed.), *The State of the Art in Numerical Analysis*, Academic Press, London, 1977, Chapter 3.1.7 (by K. W. Brodie).

Jaeger, J. C., *An Introduction to Applied Mathematics*, Clarenden Press, Oxford, 1951.

Jain, M. K., *Numerical Solution of Differential Equations*, John Wiley, New York, 1979.

Jensen, J. A. and J. H. Rowland, *Methods of Computation*, Scott, Foresman and Company, Glenview, IL, 1975.

Johnson, R. A. and D. W. Wichern, *Applied Multivariate Statistical Analysis*, 4th ed., Prentice Hall, Englewood Cliffs, NJ, 1998.

Kachigan, S. K., *Multivariate Statistical Analysis: A Conceptual Introduction*, 2nd ed., Radius Press, New York, 1991.

Kahaner, D., C. Moler, and S. Nash, *Numerical Methods and Software*, Prentice Hall, Englewood Cliffs, NJ, 1989.

Kammer, W. J., G. W. Reddien, and R. S. Varga, "Quadratic Splines," *Numerische Mathematik*, vol. 22, pp. 241–259, 1974.

Kaps, P. and Rentrop, P., "Generalized Runge-Kutta methods of order four with stepsize control for stiff ordinary differential equations," *Numerische Mathematik*, vol. 33, pp. 55–68, 1979.

Keller, H. B., *Numerical Methods for Two-point Boundary-value Problems*, Blaisdell, Waltham, MA, 1968.

Kollerstrom, N., "Thomas Simpson and 'Newton's Method of Approximation': An Enduring Myth," *British Journal for the History of Science*, v. 25 (1992), pp. 347–354.

Kolman, B., *Introductory Linear Algebra with Applications*, 6th ed., Prentice Hall, Upper Saddle River, NJ, 1997.

Lancaster, P. and K. Salkauskas, *Curve and Surface Fitting: An Introduction*, Academic Press, Boston, 1986.

Larsen, R. W., *Introduction to Mathcad*, Prentice Hall, Upper Saddle River, NJ, 1999.

Leon, S. J., *Linear Algebra with Applications*, 5th ed., Prentice Hall, Upper Saddle River, NJ, 1998.

Leon, S. J., E. Herman, and R. Faulkenberry, *ATLAST: Computer Exercises for Linear Algebra*, Prentice Hall, Upper Saddle River, NJ, 1996.

Lorenz, E., "Deterministic Nonperiodic Flows," *Journal of Atmospheric Sciences*, vol. 20, pp. 130–141, 1963.

Mardia, K. V., *Multivariate Analysis*, Academic Press, London, 1980.

Marquardt, D. W., *Journal of the Society for Industrial and Applied Mathematics*, vol. 11, pp. 431–441, 1963.

Mathcad2000 Reference Manual, Mathsoft, Cambridge, MA, 1999.

Mathcad2000 User's Guide, Mathsoft, Cambridge, MA, 1999.

Meis, T. and U. Marcowitz, *Numerical Solution of Partial Differential Equations*, Springer-Verlag, New York, 1981.

Mitchell, A. R., *Computational Methods in Partial Differential Equations*, John Wiley & Sons, London, 1969.

Mitchell, A. R. and R. Wait, *The Finite Element Method in Partial Differential Equations*, John Wiley & Sons, London, 1977.

Morrison, D. F., *Multivariate Statistical Methods*, 3rd ed., McGraw-Hill, New York, 1990.

Nicolis, G. and I. Prigogine, *Self-Organization in Nonequilibrium Systems*, John Wiley & Sons, New York, 1977.

Nussbaumer, H. J., *Fast Fourier Transform and Convolution Algorithms*, Springer-Verlag, New York, 1982.

Ortega, J. M., *Numerical Analysis—A Second Course*, Academic Press, New York, 1972.

Ortega, J. M. and W. G. Poole, *An Introduction to Numerical Methods for Differential Equations*, Pitman Publishing, 1981.

Piessens, R., E. de Doncker, C. W. Uberhuber, and D. K. Kahaner, *QUADPACK: A Subroutine Package for Automatic Integration*, Springer-Verlag, New York, 1983.

Polak, E., *Computational Methods in Optimization*, Academic Press, New York, 1971.

Powers, D. L., *Boundary Value Problems*, 3rd ed., Harcourt Brace Jovanovich, Orlando, FL, 1987.

Press, W. H., S. A. Teukolsky, W. T. Vetterling, and B. P. Flannery, *Numerical Recipes in C, The Art of Scientific Computing*, 2nd ed., Cambridge Unversity Press, 1992.

Ralston, A. and P. Rabinowitz, *A First Course in Numerical Analysis*, 2nd ed., McGraw-Hill, New York, 1978.

Reinboldt, W. C., *Methods for Solving Systems of Nonlinear Equations*, SIAM, Philadelphia, 1974.

Rice, J. R., *Numerical Methods, Software, and Analysis*, 2nd ed., Academic Press, New York, 1992.

Ridders, C. J. F., *Advances in Engineering Solftware*, vol. 4, no. 2, 1982, pp. 75–76.

Ridders, C. J. F., *IEEE Transaction on Circuits and Systems,* 1979, vol. CAS-26, pp. 979–980.

Ritger, P. D. and N. J. Rose, *Differential Equations with Applications*, McGraw-Hill, New York, 1968.

Rivlin, T. J., *An Introduction to the Approximation of Functions*, Dover, New York, 1981. (Originally published by Blaisdell Publishing, 1969.)

Roberts, C. E., *Ordinary Differential Equations: A Computational Approach*, Prentice Hall, Englewood Cliffs, NJ, 1979.

Silvester, P. P. and R. L. Ferrari, *Finite Elements for Electrical Engineers*, 3rd ed., Cambridge University Press, Cambridge, UK, 1996.

Simmons, G. F., *Calculus with Analytic Geometry*, McGraw-Hill, New York, 1985.

Simmons, G. F., *Differential Equations with Applications and Historical Notes,* McGraw-Hill, New York, 1972.

Simon, W., *Mathematical Techniques for Biology and Medicine*, Dover, New York, 1986. (Originally published by MIT Press, 1977.)

Singleton, R.C., *Communications of ACM*, vol. 11, no. 11, 1986.

Singleton, R.C., *IEEE Transactions on Audio and Electroacoustics,* vol. AU-15, pp. 91–97, 1967.

Smith, D. E., *A Source Book in Mathematics*, Dover, New York, 1959.

Spong, M. W. and M. Vidyasagar, *Robot Dynamics and Control*, John Wiley & Sons, New York, 1989.

Stoer, J. and R. Bulirsch, *Introduction to Numerical Analysis*, Springer Verlag, New York, 1980.

Strang, G. *Linear Algebra and Its Applications*, 3rd ed., Harcourt Brace Jovanovich, San Diego, CA, 1988.

Strang, G. and G. Fix, *An Analysis of the Finite Element Method*, Prentice Hall, Englewood Cliffs, NJ, 1973.

Struik, D. J., *A Concise History of Mathematics*, 4th ed., Dover, New York, 1987.

Svobony, T., *Mathematical Modeling for Industry and Engineering*, Prentice Hall, Upper Saddle River, NJ, 1998.

Taha, H. A, *Operations Research, An Introduction,* 6th ed., Prentice Hall, Upper Saddle River, NJ, 1997.

Thomson, W. T., *Introduction to Space Dynamics*, Dover, New York, 1986. (Originally published by John Wiley & Sons, 1961.)

Thomson, W. T., *Theory of Vibrations with Applications*, Prentice Hall, Englewood Cliff, NJ, 1993.

Timan, A. F., C. J. Hyman, and N. I. Achieser, *Theory of Approximation*, Dover, New York, 1993.

Troutman, J. L. and M. Bautista, *Boundary Value Problems of Applied Mathematics*, Prindle, Weber & Schmidt Publishing, 1994.

Van der Pol, B., "Forced Oscillations in a Circuit with Non-linear Resistance," *Phil. Mag.*, Vol. 3, pp. 65–80, 1927.

Van Loan, C. F., *Computational Framework for the Fast Fourier Transform,* SIAM, Philadelphia, 1992.

Vargaftik, N. B., *Tables of the Thermophysical Properties of Liquids and Gases*, 2nd ed., Hemishpere, Washington D.C., 1975.

Weinstock, R., "Isaac Newton: Credit Where Credit Won't Do," *The College Mathematics Journal*, v. 25, no. 3, May 1994, pp. 179–192.

Wilkinson, J. H., The Algebraic Eigenvalue Problem, Oxford University Press, New York, 1965.

Wilkinson, J. H. and C. Reinsch, *Linear Algebra, vol II of Handbook for Automatic Computation*, Springer-Verlag, New York, 1971.

Winston, W. L., *Operations Research, Aplications and Algorithms*, 3rd ed., Duxbury Press (Wadsworth), Belmont, CA, 1994.

Young, D. M.,and R. T. Gregory, *A Survey of Numerical Mathematics*, Vol.1 and 2, Dover, New York, 1988.

Zauderer, E., *Partial Differential Equations of Applied Mathematics*, 2nd ed. John Wiler & Sons, New York, 1989.

Zienkiewicz, O. C. and R. L. Taylor, *The Finite Element Method*, 4th ed., Vol 1, McGraw-Hill, London, 1989.

Zill, D. G., *Differential Equations with Boundary-Value Problems*, Prindle, Weber, & Schmidt, Boston, 1986.

Answers to Selected Problems

(Most answers shown to 5 digits.)

Chapter 1

P1.1 $x = 1, y = 2$

P1.3 $x = 1, y = 1$

P1.5 $x = 1, y = 3$

P1.7 for $x_0 = 0.8$, $x_1 = 0.99957$, $x_2 = 0.90965$, $x_3 = 0.96928$; conditions of the theorem are satisfied

P1.9 for $x_0 = 0.5$, $x_1 = 0.75$, $x_2 = 0.4375$, $x_3 = 0.80859$; conditions of the theorem are not satisfied

P1.11 C_1: center $(1,0)$, radius $3/8$; C_2: center $(2,0)$, radius $1/2$; C_3: center $(3,0)$, radius 0; disks disjoint, so $5/8 \le \mu_1 \le 11/8$; $3/2 \le \mu_2 \le 5/2$; $\mu_3 = 3$

P1.13 C_1: center $(1,0)$, radius $1/4$; C_2: center $(2,0)$, radius $1/2$; C_3: center $(3,0)$, radius $1/4$; disks disjoint, so $3/4 \le \mu_1 \le 5/4$; $3/2 \le \mu_2 \le 5/2$; $11/4 \le \mu_3 \le 13/4$

P1.15 C_1: center $(1,0)$, radius $3/8$; C_2: center $(2,0)$, radius $1/2$; C_3: center $(3,0)$, radius $1/3$; μ_1 and μ_2 are within the union of the regions bounded by C_1 and C_2; $23/8 \le \mu_3 \le 25/8$

P1.17 a. 10.9; b. 11.0; c. rel error for a: $-9 * 10^{-3}$, rel error for b: $9 * 10^{-5}$

P1.19 a. $x_1 = 56.98$, $x_2 = 0.02$; b. $x_1 = 56.98$, $x_2 = 0.0176$; c. $x_1 = 56.9825$, $x_2 = 0.0175$

P1.21 a. 1.2; b. 1.1; c. 1.0666 (note, this is not better, the exact result is $\arctan(2) = 1.1071...$)

P1.23 a. 5; b. 4.5; c. 4.333... (exact result: $(2^2 - 2^0)/\ln(2) = 4.3281...$)

P1.25 a. 3.5343; b. 3.1474; c. 3.0184 (exact result is 3)

P1.27 (a. and b.) $0.92857 \le y \le 1.0714$; a. $1.9572 \le x \le 2.0429$; b. $1.9572 \le x \le 2.0429$

P1.29 (a. and b.) $1.9684 \le y \le 2.0316$; a. $2.9537 \le x \le 3.0463$; b. $2.9263 \le x \le 3.0737$

Chapter 2

P2.1 $x = 1.4142$

P2.3 $x = 2.6458$

P2.5 $x = 1.5874$

P2.7 $x = 0.49492$

P2.9 $x = 0.81904$

P2.11 $x = -2.8794$, $x = -0.6527$, $x = 0.53209$

P2.13 $x = -3.1055$, $x = 0.22346$, $x = 2.882$

P2.15 $x = 2.8063$

P2.17 $x = -3.324$, $x = -1.6197$, $x = 5.9437$

P2.19 x = 0, x = 1.3333, x = 2.5

P2.21 a. x alternates between 1 and -1; b. x = 0.2541; c. $x_0 = 0.5 \rightarrow x = 0.44853$; $x_0 = 0 \rightarrow f'(x_0) = 0$; Newton's method fails. d. x = 0.87055

P2.23 x = 2.029, x = 4.9132, x = 7.9787

P2.25 x = -0.70347

P2.27 a. x = 2.4781, x = 3; b. x = 2.7368, x = 2.7

P2.29 x = 0.12313

Chapter 3

P3.1 $x = [\ 1.1429,\ 1.0000,\ -0.7143]^T = [\ 8/7,\ 1,\ -5/7\ ]^T$

P3.3 $x = [\ 1, 2, 3\ ]^T$

P3.5 $x = [\ 1, 1, 1\ ]^T$

P3.7 i) $x = [\ 1, 2, 6\ ]^T$; ii) $x = [\ -3, 5, -4\ ]^T$

P3.9 $x = [\ -1,\ 2,\ 0, 1\ ]^T$

P3.11 $x = [\ 3, -4, -5\ ]^T$

P3.13 $x = [\ 2, 1, -1\ ]^T$

P3.15 $x = [\ 1, 1, 1, 1, 1, 1\ ]^T$

P3.17 a. $x = [\ 0, 3\ ]^T$; b. $x = [\ -1, 3\ ]^T$; without rounding $x = [\ -1.001,\ 3.001\ ]^T$

P3.19 a. $x = [\ 0, 5, 0\ ]^T$; b. $x = [\ 1, 2, 3\ ]^T$; without rounding $x = [\ 1.0005,\ 1.9995,\ 2.9995]^T$

P3.27 $x = [\ 1, 1, 1\ ]^T$

P3.29 $x = [\ 1, -2, 5, 8\ ]^T$

P3.31 $x = [\ 2, -5, -2, 1, -2, 2\ ]^T$

P3.33 $x = [\ -4, 3, -3, 4, 3, 3, 5, 5\ ]^T$

P3.35 $x = [\ -3, 1, -1, -2, 3, -4, -1, -1, 3, 0\ ]^T$

P3.37 a. $x = [\ 4, -4,\ 4, 1,\ 2, 3, -4, 2\ ]^T$
b. $x = [\ -2,\ 1, 1, 0,\ 3, 0, -1,\ 5\ ]^T$
c. $x = [\ -3, -3, -2, 0, 4, -3, 3, 1\ ]^T$
d. $x = [\ 5, 2, -4, -3, 2, -4, 0, -1\ ]^T$

P3.39 a. $x = [\ 1,\ 0,\ 2, -1,\ 5, -1, -2, -5, -3, -2\ ]^T$
b. $x = [\ 1, -1, -3,\ 4, -4, -4, -1,\ 2,\ 4,\ 0\ ]^T$
c. $x = [\ -2,\ 1, -5, 4,\ -1, 2, -4,\ 2, -2,\ 4\ ]^T$
d. $x = [\ 1,\ 0,\ 1, -1, -1, 4, -4, -3,\ -3, -3\ ]^T$

Chapter 4

(Results shown are exact, not values after 3 or 10 iterations.)

P4.1 $x = [\ 1, 2,\ 3\ ]^T$

P4.3 $x = [\ 2, 1, -1\ ]^T$

P4.5 $x = [\ 0.3, 1.8, 1.7\ ]^T$

P4.7 $x = [\ 6, 3, 5, 2\ ]^T$

P4.9 $x = [\ 0, 1, 0, -1\ ]^T$

P4.11 $x = [\ 1, 2, 3, 4\ ]^T$

P4.13 $x = [\ 1, -2, 3, 2, -1\ ]^T$

P4.15 $x = [\ -2, 4, -6, 2, -4, 7\ ]^T$

P4.17 $x = [\ -2, 1, 4, -6, 3, 2, -4, 7\ ]^T$

Chapter 5

P5.1 x = 0.90644, y = 2.5704

P5.3 x = 1.38, y = 1.6335

P5.5 x = 1.0784, y = 1.9259

P5.7 x = 0.56451, y = 1.1579, z = 1.5299

P5.9 x = 1.4901, y = -0.68477, z = 0.98001

P5.11 x = 0.29704, y = 0.67481, z = 0.73066

P5.13 x = 0.33016, y = 0.47535, z = 0.60284

P5.15 x = 0.57627, y = 0.36755, z = 0.70639

Chapter 6

P6.1 a. L = [1 0 0
              2 1 0
              3 4 1]

U = [1 2 3
      0 4 5
      0 0 6]

b. A^{-1} = [1.5833 -0.1667 -0.0833
               -1.5417   1.0833 -0.2083
                0.8333 -0.6667   0.1667]

c. $\det(A) = 24$

d. x = [1.3333 -0.6667 0.3333]^T
   y = [2.1944 -2.8472 1.6111]^T

P6.3 a. L = [1   0   0
                  2   1   0
                  1   2/3   1]

U = [2   1   -2
      0   -3   6
      0   0   -1]

b. A_inv = [1/6   1/6   0
              0   1   -2
            -1/3   2/3 -1]

c. $\det(A) = 6$

d. x = [1/3 -1 -2/3]^T
   y = [-1/9 1/3 -1/9]^T

P6.5 a. L = [1.0000   0   0
                  0.5000   1.0000   0
                  0.3333   1.0000   1.0000]

U = [1.0000 0.5000 0.3333
      0   0.0833 0.0833
      0   -0.0000 0.0056]

b.   9   -36   30
    -36   192 -180
     30 -180   180

c. 4.6296e-04

d. x = [3   -24   30]^T
   y = [1791 -10116 9810]^T

P6.7
a. L = [1   0   0   0
          2   1   0   0
          3   4   1   0
         -1 -3   0   1]

U = [1   1   0   3
      0 -1 -1   -5
      0   0   3   13
      0   0   0 -13]

b. A_inv = [-0.2308   0.2051   0.3333   0.1795
              0.0769   0.4872 -0.3333   0.0513
              0.0000 -0.3333   0.3333   0.3333
              0.3846 -0.2308 -0.0000 -0.0769]

c. $\det(A) = 39$

d. x = [0.4872 0.2821 0.3333 0.0769]^T
   y = [0.0703 0.0677 0.0427 0.1164]^T

P6.9 a. L = [1   0   0   0
                   0   1   0   0
                   0   0   1   0
                   1/3 -1/9   1/9   1]

U = [3   7   4   0
      0   3   13   3
      0   0   1   4
      0   0   0   -1/9]

b. A_inv = [-98   32 -24   295
              49 -16   12 -147
             -12   4   -3   36
               3   -1   1   -9]

c. $\det(A) = -1.0000$

d. x = [205   -102   25   -6]^T
   y = [-25724 12859 -3159 796]^T

P6.11
a. L = [1   0   0   0   0   0
         2   1   0   0   0   0
         3   4   1   0   0   0
         1   0   0   1   0   0
         0   1   0 -1/3 1   0
         0   0   1   1   -2   1]

U = [1   2   3   1   0   0
      0   2   4   0   1   0
      0   0   3   0   0   1
      0   0   0   3   6 12
      0   0   0   0   2   6
      0   0   0   0   0   1]

b. B = A^{-1} =
 11.0556 -14.50   5.67   1.9444 -9.167 -5.33
 -6.1944   7.75   -2.83   -0.8056   4.083   2.167
  1.8889   -2.00   0.67   0.1111 -0.667 -0.33
  3.3333   5.00   2.00   0.6667   3.00   2.00
  2.8333   -6.50   3.00   1.1667 -5.50   -3.00
 -0.67   2.00   -1.00   -0.33   2.00   1.00

c. $\det(A) = 36$

d.
x =
[-10.3333 4.1667 -0.333 4.00 -8.00 3.00]^T
y =
[-111.435 67.8565 -23.2963 35.278 -17.694 1.222]^T

P6.13

a. L = [1 0 0 0 0 0
        2 1 0 0 0 0
        3 4 1 0 0 0
        1 0 0 1 0 0
        0 1 0 -1 1 0
        2 0 1 1 -2 1]

U = [1 2 1 1 0 0
        0 2 2 0 1 0
        0 0 1 0 0 1
        0 0 0 1 6 12
        0 0 0 0 2 6
        0 0 0 0 0 1]

b. A_inv =
  52.50 -28.50 11.00 -7.50 -16.50 -10.0
 -16.25   9.75 -3.50   2.25   4.75   2.50
   9.00 -6.00   2.00 -1.00   -2.00 -1.00
 -28.00 15.00 -6.00   4.00   9.00   6.00
  12.50   -6.50   3.00 -2.50   -5.50 -3.00
  -4.00   2.00 -1.00   1.00   2.00   1.00

c. det(A) = 4

d. x = [1.00 -0.50   1.0   0.00 -2.00   1.0]^T

y = [100.75 -31.625 17.0 -53.5 26.75 -9.0]^T

P6.17 a. dd = [4 4 3]
        bb = [0 4 2]

b. L = [1 0 0
        4 1 0
        0 2 1]

U = [4 2 0
        0 4 1
        0 0 3]

P6.19 a. dd = [5 3 2]
        bb = [0 3 3]

b. L = [1 0 0
        3 1 0
        0 3 1]

U = [5 1 0
        0 3 3
        0 0 2]

P6.21 a. dd = 1 1 1 1
        bb = 0 2 2 1

b. L = [1 0 0 0
        2 1 0 0
        0 2 1 0
        0 0 1 1]

U = [1 3 0 0
        0 1 2 0
        0 0 1 5
        0 0 0 1]

P6.23 a. dd = [1 2 4 4]
        bb = [0 1 3 3]

b. L = [1 0 0 0
        1 1 0 0
        0 3 1 0
        0 0 3 1]

U = [1 4 0 0
        0 2 2 0
        0 0 4 3
        0 0 0 4]

P6.25 a. dd = [5 1 1 4]
        bb = [0 2 3 1]

b. L = [1 0 0 0
        2 1 0 0
        0 3 1 0
        0 0 1 1]

U = [5 0 0 0
        0 1 2 0
        0 0 1 4
        0 0 0 4]

P6.27 a. dd = [1 2 3 4 5 6]
        bb = [0 2 3 4 5 6]

b. L = [1 0 0 0 0 0
        1 0 0 0 0 0
        0 3 1 0 0 0
        0 0 4 1 0 0
        0 0 0 5 1 0
        0 0 0 0 6 1]

$$U = \begin{bmatrix} 1 & -5 & 0 & 0 & 0 & 0 \\ 0 & 2 & -4 & 0 & 0 & 0 \\ 0 & 0 & 3 & -3 & 0 & 0 \\ 0 & 0 & 0 & 4 & -2 & 0 \\ 0 & 0 & 0 & 0 & 5 & -1 \\ 0 & 0 & 0 & 0 & 0 & 6 \end{bmatrix}$$

P6.29 a. dd = 5 4 2 3 4 1
bb = 0 3 2 1 3 4

b. $$L = \begin{bmatrix} 1 & 0 & 0 & 0 & 0 & 0 \\ 3 & 1 & 0 & 0 & 0 & 0 \\ 0 & 2 & 1 & 0 & 0 & 0 \\ 0 & 0 & 1 & 1 & 0 & 0 \\ 0 & 0 & 0 & 3 & 1 & 0 \\ 0 & 0 & 0 & 0 & 4 & 1 \end{bmatrix}$$

$$U = \begin{bmatrix} 5 & 2 & 0 & 0 & 0 & 0 \\ 0 & 4 & 0 & 0 & 0 & 0 \\ 0 & 0 & 2 & 2 & 0 & 0 \\ 0 & 0 & 0 & 3 & 4 & 0 \\ 0 & 0 & 0 & 0 & 4 & 4 \\ 0 & 0 & 0 & 0 & 0 & 1 \end{bmatrix}$$

P6.31 $$L = \begin{bmatrix} 1.0000 & 0 & 0 \\ 0.1667 & 1.00 & 0 \\ 1.0000 & 0 & 1.00 \end{bmatrix}$$

$$U = \begin{bmatrix} 6.00 & 2.0000 & 2.0000 \\ 0 & 1.6667 & -1.3333 \\ 0 & 0 & -1.0000 \end{bmatrix}$$

P6.33 $$L = \begin{bmatrix} 1 & 0 & 0 & 0 \\ 0 & 1 & 0 & 0 \\ -1 & 0 & 1 & 0 \\ 0 & 0 & 1 & 1 \end{bmatrix}$$

$$U = \begin{bmatrix} -1 & 1 & 0 & 0 \\ 0 & 1 & -1 & 1 \\ 0 & 0 & 1 & 0 \\ 0 & 0 & 0 & -1 \end{bmatrix}$$

P6.35
$$L = \begin{bmatrix} 1.00 & 0 & 0 & 0 & 0 & 0 \\ 0 & 1.00 & 0 & 0 & 0 & 0 \\ 0 & 0 & 1.00 & 0 & 0 & 0 \\ 0 & 0 & 0 & 1.0000 & 0 & 0 \\ 0 & 0 & 0 & 0 & 1.0000 & 0 \\ 0.50 & 0.25 & 0 & 0.0833 & 0.0486 & 1 \end{bmatrix}$$

$$U = \begin{bmatrix} -2 & 6 & 4.0 & 0 & 0 & 0 \\ 0 & 4 & 8.0 & -0.50 & 0 & 0 \\ 0 & 0 & -0.5 & 3.25 & 1.50 & 0 \\ 0 & 0 & 0 & 1.50 & 1.75 & -3.00 \\ 0 & 0 & 0 & 0 & -3.00 & 13.00 \\ 0 & 0 & 0 & 0 & 0 & -0.3819 \end{bmatrix}$$

P6.37 a. Doolittle form
$$L = \begin{bmatrix} 1 & 0 & 0 \\ 2 & 1 & 0 \\ 4 & 3 & 1 \end{bmatrix}$$

$$U = \begin{bmatrix} 9 & 18 & 36 \\ 0 & 4 & 12 \\ 0 & 0 & 1 \end{bmatrix}$$

b. Cholesky form

$$L = \begin{bmatrix} 3 & 0 & 0 \\ 6 & 2 & 0 \\ 12 & 6 & 1 \end{bmatrix}$$

$$U = \begin{bmatrix} 3 & 6 & 12 \\ 0 & 2 & 6 \\ 0 & 0 & 1 \end{bmatrix}$$

c. $x1 = [\ -1\ \ 2\ \ 1\ ]^T$
$x2 = [\ \ 2\ \ 1\ -1\ ]^T$
$x3 = [\ -2\ -2\ \ 1\ ]^T$

P6.39 Pivoting is required, Cholesky and Doolittle methods fail.

$$LU = PA = \begin{bmatrix} 2 & -1 & 0 & 0 & 0 & 0 \\ -1 & 2 & -1.0000 & 0 & 0 & 0 \\ 0 & 0 & -1.0000 & 2.0000 & -1 & 0 \\ 0 & -1 & 0.6667 & -1.0000 & 0 & 0 \\ 0 & 0 & 0 & -1.0000 & 2 & -1 \\ 0 & 0 & 0 & 0 & -1 & 2 \end{bmatrix}$$

$$L = \begin{bmatrix} 1.0 & 0 & 0 & 0 & 0 & 0 \\ -0.5 & 1 & 0 & 0 & 0 & 0 \\ 0 & 0 & 1 & 0 & 0 & 0 \\ 0 & -0.6667 & 0 & 1 & 0 & 0 \\ 0 & 0 & 0 & 1 & 1.0 & 0 \\ 0 & 0 & 0 & 0 & -0.5 & 1 \end{bmatrix}$$

$$U = \begin{bmatrix} 2 & -1.0 & 0 & 0 & 0 & 0 \\ 0 & 1.5 & -1 & 0 & 0 & 0 \\ 0 & 0 & -1 & 2 & -1 & 0 \\ 0 & 0 & 0 & -1 & 0 & 0 \\ 0 & 0 & 0 & 0 & 2 & -1 \\ 0 & 0 & 0 & 0 & 0 & 1 \end{bmatrix}$$

c. solution using LU_pivot
   xx1 = 2 1 3 -1 -3 -2
   xx2 = 1 2 3 -1 -3 -2
   xx3 = 1 -1 2 -2 3 -3

Chapter 7

Results after 10 iterations

P7.1 a. m = 3.0001, v = [0.00 0.50 1.00], error = 4.5601e-05, exact = 3
         b. m = 1.0000, v = [1.00 0.00 -1.00], error = 5.8666e-05, exact = 1.

P7.3 a. m = 3.0132, v = [1.00 1.00 0.0087], error = 0.0109, exact = 3.
         b. m = 1.9768, v = [-0.9653 -0.9653 1.0], error = 0.0290, exact = 2

P7.5 a. m = 3.0240, v = [-0.0106 0.2889 1.0000], error = 0.0162, exact = 3
         b. m = 0.9990, v = [1.0000 0.9996 -0.0012], error = 0.0012, exact = 1

P7.7 a. m = 5.1521, v = [0.2367 1.0000 0.0529], error = 0.0355, exact = 5
         b. m = 3.4033, v = [0.1618 1.0000 0.7092], error = 0.3437, exact = 3

P7.9 a. m = 2.0019, v = [0.2884 1.0000 0.5730], error = 0.0012, exact = 2
         b. m = 0.9980, v = [0.2857 1.0000 0.5714], error = 5.4704e-06, exact = 1

P7.11 a. m = 4.00,   v = [1.00 0.0010 -0.9980 0.0010], error = 0.0048, exact = 4
         b. m = 1.9993, v = [0.0005 0.5002 1.00 0.5002], error = 0.0011, exact = 2

P7.13 a. m = 4.990, v = [0.9970 1.00 0.0060 0.0030],   error = 0.0172, exact = 5
         b. m = 0.0000, v = [0.5000 1.0000 0.5000 0.0000], error = 5.6445e-15, exact = 0,
(Hint: modify the inverse power method, or see hint for P7.19 below)

*P7.15 a. m = 5.00, v = [-0.6049 -0.1369 0.50 1.00], error = 2.0968, exact = 5
         after 50 iterations, m = 5.0002,
            v = [-0.0000 1.0000 0.2500 0.5001], error = 1.1922e-04
         b. m=1.00, v=[-0.0002 -0.0006 -0.00 1.00], error = 0.0013, exact = 1

P7.17 a. m = 4.0834, v = [-0.0005 -0.0592 0.4117 1.0000 0.5291], error = 0.1150, exact = 4
         b. m = 1.9876, v = [0.5002 -0.4829 -0.9663 0.5166 1.0000], error = 0.0353, exact = 2

P7.19 a. m = 3.0132, v = [-0.0087 0.4869 1.0000 0.7566 0.5044], error = 0.0191, exact = 3; after 30 iterations,
         m = 3.0000, v = [-0.000 0.500 1.000 0.750 0.50], error = 5.6172e-06
         b. m = -5.7135e-10, v = [1.00 1.00 -0.00 -0.00 -0.00], error = 7.7881e-07, exact = 0; converged after 12
         iterations, m = -7.0636e-12, error = 4.8669e-08. (Hint: Apply inverse power method to B = A - 0.2I
         and then add 0.2 to the eigenvalue.)

P7.21 a. m = 8.7839, v = [-0.5198 1.00 0.2923], error = 0.1514;
         b. m = 8.8657, v = [-0.5198 1.00 0.2923], error = 0.1139;

P7.23 a. m = 16.6077, v = [-0.5306 1.0000 0.2966], error = 0.2326;
         b. m = 16.7373, v = [-0.5306 1.0000 0.2966], error = 0.1713;

P7.25 a. m = 45.0458, v = [-0.1743 1.0000 0.0506], error = 0.0013;
         b. m = 45.0454, v = [-0.1743 1.0000 0.0506], error = 0.0012;

*P7.27 a. m = -8.9669, v =[0.3685 -0.0528 1.00], error = 0.6893
b. m = -8.7071, v =[0.3685 -0.0528 1.00], error = 0.6384

P7.29 a. m = 57.2864, error =1.3067;
b. m = 58.0325, error = 0.8996;
eigenvector for parts a. and b.
v = [1.000 0.0220 0.1218 -0.6622],

P7.31 a. m = 48.2782, v = [-0.1454 1.0000 -0.1017 -0.0383], error = 0.0133;
b. m = 48.2800, v = [-0.1454 1.0000 -0.1017 -0.0383], error = 0.0132;

P7.33 a. m = 72.0000, v = [1.0000 0.3333 -0.6652 0.6681], error = 0.0746
b. m = 71.9997, v = [1.0000 0.3333 -0.6652 0.6681], error = 0.0746

P7.35 a. m = 33.3009, v = [-0.1043 0.4447 -0.5053 1.00 0.2490], error = 0.0637
b. m = 33.3290, v = [-0.1043 0.4447 -0.5053 1.0000 0.2490], error = 0.0532

P7.37 a. m = 76.3303, error = 0.0742;
b. m = 76.2918, error = 0.0502; eigenvector for parts a. and b.
v = [0.8155 0.1067 1.0000 -0.1053 0.4252 0.0853 0.3795],

P7.39 a. m = 194.6906, error = 0.7786;
b. m = 194.7387, error = 0.7723; eigenvector for parts a. and b.
v = [-0.3087 1.0000 -0.6857 -0.0823 -0.5172 0.9594 0.3572 0.8342 0.0161]

P7.41 a. the QR factorization of matrix A is
Q =   -0.2182 -0.6838 -0.6963
      -0.4364 -0.5698   0.6963
      -0.8729   0.4558 -0.1741

R =   -4.5826   3.9279 -5.2372
      -0.0000 -1.2536   1.1396
      -0.0000 -0.0000   0.5222

b. the upper triangular matrix
B =   3.0000 -0.6667 -7.4535
     -0.0000   1.0000   0.0000
      0.0000 -0.0000   1.0000
has the same eigenvalues as A (on the diagonal): 3, 1, 1.

c. the upper Hessenberg matrix with the same eigenvalues as A is
AA =   1.0000   0   0
      -4.4721   3.0000 -6.0000
       0.0000 -0.0000   1.0000

d. the upper Hessenberg matrix is converted to the upper triangular matrix

B =   3.0000   4.4721   6.0000
     -0.0000   1.0000 -0.0000
      0.0000 -0.0000   1.0000
with the eigenvalues on the diagonal.
(computational effort for part d is approximately 1/2 that needed for part b.)

P7.43 a.
Q = -0.8575 0.5145 0
      -0.5145 -0.8575 0
      0   0   1.0000

R = -5.8310 1.7150 -1.3720
      0   -1.0290 -0.3430
      0   0   2.0000

b. upper triangular matrix
B = 3.0147 -4.9970 1.4142
      0.0030 1.9853   0.0041
      0   0   2.0000

with the same eigenvalues as A (on the diagonal):
3.0147, 1.9853, 2.0000.

c. the upper Hessenberg matrix with the same eigenvalues as A is A itself.

P7.47 a. Q = 0.2040 -0.4011 0.8930
              0.9748 -0.0013 -0.2233
              0.0907 0.9160   0.3907

R = 176.4540 48.2165 -22.3741
      -0.0000   -0.4061   2.9493
      0.0000   0.0000   0.8372

b. B = 5.1175   21.0988 182.8169
        -0.0056 3.8943   7.3171
        -0.0001 -0.0014 2.9882
e =   5.1175   3.8943   2.9882

c. AA = -36.0000 -9.4939 -5.9047
          172.7426   44.9453 26.0879
          0.0000   0.0879   3.0547

d. B = 5.1175   -23.5183 182.5214
        0.0056   3.9911   -7.3038
        -0.0000   0.0147   2.8914
e =   5.1175   3.9911   2.8914

P7.51 a. Q = -0.4339 -0.7071 -0.0638 0.5547
               0.4339   -0.7071 0.0638   -0.5547
               0.7593   0.0000 -0.3403 0.5547
               0.2169   0.0000 0.9360   0.2774

R = -9.2195 9.2195   -4.4471 4.0132
      -0.0000 -5.6569 2.8284   -1.4142
      -0.0000 0   -1.1061 2.8504
      0.0000   0   0   1.6641

b. after 10 iterations, the matrix is almost upper triangular
B = 3.9894   -0.0020 -0.3496 13.9731
      -0.0028 3.9995   -0.0908 -2.5377
      -0.0295 -0.0056 3.0302   3.6283
      -0.0001 -0.0000 -0.0048 1.9810
e = 3.9894 3.9995   3.0302   1.9810

c. Upper Hessenberg
AA = 4.0000 0   0   0
       8.3066 1.8406   0.8521   -12.0333
       0.0000 -0.2169 3.1594   0.2408
       0.0000 0.0000   -0.0000 4.0000

d. B = 3.9894 0.2795   -8.1193 -11.3740
        0.0296   3.0462   2.0218   2.6984
        0.0000   -0.0173 1.9644   -2.8648
        0.0000   -0.0000 -0.0000   4.0000
e =   3.9894   3.0462   1.9644   4.0000

P7.55 a. Q = 0.1414 -0.9899 0   0
                0.9899   0.1414   0   0
                0   0   -0.5300 -0.8480
                0   0   -0.8480 0.5300

R = -21.2132 10.1823 -20.9304 0
      -0.0000   -0.5657 5.0912   0
      0   0   -9.4340   -0.8480
      0   0   -0.0000   0.5300

b. B = 4.0605 22.9972 9.5804   -19.1608
        0.0028   2.9395   -1.0081   2.0161
        0   0   5.0000   -8.0000
        0   0   0.0000   1.0000
e =   4.0605   2.9395   5.0000   1.0000

c. A is upper Hessenberg.
d. Householder function fails

P7.57
a. Q =
0.2626 -0.6607 0.4048   -0.4131 0.400
0.3939 -0.5919 -0.4048   0.4131 -0.400
0.2626   0.1376 -0.7625 -0.4131 0.400
0.6565 -0.3441 -0.2353 -0.4849 -0.400
0.5252 -0.2753 -0.1883   0.5029 0.600

R =
7.6158   6.3027   -7.6158   8.1410   -2.3635
0.0000   -2.5052   1.5967 -2.6703   0.4405
0.0000   -0.0000 -2.3346 -1.8263   0.3012
0.0000   0.0000 -0.0000 -2.6941   1.8679
0.0000   -0.0000   0.0000   0   0.8000

b.
B =
4.0552	-0.0524	0.5977	1.0820	11.5716
-0.0266	2.0007	-0.0077	1.2381	3.5007
-0.0591	0.0015	1.9828	2.7926	7.8998
-0.0206	0.0005	-0.0060	2.9631	2.7234
-0.0002	0.0000	-0.0001	-0.0003	1.9982

e = 4.0552 2.0007 1.9828 2.9631 1.9982

c.
AA =
2.0000	0	0	0	0
-7.3485	3.8889	-0.0994	-9.7992	-8.4256
-0.0000	-0.9938	3.1111	0.0369	2.2768
-0.0000	-0.0000	0.0000	2.0000	0.0000
-0.0000	0.0000	-0.0000	-0.0000	2.0000

d.
B =
4.0552	0.8587	5.2410	7.0225	7.6099
-0.0680	2.9786	5.1388	6.8344	4.2734
-0.0000	-0.0064	1.9662	-0.0450	-0.0288
0.0000	-0.0000	0.0000	2.0000	0.0000
-0.0000	-0.0000	-0.0000	-0.0000	2.0000

e = 4.0552 2.9786 1.9662 2.0000 2.0000

P7.59 a. QR_factor function fails
b. QR_eigen fails (since QR factors are not available)

c.
AA =
-8.0000	-5.7585	-4.9735	-7.0216	-3.5777
13.8924	11.3316	7.5722	14.1866	9.1423
0.0000	-1.5418	1.4703	-3.9991	-3.0447
0.0000	-0.0000	-0.1684	1.1980	-1.7744
-0.0000	-0.0000	0.0000	-0.0000	3.0000

d.
B =
3.0127	-0.7503	-4.6326	-3.7316	-17.6593
0.0172	1.9878	0.8999	1.3603	22.4197
0.0000	-0.0005	0.9995	-1.7889	3.3341
0.0000	0.0000	-0.0000	3.0000	0.7456
0.0000	-0.0000	-0.0000	-0.0000	-0.0000

e = 3.0127 1.9878 0.9995 3.0000 -0.0000

P7.61 a. Q =
-0.9370	-0.1735	0.3030
0.1562	-0.9844	-0.0808
0.3123	-0.0284	0.9495

R =
-6.4031	2.4988	2.6550
-0.0000	-7.7302	-0.6941
0.0000	-0.0000	1.2122

b. B =
8.8697	-0.0970	-0.0000
-0.0970	6.0033	-0.0000
-0.0000	-0.0000	1.1270

e = 8.8697 6.0033 1.1270

c. AA =
6.0000	2.2361	-0.0000
2.2361	4.0000	3.0000
-0.0000	3.0000	6.0000

d. B =
8.8697	0.0970	0.0000
0.0970	6.0033	0.0000
0.0000	0.0000	1.1270

e = 8.8697 6.0033 1.1270

P7.65 a. Q =
-0.5946	-0.2057	0.7772
0.4460	-0.8887	0.1060
0.6690	0.4097	0.6202

R =
-13.4536	23.1908	14.7172
0.0000	-37.8707	7.5864
-0.0000	-0.0000	1.6879

b. B =
45.0454	-0.0239	-0.0000
-0.0239	20.0000	-0.0000
-0.0000	-0.0000	0.9546

e = 45.0454 20.0000 0.9546

c. AA =
8.0000	10.8167	-0.0000
10.8167	23.2308	13.8462
-0.0000	13.8462	34.7692

d. B =
45.0454	0.0239	-0.0000
0.0239	20.0000	0.0000
-0.0000	0.0000	0.9546

e = 45.0454 20.0000 0.9546

P7.69 a. Q =
-0.9742	-0.0262	0.2071	0.0853
-0.0585	-0.5679	-0.0084	-0.8210
-0.0390	-0.7166	-0.4812	0.5034
0.2143	-0.4041	0.8517	0.2555

R =
-51.3225	-2.7084	-4.0918	19.5821
-0.0000	-13.8081	-17.0130	-16.6594
0.0000	0	-11.7395	35.3765
0.0000	0	0	3.6758

b. B =
58.0976	-0.1146	0.0060	-0.0000
-0.1146	35.5004	-0.5341	0.0000
0.0060	-0.5341	23.7782	-0.0000
-0.0000	0.0000	-0.0000	0.6238

e = 58.0976 35.5004 23.7782 0.6238

c. AA =   0.0000 -11.5758 -0.0000 -0.0000
         -11.5758 38.7463 11.6143   0.0000
          -0.0000 11.6143   8.0082   8.5860
          -0.0000   0.0000   8.5860 21.2456

d. B = 58.0976 -0.1147 -0.0000 0.0000
      -0.1147 35.5242   0.0775 0.0000
      -0.0000   0.0775 23.7545 0.0000
       0.0000   0.0000   0.0000 0.6238
   e = 58.0976 35.5242 23.7545 0.6238

P7.73 a. Q = -0.7894   0.1478 -0.2024 -0.5604
                  -0.2046 -0.9784   0.0096   0.0267
                   0.4093 -0.1021 -0.8578 -0.2936
                  -0.4093   0.1021 -0.4723   0.7740

R = -51.3079 -16.5764 33.1528 -33.1528
     0.0000 -17.4849   -2.0906   2.0906
    -0.0000   -0.0000 -27.0651 -20.8193
    -0.0000   -0.0000   0.0000   17.2938

b. B = 72.0000   0.0000 -0.0000   0.0000
        0.0000 18.0000   0.0000 -0.0044
       -0.0000   0.0000 36.0000   0.0000
        0.0000 -0.0044   0.0000   9.0000
   e = 72.00   18.00   36.00   9.00

c. AA = 40.5000 -31.5000   0.0000   0.0000
        -31.5000   40.5000 -0.0000 -0.0000
          0.0000   -0.0000 18.8471 -3.8118
          0.0000   0.0000 -3.8118 35.1529

d. B = 72.0000 -0.0000 -0.0000 -0.0000
       -0.0000   9.0000 -0.0000   0.0000
        0.0000 -0.0000 35.9997 -0.0791
        0.0000   0.0000 -0.0791 18.0003
   e = 72.0000   9.0000 35.9997 18.0003

Chapter 8

P8.1 a. x = 1 2 3
              y = 1 4 8
              c = 0.5000 -4.0000 4.0000

b. x = 1 2 3
   y = 1 4 8
   d = 3.0000 0.5000
       4.0000 0
   c = 1.0000 3.0000 0.5000

P8.3 a. x = 4 9 16
              y = 2 3   4
              c = 0.0333 -0.0857 0.0476

b. x = 4 9 16
   y = 2 3 4
   d = 0.2000 -0.0048
       0.1429 0
   c = 2.0000 0.2000 -0.0048

P8.5 a. x = 0 1 2
              y = 1 2 4
              c = 0.5000 -2.0000 2.0000

P8.7 a. x = -1   0   1   2
              y = 0.3333 1.0000 3.0000 9.0000
              c = -0.0556 0.5000 -1.5000 1.5000

P8.9 a. x = 0 1 2 3
              y = 0 1 0 -1
              c = 0 0.5000 0 -0.1667

P8.11 x = 0 0.6667 1.0000 2.0000
           y = 2.0000 -2.0000 -1.0000 -0.5000
           c = -1.5000 -6.7500 3.0000 -0.1875

P8.13 x = 0 0.6667 1.0000 2.0000
           y = 4.0000 -4.0000 -3.5000 -0.5000
           c = -3.0000 -13.5000 10.5000 -0.1875

P8.15 x = 0 0.5000 1.0000 1.5000
           y = 1 2 1 0
           c = -1.3333 8.0000 -4.0000 0

P8.17 x = 0 1 8 27
           y = 0 1 2 3
           c = 0 0.0055 -0.0019 0.0002

P8.19 x = -2 -1 0 1 2 3 4
           y = -14.0 0.50 3.10 0 -3.0 0 16.0
           c = -0.0194 -0.0042 0.0646 0 -0.0625 0 0.0222

Chapter 9

P9.1 y = 3.5 x -2.6667

P9.3 y = 0.1651 x + 1.4037

P9.5 y = 1.5 x + 0.8333

P9.7 a) y = 2.8 x + 1.9333
          b) y = 1.333 x^2 + 1.4667 x + 0.6

P9.9 a) y = 2.3 x + 0.3
          b) y = 0.75 x^2 + 0.05 x + 1.05

P9.11 a) y = -0.94 x + 0.4867
           b) y = 2.475 x^2 -6.055 x + 1.8067

P9.13 a) $y = -1.68 x + 0.54$
b) $y = 6.075 x^2 - 14.235 x + 3.78$

P9.15 a) $y = -0.8 x + 1.6$
b) $y = -2 x^2 + 2.2 x + 1.1$

P9.17 a) $y = 0.2299 x + 0.6207$
b) $y = -0.0132 x^2 + 0.4398 x + 0.2961$
c) $y = 0.0021 x^3 - 0.063 x^2 + 0.7115 x + 0.1344$

P9.19 a) $y = 1.12 x + 1.26$
b) $y = 0.5696 x^2 + 1.1125 x + 0.9702$
c) $y = 0.1957 x^3 + 0.5718 x^2 + 0.955 x + 0.975$

P9.21 $y = 2.9679 x + 3.5467$

P9.23 $y = -3.3164 x + 0.28$

P9.25 $y = 4.0616 x - 3.8039$

P9.27 $y = -0.8182 x^2 - 1.7372 x + 2.58$

P9.29 $y = 5.0052 x^2 - 0.5594 x + 4.3498$

P9.31 $y = 4.1402 x - 5.8261$

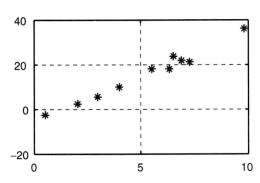

P9.33 $y = 3.9954 x^2 + 4.4814 x - 1.4743$

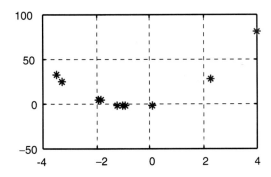

P9.35

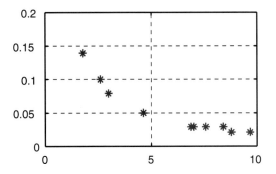

quadratic fit; err = 5.3794e-04

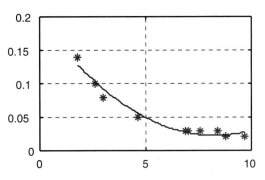

exponential (log-linear) fit; err = 8.4483e-04

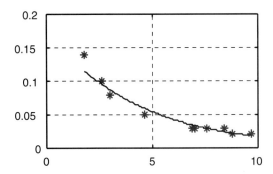

reciprocal of data

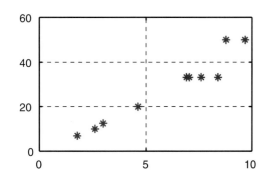

linear fit to reciprocal of data

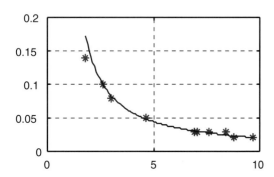

P9.37

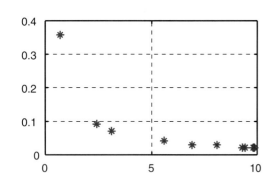

quadratic fit, err = 0.0148

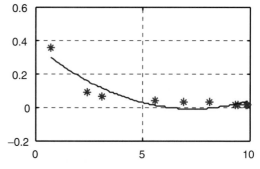

exponential (log-linear) fit, err = 0.0322

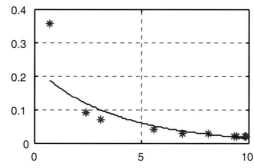

reciprocal of data

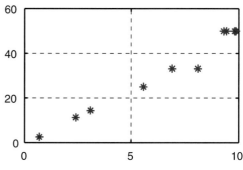

z = 1/ (5.2528 x -2.2642); error = 0.1211

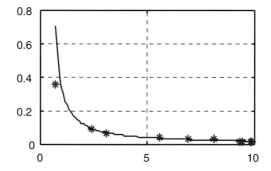

P9.39 plot data and reciprocal of data

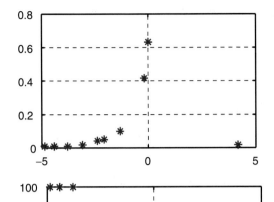

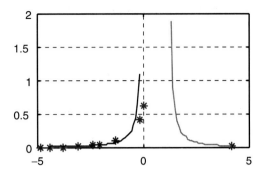

P9.41

quadratic fit to reciprocal of data

reciprocal of data

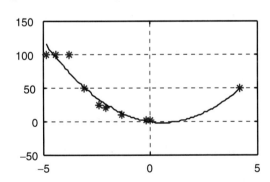

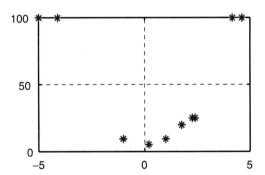

data and approximating function

Answers to Selected Problems — 685

data and best-fit reciprocal of quadratic

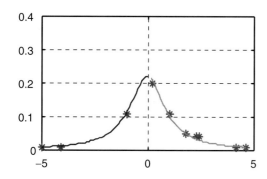

a **P9.43**
cubic fit

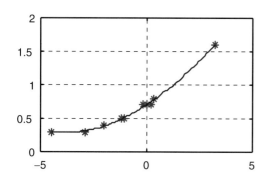

P9.45
cubic fit

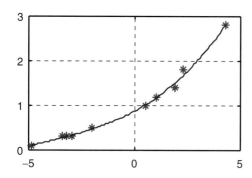

P9.47
cubic fit

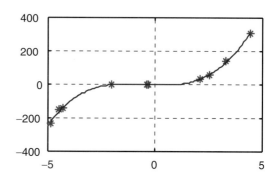

P9.49
quadratic fit

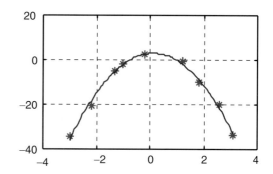

Chapter 10

P10.1 a. m = 1; a = 1/2, 1/2 b = 0, 1/2
b. m = 2; a = 1/2, 1/2, 0 b = 0, 1/2, 0

P10.3 a. m = 1; a = 1/2, -1/2 b = 0, 1/2
b. m = 2; a = 1/2, -1/2, 0 b = 0, 1/2, 0

P10.5 a. m = 1; a = 1/2,-1/3 b = 0,-1/3
b. m = 2; a = 1/2,-1/3,-1/6 b = 0,-1/3, 0

P10.7 a. m =1; a = 1/2,-4/9   b = 0,0
b. m =2; a = 1/2,-4/9,0 b = 0,0,0
c. m =4; a = 1/2,-4/9,0,-1/18 b = 0,0,0,0

P10.9 a. m = 1; a = 1/2, 1/4 ,   b = 0, 0.6036
b. m = 2; a = 1/2, 1/4, 0, b = 0, 0.6036, 0
c. m = 4; a = 1/2, 1/4, 0, 1/4, 0, b = 0, 0.6036, 0, 0.1036, 0

P10.11 a. m = 1; a = 1/2, -1/2 , b = 0, -1/2
b. m = 2; a = 1/2, -1/2, 0, b = 0, -1/2, 0

P10.13 a. m = 1; a = 1/2, 1/2 , b = 0, -1/2
b. m = 2; a = 1/2, 1/2, 0, b = 0, -1/2, 0

P10.15 a. m = 1; a = 1/2, 1/3 , b = 0, 1/3
b. m = 2; a = 1/2, 1/3, -1/6, b = 0, 1/3, 0

P10.17 a. m = 1; a = 1/2, 4/9 , b = 0, 0
b. m = 2; a = 1/2, 4/9, 0, b = 0, 0, 0
c. m = 4; a = 1/2, 4/9, 0, 1/18, b = 0, 0, 0, 0

P10.19 a. m = 1; a = 1/2, -1/4 , b = 0, -0.6036
b. m = 2; a = 1/2, -1/4, 0, b = 0, -0.6036, 0
c. m = 4; a = 1/2, -1/4, 0, -1/4, 0, b = 0, -0.6036, 0, -0.1036, 0

P10.21
g = 2.0000   1.0000+ 1.0000i 0   1.0000- 1.0000i
gg = 2.0000   1.0000- 1.0000i 0   1.0000+ 1.0000i

P10.23
g = 2.0000   -1.0000+ 1.0000i 0   -1.0000- 1.0000i
gg = 2.0000   -1.0000- 1.0000i 0   -1.0000+ 1.0000i

P10.25
g = 2.0000   -0.6667- 0.6667i   -0.6667   -0.6667+ 0.6667i
gg = 2.0000   -0.6667+ 0.6667i   -0.6667   -0.6667- 0.6667i

P10.27
g = 3.00   -1.3333   0 + 0.00i -0.3333   0.0+ 0.0i   -1.3333- 0.00i
gg = 3.00   -1.3333- 0.00i   0.00- 0.00i -0.3333+ 0.0i   0.0   -1.3333- 0.00i

P10.29 g = 4.0 1.0 +2.4142i 0 1.0 +0.4142i
0   1.0 -0.4142i   0   1.0 -2.4142i

gg = 4.0 1.0 -2.4142i 0 1.0 -0.4142i
0   1.0 +0.4142i   0   1.0 +2.4142i

Chapter 11

P11.1 a. df = 0.010 b. db = 0.010 c. dc = 0.010 d. dc2 = 0.040, dr = (4*dc -dc2)/3 = 0

P11.3 a. df = 6 b. db = 2 c. dc = 4 d. dc2 = 6.6667; dr = 3.1111

P11.5 a. df = 0.20 b. db = 0.3333 c. dc = 0.25 d. dc2 = 0.25; dr = 0.25

P11.7 a. 25 b. 21.7789 c. 22 d. 21.6411 e. 21.4043

P11.9 a. 0.4083 b. 0.4056 c. 0.4056 d. 0.4055 e. 0.4054

P11.11 a. 1.5000 b. 1.5675 c. 1.6667 d. 1.5708 e. 1.5000

P11.13 a. 1.1667 b. 1.2075 c. 1.2222 d. 1.2092 e. 1.2000

P11.15 a. 1.3090 b. 1.4172 c. 1.3727 d. 1.4241 e. 1.4434

P11.17 a. 1.6089 b. 1.6520 c. 1.5507 d. 1.6528 e. 1.7365

P11.19 a. 5.4807 b. 5.2799 c. 5.2894 d. 5.2725 e. 5.2541

Chapter 12

P12.1

c. yy = 2.00   2.2100   2.4421   2.6985   2.9818   3.2949   3.6409   4.0231   4.4456   4.9124   5.4282

d. yy = 2.00   2.2103   2.4428   2.6997   2.9836   3.2974   3.6442   4.0275   4.4511   4.9192   5.4366

P12.3

c. yy = 1.00   0.9098   0.8343   0.7704   0.7155   0.6679   0.6263   0.5895   0.5567   0.5274   0.5011

d. yy = 1.00   0.9091   0.8333   0.7692   0.7143   0.6667   0.6250   0.5882   0.555   0.5263   0.5000

P12.5

c. yy = 2.00   2.1905   2.3630   2.5194   2.6618   2.7916   2.9104   3.0194   3.1197   3.2122   3.2979

d. yy = 2.00   2.1903   2.3627   2.5191   2.6614   2.7912   2.9100   3.0190   3.1192   3.2118   3.2974

P12.7

c. yy = 2.00   1.9800   1.9212   1.8271   1.7030   1.5559   1.3933   1.2230   1.0524   0.8878   0.7343

d. yy = 2.00   1.9801   1.9216   1.8279   1.7043   1.5576   1.3954   1.2253   1.0546   0.8897   0.7358

P12.9

c. yy = 1.00   1.0008   1.0075   1.0265   1.0648   1.1310   1.2375   1.4028   1.6569   2.0505   2.6732

d. yy = 1.00   1.0010   1.0080   1.0274   1.0661   1.1331   1.2411   1.4092   1.6686   2.0730   2.7182

P12.11

d. yy = 1.00   0.7843   0.7040   0.7061   0.7564   0.8187   0.8635   0.8697   0.8242   0.7227   0.5683

P12.13

d. yy = 2.00 2.1237 2.2450 2.3643 2.4819 2.5981
2.7129 2.8266 2.9394 3.0512 3.1623

P12.15

d. yy = 5.00 4.2411 3.5621 2.9618 2.4416 2.0013
1.6411 1.3608 1.1605 1.0403 1.00

P12.17

d. yy = 3.00 2.9667 2.8724 2.7355 2.5755 2.4158
2.2757 2.1670 2.0921 2.0464 2.0218

P12.19

d. yy = -1.00 -0.8884 -0.8298 -0.802 -0.7924 -0.791
-0.7898 -0.7807 -0.7539 -0.6952 -0.5809

(P 12.21-25, Results for ABM3, n = 10)

P12.21
yy = 1.0000 1.1053 1.2236 1.3588 1.5144 1.6948
1.9047 2.1492 2.4342 2.7658 3.1513

P12.23
yy = 1.0000 1.1099 1.2409 1.3946 1.5727 1.7770
2.0097 2.2736 2.5714 2.9063 3.2820

P12.25
yy = 1.0000 1.0050 1.0203 1.0465 1.0851 1.1385
1.2101 1.3049 1.4311 1.6015 1.8376

Chapter 13

P13.1
a. uu = 2.0000 2.0000 2.0800 2.2480 2.5152 2.8963
3.4100 4.0796 4.9336 6.0068 7.3413

b. uu = 2.0000 2.0400 2.1688 2.3995 2.7491 3.2394
3.8980 4.7587 5.8638 7.2649 9.0257
exact, y= $1.5*\exp(x) + 0.5*\exp(-x) -x$;

P13.3
a. uu = 7.0000 6.2000 5.4400 4.7376 4.1050 3.5503
3.0789 2.6940 2.3975 2.1903 2.0724

b. uu = 7.0000 6.2200 5.4951 4.8350 4.2465 3.7346
3.3028 2.9534 2.6878 2.5069 2.4112
exact, y = $\exp(-x) + x \exp(-x) + x^2 - 4x + 6$

P13.5
a. uu = -2.00 -1.20 -0.7200 -0.6624 -1.1296 -2.2404
-4.1463 -7.0507 -11.2329 -17.0808 -25.1355

b. uu = -2.00 -1.360 -1.1616 -1.5471 -2.7004 -4.8751
-8.4344 -13.9072 -22.0705 -34.0707 -51.6013
exact, y = $-\exp(x) + \exp(-x) - x.^2 + 4*x - 2$

P13.7
a. uu = 2.0000 1.8000 1.6200 1.4580 1.3122 1.1810
1.0629 0.9566 0.8609 0.7748 0.6974

b. uu = 2.0000 1.8100 1.6381 1.4824 1.3416 1.2142
1.0988 0.9944 0.9000 0.8145 0.7371
exact, y = $2 \exp(-x)$

P13.9
a. uu = 1.0000 0.8000 0.6800 0.6200 0.6097 0.6446
0.7241 0.8502 1.0272 1.2615 1.5613

b. uu = 1.0000 0.8400 0.7402 0.6918 0.6912 0.7384
0.8364 0.9907 1.2094 1.5028 1.8838

P13.11
midpoint method,
uu = 1.0000 1.0000 1.0060 1.0242 1.0616 1.1271
1.2332 1.3985 1.6508 2.0322 2.6053

P13.13 Solution is similar to y = 2 cos(2x), fairly large n required for nice solution.

P13.15
midpoint method,
uu = 0 -0.6400 -1.2115 -1.7331 -2.2222 -2.6902
-3.1443 -3.5889 -4.0269 -4.4601 -4.8899

P13.17
midpoint method,
uu = 2.0000 2.0494 2.0976 2.1447 2.1908 2.2360
2.2803 2.3237 2.3664 2.4083 2.4494

P13.19
Euler's method,
uu = 5.00 6.3000 7.8200 9.5803 11.6016 13.9047
16.5106 19.4407 22.7163 26.3591 30.3906
exact, y = $2 x^2 + 3 x^3$

P13.21
Euler's method,
uu = 3.00 3.3000 3.6600 4.0745 4.5399 5.0535
5.6134 6.2183 6.8670 7.5587 8.2928

midpoint method,
uu = 3.00 3.3300 3.7147 4.1507 4.6356 5.1677
5.7456 6.3684 7.0351 7.7452 8.4982
exact, y = $2 x^2 + 1/x$

P13.23
midpoint method,
uu = 0   0.5400 0.9834 1.3495 1.6532 1.9060 2.1169 2.2929 2.4397 2.5619 2.6631

P13.25
midpoint method,
uu = 1.00 1.0000 0.9622 0.8807 0.7569 0.5975 0.4130 0.2162 0.0208 -0.1595 -0.3125

P13.27
midpoint method,
x = 0   0.1000 0.2000 0.3000 0.4000 0.5000 0.6000 0.7000 0.8000 0.9000

a = 1, uu = 1.0000 1.0950 1.1799 1.2541 1.3169 1.3668 1.4015 1.4169 1.4055 1.3505

a = 2, uu = 1.0000 1.0800 1.1168 1.1071 1.0475 0.9337 0.7605 0.5204 0.2016 -0.2192

a = 3, uu = 1.0000 1.0550 1.0126 0.8717 0.6364 0.3176 -0.0655 -0.4803 -0.8718 -1.1323

P13.29
midpoint method, on [1, 3]
k = 1, uu = 1.000 0.980 0.8999 0.7172 0.3441 -0.4089 -1.9740 -5.3898 -13.2966 -32.7878 -84.0125

k = 2, uu = 1.000 0.960 0.8006 0.4442 -0.2615 -1.6323 -4.3614 -10.0592 -22.6926 -52.6169 -128.4744

k = 3, uu = 1.000 0.940 0.7021 0.1810 -0.8179 -2.6792 -6.2099 -13.2143 -27.9828 -61.3645 -142.5102

k = 4, uu = 1.00 0.920 0.6044 -0.0724 -1.3264 -3.5583 -7.5656 -15.0492 -29.8918 -61.5631 -134.6992

Note, using larger n shows that solutions actually decay more rapidly than results above indicate.

P13.31

t	x1	x2	x3	x4
a.				
0	1.0000	0	0	0
0.1	1.2855	1.2707	1.6848	0.6997
0.2	1.5438	2.7317	3.8321	1.6441
0.3	1.7775	4.4741	6.6155	2.9190
0.4	1.9890	6.6179	10.2687	4.6398
0.5	2.1804	9.3235	15.1057	6.9627
0.6	2.3536	12.8053	21.5499	10.0981
0.7	2.5102	17.3510	30.1713	14.3305
0.8	2.6520	23.3476	41.7391	20.0436
0.9	2.7803	31.3158	57.2908	27.7552
1.0	2.8964	41.9575	78.2258	38.1647
b.				
0	0	1.0000	0	0
0.1	-0.2855	-0.2707	-1.6848	-0.6997
0.2	-0.5438	-1.7317	-3.8321	-1.6441
0.3	-0.7775	-3.4741	-6.6155	-2.9190
0.4	-0.9890	-5.6179	-10.2687	-4.6398
0.5	-1.1804	-8.3235	-15.1057	-6.9627
0.6	-1.3536	-11.8053	-21.5499	-10.0981
0.7	-1.5102	-16.3510	-30.1713	-14.3305
0.8	-1.6520	-22.3476	-41.7391	-20.0436
0.9	-1.7803	-30.3158	-57.2908	-27.7552
1.0	-1.8964	-40.9575	-78.2258	-38.1647
c.				
0	0	0	1.0000	0
0.1	0.1903	1.0803	2.5897	0.6997
0.2	0.3625	2.3692	4.6508	1.6441
0.3	0.5184	3.9557	7.3563	2.9190
0.4	0.6594	5.9585	10.9390	4.6398
0.5	0.7869	8.5365	15.7123	6.9627
0.6	0.9024	11.9029	22.0987	10.0981
0.7	1.0068	16.3442	30.6679	14.3305
0.8	1.1013	22.2463	42.1885	20.0436
0.9	1.1869	30.1290	57.6973	27.7552
1.0	1.2642	40.6932	78.5937	38.1647
d.				
0	0	0	0	1.0000
0.1	-0.0952	-0.5402	-0.7948	0.6502
0.2	-0.1813	-1.1846	-1.8254	0.1779
0.3	-0.2592	-1.9778	-3.1782	-0.4595
0.4	-0.3297	-2.9793	-4.9695	-1.3199
0.5	-0.3935	-4.2683	-7.3561	-2.4813
0.6	-0.4512	-5.9515	-10.5493	-4.0491
0.7	-0.5034	-8.1721	-14.8339	-6.1653
0.8	-0.5507	-11.1231	-20.5942	-9.0218
0.9	-0.5934	-15.0645	-28.3487	-12.8776
1.0	-0.6321	-20.3466	-38.7969	-18.0824

P13.33

t	x1	x2	x3	x4
a.				
0	1.0000	0	0	0
0.1	1.2855	1.0138	1.0659	0.2325
0.2	1.5438	2.0713	2.2897	0.5408
0.3	1.7775	3.1993	3.7161	0.9445
0.4	1.9890	4.4291	5.3993	1.4674
0.5	2.1804	5.7973	7.4047	2.1391
0.6	2.3536	7.3473	9.8117	2.9959
0.7	2.5102	9.1308	12.7170	4.0828
0.8	2.6520	11.2099	16.2382	5.4548
0.9	2.7803	13.6596	20.5188	7.1798
1.0	2.8964	6.5705	25.7336	9.3412
0	0	1.0	0	0

P13.35

t	x1	x2	x3	x4
b.				
0	0	1.0	0	0
0.1	0.3321	1.0679	-1.0418	-0.9589
0.2	0.7377	1.0345	-2.4739	-2.1003
0.3	1.2332	0.8202	-4.5103	-3.5485
0.4	1.8383	0.2996	-7.4663	-5.4783
0.5	2.5774	-0.7224	-11.809	-8.1414
0.6	3.4801	-2.5466	-18.232	-11.9044
0.7	4.5827	-5.6338	-27.7692	-17.306
0.8	5.9294	-10.6863	-41.96	-25.1414
0.9	7.5743	-18.7694	-63.096	-36.587
1.0	9.5833	-31.4941	-94.603	-53.390

P13.37

t	x1	x2	x3	x4
c.				
0	0	0	1.0000	0
0.1	-0.2673	0.9647	2.7103	0.2569
0.2	-0.3566	2.2283	4.7373	0.6605
0.3	-0.3556	3.8099	7.2624	1.2748
0.4	-0.3141	5.7966	10.5250	2.1888
0.5	-0.2593	8.3361	14.8398	3.5262
0.6	-0.2047	11.6412	20.6240	5.4580
0.7	-0.1566	16.0026	28.4345	8.2203
0.8	-0.1169	21.8114	39.0189	12.1377
0.9	-0.0856	29.5929	53.3841	17.6562
1.0	-0.0616	40.0525	72.8899	25.3870

P13.39

t	x1	x2	x3	x4
d.				
0	0	0	0	1.0000
0.1	-0.9515	-1.2072	-0.1605	1.0952
0.2	-2.1405	-2.9905	-0.6687	1.1813
0.3	-3.3964	-5.1759	-1.5203	1.2592
0.4	-4.4481	-7.4084	-2.6306	1.3297
0.5	-4.9407	-9.1499	-3.8157	1.3935
0.6	-4.4804	-9.7182	-4.7867	1.4512
0.7	-2.7098	-8.3817	-5.1684	1.5034
0.8	0.5868	-4.5112	-4.5472	1.5507
0.9	5.3622	2.2172	-2.5515	1.5934
1.0	11.1882	11.5937	1.0376	1.6321

Chapter 14

P14.1

a. xx = 0 0.0245 0.2158 0.4069 0.5984 0.7917 0.9881 1.1844 1.3808 1.5761 1.7686 1.9598 2.1512 2.3443 2.5406   2.7370 2.9333 3.1290 3.1416

yy = 1.0000 1.1602 2.3767 3.5062 4.5098 5.3554 6.0100 6.4336 6.6100 6.5334 6.2148 5.6711 4.9197 3.9798 2.8726 1.6550 0.3737 -0.9177 -1.0000

b.
  x = 0   0.3142 0.6283 0.9425 1.2566 1.5708 1.8850 2.1991 2.5133 2.8274 3.1416

yy = 1.0000 0.9527 0.8113 0.5899 0.3103 -0.0000 -0.3103 -0.5899 -0.8113 -0.9527 -1.0000

P14.3

a. xx = 0 0.0078 0.0703 0.1328 0.1953 0.2578 0.3203 0.3828 0.4453 0.5078 0.5703 0.6328 0.6953 0.7578 0.8203 0.8828   0.9453 1.0000

yy = 10.0000 9.8856 9.0118 8.2076 7.4679 6.7877 6.1629 5.5892 5.0630 4.5807 4.1393 3.7357 3.3673 3.0316 2.7263 2.4494 2.1989 2.0000

b.

x = 0   0.1   0.2   0.3   0.4   0.5   0.6   0.7   0.8   0.9   1.0

yy = 10.00 8.6195 7.4117 6.3564 5.4361 4.6352 3.9400 3.3386 2.8207 2.3772 2.00

P14.5

a. xx = 0   0.0078 0.0703 0.1328 0.1953 0.2578 0.3203 0.3828 0.4453 0.5078 0.5703 0.6328 0.6953 0.7578 0.8203 0.8828 0.9453 1.0000

   yy = 2.0000 2.0626 2.5109 2.8698 3.1444 3.3406 3.4647 3.5235 3.5236 3.4720 3.3756 3.2410 3.0747 2.8828 2.6711 2.4452 2.2099 2.0000

b.

x = 0   0.1   0.2   0.3   0.4   0.5   0.6   0.7   0.8   0.9   1.0

yy = 2.0 2.6998 3.1742 3.4469 3.5446 3.4958 3.3286 3.0709 2.7483 2.3844 2.0000

P14.7

a.

xx = 0 0.0245 0.2209 0.4172 0.6136 0.8099 1.0063 1.2026 1.3990 1.5953 1.7917 1.9880 2.1844 2.3807 2.5771 2.7734 2.9698 3.1416

yy = 0 -0.2445 -2.0309 -3.5283 -4.7511 -5.7110 -6.4173 -6.8768 -7.0940 -7.0710 -6.8075 -6.3010 -5.5466 -4.5370 -3.2625 -1.7108 0.1330 2.0000

b.

x = 0   0.3142 0.6283 0.9425 1.2566 1.5708 1.8850 2.1991 2.5133 2.8274 3.1416

yy = 0 -2.7723 -4.8234 -6.2039 -6.9479 -7.0738 -6.5847 -5.4685 -3.6976 -1.2284 2.0000

P14.9

a.

xx = 1.0000 1.0547 1.2377 1.4401 1.6452 1.8509 2.0579 2.2666 2.4770 2.6898 2.9058 3.1264 3.3428 3.5558 3.7669 3.9774 4.1884 4.4013 4.6172 4.8378 5.0655 5.2910 5.5159 5.7406 5.9637 6.1871 6.4128 6.6430 6.8803 7.1137 7.3424 7.5692 7.7962 8.0000

yy = 1.0000 0.9192 0.6588 0.3889 0.1364 -0.0923 -0.2945 -0.4664 -0.6049 -0.7074 -0.7720 -0.7975 -0.7846 -0.7379 -0.6624 -0.5627 -0.4437 -0.3101 -0.1671 -0.0199 0.1260 0.2579 0.3713 0.4617 0.5255 0.5615 0.5687 0.5465 0.4949 0.4191 0.3256 0.2191 0.1046 0

b.

x = 1.0000 1.7000 2.4000 3.1000 3.8000 4.5000 5.2000 5.9000 6.6000 7.3000 8.0000

yy = 1.000 0.1872 -0.4242 -0.6985 -0.6096 -0.2622 0.1543 0.4474 0.5007 0.3157 0

P14.11

xx = 1.0000 1.0703 1.2503 1.4541 1.7018 2.0047 2.3813 2.8604 3.3916 3.9541 4.5166 5.0791 5.6416 6.2041 6.7666 7.3291 7.8916 8.4541 9.0166 9.5791 10.0000

yy = 0   0.0456 0.1489 0.2458 0.3407 0.4306 0.5127 0.5833 0.6314 0.6592 0.6707 0.6711 0.6635 0.6504 0.6333 0.6134 0.5914 0.5680 0.5437 0.5188 0.5000

b.

x = 1.0000 1.9000 2.8000 3.7000 4.6000 5.5000 6.4000 7.3000 8.2000 9.1000 10.0000

y = 0   0.3919 0.5626 0.6360 0.6599 0.6565 0.6374 0.6090 0.5752 0.5384 0.5000

P14.13
a.
x = 1.0000 1.0078 1.0703 1.1328 1.1953 1.2578 1.3203 1.3828 1.4453 1.5078 1.5703 1.6328 1.6953 1.7578 1.8203 1.8828 1.9453 2.0000

y = 0.5000 0.5066 0.5634 0.6290 0.7049 0.7931 0.8958 1.0157 1.1557 1.3194 1.5106 1.7339 1.9941 2.2971 2.6490 3.0568 3.5284 4.0000

b.
x = 1.0000 1.1000 1.2000 1.3000 1.4000 1.5000 1.6000 1.7000 1.8000 1.9000 2.0000

y = 0.5000 0.5937 0.7117 0.8615 1.0531 1.2987 1.6136 2.0164 2.5297 3.1804 4.0000

P14.15
a.
x = 0   0.0078 0.0703 0.1328 0.1953 0.2578 0.3203 0.3828 0.4453 0.5078 0.5703 0.6328 0.6953 0.7578 0.8203 0.8828 0.9453 1.0000

y = 1.0000 1.0048 1.0438 1.0846 1.1274 1.1724 1.2199 1.2702 1.3235 1.3803 1.4408 1.5055 1.5750 1.6496 1.7300 1.8168 1.9110 2.0000

P14.17
a.
x = 1.0000 1.0078 1.0703 1.1328 1.1953 1.2578 1.3203 1.3828 1.4453 1.5078 1.5703 1.6328 1.6953 1.7578 1.8203 1.8828 1.9453 2.0000

y= 0   -0.0213 -0.1848 -0.3376 -0.4819 -0.6193 -0.7508 -0.8776 -1.0005 -1.1201 -1.2370 -1.3516 -1.4645 -1.5759 -1.6863 -1.7958 -1.9048 -2.0000

P14.19
b.
x = 0   0.0078 0.0703 0.1328 0.1953 0.2578 0.3203 0.3828 0.4453 0.5078 0.5703 0.6328 0.6953 0.7578 0.8203 0.8828 0.9453 1.0000

y = 1.0000 0.9853 0.8736 0.7712 0.6773 0.5911 0.5119 0.4393 0.3727 0.3119 0.2563 0.2058 0.1601 0.1190 0.0823 0.0499 0.0215 0.0000

P14.21
b.
x = 1.0000 1.0078 1.0703 1.1328 1.1953 1.2578 1.3203 1.3828 1.4453 1.5078 1.5703 1.6328 1.6953 1.7578 1.8203 1.8828 1.9453 2.0000

y = 1.0000 0.9922 0.9343 0.8828 0.8366 0.7950 0.7574 0.7232 0.6919 0.6632 0.6368 0.6124 0.5899 0.5689 0.5494 0.5311 0.5141 0.5000

P14.23
b.
x = 0   0.0078 0.0549 0.1019 0.1490 0.1963 0.2449 0.2974 0.3536 0.4133 0.4758 0.5383 0.6008 0.6633 0.7258 0.7883 0.8508 0.9133 0.9758 1.0000

y = 1.0000 0.9983 0.9801 0.9488 0.9046 0.8477 0.7767 0.6863 0.5751 0.4416 0.2870 0.1191 -0.0600 -0.2483 -0.4436 -0.6439 -0.8471 -1.0514 -1.2550 -1.3333

P14.25

x = 0 0.0078 0.0703 0.1328 0.1953 0.2578 0.3203 0.3828 0.4453 0.5078 0.5703 0.6328 0.6953 0.7578 0.8203 0.8828 0.9453 1.0000

y = 0 0.0180 0.1611 0.3011 0.4356 0.5631 0.6829 0.7947 0.8988 0.9958 1.0861 1.1703 1.2491 1.3229 1.3923 1.4577 1.5195 1.5708

Chapter 15

P15.1

Solution at t = 0.5

a.	0	0.0683	0.1444	0.2364	0.3523	0.50
b.	0	0.0682	0.1443	0.2363	0.3522	0.50
c.	0	3.9062	-15.6250	29.6875	-25.625	0.50 (unstable)
d.	0	0.0693	0.1461	0.2381	0.3533	0.50
e.	0	0.0682	0.1443	0.2365	0.3521	0.5000

P15.3

Solution at t = 0.5

a.	0	0.0010	0.0016	0.0016	0.0010	0
b.	0	0.0008	0.0012	0.0012	0.0008	0
c.	0	-44.2775	27.3650	27.3650	-44.2775	0 (unstable)
d.	0	0.0053	0.0086	0.0086	0.0053	0
e.	0	0.0005	0.0016	0.0016	0.0005	0

P15.5

Solution at t = 0.5

a.	0	0.0000	0.0000	0.0000	0.0000	0
b.	0	-0.0029	0.0048	-0.0048	0.0029	0
c.	0	-19809	32052	-32052	19809	0 (unstable)
d.	0	0.0000	-0.0000	0.0000	-0.0000	0
e.	0	-0.0621	0.1004	-0.1004	0.0621	0

P15.7

Solution at t = 0.5

a.	0	0.2012	0.4020	0.6020	0.8012	1.00
b.	0	0.2009	0.4008	0.6015	0.8005	1.00

c.	0	39.0625	-171.8750	334.375	-299.375	1.0 (unstable)
d.	0	0.2081	0.4129	0.6127	0.8078	1.00
e.	0	0.1985	0.3991	0.6022	0.7952	1.00

P15.9

Solution at t = 0.5

a.	0	0.0825	0.1335	0.1335	0.0825	0
b.	0	0.0724	0.1000	0.1171	0.0618	0
c.	0	90250	-120000	87790	38130	0 (unstable)
d.	0	0.4058	0.6610	0.6660	0.4141	0
e.	0	0.5248	-0.2946	0.3233	0.1623	0

P15.11

a. oscillations most easily shown graphically
b. unstable

P15.17

a.	0.00	0.0000	0.0000	0.0000	0.00
	0.00	0.0547	0.1172	0.1797	0.25
	0.00	0.1172	0.2422	0.3672	0.50
	0.00	0.1797	0.3672	0.5547	0.75
	0.00	0.2500	0.5000	0.7500	1.00

P15.19

a.	0.0000	0.0625	0.2500	0.5625	1.00
	-0.0625	0.0547	0.3047	0.6797	1.1875
	- 0.2500	-0.0703	0.2422	0.6797	1.2500
	- 0.5625	-0.3203	0.0547	0.5547	1.1875
	- 1.0000	-0.6875	-0.2500	0.3125	1.0000

P15.21

a.	1.00	1.0625	1.2500	1.5625	2.0000
	0.9375	0.9688	1.1562	1.4687	1.9375
	0.7500	0.7813	0.9688	1.2812	1.7500
	0.4375	0.4688	0.6563	0.9688	1.4375
	0.0000	0.0625	0.2500	0.5625	1.00

P15.23

a.	0.0000	0.0156	0.1250	0.4219	1.0000
	0.0156	0.1719	0.3438	0.6094	1.0781
	0.1250	0.3438	0.4844	0.6250	0.8750
	0.4219	0.6094	0.6250	0.5469	0.4844
	1.0000	1.0781	0.8750	0.4844	0.0000

P15.25

a.	0.0000	0.0000	0.0000	0.0000	0.0000
	0.2656	0.2656	0.1875	0.0157	-0.2344
	0.6250	0.6406	0.4844	0.1406	-0.3750
	1.1719	1.2031	0.9687	0.4531	-0.3281
	2.0000	2.0625	1.7500	1.0625	0.0000

Index

Numbers in parentheses indicate reference is to exercise.

A

Absolute stability region, 514
Abel, Niels Henrik, 4
Acceleration, 22–24, 41, (44), 442–444, 458–461, (473)
Accuracy, 3
Adams-Bashforth method, 502–507, 523, 552, 568
Adams-Moulton method, 508, 524
Adams Bashforth-Moulton method, 509–513, 524, (526), 552–556, (570)
Airy's equation, (570)
Al-Kashi, Jemshid, 1
Algorithm, careful application, 26
Algorithms. *See* Algorithm Index following Index
Approximate
 to n digits, 13, 14, 18
 to n decimal places, 13
Approximation, function, 349–432. *See also* Least-squares approximation, Fourier methods, Pade approximation, Taylor approximation
Arc length, 135, 157, (174–175)
Areas, 1
Archimedes, 1, 48
A-stability, 558
Assymptotic error constant, 64, 65, 81
Augmented matrix, 97

B

Babbage, Charles, 2
Babylonian mathematics, 1
Backward difference formula, 436–439, 471, (473)
Back substitution, 98, 99, 101, 107, 109, 115, 119, 121
Bank balance, (527)
Basis functions for finite-element method, 646–653, (664)
Beam, deflection, (92), 235, 245, (574), 576, 582, 598, (608)
Bernoulli equation, (528)
Bessel equation, (570)
Bessel functions, (345)
Big-Oh. *See* Order
Binary representation, 15–16
Bisection, 47, 50–54, 87, (89), (92)
Bit, 16
Blasius equation, (607)
Bouncing ball data, (391)
Boundary condition
 Dirichlet, 578
 insulated, 632
 mixed, 578, 584
 Neuman, 578
Boundary value problems 575–605, (606–608)
 finite difference method, 592–601, 605
 shooting method, 578–592, 604–605
Brent's method, 2, 85–86
Briggs, Henry, 2

Broyden-Fletcher-Goldfarb-Shanno (BFGS) method. *See* Quazi-Newton methods
Brusselator model, (199)
Building permits, (347)
Bulirsch-Stoer interpolation. *See* Rational function interpolation
Burger's equation, (665)
Butcher's method, 500
Byte, 16

C

Cancellation error, 19
Census data, (347)
Central difference formula, 436–439, 441, 471, (473), 593–594, 599, 623, 628, 634
Characteristic polynomial, 514
Chebyshev equation, (570)
Chemical processes, (199), 284, 291, 311, 332, 361, (391), (527), 531, 536, 540, 550–551, (572–573), (607–608)
Chinese mathematics, 1
Cholesky form, 201, 217–218, 227, (231)
Chopping arithmetic, 17
Clamped boundary (spline), 333, (348)
Classic Runge-Kutta method, 495–498, 523, (526), 549–551, (570). *See also* Runge-Kutta methods, fourth order
Complex conjugate, (279)

695

Composite integration methods,
    452–461, 471, (473)
  Romberg, 458–461, (473)
  Simpson, 455–457, 471, (473)
  trapezoid, 453–455, 471, (473)
Computation, efficient, 25
Computational effort, 3, 113, 119, 150.
    See also Flops
Computer representation of numbers,
    15–17, (45)
Conjugate gradient methods, 159–160
Convergence, 3
  fixed point iteration, 10–11, (43–44)
  geometric illustration of, 5, 10, 135,
    143, 144
  linear, 54, (92)
  order of, 41, 64–65, 73
  quadratic, 64, 73
  rate of, 65, 81, (281)
  tests for, 13, (92). See also Stopping
    conditions
  use of eigenvalues for, 156–157
Convergence, conditions for
  fixed-point iteration, 11
  fixed-point iteration, for systems,
    186, 196, (200)
  Gauss-Seidel, 150
  Jacobi method, 142
  Newton's method, 73
  nonlinear finite difference method,
    605
  secant method, 65
  SOR, 155
Condition of a matrix 105. See also Ill-
    conditioned matrix
Cooley, 3
Copernican revolution, 2
Crank-Nicolson method, 628–631, 659,
    (663)
Crout form, 215
Cubic equations, 2
Cubic spline interpolation, 3, 325–333,
    337, (339–341)
Cubic least squares. See Least squares
    approximation

D

Danielson, 3
Davidon-Fletcher-Powell (DFP)
    method. See Quazi-Newton meth-
    ods
DeBoor, 3
Degree of precision, 450
Derivative, approximation of, 433,
    436–444, (473). See also Backward
difference, Central difference,
    Forward difference
Derivative boundary conditions, 632
Derivative initial conditions, 633
Determinant of a matrix, 224,
    (228–229)
Diagonally dominant matrix, 104, 120,
    142, (170), (608)
Difference formulas
  first derivatives, 436–440, 471, (473),
    614
  higher derivatives, 440–441, 471, 614
  partial derivatives, 441–442,
    614–615, 620, 623, 627
Differentiation, numerical, 436–444.
    See also Richardson extrapolation
Differential equations, ordinary,
    477–608
  first-order, 477–528
  higher-order, 529–530, 532–533, 541,
    (570), 575–608
  systems, 529, 531, 534–40, 542–569,
    (571)
Differential equations, partial,
    609–662
Direct methods, 3.
    See also Gaussian elimination
Dirichlet boundary condition, 578
Discrete least-squares, 349–372,
    393–406
Discrete Fourier transform, 3, 393
Divided difference table, 295–304,
    307–312,
Disease, spread, (527)
Doolittle form, 215, (231)
Drag coefficient, (345), 478, 513, (527),
    577

E

Earnings, (284), (527–8)
Eigenvalues and eigenvectors,
    233–279. See also Power method,
    Inverse power method, QR
    method
  applications, (277–279)
  approximation of, 233–273
  bounds on, 11–12
  convergence of Jacobi method, 142
  for beam buckling, 235
  stability of finite differences for
    PDE, 621, 627–628
Electrical circuit, 94, 99, (130), 202,
    206, 220
Electrostatic potential, 543, 546–547,
    548, 580

Elliptic integrals, (346), (474)
Elliptical orbit, 49, 59, 63, 435, 457
Elliptical PDE, 609, 613, 640–661. See
    also Poisson, Laplace
Enthalpy, (345), (346)
Equations, numerical solution of, 1, 2,
    47–92
Eratosthenes, 1
Error
  cancellation, 19
  from inexact representation, 17–19
  from mathematical approximations,
    20–22
  measuring, 14–15
  relative, 14, (44)
  round-off, 3, 17–18, (44)
  total squared, 352–380, 402
  truncation, 20–22, 438, 441, (476),
    483, 502, 503, 508, 509, 620, 628
Error function, (474)
Euler's backward method, 558
Euler's method
  single ODE, 477, 479–483, 522,
    (526)
  systems of ODE, 529, 534–536,
    542–543, 567, (570–571)
Euler's modified method (Euler-
    Cauchy), 492, 522
Examples. See Example Index follow-
    ing Index
Explicit methods for PDE, 615–622,
    632–637, 659–660, (663–664)
Explicit multi-step methods for ODE,
    502–07, 523
Extrapolation. See Acceleration

F

Factorization of matrix. See LU factor-
    ization, QR factorization
False position. See Regula falsi
Fast Fourier transform. See FFT
FFT, 393, 407–427, (429–431)
Fibonacci, Leonardo, 2
Finite-difference method for ODE-
    BVP, 575
  linear, (129), 592–598, 605,
    (606–607)
  nonlinear, 599–601, 605, (607)
Finite-difference method for PDE,
    131, 132, 609, 613–644, 659–661,
    (664)
Finite-element method for elliptic
    PDE, 609, 613, 645–657, 661,
    (664)
Five-point formula, 442

Fixed-point iteration, 5, 41
  convergence theorem, 11, (43)
Fixed-point iteration for systems, 181–186
  convergence theorem, 186
Fletcher-Reeves method, 2
Floating-point, 15
Floating-point operations. See Flops
Floating sphere, 48, 53, 71
Forces on a truss, 95, 102, 111–112, (129), (169)
Flops 25, (130), (574), (666)
  basic Gaussian elimination 103–104, 113, 123
  Fourier transforms, 407
  Gauss vs. Gauss-Jordan, 123, (130)
  Householder matrix, 257
  Jacobi method, 143
  Thomas method, 119
Flow in a pipe, (91), (474)
Forward difference formula, 436–438, 471, (473), (608), 614, 615, 623
Forward elimination, 96
Fourier methods, 393–431
Fourth-order Runge-Kutta methods. See Runge-Kutta methods
Fresnel integrals, (346), (474)
Function
  orthogonal, 376–380
  rational, 316–320, 382–384
Function approximation, 3, 349–406. See also Least squares approximation, Fourier methods
Function approximation at a point, 381–384. See also Taylor approximation, Pade approximation
Functional, minimization of, 645. See also Minimum of a function

G

Gas law
  Beattie-Bridgeman, (91)
  Peng-Robinson, (91)
  van der Waals, (91)
Gauss, Carl Friedrich, 1, 3
Gauss-Jordan method, (130)
Gauss-Seidel method, 131, 134, 144–151, 166, (168–70), 640
  conditions for convergence, 150–151
Gaussian elimination, 6–7, (45). See also Thomas method 93–124, (125–130)
  basic, 96–105, 123
  for tridiagonal systems 113–121, 124
  matrix notation, 97
  row pivoting, 106–113, 123
  use by Chinese, 1
Gaussian quadrature, 462–467, 472, (473)
Geometric figures, 395, 406, 424, 427, (430–431)
Gerschgorin Circle Theorem, 41, (44, 45), 233, 621, 628
Ginzburg-Landau equation, (527)
Givens rotation, 248, 254, 257–260, 264–265, 273
Global truncation error. See Error
Gompertz growth curve, 362
Gram-Schmidt process, 376–377, (392)
Greek mathematics, 1

H

Heat capacity, (345)
Heat distribution, steady state in plane, 609. See also Laplace or Poisson equations
Heat distribution, steady state in a rod, (607)
Heat equation (temperature in a rod), 610, 614–632, (663)
Hermite equation, (570)
Hermite interpolation, 283, 310–315, 336, (341)
Hessian, 193, 194
Hessenberg, 248, 254–256, 259–261, 264–266, 267–268, 271, 273, (276)
Heun's method, 492, 493, 494, 522. See also Runge-Kutta methods, second order
Higher-order differential equation, 529–530, 532–533, 541, (570), 575–608
  conversion to system of first-order ODE, 529, 532–533, 541, (570)
Hilbert matrix, 105, 233, 375
Horner's method, 1, 25, 41, (45)
Householder, 248–257
Hyperbolic PDE, 609, 611, 613, 633–639, 660, (663–664). See also Wave equation

I

Ideal gas law. See Gas law
Ill-conditioned matrix, 105. See also Hilbert matrix
Ill-conditioned ODE, 557
Implicit methods for PDE, 623–631, 638–639, 659–660, (663–664)
Implicit multi-step methods for ODE, 508–513, 524
Inertial matrix, 234, (277–278)
Initial-value problems, 477–574
Integration. See Numerical integration
Integrals
  cosine, (474)
  Dawson, 491
  elliptic, (474)
  exponential, (474)
  Fresnel, (474)
  sine, 434, (474)
Interpolation, 283–338, (339–348). See also Polynomial, Hermite, Rational function, Cubic spline
Inverse matrix, 224–225, (228)
Inverse power method, 242–247, (274)
Iterative methods, 3. See also Jacobi, Gauss-Seidel, SOR, Bisection, Regula falsi, Secant, Newton, Muller, Fixed point Minimization, Power method, QR method for eigenvalues
Iteration, termination conditions for, 13. See also Stopping conditions

J

Jacobi method, 131, 134, 135–143, 166, (168–170)
  conditions for convergence, 142–143
Jacobian, 171, 176–179, 558

K

Khayyam, Omar, 1
Kepler, Johannes, 2, 49
Kutta, M. W., 2
Kutta's method for ODE, 499

L

Lagrange, Joseph Louis, 2
Lagrange interpolation polynomial, 283, 286–295, 336, (339), 438, 447
Laguerre equation, (570)
Lancoz, 3
Laplace (or potential) equation, 609, 612, 640–659, 661, (664). See also Elliptical PDE
Lawson's method, 500
Least squares approximation, 349–380, (389–392)
  continuous, 373–380

Least squares approximation (cont'd.)
  cubic, 364–68, (389–91)
  discrete, 352–372
  exponential, 369–370
  linear, 352–358, (389–390)
  quadratic, 359–364, (389–390)
  total, 352
Leibnitz, Gottfried Wilhelm, 2
Legendre equation, (570)
Legendre polynomials, (89), 378–380, 467
Levenberg-Marquardt method, 2, 193–194
Light intensity, 372, (392)
Linear boundary-value problems, 575, 578–584, 592–598
Linear finite-difference method. *See* Finite-difference method
Linear interpolation, 286, 295
  piecewise, 321, (339)
  use by Babylonians, 1
Linear least squares. *See* Least squares approximation
Linear shooting method. *See* Shooting method
Linear system, solving, 6
  direct method. *See* Gaussian elimination
  interative methods. *See* Jacobi, Gauss-Seidel, SOR
Local trucation error. *See* Error, truncation
Logarithms, 2
Logistic population growth. *See* Population growth
Lorenz equation, (199), (573)
Lotka-Volterra model, (574)
Lower triangular matrix, 201.
  See also LU factorization
LU factorization, 201–232
  applications, 219–225
  direct, 215–218. *See also* Doolittle, Crout, Cholesky
  Gaussian elimination,
    basic, 203–206
    pivoting, 209–214
    tridiagonal matrix, 207–208

M

Mass-spring system, (129), 555, (572)
Mathcad, introduction to 28–40
  arrays, 31–32
  graph region, 30
  programming, 38–40
  text region, 30

  toolbars, 28–29, 32–35
  variables and functions, 29–30, 32, 35–38
  workspace, 28–30
Mathcad, functions, built-in, 35–37, 81–84, 122, 131, 157–58, 192–193, 226, 270, 334–335, 385–386, 423–424, 517–520, 559–561, 602–604, 658–659
Mathcad, functions for numerical methods. *See* Mathcad function index following index
Matrix, 104–105, 113–121, 150, 156, 213–214. *See also* determinant, Hessenberg, Hilbert, Householder diagonally dominant, 104, 120, 142, (170), (608)
  ill-conditioned, 105, 233, 375
  permutation, 213–214
  positive definite, 150, 155–156, 201
  sparse, 131
  trace, 247
  tridiagonal, 113–121, 156, 592–598
  upper triangular. *See* LU, QR
Measuring error, 14–15
Middle eastern mathematics, 1
Midpoint method (ODE), 487–492, 522, (526), 537–540, 544–548, 567, (570)
Midpoint rule (integration), 450–451, 471
Minimum of a function, 187–191, 196
Motion of object with air resistance, 478, 513, 556, 577, 588
Muller's method, 47, 75–81, 87, (89), (92)
Multi-step methods for ODE. *See* Adams-Bashforth, Adams-Moulton, Adams-Bashforth-Moulton

N

Napier, John, 2
Natural cubic splines. *See* Cubic splines
Neuman boundary conditions, 578
Newton's law of cooling, (391)
Newton's method, 2, 4, 47
  function of 1 variable, 68–74, 87, (89), (92), 588–592
  systems of equations, 171, 174–180
Newton interpolation polynomial, 283, 295–309, 336, (339)
Newton-Cotes formulas, 433
  closed, 445–450, 452–457, 471
  open, 445, 450–452, 471

Newton-Raphson method, 2. *See also* Newton's method
Nine Chapters, 1
Nonlinear boundary-value problems, 575, 585–591, 599–601
Nonlinear equations, finding roots of. *See* Nonlinear functions, finding zeros of
Nonlinear finite differences. *See* Finite-difference method
Nonlinear function of one variable, finding zeros of, 4, 47–92. *See also* Bisection, Secant, Regula falsi, Newton's method, Muller's method
Nonlinear functions of several variables, finding zeros of, 171–200. *See also*, Newton's method, Fixed-point iteration, Minimization
Nonlinear pendulum. *See* Pendulum
Nonlinear shooting method. *See* Shooting method
Nonlinear system from geometry, 172, 174–175, 178
Normalized floating point system, 15
Numerical differentiation, 3, 433. *See also* Differentiation, numerical, Difference formulas, Backward difference formula, Central difference formula, Forward difference formula
Numerical linear algebra, 3, 93–170, 201–281
Numerical integration, 3, 8–9, 22–24, (44), 445–472. *See also* Trapezoid, Simpson, Gaussian quadrature, Midpoint, Romberg, Newton-Cotes
Numerical round-off, 3. *See also* Error, round-off
Nystrom method for ODE, 494

O

Oil reservoir modeling, 350, 363, 370
Operation counts. *See* Flops
Orbit. *See* Elliptical orbit
Order (big-Oh), 14. *See also* Convergence, order of; Error, truncation
Order of equations
  effect on accuracy, 26–27, 104–105
  effect on convergence, 142
Ordinary differential equations, 3, 477–608
  boundary value problems, 575–608.

See also Shooting, Finite difference
 initial-value problems 477–574. *See also* Taylor, Euler, Runge-Kutta, Adams-Bashforth, Adams-Moulton, Adams-Bashforth-Moulton
  stability, 514–516
  stiff, 557–558
Orthogonal functions, 376–380, 396
Orthonormal functions, 376
Orthogonal polynomials, 376–380. *See also* Gram-Schmidt process, Legendre polynomials
Ostrowski Theorem, 155
Over-relaxation, 151. *See also* SOR
Overflow, 17

P

Pade approximation. *See* Rational function approximation
Parabolic PDE. *See* Heat equation
Parametric curves, 285, (343–344)
Parasitic solution, 557
Partial differential equations (PDE), 3, 609–666. *See also* Heat equation, Wave equation, Poisson equation, Finite element method
Partial pivoting, 106–113, 123
Pascal, Blaise, 2
Pendulum
 nonlinear, 530, 532, 535, 539
 spherical, (572)
Permutation matrix. *See* Matrix
Persian mathematics, 1
Piecewise planar function. *See* Basis function
Piecewise polynomial interpolation
 cubic, 325–333
 linear, 321
 quadratic, 322–324
Pivot
 column, 97, 103
 element, 97, 100, 103
 row, 97
Pivoting, partial, 104–113
Planetary orbits. *See* Elliptical orbit
Poisson equation, 131, 132, 609, 640–657. *See also* Elliptical PDE
Polak-Ribiere method, 2
Polynomial, 3. *See also* Interpolation
Polynomial interpolation, 283, 286–309. *See also* Lagrange interpolation polynomial, Newton interpolation polynomial

Population growth, 351, 367–368
Potential (Laplace) equation, 131, 132, 609, 640–659, 661, (664). *See also* Elliptical PDE
Positive definite matrix, 155–156, 201
Power method, 233–247
 accelerated, 240
 basic, 236–240
 inverse, 242–247
 shifted, 241
Power spectrum, (430)
Precision, 16
Predator-prey problems, (574)
Predictor-corrector methods. *See* Adams-Bashforth-Moulton methods
Pythagorean theorem, 48

Q

QR factorization, 248–270, 272–273, (276). *See also* Householder, Givens, Hessenberg
QR method for eigenvalues, 267–270, 273, (276)
Quadratic equations, Babylonian solution, 1
Quadratic formula, effect of round-off, 19, (44)
Quadratic least squares. *See* Least squares approximation
Quadratic spline, 322–324
Quadrature, Gaussian, 462–467, 472, (473)
Quazi-Newton methods, 194–195

R

Radiation flux, (346)
Radix-2 FFT. *See* FFT
Rate of convergence. *See* Convergence
Rational function approximation, 3, 382–384, 388
Rational function interpolation, 3, 316–320, (340–341)
Rayleigh quotient, 240–241, 272, (275)
Regula falsi, 4, 47, 55, 56–59, 65, 87, (89)
Relative error. *See* Error, relative
Relaxation. *See* SOR
Riccati equation, (528)
Richardson extrapolation. *See* Acceleration
Ridder's method, 2, 85–86

Romberg integration. *See* Acceleration
Root finding. *See* Bisection, Regula falsi, Secant, Newton, Muller
Round-off error. *See* Error, round-off
Rounding, 17
Runge, Carl, 2
Runge function, 308–309, 330
Runge-Kutta methods, 2, 477, 487–501, 522–523, 537–540, 544–551
 second order, 487–493. *See also* Midpoint method
 third order, 494
 for systems, 537–540, 544–551, 567
 fourth order, 495–499, 523, (526), 549–51, (570)
 higher order, 500
Runge-Kutta-Fehlberg methods, 501, 520–521

S

Saturated oxygen, (346)
Scaling, 113
School enrollments, (347)
Secant method, 4, 47, 55, 60–65, 87, (89), (92), 585–588
Shooting method
 linear, 578–584
 nonlinear, 585–592
Significant digits, 14
Simpson's method (or rule), 2, 448–450, 455–457, 471, (473)
Solving equations of one variable. *See* Bisection, Fixed-point iteration, Muller, Newton, Regula falsi, Secant
Solving systems of linear equations. *See* Gaussian elimination, Gauss-Seidel, Jacobi, SOR
SOR method, 131, 151–157, 166, (168–70)
 conditions for convergence, 155–57
Spline interpolation, (129), 322–333, 337, (339–341)
Spring, (90)
Spring-mass system. *See* Mass-spring system
Stability
 finite difference methods for PDE, 621, 627, 628, 634, 639
 Milne's method for ODE, (528)
 weakly stable method, 515–516
 strongly stable method, 515

Steady-state heat distribution. *See* Heat equation
Steepest descent method. *See* Minimum of function
Stopping conditions, 13, 52, 58, 62, 70, 134, 146, 152, 187, 242
Stiff differential equation, 557–558
Straight-line approximations. *See* Newton, Regula falsi, Secant, Trapezoid
Successive over relaxation. *See* SOR
Successive underrelaxation, 151
Symmetric power method, 240
System of linear equations. *See* Gaussian elimination, Jacobi, Gauss-Seidel, SOR
System of nonlinear equations. *See* Newton's method, Fixed-point iteration
System of ordinary differential equations. *See* Differential equations, system

T

Taylor, Brook, 2
Taylor methods, 484–486. *See also* Euler's method
Taylor approximation, 381
Taylor polynomial, 2, 3, 15, 20–21, 493
Telegrapher's equation, (665)
Temperature conversion data, (391)
Temperature in a rod, (607). *See also* Heat equation

Termination conditions. *See* Stopping conditions
Theorems
  Fixed-point convergence, 11
  Fixed-point convergence in R^n, 186
  Gerschgorin circle, 12, 41
  Newton's method convergence, 73
  Ostrowski, 155
  Secant method convergence, 65
Thermal diffusivity, (665)
Thirteen-point formula, 442
Thomas method, 116–19, 124, (127), (130), 329, 639
Three-point difference formulas, 438–439
Torricelli's law, (527)
Trace, 247
Traffic density, (665)
Trapezoid method for ODE, 524
Trapezoid rule, 41, (44), 446–447, 453–455, 471, (473)
Tridiagonal matrix, 113–121, 156, 592–598
Trigonometric polynomial, 393, 396–428
Trigonometric approximation, 393, 396–397, 399, 400, 402–405, 428
Trigonometric interpolation, 393, 396–397, 398, 400–406, 414, 425–428
Truncation error. *See* Error

Truss, forces on, 95, 102, 111–112, (130)
Tukey, 3
Two-point boundary value problem. *See* Boundary value problem
Two-link robot arm, 173, 180, 563–566

U

Underflow, 17
Upper Hessenberg matrix. *See* Hessenberg
Upper triangular matrix. *See* LU, QR

V

Van der Pol equation, (573)
Vapor pressure, (346)
Viscosity, (345)

W

Water draining, (392), (527)
Wave equation, 609, 611, 613, 633–639, 660, (663–664). *See also* Hyperbolic PDE
Waveforms for musical instruments, 394, 426, (429)
Word, 16

Z

Zeros of a function. *See* Nonlinear function

Author Index

Abramowitz, M., (346), 472, 474
Achieser, N. I., 388
Acton, F. S., 194, 195, 197, 606
Ames, W. F., 199, 442, 472, 628, 634, 638, 639,
Anderson, R. E., 197
Ascher, U. M., 606
Atkinson, K. E., 42, 65, 73, 81, 88, 125, 142, 150, 186, 217, 295, 447, 450, 467, 472, 501, 514
Axelsson, O., 662
Ayyub, B. M., (91), (345-6), 472, (474), 569, (574), (608)
Bader, G., 569
Barker, V. A., 662
Barrett, R., 167
Bartels, R. H., 285, 338
Barsky, B. A., 285, 338
Bautista, M., 662
Beatty, J. C., 285, 338
Beauregard, R. A., 232, (277)
Becker, E. B., 662
Birkhoff, G., 661
Bloomfield, P., 428
Bjorck, A., 42
Black, W., 197
Boyce, W. E., (92), 661, (665)
Boyer, C. B., 42
Bracewell, R., 425, 428
Brent, R., 86, 88
Brezinski, C., 388
Briggs, W. L., 428
Bringham, E. O., 428
Bulirsch, R., 125, 167, 270, 273, 317, 338, 525, 606

Burden, R., 156
Carey, G. F., 662
Cash, J. R., 520, 525
Celia, M. A., 662
Cheney, E. W., 388
Clenshaw, C. W., 472
Coleman, T. F., 125
Colton, D., 661
Curtis, A. R., 472
Dahlquist. G., 42, 558
Danzig, G. B., 167
Datta, B. N., 167
Davies, A. J., 662
Davis, P. J., 338
deBoor, C., 338, (348)
deDoncker, C. W., 472
Dennis, J. E., 193, 197
Derrick, W. R., 232, 569
Deuflhard, P., 525, 569
Dillon, W. R., 197
DiPrima, R. C., (92), 661, (665)
Donato, J., 167
Dongarra, J., 125, 167, 228
Edwards, C. H., 88, 92, 478, 525, (527), 528
Eijkhout, V., 167
Elliott, D. F., 428
Eves, H., 42
Faires, J. D., 146
Farin, G., 338
Ferrari, R. L., 662
Finizio, N., 525
Fix, G., 662
Flannery, B., 42, 88, 125, 167, 197, 338, 525, 569, 606

Forsythe, G. E., 88, 125, 228, 662
Fox, L., 228, 273, 606
Fraleigh, J. B., 232
Freund, R. W., 167, (277)
Froberg, C. E., 483, 525, 528
Garcia, A. L., (129), 199, (345), 525, 527, 572, (573), 662, (665)
Gear, C. W., 525, 558
Gill, P. E., 125
Goldstein, M., 197
Golub, G. H., 104, 125, 155, 156, 167, 226, 228, 247, 271, 273, 429, 525, 662
Gray, W. G., 662
Greenbaum, A., 167
Greenberg, M. D., 88, 234, 273, 569, 572
Greenspan, D., 530
Grossman, S. I., 232, 569
Haberman, R., 606, 662
Hager, W. W., 113, 124, 167, 170, 235, 260
Hair, J. F., 197
Hall, C. A., 662
Hamming, R. W., 88
Hanna, O. T., (91), (129), 525, (527), 569, 572, 606, (607-8)
Henson, V. E., 428
Hibbeler, R. C., 88, (90), (91)
Hildebrand, F. B., 569
Himmelblau, D. M., (91)
Horn, R. A., 228
Hornbeck, R. W., 606
Hyman, C. J., 388
Inman, D. J., 569, 606

701

Isaacson, E., 42, 125, 150, 452, 472
Jacobs, D. A. H., 197
Jaeger, J. C., 576, 606
Jain, M. K., 483, 494, 500, 525
Jensen, J. A., 125, 388, 472, (475)
Johnson, R. A., 197, 228
Kachigan, S. K., 197
Kahaner, D., 333, 338, 472
Kammer, W. J., 338
Kaps, P., 562, 569
Karp, A. H., 520, 525
Keller, H. B., 42, 125, 150, 452, 472, 599, 604, 606
Kolman, B., 228, 273
Ladas, G., 525
Lancaster, P., 338
Larsen, R. W., 42
Leon, S. J., 228, 232, 273, (278), (279)
Lynch, R. E., 661
Malcolm, M. A., 88, 228
Marcowitz, U., 662
Mardia, K. V., 197
Marquardt, D. W., 197
Mattheij, R. M. M., 606
McCuen, R. H., (91), (345-6), 472, (474), 569, (574), (608)
Meis, T., 662
Mitchell, A. R., 662
Moler, C. B., 88, 125, 228, 338
Morrison, D. F., 197
Murray, W., 125
Nachtigal, N. M., 167
Nash, S., 338
Nicolis, G., 573
Nussbaumer, H. J., 429
Oden, J. T., 662
Ortega, J. M., 155, 167, 186, 197, 525, 662
Pauling, L. C., (91)

Penney, D. E., 88, (92), 478, 525, (527), 528
Piessens, R., 470, 472
Polak, E., 197
Poole, W. G., 525
Porsching, T. A., 662
Pozo, R., 167
Press, W. H., 42, 86, 88, 125, 151, 160, 167, 193, 194, 195, 197, 226, 271, 317, 338, 386, 469, 470, 520, 521, 525, 569, 604, 606
Prigogine, I., 573
Rabinowitz, P., 42, 64, 88, 125, 150, 247, 383, 429, 467, 472
Ralston, A., 42, 64, 88, 125, 150, 247, 383, 429, 467, 472
Rao, K. R., 428
Reddien, G. W., 338
Reinboldt, W. C., 197
Reinsch, C., 125, 273
Rentrop, P., 562, 569
Rice, J. R., 88, 525
Ridders, C. J. F., 86, 88, 472
Ritger, P. D., 472, (474), 525
Rivlin, T. J., 388
Roberts, C. E., 525, 606, (608)
Romine, C., 167
Rose, N. J., 472, (474), 525
Rowland, J. H., 125, 388, 472, (475)
Russell, R. D., 606
Salkauskas, K., 338
Sandall, O. C., (91), (129), 525, (527), 563, 569, (572), 606, (607-8), 559
Schnabel, R. B., 193, 197
Silvester, P. P., 662
Simmons, G. F., 42
Simon, W., (91), 531, 569, (573)
Singleton, R. C., 425, 429
Smith, D. E., 42

Spong, M. W., 563, 569
Stegun, I. A., (346), 472, 474
Stoer, J., 125, 167, 270, 273, 317, 338, 525, 606
Strang, G., 113, 124, 156, 167, 228, 273, 662
Struik, D. J., 42
Svobony, T., 662, (665)
Taha, H. A., 167
Tatham, R. L., 197
Teukolsky, S. A., 42, 125, 167, 197, 338, 525, 569, 606
Taylor, R. L., 662
Thomson, W. T., 234, 272, 569, (572)
Timan, A. F., 388
Troutman, J. L., 662
Uberhuber, C. W., 472
Van der Pol, B.,
van der Vorst, H., 167
Van Loan, C. F., 104, 125, 156, 167, 226, 228, 247, 270, 273, 429
Varga, R. S., 338
Vetterling, W. T., 42, 88, 125, 167, 197, 338, 525, 569, 606
Vargaftik, N. B., (345)
Vidyasagar, M., 563, 569
Wait, R., 662
Wasow, W. R., 662
Wichern, D. W., 197
Wilkinson, J. H., 125, 273
Winston, W. L., 167
Wright, M. H., 125
Yandl, A., 72
Young, D. M., 562,
Zauderer, E., 662, (665)
Zienkiewicz, O. C., 662
Zill, D. G., (279), 525